Table of Contents

UAV Sensors for
Environmental Monitoring

Special Issue Editors

Felipe Gonzalez Toro
Antonios Tsourdos

MDPI • Basel • Beijing • Wuhan • Barcelona • Belgrade

Special Issue Editors
Felipe Gonzalez Toro
Queensland University of Technology
Australia

Antonios Tsourdos
Cranfield University
UK

Editorial Office
MDPI AG
St. Alban-Anlage 66
Basel, Switzerland

This edition is a reprint of the Special Issue published online in the open access journal *Sensors* (ISSN 1424-8220) from 2015–2016 (available at: http://www.mdpi.com/journal/sensors/special_issues/UAV-sensors).

For citation purposes, cite each article independently as indicated on the article page online and as indicated below:

Lastname, F.M.; Lastname, F.M. Article title. *Journal Name*. **Year**. *Article number, page range.*

First Edition 2018

ISBN 978-3-03842-753-7 (Pbk)
ISBN 978-3-03842-754-4 (PDF)

About the Special Issue Editors

Felipe Gonzalez Toro, Associate Professor at the Science and Engineering Faculty, Queensland University of Technology (Australia), with a passion for innovation in the fields of aerial robotics and automation and remote sensing. He creates and uses aerial robots, drones or UAVs that possess a high level of cognition using efficient on-board computer algorithms and advanced optimization and game theory approaches that assist us to understand and improve our physical and natural world. Dr. Gonzalez leads the UAVs-based remote sensing research at QUT. As of 2017, he has published nearly 120 peer reviewed papers. To date, Dr. Gonzalez has been awarded $10.1M in chief investigator/partner investigator grants. This grant income represents a mixture of sole investigator funding, international, multidisciplinary collaborative grants and funding from industry. He is also a Chartered Professional Engineer, Engineers Australia—National Professional Engineers Register (NPER), a member of the Royal Aeronautical Society (RAeS), The IEEE, American Institute of Aeronautics and Astronautics (AIAA) and holder of a current Australian Private Pilot Licence (CASA PPL).

Antonios Tsourdos obtained a MEng on Electronic, Control and Systems Engineering, from the University of Sheffield (1995), an MSc on Systems Engineering from Cardiff University (1996) and a PhD on Nonlinear Robust Autopilot Design and Analysis from Cranfield University (1999). He joined the Cranfield University in 1999 as lecturer, was appointed Head of the Centre of Autonomous and Cyber-Physical Systems in 2007 and Professor of Autonomous Systems and Control in 2009 and Director of Research—Aerospace, Transport and Manufacturing in 2015. Professor Tsourdos was a member of the Team Stellar, the winning team for the UK MoD Grand Challenge (2008) and the IET Innovation Award (Category Team, 2009). Professor Tsourdos is an editorial board member of: Proceedings of the IMechE Part G Journal of Aerospace Engineering; IEEE Transactions of Aerospace and Electronic Systems; Aerospace Science & Technology; International Journal of Systems Science; Systems Science & Control Engineering; and the International Journal of Aeronautical and Space Sciences. Professor Tsourdos is Chair of the IFAC Technical Committee on Aerospace Control, a member of the IFAC Technical Committee on Networked Systems, Discrete Event and Hybrid Systems, and Intelligent Autonomous Vehicles. Professor Tsourdos is also a member of the AIAA Technical Committee on Guidance, Control and Navigation; AIAA Unmanned Systems Program Committee; IEEE Control System Society Technical Committee on Aerospace Control (TCAC) and IET Robotics & Mechatronics Executive Team.

Preface to "UAV Sensors for Environmental Monitoring"

The past decade has borne witness to a remarkable growth in the use of UAVs as remote sensing platforms. A wide range and selection of commercially available UAV platforms have entered the market. They can be acquired off-the-shelf at increasingly lower cost. This availability and affordability has been complemented by rapid advances in the miniaturization of cameras, sensors, instrumentation and data stores and transmission systems. UAV platforms have become smarter, more adroit, and more dexterous. The sensors they fly are lighter, cheaper, more compact, sophisticated and sensitive. Adoption of UAV technology has brought about a revolution in environmental monitoring. This transformative UAV innovation has been enthusiastically embraced by remote sensing applications and scientific disciplines.

This book highlights advances in sensor technologies, navigation systems and sensors payloads including the following: autonomous aerial refueling; calibration method using low cost MEM IMUs; low cost airborne sensing systems; vision-based detection; UAVs task and motion planning in the presence of obstacles; ground-based radio navigation systems; smart flying sensors; UAV control using 3D landmark bearing-only observations; cooperative surveillance and pursuit; new multispectral camera system with different bandpass filters; synthetic aperture radar to aid UAV navigation; formation flight of multiple UAVs; shared autonomy; dual-stack single-radio communication architectures; GNSS-based passive radar; UAV navigation in cluttered and GPS-denied environments and UAV-based photogrammetry.

The book also highlights a diverse range of new applicators including the estimation of carbon exports from heterogeneous soil landscapes; instruments to classify the water content feature of lands; wildlife monitoring and conservation; wildlife poachers detection; architectural surveys of vertical structures; environmental source localization and tracking; dust particles after blasting at open-pit mine sites; data collection from WSNs; inspection of pole-like structures; multi-UAV routing; emergency response; river hydromorphological features detection; dynamic UAV routing in traffic incident monitoring, and surveys on methods and inexpensive platforms for infrastructure inspection.

Felipe Gonzalez Toro and Antonios Tsourdos
Special Issue Editors

Article

UAV-Based Photogrammetry and Integrated Technologies for Architectural Applications—Methodological Strategies for the After-Quake Survey of Vertical Structures in Mantua (Italy)

Cristiana Achille, Andrea Adami, Silvia Chiarini, Stefano Cremonesi, Francesco Fassi, Luigi Fregonese * and Laura Taffurelli

Department of Architecture, Built Environment and Construction Engineering, ABC, Politecnico di Milano, via Ponzio, Milano 31-20133, Italy; cristiana.achille@polimi.it (C.A.); andrea.adami@polimi.it (A.A.); silvia.chiarini@polimi.it (S.C.); stefano.cremonesi@polimi.it (S.C.); francesco.fassi@polimi.it (F.F.); laura.taffurelli@polimi.it (L.T.)
* Author to whom correspondence should be addressed; luigi.fregonese@polimi.it;
  Tel.: +39-376-371-056; Fax: +39-376-371-041.

Academic Editors: Felipe Gonzalez Toro and Antonios Tsourdos
Received: 31 March 2015; Accepted: 24 June 2015; Published: 30 June 2015

Abstract: This paper examines the survey of tall buildings in an emergency context like in the case of post-seismic events. The after-earthquake survey has to guarantee time-savings, high precision and security during the operational stages. The main goal is to optimize the application of methodologies based on acquisition and automatic elaborations of photogrammetric data even with the use of Unmanned Aerial Vehicle (UAV) systems in order to provide fast and low cost operations. The suggested methods integrate new technologies with commonly used technologies like TLS and topographic acquisition. The value of the photogrammetric application is demonstrated by a test case, based on the comparison of acquisition, calibration and 3D modeling results in case of use of a laser scanner, metric camera and amateur reflex camera. The test would help us to demonstrate the efficiency of image based methods in the acquisition of complex architecture. The case study is Santa Barbara Bell tower in Mantua. The applied survey solution allows a complete 3D database of the complex architectural structure to be obtained for the extraction of all the information needed for significant intervention. This demonstrates the applicability of the photogrammetry using UAV for the survey of vertical structures, complex buildings and difficult accessible architectural parts, providing high precision results.

Keywords: tall buildings; close range photogrammetry; UAV application; automated 3D modeling techniques; terrestrial laser scanning; topographic survey; cultural heritage; emergency context survey

---

1. Introduction

Which kind of surveying strategies should be chosen in the case of complex and critical situations? This paper explains how traditional and innovative techniques in the surveying field can be applied to obtain high quality timely results in complex contexts. This case study deals with for tall buildings and critical situations like in the case of the earthquake that affected two Italian regions, the Emilia and Lombardy regions, in May 2012.

A great number of historical buildings were seriously damaged by the shocks. In particular, most of the churches located in the southern area of Mantua's province required very important structural interventions. In the city, recently declared a UNESCO World Heritage (2008) site, the Santa Barbara bell tower was heavily impacted by the effects of the earthquake and shows many problems.

To repair the damages, the first in-depth fact-finding operation is an accurate survey of the architecture and its structure. To acquire and describe complex architectural objects, researchers required a 3D database of spatial and metric information that can be used to extract three-dimensional models as well as two-dimensional representations and all the information needed for the maintenance program or restoration projects [1–5].

The bell tower of Santa Barbara has, like every tower, a structure that is quite challenging to measure correctly due to its natural vertical extension. Moreover, the dense urban pattern of the old town center makes it hard to find good positions to observe and measure all four facades from a horizontal point of view. The possible acquisition points are on the ground, very close to the tower and this creates bad measuring angles, a non-uniform resolution on the object, noisy scans, and consequently low accuracy measurements. Moreover, in this case the lantern at the top of tower cannot be seen from the ground floor, making the use of lifting equipment indispensable. An additional problem to be resolved is the presence of temporary holding structures around the top lantern that hide the structure and impose multi-angle acquisition from a close point of view.

In light of all these problems, the most suitable solution for the 3D survey of the bell tower is the integration of different sensors and methods in order the use the potential of every method correctly [6–8]. In particular, the idea for the external facades was to use a manually controlled UAV.

The solution for the 3D modeling of tall structures is the integration of laser scanner data of the internal areas with dense image matching DSM of the external façades, using a classical topographic network to georeference all data together in a single reference system. Some aspects have to be considered in order to plan the different acquisition correctly. The integration of different data requires sufficient overlap areas, same resolution and comparable accuracy.

After an opening overview on the specific architecture and its damage after the earthquake, a first analysis in this paper regards the traditional methods for the survey of tall buildings and towers. Particular attention is given to the most recent techniques of automated 3D modeling and UAV acquisition in order to highlight their pertinence in the survey of Cultural Heritage artefacts. Before describing the survey stage, some tests to verify the possibility of integration of automated image modeling with laser scanner data are described.

2. Santa Barbara Tower Bell and the Earthquake

The church of Santa Barbara is located in the old town center of Mantua and is one of the most representative buildings in the city (Figure 1).

Figure 1. Santa Barbara Church and its bell tower in the historical urban context.

Santa Barbara church was commissioned by Guglielmo Gonzaga as the family chapel inside the "Palazzo Ducale", the most important historical complex of Mantua. The architect Giovanni Battista Bertani built it in 1562–1572. The bell tower, with a height of about 49 m from the top of the dome, has a square plan (about 9.00 × 9.00 m). It is composed of six internal levels. Some ledges split the main tower in three parts; the first one is finished with plaster and the others have a brick surface. The facades have different architectural elements like niches, gables, arches and pilaster strips. In the belfry, two arches on every façade show the bells to the exterior. Above this point, there is a small round temple with a colonnade, covered by a balcony and a little dome with a lantern: this is the most characteristic part of the bell tower that characterizes the Mantua landscape.

On May 20th and 29th 2012 two big earthquake shocks were felt across a large part of the north of Italy. During that earthquake many buildings were damaged (Figure 2) and afterward it was necessary to batten down the hatches by building temporary structures for safety of the buildings and people.

In the Santa Barbara bell tower, the top was the most damaged part. The dome's lantern collapsed and in its fall broke off a part of the summit balcony, in addition to some areas of surrounding buildings. The round temple was subjected to a rotation movement and the structure had some subsidence, so many fissures appeared to the structure.

The serious damage that affected the highest part of the structure required the immediate realization of a temporary holding structure. In the initial days after the collapse, the "Protezione Civile" fire brigade secured the top of the tower with steel cables and a reticular tubular structure to prevent another collapse, before any detailed survey could be done.

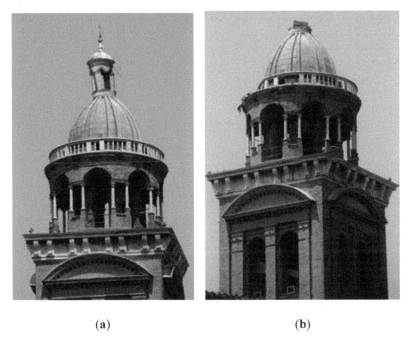

(a)   (b)

Figure 2. Santa Barbara bell tower, after the first earthquake (**a**) and the second (**b**) with the collapse of the lantern.

The project of the structure for the support of damaged parts, to save time, was based on a drawing made in 1999, during the last restoration work, and some photographs taken using a large crane. Because of the lack of complete and up-to-date drawings, the Cultural Heritage office of Mantua's Diocese asked for all the representations needed for the analysis of the present situation (e.g., material decay, tower's tilt and structural damages) and to design the restoration and complete the bell tower. Therefore, it was considered necessary to make an accurate survey of the situation after the earthquake to obtain traditional two-dimensional architectural representations (plans, elevations and sections) together with the addition of vertical profiles, facades, orthophotos and a 3D model for finite element calculation. The survey was intended at an overall scale 1:50, which requires an accuracy of less than 1 cm, with parts detailed at 1:20 scale.

3. Traditional Approach

The main problem for the operational stage of the survey was the acquisition of the highest part of the tower, due to its height, and the Northeast facade, which is set against the church's sacristy. In particular, the belfry and the round temple cannot be measured from the ground. Moreover, it was necessary to guarantee a high precision result to allow a correct and detailed design of the new structure. Before starting the survey of the bell tower, many possible solutions were considered based on traditional or well-established methods of architectural survey: topography, laser scanning and traditional digital photogrammetry.

The topographic survey is surely one of the most well-established methods. It allows all the significant points of the facades of the tower to be acquired. It requires many station points, adjusted in a topographical network, including acquiring the measurements via the method of irradiation or, better, forward intersection [9]. All this requires the possibility of moving around the tower and, the possibility of identifying raised measuring positions. In addition to the logistic difficulties, which are different case by case, the critical aspects of this method are related to time. The operation of discretization, to determine which points and parts of the architecture have to be represented, cannot be referred to the next stage of the restitution, but it must be done directly on site. This requires lengthy survey times, high logistic difficulties and moreover this methodology can no longer be used with the arrival of methods that provides high resolution measurements directly in 3D, such as laser scanner or photogrammetry, in real time or nearly real time. Topography or single point measurements methods are however necessary to register different data together and to test the results.

The photogrammetric survey shortens the acquisition time on site, as most of the information is extracted in post-processing. The capture geometry is the challenge of the photogrammetric acquisition. Acquiring images with the correct geometry is extremely necessary in order to achieve the requested precision that means, in this case, finding some suitable high places, such us balconies or roofs of surrounding buildings. Images, taken with very narrow angles, make the restitution phase more complex and, in any event, provide measurements of lower accuracy.

For laser scanning technology, in addition to the foregoing consideration, it is also necessary to analyze the different kind of instruments. Some instruments, especially the ones based on phase shift technology, have a short measurement range. This means that when the scan is made from the base of the tower or from other places around it, acquired data can be very noisy. From relevant distances, moreover, the point clouds do not have a high resolution and therefore the result cannot be very accurate.

4. The Most Recent Surveying System

The latest surveying systems deal with both the acquisition stage and the data processing. One of the most important changes in recent years regards the combination of digital photogrammetry and computer vision into new software for automated 3D modeling from images. A second innovation concerns the possibility of acquiring information (images, point clouds, IR images) from above, with remotely piloted aerial vehicles.

The integration of these two latest methods—automated 3D modeling from UAV acquired images—can solve the problems mentioned before. In fact, it is possible to obtain high resolution dense point clouds of tall buildings in a short time, even if the building is located in a high-density area.

4.1. Automated 3D Modeling

It is complex to find a single definition for this new approach to 3D modeling from images. By focusing the attention on different steps of the workflow, many definitions have been suggested: dense stereo matching (to underline the matching phase from stereo pairs), dense image matching (image as source of 3D), automated 3D image modeling (automatic approach from images).

However, this software appeared about 4 or 5 years ago and is the result of developments in the field of photogrammetry and computer vision. The most important innovation regarded the image matching process. It can be defined as the establishment of correspondences between images. The first studies in photogrammetry mainly concerned aerial images and topographic mapping problems. To foster image matching, the concepts of epipolar geometry and cross-correlation for image matching were introduced. However, it is only with the advent of digital images that research focused on the automation of procedures in order to avoid the intervention of skilled operators. Computer vision, on the other hand, focused on stereo matching but with a different aim, which was the reconstruction of 3D space, without particular attention to its accuracy. A second innovation was dealing with the possibility of automatically extracting dense point clouds from images by the use of some operators, which allows detecting and describing local features in images.

The meeting of the two sciences led to the three-dimensional reconstruction of reality, starting from images. Even with specific algorithms [10–14], these software allow the fully autonomous three-dimensional reconstruction of objects from images in different steps:

(a) Automatic extraction of characteristic features of images;
(b) Image matching to detect tie points;
(c) Computing by bundle adjustment of interior and exterior orientation;
(d) Computing of a dense point cloud.

In addition:

(e) Triangulation of points in a mesh;
(f) Texture mapping.

The ability to reconstruct three-dimensional objects from images is a common task in the Cultural Heritage field. Consequently, the possibility of using low cost systems (a simple digital camera) rather than expensive systems (laser scanners) was a relevant step. Additionally, these systems provide not only the geometry of the object, but also its texture that is metrically projected onto the 3D images. From this extracting orthophotos that are useful and widely used in this type of application is quite automatic and immediate. In addition, these determined its use not only in the surveying field, but also in virtual reconstruction or simulation. Ease of use, manageability and speed contribute to their spread. In addition, many algorithms have been developed around these methods to optimize the calculations and obtain accurate results with short time of elaboration. Several open source software packages have fostered the 3D reconstruction from images by providing an inexpensive solution and by allowing an in-depth analysis of results and algorithms.

Many researches were focused on the comparison between points-clouds from laser scanning and automated-photogrammetry [10,13,15]. Instead, one of the topics of this paper is to verify the accuracy of the method and the possibility of integrating all the data into a single reference system.

4.2. UAV Survey

The term Unmanned Aerial Vehicle (UAV), also commonly known as drones, defines a generic aircraft designed to operate with no human pilot onboard [16]. This acronym is commonly used in geomatics applications, but many other terms identify these aircraft according to their propulsion system, autonomy, maximum altitude and level of automation, giving an idea of the variety of the jobs carried out by these systems. For example drone, Remotely Operated Aircraft (ROA), Unmanned

Combat Air Vehicle (UCAV), Medium Altitude Long Endurance (MALE) UAV, Remote Piloted Aircraft System (RPAS), *etc.* Another definition describing these devices is Unmanned Aerial Systems (UAS) that comprehends the whole system, including the aerial vehicle and the ground control station.

The development of UAV solutions was primary driven by military purposes, for inspection, surveillance, and mapping of inimical areas. The quick development of civilian solutions, in particular for geomatics applications, was made possible by the spread and improvement of digital camera devices and GNSS systems, necessary for navigation and geo-referencing tasks [17].

In geomatics, the use of UAVs is a low cost alternative to traditional airborne photogrammetry for large scale topographic mapping or recording detailed 3D information of terrestrial objects and integratation with the data collected by terrestrial acquisition (for example laser scanning). UAV aerial photogrammetry cannot replace traditional photogrammetry and satellite mapping application for large territories, but they give an efficient solution for little areas and large scale surveys. This is the case of Cultural Heritage applications. Moreover, these little flying vehicles give rapid solutions to some long-standing unsolved problems, such as for example the high resolution survey of pavements and vertical surfaces such as, for example, tall towers.

The principal frames for the UAV used for geomaticsphotogrammetry applications are unpowered (balloon, kite, glider), fixed wings, copter and multi-copter platforms. Each of them have different advantages and handicaps referring to different tasks and applications. There is also a wide price range, approximately from 1000 to 50,000 Euros depending on the on-board instrumentation, flight autonomy, payload and automation capabilities. For example, low cost solutions do not allow a completely autonomous flight, and in most cases, they require human assistance for take-off and landing. Another cost-sensitive element is the engine type: in comparison with the most used electrical ones, internal combustion engines have longer endurance and permit higher payloads, but they also have higher costs and require more maintenance and pre-flight controls. The three principal frameworks of drones, used mostly for geomatics purposes, have specific characteristics:

Fixed wing: these vehicles work like a traditional aircraft with one or more propellers and fly in a straight direction. Most of these drones work in a fully autonomous mode following a pre-planned flight plan. The characteristics of this structure made the drones suitable for aero-photogrammetry works due to high autonomy and possibility of cover wider areas than a multi or single rotor configuration. The most used and portable solutions are built with light materials (less than 1 kg take-off weight) for increased portability and autonomy. In contrast, this solution has very low resistance to wind—this makes it difficult to operate in high wind conditions, and limits the payload.

Copters: these devices are basically scaled-down models of real helicopters; usually the bigger ones are equipped with gasoline engines that give very high autonomy and the possibility of carrying a much heavier payload. This type of UAV has high flexibility. On the other hand, big copter systems are also characterized by a high resistance and stability in difficult wind conditions. A problem of the systems with ICE motors is the greater need for maintenance and preparation procedures for flying.

Multi-copters: they consist of devices with three or more propellers fixed to a structure. Position and motion are controlled by managing the differential engine rotation speed of any single propeller. This type of system is very flexible in relation to the various different tasks and payloads (more propellers means more payload).

Typically, the batteries do not have a very long autonomy (15–20 min approx.) meaning batteries must be changed for long sessions of flightwork. These systems are preferred very often for the application of terrestrial photogrammetric Cultural Heritage surveys, infrastructures and civil engineering because of their high maneuverability, small dimensions, short deployment time and high stability.

The development of these systems is very fast and it is focused on increasing the flight time, accuracy of on-board navigation systems and in particular on the sensors that can be used.

5. Laboratory Tests on Data Integration

Before integrating laser scanners and topographic surveys with photogrammetric ones, made by automated 3D image modeling, it was necessary to evaluate several aspects. In particular, regarding point clouds, density of points, quality and presence of color information had to be considered.

The first aspect was strictly related to resolution, i.e., the average density of points on the surveyed surface. A homogeneous density allows the same level of detail for each part of the object to be obtained, both in photographic representations such as orthophotos and in three-dimensional models.

Even more important is the quality of measurements, which involves the concepts of accuracy and precision. The first one expresses the ability to approach the real measurement, the second the ability of repeating the measurements and obtaining a similar value. The possibility of recording RGB values is an added value as it provides more information about the object.

The first condition to integrate different data sources is that all data must respect some pre-established parameters. The immediate example regards accuracy. The scale of the survey and complexity of the object in fact determine the minimum imposed value of accuracy. If one of the two datasets does not guarantee that minimum value, they cannot be integrated. Once this first condition is guaranteed, it is necessary that the values of accuracy and precision are very similar, otherwise, the errors in the data recording stage can produce problems in the calculation of the final products.

To verify the integrability of the two different techniques, laser scanner and digital automated photogrammetry, some tests were done to verify the quality of their data. The same object was surveyed with both methods and then the results were compared. In order to have a comparison that is closer related to the case of Santa Barbara bell tower, the test involved the base section of the tower. In that position both the laser scanner and photogrammetry had no logistic problems. Some targets, for photogrammetry or laser scanner, were measured by a total station (Leica TS30, Leica Geosystems, Italy) in order to verify the registration stage of different data and to evaluate the results in a single reference system. They were located on different levels of the facade in order to have a fully 3D spatial distribution of targets. Different datasets were acquired to make a complete comparison and they included images taken with reflex cameras and laser scanner scans. In particular:

- Dataset 1: Photogrammetry—Canon 5D

  The same area was surveyed by 41 images acquired by Canon EOS 5D Mark III (Canon, Italy) with a 35 mm lens. The EOS 5D is a high-resolution camera, but it has no calibration certificate. The images were shot in jpg format and their dimensions are 5760 × 3840 pixels. The pixel size is 0.0064 m and the ground sample distance (GSD) is 0.0016 m.

- Dataset 2: Laser scanner—HDS 7000

  Single point cloud made by a Leica HDS 7000 (the same instrument used for acquisition in the after-quake context in the Santa Barbara case) overhead at a distance of 8 m from the surface of the object. It has a linearity error, as defined by the producer, ≤ 1 mm. The scan is made up of 123 million points.

The laser scanner point cloud was considered as the term of reference. From the Leica datasheet, the HDS 7000 noise standard deviation is about 0.4 mm at 10 meters distance, in the event of grey objects. The point cloud was referenced in the topographic reference system by using four black/white targets (automatically extracted in Cyclone), with an average error of 2 mm. The other eight coded targets were picked manually and used as check points. The difference between the coded targets acquired by total station and manually picked in the oriented point cloud is very small: the average error is 0.001 meter. The photogrammetric dataset was processed in Agisoft Photoscan [18] to verify their integrality with laser scanner point clouds and to verify the quality of automated 3D image modeling. The images were aligned and then were georeferenced by using coded targets, automatically detected by the software. Some coded targets were not used for registration as GCP, but as checkpoint

(CP) for accuracy verification. For an appropriate analysis work, all data were elaborated in different ways, e.g.,

Test 1 automatic image-based modeling without targets
Test 2 automatic image-based modeling with topographic targets
Test 3 automatic image-based modeling with topographic targets, optimised.

In Table 1 it is evident that by using acquired targets and the optimisation process the errors are adjusted.

Table 1. Comparison of dataset orientation.

	n Photos	Optim.	RMSE [Pixel]	GCP	CP	Mean GCP with TCRA [m]	σ_{xyz} GCP with TCRA [m]	Mean CP with TCRA [m]	σ_{xyz} CP with TCRA [m]
Test 1	41	no							
Test 2	41	no	0.145	9	4	0.000925	0.000480	0.000740	0.000203
Test 3	41	yes	0.145	9	4	0.000876	0.000415	0.001012	0.000292

Moreover, the level of accuracy achieved is very high. It means that such data, obtained by automated image modeling, are surely suitable for an architectural survey at 1:50 scale and also 1:20. By considering the accuracy level of laser scanner point cloud, it is evident that photogrammetric data can be fully integrated with the laser scanner data.

A further comparison regarded not the result of georeferencing, but the final results of the processing stage. By means of an analysis based on the Euclidean distance, the point clouds, obtained from two datasets, were compared. It was decided to carry out a comparison of point clouds because they are the first result of laser scanning and photogrammetric methods. Furthermore, this choice allowed any errors or systematic effects due to triangulation algorithms to be excluded.

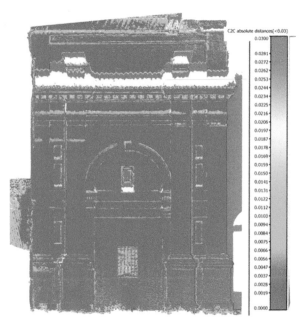

Figure 3. Comparison of photogrammetric point clouds with laser scanner one. The maximum difference, shown in red color, is 3 cm and it is present only on the borders or in missing parts.

The comparison was made in Cloudcompare [19], an open source software designed for data verification between point clouds or meshes. The point cloud of the Canon dataset, geo-referenced in the same reference system of topography and laser scanner, was compared with the laser scanner one. The differences, as shown in Figure 3, are very small and are mainly concentrated on the edges where the laser scanner data is less reliable. This test verifies that the point clouds produced for photogrammetry, both semi-metric and amateur camera, are very similar with regards to accuracy and precision and they can be integrated into a single system. In such a way, it is possible to survey complex objects with different methods that can be finally integrated to create, for example, a single 3D database.

6. The Survey of the Santa Barbara Church Bell Tower

As described before, in the survey of the bell tower all techniques, both traditional and innovative, were applied in order to obtain the best results and to solve the problem in the documentation of different parts. The acquisition of the interiors of the tower and the lowest exterior parts did not involve any kind of problem and the survey was conducted in a traditional way. The most complex part was, as expected, the upper part and in particular the exterior of the belfry and the round temple.

The first step was the realization and materialization of the topographic network to define the reference system for all the work by using the TS30 Leica Total station. The points of the network were evenly located around the tower, outside (n = 5 points), and at different levels, inside the tower (n = 16 points). From those stations 95 targets (GCP) where acquired to georeference scans and images. For the highest part of the tower, an additional ground station was created on a balcony in front of the bell tower and connected with the main topographic net on the ground.

Laser scanner acquisition was carried out with a Leica HDS 7000. It is a phase shift instrument with an onboard control system. It is a class 1 laser product with an acquisition range between 0.3 and 187 m and a 360 × 320° field of view. This instrument can acquire up to 1 million points/second at the maximum resolution. The final precision depends on several causes like instrument features, object distance from the instrument, surfaces materials, and angle of incidence of laser beam to the surfaces, *etc.* [20–22].

Many scans were necessary to acquire the total bell tower: 15 for the exterior (nine from the ground floor and six from balconies on the nearby buildings), 23 for the interior (scale, sacristy, and bells zone) and 12 for the lantern acquisition (in Figure 4 a section of the bell tower extracted from interior and exterior scans).

Figure 4. Laser scanner point clouds of interior and exterior registered in the local reference system for the elaboration of the section.

The resolutions of the interior and exterior surfaces are quite different. The interior surfaces had a more uniform resolution (3 mm at a distance of 4.5 m). Instead, for the façades, the resolution is always less, rising as the scans were done from the base of the tower, with a fixed angular step (the greater the distance, the lower the resolution). There is another effect to consider due to the height, namely the accuracy. In fact, the laser beam hit the surfaces with a very low, higher incidence angle and a bigger laser beam, resulting in a decrease of accuracy.

The first solution, trying to produce a high defined and complete acquisition of the top of the tower, was to position the laser scanner on the corner of the highest ledge, around the round temple, for the acquisition of the structure comprising arches and columns, and on the highest balcony for the acquisition of the little dome.

In this case, the main operational issue for the acquisition of a clear data was the small space surrounding the round temple. It is located at a height of 38 m from the ground, it has a diameter of about 8 m and it is placed on a 10 m square plan, which has no railing. The main problem was the presence of the temporary metal and wooden structures, placed after the earthquake, that still hold the round temple, preventing further collapses and occupying a large part of the space (Figure 5). These structures, mostly comprising metal pipes, made transportation of the laser scanner instrument to the top of the tower of the tower and its placement difficult. They also disturbed the acquisition of the architecture by reflecting the laser beam, moreover, in the test scans acquired by the extreme corner of the ledge the pipes obstructed the view of several architectural parts. The high balcony was not accessible because of the instability of the round temple stairway and the wooden supporting structure that was obstructing the only access door.

Figure 5. Some pictures of survey moments of the "round temple", on the top of the tower. The images shows the difficulties given by the presence of the metal scaffoldings and the wooden holding structure positioned after the earthquake.

6.1. Image Acquisition of the Top of the Bell Tower

One way to complete the survey of the tower was to use an UAV system. This avoided the use of cranes, which would lead to higher costs and time expenditures. Moreover, the possibility of moving a simple crane would not allow the acquisition of vertical strips without the use of *ad hoc* devices, and it would have been impossible to reach all the acquisition positions needed. Therefore the acquisition was performed in collaboration with Eos Fly (Mantua, Italy) using an octocopter (Figure 6).

Figure 6. The octocopter employed for the bell tower acquisition, and the Wi-Fi camera controller.

The choice to use a multi-copter was made by taking different aspects into account. The first consideration was the type of building: a vertical and very tall structure. UAV allows a vertical flight pattern so it permits the acquisition of vertical strips of images. Another consideration was based on the position of the building, which is in the old town center of the city, surrounded by other buildings. For this reason it was necessary to use an easy to handle vehicle.

The flight device had eight propellers fixed on the same number of arms, two gyroscopes for the flight control and the telemetry instruments (GPS and the barometric altimeter). The octocopter had a flight autonomy of about five to fifteen minutes, depending on the weight loaded on board; it was equipped with LiPo batteries (16 V 4.0 Ah). The octocopter was equipped with a reflex camera (Canon EOS 650D, APS, 18 Mpixel), the camera mount could tilt 90° vertically, from horizontal position to a zenithal one. The RC system controls both the fly operations, the camera rotation and camera trigger. The flying team included the pilot and by a photogrammetric expert able to visualize the camera view on a remotely connected monitor. This was the way to acquire images with the correct point of view and overlap.

The most relevant step was the flight plan. It is important to define the distance from the surface, the overlap between images and, as a consequence, the trajectory. To optimize the acquisition time and reduce the number of photos, the project was optimized taking into account the camera parameters, dimension and characteristics of the building and the surroundings.

The employed camera was a Canon EOS 650D with a CMOS sensor size of 5184 × 3456 pixels (22.3 × 14.9 mm) and 18 mm focal length lens. Each image was acquired with an aperture f/9 and 400 ISO. A maximum pixel size (GSD) on the object of about 3 mm was calculated, which involves an average distance of about 8 meters from the surface. An overlap of about 80% between neighboring images was expected.

The plan (Figure 7) was to acquire three vertical image-strips for each front, completed by two additional strips on the corners, which would permit the connection between adjacent fronts. For the acquisition of the round temple it was planned to realize three 360° flights around it, with a minimum of eight shots, completed with the same number of oblique shots from highest positions and a series of nadir photos.

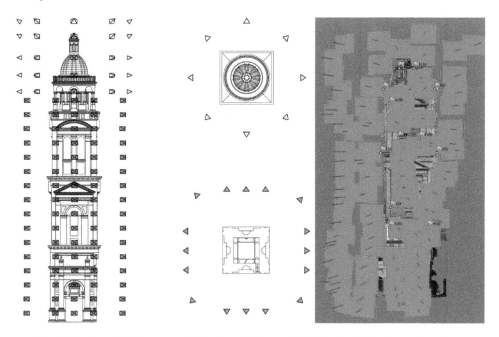

Figure 7. Design of UAV image acquisition. In red the images for the front of the bell tower, in yellow for the round temple. Result of image orientation. The difference between project and real flight are evident.

During the acquisition phase, it is recommended to have the same light conditions in order to have uniform color and illumination in each image. At the same time, shadows should be avoided. In this way, the photogrammetry texture and the orthophoto are uniform and similar in every part of the structure. For this reason, an overcast day was chosen to survey Santa Barbara allowing optimal light conditions.

Overall, 110 photos were taken for the north facade, 19 for the south one, 70 for the east, 83 for the west one and 159 for the round temple. The different numbers of images was due to the different size and position of each façade and the presence of structures. A 77% overlap was obtained with just a vertical baseline of 2.35 m and a horizontal one of 3.48 m. Only the middle strip revealed enough to acquire over the 90% of the facade width, consequently each point on the object's surface was recognized at least in eight photos (Figure 7).

6.2. Data Processing

The laser scanner and topographic surveys were completed in three days of work. As is known in survey operations, the processing phase is longer than acquisition due to the amount of data to process and the manual processing of features extraction: the only laser scanner database comprised 2.3 billion points. For the complete representation of the bell tower, the strategy was to use both a laserscanner and photogrammetric pointclouds. Photogrammetric data were integrated with laser scanner ones for the orthophotos of the lower external parts of the facades, whereas the interiors were drawn only from laser scanner point clouds.

The photogrammetric elaboration was done using Agisoft Photoscan. Due to the high number of images, it was more advisable to divide the project into sub-parts ("chunk" in Photoscan). This allows the elaboration time to be reduced, splitting it onto different PCs. The bell tower exterior model was subdivided into different parts corresponding to every single façade. Overall, five chunks were created, one for the round temple and one for each side.

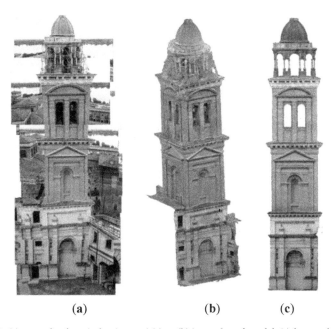

(a)   (b)   (c)

Figure 8. (a) example of vertical strip acquisition; (b) image-based model; (c) front orthophoto.

Image orientation was done using physical points on the structure whose coordinates were extracted manually on the laser scans, as previous tests validated this operation. After calculation we verified that the average GSD of all the images was about 0.003 m. The mean error in camera alignment was about 1.8 pixels meters, after optimization. The rototranslation of the model for the georeferencing in the local reference system showed an error of about 0.011 m.

As result, dense cloud and mesh models were exported from Agisoft Photoscan. The first one was useful to integrate the scanner data after noise reduction and outliers removal. Mesh models were used to extract high-resolution orthophoto (Figure 8c) even if a healing process was necessary. This optimization was done using Rapidform XOR3.

To obtain a better orthophoto in terms of image quality (coherence of lights and shadows on the facades) not all images were used. Images with strong chromatic variations and especially non-parallel to the surface images were discarded. For the orthophotos of the bell tower facades we used fewer photographs: 125 images for the South front, 83 images for the north face, 69 images for the east, 99 images for the west (from the south-west front in clockwise). The round temple orthophoto was built with 154 images.

The first operation for the laser scanner data was alignment in the local reference system made by topography. The entire database of points acquired by laser scanner is divided into internal data

13

(about 1.5 billion) and external ones (800 million only for the lower external part). The oriented point clouds were cleaned up from overlapping redundant data, noise points and outliers filtered away. The presence of a huge scaffold structure made this kind of work extremely time consuming because it was necessary to isolate the building point data from the scaffold data.

Point clouds and orthophotos were used together to realize the documentation of the actual post-earthquake situation [9]. According to the Cultural Heritage office of Mantua Diocese, some traditional architectural representations were drawn at a scale of 1:50: a plant for each tower floor with three additional horizontal sections in correspondence to the most damaged parts, two vertical sections of the complete structure and fronts integrated with orthophotos (Figures 9–11).

Figure 9. Comparison between the Santa Barbara bell tower's top laser scanner data and the respective surface orthophoto elaborated from the photogrammetric 3D model.

Laser scanner point clouds were used to draw the plants at different floors and the interior part of the sections. According to the known pipeline, point clouds were imported in AutoCad and used as the basis for the drawings. The remaining part of the sections and the exteriors façades were drawn by using both laser scanner and photogrammetric point clouds. After bell tower restitutions the front's drawings were overlapped with orthophotos to integrate vector information with color data.

Figure 10. 3D model of the round temple for FEM analysis.

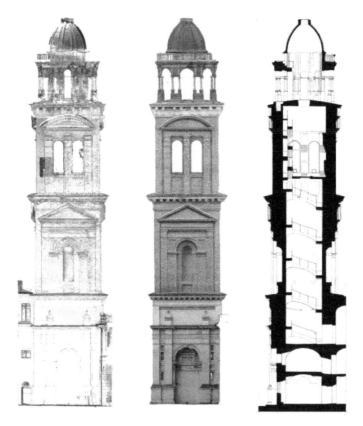

Figure 11. Santa Barbara bell tower. Laser scanning and photogrammetry data and a final elaboration of a vertical section obtained using both the technologies.

For the round temple and the lantern, the most damaged parts of the belfry, in addition to the 3D drawings (Figure 9) the Diocese also asked for a 3D model. The project of restoration and seismic retrofitting was very complex, therefore it was necessary to provide more information. The construction engineers needed a 3D model for FEM analysis.

The 3D model was built by using laser scanner and photogrammetric data together. According to the requirements of structural engineers, the 3D model was built as a solid model using simple geometric primitives modified with Boolean operations (Figure 10).

7. Conclusions

This paper focuses on the survey of vertical structures in dense urban contexts in the case of emergency conditions. This experience proves that UAV acquisition and automated 3D image modeling are an effective way to achieve fast and precise results. To reach good results a correct flight project for the UAV acquisition is fundamental, in order to obtain a photogrammetric model with good image overlap and the optimal number of images to acquire all the architecture information avoiding overabundances. This technology is cheaper than TLS surveys thanks to the use of non-metric cameras and the use of automated software for the elaboration process, which minimizes human intervention. Above all, this approach allows the problem of surveying tall buildings in old town centers or congested areas to be solved effectively.

This methodology is very promising also because it can be fully integrated with other kinds of technology and instruments that proved their great efficiency in other similar cases of emergency (TLS and topography). The results of accuracy tests allow data from UAV automated photogrammetry to be fully integrated with laser scanner point clouds. From the database, based on all data, many representations can be extracted, both vectorial (plans, sections) and raster (facade orthophoto).

This integration of sensors ensures not only the quality of results, but also a safe working environment for the surveyor, even in post-earthquake cases. This operational methodology is fully integratable into the architectural process of damage assessment and planning of restoration or maintenance.

Acknowledgments: The authors would like to thank Monsignor Giancarlo Manzoli, episcopal delegate for Cultural Heritage of the Diocese of Mantua, Arch. Alessandro Campera, Office for the Ecclesiastical Cultural Heritage of the Diocese, for their willingness and courtesy.

Author Contributions: The work presented here was carried out in collaboration among all authors. All authors have contributed to, seen and approved the manuscript.

Conflicts of Interest: The authors declare no conflict of interest.

References

1. Manfredini, A.M.; Remondino, F. A Review of Reality-Based 3D Model Generation, Segmentation and Web-Based Visualization Methods. *Int. J. Herit. Digit. Era* **2012**, *1*, 103–123. [CrossRef]
2. Achille, C.; Fassi, F.; Fregonese, L. 4 Years history: From 2D to BIM for CH: The main spire on Milan Cathedral. In Proceedings of the 2012 18th International Conference on ISEE Virtual Systems and Multimedia [VSMM], Milan, Italy, 2–6 September 2012; pp. 377–382.
3. Fassi, F.; Achille, C.; Fregonese, L. Surveying and modeling the main spire of Milan Cathedral using multiple data sources. *Photogramm. Rec.* **2011**, *26*, 462–487. [CrossRef]
4. Remondino, F.; Girardi, S.; Rizzi, A.; Gonzo, L. 3D Modeling of Complex and Detailed Cultural Heritage Using Multi-Resolution Data. *ACM J. Comput. Cult. Herit.* **2009**, *2*. [CrossRef]
5. Fregonese, L.; Scaioni, M.; Taffurelli, L. Generation of a Spatial Information System for architecture with laserscanning data. In Proceedings of the International Archives of the Photogrammetry, Remote Sensing And Spatial Information Sciences, Paris, France, 1–2 September 2009; Volume XXXVIII-3/W8, pp. 87–92.
6. Fregonese, L.; Barbieri, G.; Biolzi, L.; Bocciarelli, M.; Frigeri, A.; Taffurelli, L. Surveying and Monitoring for Vulnerability Assessment of an Ancient Building. *Sensors* **2013**, *13*, 9747–9773. [CrossRef] [PubMed]
7. Barbieri, G.; Biolzi, L.; Bocciarelli, M.; Fregonese, L.; Frigeri, A. Assessing the seismic vulnerability of a historical building. *Eng. Struct.* **2013**, *57*, 523–535. [CrossRef]
8. Guarnieri, A.; Milan, N.; Vettore, A. Monitoring of Complex Structure for Structural Control Using Terrestrial Laser Scanning (Tls) and Photogrammetry. *Int. J. Arch.* **2013**, *7*, 54–67. [CrossRef]
9. Pilot, L.; Monti, C.; Balletti, C.; Guerra, F. Il Rilievo del Torrazzo di Cremona. In Proceedings of the Atti della 2° Conferenza Nazionale Della Federazione Della ASITA "Rilevamento, Rappresentazione e Gestione dei Dati Territoriali e Ambientali", Bolzano, Italy, 24–27 November 1998.
10. Fassi, F.; Fregonese, L.; Ackermann, S.; de Troia, V. Comparison between laser scanning and automated 3D modeling techniques to reconstruct complex and extensive cultural heritage areas. In Proceedings of the International Archives of the Photogrammetry, Remote Sensing and Spatial Information Sciences, Trento, Italy, 25–26 February 2013; Volume 40-5/W1, pp. 73–80.
11. Remondino, F.; del Pizzo, S.; Kersten, T.P.; Troisi, S. Low-Cost and Open-Source Solutions for Automated Image Orientation—A Critical Overview. In Proceedings of the EuroMed 2012, Limassol, Cyprus, 29 October–3 November 2012; pp. 40–54.
12. Guidi, G.; Remondino, F. 3D modeling from real data. In *Modeling and Simulation in Engineering*; Alexandru, C., Ed.; InTech: Rijeka, Croatia, 2012; pp. 69–102.
13. Rinaudo, F.; Chiabrando, F.; Lingua, A.; Spanò, A. Archaeological site monitoring: UAV photogrammetry can be an answer. In Proceedings of the International Archives of the Photogrammetry, Remote Sensing and Spatial Information Sciences, Melbourne, Australia, 25 August–1 September 2012; Volume 39-B5, pp. 383–388.

14. Remondino, F.; Spera, M.G.; Nocerino, E.; Menna, F.; Nex, F. State of the art in high density image matching. *Photogramm. Rec.* **2014**, *29*, 144–166. [CrossRef]

15. Fiorillo, F.; Jiménez Fernàndez-Palacios, B.; Remondino, F.; Barba, S. 3D Surveying and modeling of the archeological area of Paestum, Italy. *Virtual Archaeol. Rev.* **2013**, *4*, 55–60.

16. UVS International. Available online: http://www.uvs-international.org/ (accessed on 4 February 2015).

17. Remondino, F.; Nex, F.; Sarazzi, D. Piattaforme UAV per applicazioni geomatiche. *GEOmedia* **2011**, *6*, 28–32.

18. Agisoft Photoscan Software. Available online: http://www.agisoft.com/ (accessed on 4 February 2015).

19. Cloudcompare Software. Available online: http://www.danielgm.net/cc/ (accessed on 4 February 2015).

20. Fassi, F. 3D modeling of complex architecture integrating different techniques—A critical overview. In Proceedings of the International Archives of the Photogrammetry, Remote Sensing and Spatial Information Sciences, Zurich, Switzerland, 12–13 July 2007; Volume 36-5/W47, p. 11.

21. Fassi, F.; Achille, C.; Gaudio, F.; Fregonese, L. Integrated strategies for the modeling of very large, complex architecture. In Proceedings of the International Archives of the Photogrammetry, Remote Sensing and Spatial Information Sciences, Trento, Italy, 2–4 March 2011; Volume 38-5/W16, pp. 105–112.

22. Lingua, A.; Piumatti, P.; Rinaudo, F. Digital photogrammetry: A standard approach to cultural heritage survey. In Proceedings of the International Archives of the Photogrammetry, Remote Sensing and Spatial Information Sciences, Ancona, Italy, 1–3 July 2003; Volume 34-5/W12, pp. 210–215.

Article

Towards the Development of a Low Cost Airborne Sensing System to Monitor Dust Particles after Blasting at Open-Pit Mine Sites

Miguel Alvarado [1],*, Felipe Gonzalez [2], Andrew Fletcher [3] and Ashray Doshi [4]

1   Centre for Mined Land Rehabilitation, Sustainable Mineral Institute, The University of Queensland, Brisbane 4072, Australia

2   Science and Engineering Faculty, Queensland University of Technology (QUT), Brisbane 4000, Australia; felipe.gonzalez@qut.edu.au

3   Centre for Mined Land Rehabilitation, Sustainable Mineral Institute, The University of Queensland, Brisbane 4072, Australia; a.fletcher@cmlr.uq.edu.au

4   Faculty of Engineering, Architecture and Information Technology, School of Information Technology and Electrical Engineering, The University of Queensland, St. Lucia 4072, Australia; ashraydoshi@gmail.com

*   Author to whom correspondence should be addressed; m.alvaradomolina@uq.edu.au; Tel.: +61-7-3346-4027.

Academic Editor: Vittorio M. N. Passaro

Received: 26 May 2015; Accepted: 6 August 2015; Published: 12 August 2015

Abstract: Blasting is an integral part of large-scale open cut mining that often occurs in close proximity to population centers and often results in the emission of particulate material and gases potentially hazardous to health. Current air quality monitoring methods rely on limited numbers of fixed sampling locations to validate a complex fluid environment and collect sufficient data to confirm model effectiveness. This paper describes the development of a methodology to address the need of a more precise approach that is capable of characterizing blasting plumes in near-real time. The integration of the system required the modification and integration of an opto-electrical dust sensor, SHARP GP2Y10, into a small fixed-wing and multi-rotor copter, resulting in the collection of data streamed during flight. The paper also describes the calibration of the optical sensor with an industry grade dust-monitoring device, Dusttrak 8520, demonstrating a high correlation between them, with correlation coefficients (R^2) greater than 0.9. The laboratory and field tests demonstrate the feasibility of coupling the sensor with the UAVs. However, further work must be done in the areas of sensor selection and calibration as well as flight planning.

Keywords: PM10; monitoring; blasting; fixed-wing UAV; quadcopter; optical sensor

1. Introduction

The mining and coal seam gas industries in Australia and around the world are important economic activities. Coal exports from Queensland from March 2013 to March 2014 totaled more than $24.5b [1]. These activities generate particles and gases such as methane (CH_4), carbon dioxide (CO_2), nitrogen oxides (NO_x), and sulfur oxides (SO_x) that have potentially dangerous environmental and health impacts.

Blasting in particular includes effects such as airblast, ground vibration, flyrock, toxic gases and particulate matter [2,3]. Particulate matter, aerosols, ammonia, carbon dioxide (CO_2), nitrogen, nitrogen oxides (NO_x) and sulfur oxides (SO_x) are the primary residues produced by blasting events at mining sites. In an ideal situation, the exothermic reaction produces CO_2, water vapor and molecular nitrogen (N_2); however, due to environmental and technical factors, other noxious gases are often produced in a range of concentrations [4].

In this paper, we propose the use of small unmanned aerial vehicles (UAV) carrying air quality sensors to allow precise characterization of blasting plumes in near-real time. This approach may lead to actionable data for harm avoidance or minimization. Most pollution dispersion models use predefined estimates of pollution sources and atmospheric conditions; near-real time information from within the plume has been practically impossible to collect. Flight instrument data transmitted as telemetry from the UAV provides high resolution instantaneous micrometeorological data that can assist interpretation of concentrations detected by on-board air quality sensors. In addition, this information including location, micrometeorological data and air quality, can be delivered in real time to analytical software. The data stream may therefore be used to feed flight path-planning algorithms or atmospheric dispersion models in near-real time.

In order to assess this approach, fixed-wing and multi-rotor UAVs were used. These UAVs were developed at The University of Queensland for ecological investigations. The platforms were capable of autonomous predetermined flight path planning or semi-autonomous direction. These platforms have weight restrictions and require sensors with high temporal sampling resolution (<1 s that can be digitally sampled but allow air quality sensors to be integrated and tested. In this paper we tested light-emitting diode (LED)-based optical sensors due to the combination of essential characteristics including rapid response, light weight and ease of data digitization. To date, two dust sensors have been tested with the UAV (SHARP GP2Y10 and Samyoung DSM501A) [5,6].

Characterizing blasting plumes and predicting dispersion using this approach requires integration of a number of factors:

- Development or modification of micro UAV platforms that can be safely operated near active mine blasts.
- Identification of sensors with necessary sampling rates (<1 s), weight (<500 g), data output format and sufficient sensitivity (1 mg/m^3 PM10).
- System endurance sufficient to capture plume evolution and dynamics (>20 min).
- Integration and formatting of data streams necessary for mathematical predictive models via live telemetry.

This project aims to develop tools that inform, cross calibrate and validate plume models for particulate and gaseous pollutants associated with blasting activities.

This paper is organized as follows: Section 2 reviews current methods to monitor blasting plumes, dust and gases after blasting at open-cast mine sites, and the use of UAVs for environmental monitoring and modeling approaches; Section 3 describes the current sensing system that has been developed; Section 4 describes progress in the integration of the dust sensor system with UAVs and flight testing; and, Section 5 outlines current conclusions and further work.

2. Blast-Associated Air Sampling

2.1. Methods to Monitor Blasting Plumes

Blast-associated dust is a significant potential hazard, and novel monitoring methods are continuously explored. Roy *et al.* developed a multi-platform system using ground-based dust samplers in combination with balloon-carried samplers near open pit mines. The data collected informed multiple regression and neural network models how to monitor and predict the drifting of blast plumes [7,8]. As samplers were static during blasting, this approach required detailed site-blasting plans and favorable weather conditions to determine their interconnectivity. Under this configuration, neural network models performed better than multiple regression models in predicting outcomes [9].

Furthermore, fugitive NO$_2$ and PM10 emissions of coal mining in the Hunter Valley, Australia have been examined using gravimetric and LIDAR methods. LIDAR provided long-path laser-integrated concentration signal with very low limit of detection, but required a fixed location [10]. Attalla *et al.* used a different approach by implementing NDIR (non-dispersion infrared) and mini-DOAS (differential optical absorption spectroscopy) for prediction of NOx and other pollutant gases. This method also required a fixed-location ground-based sensing apparatus. Modelling in AFTOX (Air Force toxics model) resulted in overestimation of plume concentrations at a distance [4].

Richardson (2013) assessed particulate fractions using a scintillation probe dust sensor (Environmental Beta Attenuation Monitors—EBAMs) and a real-time laser photometer (Dusttrak) in Hunter Valley and Central Queensland (Goonyella Riverside) and confirmed that PM2.5 is a small fraction of the overall suspended blast-associated particles, while PM10 is dominant [11].

2.2. Dust Sampling Sensors

The method and type of sensors used to measure contaminant gas or dust emissions will vary according to the type of emission, concentration range of concern, and required response time. Sensors are commonly based on ultrasound, optical, and electrochemical sensing elements [12–14]. These sensors can either be handheld, installed in vehicles, or form ground-based network systems. Table 1 shows different examples of sensors and their characteristics classified by the way they are implemented. Network systems are very useful when specific receptors or areas are to be monitored [15,16]. However, effective monitoring diameter, costs of installation, operation and maintenance are important considerations that may limit their use and procurement.

Table 1. Example of sensing technology used for monitoring gases in the mining, oil and gas industries.

Instrument	Description	Gases/Particles	Characteristics
Handheld			
Dräger X-am 5600 [17]	Compact instrument for the measurement of up to 6 gases; complies with standard IP67; IR sensor for CO_2 and electrochemical for other gases.	O_2, Cl_2, CO, CO_2, H_2, H_2S, HCN, NH_3, NO, NO_2, PH_3, SO_2, O_3, Amine, Odorant, $COCl_2$ and organic vapors.	Dimensions: $4.7 \times 13.0 \times 4.4$ cm Weight: 250 g
Installed in ground vehicles			
Picarro Surveyor [18,19]	Cavity ring-down spectroscopy (CRDS) technology, sensitivity down to parts-per-billion (ppb); survey gas at traffic speeds and map results in real time; real-time analysis to distinguish natural gas and other biogenic sources.	CO_2, CO, CH_4, and water vapor	Dimensions: Analyzer $43.2 \times 17.8 \times 44.6$ cm; external pump $19 \times 10.2 \times 28.0$ cm Weight: 24 kg + vehicle Power: 100–240 VAC
Stationary			
Tapered Element Oscillating Microbalance (TEOM) [20,21].	Continuous particle monitoring. The tapered element consists of a filter cartridge installed on the tip of a hollow glass tube. Additional weight from particles that collect on the filter changes the frequency at which the tube oscillates.	Total suspended particles (TSP), PM10, PM2.5	Dimensions: $43.2 \times 48.3 \times 127.0$ cm) Weight: 34 kg Power: 100–240 VAC
Networks			
AQMesh [22]	Wireless monitor; high sensitivity (levels to ppb); designed to work through a network of arrayed monitors.	NO, NO_2, O_3, CO, SO_2, humidity and atmospheric pressure.	Dimensions: $17.0 \times 18.0 \times 14.0$ cm Weight: <2 kg Power: LiPo batteries
Airborne			
Yellow scan [23]	LIDAR technology with a total weight of 2.2 kg; 80,000 shots/s; resolution of 4 cm; class 1 laser at 905 nm.	Dust and aerosols.	Dimensions: $17.2 \times 20.6 \times 4.7$ cm Weight: 2.2 kg Power: 20 W

A complex criteria matrix must be considered when selecting an airborne sensor to monitor blast plumes. Factors include dimensions (weight and size); tolerance of vibration and movement given mounting on a UAV platform (up to 15 m/s); concentration range of sensor as well as the accuracy and limitations of the sensor (e.g., response time, mean square deviation, calibration, interference of other gases, humidity and temperature).

Optical LED particulate sensors are potentially suitable devices that could be used to explore the proposed system. LED sensors have the advantages of low power consumption, high durability, compact size, and easy handling and have been tested as a reliable source light for DOAS [24]. They also have demonstrated the ability to reduce internal stray light and can be used as a light source for applications requiring numerous kilometers of total light path [25]. Several other authors have also highlighted their advantages over other types of sensors [26,27].

2.3. Use of Unmanned Aerial Vehicles (UAVs) for Environmental Monitoring

Researchers are identifying advantages of UAVs to undertake investigations in difficult terrain/landscape areas, where health and safety risks exist, or where there is a lack of resources (human and/or economic). Gas sensing with small-micro UAVs using electrochemical and optical sensors is still not well established due to rapid development of sensors and UAVs. Sensing of CO_2, CH_4 and water vapor [28], NO_2 and NH_3 [29], ethanol and CH_4 [30,31], have been conducted using rotary-winged platforms. Watai *et al.* [32], used a kite plane to monitor CO_2 using a NDIR gas analyzer which had a response time of 20 s. A spectrum-specific video camera has been developed to visualize SO_2 emissions from volcanoes by Brown *et al.* [33]. Lega *et al.* integrated a multi-rotor-sensing platform that monitors air pollutants in real time and provides 3D visualization [34]. Several variations of this platform, StillFly and BiLIFT, detect gases like CO, C_6H_6, NO_2, O_3, SO_2, NO_X and PM10, as well as thermal IR images to detect sewage discharges along the coastline of Italy [34,35]. Fixed-wing systems capable of achieving real time monitoring and providing indexed-linked samples are also currently possible [34]. Target sampling locations and source scales will be important to platform and sensor selection due to the fundamental differences between fixed-wing and multi-rotor UAVs, such as hovering capacity, endurance, and flight envelope.

Other approaches have been taken to characterize and track fugitive emission contamination plumes and register their concentrations [4,7,29,36–39]. However, to the authors' knowledge, UAVs have never been used to understand dust or gas emission associated with mine blasting. Small UAV platforms and real-time air quality sampling impose a number of novel and complex sampling requirements that still need to be addressed:

- Rapid sensor response time is important in mobile sensing platforms that move relative to both air and ground.
- Limited power requires flight efficiency and sensors with low power consumption.
- Sensors that require extended duration equilibration times (~20 s) require sampling chambers, and delayed response times result in difficult or impossible flight performance required to return to the estimated plume location.
- Multi-rotor platforms operate via GPS, and thus approximate a ground-based sampler independent of the fluid it is sampling. Fixed-wing platforms move through a defined volume of air in a given time regardless of ground location.
- Moving platforms require flow control for sample chambers to ensure calibrated values can be reported.
- The use of mathematical models is an essential element when monitoring air quality and atmospheric contamination. Defining appropriate models given the type of data collected is important. There are several approaches commonly used for air (emission factors, Gaussian, Lagrangian, Eulerian, *etc.*), each of which have limitations to their performance [40–42].
- Improve data visualization since current attempts to map pollutants in the atmosphere are presented as snapshots, not as a dynamic environment with concentration measurements that change before and after the moment a reading is produced.

3. Design of Sensing System

The sensor system consists of a gas-sensing node, the UAV, and a data integration and visualization interface.

3.1. System Architecture

The system architecture for the fixed-wing UAV with integrated dust sensor (Figure 1) and a multi-rotor carrying a telemetered dust sensor (Figure 2) are necessarily different due to the use of different autopilots. Micro meteorological data deduced from UAV platform flight control is a novel and detailed source of data for interpretation of air quality measurements. However, it requires integration with gas sensor data to allow a meaningful application.

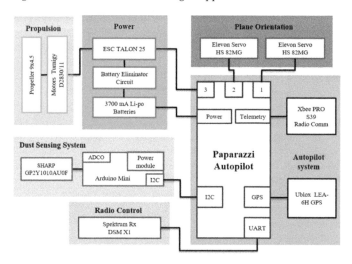

Figure 1. System architecture for the fixed-wing UAV with dust sensor.

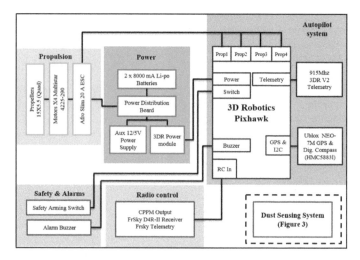

Figure 2. System architecture for quadcopter UAV with independent gas-sensing system.

3.2. Gas-Sensing Node

Two sensors were experimentally assessed to date: GP2Y10 (SHARP) and DSM501A (Samyoung) for PM10. The SHARP and Samyoung sensors tested were connected through an Arduino microcontroller to integrate sensor-telemetry data streams. The sensor and associated electronics are low weight and constrained size to allow simple installation in other multi-rotor or fixed-wing platforms. The dust-sensing module was placed on the top side of the quadrotor platform to minimize high velocity air flow that is fundamental to similar quadcopters [43,44]. Figure 3 shows the system architecture for the modular dust sensor in detail.

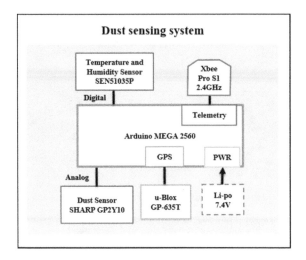

Figure 3. System architecture for the modular dust sensor.

The system is constructed around an Arduino MEGA 2560, powered by a 7.4 V lithium polymer battery, data telemetry is via XBee Pro S1 (2.4 GHz) radio transmitter while a GP-635T provides a timestamp for serial port data. Sensors include a SEN51035P temperature and humidity sensor and GP2Y10 SHARP dust sensor (Figure 3). All data was transmitted and logged on a ground station which displays received raw values and PM10 concentration readings in real time.

3.3. UAV Platforms

Both a fixed-wing and a multi-rotor UAV were selected to develop the sensing system. These aerial vehicles operate in fundamentally different ways with fixed-wing UAVs traversing a set volume of air in a given time while the hovering ability of rotary-winged UAVs allow collection of data at specific locations in space and time; however, they experience wind.

The suitability of three fixed-wing platforms was considered for integration with the optical and/or electrochemical sensors. All UAVs are constructed of expanded polypropylene (EPP) and composite materials that have been demonstrated as safe and robust platforms in the mining industry environment. Specifications for the models considered during this investigation are provided in Table 2 and Figure 4.

Table 2. Characteristics of UAVs identified as feasible platforms for this investigation.

Model	Wingspan (mm)	Length (mm)	Flying Weight (g)	Endurance (min)	Approx. Payload [3] (g)
Teklite [1]	900	575	900–950	45	200
GoSurv [2]	850	350	900–1200	50	>300
Swamp Fox [45] [1]	1800	1000	4500	40	1000

[1] Commercially available platform; [2] Fixed-wing platform designed at UQ SMI-CMLR; [3] Determined through experimental procedures.

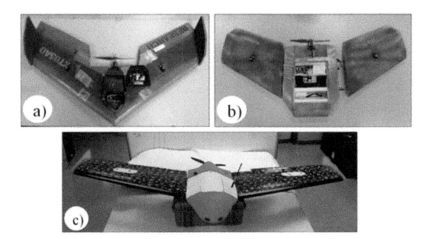

Figure 4. Fixed-wing UAV platforms, (a) Teklite; (b) GoSurv; and (c) Swamp Fox.

All UAVs listed in Table 2 have low kinetic energy (<50 joules) and low air speed (<60 km/h). Low kinetic energy and speed improve safety and simplifies data acquisition performance but require reasonably calm conditions to operate. All have a pusher-propeller design providing access to clean airflow for sensors. The Paparazzi autopilot used on Teklite and GoSurv records altitude, platform coordinates, speed and direction. They also estimate wind speed and direction by response difference. The autopilot of the Swampfox platform records airspeed using a pitot tube, as well as speed and direction with the GPS. Air speed and geolocation data are drawn from the autopilot telemetry that is integral to all small UAV operations.

The Teklite was selected as the best platform for the type of test to be conducted due to its portability, ease of integration of sensors, successful flight testing, light weight and low (<100 ft) target flight altitude. The UAV is controlled using a ground control station. The flight plan is preloaded from the ground station that displays the flight parameters of the UAV, flight route and atmospheric pollution readings in real time. The flight plan can also be modified manually using a handheld radio transmitter and/or by altering the parameters through the ground station.

The UAV can be flown from as far as 1.5 km from the ground control station. Weather conditions (wind speed and direction, temperature, *etc.*) are used to pre-plan the UAV flight path to follow and characterize the blasting plume. If required, a flight path can be modified and uploaded into the ground station based on post-blasting observations. The UAV is restricted to fly more than 35 m above ground level as a safety factor to avoid collision with trees or infrastructure.

A multi-rotor platform was used to record readings below 35 m above ground level. The system was designed for agricultural and air-monitoring surveys. Figure 5 shows the quadcopter integrated with the modular dust sensor. The multi-rotor platform has an average flight time of 20 min and a total weight of 2.5 kg (with batteries). The modular dust sensor had a total weight of 150 g and was placed inside a plywood case.

Figure 5. Quadcopter and modular gas-sensor system integrated.

4. Bench Testing of Optical Sensors

A gas chamber (see Figure 6) based on the work of Budde *et al.* [46] was constructed in order to expose the sensor node to different concentrations of particles and compare the readings against a calibrated dust-monitoring device—Dusttrak 8520. The Dusttrak has a response time of 1 s, a resolution of 0.001 mg/m^3 and is capable of monitoring PM10 and PM2.5. Smoke from standard incense sticks was used as an airborne particulate source. The Samyoung sensor produced a low correlation coefficient (R^2) of 0.5 and was therefore deemed an unsuitable option for the integration of the dust-monitoring module and UAVs.

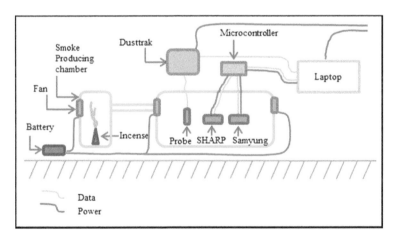

Figure 6. Gas chamber for sensor testing and calibration.

Results for SHARP (GP2Y10)

Tests for PM10 and PM2.5 where undertaken for the SHARP dust sensor. An initial data collection test was used to correlate the raw values obtained from the sensor, which is the voltage modified by the light absorption of the receiver, with the values registered by the Dusttrak (see Figure 7). A linear and second-degree calibration equation, with correlation coefficients greater than 0.9, were obtained and applied to the sensor data.

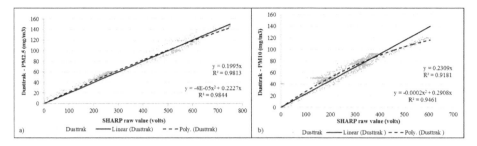

Figure 7. Correlation of raw values obtained with SHARP sensor for (**a**) PM2.5 and (**b**) PM10 *vs.* readings collected with Dusttrak (mg/m³).

The original algorithm of M. Chardon and Trefois [47] developed to use the SHARP sensor with an Arduino board was modified to take readings every second. The objective of this test was to check that the data collected by the SHARP sensor was comparable to the Dusttrak readings, results are shown in Figure 8. The offset observed in the initial test was reduced having a satisfactory match between sensors. Percentage errors were calculated obtaining 38.0% and 13.6% for PM10 linear and quadratic fits respectively. PM2.5 errors were 11.9% and 9.96% for linear and quadratic fits respectively.

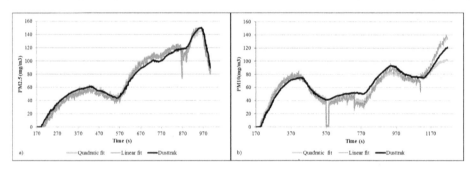

Figure 8. Linear and quadratic linear fit for raw SHARP values of (**a**) PM2.5 and (**b**) PM10 particle concentrations.

A third test was done using two SHARP sensors and the Dusttrak (Figure 9). An offset between the SHARP sensors was also observed; however this error was reduced after correlating data with the linear equation fitted previously. For this test the linear fit produced a lower percentage error for SHARP A and B, of 19.3% and 12.5% respectively, however the second degree fit produced very similar results with errors of SHARP A:21.5% and SHARP B:14.9%.

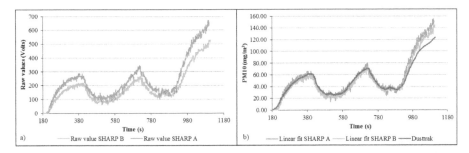

Figure 9. Dual SHARP and Dusttrak test showing (**a**) raw values data and (**b**) corrected particle measurements against Dusttrak readings.

Sensor variability due to temperature changes [5,30], was not considered for the experiment, however it will be undertaken in further tests.

5. Flight Test

5.1. SHARP Sensor Integrated to Fixed-Wing UAV

The air intake and discharge were modified to produce a continuous flow inside the SHARP sensor chamber (see Figure 10). Air sampling intake was through a carbon fiber scooped cowl on the top surface of the wing directly over the sensor inlet. Sample exhaust was through a 4 mm tube attached to the sensor outlet and extended through the lower surface of the platform.

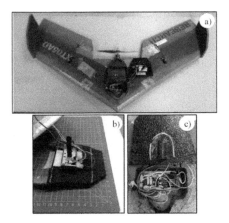

Figure 10. Modifications made to Teklite and SHARP sensor for flight, (**a**) Teklite UAV and SHARP sensor; (**b**) Air outlet for SHARP sensor; (**c**) Air intake for SHARP sensor.

5.2. Test 1: Sensor Integration

Several flights were made to test the feasibility of integrating the SHARP sensor with the Teklite platform. The first test was conducted on 6 June 2014 in order to evaluate the integration of the system. The test used a fire in an open area as an airborne particulate source. The UAV was programmed to fly around the fire for approximately 30 min. Data collected from the UAV and the air quality sensor are shown in Figures 11 and 12. The data did not report variations in particulate matter concentration in the atmosphere, as it is observed in Figure 12. Analysis of the data indicated that electrical noise caused by motor and onboard electronics was interfering with the output.

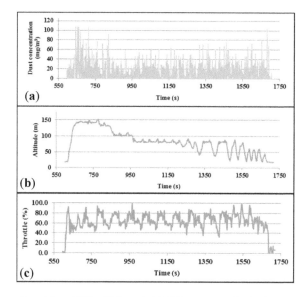

Figure 11. Data collected from Teklite flight with SHARP sensor attached. (**a**) Dust concentration; (**b**) Altitude; (**c**) Throttle.

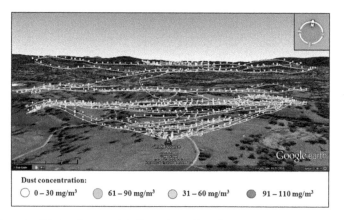

Dust concentration:

○ 0 – 30 mg/m³   ◐ 61 – 90 mg/m³   ◔ 31 – 60 mg/m³   ● 91 – 110 mg/m³

Figure 12. 3D visualization of the Test 1 data collected with Teklite- SHARP sensor.

High frequency noise consistent with electrical switching of motors and servos was filtered by installing a 50 V (0.1 μF) capacitor to the power source.

5.3. *Test 2: PM10 Monitoring*

A second field test was conducted on 13 October 2014 using "talcum powder" lifted into the atmosphere using a petrol-powered leaf blower (STIHL BG 56 Blower—max of 730 m³/h). Talcum powder was used due to its safe handling and availability. Talcum powder is composed of 0.2–0.3 mass fraction with a particle diameter no greater than 10 μm [48,49].

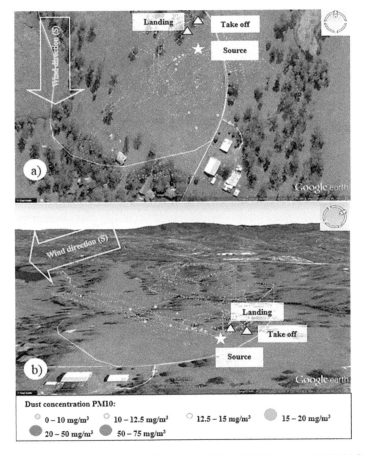

Figure 13. 3D visualization of Test 2 data collected with Teklite-SHARP sensor for PM10 (**a**) Overview and (**b**) Side view.

In order to determine PM10 concentrations measured during the flight, the data was processed using the particle correlation (Figure 7) obtained from laboratory testing of the SHARP sensors. Figure 13 shows the distribution of PM10 concentrations in the atmosphere registered by the optical sensor by using top and side 3D visualization of the particulate plume. The wind direction was towards the west−southwest and concentrations ranged from 15 mg/m^3 to 66 mg/m^3, describing the shape of the plume when dispersed by the wind.

Tests 1 and 2 demonstrated the functionality and feasible integration of the system; however, the need for systematic characterization of a particulate plume of known composition and size remained. This is required to demonstrate the ability to calculate particulate emission rates, as most parameters can be independently measured using a constant powder emission, constant emission rate, known atmospheric conditions and particle size distribution of the source.

To achieve a systematic plume characterization, it was necessary that the UAV reproduced a fixed experimental flight pattern to aid spatial calculations and also exclude biased measurements that could easily be made when flying manually into the visible plume produced by the powder ejected. A flight path consisting of concentric circles at different heights and radius was planned for Test 3. The flight path ensures the UAV covers the designated area around the source. This ensures that the sensor intersects the plume and tests the ability of the data to describe the behavior of the plume in the air space surrounding it.

5.4. Test 3: Mixed Fixed and Rotary Wings

Test 3 was undertaken on 3 March 2015. The setup for the field experiment was based on Test 2, incorporating modifications to satisfy UAV flight and rigorous plume modelling requirements. The fixed-wing and multi-rotor UAVs were able to fly following the patterns programmed for the tests. Table 3 shows the radius and heights used for the test. These parameters were defined according to the capabilities of each UAV and to collect complementary datasets at two spatial scales.

Table 3. Programmed flight parameters and UAV capabilities.

Parameters	Quadcopter	Fixed-Wing
Max. Height *	120 m	120 m
Max. Radius	100 m	200 m
Programmed Heights (MAGL)	7, 14, 21	35, 45, 55
Programmed Radius	5, 15, 35	45, 55, 65, 75, 85

* Determined by UAV height flight restrictions [50].

The talcum powder plume was generated using a petrol-powered fan connected to a 5.5 m long and 0.05 m diameter PVC stack. The powder was loaded into the airstream through an intersection custom made for the powder containers at an approximate rate of 300 g/min (Figure 14).

Figure 14. Powder ejection system setup.

For Test 3, the SHARP sensors where recalibrated due to the different characteristics including color and particle diameter that smoke and talcum powder have. The calibration procedure previously used for smoke particles was repeated for the talcum powder. A correlation equation was calculated using a polynomial fit by processing the data obtained with the Dusttrak 8520 and with the SHARP sensor (Figure 15). Integrated datasets from each platform were post-processed to visualize the concentrations measured by the fixed-wing and quadcopter during experimental flights (Figure 16).

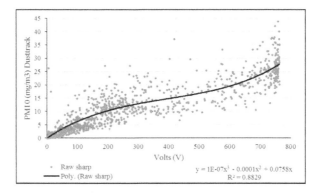

Figure 15. Correlation between talcum powder particles and raw value readings from the SHARP sensor and Dusttrak.

PM10 concentration values ranged from 0.5 mg/m^3 to 19 mg/m^3 and their distribution described the path followed by the powder plume to the west, downwind from the source (Figure 16a,b). Mid-range concentrations to the east (downwind) and north of the source are likely the result of petrol motor exhaust particles and potentially spilled talcum powder. Experimental equipment modification using battery-powered fans and venture effect powder loading are being developed.

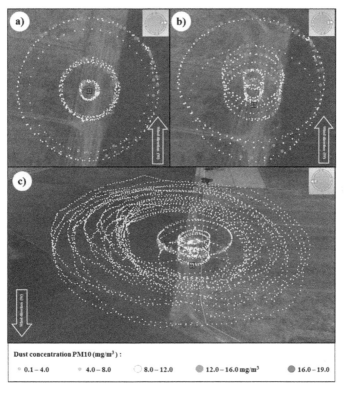

Figure 16. Flight path and PM10 concentrations monitored with the UAV quadcopter (**a**) top view and (**b**) side view; and (**c**) fixed-wing and quadcopter (overlapped flights).

Future tests will also include measurements of background levels during flight monitoring periods to determine their influence in the UAV readings.

For safety reasons and the complexity involved in flying two UAVs simultaneously, the quadcopter and the fixed-wing UAVs were not flown simultaneously. The fixed-wing UAV was flown after the quadcopter and recorded maximum concentrations of 2.0 mg/m^3 without an observable pattern (Figure 16c). Weather conditions with wind speed ranging from 7 to 9 m/s prevented the powder plume rising to the minimum programmed height of 35 m; therefore, it is unlikely that detectable particulates associated with the plume were present.

Figure 17 shows the contour plots of the powder distribution at a height of 18 m above ground level and 30 m to the west of the source. The contour plots together with the volume rendering produced with the software Voxler aid in the interpretation of the data. They produce a model of the plume which can be challenging to interpret when plotting all readings independently, due to the high density of information. Higher concentrations of PM10 particles are shown in red color which are located in the western side of the source located at the center of the plot.

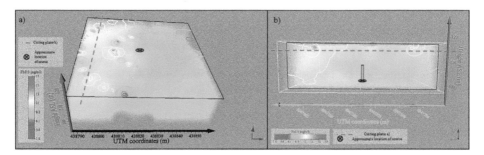

Figure 17. Volume rendering and contour plots created with quadcopter dataset (**a**) top view 18 m above ground level (from the East) and (**b**) side view 30 m away from the source (from the west).

6. Conclusions and Further Work

The sensor systems developed to date are technically capable of delivering data comparable to industrial quality dust-monitoring devices but require individual calibration equations for each sensor used to characterize dust plumes. The use of talcum powder is primarily a detection exercise at this stage as most particulate matter in this product has a diameter greater than 10 μm. System testing at PM2.5 will require a chemical source such as a smoke generator.

The tests described in this paper only measured concentrations with a precision of 1 mg/m^3; more precise readings of smaller concentrations will require the use of a different optical sensors and reference calibration with more precise equipment. Cross-contamination sources will require to be controlled in further experiments, and background levels will need to be measured to determine their content percentage in the final concentrations. These measurements will allow the programming of different flight patterns which could be focused in the intersection of the plume and will provide additional flight time.

Current experimental work indicates that integration of air quality sensor and autopilot data is feasible and will characterize airborne particulates in time and space.

Further work will be focused on the analysis of near-real time data to feed atmospheric modeling software and for flight path-planning algorithms.

Acknowledgments: The first author would like to acknowledge the support of the Centre for Mine Land Rehabilitation (CMLR) at UQ SMI, the Science and Engineering Faculty at QUT, and from the UQ eGatton program to Pfsr. Kim Bryceson, Armando Navas Borrero and Sotiris Ioannou.

Sensors **2015**, *15*, 19667–19687

Author Contributions: Miguel Angel Alvarado Molina was responsible for conducting the literature review, design and coordination of experimental procedures. He had the main role in the writing of this paper.

Felipe Gonzalez provided advice on experimental design, sensor integration and data interpretation. He also provided guidance on the structure and editing of the research article.

Andrew Fletcher provided overall project guidance and advice on experimental design, sensor integration and data interpretation. He also provided guidance regarding the formatting and editing of the research article.

Ashray Doshi assisted in the experimental design and sensor integration. He also provided advice in codes development for operation of the modular dust-sensing system and coordinated the integration of the dust sensor to UAV platforms.

Conflicts of Interest: The authors declare no conflict of interest.

References

1. Department of Natural Resources and Mines (DNRM). *Queensland Monthly Coal Report*; Queensland Government, The State of Queensland: Brisbane, Australia, 8 July 2014.
2. Raj, R. Sustainable mining systems and technologies. In *Sustainable Mining Practices*; Taylor & Francis: Oak Brook, USA, 2005; pp. 91–178.
3. NSWEPA. EPA Investigating Reports of Blasting Fumes from Wambo Coa. Available online: http://www.epa.nsw.gov.au/epamedia/EPAMedia14051501.htm (accessed on 29 July 2014).
4. Attalla, M.; Day, S.; Lange, T.; Lilley, W.; Morgan, S. *NOx Emissions from Blasting in Open Cut Coal Mining in the Hunter Valley*; Australian Coal Industry's Research Program, ACARP: Newcastle, Australia, 2007.
5. Sharp. Opto-Electronic Devices Division Electronic Components Group. Available online: http://www.dema.net/pdf/sharp/PC847XJ0000F.pdf (accessed on 10 August 2015).
6. SYhitech. DSM501A Dust Sensor Module. Available online: http://i.publiclab.org/system/images/photos/000/003/726/original/tmp_DSM501A_Dust_Sensor630081629.pdf (accessed on 8 August 2014).
7. Roy, S.; Adhikari, G.; Renaldy, T.; Singh, T. Assessment of atmospheric and meteorological parameters for control of blasting dust at an Indian large surface coal mine. *Res. J. Environ. Earth Sci.* **2011**, *3*, 234–248.
8. Roy, S.; Adhikari, G.R.; Singh, T.N. Development of Emission Factors for Quantification of Blasting Dust at Surface Coal Mines. *J. Environ. Protect.* **2010**, *1*, 346–361. [CrossRef]
9. Roy, S.; Adhikari, G.; Renaldy, T.; Jha, A. Development of multiple regression and neural network models for assessment of blasting dust at a large surface coal mine. *J. Environ. Sci. Technol.* **2011**, *4*, 284–299. [CrossRef]
10. Bridgman, H.; Carras, J.N. *Contribution of Mining Emissions to NO₂ and PM10 in the Upper Hunter Region*; ACARP: NSW, Australia, 2005.
11. Richardson, C. *PM2.5 Particulate Emission Rates From Mining Operations*; Australian Coal Industry's Research Program, ACARP: Castle Hill, Australia, March 2013.
12. Koronowski, R. FAA Approves Use of Drones by ConocoPhillips to Monitor Oil Drilling Activities in Alaska. Available online: http://thinkprogress.org/climate/2013/08/26/2524731/drones-conocophillips-alaska/ (accessed on 22 January 2013).
13. Fernandez, R. Methane Emissions from the U.S. In *Natural Gas Industry and Leak Detection and Measurement Equipment*. Available online: http://arpa-e.energy.gov/sites/default/files/documents/files/Fernandez_Presentation_ARPA-E_20120329.pdf (accessed on 10 August 2010).
14. Nicolich, K. High Performance VCSEL-Based Sensors for Use with UAVs. Available online: http://www.princeton.edu/pccmeducation/undergrad/reu/2012/Nicolich.pdf (accessed on 4 May 2014).
15. DECCW. Upper Hunter Air Quality Monitoring Network. Available online: www.environment.nsw.gov.au/aqms/upperhunter.htm (accessed on 10 October 2013).
16. DEHP. Air Quality. Available online: http://www.ehp.qld.gov.au/air/ (accessed on 10 August 2014).
17. Dräger X-am® 5600. Drägerwerk AG & Co. KGaA: Lubeck, Germany, 2014. Available online: http://www.draeger.com/sites/assets/PublishingImages/Products/cin_x-am_5600/UK/9046715_PI_X-am_5600_EN_110314_fin.pdf (accessed on 10 August 2015).
18. Picarro. *PICARRO Surveyor*; Picarro: Santa Clara, CA, USA, 2014; Available online: https://picarro.app.box.com/s/mtmyqr0k2kfotg2uf40z (accessed on 10 August 2015).
19. Picarro. *Picarro G2401 CO₂ + CO + CH₄ + H₂O CRDS Analyzer*; Picarro, Ed.; Picarro: Santa Clara, CA, USA, 2015.
20. QLDGov. Tapered Element Oscillating Microbalance. Available online: https://www.qld.gov.au/environment/pollution/monitoring/air-pollution/oscillating-microbalance/ (accessed on 20 August 2014).

21. ThermoScientific. Thermo Scientific TEOM® 1405-DF. Available online: http://www.thermo.com.cn/Resources/200802/productPDF_3275.pdf (accessed on 10 August 2015).
22. Geotech. AQMesh Operating Manual. Available online: http://www.geotechuk.com/media/215152/aqmesh_operating_manual.pdf (accessed on 10 August 2015).
23. LAvionJaune. Ultra-Light, Standalone Lidar System for UAVs (Laser Scanner, IMU, RTKGPS, Processing Unit). Available online: http://yellowscan.lavionjaune.com/data/leafletYS.pdf (accessed on 10 August 2015).
24. Sihler, H.; Kern, C.; Pöhler, D.; Platt, U. Applying light-emitting diodes with narrowband emission features in differential spectroscopy. *Opt. Lett.* **2009**, *34*, 3716–3718. [CrossRef] [PubMed]
25. Kern, C.; Trick, S.; Rippel, B.; Platt, U. Applicability of light-emitting diodes as light sources for active differential optical absorption spectroscopy measurements. *Appl. Opt.* **2006**, *45*, 2077–2088. [CrossRef] [PubMed]
26. Choi, S.; Kim, N.; Cha, H.; Ha, R. Micro Sensor Node for Air Pollutant Monitoring: Hardware and Software Issues. *Sensors* **2009**, *9*, 7970–7987. [CrossRef] [PubMed]
27. Thalman, R.M.; Volkamer, R.M. Light Emitting Diode Cavity Enhanced Differential Optical Absorption Spectroscopy (led-ce-doas): A Novel Technique for Monitoring Atmospheric Trace Gases. *Proc. SPIE* **2009**, *7462*. [CrossRef]
28. Khan, A.; Schaefer, D.; Roscoe, B.; Kang, S.; Lei, T.; Miller, D.; Lary, D.J.; Zondlo, M.A. Open-path greenhouse gas sensor for UAV applications. In Proceedings of the 2012 Conference on Lasers and Electro-Optics (CLEO), San Jose, CA, USA, 6–11 May 2012; pp. 1–2.
29. Malaver, A.; Gonzalez, F.; Motta, N.; Depari, A.; Corke, P. Towards the Development of a Gas Sensor System for Monitoring Pollutant Gases in the Low Troposphere Using Small Unmanned Aerial Vehicles. In Proceedings of Workshop on Robotics for Environmental Monitoring, Sydney University, Sydney, Australia, 11 July 2012.
30. Neumann, P.P.; Hernandez Bennetts, V.; Lilienthal, A.J.; Bartholmai, M.; Schiller, J.H. Gas source localization with a micro-drone using bio-inspired and particle filter-based algorithms. *Adv. Robot.* **2013**, *27*, 725–738. [CrossRef]
31. Bennetts, V.H.; Lilienthal, A.J.; Neumann, P.P.; Trincavelli, M. Mobile robots for localizing gas emission sources on landfill sites: Is bio-inspiration the way to go? *Front. Neuroeng.* **2011**, *4*, 735–737.
32. Watai, T.; Machida, T.; Ishizaki, N.; Inoue, G. A Lightweight Observation System for Atmospheric Carbon Dioxide Concentration Using a Small Unmanned Aerial Vehicle. *J. Atmos. Ocean. Technol.* **2005**, *23*, 700–710. [CrossRef]
33. Brown, J.; Taras, M. *Remote Gas Sensing of SO$_2$ on a 2D CCD (Gas. Camera)*; Resonance LTD: Barrie, ON, Canada, 2008.
34. Lega, M.; Napoli, R.M.A.; Persechino, G.; Kosmatka, J. New techniques in real-time 3D air quality monitoring: CO, NO$_x$, O$_3$, CO$_2$, and PM. In Proceedings of the NAQC 2011, San Diego, CA, USA, 7–11 March 2011.
35. Lega, M.; Kosmatka, J.; Ferrara, C.; Russo, F.; Napoli, R.M.A.; Persechino, G. Using Advanced Aerial Platforms and Infrared Thermography to Track Environmental Contamination. *Environ. Forensics* **2012**, *13*, 332–338. [CrossRef]
36. Saghafi, A.; Day, S.; Fry, R.; Quintanar, A.; Roberts, D.; Williams, D.; Carras, J.N. Development of an Improved Methodology for Estimation of Fugitive Seam Gas. Emissions from Open Cut Mining. Available online: http://www.acarp.com.au/abstracts.aspx?repId=C12072 (accessed on 10 August 2015).
37. Gonzalez, F.; Castro, M.P.; Narayan, P.; Walker, R.; Zeller, L. Development of an autonomous unmanned aerial system to collect time-stamped samples from the atmosphere and localize potential pathogen sources. *J. Field Robot.* **2011**, *28*, 961–976. [CrossRef]
38. Gonzalez, L.F.; Castro, M.P.; Tamagnone, F.F. Multidisciplinary design and flight testing of a remote gas/particle airborne sensor system. In Proceedings of the 28th International Congress of the Aeronautical Sciences, Optimage Ltd., Brisbane Convention & Exhibition Centre, Brisbane, QLD, Australia, 23 September 2012; pp. 1–13.
39. Malaver, A.; Motta, N.; Corke, P.; Gonzalez, F. Development and Integration of a Solar Powered Unmanned Aerial Vehicle and a Wireless Sensor Network to Monitor Greenhouse Gases. *Sensors* **2015**, *15*, 4072–4096. [CrossRef] [PubMed]

40. Reed, W.R. *Significant Dust Dispersion Models for Mining Operations*; Department of Health and Human Services: Pittsburgh, PA, USA, September 2005.
41. Stockie, J.M. The Mathematics of Atmospheric Dispersion Modeling. *SIAM Rev.* **2011**, *53*, 349–372. [CrossRef]
42. Visscher, A.D. An Air Dispersion Modeling Primer. In *Air Dispersion Modeling*; John Wiley & Sons, Inc.: Hoboken, NJ, USA, 2013; pp. 14–36.
43. Roldán, J.J.; Joossen, G.; Sanz, D.; del Cerro, J.; Barrientos, A. Mini-UAV Based Sensory System for Measuring Environmental Variables in Greenhouses. *Sensors* **2015**, *15*, 3334–3350. [CrossRef] [PubMed]
44. Haas, P.; Balistreri, C.; Pontelandolfo, P.; Triscone, G.; Pekoz, H.; Pignatiello, A. Development of an unmanned aerial vehicle UAV for air quality measurements in urban areas. In Proceedings of the 32nd AIAA Applied Aerodynamics Conference; American Institute of Aeronautics and Astronautics, Atlanta, GA, USA, 16–20 June 2014.
45. Skycam. Swamp Fox UAV. Available online: http://www.kahunet.co.nz/swampfox-uav.html (accessed on 13 June 2014).
46. Budde, M.; ElMasri, R.; Riedel, T.; Beigl, M. Enabling Low-Cost Particulate Matter Measurement for Participatory Sensing Scenarios. In Proceedings of the 12th International Confrence on Moile and Ubiquitous Multimedia MUM, Lulea, Sweden, 2–5 December 2013; ACM: Lulea, Sweden, 2013; p. 19.
47. M.Chardon, C.; Trefois, C. *Standalone Sketch to Use with a Arduino Fio and a Sharp Optical Dust Sensor GP2Y1010AU0F*; Creative Commons: San Francisco, CA, USA, 2012.
48. Fiume, M.M. *Safety Assessment of Talc As Used in Cosmetics*; Cosmetic Ingredient Review: Washington, DC, USA, 12 April 2013.
49. Klingler, G.A. *Digital Computer Analysis of Particle Size Distribution in Dusts and Powders*; Aerospace Research Laboratories, Office of Aerospace Research, United States Air Force: Wright-Patterson Air Force Base, OH, USA, 1972.
50. CASA. Civil Aviation Safety Regulations 1998. In *Unmanned Air and Rockets*; Australian Government ComLaw: Canberra, Australia, 1998.

Article

Multi-UAV Routing for Area Coverage and Remote Sensing with Minimum Time

Gustavo S. C. Avellar, Guilherme A. S. Pereira *, Luciano C. A. Pimenta and Paulo Iscold

Escola de Engenharia, Universidade Federal de Minas Gerais, Av. Antônio Carlos 6627,
Belo Horizonte 31270-901, MG, Brazil; E-Mails: gustavoavellar@ufmg.br (G.S.C.A.);
lucpim@cpdee.ufmg.br (L.C.A.P.); iscold309@gmail.com (P.I.)
* E-Mail: gpereira@ufmg.br; Tel.: +55-313-409-6687; Fax: +55-313-409-4810.

Academic Editor: Felipe Gonzalez Toro
Received: 27 June 2015 / Accepted: 27 October 2015 / Published: 2 November 2015

Abstract: This paper presents a solution for the problem of minimum time coverage of ground areas using a group of unmanned air vehicles (UAVs) equipped with image sensors. The solution is divided into two parts: (i) the task modeling as a graph whose vertices are geographic coordinates determined in such a way that a single UAV would cover the area in minimum time; and (ii) the solution of a mixed integer linear programming problem, formulated according to the graph variables defined in the first part, to route the team of UAVs over the area. The main contribution of the proposed methodology, when compared with the traditional vehicle routing problem's (VRP) solutions, is the fact that our method solves some practical problems only encountered during the execution of the task with actual UAVs. In this line, one of the main contributions of the paper is that the number of UAVs used to cover the area is automatically selected by solving the optimization problem. The number of UAVs is influenced by the vehicles' maximum flight time and by the setup time, which is the time needed to prepare and launch a UAV. To illustrate the methodology, the paper presents experimental results obtained with two hand-launched, fixed-wing UAVs.

Keywords: coverage path planning; UAVs; vehicle routing problem

---

1. Introduction

The world is on the verge of a major breakthrough, as we reach a moment in history when UAV flights become regulated in many countries around the world. Companies from different fields are currently using UAV and sensor technologies to acquire information of ground regions and to reduce the time and costs of operation. Applications, such as environment monitoring, search and rescue, precision agriculture and surveillance, may benefit from this usage of UAVs with onboard sensors for spatial coverage [1–4].

This work, which was motivated and mainly financed by FINEP (Funding Agency for Studies and Projects), a funding agency of the Brazilian government, deals with one of the most common uses of aerial robot technologies, which is the one for obtaining a series of overlapping aerial images from the ground. These images are usually post-processed for the extraction of desired information, such as digital terrain maps and vegetation indexes. In this context, efficient UAV path planning algorithms are of great importance, since the operation time, costs and the quality of the information extracted from the images are directly related to the quality of such a planning. We propose an area coverage path planning strategy to obtain images of the ground considering a multi-UAV scenario.

Several area coverage strategies have been proposed in the literature. A comprehensive recent survey of methods can be found in [5]. The large majority of strategies rely on decomposing the target area into cells that must be visited and covered. Choset [6], for example, proposed a method of exact cellular decomposition dedicated to coverage tasks. The method divides the space in convex regions that must be covered by a sequence of back and forth movements. The cells are modeled as nodes in an adjacency graph in which edges represent the existence of a common boundary between two cells. The coverage problem is then solved by first executing a graph search to determine in which order the cells should be covered. Second, the robot moves from cell to cell according to the specified sequence, performing the back and forth movements inside each cell to cover it entirely. Acar *et al.* [7] adapted this method to work on smooth and polygonal workspaces. Instead of searching for vertices, the proposed algorithm looks for connectivity changes in the workspace. Acar *et al.* [8] uses this method and improves it by describing each cell as narrow or vast. In vast cells, the back and forth movements are executed as usual, with the width of each coverage row proportional to the size of the footprint associated with the sensor carried by the robot. On the other hand, in narrow cells, the robot changes its behavior and uses its sensor to follow the associated generalized Voronoi diagram (GVD) to cover the cell.

The only optimization involved in the previously mentioned methods is in the choice of the order of cells to be followed. A simple form of path optimization is to choose the direction of the back and forth movements in each cell. This may reduce the number of turns in the path, thus reducing the effects of vehicle deceleration and acceleration due to each turn [9]. Li *et al.* [10] takes this method one step further by also addressing the connection between cells. In [10], the authors were able to prove that, for UAVs, a path with less turns is more efficient in terms of route length, duration and energy. The optimization of the sweep direction was also applied in the context of multi-UAV systems in [11]. In this case, the target area was partitioned into convex polygons, and each one of these polygons was assigned to a UAV, which had the responsibility of covering the area by following a back and forth pattern according to the optimal sweep direction.

Xu *et al.* [12] presented an area coverage strategy that combines the ideas of the exact cell decomposition of Choset [6] with the direction of coverage optimization suitable for single fixed-wing UAV operation. As the authors use a graph representation (Reeb graph), where the cells are not modeled as nodes, but as graph edges, the sequence of cells is found by solving the so-called Chinese postman problem (CPP), which is a routing problem in which the objective is to find the shortest tour that visits every edge at least once.

Some researchers have focused on solutions that take into account multiple robots, such as the work in [11] that was previously mentioned. The use of multi-robot systems has several advantages, such as the reduction of mission time due to workload division and the introduction of fault tolerance, as one robot may cover the region initially assigned to another robot in case of failure. In this context, a common approach is to consider the coverage problem as a vehicle routing problem (VRP).

In general, a VRP is the problem of finding a set of routes to be executed by a set of vehicles that must visit a set of customers at different geographical locations, given a transportation road network. These routes must fulfill all of the customers' demands of goods, satisfy the operational constraints and minimize an objective function that reflects the global transportation cost [13]. The transformation of the coverage problem into a VRP is usually carried out by building a graph in such a way that coverage is reached when a set of nodes or edges of this graph is visited at least once by a robot. The different solutions in this line vary in at least one of the following aspects: the form of constructing this graph, the manner of obtaining coverage, *i.e.*, which nodes or which edges must be visited, the operational constraints imposed on the robots and the objective function to be minimized.

In [14], the authors consider a multi-UAV routing problem with the possibility of modeling periods of loitering and also dealing with general relative timing constraints. In their formulation, it is possible to model scenarios in which a waypoint has to be visited more than once with given timing constraints. The problem of boundary coverage by a multi-robot system is addressed in [15]. In this case, the objective is to generate balanced inspection routes that cover a subset of the graph edges instead of the graph nodes. The work in [16] solves the problem of planning routes to multiple UAVs with the collected amount of information from desired regions being the objective function to be maximized. The information is acquired by down-facing cameras installed on the UAVs, and the

Sensors **2015**, *15*, 27783–27803

computation of the information takes into account the variation of the resolution at different parts of the captured image. Dynamic vehicle routing with the objective of spatial and temporal coverage of points of interest, modeled as nodes in a graph, is the focus of [17]. The spatial coverage is related to the fact that the points are distributed over the area, and the temporal coverage is associated with the existence of time constraints determining when the points have to be covered. The environment is dynamic in the sense that the targets evolve spatially and temporally. Three conflicting objectives are considered simultaneously: (i) minimization of the distance traveled; (ii) maximization of satisfaction, which models the necessity of covering the targets within given time windows; and (iii) minimization of the number of UAVs. A vehicle routing problem with multiple depots is the model used in [18]. In the VRP nomenclature, the depot is the place where the vehicles start and finish their tasks. The objective in [18] is the minimization of the longest tour performed by every UAV, which is equivalent to the minimization of the total mission time.

This paper presents a methodology for optimal time coverage of ground areas using multiple fixed-wing UAVs. Similar to other works, we solve the coverage problem by creating a graph and transforming the original problem into a vehicle routing problem. The main contribution of this work is the incorporation of specific features that are relevant in a real-world deployment. We assume the common scenario in which the number of human operators responsible for launching and retrieving the UAVs is smaller than the number of vehicles. This is incorporated in the method by defining a so-called setup time. In some situations, the setup time prevents two UAVs from being launched within small intervals of time. This means that, since one operator cannot prepare more than one UAV at the same time, the setup time of each UAV is cumulative. For example, consider a mission with one operator, two UAVs and a setup time of 4 min. The pre-flight tasks for the first UAV will take 4 min and will take 8 min for the second UAV, 4 min of which the operator was working on the first UAV while the second was idle. Similar to [18], in this work, we aim to solve the multi-UAV coverage task in minimum time, and in the same spirit of [17], it is also our goal to use a reduced number of vehicles, if possible. Our method finds the optimal routes and number of necessary UAVs automatically considering the number of human operators available. Given the constraint on the number of operators, there are scenarios in which launching a large number of UAVs may have a negative impact in the total mission time, due to the influence of the cumulative setup time. Furthermore, in order to reduce the number of turns during the mission, we also optimize the sweep direction as in [9]. We evaluate the method in a real-world experiment with two actual aerial vehicles.

This paper is organized as follows. Section 2 presents the problem statement. The proposed solution is presented in Section 3. Experimental results in simulated and real environments using multiple UAVs are shown in Sections 4 and 5, respectively. The conclusions and some perspectives for future works are presented in Section 6.

2. Problem Definition

In this paper, it is assumed that a group of M fixed-wing UAVs has the mission to cover, as quickly as possible, a polygonal convex area represented by a set P of vertices in $\mathbb{R}^2$. If a non-convex area is to be covered, it is assumed that P represents the convex hull of the area. It is considered that the maximum time of flight of each UAV is finite and known in advance. Each UAV is equipped with an on-board camera pointing down. The mission of the UAVs is to sense, using the on-board cameras, the entire region specified by P. The altitude of the UAVs flight is constant and is carefully chosen so that the resolution of the camera allows the observation of the characteristics of interest on the ground. We assume that all UAVs are identical in terms of hardware and power, although this is not a limitation to our methodology. Related to the operation of the UAVs, we assume the necessity of a setup time before the flight of each UAV. This setup time includes the connection of the batteries, the GPS fixing and the launching itself, among other tasks.

Given this, the specific problems we are dealing with in this paper are: (i) to discover the number $m \leq M$ of UAVs that minimize the time taken to cover the area represented by P; and (ii) to specify the paths for each UAV so that the mission is completed in minimum time. Notice that, given the setup time, the ideal number of UAVs cannot be trivially chosen to be M. In the same way, the setup time and the dynamic constraints of the UAV also prevent the area from being simply divided among the available UAVs. The next section will present our solution to these problems.

3. Methodology

Our strategy for solving the problem presented in the previous section is divided into two parts. In the first part, we decompose the area to be covered as a set of sweeping rows, using a methodology similar to the one proposed in [9]. These rows form the edges of a graph that is used in the second part of the method, which is based on a vehicle routing problem solution. The next subsections detail each part of the method.

3.1. Area Decomposition

In this work, we assume that the UAVs will fly over the area to be covered executing a back and forth motion in rows perpendicular to a given sweep direction, as shown in Figure 1. While following the rows, the UAV is leveled (which means that its camera is pointing down), but at the end of the row, it makes a curve outside the area to return to the next row. During such a curve, the camera is generally not pointing to the ground. Furthermore, as pointed out by Huang [9], the number of turns is directly related to the time of coverage of a given region. Therefore, as suggested in [9], the first step of our method is to find the optimal direction of coverage, which is perpendicular to the smallest height of the polygonal area. In this direction, the area can be covered with the smallest number of rows, thus with the smallest number of curves. This can be observed in Figure 1. As pointed out by [10], a path with less turns is more efficient in terms of route length, duration and energy. However, notice that the sweep direction may be chosen in different ways. It is possible, for example, to choose the direction of the coverage rows as a function of the wind, since flying against the wind may destabilize the vehicle [12].

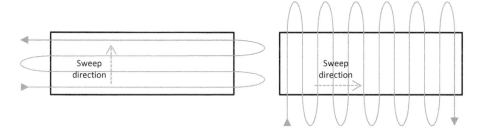

Figure 1. Coverage strategy used in this work. A rectangular area is covered using a back and forth motion along lines perpendicular to the sweep direction. Notice that the sweep direction highly influences the number of turns outside the area to be covered, thus affecting the coverage time. The optimal sweep direction is parallel to the smallest linear dimension of the area [9].

To find the optimal direction of coverage of a given polygon, a simple search procedure may be used. As shown in Figure 2, the polygon is rotated over a surface, and its height is measured. The best orientation is the one that yields the smallest height, h_{min}.

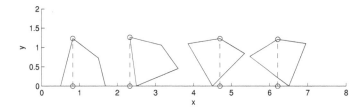

Figure 2. Procedure to search the optimal direction of coverage. The area is rotated until the smallest height is found.

Once the optimal sweep direction is found, it is then possible to distribute the rows over the area. In our approach, the distance between two rows is chosen as a function of the footprint of the on-board cameras on the ground. As shown in Figure 3, assuming that the image sensor is parallel to the ground plane (*i.e.*, the UAV is leveled), by knowing the width of the image sensor, l, the focal distance of the camera's lens, f, both in millimeters, and the distance between the camera and the ground, H (the flying height), in meters, it is possible to compute the width L of the camera's footprint, in meters, as:

$$L = H\frac{l}{f} \tag{1}$$

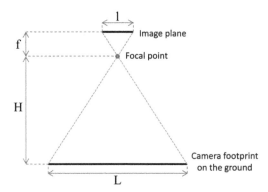

Figure 3. Relation between the size of the camera sensor, the height of flight and the camera footprint on the ground.

The number of coverage rows is then computed as:

$$N_l = \left\lceil \frac{h_{min}}{L(1-s)} \right\rceil \tag{2}$$

while the distance, in meters, between two rows is:

$$d_l = \frac{h_{min}}{N_l} \tag{3}$$

where $s \in (0, 1)$ represents the fraction of overlap between two images. This overlap is generally necessary to concatenate the images to compose an aerial map.

Assuming that the polygon that represents the region is rotated in a way that the optimal direction of coverage is parallel to the x axis of the global reference frame, the coverage rows can be defined by two planar points (x, y) with identical y coordinates given by:

$$y_i = i \times d_l - \frac{d_l}{2}, \quad i = 1, \dots, N_l \quad (4)$$

and x coordinates defined by the points where the horizontal straight line with coordinate y_i intercepts the borders of the area to be covered. Once the points are computed, they are rotated back to the original orientation.

The extreme points of the coverage rows, along with the coordinates of the UAV launch position, called the base or depot, are considered to be the set of nodes V of a graph $G = (V, E)$. Each node of the graph is numbered so that the base receives Number 1, the nodes related to the first coverage row receives Numbers 2 and 3, the ones associated with the second row are labeled 4 and 5, and so on. At the end, each coverage row is associated with subsequent even and odd nodes. The edge set, E, is composed of all lines connecting the N nodes of the graph, thus forming a complete graph, as shown in Figure 4.

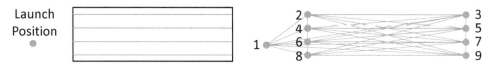

Figure 4. Graph representing the coverage problem. On the left, a rectangular region to be covered, the launch position and the covered rows. The nodes of the graph on the right are composed of the launch position and the intersection points between the coverage rows and the borders of the region. All nodes are connected by edges, forming a complete graph.

Mathematically, graph G may be represented by an $N \times N$ cost matrix C whose elements, C_{ij}, are given by the Euclidean distance between the spatial coordinates of nodes i and j. It is important to notice that C is time invariant, symmetric, *i.e.*, $C_{ij} = C_{ji}, \forall (i, j) \in E$, and its elements satisfy the triangular inequality, *i.e.*, $C_{ij} + C_{jk} \geq C_{ik}, \forall (i, j), (j, k), (i, k) \in E$. The next subsection will describe how the graph represented by C will be used in an optimization problem to allow the coverage of an area in minimum time using multiple UAVs.

3.2. Routing Strategy

Once a graph associated with the region to be covered is created, the coverage problem can be posed as a vehicle routing problem (VRP) [13]. In this class of problems, a set of customers must be visited by a set of vehicles. To transform the problem proposed in Section 2 into a VRP, each UAV will be modeled as a vehicle and each extreme point of the coverage rows as a customer. Furthermore, by proposing new constraints, it is possible to enforce the vehicles to use some pre-specified edges of the graph in their routes, so that the coverage rows will be certainly followed by one of the launched UAVs. Finally, by solving the VRP, one can obtain the set of routes that each UAV will have to perform.

Before presenting the mathematical formulation of the routing problem, we will define the constants and variables necessary for this formulation. As defined in the previous section, constant C_{ij} represents the traversing cost of the edge (i, j) between nodes i and j. To indicate whether or not the k-th UAV is going to fly from vertex i to vertex j, the binary variable $X_{ij}^k \in \{0, 1\}$ is used. Furthermore, let constant $V_{ij}^k \in \mathbb{R}$ represent the flight speed of the k-th UAV while flying from vertex i to vertex j, constant $t_s \in \mathbb{R}$ be the individual setup time and L_k be the battery duration of UAV number k. Let also $m \in \mathbb{N}$ be the variable that represents the number of UAVs designed for a mission, $M \in \mathbb{N}$ be the total

Sensors **2015**, *15*, 27783–27803

number of UAVs available, $O \in \mathbb{N}$ be the constant number of UAV operators and $N \in \mathbb{N}$ be the number of nodes of the graph. Finally, the variable d_k represents the extra time necessary to launch UAV k.

Based on the variables previously defined, the time spent by UAV k to fly its route is mathematically given by:

$$T_k = \sum_{i=1}^{N} \sum_{j=1}^{N} \frac{C_{ij}}{V_{ij}^k} X_{ij}^k + d_k$$

Our main objective is to minimize the mission time. This can be accomplished by minimizing the time of the longest route among the routes of all UAVs. Therefore, our problem is in fact a min-max problem in which we want to minimize the maximum T_k. To transform the min-max problem into a linear problem, we introduce an extra variable V, which represents the longest UAV route. The basic optimization problem is then written as:

$$\min(V) \tag{5}$$

subject to

$$\sum_{i=1}^{N} \sum_{j=1}^{N} \frac{C_{ij}}{V_{ij}^k} X_{ij}^k + d_k \leq V, k = 1, \ldots, M \tag{6}$$

$$t_s \left\lceil \frac{k}{O} \right\rceil \sum_{j=1}^{N} X_{1j}^k = d_k, k = 1, \ldots, M \tag{7}$$

As previously shown, constraint in Equation (6) accounts for the individual cost (time) of UAV k. This corresponds to T_k. By defining this constraint along with the objective function in Equation (5) using variable V, we are posing a linear version of the min-max problem, where the maximum cost among the ones of all UAVs must be minimum. By doing this, we are, in practice, minimizing the time taken to cover the complete area.

In Equation (6), the term d_k, detailed in Equation (7), accounts for the setup time, t_s, which is an extra cost that corresponds to the time spent by a human operator to prepare and launch the UAV for the mission. In a mission where only one person is operating the whole team of UAVs, each UAV will have the extra time d_k, which will be cumulative. In a system with two UAVs, for example, while UAV 1 is prepared, nothing is happening with UAV 2. After the launch of UAV 1, UAV 2 is prepared to be launched. Notice in Equation (7) that d_k is null when the UAV is not used, once X_{1j}^k is zero, indicating that UAV k did not leave Node 1, which represents the launching position.

Since the setup time is one of the main contributions of this work, before proceeding with the additional constraints necessary to guarantee a coverage solution, we will explore the effect of constraint in Equation (7) with three examples. For the first example, assume a team of $M = 3$ UAVs with individual setup time $t_s = 10$ min and a single operator ($O = 1$). The team is supposed to cover a rectangular area with 8 coverage rows. Each row can be covered in 2.5 min, and the time to reach the region and to change between rows is considered to be negligible. Given this, the cumulative setup time of each UAV, as computed by Equation (7), is given by:

$$t_s \left\lceil \frac{k}{O} \right\rceil \sum_{j=1}^{N} X_{1j}^k = d_k, k = 1, \ldots, M$$
$$10 \times \lceil 1/1 \rceil \times 1 = 10 = d_1$$
$$10 \times \lceil 2/1 \rceil \times 1 = 20 = d_2$$
$$10 \times \lceil 3/1 \rceil \times 1 = 30 = d_3$$

Notice that the cumulative setup time for UAV 3, $d_3 = 30$ min, is equal to the time a single UAV would expend to cover the region ($8 \times 2.5 + 10$ min). This indicates that the use of 3 UAVs for this mission is not worth it. In this way, the best solution would make $X^3_{1j} = 0$, indicating that UAV 3 would not be used in this mission. Two UAVs would be launched, and the best coverage time would be 25 min. For this solution, UAV 1 would cover six coverage rows and UAV 2 only two rows. As a remark, if three UAVs were considered, the best coverage time would be 32.5 min.

The second example explores the case when the number of operators is larger than one, but is smaller than the number of UAVs available. Suppose a team of $M = 5$ UAVs with individual setup time $t_s = 10$ min and a number of $O = 2$ operators. Using Equation (7), we have:

$$10 \times \lceil 1/2 \rceil \times 1 = 10 = d_1$$
$$10 \times \lceil 2/2 \rceil \times 1 = 10 = d_2$$
$$10 \times \lceil 3/2 \rceil \times 1 = 20 = d_3$$
$$10 \times \lceil 4/2 \rceil \times 1 = 20 = d_4$$
$$10 \times \lceil 5/2 \rceil \times 1 = 30 = d_5$$

As can be seen, UAVs 1 and 2 have the same setup time, because the two operators will prepare them simultaneously. After the takeoff of these UAVs, UAVs 3 and 4 will be prepared. Their setup time is 20 min, composed of 10 min of waiting for UAV 1 and 2 to be prepared and 10 min of their own preparation. For UAV 5, the same reasoning can be applied.

In a third example, consider the case where the number of operators is equal to the number of UAVs. In this case, since k/O in Equation (7) will be equal to one for all UAVs, the setup time for each UAV would be t_s.

To complete the optimization problem and to guarantee its solution, a solution to the problem posed in Section 2 is indeed, and other constraints need to be incorporated into the the basic problem. The first of these constraints is given by:

$$\sum_{i=1}^{N} \sum_{j=1}^{N} \frac{C_{ij}}{V^k_{ij}} X^k_{ij} \le L_k, k = 1, \ldots, M \tag{8}$$

which limits the maximum time of flight of UAV k by its battery duration, L_k. We consider that the charge of the battery does not decrease during the setup time or that charge decreasing is negligible. In this way, the time of flight of each UAV is simply the total time in Equation (6) subtracted by the setup time d_k. It is important to mention that constraint in Equation (8) can make the problem infeasible. This is expected if, for example, a large area is to be covered by a very small team of UAVs. The only solution in this case would be to increase the number of agents. In another situation, if the time to cover a single row is larger than the battery duration of a single UAV, the problem would also have no solution. In this case, one could go back to the first step of the methodology and try to reduce the length of the rows by changing the sweep direction (which would increase the number of turns). A more complex alternative for both situations would be to change the optimization problem to consider battery recharging. These solutions are left as suggestions for future research.

To guarantee that each node of the graph is visited only once by a single UAV, two other constraints are necessary:

$$\sum_{k=1}^{M} \sum_{i=1}^{N} X^k_{ij} = 1, j = 2, 3, 4 \ldots, N \tag{9}$$

$$\sum_{i=1}^{N} X^k_{ip} - \sum_{j=1}^{N} X^k_{pj} = 0, p = 1, 2, 3 \ldots, N, k = 1, 2, 3 \ldots, M \tag{10}$$

Notice that constraint in Equation (9) enforces that each node, except the base (represented by Node 1) is visited by only one UAV. On the other hand, constraint in Equation (10) guarantees that the UAV that arrives at a given node is the same one that leaves this node.

Sensors **2015**, *15*, 27783–27803

To enforce that each UAV path starts and finishes at the base (Node 1) and to guarantee that the path has no internal cycles, a standard sub-tour elimination constraint [19] is used:

$$u_i - u_j + N \sum_{k=1}^{M} X_{ij}^k \leq N - 1, i, j = 2, 3, 4 \ldots, N \tag{11}$$

where $u_i \in \mathbb{Z}, i = 2, 3, 4, \ldots, N$.

To make sure the VRP solution will make the UAVs cover the area modeled by graph G, the following constraint is also necessary:

$$\sum_{k=1}^{M} X_{i,i+1}^k + \sum_{k=1}^{M} X_{i+1,i}^k = 1, i = 2, 4, 6 \ldots, N \tag{12}$$

This constraint enforces that each UAV, having visited one of the nodes of a coverage row, must visit the other node of that row. This is possible given the way the nodes were numbered (see Figure 4). Constraint in Equation (12) guarantees that each UAV that visits an even node also visits the next odd node. Furthermore, a UAV that visits an odd node must visit the previous even node. Therefore, this constraint is essential to make the problem solution an actual coverage solution.

To avoid the UAVs crossing the coverage area following an edge that is not parallel to the coverage rows, two optional constraints can be added to the optimization problem:

$$\sum_{k=1}^{M} X_{i,i+1}^k = \sum_{k=1}^{M} \sum_{j=\{1,3,\ldots\}\backslash\{i+1\}}^{N} X_{i+1,j}^k, i = 2, 4, 6 \ldots, N \tag{13}$$

$$\sum_{k=1}^{M} X_{i,i-1}^k = \sum_{k=1}^{M} \sum_{j=\{1\}\cup\{2,4,\ldots\}\backslash\{i-1\}}^{N} X_{i-1,j}^k, i = 3, 5, 7 \ldots .N \tag{14}$$

In practice, constraints in Equations (13) and (14) can avoid the UAVs executing sharp turns and also photos being taken in different directions. If these issues are not important for a given task, these two constraints can be simply ignored without compromising the execution of the task.

Finally, to allow the number of UAVs used in the mission, m, to be smaller than the maximum number of UAVs available, M, we introduce the following constraints:

$$\sum_{k=1}^{M} \sum_{j=1}^{N} X_{1j}^k = m \tag{15}$$

$$m \leq M \tag{16}$$

It is important to mention that constraints in Equations (15) and (16) represent an important contribution of this work in relation to others that were previously published, such as [19], where the number of UAVs is kept constant. In this work, the number of UAVs will be chosen as a function of the minimization of $\mathcal{V}$.

By solving the problem represented by the objective function in Equation (5) and constraints in Equations (6)–(16), one would expect to obtain a solution in which the minimum number of UAVs would follow the shortest possible paths, covering in minimum time the area modeled by graph G. However, although the mission time will indeed be minimum, the objective function does not explicitly take into account the number of UAVs, which may cause the computed solution to consider a number of UAVs that is not optimal. Moreover, since our problem minimizes the longest path in terms of mission time, once the best longest path is found, there is no guarantee that the paths for the other UAVs are minimized. To make sure that the optimal number of UAVs, m, is chosen and that the paths for all UAVs are minimum, two strategies were devised.

The first one is a heuristic and depends on the adjustment of some constants. The idea is to change the utility cost function, so that it explicitly takes into account the number of UAVs and/or a combination of all UAV path costs. For example, the cost function $V + \rho m$ explicitly considers the number of UAVs. A correct choice of constant ρ will make the optimizer find the best m. In the same way, if T_k is the individual coverage time for UAV k, the cost function $V + \rho \, mean(T_k)$ would minimize the paths of all UAVs if ρ is chosen properly.

The second strategy, which will generate the optimal solution in terms of the number of UAVs and individual paths independently of parameter choices, is an iterative solution that consists of solving the optimization problem more than once. In the first iteration, the original graph is used, and the optimization problem to be solved is exactly the one previously presented. In the second iteration, the problem is reduced by removing the UAV assigned to the longest path in the first iteration, which is already optimal, and all nodes (except for the base) and associated edges that belong to its path. This procedure is repeated until all nodes, but the base, are removed from the graph. With this procedure, we have a better use of the available resources. It is then guaranteed that the optimal number of UAVs is found and that the paths for all UAVs are optimal without compromising the primary objective, which is to minimize the cost of the route with the highest cost.

In the next section, we present simulations that illustrate our methodology and the role of the constraints in the optimization problem.

4. Simulations

This section intends to illustrate the proposed methodology using a series of simulations. All simulations were executed in MATLAB on an Intel Core i5 1.7 GHz computer with 4 GB of RAM. The optimization problem was solved using the Yalmip [20] toolbox in the front-end and the Gorubi [21] solver in the back-end.

In our first simulation, we explore the cost function in Equation (5) and some of the issues mentioned before, related to the optimization of the paths of all UAVs. This simulation consists of two UAVs with a setup time of 2 min being deployed on a mission to cover a triangular-shaped area. Figure 5a shows that the green path, which corresponds to a mission of 8.16 min, was optimized in the first iteration. The red path, corresponding to a mission of 8.13 min, has unnecessary cycles and is not the minimum. This problem is solved by running the optimization algorithm once again with the removal of the first UAV and its path, as shown in Figure 5b. In this case, the UAV with the red path takes the reduced time of 8.03 min to finish its mission.

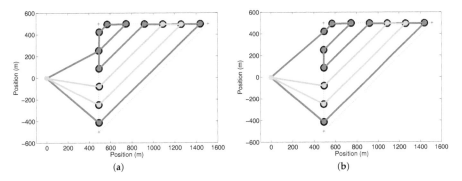

Figure 5. Effect of the cost function. (**a**) The mission represented by the green path is optimized, but the other (red) is not the minimum; (**b**) By removing the largest path and running the optimizer again, all paths are optimized.

Our second set of simulations explores the activation of constraint in Equation (12). To show the effect of this constraint, we have used two UAVs to cover a square-shaped area with no setup time, as shown in Figure 6. Figure 6a shows, in the solid line, the paths of the UAVs when constraint in Equation (12) is removed. The paths when this constraint is considered are shown in Figure 6b. Notice that the result shown in Figure 6a is a solution for a multiple traveling salesman problem (mTSP), but is not an area coverage solution.

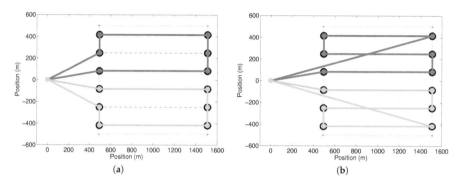

Figure 6. Effect of constraint in Equation (12). The solid lines represent the paths for the two UAVs used in the simulation. Optimization results without (**a**) and with (**b**) the constraint.

The next simulation shows the effect of constraints in Equations (13) and (14). In the result shown in Figure 7a, these constraints were removed. If we compare this result with Figure 7b, where the constraints are in place, it is possible to see that the constraints do not allow the UAV to move from one node to another, crossing the coverage region, except when the UAV is moving from or to the base. It is important to mention that the path obtained without the constraints takes 7.11 min to be followed by each UAV, while the path found with the constraints is longer and can be followed in 7.26 min. On the other hand, the problem can be computed in 4.59 s without the constraints and in 0.8 s with the constraints. This happens because these constraints act as if they are removing the diagonal edges of the graph, which, in fact, reduces the size of the problem. Thus, observe that the use of constraints in Equations (13) and (14) can be avoided if the diagonal edges are removed from the original graph.

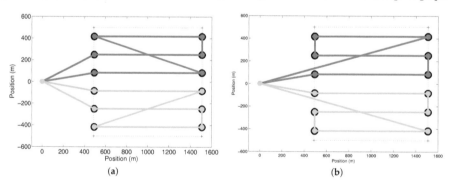

Figure 7. Effect of constraints in Equations (13) and (14). The solid lines represent the paths for the two UAVs used in the simulation. Optimization results without (**a**) and with (**b**) the constraints.

In what follows, we show the effects of constraint in Equation (7). For this set of simulations, a non-convex area is selected, and the methodology is applied to the area's convex hull. Results shown in Figure 8 represent trivial solutions for this area, when the setup time is not important. Figure 8a shows the optimization results for a mission when only one UAV was available ($M = 1$). In this case, the UAV took 20.8 min to complete the coverage, including 4 min of setup time. Figure 8b shows the optimization results for a mission when four UAVs were available ($M = 4$), each UAV with its own operator. This means that there is no cumulative setup time, and each UAV has only the original 4 min of setup time. Notice that the optimizer tries to equally divide the coverage area among the UAVs. In this simulation, the four UAVs took, respectively, 8.21, 8.36, 11.10 and 8.64 min to perform their tasks, including the setup time. A similar result would be found if the setup time were set to be null.

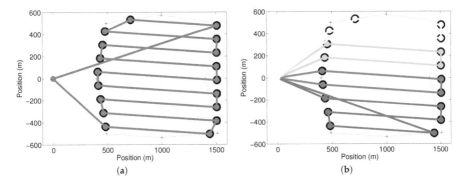

Figure 8. Two trivial solutions for the optimization problem. (**a**) The path for the only UAV available; (**b**) the paths for the four UAVs available when the setup time is not considered or each UAV has it is own operator (the setup time is noncumulative). In those cases, the area is simply divided among the available vehicles.

For the same coverage area and number of UAVs ($M = 4$), but only one human operator, the cumulative setup time in Equation (7) must be considered. Figure 9a shows that the proposed methodology, in this case, found that the problem is best solved with two UAVs ($m = 2$). In this result, the UAVs take 14.73 and 15.10 min to complete their missions, including 4 and 8 min of setup time, which is the optimal solution in this case. To show that this result is correct, we force the use of a third UAV by exploring the effect of constraint in Equation (8), which is related to the duration of the batteries. We have reduced the UAVs' maximum flight time from to 30 down to 10 min. This means that, if the previous solution is considered, none of the UAVs would be able to complete their missions. Using Equation (8), the optimizer automatically selects a third UAV, making $m = 3$, to ensure that the mission will be completed. This solution is shown in Figure 9b, where the mission times for each UAV are 11.54, 15.51 and 16.03 min, including 4, 8 and 12 min of setup time. Notice that this result is worse than the one with two UAVs, which cannot be used in practice due to the reduced flight time. Furthermore, observe that the times of flight of each UAV, computed as the mission times minus the setup times, are always smaller than the battery life (10 min).

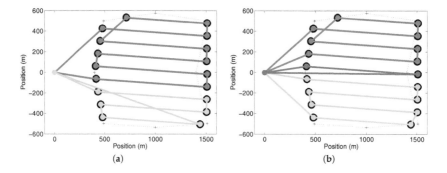

Figure 9. (**a**) Optimal solution for the problem in Figure 8 when the four UAVs available are operated by a single person and the cumulative setup time is considered. Only two UAVs were selected by the method; (**b**) Effect of constraint in Equation (8) when the UAVs' battery life is reduced and a third UAV is necessary to complete the task.

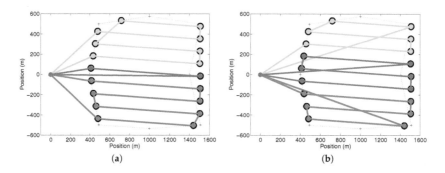

Figure 10. (**a**) Optimal solution for the problem in Figure 8 when the four UAVs available are operated by a team of two operators. Only three UAVs were selected by the method; (**b**) Optimal solution for the problem in Figure 8b when the four UAVs available are operated by a team of three operators. Again, only three UAVs were selected by the method.

For the last set of simulations, we explore the effect of a number of operators larger than one and smaller than the number of available UAVs, using the same coverage area and number of UAVs ($M = 4$) of the previous simulations, but returning the UAVs' maximum flight time to 30 min. Two simulations were made, with teams of two and three operators. In both cases, the number of UAVs selected by the method was $m = 3$. Figure 10a shows that the proposed methodology found a result similar to the one in Figure 9b when two operators were available. In this result, the UAVs take 11.51, 11.54 and 12.03 min to complete their missions, including 4 min of setup time for the first two UAVs and 8 min of setup time for the third one. Despite having the same flight path, mission times are smaller than those in Figure 9b, because the first two UAVs are simultaneously prepared and launched. The third UAV became more productive, having more flight time, when a third operator was included for the team, as shown in Figure 10b, where the mission times for each UAV are 11.10, 11.19, and 11.26 min, including 4 min of setup time for each one of them. Notice that this result is one of the best shown in this paper for this specific area. In the best result, shown in Figure 8b, the longest UAV route took 11.10 min, but this solution employed a fourth UAV and fourth operator, which probably would not be worth a gain of 0.16 min or 9.60 s.

5. Real-World Experiments

We have tested our methodology in practice using two fixed-wing UAVs controlled by the 2128^g Micropilot's autopilot. A picture of one of these UAVs is shown in Figure 11. Each UAV was equipped with a Canon Powershot ELPH130 camera with sensor of 6.17 mm in width and focal length of 5.0 mm. The camera was pointing down. This testbed is described in detail in [22]. Considering that the task would be executed at a height of 120 m and that an overlap of 30% between two consecutive images is required, we use Equation (1) to find the distance between coverage rows to be 105 m. The area to be covered was chosen to have approximate dimensions of 900 m × 1600 m, as shown in Figure 12, where the area and the coverage rows are overlaid on a Google Earth satellite image.

Figure 11. Picture of one of the UAVs used in the experiments.

Figure 12. Coverage area used in the experiments overlaid on a satellite image.

In our first experiment, we use only one UAV ($M = 1$) in order to establish a time reference to cover the area. Figure 13a shows the path followed by the UAV, while Figure 13b shows its altitude. In this second figure, it is possible to see the time when the UAV is effectively executing the mission and the setup time, when it is on the ground. The mission was completed in 27.7 min, including 8 min of setup time. In Figure 13a, a small overshoot can be noticed after each curve made by the UAV. This is due to the fact that the distance between the coverage rows is smaller than the minimum curvature radius of the UAV, which is 115 m. The turn radius is a function of the maximum roll angle, which was limited to avoid the loss of the GPS signal on the curves, once the GPS is on one of the UAV wings.

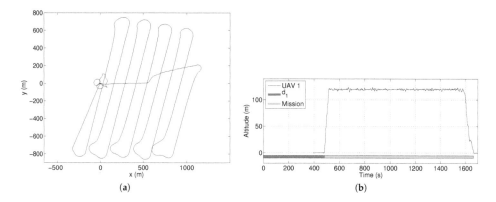

Figure 13. Coverage of the area in Figure 12 with a single UAV. (**a**) The UAV's path projected on a cartographic reference system; (**b**) the UAV's altitude. In this figure, d_1 is the UAV setup time.

In the second experiment, we used two UAVs ($M = 2$). The optimizer distributed seven coverage rows for UAV 2 and only two rows for UAV 1. The mission time was reduced to 24.2 min, including the setup time for each UAV. Notice that the mission time corresponds to the time spent by UAV 1 to cover the area, plus 8 min of setup time. The results for this mission are shown in Figure 14. A video illustrating this experiment can be seen at [23].

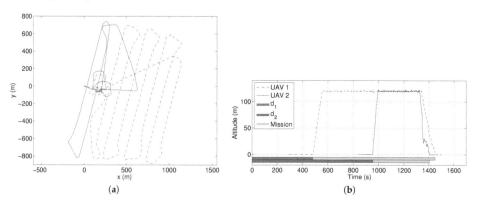

Figure 14. Coverage of the area in Figure 12 with two UAVs. (**a**) The UAVs' paths projected on a cartographic reference system; (**b**) the UAVs' altitude. In this figure, d_1 and d_2 are the cumulative setup times of UAVs 1 and 2, respectively.

For this specific area, it can be noticed that the gain in time to include an extra UAV was only 3 min. Although one can think of a mission where 3 min would make a difference, such as search and rescue operations, we can also see that in most situations, this reduction is not worth the cost of a second UAV. One way to reduce the coverage time even more is reducing the setup time by, for example, increasing the number of human operators working on a single UAV. We tested this strategy with a setup time of only two minutes. In this case, UAV 1 was assigned to five coverage rows, while UAV 2 was assigned to four rows. The mission time was reduced to 18.4 min, which corresponded to a much larger economy of time if compared to the initial 27.7 min. Data from this experiment are in Figure 15.

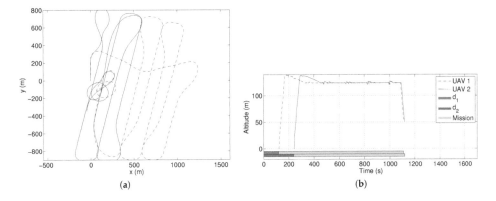

Figure 15. Coverage of the area in Figure 12 with two UAVs and a smaller setup time. (**a**) The UAVs' paths projected on a cartographic reference system; (**b**) The UAVs' altitude. In this figure, d_1 and d_2 are the cumulative setup times of UAVs 1 and 2, respectively.

6. Conclusions and Future Work

This paper presented a methodology for the coverage and sensing of ground areas using fixed-wing UAVs. The main contribution of the method is that it explicitly and formally considers some practical problems that only appear during the deployment of the actual vehicles. The number of UAVs used in the task, for instance, is chosen as a function of the size and format of the area, the maximum flight time of the vehicles and, more importantly, the time needed to prepare and launch the UAV, which we call setup time. This time was never considered before in the solutions for this kind of problem, which frequently resulted in trivial solutions where the time of coverage is inversely proportional to the number of UAVs used. This is certainly not true in practice if the number of human operators is smaller than the number of UAVs to be launched.

It is important to mention that our methodology would not find a solution to the coverage problem in some situations, which include the ones where the number of UAVs and their battery life are small given the size of the area to be covered. In such a situation, it would be interesting to have a methodology that allows the UAVs to land, recharge and take-off again to complete the mission. This strategy is considered as future work.

It is also important to say that the UAV routing strategy proposed in this paper does not take into account possible collisions among the UAVs. To avoid collisions when the planned paths intersect, a velocity planner, such as the one proposed in [24], would be necessary. The addition of such a step to our architecture is left as a future development.

Acknowledgments: The authors thank FINEP/Brazil, FAPEMIG/Brazil (Funding Agency for Research in Minas Gerais), and CNPq/Brazil (National Counsel of Technological and Scientific Development) for the financial support. The authors hold scholarships from CNPq/Brazil.

Author Contributions: Gustavo Avellar was the main developer of the proposed methodology. He also performed experiments, analyzed the results and helped with writing this manuscript. Guilherme Pereira contributed with the methodology and was responsible for writing the manuscript. Luciano Pimenta contributed with the methodology and in the writing of the manuscript. Paulo Iscold participated in the experiments, analyzed the results and reviewed the manuscript.

Conflicts of Interest: The authors declare no conflict of interest.

References

1. Dunbabin, M.; Marques, L. Robots for environmental monitoring: Significant advancements and applications. *IEEE Robot. Autom. Mag.* **2012**, *19*, 24–39.

2. Waharte, S.; Trigoni, N. Supporting search and rescue operations with UAVs. In Proceedings of the International Conference on Emerging Security Technologies, Canterbury, UK, 6–7 September 2010; pp. 142–147.

3. Barrientos, A.; Colorado, J.; del Cerro, J.; Martinez, A.; Rossi, C.; Sanz, D.; Valente, J. Aerial remote sensing in agriculture: A practical approach to area coverage and path planning for fleets of mini aerial robots. *J. Field Robot.* **2011**, *28*, 667–689.

4. Nex, F.; Remondino, F. UAV for 3D mapping applications: A review. *Appl. Geomat.* **2013**, *6*, 1–15.

5. Galceran, E.; Carreras, M. A survey on coverage path planning for robotics. *Robot. Auton. Syst.* **2013**, *61*, 1258–1276.

6. Choset, H. Coverage of Known Spaces: The Boustrophedon Cellular Decomposition. *Auton. Robot.* **2000**, *9*, 247–253.

7. Acar, E.U.; Choset, H.; Rizzi, A.A.; Atkar, P.N.; Hull, D. Morse Decompositions for Coverage Tasks. *Int. J. Robot. Res.* **2002**, *21*, 331–344.

8. Acar, E.U.; Choset, H.; Lee, J.Y. Sensor-based coverage with extended range detectors. *IEEE Trans. Robot.* **2006**, *22*, 189–198.

9. Huang, W. Optimal line-sweep-based decompositions for coverage algorithms. In Proceedings of the IEEE International Conference on Robotics and Automation, Seoul, Korea, 21–26 May 2001; pp. 27–32.

10. Li, Y.; Chen, H.; Joo Er, M.; Wang, X. Coverage path planning for UAVs based on enhanced exact cellular decomposition method. *Mechatronics* **2011**, *21*, 876–885.

11. Maza, I.; Ollero, A. Multiple UAV cooperative searching operation using polygon area decomposition and efficient coverage algorithms. In *Distributed Autonomous Robotic Systems 6*; Springer: City, Japan, 2007; pp. 221–230.

12. Xu, A.; Viriyasuthee, C.; Rekleitis, I. Efficient complete coverage of a known arbitrary environment with applications to aerial operations. *Auton. Robot.* **2014**, *36*, 365–381.

13. Toth, P.; Vigo, D. *The Vehicle Routing Problem*; Society for Industrial and Applied Mathematics: Philadelphia, PA, USA, 2002.

14. Alighanbari, M.; Kuwata, Y.; How, J.P. Coordination and control of multiple UAVs with timing constraints and loitering. In Proceedings of the American Control Conference, Denver, CO, USA, 4–6 June 2003; pp. 5311–5316.

15. Easton, K.; Burdick, J. A Coverage Algorithm for Multi-robot Boundary Inspection. In Proceedings of the 2005 IEEE International Conference on Robotics and Automation, Barcelona, Spain, 18–22 April 2005; pp. 727–734.

16. Ergezer, H.; Leblebicioglu, K. 3D Path Planning for Multiple UAVs for Maximum Information Collection. *J. Intell. Robot. Syst.* **2014**, *73*, 737–762.

17. Guerriero, F.; Surace, R.; Loscrí, V.; Natalizio, E. A multi-objective approach for unmanned aerial vehicle routing problem with soft time windows constraints. *Appl. Math. Model.* **2014**, *38*, 839–852.

18. Kivelevitch, E.; Sharma, B.; Ernest, N.; Kumar, M.; Cohen, K. A Hierarchical Market Solution to the Min-Max Multiple Depots Vehicle Routing Problem. *Unmanned Syst.* **2014**, *2*, 87–100.

19. Christofides, N.; Mingozzi, A.; Toth, P. Exact algorithms for the vehicle routing problem, based on spanning tree and shortest path relaxations. *Math. Program.* **1981**, *20*, 255–282.

20. Lofberg, J. YALMIP: A toolbox for modeling and optimization in MATLAB. In Proceedings of the IEEE International Conference on Robotics and Automation, Taipei, Taiwan, 4 September 2004; pp. 284–289.

21. Gurobi Optimization Inc. Gurobi Optimizer Reference Manual. 2015. Available online: http://www.gurobi.com (accessed on 28 October 2015).

22. Avellar, G.S.C.; Thums, G.D.; Lima, R.R.; Iscold, P.; Torres, L.A.B.; Pereira, G.A.S. On the development of a small hand-held multi-UAV platform for surveillance and monitoring. In Proceedings of the International Conference on Unmanned Aircraft Systems, Atlanta, GA, USA, 28–31 May 2013; pp. 405–412.

23. Avellar, G.S.C.   Area Coverage with Multiple UAVs.   Available online: https://youtu.be/07Kc6nIVzwA (accessed on 28 October 2015).
24. Gonçalves, V.M.; Pimenta, L.C.A.; Maia, C.A.; Pereira, G.A.S. Coordination of multiple fixed-wing UAVs traversing intersecting periodic paths. In Proceedings of the IEEE International Conference on Robotics and Automation, Karlsruhe, Germany, 6–10 May 2013; pp. 841–846.

Article

UAV Deployment Exercise for Mapping Purposes: Evaluation of Emergency Response Applications

Piero Boccardo [1], Filiberto Chiabrando [2,*], Furio Dutto [3], Fabio Giulio Tonolo [4] and Andrea Lingua [2]

[1]   Politecnico di Torino—Interuniversity Department of Regional and Urban Studies and Planning (DIST), Viale Mattioli 39, 10125 Torino, Italy; piero.boccardo@polito.it
[2]   Politecnico di Torino—Department of Environment, Land and Infrastructure Engineering (DIATI), C.so Duca degli Abruzzi 24, 10129 Torino, Italy; andrea.lingua@polito.it
[3]   Città Metropolitana di Torino—Servizio Protezione Civile, Via Alberto Sordi 13, 10095 Grugliasco (TO), Italy; furio.dutto@cittametropolitana.torino.it
[4]   Information Technology for Humanitarian Assistance, cooperation and Action (ITHACA), Via P.C. Boggio 31, 10138 Torino, Italy; fabio.giuliotonolo@ithaca.polito.it
*   Author to whom correspondence should be addressed; filiberto.chiabrando@polito.it; Tel.: +39-11-0904380; Fax: +39-11-0904399.

Academic Editor: Felipe Gonzalez Toro
Received: 14 May 2015; Accepted: 23 June 2015; Published: 2 July 2015

Abstract: Exploiting the decrease of costs related to UAV technology, the humanitarian community started piloting the use of similar systems in humanitarian crises several years ago in different application fields, *i.e.*, disaster mapping and information gathering, community capacity building, logistics and even transportation of goods. Part of the author's group, composed of researchers in the field of applied geomatics, has been piloting the use of UAVs since 2006, with a specific focus on disaster management application. In the framework of such activities, a UAV deployment exercise was jointly organized with the Regional Civil Protection authority, mainly aimed at assessing the operational procedures to deploy UAVs for mapping purposes and the usability of the acquired data in an emergency response context. In the paper the technical features of the UAV platforms will be described, comparing the main advantages/disadvantages of fixed-wing *versus* rotor platforms. The main phases of the adopted operational procedure will be discussed and assessed especially in terms of time required to carry out each step, highlighting potential bottlenecks and in view of the national regulation framework, which is rapidly evolving. Different methodologies for the processing of the acquired data will be described and discussed, evaluating the fitness for emergency response applications.

Keywords: UAV; orthophoto; emergency response; photogrammetry; civil protection

1. Introduction

An unmanned aerial vehicle (UAV), commonly known as a drone and also referred to as an unpiloted aerial vehicle or a remotely-piloted aircraft (RPA) by the International Civil Aviation Organization (ICAO) is an aircraft without a human pilot aboard [1].

As highlighted in Office for the Coordination of Humanitarian Affairs (OCHA) [2], UAVs, previously mainly associated with military applications, are increasingly being adopted for civilian uses. UAVs are exploited in civilian (and also commercial) applications, like agriculture, surveying, video making and real estate. Exploiting the decrease of costs related to UAV technology, also the humanitarian community started piloting the use of similar systems in humanitarian crises several years ago in different application fields, *i.e.*, disaster mapping and information gathering, community capacity building, logistics and even transportation of goods.

Part of the author's group, composed of researchers in the field of applied geomatics, has been piloting the use of UAVs since 2006 [3], with a specific focus on disaster management applications [4]. The present paper is therefore focused on the exploitation of UAVs in the emergency mapping domain, defined as "creation of maps, geo-information products and spatial analyses dedicated to providing situational awareness emergency management and immediate crisis information for response by means of extraction of reference (pre-event) and crisis (post-event) geographic information/data" [5]. Emergency mapping can be adopted in all of the phases of the emergency management cycle (*i.e.*, "the organization and management of resources and responsibilities for addressing all aspects of emergencies, in particular preparedness, response and initial recovery steps" [6], but the goal of the research is mainly to investigate potential applications supporting the immediate emergency response phase ("the provision of emergency services and public assistance during or immediately after a disaster in order to save lives, reduce health impacts, ensure public safety and meet the basic subsistence needs of the people affected" [6]).

As far as emergency mapping UAV applications are concerned, several organizations have already begun using the consolidated capability of UAVs to produce maps and provide high-resolution imagery as part of disaster response or disaster risk reduction programming. The main interest is in damage assessment information and in population count estimation [7].

Furthermore, taking the operational status of the current UAV technology into account, Politecnico di Torino decided to establish the Disaster Recovery Team (DIRECT [8]), a team of university students in the field of architecture and engineering with the goal to support disaster management activities through the application of geomatics techniques (3D surveying, remote sensing, mapping, WebGIS). In the framework of the DIRECT activities, a UAV deployment exercise was jointly organized with the Regional Civil Protection authority in July 2014 mainly aimed at assessing the effectiveness of the operational procedures in place to deploy UAVs for mapping purposes and the usability of the acquired data in an emergency response context.

The paper is structured in two main parts, focusing respectively on UAV deployment operational procedures (Sections 3 and 4) and image post-processing (Section 5). In detail, Section 3 addresses the main phases of the adopted operational procedure, especially in terms of time required to carry out each step, highlighting potential bottlenecks, and in view of the national regulation framework, which is rapidly evolving. Furthermore the technical features of the UAVs platforms employed during the exercise will be described in Section 4. Different methodologies for the processing of the acquired data aimed at generating and distributing in near-real-time up-to-date orthoimages will be described and discussed in Section 5, with a specific focus on the positional accuracy and the fitness for emergency response applications.

2. UAV Deployment Exercise

In the framework of the DIRECT activities and in collaboration with the Provincial Civil Protection authority (Protezione Civile della Provincia di Torino), a UAV deployment exercise was organized in July 2014 with a two-fold aim: firstly, to test operationally UAV deployment procedures and secondly to assess the usability of the acquired high resolution imagery in an emergency response context. The present section will provide a detailed description of the deployment exercise setup.

A specific area of interest (covering a surface of approximately 1.5 km^2; Figure 1) to be mapped at different levels of detail was defined in advance, with the goal to simulate an area that could have been impacted by a crisis event, e.g., a plain flood.

Figure 1. Deployment exercise location (**left**) and detail of the area to be mapped (**right**).

Two different types of UAV platforms were planned to be tested, specifically a multi-rotor system and a fixed-wing airframe (described in Section 4), with the goal to evaluate the fitness for the purpose of similar platforms in an emergency context and possible specific limitations or advantages.

From an operation perspective and taking the immediate response phase into account, the main goal is to image and map the affected areas in the shortest time possible from the mobilization request, in order to provide to civil protection authorities updated imagery depicting the situation after the event. For this purpose, a specific list of operational phases was defined in advance, with the aim to evaluate the time required to carry out each step and, consequently, to identify potential bottlenecks to be further investigated and improved. The detailed description of the aforementioned steps and the outcomes of the evaluation is provided in Section 3.

As far as the orthoimage production is concerned, two different types of test were planned:

- firstly, a fast processing in the field (Section 5.2) exploiting as ground control points (GCP) pre-positioned markers to be quickly measured with low-cost devices and low positional accuracy (*i.e.*, mobile phones; Section 5.1);
- secondly, a rigorous processing (to be carried out off-line; Section 5.3) based on the same GCPs used in the previous test, but measured with a higher accuracy with topographical instrumentation, to be mainly used as a reference in the positional accuracy assessment phase (Section 5.4).

Lastly, it was also planned to test the possibility to upload the processed image on a web mapping server directly from the field, exploiting the satellite connectivity offered by the fully-equipped mobile unit made available by the Regional Civil Protection authority (Figure 2).

Figure 2. The Regional Civil Protection Mobile Unit, fully equipped with a complete "office" package (including web satellite connectivity): on its way to the test site (**left**) and deployed in the field (**right**).

3. UAV Deployment Operational Phases: Description and Discussion

As reported in the previous section, a complete list of the operational phases required to correctly carry out a UAV survey was defined in detail. Each phase and the related aim are described in this section: furthermore, the average time required to carry out the related tasks during the exercise is provided and discussed.

Initial briefing: This phase follows a UAV service request to support emergency management activities (e.g., request to map damages to infrastructures in an industrial zone affected by a flood event). Preliminarily, it is necessary to get information on the location of the affected areas, mainly required to plan the deployment of the team and the instruments in the field and to identify an area suitable to be used as a coordination center in the field (allowing UAV's taking-off and landing operations). Secondly, the requirements of the users requesting the service are analyzed, these being strictly connected to the technology to be adopted (e.g., mapping of a single building with rotor platforms *versus* mapping wide areas with fixed-wing airframes) and the technical details of the acquisitions (e.g., spatial/spectral resolution of the sensors installed on the UAV, average flight height, *etc.*). Lastly, a thorough assessment of the request is carried out, also taking local and/or international regulations into account, for a final decision of the feasibility of a UAV deployment.

This phase took about 30–45 min during the deployment exercise in the field, but it has to be considered that this was made possible mainly by the controlled environment of a simulated exercise. According to the authors' experience, this phase could be one of the most time-consuming tasks, especially considering the specific planning activities and risk analyses to be carried out on a case by case basis according to the (evolving) Italian UAV regulation provided by Ente Nazionale Aviazione Civile (ENAC [9]).

Deployment of the team in the affected area: This phase is required to allow the mapping team (and the related hardware instrumentation, *i.e.*, UAV platforms, ground station, mobile devices for GCP collection, markers, *etc.*) to reach the selected operation base.

The time required to carry out this operation is obviously strictly related to the organization of the institution managing the UAV deployment and the possible pre-positioning of UAV teams in areas at risk or allowing a complete local/regional/national coverage: the time span could therefore range from a few hours (as in the exercise case) to days.

Remotely-piloted aircraft system setup: This phase is aimed at correctly setting up the hardware and software required for the UAV mission and carrying out the flight operations, as well as the subsequent image downloading and related post-processing (if required). During the exercise, this preliminary phase took about 20 min.

Marker pre-positioning and survey: In order to grant an adequate positional accuracy of the processed orthophoto, it is required to identify on the acquired images features of interest whose coordinates with respect to a defined reference system are known with a suitable accuracy. Considering that most often in operational cases, large-scale maps (with a sufficient level of detail) are not available or the accuracy achievable exploiting global map services (e.g., Google Maps, OpenStreetMap) is not considered sufficient for the aim of the survey, it may be necessary to pre-position *ad hoc* markers that will be clearly visible in the acquired imagery. During the positioning operations, which should be carefully planned in the case of large areas to be surveyed to grant a homogeneous distribution of the markers and to speed up the process, the coordinates of the marker should be measured at the same time for efficiency purposes (see Section 5.1 for details).

The positioning and measuring of 16 markers during the exercise took approximately 1 h.

Flight plan setup: This task is crucial to plan the flight operations, and it is strictly related to the required technical features of the final map products, mainly in terms of spatial resolution and required overlapping percentages between images (to allow a proper photogrammetric 3D processing). UAV systems include *ad hoc* software to semi-automatically generated flight/acquisition plans to be uploaded onto the autopilot system. A skilled operator should be able to carry out this task in a few minutes: this was the case during the exercise.

UAV flight operations: This phase includes all of the tasks required to operate the UAV and to properly acquire the required data, including taking-off, landing and data downloading operations. The flight tests performed during the exercises took an average time of 15 min, ranging from 9 min (at a higher flight height) with rotor platforms to 17 min (covering a larger area) with fixed-wing platforms. It has to be noted that fixed-wing platforms are generally capable of covering larger areas in a shorter time than rotor platforms. It is also stressed the need to have redundant batteries (or even platforms) to optimize the mission time, limiting as much as possible the overall downtime.

Data processing (orthophoto generation): Once the raw images are downloaded, it is required to process the data in order to generate the required value-added products, including pre-processing tasks as data resampling or image selection (to limit the processing only to the minimum set of images actually required to cover the affected areas). The average time required to carry out this step during the exercise was about 2 h.

Data dissemination: This is the last phase of the activities, which is aimed to make the generated products (e.g., orthophoto) available to the users (which may be in remote locations, e.g., national civil protection headquarters). During the exercise, this step was carried out exploiting an *ad hoc* web map server, making the final orthoimage accessible as a standard OGC (Open Geospatial Consortium) WMS (Web Map Service). Image uploading and service setup required about 10 min; clearly, the actual time is strictly related to the data file size and the available upload bandwidth.

It has to be highlighted that some of the aforementioned operations can be carried out in parallel in order to minimize the overall acquisition time, obviously depending on the number of people composing the team.

4. Employed UAV Platforms and Sensors

The UAV deployment exercise was planned to test two different platforms: a multi-rotor RPAS (Remotely Piloted Aircraft System) and a fixed-wing system.

4.1. Rotor Platform and Sensor

The first test was performed over an area of 400 m × 400 m using a Hexakopter by Mikrokopter (Figure 3).

Figure 3. Hexakopter multi-rotor platform (**left**) and take-off and flying phases (**center** and **right**).

The system (technical details are available at [10]) is composed of six motors and the electronic equipment required for both the remote control and the automatic flight (*i.e.*, one flight control adaptor card, one remote control, one navigation control, one GPS receiver, a three-axis magnetometer, one wireless connection kit) as well as one computer serving as the ground control station.

As far as the image acquisition sensor is concerned, the multi-rotor platform is equipped with a commercial off-the-shelf (COTS) Sony Nex 5 digital mirror less camera. The digital camera is mounted on a servo-assisted support that grants electronically-controlled rotations along two directions (the x- and h-axis) with the goal to acquire vertical imagery (lens axis pointing at nadir).

In Table 1, the main technical features of the employed multi-rotor platform are reported.

Table 1. Main technical features of the employed multi-rotor platform.

Hardware	
Weight	2.5 kg
Payload	1 kg
Propulsion	Electric
Camera (focal length = 16 mm)	Sony Nex 5 (pixel size 5.22 μm)
Operation	
Maximum flight time	12–15 min
Nominal cruise speed	3–5 m/s
Ground Sampling distance at 100 m	0.032 m
Linear landing accuracy	Approximately 1 m

In order to have two different sets of test images and to simulate operations in the two volumes of space mentioned by the Italian RPAS regulation, two different flights were planned and carried out. One flight with a height of 70 m (the maximum height above the ground allowed by volume of space V70 according to the ENAC regulation) and a second flight performed at 150 m (the maximum height above the ground allowed by volume of space V150 according to the ENAC regulation). In the first case, the area was covered with eight stripes and 190 images characterized by a ground sample distance (GSD) of 0.022 m. The second flight allowed the coverage of the area with four stripes and 120 images with a GSD of 0.05 m. The longitudinal and lateral overlapping were respectively set at 80% and 30% in both cases. Using the Hexakopter, the flight plan is managed by the Mikrokopter tool OSD that connects the platform to the ground station. The tool is exploited to set all of the parameters of the flight plan, using as a reference map of the area one of the images made available by several on-line map servers (e.g., Google Maps, Bing Maps, *etc.*). The flight planning approach is the usual one based on the definition of waypoints (a reference point in the physical space used for the navigation, usually placed at the beginning and at the end of the stripes).The RPAS aims to follow the direction between the starting and ending points according to the defined azimuth. In Figure 4, two screen shots of the planned flight plans are shown. A good practice is also to insert the last waypoint close to the take-off and landing area (Figure 4 (left), waypoint P17), in order to simplify the manual landing operation.

Figure 4. The multi-rotor RPAS planned flight plans: 70-m flight height above the ground (**left**) and 150-m flight height above the ground flight (**right**).

Using the Mikrokopter tool, several other parameters could be set, such as flight speed, elevation, shooting time, *etc.* According to the aim of the test, the time required to carry out the relevant phases of these flights was recorded. In the first case (70 m flight height), 10 min were required for the planning phase and 13 min for flying over the area. The second flight (150 m flight height) was planned in 5 min, and the images were acquired in 9 min. Finally, the images stored in the SD card were downloaded on a laptop (for the post-processing phase) in about 5 min.

4.2. Fixed-Wing Platform and Sensor

A fixed-wing RPAS was also evaluated. The platform employed in the test was a consumer product of the SenseFly company (technical details are available at [11]), namely the eBee autonomous flying drone (Figure 5).

Figure 5. The eBee fixed-wing platform (**left**) and take-off phase (**right**).

During the UAV deployment exercise, the system was used with an RGB Canon Ixus COTS camera to acquire visible imagery. Other sensors, tailored to the system, are also available, *i.e.*, a near-infrared camera (to acquire false color imagery), the multi-SPEC 4C camera that acquires multi-spectral images (green, red, red-edge and near-infrared) and the thermal sensor that produces grey scale images and video with high pixel density and thermal resolution.

The main technical features of the platform are reported in Table 2.

Table 2. Main technical features of the employed fixed-wing RPAS.

Hardware	
Weight (camera included)	0.700 kg
Wingspan	96 cm
Propulsion	Electric
Camera (focal length = 4.3 mm)	16 Mp Ixus/ELPH (pixel size 1.33 µm)
Operation	
Maximum flight time	50 min
Nominal cruise speed	11–25 m/s
Ground Sampling distance at 100 m	0.031 m
Linear landing accuracy	Approximately 5 m

The area selected for the test of the fixed-wing platform flight was approximately 1 km². In order to have a GSD suitable for producing an orthophoto fitting a map scale of about 1:500, the flight height was set at 150 m. Sixteen stripes and 160 images were acquired with a 70% longitudinal and 30% lateral overlapping (Figure 6). The GSD (of the acquired images according to the aforementioned parameters was about 0.05 m.

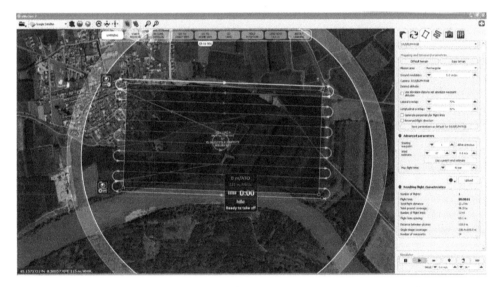

Figure 6. The fixed-wing RPAS flight plans: 150-m flight height above the ground.

The eBee flight plan is managed through the eMotion software by SenseFly, which is based on an automatic photogrammetric approach. Using eMotion, the initial parameters to set up are the ones related to the area of interest, the GSD and the overlapping percentages between the images (lateral and longitudinal). Once these parameters are set, the software automatically calculates the number of stripes required to cover the area of interest, as well as the flight height. The second planning step is focused on take-off and landing settings. This step is obviously crucial, with RPAS automatically managing the landing phase, since an area without vertical obstacles in a certain range is absolutely necessary. The eBee performance during the landing phase is excellent, considering that (exploiting the onboard GNSS and ultrasound proximity sensors) the platform is able to land in a predefined area with an accuracy of about 5 m.

As far as the time stamps are concerned, the flight planning was carried out in about 10 min, while the flight time was 17 min. The raw images' download was carried out in about 5 min.

4.3. Summary of the Time Required for Each Flight Step

Table 3 summarizes the time required for flight planning, flight execution and image downloading in each performed tests.

Table 3. Time required for each flight step for the different test configurations.

Platform and Flight Height	Planning (min)	Flight Time (min)	Data Download (min)	Acquired Area (km²)
Microkopter (70 m)	10	13	5	0.15
Microkopter (150 m)	5	9	5	0.20
eBee (150 m)	10	17	5	1

5. Orthophoto Generation and Accuracy Assessment

One of the aims of the test was the generation of a cartographic product able to document from the geometric and thematic perspective the surveyed area. An orthophoto is a geospatial dataset meeting these requirements, since it is a cartographic product able to provide both geometrically-corrected information (with uniform scale and corrected distortions) and radiometric information. In order to perform the orthorectification process of a set of images (covering the surveyed area from different points of view) it is necessary to: (i) estimate the position and attitude of the images (external orientation parameters) in a defined reference system (cartographic or local); and (ii) have the elevation model of the area, generally extracted during the processing as a digital surface model (DSM).

The aforementioned process is based on the photogrammetric approach. Currently, a fast and cost-effective approach for the estimation of the interior and external orientation, the extraction of the DSM and the production of the orthophoto is based on the SfM (structure from motion [12]) methodology, implementing the usual photogrammetric workflow through computationally-efficient computer vision algorithms.

In the framework of this test, Photoscan software (by Agisoft) was employed, which is one of the most popular tools used for 3D reconstruction purposes in the international community [13–16]. Being COTS software, very few algorithm details are available. The tie points extraction is performed using a modified SIFT (scale-invariant feature transform, [17]) approach. The external orientation (bundle block adjustment) is performed using the most common algorithm of the computer vision community, namely the Gauss–Markov approach [18]. The image matching seems to be realized using a semi global matching-like approach [19]. After the point cloud extraction step, as final products, a 3D model (Triangulated Irregular Network—TIN format), textures, a DSM and, eventually, the orthophoto are generated (Figure 7).

Figure 7. Oriented images with the number of overlaps (**left**), DSM (**center**) and orthoimagery (**right**).

In the following section, the strategy used for the realization of the orthophoto is reported.

In order to evaluate different scenarios, the data processing steps were grouped according to the characteristic of the flights and the number of GCPs used for the bundle block adjustment.

Naturally, all of the processing steps were evaluated in terms of time (starting from the ground control point acquisition up to the web upload through a map server) and the positional accuracy of the realized cartographic product.

5.1. GCP Acquisition

The correct workflow to generate a geospatial product with controlled positional accuracy requires the coordinates of several ground control points (GCPs) to be used in the georeferencing process (unless a direct georeferencing approach is adopted, as in one of the eBee-related tests) and for the accuracy evaluation. For this purpose, at least four GCPs are required, but a redundant number of points are obviously measured. Usually, a good compromise is ten points in a square kilometer [20,21].

In certain conditions, it is not possible to easily identify natural features (to be used as GCP) on the images; therefore, it is necessary to pre-position *ad hoc* markers before the UAV flight. Those markers should homogeneously cover the whole area of interest and should be easily pin-pointed on the images. During the deployment exercise, 16 square wood panels (40 cm × 40 cm) with a black circle on a white background were use as markers (Figure 8c,d).

The markers were then measured with two different approaches: firstly using a user-friendly device (e.g., a common GPS-equipped smartphone with a suitable app with the aim to test the results achievable with a user-friendly and low-cost technology) and then by means of a real-time kinematic (RTK) GPS survey, to have higher precision to be used as a reference in the accuracy assessment phase.

In the first case, during the test, the GCPs were measured using two different mobile applications available for Android (U-Center) and Windows (GPS-GPX logger), respectively installed on a Samsung Galaxy S5 and a Lumia 1020, involving two different teams (three persons each).

With the employed smartphones, the obtained precision according to the results reported in the employed apps was about 3 m, as expected with mobile GPS embedded sensors [22]. The GCP survey phase took about 45 min. Figure 8 shows a screenshot of one of the employed apps (a, b) and a detail of the marker (c, d).

Figure 8. GPS coordinates measured by Samsung S4 with the GPS-GPX logger app (**a**), with the Nokia Lumia app (**b**); Details of the measured markers (**c**,**d**).

Concerning the real-time kinematic (RTK) GPS survey, as expected with a GNSS RTK survey, the coordinates of the GCPs were measured with a horizontal accuracy of 0.02 m and a vertical accuracy of about 0.04 m.

In this case, the survey was quite time consuming compared to the previous one: the points were surveyed in about 55 min with a Leica GNSS system 1200 using a virtual reference station (VRS) approach [23].

Table 4 summarizes the number of acquired points, the acquisition time and the mean accuracy of the two methods.

Table 4. Time required for ground control point (GCP) survey and average estimated accuracy. RTK, real-time kinematic.

Survey System	GCPs (No.)	Acquisition Time (min)	Average Planimetric Accuracy (m)	Average Elevation Accuracy (m)
Smart phone	16	45	3	3
GNSS-RTK	16	55	0.15	0.20

5.2. Orthophoto Generation: Fast Processing in the Field

The acquired images and GCPs coordinates were then exploited to generate an orthophoto with Photoscan. The software workflow is based on: photo alignment, dense cloud generation, georeferencing, mesh generation, texture generation and orthophoto export.

To speed up all of the operations (the goal being processing in the field to limit the processing time as much as possible), the accuracy of each step was set at the lower level possible. The image processing steps were carried out on a laptop with the following technical features: Pentium i7 2.40 GHz, 16 GB RAM with an Nvidia GeForce GTX 670 2 GB. The data acquired during the two multi-rotor flights (70-m and 150-m flight height) were processed using the GCPs' coordinates with lower accuracy (the ones being quickly available in the field). Figure 9 shows the output orthophoto.

Figure 9. Orthophoto based on the images acquired with the multi-rotor UAV: 70-m flight height (**left**) and 150-m flight height (**right**).

The eBee data were processed with a direct georeferencing approach exploiting the GPS/IMU position/attitude measurements as initial approximate values [24]. The sensors installed on the eBee platform are a GPS chip (U-Blox chipset), which provides a position based on C/A (Coarse/Acquisition) code at 1 Hz (no raw code data is recorded), and an attitude sensor, which provides the three attitude angles (roll, pitch and heading). In order to import those parameters in Photoscan, the export option of the eMotion software was exploited to get an ASCII file with the coordinates (latitude, longitude, altitude) and attitude (heading, pitch and roll) of each acquired image (Figure 10).

```
#Rev: sensefly/geoinfo/2

fileName      latitude      longitude     altitude_amsl     altitude_wgs84     heading       pitch          roll
IMG_0653.JPG  45.1584770000 8.3698379000  279.9995117188    330.2305297852     85.2294235229 2.4848692417   -1.4296081066
IMG_0654.JPG  45.1584861000 8.3704315000  281.0472717285    330.5222473145     87.5284271240 3.2318589687   1.1986414194
IMG_0655.JPG  45.1585244000 8.3710128000  282.5871276855    332.5290832520     69.8725891113 1.5545492172   -0.0942675471
IMG_0656.JPG  45.1585208000 8.3716217000  283.3697814941    333.5429992676     90.2204895020 3.4727685452   -4.9474272728
IMG_0657.JPG  45.1584984000 8.3721966000  283.1581420898    333.3637695312     82.2723999023 0.9797640443   -2.0585665703
IMG_0658.JPG  45.1585139000 8.3728102000  281.2704467773    331.3168945312     88.5572204590 2.6897964478   4.5214247704
IMG_0659.JPG  45.1585317000 8.3733939000  281.6551818848    331.0068054199     82.5366973877 3.7361104488   6.9712395668
IMG_0660.JPG  45.1585240000 8.3739759000  282.1534729004    331.5271911621     89.9098205566 -2.0950198174  2.1271202564
```

Figure 10. Example of the external orientation parameters as measured by the eBee platform, required for the direct georeferencing approach.

All of the processing steps carried out with the images acquired by the multi-rotor platform were applied also for the eBee data, obviously excluding the GCP step. Figure 11 shows the final orthophoto.

Figure 11. Orthophoto based on the images acquired with the fixed-wing UAV (direct georeferencing approach).

For comparison purposes, the same data were re-processed also using as input the GCPs measured with smartphones. Table 5 summarizes the main settings of the different data processing in the field.

Table 5. Main settings of the image processing step in the field.

Platform	Flight Height (m)	Employed GCPs Measured with Smartphones (No.)	Processed Images (No.)	GSD (m)	Processing Time (h:min:s)
Multi-rotor	70	6	190	0.044	1:43:00
Multi-rotor	150	10	120	0.091	1:40:00
eBee	150	N/A	163	0.040	1:50:00
eBee	150	10	163	0.040	1:55:00

In conclusion, the test proved multi-rotor platforms to be more flexible, allowing one to carry out the take-off and landing operations in smaller areas even with vertical obstacles. Furthermore, their flight height can be lower than the one required by a fixed-wing platform, providing images with higher GSD and more flexibility in setting the required photogrammetric parameters (e.g., stereo pair base to optimize the estimated vertical accuracy). These platforms are usually the first choice for surveying small areas or isolated buildings.

On the other hand, if large areas need to be mapped, the preferred approach is based on fixed-wing UAVs, allowing speeding up of the acquisition process (by flying at a 150-m height, it is possible to cover 1 km^2 in about 15 min). The main disadvantage of these platforms is the need for relatively large landing areas (at least 50 m long) without vertical obstacles to allow the platform to correctly approach the landing point (although the eBee platform has also a circular landing mode to cope with this issue).

5.3. Orthophoto Generation: Rigorous Off-Line Processing

In order to evaluate the possible differences in terms of accuracy and processing time between a fast processing approach (GCPs measured by smartphones and low accuracy processing setting, standard performance laptop) and a more rigorous approach (GCPs measured by GNSS RTK and medium/high accuracy processing setting, high performance desktop computer), all of the acquired data were re-processed in the Geomatics Laboratory of the Politecnico di Torino with a higher performance computer (Pentium i7 3.70 GHz, 16 GB RAM with two Nvidia GeForce GTX 7502 GB).

The Photoscan workflow was therefore set to exploit medium/high accuracy settings during the processing; specifically, a medium accuracy value was set for the tie point extraction step and a high accuracy value for the mesh generation phase. The time required for processing was shorter than the time required in the field, despite the more rigorous setting, thanks to the higher performance of the adopted desktop computer; on the other hand, the final results in terms of radiometric quality were comparable to the outputs generated in the field, demonstrating that even a fast processing in the field fulfils the emergency response requirements.

As expected, the GCPs measured with the RTK survey allowed improving of the positional accuracy of the orthophoto (the details of the accuracy assessment phase are provided in Section 5.4).

Table 6 summarizes the main settings of the different data off-line processing.

Table 6. Main settings of the off-line image processing step.

Platform	Flight Height (m)	Employed GCPs Measured with GNSS RTK (No.)	Processed Images (No.)	GSD (m)	Processing Time (h:min:s)
Multi-rotor	70	6	190	0.044	1:30:00
Multi-rotor	150	10	120	0.091	1:25:00
eBee	150	N/A	163	0.040	1:40:00
eBee	150	10	163	0.040	1:50:00

5.4. Accuracy Assessment Results

According to the typical geomatics researcher mind-set, all of the products were compared in order to assess the geometric accuracy, with the final goal to estimate the possible nominal map scale of the orthophoto (according to the Italian national mapping standards).

The accuracy was evaluated using as check points (CP) the most accurate surveyed coordinates (by means of a GNSS RTK survey). For the orthophoto based on GCPs measured by smartphones and on the direct georeferencing approach, all 16 points measured by the GNSS RTK survey were used as CP. For the orthophoto based on GCPs measured by the GNSS RTK survey, the accuracy evaluation was performed using as CP six points not employed in the photogrammetric process.

Table 7 summarizes the average positional accuracy and the estimated nominal map scale of the orthophoto products.

Table 7. Results of the accuracy assessment phase of the orthophoto products.

Product ID	Platform	Flight Height (m)	Employed Strategy	Easting Average Accuracy (m)	Northing Average Accuracy (m)	Elevation Average Accuracy (m)	Nominal Scale (According to the Planimetric Error)
1	Multi-rotor	70	Smartphone GCPs	1.24	1.03	2.65	1:5000
2	Multi-rotor	70	RTK GCPs	0.06	0.03	0.04	1:500
3	Multi-rotor	150	Smartphone GCPs	1.10	0.78	6.45	1:5000
4	Multi-rotor	150	RTK GCPs	0.10	0. 25	0.36	1:1000
5	eBee	150	Direct geo	1.48	1.55	12.01	1:10,000
6	eBee	150	RTK GCPs	0.08	0.10	0.48	1:1000

According to the results summarized in Table 7, it is clear that the usage of GCPs measured by traditional GNSS survey techniques leads to higher positional accuracy that fits very large nominal map scales (Products 2, 4 and 6). On the other hand, it is also evident that the accuracies of orthophotos based on smartphone-measured GCPs (Products 1, 3), having displacements lower than 1.5 m, fit medium nominal scales that are indeed suitable for emergency mapping purposes. The more interesting consideration is the fact that the same level of accuracy can be obtained with a direct georeferencing approach (Product 6), without any need to measure GCPs in the fields, therefore reducing the overall time required to get usable post-event products and allowing technicians to carry out surveys also in areas with limited accessibility (where a proper GCP survey would not be feasible). It has to be highlighted that the direct georeferencing approach could be further improved using a more accurate GNSS sensor, such as the one embedded in the recently released RTK version of the eBee platform: the first tests highlight that centimeter accuracy can be obtained without any control points [25].

6. Example of Value-Added Application in the Emergency Mapping Domain

An up-to-date, very high resolution orthophoto is one of the main data sources exploited for emergency mapping activities, with the main goal to identify the most affected areas and to assess the damages to the main infrastructures, providing value-added products to end users. Typical value-added products are in the form of cartographic products, and generally, the underlying datasets are also provided to end users as vector layers.

Most often, the aforementioned input imagery is acquired from satellite platforms, due to the possibility to trigger a wide range of satellite sensors (both optical and SAR) with the goal to get usable images in the shortest time frame possible. Nevertheless, in some circumstances, satellite platforms may not be able to provide adequate imagery for several reasons: type of post-event analysis requiring a spatial resolution larger than the one offered by satellite sensors (*i.e.*, 0.3 m as of May 2015), persistency of cloud coverage and need for optical imagery, illumination conditions of the affected areas at the time of satellite passage (at very high latitudes), possible acquisition conflicts leading to repeated acquisition cancellations, orography of the affected areas leading to heavy geometrical distortions. In those cases, aerial imagery acquired by UAV platforms is a possible alternative data source to be used for the post-event analysis. Generally, UAV imagery is also considered an *in situ* dataset that can complement satellite-based analysis: the European Commission is already assessing the potential role of UAVs, using the collected imagery "as an alternative and/or complementary source of post-event imagery in emergency situations and in a rapid response and mapping context" [26].

When very high resolution optical imagery is available, the typical workflow to assess the damages to infrastructure is based on computer-aided photo interpretation (CAPI) techniques carried out by skilled image interpreters [27], which certainly benefits from the typical very high spatial resolution of UAV imagery.

Figure 12 shows an example of a value-added map produced during the response phase to an earthquake that struck Northern Italy on May 2012 (in the framework of the European Commission Copernicus Emergency Management Service Mapping) assessing damages to buildings on the basis of very high resolution optical imagery acquired from aerial and satellite platforms.

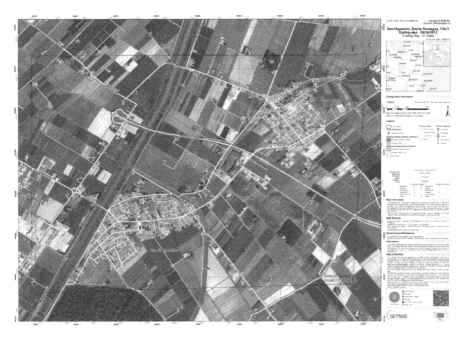

Figure 12. Example of post-event damage assessment map based on both aerial and satellite imagery (Copernicus Emergency Management Service–Mapping [26]).

7. Discussion and Conclusions

As described in Section 2, the UAV deployment exercise had a two-fold aim: to operationally test UAV deployment procedures with a focus on the timeliness of the service and, secondly, to assess the usability of the acquired high resolution imagery in an emergency response context, especially in terms of spatial accuracy.

As far as the UAV deployment procedures are concerned, the evaluation of the time required to carry out each operational task highlighted that the more demanding steps are the ones required to initially assess the request and to deploy the team in the field. The team deployment operation can reasonably last from a few hours to days, especially considering possible accessibility issues most often characterizing disaster-affected areas. Once in the field, each technical procedure related to the flight survey and the image processing takes from a few minutes (e.g., flight plan planning) to a couple of hours maximum (e.g., marker GPS survey and image processing). The outcomes of the tests clearly demonstrate that an aerial survey of the affected areas (limited to a few square kilometers) can be technically carried out in a time frame fulfilling emergency management requirements (*i.e.*, to get up-to-date post-event information in the shortest time possible). The main bottleneck from the operational procedure perspective is related to the need to deploy all of the required instrumentation in the affected areas: in the framework of an operational service, the authors strongly believe that only pre-positioned stand-by UAV teams covering specific geographical areas may grant the possibility to reach the target area in a reasonable and limited timeframe (a few hours possibly). In the case of international operations, the time required to get valid visas (if necessary) has also to be taken into account. Furthermore, the existing (and rapidly evolving) national (and international) regulations should be thoroughly evaluated and addressed well in advance, to be sure that all the regulation-related tasks (e.g., special flight permission requests or the request to issue a NoTAM, Notice To AirMen) can be carried out quickly and correctly. A discussion with the relevant aviation authorities focused on the specific domain of RPAS for emergency management purposes should be also encouraged.

Adverse weather conditions (scarce visibility, heavy rains, strong winds) can also play an important role in possible delays of flight operations once in the field, due to the platforms' technical limitations (although waterproof platforms exist, the capability to withstand strong winds is limited to certain air speed thresholds).

Concerning the extent of the area to be surveyed, multi-rotor platforms are usually the first choice when small areas or isolated buildings must be surveyed, therefore fitting the requirements of emergencies like industrial accidents (including potential needs for indoor flights in case of search and rescue applications [28]) and landslides.

If larger areas need to be mapped, which is generally the case in disasters like floods and wildfires, the preferred approach is based on fixed-wing UAVs, allowing speeding up of the acquisition process (at a 150-m height, 1 km^2 can be covered in about 15 min). The use of UAV fleets, covering in parallel different portions of the affected area, could be planned in case of events that involve vast regions as an alternative to a typical mission based on several subsequent flights of the same platform. Furthermore, it has to be highlighted that according to the authors' experience, the use of a UAV-based approach fits the need to get detailed information over the most impacted areas: the overview of the situation over the whole affected area is generally based on satellite imagery. Concerning the image processing phase, the number of images to be processed and the need for almost completely automated algorithms lead to time-consuming operations to get a usable orthophoto (*i.e.*, a couple of hours of processing in the field to cover about 1 km^2). The adoption of proper tools aimed at preliminary extracting the minimum number of images among the collected ones (downsampling them to the spatial resolution actually required to allow a proper post-event analysis) is therefore encouraged.

Rigorous positioning accuracy assessment tests (Section 5.4) clearly demonstrate that even using a direct georeferencing approach (*i.e.*, no need for measured GCP), it is possible to obtain horizontal accuracies of about 1.5 m. Similar values are obviously suitable for the immediate response phase, allowing the responders to uniquely identify the affected features and to assess their damage grade in the post-event imagery.

The availability of the satellite connectivity offered by the Regional Civil Protection Mobile Unit allowed the team to test also the upload of the processed image on a web mapping server directly from the field. The test was indeed successful, and it demonstrated that the data can be made available in a few minutes (the actual time is strictly related to the available bandwidth and the size of the imagery) to the broader humanitarian community, e.g., to crowd mapping volunteers, which can definitively speed up the post-event analysis (*i.e.*, identification of the affected areas and, when possible, the assessment of damages to main infrastructures).

Acknowledgments: The authors would like to especially thank the researchers, operators and students involved in the UAV data acquisition and processing, namely: Antonia Spanò, Paolo Maschio, Marco Piras, Francesca Noardo, Irene Aicardi, Nives Grasso, the team DIRECT students, the civil protection volunteers and the municipality of Morano sul Po. Furthermore the authors would like to thank Luciana Dequal for the English language proof reading.

Author Contributions: All authors conceived, designed and carried out the operational activities in the field related to the UAV deployment exercise. All authors have made significant contributions to the paper, specifically: Filiberto Chiabrando and Andrea Lingua processed the imagery; Piero Boccardo and Fabio Giulio Tonolo contributed to the comparison analysis; Filiberto Chiabrando and Fabio Giulio Tonolo conceived the paper structure and wrote the first draft, which was revised and finalized with the contribution of Piero Boccardo, Andrea Lingua and Furio Dutto.

Conflicts of Interest: The authors declare no conflict of interest.

References

1. International Civil Aviation Organization (ICAO). Available online: http://www.icao.int/ (accessed on 20 March 2015).
2. OCHA Policy Development and Studies Barnch—PDSB (2014), Unmanned Aerial Vehicles in Humanitarian Response. Available online: https://docs.unocha.org/sites/dms/Documents/Unmanned%20Aerial%20Vehicles%20in%20Humanitarian%20Response%20OCHA%20July%202014.pdf (accessed on 6 March 2015).
3. Bendea, H.; Boccardo, P.; Dequal, S.; Giulio Tonolo, F.; Marenchino, D. New technologies for mobile mapping. In Proceedings of the 5th international symposium in Mobile Mapping Technology, Padova, Italy, 28–31 May 2007.
4. Bendea, H.; Boccardo, P.; Dequal, S.; Giulio Tonolo, F.; Marenchino, D.; Piras, M. Low cost UAV for post-disaster assessment. In Proceedings of The XXI Congress of the International Society for Photogrammetry and Remote Sensing, Beijing, China, 3–11 July 2008; Volume XXXVII Part B8, pp. 1373–1380.
5. International Working Group on Satellite-based Emergency Mapping (IWG-SEM). Emergency Mapping Guidelines. Available online: http://www.un-spider.org/sites/default/files/IWG_SEM_EmergencyMappingGuidelines_A4_v1_March2014.pdf (accessed on 26 February 2015).
6. The United Nations Office for Disaster Risk Reduction (UNISDR) Terminology on Disaster Risk Reduction. Available online: http://www.unisdr.org/files/7817_UNISDRTerminologyEnglish.pdf (accessed on 19 February 2014).
7. Gilman, D.; Meier, P. Humanitarian UAV Network—Experts Meeting, un Secretariat. Available online: https://docs.unocha.org/sites/dms/Documents/Humanitarian%20UAV%20Experts%20Meeting%20Summary%20Note.pdf (accessed on 6 March 2015).
8. DIRECT. Available online: http://www.polito.it/direct (accessed on 23 March 2015).
9. ENAC. Remotely Piloted Aerial Vehicles Regulation. Available online: https://www.enac.gov.it/repository/ContentManagement/information/N1220929004/Reg%20SAPR%20english_022014.pdf (accessed on 10 March 2015).
10. MikroKopter. Available online: http://www.mikrokopter.de (accessed on 23 March 2015).
11. SenseFly Company. Available online: https://www.sensefly.com/drones/ebee.html (accessed on 15 April 2015).
12. Ullman, S. The interpretation of structure from motion. *Proc. R. Soc. Lond.* **1999**, *B203*, 405–426. [CrossRef]
13. Kersten, T.P.; Lindstaedt, M. Automatic 3D Object Reconstruction from Multiple Images for Architectural, CulturalHeritage and Archaeological Applications Using Open-Source Software and Web Services. *Photogramm. Fernerkund. Geoinf.* **2012**, *6*, 727–740. [CrossRef]
14. Koutsoudis, A.; Vidmar, B.; Ioannakis, G.; Arnaoutoglou, F.; Pavlidis, G.; Chamzas, C. Multi-image 3D reconstruction data evaluation. *J. Cult. Herit.* **2014**, *15*, 73–79. [CrossRef]
15. Chiabrando, F.; Lingua, A.; Noardo, F.; Spanò, A. 3D modelling of trompe l'oeil decorated vaults using densematching techniques. In Proceedings of the ISPRS Annals of the Photogrammetry, Remote Sensing and Spatial Information Sciences, Riva del Garda, Italy, 23–25 June 2014; Volume II-5. 2014ISPRS Technical Commission V Symposium.
16. Remondino, F.; del Pizzo, S.; Kersten, T.; Troisi, S. Low-cost and open-source solutions for automated image orientation—A critical overview. In Proceedings of the 4th International Conference, EuroMed 2012 (LNCS 7616). Limassol, Cyprus, 29 October–3 November 2012; pp. 40–54.
17. Lowe, D. Distinctive image features from scale invariant keypoints. *Int. J. Comput. Vis.* **2004**, *60*, 91–110. [CrossRef]
18. Triggs, B.; McLauchlan, P.F.; Hartley, R.I.; Fitzgibbon, A.W. Bundle Adjustment A Modern Synthesis. In *Vision Algorithms: Theory and Practice Lecture Notes in Computer Science*; Springer: Berlin/Heidelberg, Germany, 2000; pp. 298–372.
19. Hirschmuller, H. Stereo processing by semiglobal matching and mutual information. *IEEE Trans. Pattern Anal. Mach. Intell.* **2008**, *30*, 328–342. [CrossRef]

20. McGlone, J.C.; Mikhail, E.M.; Bethel, J.; Mullen, R. *Manual of Photogrammetry*, 5th ed.; American Society of Photogrammetry and Remote Sensing: Bethesda, MA, USA, 2004.

21. Krauss, K. *Photogrammetry, Volume 2: Advanced Methods and Applications*, 4th ed.; Dummler's Verlag: Bonn, Germany, 1997.

22. Zandbergen, P.A.; Barbeau, S.J. Positional accuracy of assisted gps data from high-sensitivity gps-enabled mobile phones. *J. Navig.* **2011**, *64*, 381–399. [CrossRef]

23. Landau, H.; Vollath, U.; Chen, X. Virtual reference station systems. *J. Glob. Position. Syst.* **2002**, *1*, 137–143. [CrossRef]

24. Cramer, M.; Stallmann, D.; Haala, N. Direct georereferencing using GPS/Inertial exterior orientations for photogrammetric applications. *Int. Arch. Photogramm. Remote Sens.* **2000**, *33*, 198–205.

25. Roze, A.; Zufferey, J.-C.; Beyeler, A.; McClellan, A. eBee RTK Accuracy Assessment. White Paper Sense Fly 2014. Available online: https://www.sensefly.com/fileadmin/user_upload/documents/eBee-RTK-Accuracy-Assessment.pdf (accessed on 15 May 2015).

26. Copernicus Emergency Management Service—Mapping. Available online: http://emergency.copernicus.eu/mapping/ems/new-phase-brief (accessed on 7 May 2015).

27. Tomic, T.; Schmid, K.; Lutz, P.; Domel, A.; Kassecker, M.; Mair, E.; Grixa, I.L.; Ruess, F.; Suppa, M.; Burschka, D. Toward a Fully Autonomous UAV: Research Platform for Indoor and Outdoor Urban Search and Rescue. *IEEE Robot. Autom. Mag.* **2012**, *19*, 46–56. [CrossRef]

28. Ajmar, A.; Boccardo, P.; Disabato, F.; Giulio Tonolo, F. Rapid Mapping: Geomatics role and research opportunities. *Rendiconti Lincei* **2015**. [CrossRef]

Article

Automated Identification of River Hydromorphological Features Using UAV High Resolution Aerial Imagery

Monica Rivas Casado [1,†,*], Rocio Ballesteros Gonzalez [2,†], Thomas Kriechbaumer [1] and Amanda Veal [3]

1 School of Energy, Environment and Agrifood, Cranfield University, Cranfield MK430AL, UK; t.kriechbaumer@cranfield.ac.uk

2 Regional Centre of Water Research Centre (UCLM), Ctra. de las Peñas km 3.2, Albacete 02071, Spain; rocio.ballesteros@uclm.es

3 Hydromorphological Team, Environment Agency, Manley House, Kestrel Way, Exeter, Devon EX27LQ, UK; amanda.veal@environment-agency.gov.uk

* Author to whom correspondence should be addressed; m.rivas-casado@cranfield.ac.uk; Tel.: +44-1234-750111 (ext. 2706).

† These authors contributed equally to this work.

Academic Editors: Felipe Gonzalez Toro and Antonios Tsourdos
Received: 29 July 2015; Accepted: 28 October 2015; Published: 4 November 2015

Abstract: European legislation is driving the development of methods for river ecosystem protection in light of concerns over water quality and ecology. Key to their success is the accurate and rapid characterisation of physical features (*i.e.*, hydromorphology) along the river. Image pattern recognition techniques have been successfully used for this purpose. The reliability of the methodology depends on both the quality of the aerial imagery and the pattern recognition technique used. Recent studies have proved the potential of Unmanned Aerial Vehicles (UAVs) to increase the quality of the imagery by capturing high resolution photography. Similarly, Artificial Neural Networks (ANN) have been shown to be a high precision tool for automated recognition of environmental patterns. This paper presents a UAV based framework for the identification of hydromorphological features from high resolution RGB aerial imagery using a novel classification technique based on ANNs. The framework is developed for a 1.4 km river reach along the river Dee in Wales, United Kingdom. For this purpose, a Falcon 8 octocopter was used to gather 2.5 cm resolution imagery. The results show that the accuracy of the framework is above 81%, performing particularly well at recognising vegetation. These results leverage the use of UAVs for environmental policy implementation and demonstrate the potential of ANNs and RGB imagery for high precision river monitoring and river management.

Keywords: Unmanned Aerial Vehicle; photogrammetry; Artificial Neural Network; feature recognition; hydromorphology

1. Introduction

Environmental legislation [1–3] aiming to improve the quality of riverine ecosystems has driven the development of a vast number of methods for the hydromorphological assessment of rivers [4]. Within this context, hydromorphology refers to the physical characteristics of the shape, boundaries and content of a river [1]. There are currently over 139 different hydromorphological assessment methods used to characterise both physical in-stream and riparian habitat, river channel morphology, hydrological regime alteration or longitudinal river continuity. In Europe alone, the implementation of the Water Framework Directive (WFD [1]) has led to the development and use of over 73 methodologies [4] such as the LAWA method in Germany [5] or the CARAVAGGIO assessment in Italy [6]. In the United Kingdom (UK), the key methods adopted are the River Habitat Survey (RHS) [7,8] and the River-MImAS [9]. The comparison of results obtained from different assessment methods is convoluted, this highlighting the need for an unbiased and standardised protocol for hydromorphological characterisation.

The existing approaches are commonly implemented via *in-situ* mapping [8] or aerial imagery assessment [10]. The former relies on the expertise of the surveyor identifying hydromorphological features and does not allow for the objective re-assessment of records after survey completion. Moreover, due to practical time and cost constraint these surveys are difficult to repeat at high frequencies and are limited to accessible reaches [11–14]. Therefore, such assessments lack of spatial detail and do not capture the spatio-temporal variability within river reaches. In contrast, approaches based on aerial imagery rely on the visual or automated identification of key river characteristics from off-the-shelf imagery of generally 12.5 cm or 25 cm resolution. Here, the quality of the assessment depends upon the accuracy of the classification approach and the characteristics of the imagery, such as resolution and wavelength bands.

There are three types of image classification techniques [15]: object-based image analysis, unsupervised and supervised image classification. Object-based techniques rely on multi-resolution segmentation and are able to simultaneously generate objects of different shapes and scales by grouping pixels of similar characteristics. Unsupervised classification groups pixels based on their reflectance properties whereas supervised classification is based in the concept of segmenting the spectral domain into areas that can be associated with features of interest. The later method requires a training process by which representative samples of features of interest are identified and used to classify the entire image. There is a large array of algorithms for the task [15] such as maximum likelihood, Gaussian mixture models, minimum distance and networks of classifiers. Amongst all the existing supervised classification approaches, methods based on Artificial Neural Networks (ANNs) have been shown to enable image pattern recognition at particularly high precision with both coarse [16,17] and fine resolution imagery [18,19]. For example, [20–22] used ANNs to identify green canopy cover from background soil and shadows. ANNs have also been used successfully to map water bodies [23] and flood extent [24]. However, to the authors' knowledge the usefulness of ANNs in the classification of river features as part of hydromorphological assessment has not been tested yet.

The use of image classification techniques for river mapping is well documented and has been applied successfully on hyperspectral imagery for the identification of hydraulic and habitat patterns [25], woody debris [14], channel substrate [26] and riparian vegetation [26]. Although hyper and multispectral bands are the preferred wavelength bands to classify hydromorphological features, they require exhaustive data processing algorithms and post-interpretation [26]. Spaceborne hyperspectral imagery (30 to 50 m ground resolution) does not offer the required resolution for detail river feature identification and fails to provide global spatio-temporal coverage. It may therefore be difficult to obtain off-the-shelf data for a given location and event (time) of interest. Airborne hyperspectral imagery can address this limitation but data capture can be expensive as it requires hired hyperspectral flights tailored to the objectives of the research. The spatial (ground) resolution of airborne hyperspectral imagery (typically above meter resolution) is significantly larger than typical high-resolution RGB photographs as a result of a lower number of photons per channel than imaging spectrometers [26]. Hyperspectral imagery may also be highly sensitive to imaging geometry (e.g., differences from the middle to the edge of the flight path) and environmental conditions (e.g., water vapour). In addition, hyperspectral imagery does not represent a snapshot of the area as pixel's data is collected consecutively—*i.e.*, there is a lag between acquisitions of consecutive pixels. For airborne imagery, this translates into unequal pixel geometries and major issues in image rectification [26].

The classification of key features from RGB imagery has already been proved to be an efficient method for the automated identification of macrophytes [27]. This also holds true for vision based classification techniques for both geomorphic and aquatic habitat features [28] and is consistent with the excellent results shown for the characterisation of fluvial environments [29,30]. Recent studies have attempted to improve the results of existing classification methods by using Unmanned Aerial Vehicle (UAV) high resolution RGB aerial imagery. For example, [31] used UAV and ultra-light aerial vehicles imagery of resolutions between 3.2 cm and 11.8 cm to assess rates of vegetation recruitment and survival on braided channels in the Alps and [32] used UAV aerial imagery with resolutions from 1 cm to 10 cm to quantify the temporal dynamics of wood in large rivers.

The combination of UAV high resolution imagery and automated classification techniques for the identification of features within river environments has been documented by several authors [33,34]. In [33] supervised machine learning approaches were used to identify different types of macrophytes whereas in [34] standing dead wood presence in Mediterranean forests was mapped combining 3.2 to 21.8 cm resolution imagery and object oriented classification approaches.

The increased availability of low cost, vertical take-off UAV platforms and the increased legal requirement for improved river monitoring protocols make particularly attractive the used of RGB UAV high resolution aerial imagery for the development of a plausible, transferable and standardised framework for hydromorphological assessment. This solution could offer both timely (on-demand) and detailed (higher resolution) information than remote sensing imagery. Here, we extend the combined use of UAV high resolution RGB aerial imagery and ANNs to automatically classify all existing hydromorphological features along a 1.4 km river reach. The aim is to develop and assess a framework combining UAVs, high resolution imagery and ANNs for the unbiased characterisation of river hydromorphology. This is achieved through the following three core objectives:

i.   To adapt existing ANN classification software for the identification of hydromorphological features,
ii. To assess the suitability of UAV high resolution aerial imagery for (i),
iii. To quantify the accuracy of the operational framework derived from (i) and (ii).

2. Experimental Section

2.1. Study Site

The case study area is a 1.4 km reach in the upper catchment of the river Dee near Bala dam, Wales, UK (Figure 1a). The river Dee flows North-East from its origin in Dduallt (Snowdonia) into Bala lake to descend East to Chester and discharge in an estuary between Wales and the Wirral Peninsula in England. It defines the boundary between Wales and England for several miles from Bangor-on-Dee to Aldford. The catchment area of the 110 km long river covers 1816 km^2, with the study site located approximately at 30 km from its origin. The fieldwork took place from the 20th to the 25th of April 2015 under low flow conditions and with a constant volumetric flow rate of 4.8 m^3·s^{-1}. The UAV imagery was collected on the 21 April 2015.

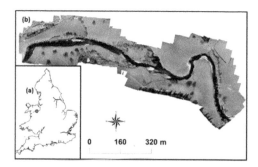

Figure 1. (**a**) Location of the study site along the river Dee near Bala, Wales, UK; (**b**) Detailed view of the study area.

2.2. Sampling Design

A total of 60 1 m × 1 m Ground Control Points (GCPs) were distributed uniformly within the flying area (Figure 2) to obtain parameters for external orientation [28,35]. The centroid of each 1 m × 1 m white 440 g PVC GCP was established via its square diagonals. Opposite facing triangles were painted in black to facilitate centroid identification (Figure 2). GCPs were pinned with pegs to the ground through four metallic eyelets. The locations of the GCP centroids were obtained from a Leica GS14 Base and Rover Real Time Kinematic (RTK) GPS with a positioning accuracy of 1–2 cm in the X, Y and Z dimensions. Further 25 yellow and white check points (XPs, Figure 2) were set to quantify image coregistration model errors [35] using the same deployment strategy as for the GCPs. Velocity and depth measurements within the channel were obtained using a SonTek RiverSurveyor M9 Acoustic Doppler Current Profiler (ADCP) mounted on an ArcBoat radio control platform [36] (Figure 3). The reach was sampled following a bank to bank zig-zag pattern to capture the spatial variability in channel depth and water velocity.

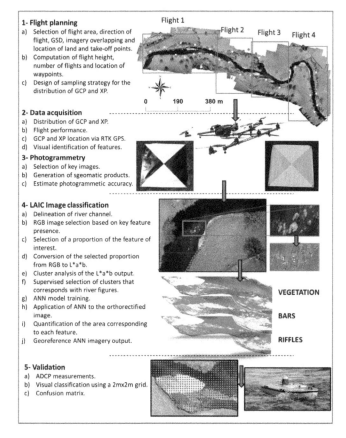

Figure 2. Workflow summarising the steps followed in the photogrammetry using Photoscan Pro and the image classification using the Leaf Area Index Calculation (LAIC) software, based on the workflows presented by [21,37], respectively. GDS, GCP and XP stand for Ground Sampling Distance, Ground Control Point (red points) and Check Point (yellow points), respectively.

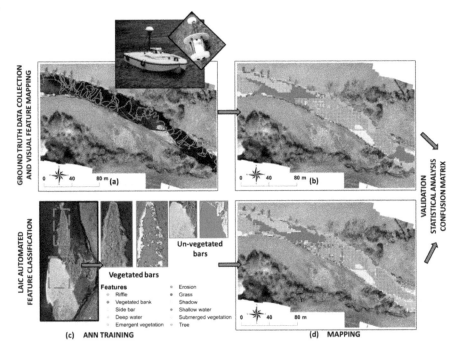

Figure 3. Detailed diagram of the workflow for the Leaf Area Index Calculation (LAIC) image classification and validation based on [18] (**a–d**). (**a**) 300 m section within the reach showing the ADCP measurements obtained along with a detailed image of the radio control boat and ADCP sensor used; (**b**) Map showing the hydromorphological features obtained from visual identification on a 2 m × 2 m regular grid; (**c**) Examples of sections selected for and outputs obtained from the Artificial Neural Network (ANN) training; (**d**) Map showing the hydromorphological feature classification obtained with ANN on a 2 m × 2 m regular grid.

2.3. UAV and Sensor

Aerial imagery in the visible spectrum was acquired via an AsTec Falcon 8 octocopter (ASCTEC, Krailling, Germany) equipped with a Sony Alpha 6000 camera (Sony Europe Limited, Weybridge, Surrey, UK) and a u-blox LEA 6S GPS. The 0.77 m × 0.82 m × 0.12 m octocopter has a vertical take-off weight of 1.9 kg, where the sensor payload accounts for 0.34 kg. Fully charged Lithium Polymer (LiPo) batteries (6250 mAh) provided a maximum flight time of 22 min. The Falcon 8 can tolerate wind speeds of up to 15 m·s^{-1}—a threshold that was never exceed during data collection. The weather conditions during the flight, based on Shawbury meteorological aerodrome report (METAR), presented surface winds of speeds between 1 m·s^{-1} and 3 m·s^{-1} and directions varying from 60° to 350°, with prevailing visibility up to 10,000 m AMSL and a few clouds at 12,800 m AMSL.

The study area was surveyed through four consecutive flight missions (Figure 2). Full spatial coverage was ensured through the combination of longitudinal and cross-sectional multipasses. The flight was pre-programmed with the AscTec Navigator software version 2.2.0 for a number of waypoints to achieve 60% along track and 80% across track image overlap for the camera parameters described in Table 1 and a flight height of 100 m. The resulting ground sample distance (GSD) was 2.5 cm. Each waypoint represented the centre of a frame and had associated GPS coordinates as well as yaw, pitch and roll information. The UAV was held stationary at waypoints and transited between them at a speed of 3 m·s^{-1}. The operator was a fully qualified RPQ-s (Small UAV Remote

Pilot Qualification) pilot and followed Civil Aviation Authority (CAA) legislation CAP393 [38] and CAP722 [39].

The Sony Alpha 6000 camera (Table 1) relies on complementary metal oxide semiconductor (CMOS) image sensor technology. The APS-C 2.82 cm (1.11 inches) diameter CMOS sensor provides images of 24.3 effective megapixels (6000 × 4000 pixels). The colour filter type used was RGB.

Table 1. Key characteristics for the Sony Alpha 6000 complementary metal oxide semiconductor (CMOS) sensor.

Characteristics	Sony Alpha 6000
Sensor (Type)	APS-C CMOS Sensor
Million Effective Pixels	24.3
Pixel Size	0.00391 mm
Image size (Columns and Rows)	6000 × 4000
Lens	24–75 mm (35 mm)
Focal	3.5–5.5
ISO range	100–51,200

2.4. Photogrammetry

A total of 394 frames out of 746 were selected for the photogrammetric analysis. The selection was based upon image quality and consecutive spatial coverage. Photoscan Pro version 1.1.6 (Agisoft LLC, St. Petersburg, Russia) was used to generate an orthoimage. Figure 2 summarises the workflow adapted from [37]. The coordinates for each of the GCPs were used to georeference (scale, translate and rotate) the UAV imagery into the coordinate system defined by the World Geodetic System (WGS84) and minimise geometric distortions. Image coregistration errors were estimated at each GCP as the difference between the positions measured through RTK GPS and the coordinates derived from the imagery. A combined measure of error for x and y is obtained from Equation (1):

$$\text{RMSE} = \sqrt{\frac{\sum_{j=1}^{N}\left[\left(\hat{x}_j - x_j\right)^2 + \left(\hat{y}_j - y_j\right)^2\right]}{N}} \tag{1}$$

where RMSE is the Root Mean Squared Error, $\hat{x}$ and $\hat{y}$ are the image derived coordinates at location j, x and y are the associated RTK GPS positions and N is the number of points assessed. If the XPs are assessed independently, their RMSE becomes an unbiased validation statistic.

The overall process to obtain the geomatic products (*i.e.*, point cloud, orthoimage and digital terrain model) required 12 h of processing time based on the performance of a computer with an Intel Core i7-5820K 3.30 GHz processor, 32 Gb RAM and 2 graphic cards (NVIDIA Geoforce GTX 980 and NVIDIA Qadro K2200).

2.5. Image Classification

The hydromorphological feature classification was implemented with the Leaf Area Index Calculation (LAIC) software (Figures 2 and 3). LAIC is a MATLAB-based supervised ANN interface designed to discriminate green canopy cover from ground, stones and shadow background using high resolution UAV aerial imagery [20,21]. In brief, LAIC relies on clustering techniques to group the pixels from high resolution aerial imagery based on CIELAB characteristics. From the three parameters describing the CIELAB space [40,41] (*i.e.*, lightness (*L*), green to red scale (*a*) and blue to yellow scale (*b*)) only *a* and *b* are taken into account by the clustering algorithm. For this purpose, RGB collected at representative waypoints was transformed to *L* a* b** colour-space. A k-means clustering algorithm of the RGB levels on the *L* a* b** colour-space transformed imagery was then implemented. The analysis groups pixels into k-clusters (*k*) with similar red/green (*a*) and yellow/blue (*b*) values. The number of clusters (*k*) depends on the feature being identified and was determined following an iterative process that increased k by one up to a maximum of ten clusters until visually satisfactory results were obtained. Within this context, visually satisfactory results required the image outputs (Figure 3) to show that the feature of interest (Table 2) had been adequately identified. This supervised method was used as a basis to calibrate a Multilayer Perceptron (MP) ANN, which was applied to the remaining images. The calibration process is highly-time consuming and therefore, only a small section of the imagery can be used for this purpose. Key to the success of the ANN was the adequate selection of the small proportion of imagery used for the ANN calibration and training process. These were chosen based on the presence of such features in at least 50% of the selected area. Here, we looked at the clarity of the colours as well as the contrast in a and b. Images with shadows or including features that could be confused with the one of interest were not selected.

The basic MP ANN is a simple, binary linear classifier that, once trained places patterns into one of the two available classes by checking on which side of the linear separating surface they lay [15]. In this study, we used a more complex form of MP ANN [16] based on three consecutive layers named input, hidden and output, hereafter. Each layer is composed of inter-connected nodes, also known as neurons. The results from the cluster analysis are input into the first layer which performs linear combinations of the input parameters to give a set of intermediate linear activation variables. In turn, these variables are transformed by non-linear activation functions in the hidden layer where a secondary layer of weights and biases provides a set of activation values. These are then fed into the output layer to obtain the final output values. The weights were adjusted via an iterative back propagation process based on the comparison of the outputs with target features characterised in the training process. The network is initialised with a set of weights. For each training pixel, the output of the network is estimated using the structure beforehand mentioned. The weights are then corrected based on the resulting outcome. This iteration process stops once an optimisation algorithm has been satisfied. In this study, a quasi-Newton optimisation algorithm aiming at minimising the estimation error [16] was used in the training process to ensure the non-linear structure of the MP ANN was accounted for [20].

The number of outputs nodes depends on how the outputs are used to represent the features. In this study, the number of output processing elements is the same as the number of training classes. Each class was trained separately and therefore, there was only one output node—with value one when the RGB values corresponded to the selected feature and zero for any other instances [20]. The software outcomes provided (i) a classified map of the study area and (ii) estimates of the areas allocated to each of the classified features. The ANN was used to recognise the key features described in Table 2. These features are used by Governmental agencies [8] and environmental scientists [35,37] alike to describe homogeneous areas of substrate, water and vegetation within the reach.

Table 2. Hydromorphological features identified within the study area based on [8].

	Feature	Description
Substrate Features	Side Bars	Consolidated river bed material along the margins of a reach which is exposed at low flow.
	Erosion	Predominantly derived from eroding cliffs which are vertical or undercut banks, with a minimum height of 0.5 m and less than 50% vegetation cover.
Water Features	Riffle	Area within the river channel presenting shallow and fast-flowing water. Generally over gravel, pebble or cobble substrate with disturbed (rippled) water surface (*i.e.*, waves can be perceived on the water surface). The average depth is 0.5 m with an average total velocity of 0.7 m·s^{-1}.
	Deep Water (Glides and Pools)	Deep glides are deep homogeneous areas within the channel with visible flow movement along the surface. Pools are localised deeper parts of the channel created by scouring. Both present fine substrate, non-turbulent and slow flow. The average depth and is 1.3 m and the average total velocity is 0.3 m·s^{-1}.
	Shallow Water	Includes any slow flowing and non-turbulent areas. The average depth is 0.8 m with an average total velocity of 0.4 m·s^{-1}.
Vegetation Features	Tree	Trees obscuring the aerial view of the river channel.
	Vegetated Side Bars	Side bar presenting plant cover in more than 50% of its surface area.
	Vegetated Bank	Banks not affected by erosion.
	Submerged Free Floating Vegetation	Plants rooted on the river bed with floating leaves.
	Emergent Free Floating Vegetation	Plants rooted on the river bed with floating leaves on the water surface.
	Grass	Present along the banks as a result of intense grazing regime.
Shadows		Includes shading of channel and overhanging vegetation.

To simplify the complexity of the ANN implementation, only the imagery within the area defined by the bank channel boundary (46,836 m^2) was considered in the overall classification. The 2.5 cm orthoimage was divided into 20 tiles for processing via LAIC. This decreased the central processing unit (CPU) demands during feature recognition. The tiles were georeferenced and mosaicked together to obtain a complete classified image for the study area. The overall classification process with LAIC was undertaken in less than 7 h.

2.6. Radio Control Boat and ADCP Sensor

Water depths and velocities were obtained with an ADCP and used to define an informed threshold between deep and shallow waters (Table 2). Water depths were used in the validation process whereas water velocities were only used for descriptive purposes in Table 2. ADCPs are hydro-acoustic sensors that measure water velocities and depth by transmitting acoustic pulses of specific frequency [42]. ADCPs deployed from a moving vessel provide consecutive profiles of the vertical distribution of 3D water velocities and depth along the vessel trajectory. The ADCP used in this study was a RiverSurveyor M9 ADCP with a SonTek differentially corrected GPS [43]. The ADCP data were collected at a frequency of 1 Hz and an average boat speed of 0.35 m·s^{-1}.

2.7. Validation

Hydromorphological features (Table 2) were mapped along the reach during a "walk-over" survey. A 2 m × 2 m grid was overlaid onto the othoimage in a Geographical Information System (GIS) environment (ArcGIS 10.3, Redlands, CA, USA). The hydromorphological feature classes were visually assigned to each of the 13,085 points defined by the regular grid. The visual classification was aided by 118 documented colour photographs and 470 RTK GPS measurements providing the exact locations of the hydromorphological features described in Table 2. The visual point class outputs were compared to those obtained from the ANN classified image via a confusion matrix [44], where the visual classification was considered to be the ground truth. Measures of accuracy (AC), true positive ratios (TPR), true negative ratios (TNR), false negative ratios (FNR) and false positive ratios (FPR) were derived for each hydromorphological feature (i) as follows:

$$AC = \frac{TN + TP}{TN + TP + FN + FP} \tag{2}$$

$$TPR_i = \frac{TP_i}{FN_i + TP_i} \tag{3}$$

$$TNR_i = \frac{TN_i}{TN_i + FP_i} \tag{4}$$

$$FNR_i = \frac{FN_i}{FN_i + TP_i} \tag{5}$$

$$FPR_i = \frac{FP_i}{TN_i + FP_i} \tag{6}$$

where TP (true positives) is the number of points correctly identified as class i, FN (false negatives) is the number of points incorrectly rejected as class i, TN (true negatives) is the number of points correctly rejected as class i and FP (false positives) is the number of points incorrectly identified as class i.

TPR, TNR, FNR and FPR are estimated for each of the features of interest whereas AC is a single value of overall classification performance. AC as well as all the ratios beforehand mentioned range from 0 to 1 or 0% to 100% when reported in percentages. Both true positives (TP) and true negatives (TN) estimate the number of points that have been correctly identified or rejected to fall within a particular class. Therefore, TPR and TNR quantify the power of LAIC at classifying features correctly when compared to the ground truth. Both false negatives (FN) and false positives (FP) estimate the number of points that have been falsely rejected or falsely identified to fall within a particular class. Hence, FNR and FPR show the rates of misclassification when compared to the ground truth values.

3. Results

The image coregistration model errors estimated from the GCP and XP positions were consistent and within the proposed thresholds reported by [35]. The average error was below 1.7 cm (X = 1.1 cm, Y = 1.0 cm and Z = 1.6 cm) for the GCP and below 2 cm (X = 0.4 cm, Y = 2.0 cm and Z = 0.53 cm) for XP. The RMSEs were 1.5 cm and 0.09 cm for GCP and XP, respectively. The accuracy of the ANN classification (Equation (2)) was 81%, meaning that a total of 10,662 points out of 13,085 were classified correctly, with the majority of classes showing a TPR above 85% (Tables 3 and 4). The ANN reached a solution for the backpropagation process under 60 iterations.

Table 3. Confusion matrix of visual classification (VC) *versus* Artificial Neural Network (ANN) classification. Feature codes have been abbreviated as follows: side bars (SB), erosion (ER), riffle (RI), deep water (DW), shallow water (SW), tree (TR), shadow (SH), vegetation (VG), vegetated bar (VB), vegetated bank (VK), submerged vegetation (SV), emergent vegetation (EV) and grass (GR). GE stands for georeferencing error.

Feature	VC	ANN Classification									Total
		SB	ER	RI	DW	SW	TE	SH	VG	GE	
SB	1334	1097	-	8	-	2	-	10	214	3	1334
ER	287	-	22	13	1	3	-	10	238	-	287
RI	3339	-	1	2717	-	318	-	219	76	8	3339
DW	2082	-	-	60	1927	54	-	8	29	4	2082
SW	2573	-	-	262	80	1514	-	493	217	7	2573
TR	1755	-	-	76	1	29	496	135	1013	5	1755
VB	299	-	-	-	-	-	-	-	299	-	299
VK	313	-	10	-	6	-	-	15	281	1	313
SV	468	-	-	160	-	125	-	46	135	2	468
EV	71	-	1	9	-	2	-	1	58	-	71
GR	344	-	-	-	-	-	-	-	343	1	344
SH	220	-	4	-	-	-	-	180	31	-	220
Total	13,085	1097	38	3305	2015	2052	496	1117	2934	31	13,085

The georeferencing errors derived from tile misalignment of LAIC outputs accounted for 0.2% of the totality of points. The TPR of shadow identification was above 80% with no significant misclassification results. FNR and FPR were below 20% in all cases except for erosion, shallow water and submerged vegetation feature classes. These values are very low and can be explained by misclassification errors described in the following sections. The overall study area (46,836 m^2) was dominated by a combination of riffle (31%) and deep water (24%) features (Table 5). Figure 4 presents an example classification output for some of the features identified within a selected tile.

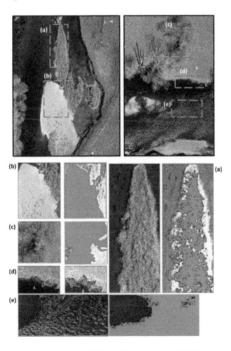

Figure 4. Example of trained outputs for (**a**) Vegetation in bars; (**b**) Side bars with no vegetation; (**c**) Trees; (**d**) Erosion and (**e**) Riffle. The outputs portray the portion of the imagery selected for analysis and the pixels selected (pink) by the cluster algorithm.

Table 4. True positive ratio (TPR), true negative ratio (TNR), false negative ratio (FNR) and false positive ratio (FPR) for each of the class features identified by the Artificial Neural Network (ANN) within the river reach.

Feature Identification (ANN)		TPR	TNR	FNR	FPR
Substrate	Bars	0.822	0.765	0.178	0.000
Features	Erosion	0.077	0.786	0.923	0.001
	Riffle	0.814	0.756	0.074	0.060
Water Features	Deep Water	0.926	0.741	0.074	0.008
	Shallow Water	0.588	0.815	0.412	0.051
	Trees	0.860	0.757	0.140	0.082
	Vegetated Bar	1.000	0.765	0.000	0.082
Vegetation	Vegetated Bank	0.898	0.767	0.102	0.082
	Submerged Vegetation	0.288	0.788	0.712	0.082
	Emergent Vegetation	0.817	0.770	0.183	0.082
	Grass	0.997	0.750	0.003	0.082
Shadow		0.818	0.770	0.182	0.073

Table 5. Areas for each of the features estimated from the Artificial Neural Network (ANN) classification.

Feature	Area (m^2) ANN
Bars	4992
Erosion	338
Riffle	12,758
Deep Water	10,008
Shallow Water	7977
Vegetation	10,080
Shadow	683
Total	46,836

3.1. Substrate Features

Erosion features (Figure 4d) had the lowest TPRs, scoring only 8% (Table 3). Many of the eroded banks within the reach were vertical cliffs and not visible from a 2D planar aerial view (Figure 5c). The majority of these areas presented grass up to the edge of the cliff and vegetated sediment deposits at the bottom. The ANN classification therefore defaulted to vegetation or shadow (generated by the cliff) in the majority of observations (Table 3 and Figure 6).

Side bars (Figures 4b and 5d) had a TPR and TNR above 76% (Table 3). Misclassification occurred when vegetation appeared within the bar—the class defaulting to vegetation in all instances (Figures 5f and 6). The ANN was able to correctly identify the vegetation (*i.e.*, TPR for vegetated bar was 100% in Table 3) but unable to specify the feature where the vegetation was encountered (e.g., bank, bar or tree spring shot). If these FN were considered TP, the TPR ratio for side bars increased to 98%.

3.2. Water Features

The ANN based classification achieved a TPR and TNR above 75% when identifying riffles. However, for LAIC to correctly classify a point as a riffle (Figure 4e), this feature had to coincide with shallow water and a rippled water surface (Figure 6). Confusion of riffles for shallow water occurred when rippled surfaces were not present within the riffle. All the points thus misclassified fell within close proximity of riffles and could be considered as TP, increasing TPR to 91%. Misclassification as shadow occurred in (i) areas where mossy submerged vegetation presented a deep brown colour and could not be distinguished from the channel bed or (ii) shallow water areas presenting a darker colour due to sedimentation.

Deep water (Figure 5e) showed the highest TPR at 92% with the majority of misclassified features falling under riffles and occurring along the transition from riffles to pools. Similarly, confusion with shallow waters occurred in the transition zone from deep to shallow areas (Figures 5e and 6). Deep water was classified as vegetation near the banks. This was primarily due to georeferencing errors when mosaicking LAIC tail outputs.

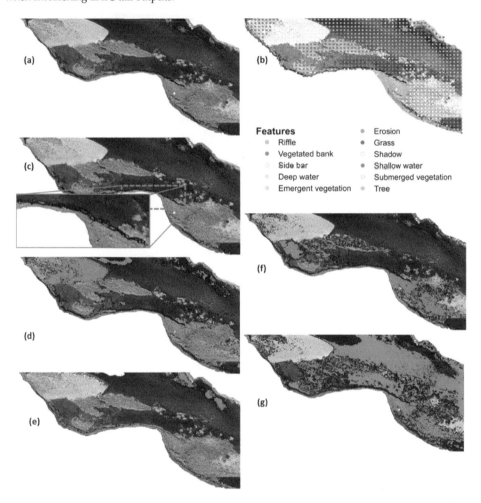

Figure 5. Example of Artificial Neural Network (ANN) classification outputs obtained with the Leaf Area Index Calculation (LAIC) for a selected portion of the orthoimage. Pixels elected within each class are shown in pink. (**a**) Original image; (**b**) Visual classification for the points defined by a 2 m × 2 m regular grid; (**c**) Erosion; (**d**) Side bars; (**e**) Deep water; (**f**) Vegetation (all classes); (**g**) Riffles. The image is not to scale.

Shallow waters presented the same errors as those already described beforehand. Here, shallow water with rippled water surface was automatically classified as riffles. This could not be identified as an error in classification but as higher resolution in feature detection than expected. When correcting for this effect, the TPR increased to 69% for shallow water, with a rate of TN equal to 81%. Misclassification of shallow waters as shadows primarily occurred in areas below trees or water areas obscured by brown submerged vegetation. In the first instance, we assumed that the visual classification was inaccurate whereas in the second case vegetation generated dark undistinguishable and even shadow patterns.

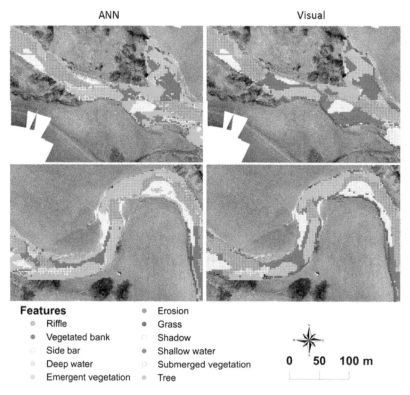

Figure 6. Classification outputs at each of the points defined by a 2 m × 2 m regular grid obtained with (**Left**) The Leaf Area Index Calculation (LAIC) Artificial Neural Network (ANN) and (**Right**) The visual identification for two sections within the study reach.

3.3. Vegetation

All vegetation classes combined resulted in a TPR of 81% with all the individual classes above 80% except for submerged vegetation (29%). Submerged vegetation was generally classified as either shallow water or riffle whereas classification as shadow primarily appeared in areas where the vegetation presented a brown colour due to sediment deposits.

Vegetated banks were primarily confused with erosion or shadows. Erosion corresponded to sediment deposits where senescent vegetation was present whereas shadows were the result of LAIC detecting features at a higher level of resolution than that obtained from visual classification. In all instances LAIC correctly identified the shadows generated by the vegetation. For the tree feature class, all misclassifications as riffles or shallow water accounted for LAIC extracting the riffle class from the water that was visible between branches. In general, trees misclassified as shadows corresponded to shadows generated by the tree branches. No significant misclassification errors were identified for emergent free floating vegetation, grass and vegetation in bars, with TRP values above 99%.

4. Discussion

This paper presented an operational UAV based framework for the identification of hydromorphological features using ANNs and high resolution RGB aerial imagery. To the authors' knowledge, this is the first study looking at the development of a framework for the automated classification of all hydromorphological features within a river reach. The framework proved to be accurate at 81% and enabled the identification of hydromorphological features at a higher level of detail than that provided by the ground truth data. For example, it allowed the separation of trees from the grass or water underneath, as well as the recognition of hydraulic units (e.g., rippled water surface).

The approach is: (i) transferable, because once the ANN has been trained, it can be applied to other river sections without direct supervision; (ii) unbiased, because it enables the objective classification of hydromorphological features and (iii) flexible, because it is able to identify the multiple hydromorphological features that can be attributed to a single point. Hydromorphological features are not discrete units but an interaction of multiple layers with fuzzy limits that reflect the spatial complexity of river environments [45]. Points can therefore fall simultaneously within multiple feature classes (e.g., riffles and submerged vegetation). LAIC has the potential to identify multiple classes for a single point based on the hydraulic, habitat or vegetation characteristics observed through the selection of different number of clusters during the ANN training.

Shadows within the imagery pose one of the primary barriers to correct feature identification. This issue has been recognised in previous work by the authors [21,22]. The overall ANN accuracy (81%) could therefore be improved through detailed flight planning that aims at minimising the presence of shadows. Thoughtful selection of the time of the flight to avoid shadows will also increase the potential for erosion identification. Flight optimisation also needs to consider (i) seasonality and (ii) flight direction. Winter flights present the advantage of exposing the totality of the river channel whereas spring and summer flights will not enable the identification of in-channel features under dense vegetation cover. However, fully developed vegetation will expose different green RGB totalities and allow LAIC to identify plant species. The optimisation of the flight direction is essential for wide area mapping (e.g., at sub-catchment scale). This will reduce the flight time, ensure all the imagery is collected under similar light conditions, minimise the number of frames and the CPU load required to build the orthoimage. Previous research with UAVs in rivers [46] have proved longitudinal multipasses to be more efficient than cross-sectional ones without compromising the quality of the photogrammetric process.

The overall approach here presented is based on near real-time RGB imagery of higher resolution than multi- and hyperspectral imagery from manned aircraft used in the past for similar purposes [14,25]. It complements existing tools for characterising rivers or fluvial features [47] such as the Fluvial Information System [48] or the "FluvialCorridor" [49]. Similarly, it provides the basis for the comparison and harmonisation of results obtained from the exhaustive list of available hydromorphological assessment methodologies [50]—a much sought outcome required by the WFD for intercalibration purposes [51]. Although time consuming, the *k*-means based ANN approach was preferred to visual photointerpretation as it provides an objective way for the classification of river environments that could be automatically applied to the national river network. Some of the time consuming steps only need to be carried out occasionally. For example, the ANN training is a one-off process that only needs to be repeated whenever significant changes in environmental conditions occur (e.g., river turbidity or weather pattern variation). Other time-consuming steps such as frame selection or GCP location can be optimised thus significantly reducing the time demand.

The technique relies on a minimum threshold of illumination and limited presence of shadows. The ideal flying time for data capture purposes is therefore solar noon. In this study, all the frames were captured on the same day under stable weather conditions and within a 6 h interval. However, these minimal changes in sun position and orientation could result in an increased number of miss-classified features along the downstream sections of the reach due to an increased presence of shadows. Illumination is not a key factor affecting the classification, as long as a minimum level of brightness is present. This is because LAIC bases the clustering technique only on the *a*b* parameters of the CIELAB space [40,41], without taking into account the luminosity (L).

It is important to note that for the adoption of the framework at national and international level, several operational limitations for small UAVs should be addressed. This includes battery life endurance, platform stability as well as standardised and flexible international airspace regulatory frameworks. The ANN approach already provides a fast processing platform for the recognition of patterns from UAV high resolution aerial imagery, with the UAV platforms presenting the main limitations for large scale mapping. Octocopters such as the one used in this study have the ability to hover over target areas to provide high resolution still imagery. Key to the UAV high performance is the design-integration of a gimbal that effectively negate pitch, yaw and roll. This camera gimbal allows the capture of nadir images to perform a highly-accurate photogrammetric workflow. Fix wing platforms can be a better alternative to cover larger areas but this may come at a cost to imagery quality and resolution. Further considerations to increase the accuracy of the ANN based feature identification relates to the reduction of georeferencing and misalignment mosaic errors (2%) from LAIC outputs. This issue can be addressed by ensuring the outputs generated are automatically georeferenced and ready to upload into a GIS platform.

5. Conclusions

The ANN based framework herein described for the recognition of hydromorphological river features relies heavily on the use of UAVs for the collection of high resolution RGB true colour aerial photography. The approach presented provides robust automated classification outputs of river hydromorphological features. The Artificial Neural Network (ANN) Leaf Area Index Calculation (LAIC) software used for pattern recognition enabled identification of hydromorphological features at a higher level of detail than that derived from visual observation. The framework leverages the use of UAVs for environmental policy implementation and demonstrates the potential of ANNs and RGB true colour imagery for precision river monitoring and management. Key advantages for its large-scale implementation rely on its flexibility, transferability and unbiased results. Time-consuming tasks within the framework can be optimised to reduce CPU demand and processing time. Further work should look at enhancing the operational thresholds of small UAVs by for example, increasing battery live or increasing overall stability under gusty conditions.

Acknowledgments: We would like to thank the Environment Agency and EPSRC for funding this project under an EPSRC Industrial Case Studentship voucher number 08002930. Special thanks go to SkyCap and Natural Resources Wales for their help and support with data collection. The authors acknowledge financial support from the Castilla-La Mancha Regional Government (Spain) under the Postdoctoral Training Scholarship Programa Operativo 2007–2013 de Castilla-La Mancha.

Author Contributions: M.R.C. is the principal investigator and corresponding author. She led and supervised the overall research and field data collection. This includes financial, contractual and health an safety management duties. R.C. structured and wrote the paper in collaboration with R.B.G. B.G. helped with the flight planning, overall data collection and processed all the UAV imagery. T.K. was in charge of processing the ADCP data, advised on the ADCP data collection, wrote parts of the paper and helped structuring and drafting the manuscript. A.V. facilitated data collection through Environment Agency and Natural Resources Wales and secured part of the funding to carry out the research. Both, B.G. and A.V. were key for the interpretation of the UAV results.

Conflicts of Interest: The authors declare no conflict of interest.

References

1. European Commission. Directive 2000/60/EC of the European Parliament and of the Council of 23 October 2000 establishing a framework for community action in the field of water policy. *Off. J. Eur. Union* **2000**, *327*, 1–72.
2. European Commission. Council Regulation (EC) No 1100/2007 of 18 September 2007 establishing measures for the recovery of the stock of European eel. *Off. J. Eur. Communities* **2007**, *248*, 17–23.

3. Russi, D.; ten Brink, P.; Farmer, A.; Badura, R.; Coates, D.; Förster, J.; Kumar, R.; Davidson, N. *The Economics of Ecosystems and Biodiversity for Water and Wetlands*; Institute for European Environmental Policy: London, UK, 2013.

4. Rinaldi, M.; Belleti, B.; van de Bund, W.; Bertoldi, W.; Gurnell, A.; Buijse, T.; Mosselman, E. *Review on Eco-Hydromorphological Methods. Deliverable 1.1, REFORM (REstoring Rivers for Effective Catchment Management)*; European Commission: Belgium, Germany, 2013.

5. LAWA. *Gewasserstrukturgutekartierung in der Bundesrepublik Deutschland—Ubersichtsverfahren. Empfehlungen Oberirdische Gewasser*; Landerarbeitsgemeinschaft Wasser: Schwerin, Germany, 2002.

6. Buffagni, A.; Erba, S.; Ciampitiello, M. *Il Rilevamento Idromorfologici e Degli Habitat Fluviali nel Contest della Direttiva Europea Sulle Acque (WFD): Principi e Schede di Applicazione del Metodo Caravaggio*; CNR IRSA: Bari, Italia, 2005; Volume 2, pp. 32–34.

7. Raven, P.J.; Fox, P.; Everard, M.; Holmes, N.T.H.; Dawson, F.H. River Habitat Survey: A new system for classifying rivers according to their habitat quality. In *Freshwater Quality: Defining the Indefinable?* Boon, P.J., Howell, D.L., Eds.; The Stationery Office: Edinburgh, UK, 1997; pp. 215–234.

8. Environment Agency. *River Habitat Survey in Britain and Ireland*; Environment Agency: London, UK, 2003.

9. UK Technical Advisory Group. *UK Environmental Standards and Conditions*; UKTAG: London, UK, 2008.

10. North Ireland Environment Agency. *River Hydromorphological Assessment Technique (RHAT)*; NIEA: Northern Ireland, UK, 2009.

11. Poole, G.C.; Frisell, G.A.; Ralph, S.C. In-stream habitat unit classification: Inadequacies for monitoring and some consequences for management. *Water Resour. Bull.* **1997**, *33*, 879–896. [CrossRef]

12. Legleiter, C.J.; Marcus, W.A.; Lawrence, R.L. Effects of sensor resolution of mapping in-stream habitats. *Photogramm. Eng. Remote Sens.* **2002**, *68*, 801–807.

13. Leuven, R.S.E.; Pudevigne, I.; Teuww, R.M. *Application of Geographical Information Systems and Remote Sensing in River Studies*; Backhuys Publishers: Leiden, Netherlands, 2002.

14. Leckie, D.G.; Cloney, E.; Cara, J.; Paradine, D. Automated mapping of stream features with high-resolution multispectral imagery: An example of capabilities. *Photogramm. Eng. Remote Sens.* **2005**, *71*, 145–155. [CrossRef]

15. Richards, J.A. *Remote Sensing Digital Image Analysis: An Introduction*, 5th ed.; Springer: London, UK, 2013.

16. Nabney, I.T. *Netlab: Algorithms for Pattern Recognition*; Springer: Birmingham, UK, 2002.

17. Bishop, A. *Neural Networks for Pattern Recognition*; Oxford University Press: Oxford, UK, 1995.

18. Ballesteros, R.; Ortega, J.F.; Hernández, D.; Moreno, M.A. Applications of georeferenced high-resolution images obtained with unmanned aerial vehicles. *Precis. Agric.* **2014**, *15*, 593–614. [CrossRef]

19. Camargo, D.; Montoya, F.; Moreno, M.; Ortega, J.; Corcoles, J. Impact of water deficit on light interception, radiation use efficiency and leaf area index in a potato crop. *J. Agric. Sci.* **2015**, *1*, 1–12. [CrossRef]

20. Córcoles, J.I.; Ortega, J.F.; Hernández, D.; Moreno, M.A. Estimation of leaf area index in onion using unmanned aerial vehicle. *Biosyst. Eng.* **2013**, *115*, 31–42. [CrossRef]

21. Ballesteros, R.; Ortega, J.F.; Hernández, D.; Moreno, M.A. Description of image acquisition and processing. *Precis. Agric.* **2014**, *15*, 579–592. [CrossRef]

22. Ballesteros, R.; Ortega, J.F.; Hernandez, D.; Moreno, M.A. Characterization of Vitis Vinifera L. canopy using unmanned aerial vehicle-based remote sensing photogrammetry techniques. *Am. J. Enol. Vitic.* **2015**, *66*, 120–129. [CrossRef]

23. Goswami, A.K.; Gakhar, S.; Kaur, H. Automatic object recognition from satellite images using Artificial Neural Network. *Int. J. Comput. Appl.* **2014**, *95*, 33–39.

24. Skakun, S. A neural network approach to flood mapping using satellite imagery. *Comput. Inform.* **2010**, *29*, 1013–1024.

25. Legleiter, C.J. Spectral driven classification of high spatial resolution, hyperspectral imagery: A tool for mapping in-stream habitat. *Environ. Manag.* **2003**, *32*, 399–411. [CrossRef]

26. Fonstad, M.J. Hyperspectral imagery in fluvial environments. In *Fluvial Remote Sensing for Science and Management*; John Wiley & Sons: West Sussex, UK, 2012; pp. 71–81.

27. Anker, Y.; Hershkovitz, Y.; Ben-Dor, E.; Gasith, A. Application of aerial digital photography for macrophyte cover and composition survey in small rural streams. *River Res. Appl.* **2014**, *30*, 925–937. [CrossRef]

28. Tamminga, A.; Hugenholtz, C.; Eaton, B.; Lapointe, M. Hyperspatial remote sensing of channel reach morphology and hydraulic fish habitat using an unmanned aerial vehicle (UAV): A first assessment in the context of river research and management. *River Res. Appl.* **2014**, *31*, 379–391. [CrossRef]

29. Husson, E.; Hagner, O.; Ecke, F. Unmanned aircraft systems help to map aquatic vegetation. *Appl. Veg. Sci.* **2014**, *17*, 567–577. [CrossRef]

30. Kaneko, K.; Nohara, S. Review of effective vegetation mapping using the UAV (Unmanned Aerial Vehicle) method. *J. Geogr. Inf. Syst.* **2014**, *6*, 733–742. [CrossRef]

31. Hervouet, A.; Dunford, R.; Piegay, H.; Belleti, B.; Tremelo, M.L. Analysis of post-flood recruitment patterns in braided channel rivers at multiple scales based on an image series collected by unmanned aerial vehicles, ultra-light aerial vehicles and satellites. *GISci. Remote Sens.* **2011**, *48*, 50–73. [CrossRef]

32. MacVicar, B.J.; Piegay, H.; Henderson, A.; Comiti, F.; Oberlin, C.; Pecorari, E. Quantifying the temporal dynamics of wood in large rivers: Field trials of wood surveying, dating, tracking, and monitoring techniques. *Earth Surf. Process. Landf.* **2009**, *34*, 2031–2046. [CrossRef]

33. Goktagan, A.H.; Sukkarich, S.; Bryson, M.; Randle, J.; Lupton, T.; Hung, C. A rotary-wing unmanned aerial vehicle for aquatic weed surveillance and management. *J. Intell. Robot. Syst.* **2010**, *57*, 467–484. [CrossRef]

34. Dunford, R.; Michel, K.; Gagnage, M.; Piegay, H.; Tremelo, M.L. Potential and constraints of unmanned aerial vehicle technology for the characterization of Mediterranean riparian forest. *Int. J. Remote Sens.* **2009**, *30*, 4915–4935. [CrossRef]

35. Vericat, D.; Brasington, J.; Wheaton, J.; Cowie, M. Accuracy assessment of aerial photographs acquired using lighter-than-air blimps: Low-cost tools for mapping river corridors. *River Res. Appl.* **2009**, *25*, 985–1000. [CrossRef]

36. Kriechbaumer, T.; Blackburn, K.; Everard, N.; Rivas-Casado, M. Acoustic Doppler Current Profiler measurements near a weir with fish pass: Assessing solutions to compass errors, spatial data referencing and spatial flow heterogeneity. *Hydrol. Res.* **2015**, in press.

37. Woodget, A.S.; Carbonneau, P.E.; Visser, F.; Maddock, I.P. Quantifying submerged fluvial topography using hyperspatial resolution UAS imagery and structure from motion photogrammetry. *Earth Surf. Process. Landf.* **2014**, *40*, 47–64. [CrossRef]

38. Civil Aviation Authority. *CAP 393 Air Navigation: The Order and Regulations*, 4th ed.; The Stationery Office: Norwich, UK, 2015.

39. Civil Aviation Authority. *CAP 722 Unmanned Aircraft System Operations in UK Airspace—Guidance*, 6th ed.; The Stationery Office: Norwich, UK, 2015.

40. Comission Internationale de L'Eclairage. *Colorimetry*; CIE Publication: Paris, France, 1978.

41. Comission Internationale de L'Eclairage. *Recommendations on Uniform Color Spaces, Color-Difference Equations, Psychometric Color Terms*; CIE Publication: Paris, France, 1978; Volume 15, pp. 1–21.

42. Mueller, D.S.; Wagner, C.R. *Measuring Discharge with Acoustic Doppler Current Profilers from a Moving Boat*; United States Geological Survey: Reston, WV, USA, 2009.

43. SonTek. *RiverSurveyor S5/M9 System Manual*; SonTek: San Diego, CA, USA, 2014.

44. Kohavi, R.; Provost, F. On applied research in machine learning. In *Editorial for the Special Issue on Applications of Machine Learning and the Knowledge Discovery Process*; Columbia University: New York, NY, USA, 1998; Volume 30, pp. 127–274.

45. Wallis, C.; Maddock, I.; Visser, F.; Acreman, M. A framework for evaluating the spatial configuration and temporal dynamics of hydraulic patches. *River Res. Appl.* **2012**, *28*, 585–593. [CrossRef]

46. Ortega-Terol, D.; Moreno, M.A.; Hernández-López, D.; Rodríguez-Gozálvez, P. Survey and classification of Large Woody Debris (LWD) in streams using generated low-cost geomatic products. *Remote Sens.* **2014**, *6*, 11770–11790. [CrossRef]

47. Schmitt, R.; Bizzi, B.; Castelletti, A. Characterizing fluvial systems at basin scale by fuzzy signatures of hydromorphological drivers in data scarce environments. *Geomorphology* **2014**, *214*, 69–83. [CrossRef]

48. Carboneau, P.; Fonstad, M.A.; Marcus, W.A.; Dugdale, S.J. Making riverscapes real. *Geomorphology* **2012**, *137*, 74–86. [CrossRef]

49. Roux, C.; Alber, A.; Bertrand, M.; Vaudor, L.; Piégay, H. "FluvialCorridor": A new ArcGIS toolbox package for multiscale riverscape exploration. *Geomorphology* **2015**, *242*, 29–37. [CrossRef]

50. Belletti, B.; Rinaldi, M.; Buijse, A.D.; Gurnell, A.M.; Mosselman, E. A review of assessment methods for river hydromorphology. *Environ. Earth Sci.* **2015**, *73*, 2079–2100. [CrossRef]
51. European Commission. Common implementation strategy for the Water Framework Directive (2000/60/EC). In *Towards a Guidance on Establishment of the Intercalibration Network and the Process on the Intercalibration Exercise;* Office for the Official Publications of the European Communities: Belgium, Germany, 2003.

Article

Autonomous Aerial Refueling Ground Test Demonstration—A Sensor-in-the-Loop, Non-Tracking Method

Chao-I Chen *, Robert Koseluk, Chase Buchanan, Andrew Duerner, Brian Jeppesen and Hunter Laux

Advanced Scientific Concepts Inc., 135 East Ortega Street, Santa Barbara, CA 93101, USA; bkoseluk@asc3d.com (R.K.); cbuchanan@asc3d.com (C.B.); aduerner@asc3d.com (A.D.); bjeppesen@asc3d.com (B.J.); hlaux@asc3d.com (H.L.)
* Author to whom correspondence should be addressed; cchen@asc3d.com;
   Tel.: +1-805-966-3331; Fax: +1-805-966-0059.

Academic Editor: Felipe Gonzalez Toro
Received: 15 October 2014; Accepted: 4 May 2015; Published: 11 May 2015

Abstract: An essential capability for an unmanned aerial vehicle (UAV) to extend its airborne duration without increasing the size of the aircraft is called the autonomous aerial refueling (AAR). This paper proposes a sensor-in-the-loop, non-tracking method for probe-and-drogue style autonomous aerial refueling tasks by combining sensitivity adjustments of a 3D Flash LIDAR camera with computer vision based image-processing techniques. The method overcomes the inherit ambiguity issues when reconstructing 3D information from traditional 2D images by taking advantage of ready to use 3D point cloud data from the camera, followed by well-established computer vision techniques. These techniques include curve fitting algorithms and outlier removal with the random sample consensus (RANSAC) algorithm to reliably estimate the drogue center in 3D space, as well as to establish the relative position between the probe and the drogue. To demonstrate the feasibility of the proposed method on a real system, a ground navigation robot was designed and fabricated. Results presented in the paper show that using images acquired from a 3D Flash LIDAR camera as real time visual feedback, the ground robot is able to track a moving simulated drogue and continuously narrow the gap between the robot and the target autonomously.

Keywords: 3D Flash LIDAR; autonomous aerial refueling; computer vision; UAV; probe and drogue; markerless

1. Introduction

In-flight aerial refueling was first proposed by Alexander P. de Seversky in 1917 and put into practice in the United States in the 1920s. The original motivation was to increase the range of combat aircraft. This process of transferring fuel from the tanker aircraft to the receiver aircraft enables the receiver aircraft to stay in the air longer and is able to take off with a greater payload. This procedure was traditionally performed by a veteran pilot due to the required maneuvering skills and fast reaction times. In recent years, more and more unmanned air vehicles (UAVs) are used in both military and civilian operations, which motivate researchers to develop solutions to achieve the goal of autonomous aerial refueling (AAR) [1–3]. The ability to autonomously transfer and receive fuel in flight will increase the range and flexibility of future unmanned aircraft platforms, ultimately extending carrier power projection [4].

There are two commonly used methods for refueling aircraft in flight: the probe and drogue (PDR) method [5], and the boom and receptacle (BRR) method [6]. The former PDR method is the focus of this paper and is the standard aerial refueling procedure for the US Navy, North Atlantic Treaty Organization (NATO) nations, Russia and China. The tanker aircraft releases a long flexible hose in the PDR method; at the end of the hose is attached a cone-shaped drogue. A receiver aircraft extends a rigid arm called a probe on one side of the aircraft. Because the tanker simply flies straight and allows the drogue to trail behind without making efforts to control the drogue, the pilot of the receiver aircraft is responsible to make sure the probe mounted on the receiver aircraft links up with the drogue from the tanker. This phase is called the approach phase. After the connection is made, the two aircraft fly in formation during which time fuel pumps from the tanker aircraft to the receiver aircraft. This phase is called the station keeping phase, because maintaining a stationary relative position between the tanker aircraft and the receiver aircraft is critical. The final separation phase is completed after the probe is pulled out of the drogue when the receiver aircraft decelerates hard enough to disconnect. One advantage of using the PDR method is that this refueling method allows multiple aircraft to be refueled simultaneously.

The boom and receptacle (BRR) method, on the other hand, utilizes a long rigid, hollow shaft boom extended from the rear of the tanker aircraft. The boom is controlled by an operator who uses flaps on the boom to supervise and direct the boom to the coupling receiver aircraft's receptacle. The workload of completing the refueling task is shared between the receiver pilot and the boom controller. This method is adapted by the US Air Force (USAF) as well as the Netherlands, Israel, Turkey, and Iran. Although boom and receptacle method provides higher fuel transfer rate and reduces the receiver pilot's workload, the modern probe and drogue systems are simpler and more compact by comparison. More detailed comparison between these two different operation methods can be found in [2].

There are two required steps in the approach phase before the connection between the receiver and tanker aircrafts can be made—The flight formatting step and the final docking step. The flight formatting step utilizes global positioning systems (GPS) and inertial navigation systems (INS) on each aircraft, combined with a wireless communication system to share measurement information. Modern Differential GPS (DGPS) systems are commonly applied to solve autonomous aerial refueling and they provide satisfactory results in guiding an aircraft to a proximate position and maintaining the close formation between the tanker and receiver [7–12]. This technique is not, however, suitable for the final docking step where a physical contact between the probe and the drogue is required. The major challenge is that some aerodynamic effects occur on the drogue and the hose as well as the receiver aircraft itself during the final docking phase. Some of these in flight effects are observed and reported [13,14]. Unfortunately, this dynamic information cannot be captured using GPS and INS sensors because neither sensor can be easily installed on a drogue, which makes the final docking step challenging. Furthermore, the update rate of the GPS system is generally considered too slow for object tracking and terminal guidance technologies that are needed in the final docking step.

Machine vision techniques are generally considered to be more suitable for the final docking task. Many vision based navigation algorithms have been developed for UAV systems [15–20]. For aerial refueling, specific developments include feature detection and matching [21–24], contour method [25], and modeling and simulation [26–30]. In addition to passive imaging methods, landmark-based approaches have also been investigated by researchers. Junkins *et al.*, developed a system called VisNav [31], which employs an optical sensor combined with structured active light sources (beacons) to provide images with particular patterns to compute the position and orientation of the drogue. This hardware-in-the-loop system has been used in several studies [32–34] and the results suggest that it is possible to provide high accuracy and very precise six degree-of-freedom position information for real-time navigation. Pollini *et al.* [28,35] also suggest this landmark-based approach and proposed placing light emitting diodes (LEDs) on the drogue and using a CCD camera with infrared (IR) filter to identify the LEDs. The captured images are then served as input information for Lu, Hager and Mjolsness (LHM) algorithm [36] to determine the relative position of the drogue. One major disadvantage of using a beacon type system in the probe and drogue refueling is that non-trivial hardware modifications on the tanker aircraft are required in order to supply electricity and support communication between the drogue and the receiver aircraft.

Martinez *et al.* [37] proposes the use of direct methods [38] and hierarchical image registration techniques [39] to solve the drogue-tracking problem for aerial refueling. The proposed method does not require the installation of any special hardware and it overcomes some drawbacks caused by partial occlusions of the features in most existing vision-based approaches. The test was carried out in a robotic laboratory facility with a unique test environment [40]. The average accuracy of the position estimation was found to be 2 cm for the light turbulence conditions and 10 cm for the moderate turbulence conditions. However, it is well known that traditional vision based technologies are susceptible to strong sunlight or low visibility conditions, such as on a dark night or in a foggy environment. As the 2D image quality declines, the accuracy of the inferred 3D information will unavoidably deteriorate.

It is still a difficult problem to reconstruct 3D information reliably from 2D images due to the inherent ambiguity caused by projective geometry [41–43] and the benefits of using 2.5D information in robotic systems for various tasks have been documented [44,45]. For probe-and-drogue autonomous aerial refueling application specifically, using a time-of-flight (ToF) based 3D Flash LIDAR system [46, 47] to acquire information of the drogue in 3D space has been proposed in Chen and Stettner's work [48]. They utilized the characteristics of the 2.5D data sensor provided and adopted a level set method (LSM) [49,50] to segment out the drogue for target tracking purposes. Because of the additional range information associated with each 2D pixel, the segmentation results become more reliable and consistent. The indoor experiments were carried out in a crowed laboratory, but the detected target showed promising results.

There are two major challenges for the final docking (or hitting the basket) step in the probe-and-drogue style autonomous aerial refueling: (1) the ability to reliably measure the orientation as well as relative position between the drogue trailed from the tanker aircraft and the probe mounted on the receiver aircraft and (2) advanced control systems to rapidly correct the approaching course of the receiver aircraft to ensure the eventual connection between the probe and the drogue. This paper offers a potential solution to the former difficult task. Although some design descriptions of the ground test robot are also presented, the intent is only to evaluate the proposed method in a more practical experimentation. We encourage readers who are interested in navigation and control aspects of the unmanned systems to consult more domain specific references, such as [51–53]. This paper employs a 3D Flash LIDAR camera as the source of the input data, but differs from [48] in that this paper suggests a sensor-in-the-loop method incorporating both hardware and software elements. In addition, a ground feasibility test was performed to demonstrate the potential for in air autonomous aerial refueling tasks.

2. Method

The method section is organized as follows. Descriptions of the sensor employed for data acquisition is first introduced in Section 2.1. Reasons for choosing this type of sensor over other sensors are discussed in depth. Section 2.2 presents characteristic analysis results of a real drogue for aerial refueling task. Section 2.3 briefly described how the 3D Flash LIDAR camera internally computes range information followed by the discussions of a more forgiving drogue center estimation method in Section 2.4.

2.1. 3D Flash LIDAR Camera

A 3D Flash LIDAR camera is an eye safe, time-of-flight (ToF) based vision system using a pulsed laser. Because the camera provides its own light source, it is not susceptible to lighting changes, which are typical challenges for traditional vision based systems. Figure 1 illustrates the comparison between a regular 2D image on the left and an image acquired from the 3D Flash LIDAR camera on the right under a strong sun light condition.

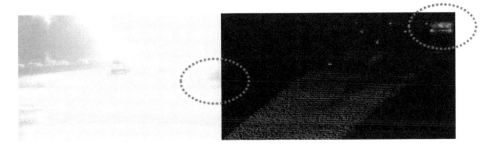

Figure 1. 2D camera *vs.* 3D Flash LIDAR camera.

This camera is, however, similar to a traditional 2D camera that uses a focal plane array with 128 × 128 image resolution. The only difference is that a 3D Flashed LIDAR camera provides additional depth information for every pixel. Each pixel triggers independently and the associated counter for the pixel will record the time-of-flight value of the laser pulse to the objects within the field of view (FOV) of the camera. Because of this similarity, the relationship between a 2D point and the 3D world can be described using a commonly used pinhole model as shown in Equations (1) and (2) below [43].

$$x = KR[\, I \mid -\tilde{C}\,]X \tag{1}$$

$$K = \begin{bmatrix} f & 0 & p_x \\ 0 & f & p_y \\ 0 & 0 & 1 \end{bmatrix} \tag{2}$$

The upper-case X is a 4-elements vector in a 3D world coordinate frame while the lower-case x is a 3-elements vector in the 2D image coordinate system. K is the internal camera parameter matrix with focal length f and principle point (p_x, p_y) information. R represents a 3×3 rotation matrix together with $\tilde{C}$ are called the external parameters which relate the camera orientation and position to the world coordinate system. Converting each pixel into 3D space is a straightforward task that requires simple geometric equations when the 3D Flash LIDAR camera is used because the depth information is available and therefore complicated mathematical inference is no longer needed. Moreover, it is possible to construct geo-reference information for every point if the global coordinates of the camera are known because the calculated 3D positions are relative to the camera center.

Another common property both types of cameras share is that they can easily change the FOV by choosing different lenses. For a fixed resolution image, it is expected to observe more details when a narrower FOV lens is selected as shown in Figure 2. The total number of pixels that will be illuminated on a known object at a certain distance can be estimated. The blue curve in Figure 2 represents the case of a 45° FOV lens, while the similar curve in red represents a 30° FOV lens. All of these 2D pixels detected by the camera can be uniquely projected back in 3D space for position estimates and the process does not require additional high quality landmarks. Figure 2 also shows rapid growth of the total number of pixels in the images as the distance between the target and the camera decreases.

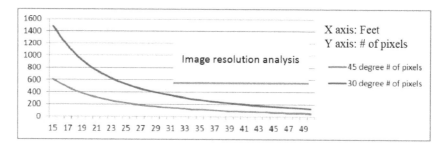

Figure 2. Range *vs.* resolution analysis, Number of pixels that will be illuminated on a 27-inch (68.58 cm) diameter object from 15 feet (4.57 m) to 50 feet (15.24 m).

Unlike a traditional scanning based lidar system, a 3D Flash camera does not have moving parts. All range values for the entire array are computed after only one shot of laser pulse. This camera is therefore capable of capturing images of a high-speed moving object without motion blur, which is particularly important for the autonomous aerial refueling application. Figure 3 shows the propeller of an airplane rotating at 220 meters per second, which is frozen by speed-of-light imaging. Figure 4 shows a seagull taking off from a roof in consecutive frames of motion. As can be seen, each snap shot is a clean image. Furthermore, as the 3D Flash LIDAR sensor shares so many common properties with conventional CCD cameras, many existing computer vision based algorithms, libraries and tools can be adapted to help solving traditionally difficult problems with comparatively minor modifications. OpenCV (Open Source Computer Vision) [54], for example, is one of the most popular libraries in the computer vision field of research and the PCL (point cloud library) [55] for both 2D and 3D point cloud data processing. As for autonomous systems, Robot Operating System (ROS) is a collection of software frameworks [56] and a useful resource for researchers since machine vision is an essential component for robots as well.

Figure 3. Propeller tips move at 220 meters per second without motion blur.

Figure 4. A seagull is taking off from a rooftop.

2.2. MA-3 Drogue

Instead of passively using the default settings in the 3D Flash LIDAR camera, experiments were carried out to explore proper settings for the drogue detections task and invaluable information was acquired using a real Navy drogue from PMA 268. The experimental results show the drogue contains retro-reflective materials and it was fortunate in terms of detecting the drogue at all needed distances. Figure 5 summarizes the experimental results. The drogue was facing up and located on the ground. A 3D Flash LIDAR camera was set up about 20 feet (6.1 m) above the drogue on our 2nd floor balcony, facing down perpendicularly. Figure 5a–c mimic images that would be observed from the receiver aircraft. Figure 5a is a regular 2D color image for visual reference purpose and Figure 5b is the intensity image captured by the 3D Flash camera. Figure 5b,c are the same images except the laser energy in Figure 5c is only 0.01% of that in Figure 5b after a neutral density filter is applied. The same strong retro reflective signals are also observed when switching the view point from the receiver aircraft to the tanker side as shown in Figure 5d–f. Although the majority of research related to the probe-and-drogue style autonomous refueling focuses on simulating scenarios of mounting sensors in the receiver aircraft. The possibility of equipping sensors on the tanker side has also been considered. This experiment is designed to help us understand what can be expected from the sensor output under different parameter settings and raise a flag if some limitations are found. Fortunately, there are no obvious show stoppers for either option in terms of received signals.

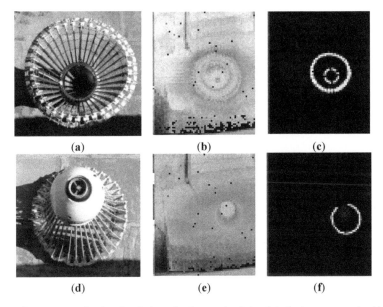

Figure 5. Strong retro-reflective signals from the drogue. (**a–c**) simulate the images perceived by the receiver aircraft; (**d–f**) simulate the images perceived by the tanker.

To limit the scope of this paper's discussion, the assumption of observing the drogue from the receiver aircraft is made. It is crucial to balance the laser power and the camera sensitivity setting to achieve the most desired signal return level. One of the challenges in the system design lies in satisfying two extreme cases in the autonomous aerial refueling application: (1) the laser must generate enough power to provide sufficient returns from the drogue when it is at the maximum required range; and (2) when the drogue is very close to the camera as expected in the final docking step, a mechanism to avoid saturating the acquired data (due to the powerful laser), is also mandatory. Based on the observation described earlier, the first extreme case does not seem to be a concern at anymore. All efforts should be dedicated to solve the second extreme scenario.

2.3. Range Calculation

How a 3D Flash camera provides ready to use range information is briefly discussed in this section. Target range measurement, based on time-of-flight of the laser pulse, is determined independently in each unit cell. With a high reflectivity target, such as retro-reflective materials in the autonomous aerial refueling application, the return amplitude can be saturated. In this saturated case, the time-of-flight can be interpolated. The saturation algorithm is certainly suitable for the close up scenario when the probe of the receiver aircraft is about to make a connection with the drogue. However, the detected signals from the drogue may not always be highly saturated when the refueling process starts from some distance away, since the laser energy follows an inverse-square law. Even a retro-reflective drogue may look like a low reflectivity target when the distance between the target and the camera is large. To include this non-saturating case, the time-of-flight can also be interpolated from the non-saturated signal. To achieve the best of both worlds, a 3D Flash LIDAR camera optimizes the non-saturating and saturating algorithms. The optimized algorithm is implemented in the camera's field-programmable gate array (FPGA) for real time output.

2.4. Drogue Center Estimation

The ability to reliably measure the relative position and orientation between the drogue trailed from the rear of the tanker aircraft and the probe equipped on the receiver aircraft is one of the main challenges in the autonomous aerial refueling application. In the previous work [48], a level-set front propagation routine is proposed for target detection and identification tasks. Together with sufficient domain knowledge to quickly eliminate unlikely target candidates, the proposed method provides satisfactory results in estimating the center of the drogue after all 3D points on the drogue are identified. Any computation related to the relative position and orientation becomes straightforward when the center point in 3D space is established.

Table 1. Domain knowledge table.

Item	How We Can Use This Information
Single object tracking	Cross over issue is not considered
Single camera	Information handling is simplified
Simple background	Only have probe, drogue, hose, and tanker
Plane movement	Drogue randomly moves in horizontal/vertical directions
Known object of interest	Highly reflective materials. Camera setting can be simplified
Bounded field of view	Use automatic target detection and recognition for each frame instead of tracking which will fail if the target is outside the FOV.

After learning more about the characteristics of the real drogue discussed in Section 2.2, the domain knowledge references are updated and summarized in Table 1. One important piece of information, which was missing in the previous work [48], is that a drogue appears to contain high reflective materials at least to the wavelength a 3D Flash LIDAR camera detects. The first rational idea for a sensor-in-the-loop approach would be taking advantage of this fact. By lowering the camera gain, a 3D Flash LIDAR camera will detect only strong signal returns from highly reflective materials such as a refueling drogue. Figure 6 shows a few snap shots from a video sequence while lowering the camera gain continuously. As can be seen, by gradually applying these changes (from left to right), a crowded lab disappears in the final frame and only the high reflectivity target remains. This simple adjustment in the camera makes the subsequent analysis much easier because there are fewer pixels left to process and the majority of these remaining pixels are on the target of interest. Therefore, less computational cost can be expected while the confidence level of the detected target increases because there is not much room for an image processing algorithm to make a mistake. This is the essence of adapting a sensor-in-the-loop approach when solving a difficult problem. Data acquisition and image processing are often coupled together but treated as two separate components in a system pipeline. While it is convenient to isolate individual components for discussion purposes, global optimization from the total system point of view most likely cannot be achieved without considering both components simultaneously.

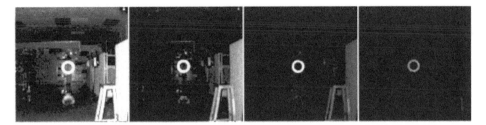

Figure 6. Adjusting camera setting.

Is object tracking is really necessary; or is the automatic target detection and recognition (ATD/R) all we need? The ATD/R module is essential for most of the automatic tracking systems and is usually engaged in either the initialization stage or the recovery stage where establishing the target of interest is required. Given the fact that a 3D Flash LIDAR camera, like a traditional 2D camera, can only image objects within its FOV, when the target drifts out of the FOV and then reappears later, an ATD/R component is required to reinitialize the target. A tracking process implicitly assumes that the target appearing in the current frame would be located somewhere close to where it was in the previous frame. Therefore, the tracking algorithm is designed to narrow the searching space to limit the computational cost. In the previous work [48], a level set front propagation algorithm is used to track the target. Existing information such as the silhouette of the target and the estimated center in the previous frame is used for seed point selection to efficiently identify the target. A tracking process does not seem to be required when the target of interest can be reliably identified with proper camera settings as shown in Figure 6.

As Figure 5c clearly illustrated, a drogue has retro-reflective materials on both the canopy (the outer ring) and the center rigid body structure (the inner ring). Although some distortion might be expected from the canopy of the drogue, it usually forms a circle thanks to aerodynamic flow. Many research papers have been published on circle fitting [57–65]. In general, the basic problem is to find a circle that best represents a collection of $n \geq 3$ points in 2D space (image coordinate system) labeled $(x_1, y_1) (x_2, y_2), \ldots, (x_n, y_n)$ with the circle equation described by $(x - a)^2 + (y - b)^2 = r^2$ and we need to determine the center (a, b) and radius r. One reasonable error measure of the fit will be given by summing the squares of the distance from the points to the circle as shown in Equation (3) below.

$$SS(a,b,r) = \sum_{i=1}^{n} \left(r - \sqrt{(x_i - a)^2 + (y_i - b)^2} \right)^2 \tag{3}$$

Coope [59] discusses numerical algorithms for minimizing SS (sum of squares) over a, b, and r. With various ways of formulating the same problem, each circle fitting algorithm results in different accuracy, convergence rate and tolerance level for noise. The goal of this paper is not to develop a new algorithm to determine the center of the circle, but to evaluate and select one algorithm that can reliably and efficiently output the center of the drogue when an image frame from 3D Flash LIDAR is presented. Utilizing these well-studied algorithms, we expect to enlarge the effective working area beyond the FOV boundary because these algorithms can estimate the center of a circle even if the circle is partially occluded.

Figure 7 summarized the evaluation results using the real target—A MA-3 drogue. The drogue was oriented vertically upward on the ground 20 feet (6.1 m) below the second floor balcony as shown in the middle intensity image of Figure 7. The 3D Flash LIDAR camera was moving in toward the balcony while facing down perpendicularly to create images of partially occluded drogue for testing as shown in the left and the right intensity images of Figure 7. These images of the drogue have been manually segmented and processed using 12 circle fit algorithms with Chernov's Matlab implementation [58,66]. Over the 100 frames of the sequence, the initial images of the drogue are occluded by the railing of the balcony, producing a partial arc. As the camera moves, the arc becomes a complete circle. The sequence ends with the camera returning to the initial origin.

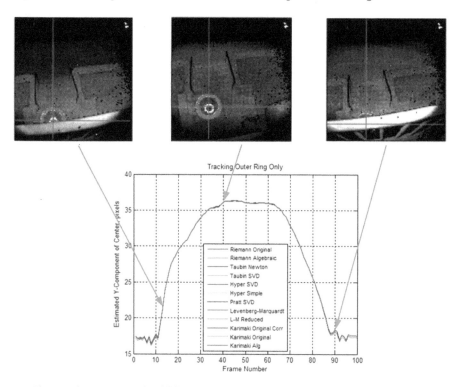

Figure 7. Comparison study of different circle fitting algorithms by using only the outer ring.

The upper plots from left to right are the intensity images acquired from the 3D Flash LIDAR for frames 13, 40 and 90. Pixels segmented for the outer ring only are shown in red, and the red cross-hair shows the estimated center using the Taubin SVD algorithm [65]. Since the primary movement is in the y-direction, the lower plot shows the estimated Y component of the ring's center *vs.* frame number for each of the 12 fit algorithms indicated in the legend. The plot shows very close agreement of all 12 algorithms with the real drogue data, including those frames where part of the ring is occluded by the railing on the balcony. While the ground truth of the actual center is not available, the estimated center in the intensity image appears subjectively to be reasonably accurate. Barrel distortion of the receiver lens is evident in the railing, but not particularly noticeable in the ring image or the estimate of the ring's center.

However, the close agreement conclusion does not hold when both inner and outer rings are used. Figure 8 shows the comparison study results using the same sequence. Again, the segmentation is performed manually. The plot shows that some of the algorithms were affected more than others having both rings present and, similarly, some algorithms were more adept at handling the appearance of the inner ring. The algorithms that were upset by the appearance of the inner ring were the Levenberg-Marquardt [67–69] and Levenberg-Marquardt Reduced algorithms (both are iterative geometric methods) and the final two Karimaki algorithms [61], lacking the correction technique. All of the other algorithms agreed closely and handled the appearance of the inner ring very well. Based on these analysis results, a Newton-based Taubin algorithm is selected and implemented in the ground navigation robot.

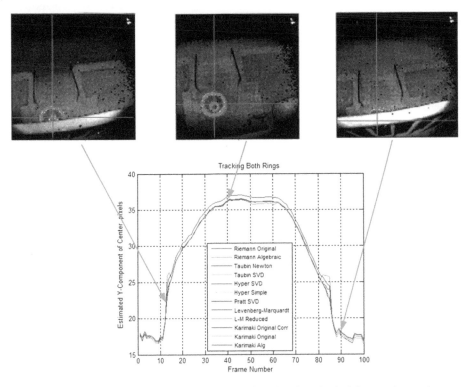

Figure 8. Comparison study of different circle fitting algorithms by using both inner and outer rings.

Three major changes improve the overall robustness of the drogue center estimation process. (1) Pixel connectivity is no longer required. The level set front propagation algorithm proposed in the previous work [48] implicitly assumes the target of interest is one connected component. If this assumption fails in practice scenarios, the analysis for subsequent segmentation and for detecting and extracting the target automatically will be unavoidably complicated. In contrast, the circle fitting algorithms, by design, perform well on disconnected segments or even on sparse input pixels; (2) Partially out of FOV cases are handled naturally. Additional domain knowledge and heuristics need to be incorporated into the previously proposed method for the segmentation routine to handle partially out of FOV cases reliably. The circle fitting algorithms, on the other hand, handle these cases without any special treatments. As can been seen in Figures 7 and 8, a small segment of an arc is all these algorithms require to predict the circle center; (3) Estimation of the target size in advance is not required. Given the FOV of a camera as well as the distance between a known object and the camera, it is possible to estimate the total number of pixels in each image that will be illuminated on this object as shown in Figure 2. Such information is very important for the previously proposed segmentation routine to quickly eliminate the unlikely candidates. It is, however, not applicable to the circle fitting algorithms because observing the entire target is not required. A reliable estimation requires careful attention to outliers. The common outlier removal algorithm, random sample consensus (RANSAC) [70], can be integrated into two different stages of the proposed method—When the center of the target in 2D image is estimated and when the final output range/depth of the target center in 3D space is estimated.

With all the benefits described above, this paper suggests a more fault tolerant drogue center estimation method, which combines camera sensitivity setting with a circle fitting algorithm. This fault tolerant capability is desirable in a practical system, which is expected to handle challenging scenarios, such as imaging a used drogue that may be covered with spilled fuel, estimating the center of the drogue when it drifts partially out of FOV, and processing images with pixel outages.

3. Ground Test Evaluation

We evaluate the proposed sensor-in-the-loop method through a ground test. The goal of this autonomous aerial refueling ground test is three fold: (1) demonstrate the proposed method that combines the sensitivity setting of a 3D Flash LIDAR camera with computer algorithms is able to successfully provide information for terminal guidance; (2) the ground test should be performed in real time, not much extra computational power is required because the camera is doing most of the range calculation; and (3) the ground test should be completed autonomously.

To achieve these three goals, a small ground navigation robot is designed and fabricated due to lack of off-the-shelf options specifically for autonomous aerial refueling evaluation. This section is organized as follows: the design idea as well as the capabilities of the robot is discussed in Section 3.1. As we mentioned earlier, the control electronics, although briefly discussed in Section 3.2, is not the focus of this paper. The mock up drogue is described in Section 3.3, followed by the pseudo codes in the robot. Finally, the experimental results and evaluation as well as discussion are shown in Sections 3.5 and 3.6, respectively.

3.1. Robot Design and Fabrication

For demonstration purposes, this robot possesses three types of simplified motions: X (left-right), Y (forward-back), and Z (up-down). At first glance, allowing the robot to move in the Y-direction is simple, only a drivetrain and a platform are required. Movement in the X-direction can be achieved by having two drive motors that each independently controls wheels on the left and right side. From the point of view of the camera, however, this does not properly simulate the motion of the refueling aircraft. Similar to changing lanes on the freeway in a ground vehicle, the aircraft most likely maintains forward-looking direction when it moves side-to-side or up-and-down with very little rotation. Rotation along Y-axis is not applicable in a ground test. To stay within the scope of the goal, the robot has a pivoting turret, allowing the camera to face forward, while the base steers left and right. This requirement adds little complexity, as a stock ring-style turntable paired with a motor and drive belt allow the camera to pivot as needed.

The Z-direction requires a mechanism that is able to raise and lower the camera with both control and stability because the actual height of the camera is important in the ground test. The robot requires the exact amount of traveling distance for each movement. The requirement also calls for a large traveling range, approximately 20 inches (50.8 cm), in the Z-direction. A scissor lift powered by a lead screw design is selected for its capability of control and stability, along with being compact and having a large traveling range. For the goal of facing forward as discussed earlier, a ring-style turntable is attached to the bottom, allowing the scissor lift to pivot freely. Motion control is achieved by adding a timing belt wrapped around the outside of the turntable and held securely with a setscrew; then a pulley attached to a motor is also mated to the belt to give the system motion. The pulley motor assembly serves an additional purpose of applying tension to the belt by being mounted on slots. Figure 9 shows the virtual model's cross sectional view of the turntable and the actual image of how the pulley and belt interact.

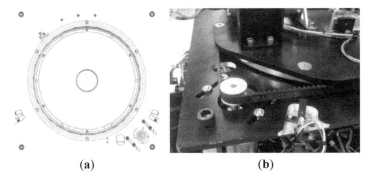

Figure 9. (a) Cross section view of the turntable in the virtual model; (b) Actual image of the interaction of pulley and belt.

The scissor lift is a crucial component for this robot, as it took up the majority of design time and fabrication cost. The end result is a functional lift that is capable of rising 20 inches (50.8 cm) in a few seconds at maximum speed and can also sit at a compact minimum position. To group all connecting wires between the 3D Flash LIDAR camera at the top of robot and control electronics at the base, E-chain is designed and fabricated as one of the essential pieces of the scissor lift. In addition to gathering all of the wires and isolating them from moving parts in the robot, the E-chain also avoids excess wire bending with its minimum bend radius, reducing wire fatigue due to the lift moving from high to low positions. Figure 10 shows the completed scissor lift assembly with the 3D Flash LIDAR camera.

Figure 10. Completed scissor lift assembly with the 3D Flash LIDAR camera at the top.

3.2. Control Electronics

Talon SR speed controllers are used to control motors of the robot when they receive command pulses from an off-the-shelf Arduino Mega micro-controller. Both Z (lifting motion) and Theta (pan motion) axes have limit switches to prevent the motor from traveling beyond the designed 180° boundary. Position feedback for each axis was provided by a rotary encoder. The Arduino micro controller uses the encoder position for speed regulation and position tracking. Unlike many expensive high-end motion controllers that apply proportional-integral-derivative (PID) control or use an S-Curve like pattern generator to provide soft starting and stopping motion by gently increasing or decreasing the speed gradually, the Arduino provides only a triangular waveform. It is, however, sufficient for this ground test demonstration if some fuzzy logic is incorporated in the applied triangular curve pattern to prevent jerky motions.

Also, to demonstrate that fairly little computational power is required to complete the task, a popular commercial off-the-shelf (COTS) single board processor, Beagle Bone Black, is chosen to handle all higher level decision making processes. The Beagle Bone Black is responsible for receiving range and intensity images from the 3D Flash LIDAR camera via Ethernet and in real time performing drogue center estimation proposed in this paper. Finally, serial commands derived from the relative drogue position need to be sent to the COTS motion controller, Arduino, to achieve the desired motions. Concurrent motions such as moving both wheels in unison can be accomplished by sending commands for both axes and then executing them simultaneously. Figure 11 shows the final ground navigation robot and its system block diagram.

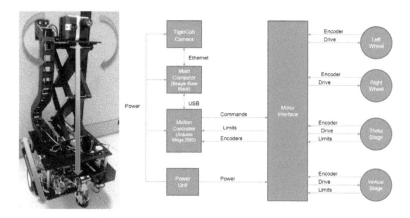

Figure 11. Ground navigation robot (**left**) and its system block diagram (**right**).

3.3. Full Size Drogue Mock-up

For the purpose of maneuvering the drogue target easily during the ground test, a full size drogue mock-up is built from cardboard and retro reflective tape strips as shown in Figure 12a. This simulated drogue consists of two concentric cardboard rings connected by three light weight wooden rods to mimic the outer parachute ring and the center rigid body portion of the real drogue as shown in Figure 12b. Figure 12b shows a real MA-3 drogue mounted on an engine stand and the outer parachute ring was expanded by stiff wires to create a profile similar to that expected during the aerial refueling task. The picture was taken on 14 November 2012 in Eureka, California, which is noted for heavy fog during the winter. One data sequence was captured earlier on that day at 3:23 a.m. The drogue was located in front of the small shed and was about 60 feet (18.29 m) away from the 3D Flash LIDAR camera.

      (a)   (b)   (c)   (d)

Figure 12. The full size drogue mock up.

The visible image in Figure 12c appears dark and blurry due to the foggy condition at the time while the intensity information acquired from a 3D Flash LIDAR camera shows two very distinct retro-reflective rings. This encouraging observation suggests that a 3D Flash LIDAR camera, together with the proposed center estimation method, has potential to provide terminal guidance information in the autonomous aerial refueling application, even in degraded visual environments (DVE) such as fog and cloud. This idea requires more rigorous experiments and the discussion is beyond the scope of this paper.

3.4. Pseudo Codes Implemented on the Ground Robot

1. Loop Until *Distance* < *Distance_Threshold*
2. Input one 128 × 128 3D Flash LIDAR image *A*
3. *Final_list_count* = 0;   //initialize
4. *Distance* = 0;   //initialize
5. *Final_list* = { };   //initialize
6. For each pixel *p* with quadruplets information—(x, y, range, intensity) in *A* {
7. if (the intensity of *p* > (range associated) minimum intensity threshold){
8.   Add *p*(x, y, 1/*range*) into *Final_list*
9.   *Final_list_count++*;
10.   *Distance* + = *p*'s range;
11.   }
12. } End For
13. if (*Final_list_count* < 10)   continue;
14. *Distance* = *Distance*/*Final_list_count*;
15. *Estimated_Center* = Taubin_Circle_Fit (*Final_list*)
16. Ground_Robot_Motion (*Estimated_Center, Distance*)
17. End loop

The above pseudo code illustrates how the ground navigation robot processes input images acquired from a 3D Flash LIDAR camera and calculates necessary information to carry out its next move. As can be seen, it is not a tracking algorithm, instead this algorithm performs center estimation and distance computation on every individual frame without using any information from previous frames. The robot will stop moving after the computed distance value is smaller than a pre-determined threshold (Line 1). All the thresholds in this pseudo code are adjustable and are expected to be changed in the flight test once the final configuration is determined.

An image from the 3D Flash LIDAR is a 128 by 128 array with co-registered range and intensity information for every pixel. To speed up the execution time, a prescreening is performed from Line 6 to Line 12. Only pixels with high enough intensity values will be kept for subsequent processes. Each selected pixel contributes its (x, y) coordinate information as well as the weight for Newton-based Taubin algorithm [65] to estimate center in Line 15. Experimental results suggest that using 1/range weighting formula in Line 8 to separate the outer canopy portion of the drogue from the center rigid body part of the drogue is beneficial. The robot is designed to keep the estimated drogue center on the center of the feedback image acquired from Flash LIDAR. Horizontal and vertical deviations in either x- or y-axes will trigger robot movements, such as turn and height adjustment, in Line 16.

3.5. Perform Ground Test Autonomously

To perform the ground test at an undisturbed location, a large conference room measuring 90 feet (27.43 m) × 44 feet (13.41 m) is used. The robot and the simulated drogue target are set up in opposite corners of the room facing each other to establish the 100 feet (30.48 m) travel distance configuration. According to the report of Autonomous Airborne Refueling Demonstration (AARD) project [71], we believe the selected venue with 100 feet (30.48 m) length is a representative and sufficient setup to evaluate the sensor. The process consisted of a Trail position, a Pre-Contact position and a Hold position. The Trail position is for the refueling aircraft to initialize the rendezvous for the closure mode and it is located at 50 feet (15.24 m) behind the Pre-Contact position. A Pre-Contact position is at 20 feet (6.1 m) behind the drogue where a closure rate of 1.5 feet (0.46 m)/s is used to capture the drogue. After the drogue is captured, the closure velocity of the receiver aircraft is reduced as the aircraft continues forward to the Hold position. The Hold position is normally 10 feet (3.05 m) ahead of the average drogue location. The normal traveling distance to complete this process is 80 feet (24.38 m).

A digital camcorder was placed on top of the 3D Flash LIDAR camera to record visible videos at the same time, as shown in Figure 13. The output images from the Flash LIDAR camera are stored in the secure digital (SD) memory card on the Beagle Bone Black processor. Please note that no careful alignment has been performed in synchronizing frames from the two cameras during this test. The main purpose of the 2D camcorder is to provide feedback for intuitive reference. Spatial alignment between 2D and 3D images is not the focus of this ground test either, because various lenses with different FOVs, 9°, 30°, and 45°, are evaluated.

Figure 14 shows some snapshots of the ground test results. Each image consists of three pictures—A regular visible 2D picture from the camcorder superimposed by two small lidar images at the bottom right corner. The left small lidar image is the original intensity image from the 3D Flash LIDAR camera given a proper sensitivity while the right small image displays pixels actually used in the center estimation process. The red cross-hair highlights the computed center in 2D image space and the range-color-coded drogue visualizes the distance. The color palette for range indication, from 0 feet (0 m) to 100 feet (30.48 m), is also included in the Figure 14 where orange represents the farthest distance of 100 feet (30.48 m).

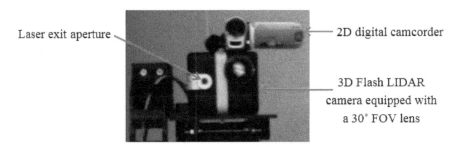

Figure 13. 2D digital camcorder and 3D Flash LIDAR camera.

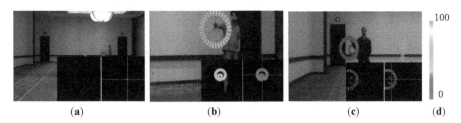

Figure 14. Ground test experimental results using various FOV lenses: (**a**) 45° FOV lens; (**b**) 30° FOV lens; and (**c**) 9° FOV lens; (**d**) Color palette from 0 feet (0 m) to 100 feet (30.48 m).

Figure 15. Autonomous aerial refueling ground test demonstration.

As can be seen in Figure 14a, the simulated drogue located at about 90 feet (27.43 m) away appears to be very small in 3D Flash LIDAR imagery when a 45° FOV lens is used. This observation suggests a narrower FOV lens would be beneficial as the analysis result concluded earlier in Figure 2. A range dependent intensity threshold was applied to quickly exclude some non-target pixels, resulting in more accurate drogue center estimation as shown in Figure 14b. The final experimental result in Figure 14c illustrates what can be expected if a 9° FOV lens and only 1% of the laser energy is supplied. As can be seen, only the retro-reflective tape is visible in this configuration. Valuable lessons are learned to better adjust the parameters in this integrated ground test system. At the end of this test, all three objectives have been successfully achieved. Figure 15 shows nine consecutive snapshots of one test run.

3.6. Evaluation and Discussion

In the probe-and-drogue refueling process, it is not uncommon for the drogue to make contact with or possibly cause damage to the receiver aircraft. For system safety analysis purposes, miss and catch criteria were imposed in the Autonomous Airborne Refueling Demonstration (AARD) project [71]. The concept of the catch criteria, as shown in Figure 16, are sensible evaluation options to be adapted in this ground test. The capture radius, R_c, suggested by the project pilot with a 90 percent success rate and was defined as being 4 inches (10.16 cm) inside the outer ring of the drogue. In a successful capture, the probe must remain within the zone with green stripes, a tube coaxial to the drogue defined by R_c, and transition into the zone with blue stripes during the hold stage.

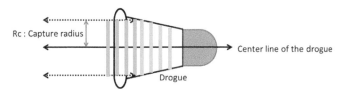

Figure 16. The catch criteria.

In the real operating scenario, the 3D Flash LIDAR camera is most likely to be rigidly mounted close to the probe on the receiver aircraft. As the 3D point clouds generated from the camera data are all relative to the focal point of the camera, the relationship describing the point clouds and the tip of the prober will be standard 3D rotation and translation matrices, like regular rigid body transformations. Without loss of generality, the following data set in Figure 17, from the same sequence that generated snap shots in Figure 15, will demonstrate a successful catch.

Figure 17 consists of seven subfigures. The center subfigure shows a pyramid shape area representing the enclosed space in 3D, which can be observed by the 3D Flash LIDAR camera with a 30° FOV lens. The tip of the pyramid is where the camera is located and the dashed arrow from the tip shows the direction this camera is pointing. Within the pyramid, six points in 3D space are shown with different colors and associated time stamped labels. Detailed information of these six points are displayed in a circle around the center subfigure. In the top left corner, a pair of intensity images captured by the camera in the beginning of the test run at T0 (0 s). The image on the left shows the original input data while the other shows the data actually feed into the center estimation routine after filtering, as described in Section 3.4 Pseudo Codes Line 7. The red cross-hair represents the estimated center with estimated range equaling 95 feet (28.96 m). The range estimation is a simple average computation as shown in Pseudo Codes Line 10 and Line 14, which guaranties a bounded number between the outer canopy and the inner of the rigid body of the drogue. The 3D point is displayed in white labeled T0 in the center subgraph.

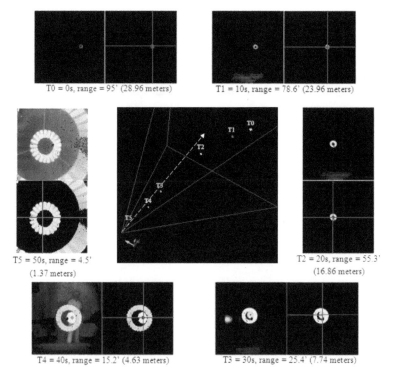

Figure 17. Experimental results from a successful catch run.

In clockwise order, the top right corner shows the image pair acquired at T1 (10 s) with estimated range equaling 78.6 feet (23.96 m). As can be seen in T2 (20 s) data with a 55.3 feet (16.86 m) range, the observed target becomes brighter due to the laser energy at the closer range (inverse-square law). With the sensor-in-the-loop approach, the camera parameters are adjusted in a way to only favor strong signals like the reflective materials on the drogue and signals returned from the carpet floor in the conference room during ground test are too weak to pass the threshold test in Pseudo Codes Line 7. Fortunately, those distractions are not expected in the flight scenarios. One may notice in T3, T4 and T5 data sets, there are perceptible defect pixels in the sensor where no range or intensity values are reported. We intentionally employed this non-perfect camera to perform this ground test to better evaluate the robustness of the center estimation module under more practical conditions. Also, in real world scenarios, a full circle may not be detected due to the drogue partially out of field of view, or occlusion by the prober, or imaging a used drogue that has spots covered with spilled fuel, resulting in lower reflected signal returns than normally expected. We are pleased to find the circle fitting based algorithm is, in fact, more forgiving, and potentially can be integrated in a deployed system.

As can be seen in the center subgraph of Figure 17, all 3D points captured at different times lie within the 30° FOV boundary because the control logic of the robot was designed to align the estimated drogue center at the center of the image while continuously shortening the estimated range to the target. Although during the test, partially outside field of view cases like Figure 14c may occur occasionally (or intentionally during evaluation) the ground robot makes a proper course correction in the next frame and brings the target back to the area close to the image center. The ground navigation robot always catches the drogue during the final docking step (or hitting the basket step) for all test runs using the miss and catch criteria defined in AARD project. The ground test demonstrated in this paper, however, is a simplified evaluation, and it is expected to have a much more sophisticated navigation and control development effort to carry out a similar test in the air. Not to mention additional turbulence conditions and aircraft generated aerodynamics complications which were all omitted from the ground test. A successful ground demonstration using an autonomous system is an encouraging step toward the logical subsequent flight evaluation. Although the algorithm implemented on the ground robot does not use any information from the previous image frame (target detection only, non-tracking), the proposed method is not limited to isolated frames. Instead, use of past information is highly recommended for trajectory prediction and course smoothing purposes, especially in the flight test.

The goal of the proposed sensor-in-the-loop approach is to consider data acquisition performed by sensor hardware and image processing carried out by software algorithms simultaneously. To optimize the system as a whole for a specific application, partitioning tasks between the hardware and software components, using their complementary strengths, is essential. This paper suggests one combination: lowest gain and highest bandwidth setting in the sensor with a circle fitting algorithm, to estimate the 3D position of the center of the drogue for the autonomous aerial refueling application. It is possible and advantageous to make a more intelligent system by adaptively varying parameters on-the-fly, such as with automatic gain control (AGC) and selecting appropriate data processing algorithms depending on the observed scenery. These interesting, yet challenging, topics deserve further research.

4. Conclusions

A sensor-in-the-loop, non-tracking approach is proposed to address the probe and drogue (PDR) style autonomous aerial refueling task. By successfully using a surrogate robot to perform the final docking stage of the aerial refueling task on the ground, the experimental results suggest that applying computer vision fault tolerant circle fitting algorithms on images acquired by a 3D Flash LIDAR camera with lowest gain and highest bandwidth settings has great potential to reliably measure the orientation and relative position between the drogue and the prober for unmanned aerial refueling applications. To the best of our knowledge, we are the first group to demonstrate the feasibility of using a camera-like time-of-flight based sensor on an autonomous system. The sensor-in-the-loop design concept seeks an optimum solution by balancing tasks between the hardware and software components, using their complementary strengths, and is well-suited to solve challenging problems for future autonomous systems. This paper concludes a successful ground test and we are looking for opportunities to further verify the proposed method in-flight and eventually deploy the solution.

Sensors **2015**, *15*, 10948–10972

Acknowledgments: This research is supported under Contract N68936-12-C-0118 from the Office of Naval Research (ONR). Any opinions, findings, and conclusions expressed in this document are those of the authors and do not necessarily reflect the view of the Office of the Naval Research.

Author Contributions: Chao-I Chen and Robert Koseluk developed image and signal processing methods for the ground test. Chase Buchanan and Andrew Duerner designed and assembled the ground navigation robot. Brian Jeppesen and Hunter Laux planned and implemented the control electronics of the ground navigation robot.

Conflicts of Interest: The authors declare no conflict of interest.

References

1. Mao, W.; Eke, F.O. A Survey of the Dynamics and Control of Aircraft during Aerial Refueling. *Nonlinear Dyn. Syst. Theory* **2008**, *8*, 375–388.
2. Tomas, P.R.; Bhandari, U.; Bullock, S.; Richardson, T.S.; du Bois, J. Advanced in Air to Air Refuelling. *Prog. Aerosp. Sci.* **2014**, *71*, 14–35. [CrossRef]
3. Nalepka, J.P.; Hinchman, J.L. Automated Aerial Refueling: Extending the Effectiveness of Unmanned Air Vehicles. In Proceedings of the AIAA Modeling and Simulation Technologies Conference, San Francisco, CA, USA, 15–18 August 2005; pp. 240–247.
4. Capt. Beau Duarte, the manager for the Navy's Unmanned Carrier Aviation Program, in the Release. Available online: http://www.cnn.com/2015/04/22/politics/navy-aircraft-makes-history/index.html (accessed on 23 April 2015).
5. Latimer-Needham, C.H. Apparatus for Aircraft-Refueling in Flight and Aircraft-Towing. U.S. Patent 2,716,527, 30 August 1955.
6. Leisy, C.J. Aircraft Interconnecting Mechanism. U.S. Patent 2,663,523, 22 December 1953.
7. Williamson, W.R.; Abdel-Hafez, M.F.; Rhee, I.; Song, E.J.; Wolfe, J.D.; Cooper, D.F.; Chichka, D.; Speyer, J.L. An Instrumentation System Applied to Formation Flight. *IEEE Trans. Control Syst. Technol.* **2007**, *15*, 75–85. [CrossRef]
8. Williamson, W.R.; Glenn, G.J.; Dang, V.T.; Speyer, J.L.; Stecko, S.M.; Takacs, J.M. Sensors Fusion Applied to Autonomous Aerial Refueling. *J. Guid. Control Dyn.* **2009**, *32*, 262–275. [CrossRef]
9. Kaplan, E.D.; Hegarty, G.J. *Understanding GPS: Principles and Applications*, 2nd ed.; Artech House: Norwood, MA, USA, 2006.
10. Khanafseh, S.M.; Pervan, B. Autonomous Airborne Refueling of Unmanned Air Vehicles Using the Global Position System. *J. Aircr.* **2007**, *44*, 1670–1682. [CrossRef]
11. Brown, A.; Nguyen, D.; Felker, P.; Colby, G.; Allen, F. Precision Navigation for UAS critical operations. In Proceeding of the ION GNSS, Portland, OR, USA, 20–23 September 2011.
12. Monteiro, L.S.; Moore, T.; Hill, C. What Is the Accuracy of DGPS? *J. Navig.* **2005**, *58*, 207–225. [CrossRef]
13. Hansen, J.L.; Murray, J.E.; Campos, N.V. The NASA Dryden AAR Project: A Flight Test Approach to an Aerial Refueling System. In Proceeding of the AIAA Atmospheric Flight Mechanics Conference and Exhibit, Providence, RI, USA, 16–19 August 2004; pp. 2004–2009.
14. Vachon, M.J.; Ray, R.J.; Calianno, C. Calculated Drag of an Aerial Refueling Assembly through Airplane Performance Analysis. In Proceeding of the 42nd AIAA Aerospace Sciences and Exhibit, Reno, NV, USA, 5–8 January 2004.
15. Campoy, P.; Correa, J.; Mondragon, I.; Martinez, C.; Olivares, M.; Mejias, L.; Artieda, J. Computer Vision Onboard UAVs for Civilian Tasks. *J. Intell. Robot. Syst.* **2009**, *54*, 105–135. [CrossRef]
16. Conte, G.; Doherty, P. Vision-based Unmanned Aerial Vehicle Navigation Using Geo-referenced Information. *EURASIP J. Adv. Signal Process.* **2009**. [CrossRef]
17. Luington, B.; Johnson, E.N.; Vachtsevanos, G.J. Vision Based Navigation and Target Tracking for Unmanned Aerial Vehicles. *Intell. Syst. Control Autom. Sci. Eng.* **2007**, *33*, 245–266.
18. Madison, R.; Andrews, G.; DeBitetto, P.; Rasmussen, S.; Bottkol, M. Vision-aided navigation for small UAVs in GPS-challenged Environments. In Proceeding of the AIAA InfoTech at Aerospace Conference, Rohnert Park, CA, USA, 7–10 May 2007; pp. 318–325.
19. Ollero, A.; Ferruz, F.; Caballero, F.; Hurtado, S.; Merino, L. Motion Compensation and Object Detection for Autonomous Helicopter Visual Navigation in the COMETS Systems. In Proceeding of the IEEE International Conference on Robotics and Autonomous, New Orleans, LA, USA, 26 April–1 May 2004; pp. 19–24.

20. Vendra, S.; Campa, G.; Napolitano, M.R.; Mammarella, M.; Fravolini, M.L.; Perhinschi, M.G. Addressing Corner Detection Issues for Machine Vision Based UAV Aerial Refueling. *Mach. Vis. Appl.* **2007**, *18*, 261–273. [CrossRef]

21. Harris, C.; Stephens, M. Combined Corner and Edge Detector. In Proceedings of the 4th Alvery Vision Conference, Manchester, UK, 31 August–2 September 1988; pp. 147–151.

22. Kimmett, J.; Valasek, J.; Junkings, J.L. Autonomous Aerial Refueling Utilizing a Vision Based Navigation System. In Proceedings of the AIAA Guidance, Navigation and Control Conference and Exhibition, Monterey, CA, USA, 5–8 August 2002.

23. Nobel, A. Finding Corners. *Image Vis. Comput.* **1988**, *6*, 121–128. [CrossRef]

24. Smith, S.M.; Bradly, J. M SUSAN—A New Approach to Low Level Image Processing. *Int. J. Comput. Vis.* **1997**, *23*, 45–78. [CrossRef]

25. Doebbler, J.; Spaeth, T.; Valasek, J.; Monda, M.J.; Schaub, H. Boom and Receptacle Autonomous Air Refueling using Visual Snake Optical Sensor. *J. Guid. Control Dyn.* **2007**, *30*, 1753–1769. [CrossRef]

26. Herrnberger, M.; Sachs, G.; Holzapfel, F.; Tostmann, W.; Weixler, E. Simulation Analysis of Autonomous Aerial Refueling Procedures. In Proceedings of the AIAA Guidance, Navigation, and Control Conference and Exhibition, San Francisco, CA, USA, 15–18 August 2005.

27. Fravolini, M.L.; Ficola, A.; Campa, G.; Napolitano, M.R.; Seanor, B. Modeling and Control Issues for Autonomous Aerial Refueling for UAVs using a Probe-Drogue Refueling System. *Aerosp. Sci. Technol.* **2004**, *8*, 611–618. [CrossRef]

28. Pollini, L.; Campa, G.; Giulietti, F.; Innocenti, M. Virtual Simulation Setup for UAVs Aerial Refueling. In Proceedings of the AIAA Modeling and Simulation Technologies Conference and Exhibit, Austin, TX, USA, 11–14 August 2003.

29. Pollini, L.; Mati, R.; Innocenti, M. Experimental Evaluation of Vision Algorithms for Formation Flight and Aerial Refueling. In Proceedings of the AIAA Modeling and Simulation Technologies Conference and Exhibit, Providence, RI, USA, 16–19 August 2004.

30. Mati, R.; Pollini, L.; Lunghi, A.; Innocenti, M.; Campa, G. Vision Based Autonomous Probe and Drogue Refueling. In Proceedings of the 14th Mediterranean Conference on Control and Automation, Ancona, Italy, 28–30 June 2006.

31. Junkins, J.L.; Schaub, H.; Hughes, D. Noncontact Position and Orientation Measurement System and Method. U.S. Patent 6,266,142 B1, 24 July 2001.

32. Kimmett, J.; Valasek, J.; Junkins, J.L. Vision Based Controller for Autonomous Aerial Refueling. In Proceedings of the 2002 IEEE International Conference on Control Applications, Glasgow, Scotland, 17–20 September 2002.

33. Tandale, M.D.; Bowers, R.; Valasek, J. Trajectory Tracking Controller for Vision-Based Probe and Drogue Autonomous Aerial Refueling. *J. Guid. Control Dyn.* **2006**, *4*, 846–857. [CrossRef]

34. Valasek, J.; Gunman, K.; Kimmett, J.; Tandale, M.; Junkins, L.; Hughes, D. Vision-based Sensor and Navigation System for Autonomous Air Refueling. In Proceeding of the 1st AIAA Unmanned Aerospace Vehicles, Systems, Technologies, and Operations Conference and Exhibition, Vancouver, BC, Canada, 20–23 May 2002.

35. Pollini, L.; Innocenti, M.; Mati, R. Vision Algorithms for Formation Flight and Aerial Refueling with Optimal Marker Labeling. In Proceeding of the AIAA Modeling and Simulation Technologies Conference and Exhibition, San Francisco, CA, USA, 15–18 August 2005.

36. Lu, C.P.; Hager, G.D.; Mjolsness, E. Fast and Globally Convergent Pose Estimation from Video Images. *IEEE Trans. Pattern Anal. Mach. Intell.* **2000**, *22*, 610–622. [CrossRef]

37. Martinez, C.; Richardson, T.; Thomas, P.; du Bois, J.L.; Campoy, P. A Vision-based Strategy for Autonomous Aerial Refueling Tasks. *Robot. Auton. Syst.* **2013**, *61*, 876–895. [CrossRef]

38. Irani, M.; Anandan, P. About Direct Methods. *Lect. Notes Comput. Sci.* **2000**, *1833*, 267–277.

39. Bergen, J.R.; Anandan, P.; Hanna, K.J.; Hingorani, R. Hierarchical Model-Based Motion Estimation. *Lect. Notes Comput. Sci.* **1992**, *588*, 237–252.

40. Thomas, P.; Bullock, S.; Bhandari, U.; du Bois, J.L.; Richardson, T. Control Methodologies for Relative Motion Reproduction in a Robotic Hybrid Test Simulation of Aerial Refueling. In Proceeding of the AIAA Guidance, Navigation, and Control Conference, Minneapolis, MN, USA, 13–16 August 2012.

41. Christopher Longuet-Higgins, H. A computer algorithm for reconstructing a scene from two projections. *Nature* **1981**, *293*, 133–135. [CrossRef]

42. Hartley, R. In Defense of the Eight-Point Algorithm. *IEEE Trans. Pattern Recogn. Mach. Intell.* **1997**, *19*, 580–593. [CrossRef]
43. Hartley, R.; Zisserman, A. *Multiple View Geometry in Computer Vision*, 2nd ed.; Cambridge University Press: Cambridge, UK, 2004.
44. Malis, E.; Chaumette, F.; Boudet, S. 2 $\frac{1}{2}$D Visual Servoing with Respect to Unknown Objects Through a New Estimation Scheme of Camera Displacement. *Int. J. Comput. Vis.* **2000**, *37*, 79–97. [CrossRef]
45. Lots, J.F.; Lane, D.M.; Trucco, E. Application of a 2 $\frac{1}{2}$D Visual Servoing to Underwater Vehicle Station-keeping. In Proceeding of the IEEE OCEANS 2000 MTS/IEEE Conference and Exhibition, Providence, RI, USA, 11–14 September 2000.
46. Stettner, R. High Resolution Position Sensitive Detector. U.S. Patent 5,099,128, 24 March 1992.
47. Stettner, R. Compact 3D Flash Lidar Video Cameras and Applications. *Proc. SPIE* **2010**, *7684*. [CrossRef]
48. Chen, C.; Stettner, R. Drogue Tracking Using 3D Flash LIDAR for Autonomous Aerial Refueling. *Proc. SPIE* **2011**, *8037*. [CrossRef]
49. Osher, S.; Sethian, J.A. Fronts Propagating with Curvature-dependent Speed: Algorithms Based on Hamilton-Jacobi Formulations. *J. Comput. Phys.* **1988**, *79*, 12–49. [CrossRef]
50. Sethian, J.A. *Level Set Methods and Fast Marching Methods: Evolving Interfaces in Computational Geometry, Fluid Mechanics, Computer Vision, and Materials Science*; Cambridge University Press: Cambridge, UK, 1999.
51. Bonin-Font, F.; Ortiz, A.; Oliver, G. Visual Navigation for Mobile Robots: A Survey. *J. Intell. Robot. Syst.* **2008**, *53*, 263–296. [CrossRef]
52. Kendoul, F. Survey of Advances in Guidance, Navigation, and Control of Unmanned Rotorcraft Systems. *J. Field Robot.* **2012**, *29*, 315–378. [CrossRef]
53. Cai, G.; Dias, J.; Seneviratne, L. A Survey of Small-Scale Unmanned Aerial Vehicles: Recent Advances and Future Development Trends. *Unmanned Syst.* **2014**, *2*, 1–25.
54. OpenCV. Available online: http://opencv.org (accessed on 23 April 2015).
55. Rusu, R.B.; Cousins, S. 3D is Here: Point Cloud Library. In Proceedings of the IEEE International Conference on Robotics and Automation (ICRA), Shanghai, China, 9–13 May 2011; pp. 9–13.
56. Robot Operating System (ROS). Available online: http://www.ros.org (accessed on 23 April 2015).
57. Chernov, N. *Circular and Linear Regression: Fitting Circles and Lines by Least Squares*, 1st ed.; Chapman & Hall/CRC Monographs on Statistics & Applied Probability Series; CRC Press: London, UK, 2010.
58. Chernov, N. Matlabe Codes for Circle Fitting Algorithms. Available online: http://people.cas.uab.edu/~mosya/cl/MATLABcircle.html (accessed on 23 April 2015).
59. Coope, I.D. Circle Fitting by Linear and Nonlinear Least Squares. *J. Optim. Theory Appl.* **1993**, *76*, 381–388. [CrossRef]
60. Gander, W.; Golub, G.H.; Strebel, R. Least squares fitting of circles and ellipses. *Bull. Belg. Math. Soc.* **1996**, *3*, 63–84.
61. Karimaki, V. Effective Circle Fitting for Particle Trajectories. *Nuclear Instrum. Methods Phys. Res. Sect. A Accel. Spectrom. Detect. Assoc. Equip.* **1991**, *305*, 187–191. [CrossRef]
62. Kasa, I. A Curve Fitting Procedure and Its Error Analysis. *IEEE Trans. Instrum. Measur.* **1976**, *25*, 8–14. [CrossRef]
63. Nievergelt, Y. Hyperspheres and hyperplanes fitted seamlessly by algebraic constrained total least-squares. *Linear Algebra Appl.* **2001**, *331*, 43–59. [CrossRef]
64. Pratt, V. Direct Least-Squares Fitting of Algebraic Surfaces. *Comput. Graph.* **1987**, *21*, 145–152. [CrossRef]
65. Taubin, G. Estimation of Planar Curves, Surfaces and Non-planar Space Curves Defined by Implicit Equations, with Applications to Edge and Range Image Segmentation. *IEEE Transit. Pattern Anal. Mach. Intell.* **1991**, *13*, 1115–1138. [CrossRef]
66. Matlab: The Language Technical Computing. Available online: http://www.mathworks.com/products/matlab/ (accessed on 23 April 2015).
67. Gill, P.R.; Murray, W.; Wright, M.H. *The Levenberg-Marquardt Method. Practical Optimization*; Academic Press: London, UK, 1981; pp. 136–137.
68. Levenberg, K. A Method for the Solution of Certain Problems in Least Squares. *Q. Appl. Math.* **1944**, *2*, 164–168.
69. Marquardt, D. An Algorithm for Least-Squares Estimation of Nonlinear Parameters. *SIAM J. Appl. Math.* **1963**, *11*, 431–441. [CrossRef]

70. Fischler, M.A.; Bolles, R.C. Random Sample Consensus: A Paradigm for Model Fitting with Application to Image Analysis and Automated Cartography. *Commun. ACM* **1981**, *24*, 381–395. [CrossRef]

71. Dibley, R.P.; Allen, M.J.; Nabaa, N. Autonomous Airborne Refueling Demonstration Phase I Flight-Test Results. In Proceedings of the AIAA Atmospheric Flight Mechanics Conference and Exhibit, Hilton Head, SC, USA, 20–23 August 2007.

Article

A New Calibration Method Using Low Cost MEM IMUs to Verify the Performance of UAV-Borne MMS Payloads

Kai-Wei Chiang [1], Meng-Lun Tsai [1], El-Sheimy Naser [2,†], Ayman Habib [3,†] and Chien-Hsun Chu [1,*]

[1] Department of Geomatics, National Cheng-Kung University, No.1, Daxue Rd., Tainan 70101, Taiwan; kwchiang@mail.ncku.edu.tw (K.-W.C.); taurus.bryant@msa.hinet.net (M.-L.T.)

[2] Department of Geomatics Engineering, University of Calgary, Calgary City, AB T2N 1N4, Canada; elsheimy@ucalgary.ca

[3] Lyles School of Civil Engineering, Purdue University, Lafayette city, IN 47907, USA; ahabib@purdue.edu

* Author to whom correspondence should be addressed; chienhsun0229@msn.com; Tel.: +886-237-3876 (ext. 857).

† These authors contributed equally to this work

Academic Editor: Antonios Tsourdos

Received: 9 January 2015; Accepted: 6 March 2015; Published: 19 March 2015

Abstract: Spatial information plays a critical role in remote sensing and mapping applications such as environment surveying and disaster monitoring. An Unmanned Aerial Vehicle (UAV)-borne mobile mapping system (MMS) can accomplish rapid spatial information acquisition under limited sky conditions with better mobility and flexibility than other means. This study proposes a long endurance Direct Geo-referencing (DG)-based fixed-wing UAV photogrammetric platform and two DG modules that each use different commercial Micro-Electro Mechanical Systems' (MEMS) tactical grade Inertial Measurement Units (IMUs). Furthermore, this study develops a novel kinematic calibration method which includes lever arms, boresight angles and camera shutter delay to improve positioning accuracy. The new calibration method is then compared with the traditional calibration approach. The results show that the accuracy of the DG can be significantly improved by flying at a lower altitude using the new higher specification hardware. The new proposed method improves the accuracy of DG by about 20%. The preliminary results show that two-dimensional (2D) horizontal DG positioning accuracy is around 5.8 m at a flight height of 300 m using the newly designed tactical grade integrated Positioning and Orientation System (POS). The positioning accuracy in three-dimensions (3D) is less than 8 m.

Keywords: UAV; IMU; GPS; MMS; DG; calibration

1. Introduction

Mobile mapping is executed by producing more than one image that includes the same object acquired from different positions, allowing the 3D positions of the object with respect to the mapping frame to be measured [1]. Multi-sensors can be mounted on a variety of platforms, such as satellites, aircraft, helicopters, terrestrial vehicles, water-based vessels, and even people. As a result, mapping has become mobile and dynamic. Mobile mapping technology enables DG by integrating GPS and INS, which makes Exterior Orientation Parameters (EOPs) of accurate images available at any given time [2]. The integration of INS and GPS improves the geo-referencing of photogrammetric data and frees it from operational restrictions that require ground reference information. Operational flexibility is greatly enhanced in all cases where a block structure is not needed [3]. Costs are considerably reduced, especially in areas where little or no ground control is available [4].

Sensors **2015**, *15*, 6560–6585

As the number of natural disasters caused by climate change increases, rapid spatial information acquisition capability using remote sensing and mobile mapping applications has received wide attention. Consequently, the development of a rapidly deployable and low cost system for collecting near-real-time spatial information without any ground reference over the disaster area should be of great interest.

The current achievable accuracy of commercial airborne MMSs is sufficient for numerous mapping applications. In addition, cost has decreased and production efficiency has increased with the use of DG-based photogrammetric platforms [4]. An airborne MMS that is relatively free of government regulations and inexpensive but maintains high mobility for small area surveys or rapid spatial information acquisition is desirable for urgent response events such as disaster relief and assessment.

Satellite images can be constrained by a number of factors, such as weather, availability of stereo coverage, temporal and geometric resolution, minimum area order and price. Thus airborne platforms such as aircraft, helicopters, kites, balloons and UAVs are good and generally cheap alternatives, especially since recent developments in small and medium format digital cameras have made great advances in automated image processing. Numerous studies have been conducted for applying UAV to photogrammetric research [1,5,6]. Nowadays despite the widespead availability of very high resolution satellite imagery, large scale photogrammetric mapping applications still primarily use aerial images because for large areas, aircraft are usually employed as a platform for acquiring aerial images. However, for small and remote area mapping, UAV is a very good and inexpensive platform and imaging alternative. It should be particularly attractive for developing countries.

Generally speaking, the main applications of UAVs may be defined as observation, maintenance, surveillance, monitoring, remote sensing and security tasks [7]. In recent years, more and more UAV-based photogrammetric platforms have been developed and their performance has been proven in certain scenarios [8]. Chiang *et al.* [9] developed a DG based UAV photogrammetric platform where an INS/GPS integrated POS system was implemented to provide the DG capability of the platform. The preliminary results show horizontal DG positioning accuracies in the East and North directions of below 10 m at a flight height of 300 m without using any GCP. The positioning accuracy in the Up direction is less than 15 m. Such accuracy is good enough for near real time disaster relief.

Rehak *et al.* [10] developed a low cost UAV for direct geo-referencing. The advantage of such a system lies in its high maneuverability, operation flexibility as well as capability to acquire image data without the need of establishing GCPs. Moreover, the precise geo-referencing has better final mapping accuracy when employing integrated sensor orientation, limiting the number and distribution of GCPs, thus saving time in their signalization and surveying.

Generally speaking, the selection of a platform is application dependent. The primary objective of developing a UAV based photogrammetric platform is to meet requirements such as small operational area, rapid deployment, low cost, high mobility, and acceptable positioning accuracy. Therefore, it is not practical to use such platforms as replacements for conventional photogrammetric applications [11].

2. Problem Statement

As indicated previously, Chiang *et al.* [9] utilized a low cost INS/GPS integrated POS system to provide the DG capability of the UAV photogrammetric platform. Figure 1 shows a DG module proposed for facilitating GCP free photogrammetry applications and INS/GPS integrated POS aided bundle adjustment photogrammetry. The EVK-6T GPS receiver from U-blox is used in the DG module. This model was chosen because of its L1 carrier phase measurements for DGPS processing, which provides sufficient positioning accuracy.

Sensors **2015**, *15*, 6560–6585

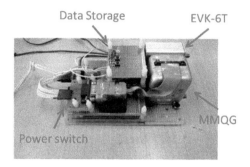

Figure 1. The DG module configuration.

The IMU used for the DG module is an MMQ-G from BEI SDID (Concord, CA, USA). This model has been chosen due to its compact size and weight. The MMQ-G IMU integrates MEMS quartz rate sensors (100 deg/h in run bias) and vibrating quartz accelerometers. The total budget of the proposed POS module is around 10,000 US dollars. MEMS inertial sensors have advanced rapidly; thus, the inclusion of MEMS inertial sensors for UAV-borne MMS applications has good potential in terms of cost and accuracy.

With the advance of MEMS inertial technology, some commercial MEMS IMUs now provide better sensor stability while significantly reducing POS cost. Therefore, the first objective of this study was to develop a new POS module using a tactical grade IMU with 6 deg/h gyro in run bias but costing only one third of the POS module proposed in [9].

The most common integration scheme used today is the Loosely Coupled (LC) integration scheme, as shown in Figure 2. The position and velocity estimated by the GPS Kalman filter (KF) are processed in the navigation KF to aid the INS, a process also known as decentralized, or cascaded, filtering. This kind of integration has the benefit of a simpler architecture that is easy to utilize in navigation systems. However, the errors in the position and velocity information provided by the GPS KF are time-correlated, which can cause degradation in performance or even instability of the navigation KF if these correlations are not compensated for [12].

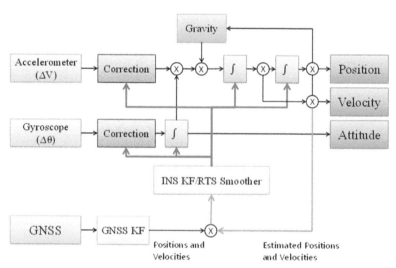

Figure 2. The LC integration scheme.

In the case of incomplete constellations, *i.e.*, fewer than four satellites in view, the output of the GPS receiver has to be ignored completely, leaving the INS unaided [13]. When a UAV flies in the open sky, the GPS signals are not obstructed or reflected by high buildings. There is no GPS outage and the user can receive data from more than four satellites, theoretically.

However, the vibration of the UAV platform and certain maneuvers, such as sharp turns or sudden movements due to strong winds, can cause a loss of the logged GPS raw measurements [14].

This problem grows worse when carrier phase measurements are applied. Thus the accuracy of the POS solutions deteriorates significantly when a low cost MEMS IMU and the LC scheme are used during partial GPS outages. Therefore the second objective of this study is to apply a robust INS/GPS integration scheme to avoid the partial GPS outages taking place in UAV scenarios.

Figure 3 and Equation (1) illustrate the general concept of the airborne DG. With this implementation, the coordinates of a mapping feature can be obtained directly through measured image coordinates.

This procedure works based on *a priori* knowledge of various systematic parameters, as shown in the following representation:

$$r_{oA}^l = r_{ob}^l(t) + R_b^l(t)\left(s_A R_c^b r_{ca}^c + r_{bc}^b\right) \tag{1}$$

In the formula, the "r" means a vector and "R" means a rotation matrix. Their superscripts and subscripts represent the frame. But the subscript of vector means start-point and end-point of this vector. r_{oA}^l is the coordinate vector of feature point (A) in the Local Level frame (LLF, l-frame); $r_{ob}^l(t)$ is the interpolated coordinate vector of the navigation sensors (INS/GPS) in the l-frame; s_A is a scale factor, determined by stereo techniques, laser scanners or a Digital Terrain Model (DTM); $R_b^l(t)$ is the interpolated rotation matrix from the navigation sensor body frame (b-frame) to the l-frame; (t) is the time of exposure, *i.e.*, the time of capturing the images, determined by synchronization; R_c^b is the rotation matrix from the camera frame (c-frame) to the b-frame, determined by calibration; r_{ca}^c is the coordinate vector of the point (a) in the c-frame (*i.e.*, image coordinate); and r_{bc}^b is the vector between the IMU center and the camera perspective center in the b-frame, determined by calibration.

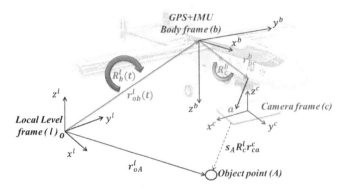

Figure 3. The concept of airborne DG.

The physical meanings of R_c^b and r_{bc}^c are given in Figures 4 and 5, respectively. Traditional calibration procedure is implemented to acquire the rotation matrix (R_c^b) between the camera and IMU by using the rotation matrix (R_b^l) provided by the IMU and the rotation matrix (R_c^l) provided by conventional bundle adjustment using the l-frame's control field during the calibration procedure using the following equation [15]:

$$R_c^b = R_l^b R_c^l \tag{2}$$

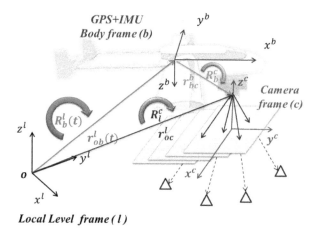

Figure 4. Concept of boresight angle calibration.

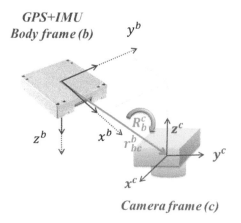

Figure 5. Concept of lever arm calibration.

The lever arm vector r^b_{GPSb} between the GPS phase center and the IMU center is determined through a surveying process. The lever arm vector r^b_{bc} between the camera and the IMU centers is determined through a two-step procedure: first, the EOPs of the images are calculated through bundle adjustment by measuring the image points when the flight mission had completed, and second, the interpolation of INS/GPS smoothed POS solutions at the image exposure time is implemented. The lever arm and boresight angle are obtained by comparing the differences of the position and the attitude between the EOP and the interpolated INS/GPS solutions using the following equation:

$$r^b_{bc} = R^b_l \begin{pmatrix} X^l_{oc} - X^l_{ob} \\ Y^l_{oc} - Y^l_{ob} \\ Z^l_{oc} - Z^l_{ob} \end{pmatrix} \tag{3}$$

where r^b_{bc} is the lever arm vector to be estimated, $(X^l_{ob}, Y^l_{ob}, Z^l_{ob})$ represents the positional vector of the INS center in the l-frame provided by INS/GPS integrated POS solutions, and $(X^l_{oc}, Y^l_{oc}, Z^l_{oc})$ represents the positional vector of the camera center in the l-frame provided by bundle adjustment. Once these parameters are well calibrated and the sensors are fixed on the platform, the proposed platform will be able to conduct GCP-free DG missions without conventional bundle adjustments for future flights.

However, in addition to those lever arms and boresight angles, the camera shutter delay that represents the time delay between the trigger time used for interpolating POS solutions and the exposure time when an image is taken should be calibrated simultaneously in kinematic mode, as explained in [9]. In practice, the trigger signal is sent to the camera (to take a picture) and to the recorder (to record the time mark) simultaneously. After this, the smoothed INS/GPS solutions can be interpolated at the time mark of each image. However, the camera exposure time will always be slightly different from the recorded time mark due to the delay caused by signal transmission time. This deviation of time leads to a systemic shift of position and attitude of each image along the forward direction. Therefore, exposure time delay compensation should be applied to estimate the magnitude of the time delay at each exposure station. To develop a system to compensate for this situation, the third objective of this study is to produce a new calibration method to solve this problem. The proposed method not only estimates the lever-arm and boresight, but estimates the deviation of time using the same measurements used by the traditional calibration method.

3. The Configuration of the Proposed Platform

The proposed UAV platform and its specifications are illustrated in Figure 6, in which it can be seen that the proposed UAV platform is designed for medium range applications. The wing span is 4 m and the payload is 40 kg. The flexible flight altitude and eight–hour maximum flight-time make the platform suitable for small area and large scale photogrammetric missions. This model is jointly developed by the Department of Geomatics, NCKU and GeoSat Informatics Technology Co. Figure 7 depicts the tactical grade DG module designed in this study to facilitate direct photogrammetry as well as INS/GPS POS aided bundle adjustment.

Wing span	3.8 m
Fuselage length	3.3 m
height	0.8 m
Payload	40 kg
Endurance	> 8 hr
Range	800 km
Flight height	4000 m
Max speed	145 km/hr

Figure 6. The proposed UAV platform.

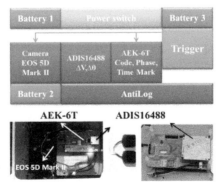

Figure 7. The configuration of DG module.

Figure 8 illustrates the specifications of the GPS receiver, AEK-6T (Thalwil city, Switzerland) from the U-blox, which is applied in the DG module. This model has been chosen because it can provide L1 carrier phase raw measurements that can be applied for differential GPS processing with single frequency carrier phase measurements to provide sufficient positioning accuracy. In addition, it supplies Pulse Per Second (PPS) output used to synchronize the time mark used to trigger the DG module's camera.

Item	AEK-6T	
Function	L1 carrier phase measurement and pseudo range	
Communication port	USB, RS232 port	
Sample rate	10 Hz	
Voltage	5V	
Dimension	74mm x 54mm x 24mm	

Figure 8. The GPS receiver of DG module.

Figure 9 illustrates the two IMUs used for the previous and new DG module, MMQ-G from BEI SDID and ADIS16488 (Newburyport city, MA, USA) from Analog Devices, respectively. These models have been chosen because of their compact size and weight. The retail price of the ADIS16488 IMU was around 1500 USD while the MMQ-G was 10,000~12,000 USD in 2008. Based on the specifications given below, the new version of the DG module is at least six times superior to previous version in terms of the quality of inertial sensors, but costs only one fifth of the original budget. A digital camera (EOS 5D Mark II, Canon, Tokyo city, Japan) is applied in this study. Figure 10 shows the picture and specifications of the camera.

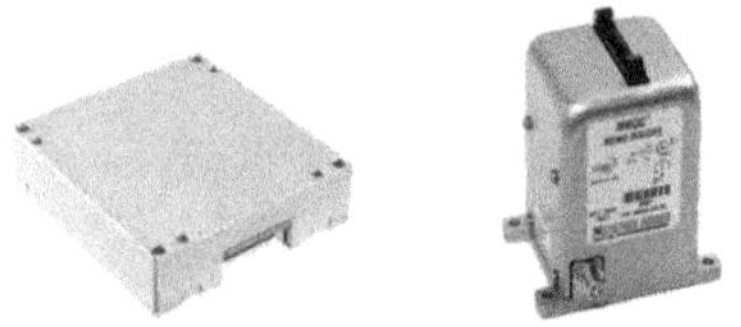

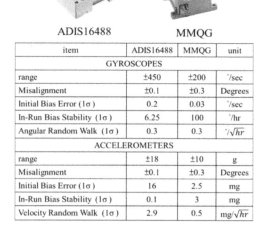

ADIS16488   MMQG

item	ADIS16488	MMQG	unit
GYROSCOPES			
range	±450	±200	°/sec
Misalignment	±0.1	±0.3	Degrees
Initial Bias Error (1σ)	0.2	0.03	°/sec
In-Run Bias Stability (1σ)	6.25	100	°/hr
Angular Random Walk (1σ)	0.3	0.3	$°/\sqrt{hr}$
ACCELEROMETERS			
range	±18	±10	g
Misalignment	±0.1	±0.3	Degrees
Initial Bias Error (1σ)	16	2.5	mg
In-Run Bias Stability (1σ)	0.1	3	mg
Velocity Random Walk (1σ)	2.9	0.5	$mg/\sqrt{hr}$

Figure 9. The IMUs for DG module.

To supply the power required for the individual sensors with various power requirements from the battery, a power switch module has been designed. An RS232 port is used to transmit the measurements collected by the MMQ-G/ADIS16488 IMU to the data storage module. Since the camera has its own power supply, it is not considered in the power supply design. The pixel size of camera is 0.0064 mm and the focal length is fixed on about 20 mm. The data storage module used to record the measurements collected by MMQ-G/ADIS16488, EVK-6T, and the synchronized time mark used to trigger the camera is an Antilog from Martelec (Alton city, UK). It was chosen due to its flexibility, low power consumption, and reliability. Since the camera has its own storage mechanization, it is not included in this module.

Canon	EOS 5D Mark II		EF 20mm f/2.8 USM	
Size(mm³)	152x113.5x71(mm³)		70.6(Diameter, mm)x77.5(mm)	
Weight (kg)	0.85+0.08(Battery，kg)		0.405(kg)	
Format	Image size (mm²)	36x24	Angle (degree)	94
	pixel(million)	2200	Focal length(mm)	20

Figure 10. Canon EOS 5D Mark II & EF 20 mm f/2.8 USM.

4. Proposed Calibration Algorithm

As mentioned previously, the lever arms and boresight angles can be obtained by a traditional calibration method. When calibrating the lever arms and boresight angles, the perspective center of each image (r_{oc}^l) is exactly known after executing the bundle adjustment; the calculation of the INS/GNSS position vector (r_{ob}^l) and rotation matrix (R_l^b) is conducted by the interpolation at the trigger time received, after which the lever arms (r_{bc}^b) can be solved using the following equation:

$$r_{bc}^b(t) = R_l^b(t)(r_{oc}^l(t) - r_{ob}^l(t)) \tag{4}$$

In terms of the boresight angles' calibration, the rotation matrix between the camera frame and the local level frame of each image (R_c^l) is also obtained through the bundle adjustment results, and the rotation matrix between the body frame and mapping frame of each image can be obtained through integrated solutions. The rotation matrix (R_c^b) can be calculated using the matrix multiplication:

$$R_c^b(t) = R_l^b(t)R_c^l(t) \tag{5}$$

However, there is another important parameter: exposure time delay (*t*). This is the time difference between the timing recorded and the actual camera exposure. The traditional calibration formula supposes that time delay is zero. Because $R_c^l(t)$ in the formula is solved from picture through photogrammetry, the "t" should be the exposure time of camera. To compensate for this gap, the DG equation has been modified as follows:

$$t_c = t_b + \Delta t \tag{6}$$

$$r_{oA}^l = r_{ob}^l(t_c) + R_b^l(t_c)\left(r_{bc}^b + sR_c^b r_{ca}^c\right) = r_{ob}^l(t_b + \Delta t) + R_b^l(t_b + \Delta t)(r_{bc}^b + sR_c^b r_{ca}^c) \tag{7}$$

where, "t_c" is the exposure time of the camera and "t_b" is the time recorded by the recorder. In the practice, the solved EOPs by bundle adjustment are at the exposure time but the proposed position and attitude of body are at recorded time. So the traditional calibration equations should be represented to kinematic mode.

$$r_{bc}^b(t_c, t_b) = R_l^b(t_b)(r_{oc}^l(t_c) - r_{ob}^l(t_b)) \tag{8}$$

$$R_c^b(t_c, t_b) = R_l^b(t_b)R_c^l(t_c) \tag{9}$$

When the platform of MMS is in kinematic mode, the accuracy of 3D positioning is significantly affected by the exposure time delay. Therefore, the proposed calibration method has been developed to reduce the impact of the exposure time delay. In the following derivation, the magnitude of the exposure time delay is assumed to be a small unknown constant. Because it is small, the IMU rotation matrix is assumed fix during delay period. The derivation of the related equation is described below:

$$
\begin{aligned}
r_{bc}^b(t_c) &= R_l^b\left(r_{oc}^l(t_c) - r_{ob}^l(t_c)\right) \\
&= R_l^b(r_{oc}^l(t_c) - (r_{ob}^l(t_b) + r_{ob}^l(\Delta t))) \\
&= R_l^b\left(r_{oc}^l(t_c) - r_{ob}^l(t_b) - \dot{r}_{ob}^l * \Delta t\right) \\
&= r_{bc}^b(t_c, t_b) - v_{ob}^b * \Delta t R_b^c(t_c) = R_l^c(t_c) * R_b^l(t_c) \\
&= R_l^c(t_c) * R_b^l(t_b) * R_b^l(\Delta t) \\
&= R_b^c(t_c, t_b) * R_b^l(\Delta t) \rightarrow R_c^b(t_c) = R_l^b(\Delta t) * R_c^b(t_c, t_b) = \dot{R}_l^b * \Delta t * R_c^b(t_c, t_b)
\end{aligned}
\tag{10}
$$

where, "r", "R" and "t" mean position vector, rotation matrix and time, respectively, and the superscript and subscript are the frame. We suppose the frames of INS/GNSS and MMS are overlaid, so "b" is the MMS body frame and also the INS/GNSS frame. The "c" and "l" are camera and local level frame, respectively.

The calibration process is carried out in the local level frame to avoid unnecessary systematic error sources due to coordinate transformation. The equation builds the relationship between the measurement and unknowns, including lever arm, boresight and exposure time delay (Δt). These are arranged and rewritten as shown below. The proposed method is implemented using the Least Square (LS) method. Before processing the LS, the rotation matrix function is re-written in Quaternions form, so the unknown items of boresight angle are [q0, q1, q2, q3]. In the coefficient matrix "A", the coefficients of boresight angle are also differential by Taylor series. Therefore, the unknown are time delay [Δt], lever arms that have three elements including [x, y, z] in the body frame and the boresight angles that have four elements including [q0, q1, q2, q3]:

$$
\left\{
\begin{aligned}
\Delta r_{bc}^b(t_c, t_b) &= r_{bc}^b(t_c) + v_{ob}^b * \Delta t \\
\Delta R_c^b(t_c, t_b) &= \Delta t * \dot{R}_b^l * R_c^b(t_c) \\
\Delta L + V &= AX \\
\Delta X &= [r_{bc}^b, R_c^b, \Delta t]_{8x1}^T \\
\Delta L &= [r_{bc}^b(t_c, t_b), R_c^b(t_c, t_b)]_{7x1}^T
\end{aligned}
\right.
\tag{11}
$$

Generally speaking, the accuracy of the calibration procedure is dominated by the quality of the INS/GNSS POS data and the bundle adjustment results. This relationship also implicitly affects the performance of the MMS. The traditional calibration method does not calibrate the exposure time delay simultaneously. If the method is applied to calibrate an MMS operating in kinematic mode, the impact of the exposure time delay will propagate to the lever arm and boresight, respectively. The proposed method can avoid this problem and provide the best estimates of lever arm, boresight and exposure time delay at the same time. On another note, the distribution of the GCPs in the image and the quality of the INS/GNSS solutions are very important during the calibration procedure. After obtaining calibration parameters, the DG task can be performed seamlessly without GCPs as long as the spatial relationships of all the sensors within this MMS module remain fixed.

5. Data Processing Strategy

For the determination of the delay, lever arm and boresight parameters, the EOPs, including position and attitude of the images, must be solved using the bundle adjustment. However, some errors will occur during the image measurements due to imperfections of cameras during production. Thus the camera calibration must be performed. The objective of camera calibration is to analyze the interior orientation parameters (IOPs), such as the lens distortion, the focal length, and the principle point. If this is done carefully, systematic errors can be diminished during the image point measurements. Therefore, in order to process the system calibration above and check the ability of DG, the establishment of the camera control field and the ground control field must be done in this research.

A circular plane is set up for camera calibration. The diameter of this plane is 240 cm, and more than two hundred artificial landmarks are distributed evenly across it. Landmarks provided by Australis software are also included. Such design can be used to calibrate various cameras with different resolution and focal length process, as shown in Figure 11 [9].

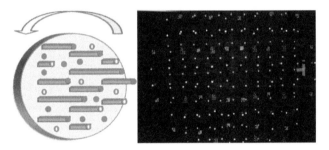

Figure 11. The camera control field.

Generally speaking, the amount of the images captured from multi-angle at the different locations could be restrained if the space of the camera control field is not enough. However, in the proposed field architecture, the relation of each landmark is fixed during the field rotation. That means its local coordinate system is also invariable. Each image can be shot at the same location but with different rotation angle. Compared to change the location of the shot, this design can overcome the restriction of the field space and provide sufficient reliability of the camera calibration, as shown in Figure 12. The camera control field is designed to acquire images with the best intersection geometry and avoid the high correlation between parameters. Thus, the calibration can be processed in the small space such as our control field (only $4 \times 4 \times 3$ m^3) with the best intersection geometry and the low correlation between parameters.

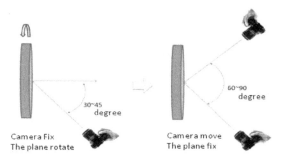

Figure 12. Relation between two situations.

The analysis of IOPs such as the focal length, the principal point, and the lens distortion is the objective of this process. The bundle method with self-calibration is proposed to determine the interior parameters CCD cameras have applied. The equation is included in the bundle adjustment [16]:

$$x_a = x_p - c\frac{r_{11}(X_A - X_O) + r_{12}(Y_A - Y_O) + r_{13}(Z_A - Z_O)}{r_{31}(X_A - X_O) + r_{32}(Y_A - Y_O) + r_{33}(Z_A - Z_O)} + \Delta x \tag{12}$$

$$y_a = y_p - c\frac{r_{21}(X_A - X_O) + r_{22}(Y_A - Y_O) + r_{23}(Z_A - Z_O)}{r_{31}(X_A - X_O) + r_{32}(Y_A - Y_O) + r_{33}(Z_A - Z_O)} + \Delta y \tag{13}$$

where:

C: The focal length;

x_p, y_p: The principal points;

x_a, y_a: The coordinates of target A in camera frame;

$r_{11} \sim r_{33}$: The rotation matrix;

X_A, Y_A, Z_A : The coordinates of target A in object frame;

$\Delta x, \Delta y$: The lens distortion.

This research adapts the commercial software, Australis [17], to solve for those parameters. It can process calibration automatically after the image is imported. A lens distortion model that includes seven parameters is enough for most kinds of cameras:

$$\Delta x = \bar{x} + \left(K_1 r^2 + K_2 r^4 + K_3 r^6\right)\bar{x} + P_1\left(r^2 + 2\bar{x}^2\right) + 2P_2 xy + b_1 x + b_2 y \tag{14}$$

$$\Delta y = \bar{y} + \left(K_1 r^2 + K_2 r^4 + K_3 r^6\right)\bar{y} + P_1\left(r^2 + 2\bar{y}^2\right) + 2P_1 xy \tag{15}$$

where: $\bar{x} = (x - x_p), \bar{y} = (y - y_p)$, $r = \sqrt{\bar{x}^2 + \bar{y}^2}$

K_1, K_2 and K_3: The radial lens distortion parameters;

P_1 and P_2: The decentric lens distortion parameters.

b_1 and b_2: The affine deformation parameters.

After obtaining proper IOPs, those parameters can be applied to enhance the accuracy of EOPs estimation of the bundle adjustment for system calibration and DG. For the determination of calibration parameters, the EOPs of each image need be known. They can be calculated using the bundle adjustment control field. So the two control fields are built for calibrating those systems applied in the study. Figure 13 illustrates the distribution of ground control points (GCPs) in two control fields which have been set up at distances of 400 and 800 m (Figure 13). The GCPs are accurately surveyed using differential-GNSS with carrier phase measurements and processed with network adjustment software. The standard deviation of GCPs is 3 mm.

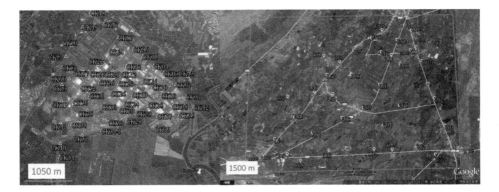

Figure 13. The distribution of GCPs in two control fields.

The image acquisition for the system calibration process is performed via flying UAV photogrammetric over the ground control field. The measurements of the image points are measured first. Second, the Australis software is used to complete the bundle adjustment to get the EOPs of each image. After performing the interpolation of INS/GPS positioning and orientation solution (POS) at trigger time, the differences between the EOPs and interpolated POS are derived for further processing. The differences are used to calculate calibration parameters using the previously mentioned calibration algorithm. After obtaining calibration parameters, the DG task can be performed exactly without using any GCP. On the other hand, traditional photogrammetric processes can use INS/GPS POS and GCPs throughout the whole area of interest to assist the conventional bundle adjustments process [18].

The grey, yellow and green scopes in Figure 14 illustrate the process of INS/GPS POS assisted AT, system calibration and DG, respectively. The INS/GPS POS helps the AT to execute AT after the three steps are finished, as tie-points and control points are measured, the IOPs of the cameras is calibrated and INS/GPS POS is interpolated. The EOPs of each image is obtained through AT, and the final products can be completed using programs like ortho-photo, Mosaic and DEM/DSM. INS/GPS POS assisted AT is included in the combination of calibration and DG processes, as shown in Figure 14.

The calibration procedure requires the EOPs of each image and the interpolated INS/GPS POS solutions. Therefore, the calibration can be executed after completing AT, after which the calibration report is generated. The DG function can provide positioning of interesting points without using GCP with interpolated INS/GPS POS and calibration report. In fact, the final products of DG are the same as AT. However, the INS/GPS POS assisted AT has to implement dense GCPs throughout the whole area under analysis before taking pictures. On the other hand, the DG mode only requires a control field for calibration purposes which is not required for every mission once it has been performed; the payload remains fixed after the last calibration.

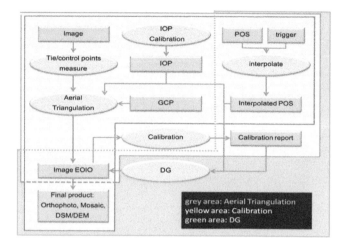

Figure 14. The process of INS/GPS POS assisted AT, system calibration and DG.

To avoid losing a lock on the GPS satellite due to the vibration of the UAV platform and certain maneuvers, this study applies the TC scheme to provide more robust POS solutions and overcome hardware limitations even when there may be frequent partial GPS outages to overcome. The TC scheme uses a single KF to integrate GPS and IMU measurements, as shown in Figure 15, which shows how the raw measurements are collected from the IMU and converted into position, velocity, and attitude measurements in the desired coordinate system using the INS mechanization algorithms. In the TC integration, the GPS pseudo range, delta range, and carrier phase measurements are processed directly in the INS KF [19]. The primary advantage of this integration is that raw GPS measurements can still be used to update the INS when fewer than four satellites are available. This is of special benefit in a complex environment, such as downtown areas where the reception of the satellite signals is difficult due to obstruction. Also, in cases when carrier phase GPS measurements are used, the IMU measurements can be used to reduce ambiguity in the resolution algorithm.

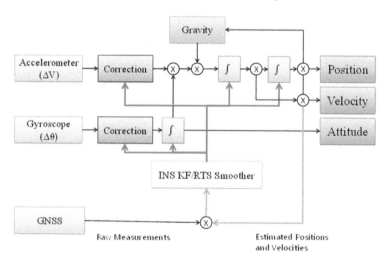

Figure 15. The TC integration scheme.

Post-mission processing, when compared to real-time filtering, has the advantage of using the data of the whole mission for estimating the trajectory. It is impossible using the whole data to filtering on real-time because only part of data is available except the last. After filtering is used in the first step, an optimal smoothing method, such as the Rauch-Tung-Striebel (RTS) backward smoother, can be applied [4]. This uses filtered results and their covariance as a first approximation which is then improved by using additional data that was not used in the filtering process. Depending on the type of data used, the improvement obtained by optimal smoothing can be considerable [20].

For a geo-referencing process which puts POS stamps on images and a measurement process that obtains three-dimensional coordinates of all important features and stores them in a Geographic Information System (GIS) database, only post-mission processing can be implemented due to the complexity of the task [21]. Therefore, most commercially available DG systems operate in real-time only for data acquisition and conduct most of the data processing and analysis in post-mission mode. Figure 16 illustrates the POS software (Pointer. POS) developed in this study, which includes the GNSS processing engine, INS mechanizations in different navigation frames, as well as the optimal estimation engine, which can perform optimal fusion in LC, TC and Hybrid Tightly Coupled (HTC) schemes.

After processing POS and bundle adjustment solutions using measurements acquired over control fields, calibration and performance verification can be achieved. First, the position and attitude of POS must be converted to [x, y, z] and the normalized quaternions form for the further processing. The smoothed POS solutions are interpolated by linear interpolation at trigger time. To keep the coordinates consistent, the POS coordinates need to be converted to coordinates of interest. This can be performed through series of transformation methods [22].

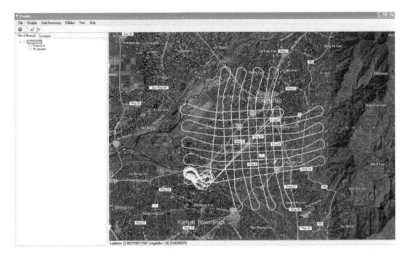

Figure 16. POS software.

The DG procedure is done by using smoothed POS solutions at trigger time and a calibration report to obtain IOPs and EOPs of each image. The three-dimensional coordinates of points of interest can be solved by conventional photogrammetric technology such as collinearity equation and intersection. The statistical analysis of MMS performance is estimated by check points and then output to the MMS performance report. Figure 17 illustrates the data processing procedure adopted in this study.

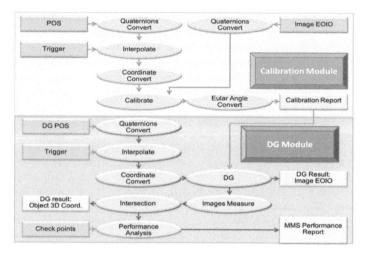

Figure 17. Data processing procedure.

6. Results and Discussion

To validate the impact of flight height on DG performance, a field test was conducted in the fall of 2011 at the first control field. The DG payload used in this scenario was the previous version and the flight altitudes set for aerial photography were set to 300 and 600 m above ground. The scope of the test zone is 3 km × 3 km, which covers the first control field, as shown in Figure 18a with the red square. The blue region illustrates the fly zone approved for this test. In addition, to compare the performance of previous and new versions of DG modules, the second test was conducted in the fall of 2013 at a second control field. The tested IMU was the ADIS16488 IMU and the flight altitude set for aerial photography was set to 300 and 600 m above ground in this test. The scope of the test zone is 3 km × 3 km, which covers the second control field shown in the red square in Figure 18b. The blue region illustrates the fly zone approved for this test.

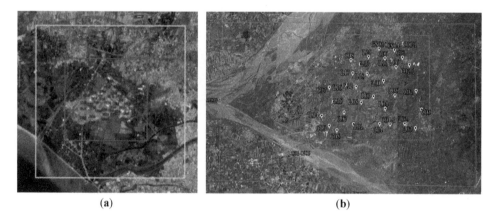

(a)   (b)

Figure 18. The scopes of the two tests.

Due to the effect of side winds, the attitude of UAV, the transversal and longitudinal overlapping were increased to 80% and 40% respectively to insure that the coverage of the stereo pair would overlap completely during the test flight. Although more images will have to be processed, this method guarantees complete coverage by the stereo pair. Figure 19 illustrates the flight trajectories of the first test at flight heights of 600 and 300 m which calls UAV-MMQG. Figure 20 depicts the trajectory of the second test which calls UAV-ADIS. The ground sample distances (GSD) of 600 m and 300 m of flight heights are about 20 cm and 10 cm.

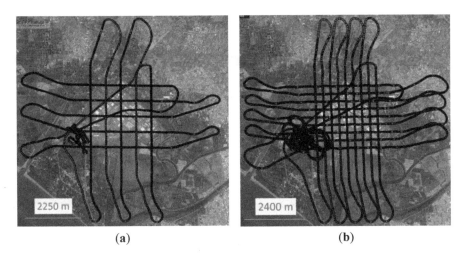

(a)   (b)

Figure 19. The trajectories of the first test flight. (a) UAV-MMQG-600; (b) UAV-MMQG-300.

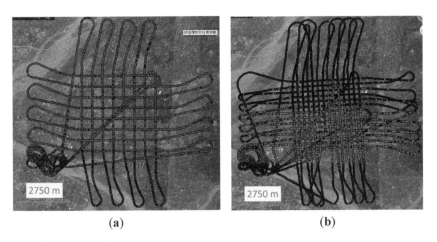

(a)   (b)

Figure 20. The trajectories of the second test flight. (a) UAV-ADIS-600; (b) UAV-ADIS-300.

6.1. Calibration Results

Traditional and proposed calibration procedures are implemented in this study to estimate calibration parameters of each camera for further study. The proposed software, as shown in Figure 21, was developed using Visual Studio 2008 C++, QT, OpenCV and OpenGL for system calibration and DG verification.

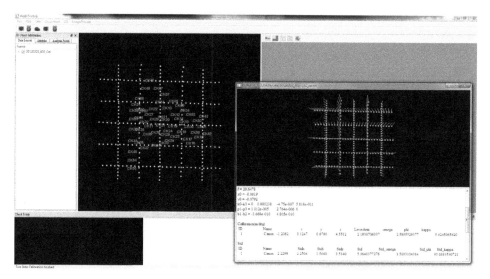

Figure 21. The calibration operation of the program.

The EOPs of the images are calculated first with Australis software through the bundle adjustment by measuring the image points when the flight mission has been completed. Then the trajectories of INS/GPS integrated POS are obtained through the use of TC schemes with the RTS smoother. The interpolation of INS/GPS smoothed POS solutions at the image exposure time is then performed. The lever arm and boresight angle for each epoch are applied by comparing the differences of the position and the attitude between the exterior orientation parameters and the interpolated INS/GPS solutions. The proposed software solves these calibration parameters using the methods described, and generates a calibration report, as shown in Table 1.

The center of POS and the camera is roughly overlaid along the x and y axis when they are assembled in the payload frame. As shown in Table 1, the relative accuracy of the proposed method is better than that of the traditional calibration method. The most probable values and standard deviation of DG modules with the 600 and 300 m flight heights are compared based on those calibration methods. As illustrated in Table 2, the standard deviation and the most probable values of the 300 meter flight height scenario is much better than those of the 600 m flight height scenario for both calibration methods. This finding illustrates the fact that the calibration flight test should be conducted at height of 200–300 m for the UAV used in this study. In addition, the relative accuracy and the most probable values for the new DG module illustrate results superior to those of the previous DG module, showing that the calibration results can be improved significantly with a better POS module. In addition, the accuracy of the proposed calibration method is superior to the traditional calibration method because the bias of exposure time delay contaminates lever-am and boresight parameters in the traditional calibration method. Figure 22 illustrates the impact of exposure time delay on lever arm parameters at each epoch. Because the center of IMU is designed on overlaying the center of camera, the lever-arm should be near zero in the level direction. The delay leads to a bias which depends on the velocity of the platform in the forward direction.

Table 1. The results of two calibration methods.

		(s)	Lever-Arm (m)			Boresight (deg)		
		Delay	X	Y	Z	Omega	Phi	Kappa
			Traditional Calibration					
UAV MMQG 600	most probable value		−1.2062	3.1247	0.6790	3.79797404	5.20251129	−2.24883285
	Standard deviation		2.0572	1.1891	2.7717	3.69862986	3.49680553	5.86546832
			Proposed Calibration					
	most probable value	−0.1072	−1.1205	−0.001	1.0557	4.61012742	5.21063752	−1.79112843
	Standard deviation	0.0036	0.0714	0.1317	0.0858	0.54718413	0.37337511	0.22203555
			Traditional Calibration					
UAV MMQG 300	most probable value		0.0141	3.8105	0.1513	4.40651630	0.51021908	−0.00873761
	Standard deviation		0.4595	1.1863	1.1839	4.37609122	5.47624647	3.77899241
			Proposed Calibration					
	most probable value	−0.1272	−0.0548	0.1523	0.3580	4.38480491	0.53859868	0.25962901
	Standard deviation	0.0043	0.0767	0.1125	0.0771	1.53376932	1.63520332	2.09804173
			Traditional Calibration					
UAV ADIS 600	most probable value		−2.0806	6.1863	1.3527	1.26444091	0.500277771	−0.193128818
	Standard deviation		2.2299	2.2504	1.5660	2.61737788	1.99383288	3.03577883
			Proposed Calibration					
	most probable value	−0.2272	−2.1138	−0.2578	1.8353	1.34584366	0.51119238	−0.28419414
	Standard deviation	0.0050	0.0804	0.1293	0.0811	0.15690974	0.14208760	0.14200760
			Traditional Calibration					
UAV ADIS 300	most probable value		−0.8034	3.3460	−0.0718	2.464419253	4.499372566	−1.291911667
	Standard deviation		0.6815	0.7948	0.7347	4.31193476	1.05151768	1.70347184
			Proposed Calibration					
	most probable value	−0.1394	−0.6251	−0.3909	0.1992	2.37524174	4.51749719	3.28462927
	Standard deviation	0.0026	0.0425	0.0703	0.0430	0.99552547	0.07104126	0.06245981

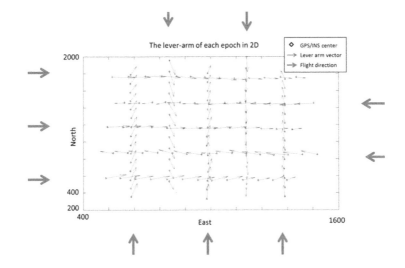

Figure 22. The lever-arm of each epoch.

Table 2. The statistical analysis of DG based on traditional calibration method.

(m)	Traditional Calibration					Proposed Calibration				
	E	N	U	2D	3D	E	N	U	2D	3D
					UAV-MMQG-600(48)					
AVG	0.0560	−2.8740	−0.3470	2.8745	2.8954	−1.5370	−4.6440	1.1730	4.8917	5.0304
STD	11.2300	8.3610	15.4680	14.0007	20.8633	10.4340	7.9770	13.8660	13.1340	19.0989
RMS	11.1240	8.7660	15.3250	14.1628	20.8672	10.4390	9.1590	13.7710	13.8874	19.5576
					UAV-MMQG-300(51)					
AVG	−0.5090	0.0550	−3.3960	0.5120	3.4344	0.1880	−1.6330	−1.2860	1.6438	2.0871
STD	8.9410	6.5850	14.9560	11.1042	18.6276	8.2970	6.1900	11.4760	10.3516	15.4549
RMS	8.8680	6.5210	15.1930	11.0075	18.7615	8.2190	6.3440	11.4380	10.3826	15.4475
					UAV-ADIS-600(32)					
AVG	−0.1040	−0.9900	−1.2560	0.9954	1.6026	0.0480	0.9050	−1.3990	0.9063	1.6669
STD	5.9380	7.3070	11.2860	9.4155	14.6978	6.0310	6.9060	10.0350	9.1687	13.5929
RMS	5.8440	7.2490	11.1790	9.3113	14.5489	5.9370	6.8680	9.9750	9.0784	13.4877
					UAV-ADIS-300(24)					
AVG	0.4880	−1.3150	−0.0100	1.4026	1.4027	−0.4850	−0.3880	−1.4470	0.6211	1.5747
STD	4.6590	5.5380	5.2160	7.2371	8.9209	3.8810	4.5650	5.2340	5.9918	7.9559
RMS	4.5870	5.5780	5.1060	7.2218	8.8445	3.8300	4.4850	5.3240	5.8978	7.9454

6.2. Verification of the DG Capability of the Proposed UAV Photogrammetric Platform

The software developed in this study can also perform the DG verification using a collinearity equation and intersection to calculate the coordinates of the check point, as shown in Figure 23, which presents the relevant information, including the coordinates of the control points, POS, calibration report and trigger file—which have been imported into the software which calculates the EOPs for each image using the DG function. Users can perform image point measurements of the check points which appear in the different images. The results of the space intersection of check points are obtained from these images, after which their coordinates, derived through GCP free mode, are then compared with the already-known coordinates. The reference coordinates of the check points are obtained through the precise control survey with GNSS RTK technology and network adjustment. Therefore, the DG coordinates of those check points can then be compared with their reference coordinates.

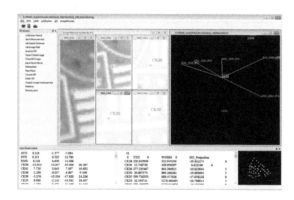

Figure 23. The DG program.

Table 2 illustrates the statistical analysis of DG based on different flights height, DG modules and calibration methods. Figures 24–27 illustrate the DG performance for the scenarios with those check points. It can clearly be seen that the newly developed DG payload is significantly better than the previous version of the DG payload. The horizontal positioning accuracy of new DG is best at about 5.8 m in 2D and 7.9 m in 3D. On the other hand, Figure 28 and Table 3 illustrate the positional errors with traditional photogrammetry using data acquired by ADIS 16488 with 300 m flying height. The following are several comparisons of three factors which are hardware, flight height and calibration method.

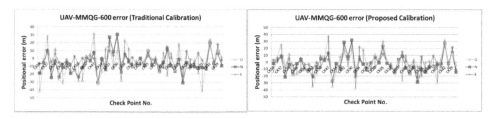

Figure 24. DG error based on MMQG with 600 m.

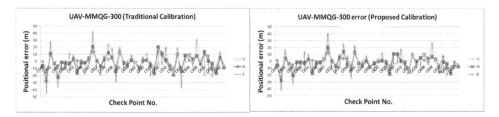

Figure 25. DG error based on MMQG with 300 m.

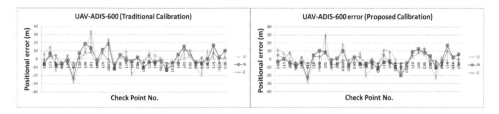

Figure 26. DG error based on ADIS 16488 with 600 m.

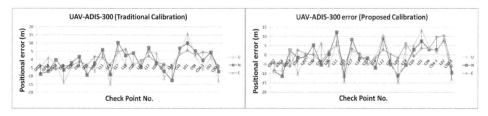

Figure 27. DG error based on ADIS 16488 with 300 m.

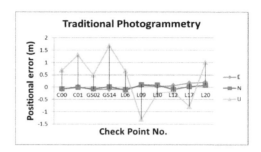

Figure 28. The positional errors of traditional photogrammetry based on ADIS 16488 with 300 m.

Table 3. The statistical analysis of traditional photogrammetry based on ADIS 16488 with 300 m.

meter	E	N	U
AVG	0.028	0.005	0.333
AVG	0.105	0.079	0.943
RMS	0.104	0.075	0.954

Tables 4 and 5 illustrate the improvement rates analysis for these scenarios. The first scenario is the relationship of the low cost POS to the flight height. The accuracy is based on MMQG, with the 300 m flight improved by about 28.7% and 11.2% in term of 2D and 3D absolute positional errors, respectively, as compared with the 600 meter flight height using the traditional calibration method. At the same time, the accuracy of the ADIS 16488 also improves by 28.9% and 64.5% in terms of 2D and 3D. The statistical numbers of MMQG and ADIS16488 IMU improve by 33.8%, 26.6% and 53.9%, 69.8% using the proposed method.

Table 4. The improvement rate of DG accuracy based on different flight and DG modules.

%			MMQG-300	ADIS16488-300
Traditional Calibration	MMQG-600	2D	28.67	96.11
		3D	11.22	135.93
	MMQG-300	2D	0.00	52.42
		3D		112.12
	ADIS16488-600	2D	−18.22	28.93
		3D	−28.95	64.50
Proposed Calibration	MMQG-600	2D	33.76	135.47
		3D	26.61	146.15
	MMQG-300	2D	0.00	76.04
		3D		94.42
	ADIS16488-600	2D	−14.37	53.93
		3D	−14.53	69.76

The second scenario compares the performance of two DG modules. Comparing the two results when using MMQG-600 and ADIS16488-300, the results of ADIS16488 IMU are superior to those of MMQG. It improves by 96.1% and 135.9% in terms of 2D and 3D, respectively, for absolute positional errors using the traditional calibration method, and improves by 135.5% for 2D and 146.2% for 3D in terms of absolute positional errors using the proposed method.

The third scenario is the relationship between two IMUs. The improvements of proposed IMU are about 52.4% and 112.1% with 300 meter flight height and 76.0% and 94.4% with 600 meter flight height compared to MMQG.

The last analysis is the improvement rate based on the new calibration method. The proposed method has proven effective in all scenarios to which it has been applied in this study. Based on MMGQ, it improves 2.0% and 6.7% in terms of 2D and 3D, respectively, in regards to absolute positional errors for the 600 m flight and 6.0% and 21.5% in terms of 2D and 3D, respectively, for absolute positional errors for the 300 m flight. The DG accuracy levels provided by the proposed DG modules with the proposed method reach 2.6% for 2D and 7.5% for 3D in absolute positional errors for the 600 m flight and 22.5% and 11.3% for the 300 m flight.

Table 5. The improvement rate of DG accuracy with the proposed method.

%	Proposed Calibration	
	2D	3D
MMQG-600	1.98	6.70
MMQG-300	6.02	21.45
ADIS16488-600	2.62	7.52
ADIS16488-300	22.45	11.32

The approximate error budgets of the proposed tactical grade DG module for flight heights of 600 and 300 m are given in Table 6. (Table 3 also looked at how the primary DG positional error sources are related to the quality of the gyroscopes used within the IMU.) The proposed DG module improves the kinematics positioning accuracy of trajectory to within less than 1 m by using single frequency carrier phase measurements. In addition, the remaining positional error sources can be mitigated by replacing the current IMU with superior gyroscopes.

Table 6. Error budgets of the new DG system.

Error source	ADIS 16488 with 600 Flight Height		ADIS 16488 with 300 Flight Height	
	Magnitude	Impact on (DG Error)	Magnitude	Impact on (DG Error)
INS/GNSS Positional error	0.1–0.2 m	0.1–0.2 m	0.1–0.2 m	0.1–0.2 m
INS/GNSS Orientation error	0.15–0.25 degree	1.6–2.5 m	0.15–0.25 degree	1.6–2.5 m
Calibration error -Boresight	0.15–0.25 degree	1.6–2.5 m 0.1–0.2 m	0.15–0.25 degree	0.8–1.3 m 0.1–0.2 m
-Lever-arm	0.1–0.2 m		0.1–0.2 m	
Synchronization error -Position -Orientation	1–2 ms	120 km/h fly speed0.036–0.072 m0.3–0.6 m	1–2 ms	120 km/h fly speed 0.036–0.072 m 0.3–0.6 m

The primary contribution of this study is the implementation of a UAV based photogrammetric platform with DG ability and the verification of its performance in terms of DG accuracy for various situations using a low cost tactical grade IMU. In addition, the preliminary results indicate that the DG accuracy in GCP free mode can meet the requirements for rapid disaster mapping and relief applications.

The total cost of the proposed POS module is below 2000 US dollars, making it suitable for rapid disaster relief deployment to provide near real-time geo-referenced spatial information. The data processing time for the DG module, including POS solution generalization, interpolation, EOP generation, and feature point measurements, is less than one hour.

7. Conclusions

This study develops a long endurance DG based fixed-wing UAV photogrammetric platform in which a low cost tactical grade integrated Positioning and Orientation System (POS) is developed. In

Sensors **2015**, *15*, 6560–6585

addition, a novel kinematic calibration method including lever arms, boresight angles and camera shutter delay is proposed and compared with traditional calibration method. Furthermore, the performance of DG is also analyzed based on the two methods with different flights and two DG modules. The results presented in this study indicate that the accuracy of DG can be significantly improved by lower flight heights and hardware with superior specifications. The proposed method improves the accuracy of DG by about 10%.

The preliminary results show that horizontal DG positioning accuracies in two-dimension (2D) are around 8 m at a flight height of 600 m with the newly designed tactical grade integrated Positioning and Orientation System (POS). The positioning accuracy in three-dimensions (3D) is less than 12 m. Such accuracy is good for near real-time disaster relief.

The DG ready function of the proposed platform guarantees mapping and positioning capability even in GCP free environments, which is very important for rapid urgent response for disaster relief. Generally speaking, the data processing time for the DG module, including POS solution generalization, interpolation, Exterior Orientation Parameters (EOP) generation, and feature point measurements, is less than one hour.

Acknowledgments: The author would acknowledge the financial supports provided by the National Science Council of Taiwan NSC (102-2221-E-006-137-MY3).

Author Contributions: Chien-Hsun Chu conceived the new calibration method and performed it. Futhermore, he finished solve data and wrote software to analysis the result of DG and calibration. Meng-Lun Tsai finished do the experiments, collection of data and helped solve data. Kai-Wei Chiang provided the direction of this study. After finishing the draft of the study, he commented and corrected this study. El-Sheimy Naser and Ayman Habib are professonal professors in the geomatics field. They provide the comments for this study after finishing the first verson.

Conflicts of Interest: The authors declare no conflict of interest.

References

1. Bachmann, F.; Herbst, R.; Gebbers, R.; Hafner, V.V. Micro UAV Based Georeferenced Orthophoto Generation in VIS + NIR for Precision Agriculture. *Int. Arch. Photogramm. Remote Sens. Spat. Inf. Sci.* **2013**, *XL-1/W2*, 11–16. [CrossRef]
2. Gibson, J.R.; Schwarz, K.P.; Wei, M.; Cannon, M.E. GPS-INS data integration for remote sensing. In Proceedings of IEEE Position Location and Navigation Symposium (PLANS '92), Monterey, CA, USA, 23–27 March 1992. [CrossRef]
3. Schwarz, K.P.; Chapman, M.E.; Cannon, E.; Gong, P. An integrated INS/GPS approach to the georeferencing of remotely sensed data. *Photogramm. Eng. Remote Sens.* **1993**, *59*, 1667–1674.
4. Chiang, K.W.; Noureldin, A.; El-Sheimy, N. A new weight updating method for INS/GPS integration architectures based on neural networks. *Meas. Sci. Technol.* **2004**, *15*, 2053–2061. [CrossRef]
5. Eisenbeiss, H. The autonomous mini helicopter: A powerful platform for mobile mapping. *Int. Arch. Photogramm. Remote Sens. Spatial Inf. Sci.* **2008**, *XXXVII*, 977–983.
6. Haubeck, K.; Prinz, T. A UAV-Based Low-Cost Stereo Camera System for Archaeological Surveys—Experiences from Doliche (Turkey). *Int. Arch. Photogramm. Remote Sens. Spat. Inf. Sci.* **2013**, *XL-1/W2*, 195–200. [CrossRef]
7. Grant, M.S.; Katzberg, S.J.; Lawrence, R.W. GPS remote sensing measurements using aerosonde UAV. *Am. Instit. Aeronaut. Astronaut.* **2005**, *2005–7005*, 1–7.
8. Nagai, M.; Shibasaki, R. Robust trajectory tracking by combining GPS/IMU and continual CCD images. *Int. Symp. Space Technol. Sci.* **2006**, *25*, 1189–1194.
9. Chiang, K.W.; Tsai, M.L.; Chu, C.H. The Development of an UAV Borne Direct Georeferenced Photogrammetric Platform for Ground Control Point Free Applications. *Sensors* **2012**, *12*, 9161–9180. [CrossRef] [PubMed]
10. Rehak, M.; Mabillard, R.; Skaloud, J. A Micro-UAV with the Capability of Direct Georeferencing. *Int. Arch. Photogramm. Remote Sens. Spatial Inf. Sci.* **2013**, *XL-1/W2*, 317–323. [CrossRef]

11. Grenzdorffer, G.J.; Engel, A.; Teichert, B. The photogrammetric potential of low-cost UAVs in forestry and agriculture. *Int. Arch. Photogramm. Remote Sens. Spatial Inf. Sci.* **2008**, *XXXVII*, 1207–1213.

12. Wendel, J.; Trommer, G.F. Tightly coupled GPS/INS integration for missile applications. *Aerospace Sci. Technol.* **2004**, *8*, 627–634. [CrossRef]

13. Lewantowicz, Z.H. Architectures and GPS/INS integration: Impact on mission accomplishment. In Proceedings of IEEE Position Location and Navigation Symposium (PLANS '92), Monterey, CA, USA, 23–27 March 1992. [CrossRef]

14. Tsai, M.L.; Chiang, K.W.; Huang, Y.W.; Lo, C.F.; Lin, Y.S. The development of a UAV based MMS platform and its applications. In Proceedings of the 7th International Symposium on Mobile Mapping Technology, Cracow city, Poland, 13–16 June 2011.

15. Fraser, C.S. Digital camera self-calibration. *ISPRS J. Photogramm. Remote Sens.* **1997**, *52*, 149–159. [CrossRef]

16. Li, Y.H. The Calibration Methodology of a Low Cost Land Vehicle Mobile Mapping System. In Proceedings of the 23rd International Technical Meeting of The Satellite Division of the Institute of Navigation (ION GNSS 2010), Portland, OR, USA, 21–24 September 2010; pp. 978–990.

17. Cronk, S.; Fraser, C.S.; Hanley, H. Hybrid Measurement Scenarios in Automated Close-Range Photogrammetry. *Int. Arch. Photogramm. Remote Sens. Spatial Inf. Sci.* **2006**, *XXXVII Pt B3b*, 745–749.

18. Tao, V.; Li, J. Advances in Mobile Mapping Technology. In *International Society for Photogrammetry and Remote Sensing (ISPRS) Book Series 2007*; Taylor and Francis Group: London, UK.

19. Scherzinger, B.M. Precise robust positioning with Inertial/GPS RTK. In Proceedings of the ION-GNSS 2000, Salt Lake City, UT, USA, 19–22 September 2000; pp. 155–162.

20. Gelb, A. *Applied Optimal Estimation*; The MIT Press: Cambridge, MA, USA, 1974.

21. El-Sheimy, N. Introduction to Inertial Navigation. ENGO 699.71 Lecture Notes; Department of Geomatics Engineering, University of Calgary: Alberta, Canada, 2002.

22. Skalouda, J.; Legatb, K. Theory and reality of direct georeferencing in national coordinates. *ISPRS J. Photogramm. Remote Sens.* **2008**, *63*, 272–282. [CrossRef]

 *sensors*

Article

Adaptive Environmental Source Localization and Tracking with Unknown Permittivity and Path Loss Coefficients [†]

Barış Fidan * and Ilknur Umay

Department of Mechanical and Mechatronics Engineering, University of Waterloo, 200 University Ave W, Waterloo, ON N2L 3G1, Canada; iumay@uwaterloo.ca
* Correspondence: fidan@uwaterloo.ca; Tel.: +1-519-888-4567 (ext. 38023)
† This paper is an extended version of our paper published in Fidan, B.; Umay, I. Adaptive source localization with unknown permittivity and path loss coefficients. In Proceedings of the 2015 IEEE International Conference on Mechatronics, Nagoya, Japan, 6–8 March 2015; pp. 170–175.

Academic Editors: Felipe Gonzalez Toro and Antonios Tsourdos
Received: 16 August 2015; Accepted: 3 December 2015; Published: 10 December 2015

Abstract: Accurate signal-source and signal-reflector target localization tasks via mobile sensory units and wireless sensor networks (WSNs), including those for environmental monitoring via sensory UAVs, require precise knowledge of specific signal propagation properties of the environment, which are permittivity and path loss coefficients for the electromagnetic signal case. Thus, accurate estimation of these coefficients has significant importance for the accuracy of location estimates. In this paper, we propose a geometric cooperative technique to instantaneously estimate such coefficients, with details provided for received signal strength (RSS) and time-of-flight (TOF)-based range sensors. The proposed technique is integrated to a recursive least squares (RLS)-based adaptive localization scheme and an adaptive motion control law, to construct adaptive target localization and adaptive target tracking algorithms, respectively, that are robust to uncertainties in aforementioned environmental signal propagation coefficients. The efficiency of the proposed adaptive localization and tracking techniques are both mathematically analysed and verified via simulation experiments.

Keywords: localization; sensor networks; path loss coefficient; RSS; TOF

1. Introduction

There has been significant research interest in the use of mobile sensory units and wireless sensor networks (WSNs) in various application areas, including environmental monitoring, especially in the last two decades. Typical mobile sensory units for environmental monitoring are autonomous vehicles (AVs) with certain types of sensor loads, and typical environmental monitoring WSNs are coordinated teams of such AVs, as well as sensor arrays on individual AVs. The main tasks of the environmental monitoring AVs and WSNs are localizing and state-observing various target objects, including animals, fire sources, fire fighter units, radioactive and biochemical emission sources and electromagnetic signal sources [1–4]. A key component in AV-based environmental monitoring is sensor instrumentation and localization algorithms utilizing these sensors. For localization, sensors are often used in sensor array or WSN forms. Use of such WSNs on AVs, e.g., UAVs, have various environmental applications, such as motion tracking, precision agriculture, coastline monitoring, rescue tasks, detecting and tracking fire, chemical and radioactive sources and pollutants [3–12].

Accurate signal-source and signal-reflector target localization via the aforementioned sensory units and WSNs requires precise knowledge of specific signal propagation properties of the environment. Such properties can be modelled by certain diffusion or propagation formulas, which involve some environmental coefficients, which are specific to the particular setting and which may be constant or time/space dependent. Environmental coefficients for radiation tracking and fire positioning are, respectively, radioactive sensor detection count rate and fire propagation velocity [6,8,9]. For electromagnetic signal source or reflector localization, typically, received signal strength (RSS) and time-of-flight (TOF)-based range sensors are used. Modelling of electromagnetic signal propagation for use by such sensors is more advanced [1,2]. The corresponding environmental coefficients are the path loss coefficient (η) for RSS and the signal permittivity coefficient (ε) for TOF.

In the literature, some preliminary studies for position tracking of radioactive and fire sources based on environmental coefficients are introduced in [6,8,9]. However, these studies either provide some rough data or assume *a priori* data from measurements on the environmental coefficients, which are the count rate for a radioactive source and the velocity of the hot gasses for fire localization. Electromagnetic signal source localization has various environmental monitoring applications, including surveillance of environmental (UAV, fire-fighter, robot) assist units, surveillance of objects tagged by electromagnetic signal sources or reflectors, surveillance of environmental intruders and positioning for rescue tasks [4,5,13,14]. A particular application is in fire-rescue systems, aiming at recognition and localization of the fire fighters [13,14].

Since localization algorithms are vulnerable to inaccuracies in the knowledge of the environmental coefficients, various approaches are proposed in the literature for the estimation of these coefficient or the compensation of uncertainties in the algorithms [15–19]. These approaches in general have a recursive nature and either still carry a significant amount of inaccuracy or require significant computational complexity for training and iteration of the estimation algorithms.

In this paper, we propose a more direct and static calculation technique for estimating the environmental coefficients, the path loss coefficient (η) for RSS and the signal permittivity coefficient (ε) for TOF, using a range sensor triplet during adaptive localization and tracking of a signal source by a mobile agent equipped with this sensor triplet. The triplet is designed to have a fixed rigid geometry where the z-coordinates of the sensors are equidistant. The proposed environmental coefficient estimation technique is integrated with a recursive least squares (RLS)-based adaptive localization scheme and an adaptive motion control law, to construct adaptive target localization and adaptive target tracking algorithms, respectively, that are robust to uncertainties in the aforementioned environmental signal propagation coefficients. The efficiency of the proposed adaptive localization and tracking techniques is both mathematically analysed and verified via simulation experiments.

Although the focus of this paper is on the localization of electromagnetic signal sources and reflectors in the environment and monitoring of objects based on such localization, the techniques studied in the paper have potential to be applied to the localization of the aforementioned fire source, radioactive emission source or biochemical source applications, as well.

The rest of the paper is organized as follows: The target localization and tracking problems of interest are defined, and the TOF and RSS-based range measurement and localization methods are briefly explained in Section 2. The details of the proposed environmental coefficient estimation technique are provided in Section 3. Sections 4 and 5 present, respectively, the adaptive localization and adaptive tracking control designs. Simulation test results are provided in Section 6. The paper is concluded with some final discussions and remarks provided in Section 7.

2. Distance-Based Localization and Tracking

In this section, we formally state the source localization and tracking problems of interest and present the considered sensor instrumentation setting. The main principles and mathematical modelling of RSS and TOF-based distance measurement techniques are presented in Sections 2.2 and 2.3. The effect of the environment on these techniques is briefly discussed in Section 2.4. For the methodology, we propose later to overcome the environmental uncertainties, use of a single sensor unit on the UAV is not sufficient; a sensor triplet, as a minimal sensor array, is required to be used. Hence, the problem definition in the next subsection assumes the use of a sensor triplet.

2.1. Localization and Tracking Problems

Consider a moving UAV equipped with sensor triplet $S = (S_1, S_2, S_3)$, where the sensors are identical and sense the intensity of the signal emitted by a target source located at some unknown position

$$p_T = [x_T, y_T, z_T]^T \qquad (1)$$

Note that p_T may be time varying. Denote the position of the UAV at time instant $t = kT_s$ for $k = 0, 1, 2, \ldots$, where T_s is the common sampling time used by the UAV sensors and processors, by

$$p[k] = [x[k], y[k], z[k]]^T \qquad (2)$$

and the position vector of each sensor S_i by

$$p_i[k] = [x_i[k], y_i[k], z_i[k]]^T \qquad (3)$$

Assume that $p_i[k]$ and the target-sensor distance

$$d_i[k] = \|p_T - p_i[k]\| \qquad (4)$$

for each sensor S_i are available to the processing unit of the sensory UAV. For simplicity, let the UAV position (body reference point) be defined as that of S_2, *i.e.*, let

$$p[k] = p_2[k] \qquad (5)$$

and hence, the target-UAV distance is defined as

$$d[k] = \|p_T - p[k]\| \qquad (6)$$

The 3D Localization Problem is to generate on-line estimate $\hat{p}_T[k]$ of p_T using the measurements of $d_i[k]$ and $p_i[k]$. An illustration of the localization task setting is given in Figure 1.

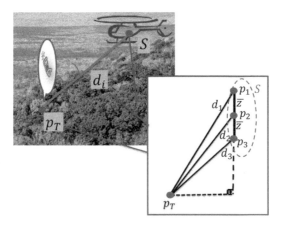

Figure 1. An illustration of the localization task setting and the proposed sensor array geometry, mobile sensor triplet unit (MSTU).

In many practical cases, the UAV altitude with respect to the target T, e.g., when T is a ground target, is maintained constant and/or available for measurement, and the practical localization is to find the x and y coordinates of T. Accordingly, the (reduced order) Lateral Localization Problem is to generate on-line estimate $\hat{p}_{TL}[k]$ of $p_{TL} = [x_T, y_T]^T$ using the measurements of $d_i[k]$ and $p_i[k]$ and knowledge of z_T.

The Lateral Tracking Problem is to produce the control input for the UAV, using d_i and p_i measurements, such that $p_L[k] = [x[k], y[k]]^T$ asymptotically converges to p_{TL}. For brevity, we skip the low level dynamic control design, assume perfect tracking of a velocity command and focus on the generation of the lateral velocity input

$$v_L[k] = \dot{p}_L[k] = \left.\frac{dp_L}{dt}\right|_{t=kT_s} \tag{7}$$

as the high level kinematic control input only.

Note that, in both of the Localization and Lateral Tracking Problems defined above, p_T may be time varying. Even though it is treated as constant in adaptive localization and tracking control scheme designs, simulation scenarios with time-varying p_T are successfully tested, as demonstrated in Section 6.

2.2. RSS-Based Techniques

RSS is a distance measurement technique based on the signal power (or strength) measured by a receiver located at the sensor [1,20]. In a generic RSS setting, the target signal source, which is required to be localized, emits a signal with original power P_T. The power P_S received by S follows an exponential decay model, which is a function of P_T, the distance d_T between S and T and a coefficient η modelling the signal propagation behaviour in the corresponding environment, called the path loss coefficient (exponent). The widely-used corresponding mathematical model is

$$P_S = K_l P_T d_T^{-\eta} \tag{8}$$

where K_l represents other factors that include the effects of antenna height and antenna gain. K_l is often considered to be log-normal and is often ignored in algorithm design leading to the simplified model

$$P_S = P_T d_T^{-\eta} \tag{9}$$

The RSS technique often provides cost savings over deploying localization-specific hardware, and all current standard radio technologies, such as Wi-Fi and ZigBee, provide RSS measurements. However, RSS can have multi-path effects that include shadowing, reflection, diffraction and refraction due to unpredictable environmental conditions, particularly for indoor applications [21]. In modelling, these effects are also lumped and included in the coefficient K_l of Equation (8).

2.3. TOF-Based Techniques

In TOF-based techniques, each sensor is composed of a transmitter unit, a receiver unit and a precision timer. The transmitter emits a signal, which is reflected by the target T and received by the receiver; and the time of flight, *i.e.*, the time elapsed between the signal's emission and receiving of its reflection, is used to deduce the distance between the sensor and the target T. The environmental characteristics are summarized in the electromagnetic (e.g., radio-frequency (RF)) signal propagation velocity

$$v = \frac{c}{\sqrt{\varepsilon}} \tag{10}$$

where c is the speed of light and ε denotes the (relative) permittivity coefficient.

Range is calculated by multiplying this propagation velocity and the measured TOF value. The corresponding mathematical model [22] can be formulated as

$$t_F = \frac{2d_T}{v_{ave}} = d_T\sqrt{\bar{\varepsilon}} \qquad (11)$$

where:

$$\bar{\varepsilon} = \frac{4\varepsilon}{c^2} = \frac{4}{v_{ave}^2}$$

Here, it is assumed that the TOF sensor emits a signal at $t_D = kT_S$ with a sampling period $T_S > 0$ and stores the TOF t_F value when the signal is received back at time $t[k] = kT_S + t_F[k]$. T_S is chosen large enough to enable TOF measurements to satisfy $t_F[k] \ll T_S$ for any k.

The value of the TOF t_F above can be measured using the phase of the received narrow-band carrier signal or via direct measurement of the arrival time of a wide-band narrow pulse [23]. The TOF-based technique, in general, requires strict time synchronization for the target and the receiver(s) [1].

2.4. Effect of the Environment

Information about the path loss exponent η for RSS-based techniques and the relative permittivity ε for TOF-based techniques have a vital effect on the measurement [15]. In many practical settings, these parameters are unknown and even variable in some due to the influences of variances on the weather conditions, human behaviour and the actuator effect at the anchor nodes. It is shown that using the wrong data on the path loss coefficient, η, has a huge effect on the accuracy of the position estimate [18].

Finding accurate estimation of these parameters is studied in the literature [16–19]. Most of the relevant works follow recursive algorithms involving training by data off-line or two-step on-line coefficient estimation and localization based on the estimate coefficients [16,17]. The off-line identification approaches require a large amount of training data for producing accurate estimates of the coefficients. The two-step on-line iterative approaches, on the other hand, may not lead to a successful level of accuracy during the joint coefficient estimation and localization process. The work in [15,18], for example, proposes iterative methodologies for the RSS case to obtain the unknown target location and path loss coefficient of the environment simultaneously in 2D with lower complexity. The iteration in these methodologies has the iterative steps of estimating the position of the target for the latest estimate of the path loss coefficient, calculating the corresponding RSS estimate and RSS estimate error and the application of an LS-based search for iterating the path loss coefficient estimate in order to minimize the RSS estimate error. During this process, upper and lower bounds for the path loss coefficient are assumed to be known.

A more systematic RLS procedure to simultaneously estimate the target position and the environmental coefficient is proposed in [22], for a TOF setting. In this work, a linear parametric formulation of the estimation problem, having a separate lumped parameter vector for both the unknown position and the unknown environmental coefficient (average signal propagation velocity during TOF), is derived, and an RLS algorithm is designed for this formation. The RLS algorithm proposed in [22] is an automatic recursive algorithm not requiring any heuristic search, has guaranteed convergence properties and has tunable design coefficients for tuning transient performance trade-off between faster convergence and reduced sensitivity to measurement noise, as superior properties compared to [15,18]. Yet, since it solves the same essential simultaneous minimization problem, there is no loss of estimation accuracy.

In an attempt to separate the target position estimation and environmental coefficient estimation problems and to advance the estimation accuracy level, henceforth, we propose a geometric sensor array technique for the environmental coefficient estimation problem in this paper. In Section 3, we present this technique, which overcomes the aforementioned issues via static or instantaneous calculation based on certain geometric relations. The required additional cost is the use of triplets of sensors at the nodes of the WSN or the sensory mobile agent of interest in place of single sensor units. We later provide comparative simulations in Section 6, to demonstrate the performance of the methodology, compared to that of [22], noting the relation with the works [15,18] mentioned above.

3. The Coefficient Estimation Technique

Consider the 3D and Lateral Localization Problems defined in Section 2.1. These problems are defined assuming the availability of distance measurements $d_i[k]$, bypassing how $d_i[k]$ are produced processing the actual measurements of RSS or TOF by the sensor triplet $S = (S_1, S_2, S_3)$. In this section, we present our proposed geometric technique to produce the estimates of $d_i[k]$ using the available RSS or TOF measurements, which is equivalent to estimation of the path loss coefficient η for RSS or the signal permittivity coefficient ε, noting the model Equations (8)–(11).

In our design, we assert the geometric formation of the sensor triplet S to be maintained as rigid, such that S_1, S_2 and S_3 are aligned in the z direction with constant spacing $\bar{z}$, as depicted in Figure 1. That is, let the position of S_i at step k for $i = 1, 2, 3$ be given by

$$p_i[k] = [x[k], y[k], z_i[k]]^T \tag{12}$$

where $z_1[k] = z[k] + \bar{z}$, $z_2[k] = z[k]$ and $z_3[k] = z[k] - \bar{z}$ for some constant $\bar{z}$. Note that the spacing $\bar{z}$ is known, since it is a design constant.

At each step k, note that

$$d_1^2 - d_2^2 = (z + \bar{z} - z_T)^2 - (z - z_T)^2 = \bar{z}^2 + 2\bar{z}(z - z_T) \tag{13}$$

$$d_3^2 - d_2^2 = (z - \bar{z} - z_T)^2 - (z - z_T)^2 = \bar{z}^2 - 2\bar{z}(z - z_T) \tag{14}$$

Adding Equations (13) and (14), we obtain

$$d_1^2[k] - 2d_2^2[k] + d_3^2[k] = 2\bar{z}^2 \tag{15}$$

We propose the use of Equation (15) for the estimation of the environmental coefficient, $\eta[k]$ for RSS or $\bar{\varepsilon}[k]$ for TOF. The time dependence of these coefficients comes mainly from time variations in the position of S (and p_T if the target is not stationary) and, hence, the time variation in the environment between T and S.

More specifically, in the case of TOF, using Equation (11), Equation (15) can be rewritten as

$$\frac{t_{F1}^2[k]}{\bar{\varepsilon}[k]} - 2\frac{t_{F2}^2[k]}{\bar{\varepsilon}[k]} + \frac{t_{F3}^2[k]}{\bar{\varepsilon}[k]} = 2\bar{z}^2 \tag{16}$$

and, hence

$$\bar{\varepsilon}[k] = \frac{t_{F1}^2[k] - 2t_{F2}^2[k] + t_{F3}^2[k]}{2\bar{z}^2} \tag{17}$$

Similarly, in the case of RSS, using Equation (9), for each sensor S_i we have

$$\frac{P_T}{P_{Si}} = d_i^\eta \tag{18}$$

where P_{Si} denotes the signal power received by S_i. Hence, Equation (15) can be rewritten as

$$f(\bar{\eta}) = \zeta_1^{\bar{\eta}} - 2\zeta_2^{\bar{\eta}} + \zeta_3^{\bar{\eta}} = 2\bar{z}^2 \tag{19}$$

where $\zeta_i = \frac{P_T}{P_{Si}}$ and $\bar{\eta} = \frac{2}{\eta}$.

In the RSS case, although we cannot obtain a closed form solution for the coefficient η (or $\bar{\eta}$) similar to Equation (17), pre-calculated look-up tables for Equation (19) can be used (if preferred, together with some iterative accuracy fine-tuning methods) to solve Equation (19) for $\bar{\eta}$. A detailed formal study of such a design is out of the scope of this paper. However, for an *ad hoc* solution, one can observe that, in Equation (19), $\zeta_1, \zeta_2, \zeta_3, 2\bar{z}^2$ are known/measured numbers, and $\bar{\eta}$ is the only unknown. For a UAV tracking a ground target, we have $P_{S1} < P_{S2} < P_{S3} < P_T$, since $d_1 > d_2 > d_3$. Hence, we have $\zeta_1 > \zeta_2 > \zeta_3$. Further, typically, $2 \leq \eta \leq 5$. Therefore, $0.4 \leq \bar{\eta} \leq 1$. For typical settings, $f(\bar{\eta})$ in Equation (19) is monotonic with no local minimum in the interval $0.4 \leq \bar{\eta} \leq 1$. Applying a three-step grid search, with six equal intervals of a size of 0.1 in the first step and 10 equal intervals of sizes 0.01 and 0.001 in the second and third steps, respectively, $\bar{\eta}$ can be calculated with an error tolerance of ± 0.001. Such search is real-time implementable, and better results can be obtained using more steps.

4. Adaptive Source Localization Scheme

In this section, an RLS-based adaptive source localization scheme is designed to perform the target localization tasks of the 3D Localization Problem and the Lateral Localization Problem defined in Section 2.1. The adaptive localization scheme is to generate the estimate $\hat{p}_T[k]$ of p_T using the information of $p_i[k]$, which is obtained by the self-positioning system of the UAV together with the geometric relation Equation (12) and $d_i[k]$, which is obtained using the proposed technique in Section 3. Similarly to [22], the unknown target position vector, $\hat{p}_T$, is treated as constant in the design, and the influence of the drifting of the target is analysed later. The adaptive localization scheme is designed as an RLS algorithm with a forgetting factor [22,24] based on a linear parametric model, separately derived for each of the 3D Localization and Lateral Localization Problems, in the sequel.

4.1. 3D Localization

We first study solution of the 3D Localization Problem. To derive a linear parametric model for this problem, from Equations (4) and (5), we have

$$
\begin{aligned}
d^2[k] &= (p[k] - p_T)^T (p[k] - p_T) \\
&= \|p[k]\|^2 + \|p_T\|^2 - 2p_T^T p[k]
\end{aligned}
\tag{20}
$$

Evaluating Equation (20) at steps k and $k-1$ and taking the difference, we obtain

$$
d^2[k] - d^2[k-1] = \|p[k]\|^2 - \|p[k-1]\|^2 - 2p_T^T (p[k] - p[k-1])
\tag{21}
$$

which can be written in the linear parametric model form

$$
\zeta[k] = p_T^T \phi[k]
\tag{22}
$$

where $\phi[k]$ and $\zeta[k]$ are defined as

$$
\phi[k] = p[k] - p[k-1]
\tag{23}
$$

$$
\zeta[k] = \frac{1}{2} \left(\|p[k]\|^2 - \|p[k-1]\|^2 - (d^2[k] - d^2[k-1]) \right)
\tag{24}
$$

Based on the linear parametric model Equations (22)–(24), various estimators can be designed to produce the estimate $\hat{p}_T$ of p_T. Next, we design an on-line RLS estimator based on the parametric model Equations (22)–(24). Following the design procedure in [24], we obtain the following RLS adaptive law with a forgetting factor and parameter projection:

$$\hat{p}_T[k] = Pr\left(\hat{p}_T[k-1] + \Gamma[k]\phi[k]\epsilon[k]\right) \tag{25}$$

$$\epsilon[k] = \zeta[k] - \hat{p}_T^T[k-1]\phi[k] \tag{26}$$

$$\Gamma[k] = \frac{1}{\beta_f}\left(\Gamma[k-1] - \frac{\Gamma[k-1]\phi[k]\phi[k]^T\Gamma[k-1]}{\beta_f + \phi[k]^T\Gamma[k-1]\phi[k]}\right) \tag{27}$$

where $\epsilon[k]$ is the (measurable) output estimate error, $\Gamma[k]$ is the 3×3 dynamic adaptive gain matrix (called the covariance matrix), $0 < \beta_f < 1$ is the forgetting factor coefficient and $Pr(.)$ is the parameter projection operator used to satisfy $\hat{p}_{T3} = \hat{z}_T \in R_z$, where the target attitude is assumed to be known *a priori* to lie in the range $R_z = [z_{T,min}, z_{T,max}]$. Initial covariance matrix $\Gamma[0] = \Gamma_0$ is selected to be positive definite, which guarantees together with Equation (27) that $\Gamma[k]$ is positive definite for all k.

4.2. Lateral Localization

The Lateral Localization Problem is a relaxed form of the 3D Localization Problem, reducing its parametric model order by one. To derive the reduced order linear parametric model, we rewrite Equation (22) as

$$\zeta[k] = p_T^T\phi[k] = p_{TL}^T\phi_L[k] + z_T\phi_z[k] \tag{28}$$

where $\phi_L[k]$ and $\phi_z[k]$ are defined as

$$\phi_L[k] = p_L[k] - p_L[k-1] \tag{29}$$

$$\phi_z[k] = z[k] - z[k-1] \tag{30}$$

Using the available information of $\phi_z[k]$, we obtain the reduced order linear parametric model

$$\zeta_L[k] = p_{TL}^T\phi_L[k] \tag{31}$$

where $\zeta_L[k]$ is defined as

$$\zeta_L[k] = \frac{1}{2}\left(\|p[k]\|^2 - \|p[k-1]\|^2 - (d^2[k] - d^2[k-1])\right) - z_T(z[k] - z[k-1]) \tag{32}$$

In the design of on-line RLS estimator for the reduced order parametric model Equation (31), we do not need the parameter projection for the z-coordinate any more. Further, the model order is two instead of three. Hence, the RLS adaptive law for this case is redesigned as follows:

$$\hat{p}_{TL}[k] = \hat{p}_{TL}[k-1] + \Gamma[k]\phi_L[k]\epsilon_L[k] \tag{33}$$

$$\epsilon_L[k] = \zeta_L[k] - \hat{p}_{TL}^T[k-1]\phi_L[k] \tag{34}$$

$$\Gamma_L[k] = \frac{1}{\beta_f}\left(\Gamma_L[k-1] - \frac{\Gamma_L[k-1]\phi_L[k]\phi_L[k]^T\Gamma_L[k-1]}{\beta_f + \phi_L[k]^T\Gamma_L[k-1]\phi_L[k]}\right) \tag{35}$$

where $0 < \beta_f < 1$ is the forgetting factor coefficient, as before, and the adaptive gain (covariance) matrices $\Gamma_L[0] = \Gamma_{L0}$ and, hence, $\Gamma_L[k]$, for all $k > 0$ are 2×2 and positive definite.

4.3. Stability and Convergence of the Adaptive Localization Laws

The adaptive localization law Equations (25) and (33) are discrete-time RLS algorithms with a forgetting factor (and parameter projection). Such algorithms are studied in detail in [24]. It is also established there and in the references therein that, for 3D localization, if $\phi[k]$ is persistently exciting (PE), viz., if it satisfies

$$\lim_{K \to \infty} \lambda_{min} \left(\sum_{k=0}^{K} \phi[k]\phi^T[k] \right) = \infty \tag{36}$$

or if the 3×3 matrix

$$\sum_{k=k_0}^{k_0+l-1} \phi[k]\phi^T[k] - \alpha_0 l I \tag{37}$$

where I is the identity matrix and $\lambda_{min}(\cdot)$ denotes the minimum eigenvalue, is positive semi-definite for some $\alpha_0 > 0$, $l \geq 1$ and for all $k_0 \geq 1$, then $\hat{p}_T[k] \to p_T$ as $k \to \infty$. The geometric interpretation of the above PE condition is that the UAV is required to avoid converging to planar motion, i.e., to avoid $p[k]$ converging to a certain fixed 2D plane.

Similarly, for 2D localization, if $\phi_L[k]$ is PE, then $\hat{p}_{TL}[k] \to p_{TL}$ as $k \to \infty$; with the geometric interpretation that the UAV is required to avoid converging to linear motion, i.e., to avoid $p_L[k]$ converging to a certain fixed 1D line.

5. Adaptive Tracking Control

In this section, our proposed control scheme for the Lateral Tracking Problem defined in Section 2.1 is presented. Similarly to Section 4, the required information of $p_i[k]$ and $d_i[k]$ is obtained on-line using the self-positioning system of the UAV together with the geometric relation Equation (12) and the proposed environmental coefficient estimation technique in Section 3, respectively. The adaptive tracking control scheme is designed following a discrete-time version of the approach in [25].

The lateral tracking objective of Section 2.1 is considered as assigning a tracking control law to generate the lateral velocity $v_L[k]$ based on estimate $\hat{p}_T[k]$ of the unknown target position to achieve

$$\lim_{k \to \infty} d_L[k] = 0 \tag{38}$$

where

$$d_L[k] = \|p_L[k] - p_{LT}\| = \left(d^2[k] - (z[k] - z_T)^2 \right)^{1/2} \tag{39}$$

is available for measurement and, hence, can be used as a variable in the control law. In the design of the proposed adaptive target tracking control scheme, with the block diagram provided in Figure 2, we follow a certainty equivalence approach similar to [25], integrating three modular tools:

(i) The adaptive localization scheme of Section 4 to produce on-line estimate $\hat{p}_T[k]$ of target position p_T.
(ii) A motion control law fed by the estimate $\hat{p}_T[k]$ in place of unknown p_T to generate the lateral velocity $v_L[k]$, aiming to drive the estimated lateral distance $\|p_L[k] - \hat{p}_{LT}[k]\|$ to zero.
(iii) A low amplitude periodic auxiliary control signal to be augmented to the motion control law to satisfy the PE condition needed for guaranteeing the convergence of the location estimate $\hat{p}_T[k]$ to p_T.

In the design of Modules (ii) and (iii), we adopt and discretize the continuous-time adaptive target pursuit design in [25] to form the following discrete (augmented) motion control law:

$$v_L[k] = \frac{\hat{p}_{TL}[k] - \hat{p}_{TL}[k-1]}{T_s} - \beta_c(p_L[k] - \hat{p}_{TL}[k]) + f(d_L[k])v_a[k] \tag{40}$$

to be applied to the motion dynamics, using zero order hold, as

$$\dot{p}_L(t) = v_L(t) = v_L[k], \text{ for } kT_s \leq t < (k+1)T_s \tag{41}$$

144

where $\beta_c > 0$ is the proportional control gain,

$$v_a[k] = v_a(kT_s) = a_\sigma \begin{bmatrix} \sin a_\sigma kT_s \\ \cos a_\sigma kT_s \end{bmatrix} \qquad (42)$$

is the periodic auxiliary control signal with frequency a_σ, and $f(\cdot)$ is a strictly increasing and bounded function that is zero at zero and satisfies $f(d_F) \leq d_F$, $\forall d_F > 0$. The function $f(\cdot)$ is used to attenuate the auxiliary signal amplitude as the UAV gets closer to the target T.

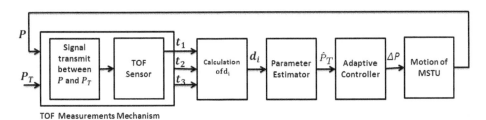

TOF Measurements Mechanism

Figure 2. Block diagram of the proposed adaptive lateral target tracking control scheme.

Based on the analysis provided in [25], we observe the properties of $v_a(t)$ summarized in the following lemma:

Lemma 1. *The auxiliary signal v_a defined in Equation (42) satisfies the following:*

(i) *There exist positive $T_1, \alpha_i > 0$, such that for all $t \geq 0$, there holds:*

$$\alpha_1 \|v_a(0)\|^2 I \leq \int_t^{t+T_1} v_a(\tau) v_a(\tau)^\top d\tau \leq \alpha_2 \|v_a(0)\|^2 I.$$

(ii) *For every $\theta \in \Re^2$, and every $t > 0$, there exists $t_1(t, \theta) \in [t, t+T_1]$, such that $\theta^\top v_a(t_1(t, \theta)) = 0$.*
(iii) *For all $t \geq 0$, $\|v_a(t)\| = \|v_a(0)\| = a_\sigma$.*
(iv) *There exists a design constant a_σ, such that the discrete time signal $v_a[k] = v_a(kT_s)$ is PE.*

Proof. (i) is a direct corollary of Lemma 8.1 of [26]. (ii) and (iii) are direct corollaries of, respectively, Theorem 5.1 and Lemma 3.1 of [25]. (iv) follows from (i) and (iii). □

Lemma 1 and classical arguments of the discretization of continuous-time dynamic systems lead to the validity of the boundedness and convergence results in Theorems 4.1 and 4.2 of [25] for our case, as well, as summarized in the following proposition:

Proposition 1. *Consider the closed-loop adaptive tracking control system composed of the adaptive law Equation (33), the motion control law Equation (40) and the motion dynamics Equation (41). Assume that $\beta_c > \bar{\sigma}'$ for the upper bound $\bar{\sigma}'$ defined in Lemma 1. Then, there exists a sufficiently small sampling time T_s, such that all of the closed-loop signals are bounded and $p_L[k]$ asymptotically converges to p_{TL}.*

6. Simulations

In this section, we perform simulation testing of the proposed adaptive localization and target tracking schemes. First, we consider a scenario where the UAV is equipped with a TOF-based range sensor triplet. In all of the simulations, the actual average permittivity is taken to be $\varepsilon_{ave} = 5$ considering solid objects and air humidity in the signal propagation paths for TOF sensors. The vertical spacing for the sensor triplet $S = (S_1, S_2, S_3)$ is chosen as $\bar{z} = 10$ cm, and the common sampling time is selected as $T_s = 1$ s.

The task of the UAV is to estimate the location p_T of (and track) a certain target T. For this task, the UAV uses the localization algorithm Equation (25), and in order to guarantee estimation convergence per the discussions at the end of Section 4, it follows a PE path, *i.e.*, a path satisfying ϕ to be PE. As such a PE path, we consider the following path, whose x and y coordinate components are plotted in Figure 3:

$$x(t) = 500\cos(0.1t) + 50 \text{ m} \tag{43}$$

$$y(t) = 300\cos(0.2t) \text{ m} \tag{44}$$

$$z(t) = 5\sin(0.1t) + 39 \text{ m} \tag{45}$$

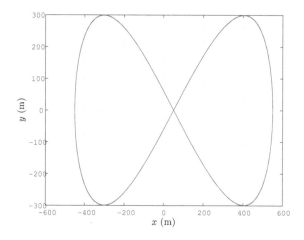

Figure 3. Lateral trajectory $(x(t), y(t))$ (m) (or $(x[k], y[k])$) of the UAV.

We consider the following design parameter selections for the localization algorithm Equation (25):

$$\beta_f = 0.9 \tag{46}$$

$$\Gamma[0] = I \tag{47}$$

$$\hat{p}_T[0] = [0, 0, 0]^T \text{ m} \tag{48}$$

We consider two cases for the target position p_T in the following two subsections. As a continuation of the discussion at the end of Section 2.4, we compare the results using our prosed algorithm with the results using the simultaneous location and permittivity coefficient estimation scheme of [22] (for the same setting). As opposed to the proposed approach in Section 3, which gives very accurate results instantaneously, in the simulation of the scheme of [22], permittivity coefficient estimation is done recursively together with the target location estimation. The initial permittivity coefficient estimate for this recursive estimation is chosen as $\hat{\varepsilon}_{ave}[k] = 10$.

6.1. Stationary Target Localization

First, we consider a stationary target located at

$$p_T = [100, 75, 8]^T \text{ m} \tag{49}$$

The localization results for this case are plotted in Figures 4 and 5. In Figure 4, we compare the results using our prosed algorithm with the results using the simultaneous location and permittivity coefficient estimation scheme of [22] (for the same setting).

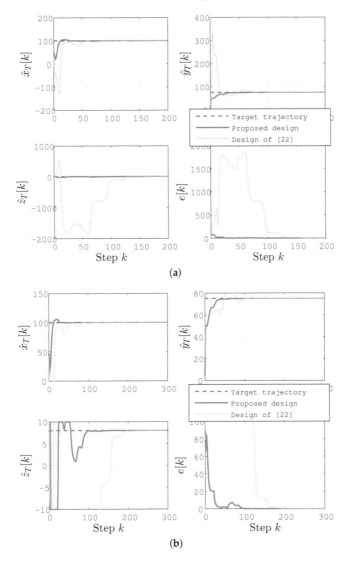

Figure 4. Location estimate $\hat{p}_T[k]$ and estimation error $e[k] = \|\hat{p}_T[k] - p_T\|$ (m) for the stationary target case. (b) The scaled version of (a) to provide the details of the convergence characteristics for the proposed scheme.

It is clearly seen in these figures that, using the prosed design, all of the coordinates of the position estimate $\hat{p}_T[k]$ rapidly converge to their actual values, leading the estimation error $e[k] = \|\hat{p}_T[k] - p_T\|$ to converge to zero. The estimates converge significantly faster than those using the design of [22], with significantly smaller overshoot/undershoot. One can further enhance the performance of localization by fine-tuning the design parameters given above.

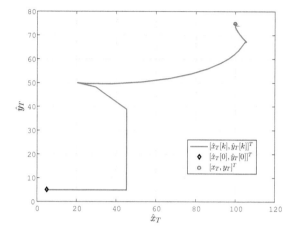

Figure 5. Lateral coordinate estimates $(\hat{x}_T[k], \hat{y}_T[k])$ (m) for the stationary target case.

6.2. Drifting Target Localization

As a second scenario, we consider a slowly drifting target T with position

$$x_T(t) = 0.1t + (2\sin(0.05t) + 100) \text{ m}$$
$$y_T(t) = 0.05t + (2\sin(0.05t) + 75) \text{ m}$$
$$z_T(t) = (0.5\sin(0.01t) + 8) \text{ m}$$

and hence, velocity

$$V_T \;=\; [(0.1 + 0.1\cos(0.05t)), (0.05 + 0.1\cos(0.05t)) \\ (0.005\cos(0.01t))]^T \text{ m/s}$$

The localization results for this case are plotted in Figures 6 and 7.

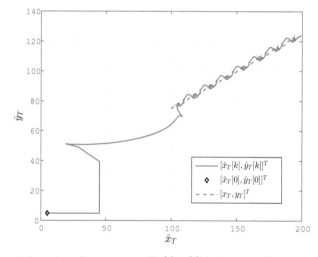

Figure 6. Lateral coordinate estimates $(\hat{x}_T[k], \hat{y}_T[k])$ (m) for the drifting target case.

As can be seen in these figures, convergence characteristics are comparable to the stationary target case; however, due to the motion of the target, perfect convergence is impossible as long as the velocity v_T of the target is not known *a priori*. The estimation accuracy, however, is significantly better, and the estimates converge significantly faster than those using the design of [22], with significantly smaller overshoot/undershoot.

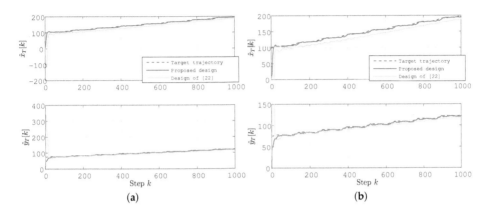

Figure 7. Location estimate $\hat{p}_T[k]$ for the drifting target case. (**b**) The scaled version of (**a**) to provide the details of the convergence characteristics for the proposed scheme.

6.3. Drifting Target Tracking

Next, we consider the tracking problem for the scenario considered in the previous subsection. The target motion is the same. The motion control law designs are selected as $\beta_c = 3$, $a_\sigma = 0.01$. The simulation results are shown in Figure 8. We can easily see that the $p_L[k]$ values converge to $\hat{p}_{TL}[k]$, as well as $p_{TL}[k]$ values. The simulation results demonstrate that the tracking task of the target is achieved. Better tracking performance can be obtained by fine-tuning the adaptive localization and target tracking control design terms.

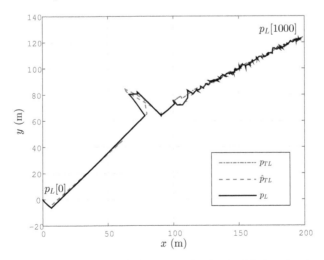

Figure 8. Lateral tracking control for a drifting target for the simplified motion dynamics model Equation (41).

To further examine the ignored actuator dynamics and disturbance effects, the simulation above is performed for the following modified version of the motion dynamics model Equation (41):

$$\dot{p}_L = \frac{1}{\tau_a s + 1}[v_L] + w_v \tag{50}$$

$$v_L(t) = v_L[k], \text{ for } kT_s \le t < (k+1)T_s \tag{51}$$

where $\frac{1}{\tau_a s + 1}$ is the transfer function of the actuator dynamics with time coefficient considered to be $\tau_a = 0.2$ (s), and w_v is a band limited white noise with power 0.1, representing further motion control disturbances. The simulation results shown in Figure 9 demonstrate that the results are comparable to those in Figure 8.

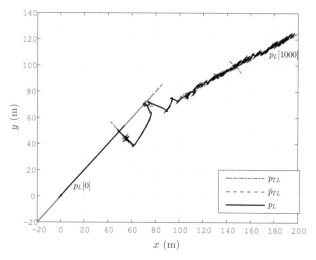

Figure 9. Lateral tracking control for a drifting target for the detailed motion dynamics model Equation (50).

7. Conclusions

In this paper, a geometric cooperative technique has been proposed to instantaneously estimate permittivity and path loss coefficients in electromagnetic signal source and reflector localization and tracking tasks, focusing on environmental monitoring applications. The details of the technique are provided for RSS and TOF-based range sensor settings. The use and performance of the technique are analysed and demonstrated on its integration with a discrete time RLS-based adaptive localization and target tracking control schemes. A set of UAV-based target localization and tracking simulation scenarios are provided to demonstrate the effectiveness of the integration of the adaptive localization and tracking schemes and the proposed coefficient estimation technique. The proposed technique involves only static instantaneous calculation based on a certain geometric relation and, hence, provides a computationally efficient way to solve the localization problems in environments with unknown permittivity and path loss coefficients, compared to the other relevant techniques proposed in the literature.

Ongoing and future follow up research directions include more formal analysis of various localization and tracking schemes using the proposed coefficient estimation technique, applications in other domains, such as biomedical monitoring [27], and real-time implementation and experimental testing. Implementation and testing of such a system can be considered in two layers, the hardware layer and the software layer. For the hardware layer implementation, a sensor triplet unit, as described in Sections 2.2 and 2.3, together with onboard CPU and communication (to broadcast the on-line localization/tracking information) units at the mobile agent, which can be mounted on the surveillance UAV. The software layer implementation will include the low level coding of the proposed algorithms, which are all real-time implementable, and further embedded software for the CPU interface with sensor and communication units. The setup for this architecture can be constructed using standard hardware units and software, such as those used for the experiments in [28,29].

Acknowledgments: The authors thank Fatma Kiraz for her assistance in the literature review.

Author Contributions: Barış Fidan proposed the base sensor array technique, as well as the general structure of the localization and tracking schemes and managed the manuscript preparation. Both authors contributed to the detailed design, fine-tuning and mathematical analysis of the schemes, the interpretation of the simulation test results and the preparation of the manuscript.

Conflicts of Interest: The authors declare no conflict of interest.

References

1. Mao, G.; Fidan, B. *Localization Algorithms and Strategies for Wireless Sensor Networks*; IGI Global—Information Science Publishing: Hershey, PA, USA, 2009.
2. *Wireless Body Area Networks Technology, Implementation and Applications*; Yüce, M.R., Khan, J.Y., Eds.; Pan Stanford Publishing: Singapore, 2012.
3. Lomax, A.S.; Corso, W., Etro, J.F. Employing unmanned aerial vehicles (UAVs) as an element of the Integrated Ocean Observing System. *MTS/IEEE Proc.* **2005**, *1*, 184–190.
4. Susca, S.; Bullo, F.; Martinez, S. Monitoring environmental boundaries with a robotic sensor network. *IEEE Trans. Control Syst. Technol.* **2008**, *16*, 288–296.
5. Arnold, T.; de Biasio, M.; Fritz, A.; Leitner, R. UAV-based multispectral environmental monitoring. In Proceedings of the 2010 IEEE Sensors Conference, Kona, HI, USA, 1–4 November 2010; pp. 995–998.
6. Wang, S.; Berentsen, M.; Kaiser, T. Signal processing algorithms for fire localization using temperature sensor arrays. *Fire Saf. J.* **2005**, *40*, 689–697.
7. Liu, Z. A supervisory approach for hazardous chemical source localization. In Proceedings of the 2013 IEEE International Conference on Mechatronics and Automation (ICMA), Takamatsu, Japan, 4–7 August 2013; pp. 476–481.
8. Howse, J.W.; Ticknor, L.O.; Muske, K.R. Least squares estimation techniques for position tracking of radioactive sources. *Automatica* **2011**, *37*, 1727–1737.
9. Brennan, S.M.; Mielke, A.M.; Torney, D.C.; Maccabe, A.B. Radiation detection with distributed sensor networks. *IEEE Comput.* **2004**, *37*, 57–59.
10. Campbell, M. *Sensor Systems for Environmental Monitoring*; Springer Science and Business Media: Berlin, Germany, 2012.
11. Lin, J.; Xiao, W.; Lewis, F.L.; Xie, L. Energy-efficient distributed adaptive multisensor scheduling for target tracking in wireless sensor networks. *IEEE Trans. Instrum. Meas.* **2009**, *58*, 1886–1896.
12. Yoon, S.H.; Ye, W.; Heidemann, J.; Littlefield, B.; Shahabi, C. SWATS: Wireless sensor networks for steam-flood and water-flood pipeline monitoring. *IEEE Netw.* **2011**, *25*, 50–56.
13. Liu, Y.; Yuan, X.; Chen, Y.; Lin, Y. Dynamic localization research for the fire rescue system. *Procedia Eng.* **2011**, *15*, 3282–3287.
14. Ge, Q.; Wen, C.; Duan S. Fire localization based on range-range-range model for limited interior space. *IEEE Trans. Instrum. Meas.* **2014**, *63*, 2223–2237.
15. Wang, G.; Chen, H.; Li, Y.; Jin, M. On received-signal-strength based localization with unknown transmit power and path loss exponent. *IEEE Wirel. Commun. Lett.* **2012**, *1*, 536–539.
16. Gezici, S. A survey on wireless position estimation. *Wirel. Pers. Commun.* **2008**, *44*, 263–282.
17. Çamlıca, A.; Fidan, B.; Yavuz, M. Implant localization in the human body using adaptive least square based algorithm. In Proceedings of the ASME 2013 International Mechanical Engineering Congress and Exposition, San Diego, CA, USA, 15–21 November 2013.
18. Salman, N.; Kemp, A.H.; Ghogho, M. Low complexity joint estimation of location and path-loss exponent. *IEEE Wirel. Commun. Lett.* **2012**, *1*, 364–367.

19. Mao, G.; Anderson, B.D.O.; Fidan, B. Path loss exponent estimation for wireless sensor network localization. *Comput. Netw.* **2007**, *51*, 2467–2483.

20. So, H.C. Source localization: Algorithms and analysis. In *Handbook of Position Location: Theory, Practice and Advances*; Wiley-IEEE Press: Hoboken, NJ, USA, 2011; pp. 813–836.

21. *Handbook of Position Location: Theory, Practice and Advances*; Zekavat, S.A., Buehrer, R.M., Eds.; Wiley-IEEE Press: Hoboken, NJ, USA, 2011.

22. Fidan, B.; Çamlıca, A.; Güler, S. Least-squares-based adaptive target localization by mobile distance measurement sensors. *Int. J. Adapt. Control Signal Proc.* **2015**, *29*, 259–271.

23. Pahlavan, K.; Li, X.; Makela, J.P. Indoor geolocation science and technology. *IEEE Commun. Mag.* **2002**, *40*, 112–118.

24. Ioannou, P.A.; Fidan, B. *Adaptive Control Tutorial*; Society for Industrial and Applied Mathematics (SIAM): Philadelphia, PA, USA, 2006.

25. Fidan, B.; Dasgupta, S.; Anderson, B.D.O. Adaptive range measurement-based target pursuit. *Int. J. Adapt. Control Signal Proc.* **2013**, *27*, 66–81.

26. Shames, I.; Dasgupta, S.; Fidan, B.; Anderson, B.D.O. Circumnavigation from distance measurements under slow drift. *IEEE Trans. Autom. Control* **2012**, *57*, 889–903.

27. Umay, I.; Fidan, B.; Yüce, M.R. Wireless capsule localization with unknown path loss coefficient and permittivity. In Proceedings of the 2015 IEEE/RAS International Conference on Advanced Robotic, Istanbul, Turkey, 27–31 July 2015; pp. 224–229.

28. Whitehouse, K.; Karlof, C.; Culler, D. A practical evaluation of radio signal strength for ranging-based localization. *ACM SIGMOBILE Mob. Comput. Commun. Rev.* **2007**, *11*, 41–52.

29. Fabresse, F.R.; Caballero, F.; Ollero, A. Decentralized simultaneous localization and mapping for multiple aerial vehicles using range-only sensors. In Proceedings of the 2015 IEEE International Conference on Robotics and Automation, Seattle, WA, USA, 26–30 May 2015; pp. 6408–6414.

Article

Vision-Based Detection and Distance Estimation of Micro Unmanned Aerial Vehicles

Fatih Gökçe *, Göktürk Üçoluk, Erol Şahin and Sinan Kalkan

Department of Computer Engineering, Middle East Technical University, Üniversiteler Mahallesi, Dumlupınar Bulvarı No. 1, 06800 Çankaya Ankara, Turkey; ucoluk@ceng.metu.edu.tr (G.Ü.); erol@ceng.metu.edu.tr (E.Ş.); skalkan@ceng.metu.edu.tr (S.K.)
* Author to whom correspondence should be addressed; fgokce@ceng.metu.edu.tr;
  Tel.: +90-312-210-5545; Fax: +90-312-210-5544.

Academic Editor: Felipe Gonzalez Toro
Received: 3 June 2015 / Accepted: 31 August 2015 / Published: 18 September 2015

Abstract: Detection and distance estimation of micro unmanned aerial vehicles (mUAVs) is crucial for (i) the detection of intruder mUAVs in protected environments; (ii) sense and avoid purposes on mUAVs or on other aerial vehicles and (iii) multi-mUAV control scenarios, such as environmental monitoring, surveillance and exploration. In this article, we evaluate vision algorithms as alternatives for detection and distance estimation of mUAVs, since other sensing modalities entail certain limitations on the environment or on the distance. For this purpose, we test Haar-like features, histogram of gradients (HOG) and local binary patterns (LBP) using cascades of boosted classifiers. Cascaded boosted classifiers allow fast processing by performing detection tests at multiple stages, where only candidates passing earlier simple stages are processed at the preceding more complex stages. We also integrate a distance estimation method with our system utilizing geometric cues with support vector regressors. We evaluated each method on indoor and outdoor videos that are collected in a systematic way and also on videos having motion blur. Our experiments show that, using boosted cascaded classifiers with LBP, near real-time detection and distance estimation of mUAVs are possible in about 60 ms indoors (1032×778 resolution) and 150 ms outdoors (1280×720 resolution) per frame, with a detection rate of 0.96 F-score. However, the cascaded classifiers using Haar-like features lead to better distance estimation since they can position the bounding boxes on mUAVs more accurately. On the other hand, our time analysis yields that the cascaded classifiers using HOG train and run faster than the other algorithms.

Keywords: UAV; micro UAV; vision; detection; distance estimation; cascaded classifiers

1. Introduction

Advances in the development of micro unmanned aerial vehicles (mUAVs), which are UAVs less than 5 kg [1], have led to the availability of highly capable, yet inexpensive flying platforms. This has made the deployment of mUAV systems in surveillance, monitoring and delivery tasks a feasible alternative. The use of mUAVs in monitoring the state of forest fires where the mission spreads over a large region and flying over the fire is dangerous [2] or in delivering packages in urban areas [3] as a faster and less expensive solution is being explored. Moreover, the widespread interest in the public has also resulted in mUAVs, which are often referred to as drones, showing up in places, such as the White House, where conventional security measures are caught unprepared [4] or in traffic accidents or in fires where the presence of mUAVs, flown by hobbyists to observe the scene, posed a danger to police and fire-fighter helicopters and resulted in delays in their deployment [5]. In all of these cases, the need for the automatic detection and distance estimation of mUAVs, either from the ground or from a flying platform (which can be another mUAV or a helicopter) against a possibly cluttered background is apparent.

The main objective of our study is the evaluation of vision as a sensor for detection and distance estimation of mUAVs. This problem poses a number of challenges: First, mUAVs are small in size and often do not project a compact and easily segmentable image on the camera. Even in applications where the camera is facing upwards and can see the mUAV against a rather smooth and featureless sky, the detection poses great challenges. In multi-mUAV applications where each platform is required to sense its neighbors and in applications where the camera is placed on a pole or on a high building for surveillance, the camera is placed at a height that is the same or higher than the incoming mUAV, and the image of the mUAV is likely to be blended against feature-rich trees and buildings, with possibly other moving objects in the background, so the detection and distance estimation problem becomes challenging. Moreover, in multi-mUAV applications, the vibration of the platform, as well as the size, power, weight and computational constraints posed on the vision system also need to be considered.

Within this paper, we report our work towards the development of an mUAV detection and distance estimation system. Specifically, we have created a system for the automatic collection of data in a controlled indoor environment, proposed and implemented the cascaded approach with different features and evaluated the detection performance and computational load of these approaches with systematic experiments on indoor and outdoor datasets.

For the cooperative operation of mUAVs and for also sensing and avoiding purposes, relative localization in 3D space, which requires the estimation of relative bearing, elevation and distance, is critical. Relative bearing and elevation can be estimated easily by detecting an mUAV in an image. However, for distance estimation, additional computation is needed. Due to the scale estimation problem in monocular vision and the excessive variability of the possible appearances of an mUAV for the same distance, the problem is challenging. Considering the demand for the distance information, we also developed a method to estimate the relative distance of a detected mUAV by utilizing the size of the detection window. We have performed indoor experiments to evaluate the performance of this approach in terms of both distance and time-to-collision estimation.

2. Related Studies

In this section, we discuss the relevant studies in three parts. In the first part, general computer vision approaches related to object detection and recognition are reviewed. The second and third parts summarize the efforts in the robotics literature to detect and localize mUAVs using computer vision and other modalities, respectively.

2.1. Object Detection and Recognition Approaches with Computer Vision

In computer vision and pattern recognition (CVPR), object detection and recognition has been extensively studied (see [6,7] for comprehensive reviews), with applications ranging from human detection, face recognition to car detection and scene classification [8–13]. The approaches to detection and recognition can be broadly categorized into two: keypoint-based approaches, and hierarchical and cascaded approaches.

2.1.1. Keypoint-Based Approaches

In keypoint-based methods, CVPR usually detects salient points, called interest points or keypoints, in the "keypoint detection" phase. In this phase, regions in the image that are likely to have important information content are identified. The keypoints should be as distinctive as possible and should be invariant, *i.e.*, detectable under various transformations. Popular examples of keypoint detectors include fast corner detection (FAST) [14,15], Harris corner detection (HARRIS) [16], maximally stable extremal region extractor (MSER) [17] and good features to track (GFTT) [18] (see [19] for a survey of local keypoint detectors).

In the next phase of keypoint-based approaches, intensity information at these keypoints is used to represent the local information in the image invariant to transformations, such as rotation, translation, scale and illumination. Examples of the keypoint descriptors include speeded-up robust features (SURF) [20], scale-invariant feature transform (SIFT) [21], binary robust independent elementary features (BRIEF) [22], oriented FAST and rotated BRIEF (ORB) [23], binary robust invariant scalable keypoints (BRISK) [24] and fast retina keypoint (FREAK) [25].

Extracted features are usually high dimensional (e.g., 128 in the case of SIFT, 64 in SURF, *etc.*), which makes it difficult to use distributions of features for object recognition or detection. In order to overcome this difficulty, the feature space is first clustered (e.g., using k-means), and the cluster labels are used instead of high-dimensional features for, e.g., deriving histograms of features for representing objects. This approach, called the bag-of-words (BOW) model, has become very popular in object recognition (see, e.g., [26–28]). In BOW, histograms of cluster labels are used to train a classifier, such as a naive Bayes classifier or a support vector machine (SVM) [29], to learn a model of the object.

In the testing phase of BOW, a window is slid over the image, and for each position of the window in the image, a histogram of the cluster labels of the features in that window is computed and tested with the trained classifier. However, the scale of the window imposes a severe limitation on the size of the object that can be detected or recognized. This limitation can be overcome to only a certain extent by sliding windows of different scales. However, this introduces a significant computational burden, making it unsuitable for real-time applications.

2.1.2. Hierarchical and Cascaded Approaches

A better approach in CVPR is to employ hierarchical and cascaded models into recognition and detection. In such approaches, shape, texture and appearance information at different scales and complexities is processed, unlike the regular keypoint-based approaches. Processing at multiple levels has been shown to perform better than the alternative approaches (see, e.g., [30]).

In hierarchical approaches, such as the deep learning approaches [31], features of varying scale are processed at each level: in lower levels of the hierarchy, low-level visual information, such as gradients, edges, *etc.*, are computed, and with increasing levels in the hierarchy, features of the lower levels are combined, yielding corners or higher-order features that start to correspond to object parts and to objects. At the top of the hierarchy, object categories are represented hierarchically. For detecting an object in such an approach, the information needs to pass through all of the hierarchies to be able to make a decision.

An alternative approach is to keep a multi-level approach, but prune processing as early as possible if a detection does not seem likely. Such cascaded approaches, which are inspired, especially, from ensemble learning approaches [32] in machine learning, perform quickly, but coarse detection at early stages and only candidates passing earlier stages pass on to higher stages where finer details undergo computationally-expensive detailed processing. This way, these approaches benefit from speed ups by processing candidate regions that are highly likely to contain a match [33]. A prominent study, which also forms the basis of this study, is the approach by Viola and Jones [10,34], which builds cascades of classifiers at varying complexities using Haar-like features and adopting the AdaBoost learning procedure [35]. Viola and Jones [10,34] applied their method to face detection and demonstrated high detection rates at high speeds. The approach was later extended to work with local binary patterns for face recognition [36] and histogram of oriented gradients for human detection [37], which are more descriptive and faster to compute than Haar-like features.

2.2. Detection and Localization of mUAVs with Computer Vision

With advances in computational power, vision has become a feasible modality for several tasks with mUAVs. These include fault detection [38], target detection [39] and tracking [40], surveillance [41,42], environmental sensing [43], state estimation and visual navigation [44–49], usually combined with other sensors, such as GPS, an inertial measurement unit (IMU), an altimeter or a magnetometer.

Recently, vision has been used for mUAV detection and localization by recognizing black-and-white special markers placed on mUAVs [50,51]. In these studies, circular black-white patterns are designed and used for detection and distance estimation, achieving estimation errors less than 10 cm in real time. However, in some applications where it is difficult to place markers on mUAVs, such approaches are not applicable, and a generic vision-based detection system, such as the one proposed in the current article, is required.

In [52], leader-follower formation flight of two quadrotor mUAVs in an outdoor environment is studied. Relative localization is obtained via monocular vision using boosted cascaded classifiers of Haar-like features for detection and Kalman filtering for tracking. In order to estimate distance, they used the width of the leader with the camera model. They tested their vision-based formation algorithm in a simulation and with real mUAVs. Results for only the real-world experiments are provided where the follower tries to keep a 6-m distance from the leader flying up to a speed of 2 m/s. Their results present only the relative distance of the mUAVs during a flight where the distance information is obtained probably (not mentioned clearly) from GPS. Although they claim that the tracking errors converge to zero, their results indicate that the errors always increase while the leader has a forward motion. Only when the leader becomes almost stationary after 35 s of the total 105 s flight do the errors start to decrease.

In [53], the 2D relative pose estimation problem is studied by extending the approach in [52]. Once the mUAV is detected via a cascaded classifier, its contours are extracted, and for these contours, the best matching image from a set of previously collected images for different view angles is determined. Then, the orientation is estimated by computing the best fitting affine transformation via least squares optimization. Their experimental results are not sufficient to deduce the performance of pose estimation. Furthermore, they use the estimated pose to enhance the relative distance estimation method applied in [52]. According to the results given for only 50 frames, there seems to be an improvement; however, the error is still very high (up to three meters for a 10-m distance with a variance of 1.01 m), and GPS is taken as the ground truth whose inherent accuracy is actually not very appropriate for such an evaluation.

Both studies [52,53] mentioned above use boosted cascaded classifiers for mUAV detection; however, they provide no analysis about the detection and computational performance of the classifiers. The methods are tested only outdoors, and the results for the tracking and pose estimation are poor for evaluating the performances of the methods. They use Haar-like features directly without any investigation. Moreover, no information is available about the camera and processing hardware used. The detection method is reported to run at 5 Hz.

In [54], the collision detection problem for fixed-winged mUAVs is studied. A morphological filter based on the close-minus-open approach is used for the preprocessing stage. Since morphological filters assume a contrast difference between the object and the background, once the image is preprocessed, the resulting candidate regions should be further inspected to get the final estimation. This is very crucial, as the morphological filters produce a large amount of false positives, which have to be eliminated. For this purpose, they combined the morphological filtering stage with two different temporal filtering techniques, namely, Viterbi based and hidden Markov model (HMM) based. The impact of image jitter and the performance of target detection are analyzed by off-board processing of video images on a graphical processing unit (GPU). For jitter analysis, videos recorded using a stationary camera are used by adding artificial jitter at three increasing levels, low, moderate and extreme. Both temporal filtering techniques demonstrate poor tracking performances in the case of extreme jitter where inter-frame motion is greater than four pixels per frame. Some failure periods are also observed for the HMM filter in the moderate jitter case. Target detection performance experiments are performed on videos captured during three different flights with an onboard camera mounted on a UAV. Two of these include head-on maneuvers, and in the third one, UAVs fly at right angles to each other. A detection distance between 400 and 900 m is reported allowing one to estimate a collision before 8–10 s of the impact.

There are also studies for detecting aircraft via vision [55–57]. Although we include mainly the literature proposed for mUAVs in this section, these studies are noteworthy, since they are potentially useful for mUAVs, as long as the size, weight and power (SWaP) constraints of mUAVs are complied with. In [55], aircraft detection under the presence of heavily cluttered background patterns is studied for collision avoidance purposes. They applied a modified version of boosted cascaded classifiers using Haar-like features for detection. Temporal filtering is also integrated with the system to reduce false positives by checking the previous detections around a detection before accepting it as valid. Their method does not estimate the distance. Experimental results presented on videos recorded via a camera mounted on an aircraft and having a collision course and crossing scenarios indicate a detection rate of around 80% with up to 10 false positives per frame. No distance information is available between target and host aircraft. Looking at the images, the distance seems to be on the order of some hundred meters. The performance of the system in close distances is also critical, which is not clearly understood from their experiments. They report that their method has a potential of real-time performance; however, no information is available about the frame size of the images and the processing hardware.

The work in [56,57] presents another approach for aircraft detection for sensing and avoiding purposes. They propose a detection method without distance estimation consisting of three stages, which are: (1) morphological filtering; (2) SVM-based classification of the areas found by Stage 1; and (3) tracking based on the similarity likelihoods of matching candidate detections. They tested their method on videos recorded using stationary cameras of various imaging sensor, lens and resolution options. These videos include aircraft flying only above the horizon; therefore, the background patterns are less challenging than the below horizon case, which is not investigated in the study. A detection rate of 98% at five statute miles with one false positive in every 50 frames is reported with a running time of 0.8 s for a four megapixel frame.

2.3. Detection and Localization of mUAVs with Other Modalities

There are many alternative sensing methods that can be used for relative localization among mUAVs. One widely-used approach is the Global Positioning System (GPS): in a cooperative scenario, each mUAV can be equipped with GPS receivers and shares its position with other agents [58]. However, GPS signals could be affected by weather, nearby hills, buildings and trees. The service providers may also put limitations on the availability and accuracy of the GPS signals. Moreover, the accuracy of GPS signals is not sufficient for discriminating between close-by neighboring agents unless a real-time kinematic GPS (RTK-GPS) system is used [59]. However, RTK-GPS systems require additional base station unit(s) located in the working environment.

Alternative to GPS, modalities, such as (1) infrared [60–65]; (2) audible sound signals [66,67] and (3) ultrasound signals [68–70] can be used; however, they entail certain limitations on the distance between the mUAVs and the environments in which they can perform detection. The infrared tends to be negatively affected by sunlight, hence not very suitable for outdoor applications. Sound can be a good alternative; yet, when there are close-by agents, interference becomes a hindrance for multi-mUAV systems, and audible sound signals are prone to be affected by external sound sources. Multipath signals can disturb the measurements severely. The speed of the sound limits the achievable maximum update rate of the system. Moreover, current ultrasound transducers provide limited transmission and reception beam angles, complicating the design of a system with omni-directional coverage.

An alternative modality commonly used by planes is radio waves (*i.e.*, radar). The limitation with radar, however, is that the hardware is too heavy and expensive to place on an mUAV. Recently, there has been an effort to develop an X-band radar to be used on mUAVs [71,72].

Ultra-wide band (UWB) radio modules, which allow two-way time-of-flight and time-difference-of-arrival measurements, and the signal strength between radio frequency (RF) devices could be thought of as other alternatives. However, both techniques need anchor units placed in the environment. The use of UWB modules without beacon units could be considered an aiding method to enhance the performance of localization systems that depend on other modalities. Signal strength between RF devices does not allow one to design an accurate system due to the uncertainties arising from antenna alignment and the effects of close objects.

2.4. The Current Study

As reviewed above, there is an interest in detection and distance estimation of aerial vehicles via vision for various purposes, such as cooperation and collision avoidance. Table 1 summarizes these studies in terms of various aspects. Looking at this comparison table and the above explanations, our study fills a void with regard to the comprehensive and systematic analysis of cascaded methods with videos, including very complex indoor and outdoor scenes, providing also an accurate distance estimation method.

Table 1. Comparison of the studies on the visual detection of aerial vehicles.

Study	Vehicle	Detection Method	Detection Performance	Motion Blur	Training Time	Testing Time	Background Complexity	Environment	Distance Estimation
Lin *et al.*, 2014	mUAV	Boosted cascaded classifiers with Haar-like features	No	No	No	No	Medium	Outdoor	Yes (low accuracy)
Zhang *et al.*, 2014	mUAV	Boosted cascaded classifiers with Haar-like features	No	No	No	No	Medium	Outdoor	Yes (low accuracy)
Petridis *et al.*, 2008	Aircraft	Boosted cascaded classifiers with Haar-like features	Yes	No	No	No	High	Outdoor	No
Dey *et al.*, 2009; 2011	Aircraft	Morphological filtering	Yes	No	NA	No	Low	Outdoor	No
Lai *et al.*, 2011	mUAV	Morphological filtering	Yes	Yes	NA	Yes	High	Outdoor	No
Current study	mUAV	Boosted cascaded classifiers with Haar-like, LBP and HOG features	Yes	Yes	Yes	Yes	High	Indoor and Outdoor	Yes

The main contribution of the article is a systematic analysis of whether an mUAV can be detected using a generic vision system under different motion patterns both indoors and outdoors. The tested indoor motion types include lateral, up-down, rotational and approach-leave motions that are precisely controlled using a physical platform that we constructed for the article. In the outdoor experiments, we tested both calm and agile motions that can also include a moving background. Moreover, the effect of motion blur is also analyzed in a controlled manner. To the best of our knowledge, this is the first study that presents a comprehensive and systematical investigation of vision for detection and distance estimation of mUAVs without special requirements, e.g., the markers used by [50,51].

Besides detecting the quadrotor, our study also integrates a distance estimation method in which a support vector regressor estimates the distance of the quadrotor utilizing the dimensions of the bounding box estimated in the detection phase.

Since it is faster than the alternatives and does not require a large training set, we use cascaded classifiers for detection, which consist of multiple (classification) stages with different complexities [10,34,36,37]. The early (lower) stages of the classifier perform very basic checks to eliminate irrelevant windows with very low computational complexity. The windows passing the lower stages are low in number and undergo heavier computations to be classified as mUAV or background. In order to train a cascaded classifier, we use different feature types proposed in the literature and compare their performances.

3. Methods

In this section, we describe the cascaded detection methods used in this paper; namely, the method of Viola and Jones [10,34] and the ones that extend it [36,37].

3.1. A Cascaded Approach to mUAV Detection

Cascaded classifiers are composed of multiple stages with different processing complexities [10,34,73]. Instead of one highly complex single processing stage, cascaded classifiers incorporate multiple stages with increasing complexities, as shown in Figure 1.

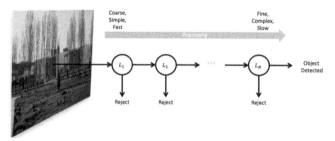

Figure 1. The stages of processing in a cascaded approach. At each stage, a decision to reject or to continue processing is made. If all stages pass, then the method declares detection of the object.

The early stages of the classifier have lower computational complexities and are applied to the image to prune most of the search space quickly. The regions classified as mUAV by one stage of the classifier are passed to the higher stages. As the higher level of stages are applied, the classifier works on a smaller number of regions at each stage to identify them as mUAV or background. At the end of the last stage, the classifier returns the regions classified as mUAV.

In the method proposed by [10,34], which relies on using AdaBoost learning, combinations of weak classifiers are used at each stage to capture an aspect of the problem to be learned. A weak classifier, $h_f(\mathbf{x})$, simply learns a linear classification for feature f with a threshold θ_f:

$$h_f(\mathbf{x}) = \begin{cases} 1 \text{ if } pf(\mathbf{x}) < p\theta_f \\ 0 \text{ otherwise} \end{cases} \tag{1}$$

where p is the polarity indicating the inequality direction. The best performing weak classifiers are combined linearly to derive a stronger one (at a stage of the cascade); see Algorithm 1.

Algorithm 1: AdaBoost Learning.

input : The training samples: $\{(\mathbf{x}_i, l_i)\}$, $i = 1, ..., N$, where $l_i = 1$ for positive and $l_i = 0$ for negative samples. $N = m + o$, where m and o are the number of positive and negative samples, respectively.

output: Strong classifier, $h(\mathbf{x})$, as a combination of T weak classifiers.

1 - Initialize the weights for samples:
$$w_{1,i} = \tfrac{1}{2m} \text{ for positive samples and } w_{1,i} = \tfrac{1}{2o} \text{ for negative samples.}$$

2 **for** $t = 1$ *to* T **do**

3   - Normalize weights so that w_t add up to one:
$$\hat{w}_{t,i} = \frac{w_{t,i}}{\sum_{j=1}^{n} w_{t,j}} \tag{2}$$

    for *each feature* $f \in \mathcal{F}$, *the set of all features* **do**

4   - Train a weak classifier h_f for learning from only feature f.

5   - Calculate the error of classification:
$$\epsilon_f = \sum_{i=1}^{n} \hat{w}_{t,i} |h_f(\mathbf{x}_i) - l_i| \tag{3}$$

6   - Among the weak classifiers, h_f, $\forall f \in \mathcal{F}$, choose the one with the lowest error (ϵ_t):
$$h_t = \arg\min_{f \in \mathcal{F}} \epsilon_f \tag{4}$$

    - Update the weights:
$$w_{t+1,i} = \hat{w}_{t,i} \left(\frac{\epsilon_t}{1 - \epsilon_t} \right)^{e_i} \tag{5}$$

    where $e_i = 1$ if $\mathbf{x}_i$ is classified correctly and zero if it is not.

7 - The final classifier is then the combination of all of the weak ones found above:
$$h(\mathbf{x}) = \begin{cases} 1 & \text{if } \sum_{t=1}^{T} \alpha_t h_t(\mathbf{x}) \geq \frac{1}{2} \sum_{t=1}^{T} \alpha_t \\ 0 & \text{otherwise} \end{cases} \tag{6}$$

where $\alpha_t = \log \frac{1-\epsilon_t}{\epsilon_t}$.

In the approach of Viola and Jones [10,34], the AdaBoost algorithm is used to learn only one stage of the cascade of classifiers: in the cascade, simpler features are used in the earlier stages, whereas bigger and more complex features are only processed if the candidate window passes the earlier stages. The method constructs the cascade by simply adding a new stage of the AdaBoost classifier when the current cascade does not yield the desired false positive and detection rates; see Algorithm 2 and Figure 1.

Algorithm 2: Learning A Cascade of Classifiers (adapted from [34]).

---

input : Positive and negative training samples: $\mathcal{P} = \{\mathbf{x}_1^+, \mathbf{x}_2^+, ..., \mathbf{x}_L^+\}$, $\mathcal{N} = \{\mathbf{x}_1^-, \mathbf{x}_2^-, ..., \mathbf{x}_M^-\}$
output: The cascade of classifiers

1 **initialize:**

    $i = 0$: The stage number
    $F_i = 1.0$: False positive rate of the current cascaded classifier
    $D_i = 1.0$: Detection rate of the current cascaded classifier
    $\mathcal{N}_i = \mathcal{N}$: Negative samples for the current cascaded classifier
    f : User defined maximum acceptable false positive rate per layer
    d : User defined minimum acceptable detection rate per layer

  while $F_i > F_{target}$ **do**

2     $i \leftarrow i + 1$

3     $n_i = 0$

4     $F_i \leftarrow F_{i-1}$

5   **while** $F_i > f \times F_{i-1}$ **do**

6         $n_i \leftarrow n_i + 1$

7   - Train a classifier h_{n_i} on $\mathcal{P}$ and $\mathcal{N}_i$ with n_i features using AdaBoost (see Algorithm 1)

8   - Determine F_i and D_i using the current cascaded detector

9   - Decrease threshold θ_i for h_{n_i} until $D_i > d \times D_{i-1}$

10   **if** $F_i > F_{target}$ **then**

11   - Run the current cascaded detector with θ_i on negative images

12   - Put any false negative windows into $\mathcal{N}_{i+1}$

Such an approach can only become computationally tractable if the features can be extracted in a very fast manner. One solution is using integral images, as proposed by Viola and Jones. In Section 3.1.1, we will describe them.

The cascaded detectors are usually run in multiple scales and locations, which lead to multiple detections for the same object. These are merged by looking at the amount of overlap between detections, as a post-processing stage.

3.1.1. Integral Images

In order to speed up the processing, the computation of each feature in a window is performed using the integral images technique. In this method, for a pixel (i, j), the intensities of all pixels that have a smaller row and column number are accumulated at (i, j):

$$II(i, j) = \sum_{c=1}^{i} \sum_{r=1}^{j} I(c, r) \tag{7}$$

where I is the original image and II the integral image. Note that II can be calculated incrementally from the II of the neighboring pixels more efficiently.

Given such an integral image, the sum of intensities in a rectangular window can be calculated easily by accessing four values. See Figure 2 for an example: the sum of intensities in Window A can be calculated as $II_4 + II_1 - (II_2 + II_3)$ [10].

Figure 2. The method of integral images for the efficient computation of sums of intensities in a window. The sum of intensities in Window A can be calculated as $II_4 + II_1 - (II_2 + II_3)$.

3.2. Cascaded Detection Using Haar-like Features (C-HAAR)

Haar-like features [74] are extensions of Haar wavelets to images. They can be used to extract meaningful information about the distribution of intensities in the form of various configurations of ON and OFF regions in an image window, as shown in Figure 3. Combined with integral images, calculating the responses of Haar-like features at a pixel can be extremely sped up, making it a suitable candidate for the cascaded approach.

Figure 3. Sample Haar-like features used in our study.

In this paper, we are using the extended set of Haar-like features described in [73]. The detector window is run over the image at multiple scales and locations.

3.3. Cascaded Detection Using Local Binary Patterns (C-LBP)

In LBP [75], a widely-used method for feature extraction, a window is placed on each pixel in the image within which the intensity of the center pixel is compared against the intensities of the neighboring pixels. During this comparison, larger intensity values are taken as one and smaller values as zero. To describe it formally, for a window $\Omega(x_c, y_c)$ at pixel (x_c, y_c) in image I, LBP pattern L_p is as $L_p(x_c, y_c) = \otimes_{(x,y) \in \Omega(x_c, y_c)} \sigma(I(x,y) - I(x_c, y_c))$, where $\otimes$ is the concatenation operator and $\sigma(.)$ is the unit step function:

$$\sigma(x) = \begin{cases} 0 & \text{if } x < 0 \\ 1 & \text{otherwise} \end{cases} \quad (8)$$

The concatenation of ones and zeros can be converted to a decimal number, representing the local intensity distribution around the center pixel with a single number:

$$L_2(x_c, y_c) = \sum_{i=0}^{|\Omega(x_c, y_c)|} 2^i \times L_p^i(x_c, y_c) \quad (9)$$

The cascaded approach of Viola and Jones [10,34] has been extended by Liao *et al.* [36] to use a statistically-effective multi-scale version of LBP (SEMB-LBP) features. In multi-scale LBP, instead of comparing the intensities of pixels, the average intensities of blocks in the window are compared to the central block; see Figure 4. Then, SEMB-LBP at scale s is defined as follows:

$$SEMB - LBP_s = \{\iota$$

(10)

where $rank(H_s)$ is the rank of H_s after a descending sort; N is the number of uniform patterns, *i.e.*, LBP binary strings where there are at most two $0 \rightarrow 1$ or $1 \rightarrow 0$ transitions in the string; and H_s is the histogram at scale s:

$$H_s(\iota) = 1_{[f_s(x,y)=\iota]}, \iota = 0, ..., L - 1$$

(11)

where $f_s(x,y)$ is the outcome of the multi-scale LBP at pixel (x,y). In the current article, we test C-LBP with scales $(3 \times u, 3 \times v)$, where $u = 1, ..., 13$ and $v = 1, ..., 7$, and N is set to 63, as suggested by [36]. In order to speed up the computation, the integral image method is used on each bin of the histogram.

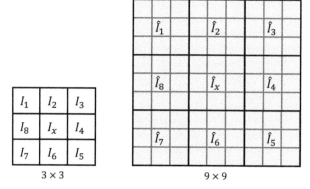

Figure 4. In LBP, the center pixel is compared to the others usually in a 3×3 window (**Left**). In the multi-block version (**Right**), average intensities in the blocks are compared instead.

3.4. Cascaded Detection Using Histogram of Oriented Gradients (C-HOG)

Histogram of oriented gradients (HOG) computes a histogram of gradient occurrences in local grid cells [11]. HOG has been demonstrated to be very successful in human detection and tracking. HOG of an image patch P is defined as follows:

$$HOG(k) = \sum_{p \in P} \delta\left(\left\lceil \frac{\theta^p}{L} \right\rceil - k \right)$$

(12)

where $\delta(\cdot)$ is the Kronecker delta which evaluates to one if and only if its input is zero, L is a normalizing constant and θ^p is the image gradient orientation at point p. $HOG(k)$ is the value of the k-th bin in a K-bin histogram. In the experiments, we set K to 9 which makes the value of L equal to $180/K = 20$ [11].

Zhu *et al.* [37] extended HOG features so that the features are extracted at multiple sizes of blocks at different locations and aspect ratios. This extension enables the definition of an increased number of blocks on which AdaBoost-based cascaded classification (Section 3.1) can be applied to choose the best combination. The integral image method is used on each bin of the histogram to speed up the computation.

3.5. Distance Estimation

Having detected the rectangle bounding an mUAV using one of the cascaded approaches introduced above, we can estimate its distance to the camera using the geometric cues. For this, we collect a training set of $\{(w_i, h_i), d_i\}$, where w_i, h_i are the width and the height of the mUAV bounding box, respectively, and d_i is the known distance of the mUAV. Having such a training set, we train a support vector regressor (SVR [76]). Using the trained SVR, we can estimate the distance of the mUAV once its bounding box is estimated.

4. Experimental Setup and Data Collection

The experimental setup, shown in Figure 5, consists of the following components:

- mUAV: We used a quadrotor platform shown in Figure 6a. Open-source Arducopter [77] hardware and software are used as the flight controller. The distance between the motors on the same axis is 60 cm. Twelve markers are placed around the plastic cup of the quadrotor for which we define a rigid body. The body coordinate frame of the quadrotor is illustrated in Figure 6a. The x_Q-axis and y_Q-axis are towards the forward and right direction of the quadrotor, respectively. The z_Q-axis points downwards with respect to the quadrotor.
- Camera: We use two different electro-optic cameras for indoors and outdoors due to varying needs in both environments. For indoors, the synchronization property of the camera is vital, since we have to ensure that the 3D position data obtained from the motion capture system and the captured frames are synchronized in time. Complying with this requirement, we use a camera from Basler Scout (capturing 1032×778 resolution videos at 30 fps in gray scale) mounted on top of the motion capture system. It weighs about 220 g, including its lens, whose maximum horizontal and vertical angle of views are $93.6°$ and $68.9°$, respectively. The power consumption of the camera is about 3 W, and it outputs the data through a Gigabit Ethernet port. The body coordinate frame of the camera is centered at the projection center. The x_C-axis is towards the right side of the camera; the y_C-axis points down from the camera; and the z_C-axis coincides with the optical axis of the camera lens, as depicted in Figure 6b.

  Due to difficulties in powering and recording of the indoor camera outdoors, we use another camera (Canon$^{(r)}$ PowerShot A2200 HD) to capture outdoor videos. This camera is able to record videos at a 1280×720 resolution at 30 fps in color. However, we use gray scale versions of the videos in our study.

  Although we needed to utilize a different camera outdoors due to logistic issues, we should note that our indoor camera is suitable to be placed on mUAVs in terms of SWaP constraints. Moreover, alternative cameras with similar image qualities compared to our cameras are also available on the market, even with less SWaP requirements.
- Motion capture system (used for indoor analysis): We use the Visualeyez II VZ4000 3D real-time motion capture system (MOCAP) (PhoeniX Technologies Incorporated) that can sense the 3D positions of active markers up to a rate of 4348 real-time 3D data points per second with an accuracy of 0.5~0.7 mm RMS in ~190 cubic meters of space. In our setup, the MOCAP provides the ground truth 3D positions of the markers mounted on the quadrotor. The system provides the 3D data as labeled with the unique IDs of the markers. It has an operating angle of $90°$ ($\pm45°$) in both pitch and yaw, and its maximum sensing distance is 7 m at minimum exposure. The body coordinate frame of the MOCAP is illustrated in Figure 6c.
- Linear rail platform (used for indoor analysis): We constructed a linear motorized rail platform to move the camera and the MOCAP together in a controlled manner, so that we are able to capture videos of the quadrotor only with single motion types, *i.e.*, lateral, up-down, rotational and approach-leave motions. With this platform, we are able to move the camera and MOCAP assembly on a horizontal line of approximately 5 m up to a 1-m/s speed.

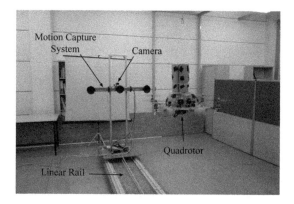

Figure 5. The setup used in indoor experiments. The rail was constructed in order to be able to move the camera with respect to the quadrotor in a controlled manner. This allows analyzing the performance of the methods under different motion types.

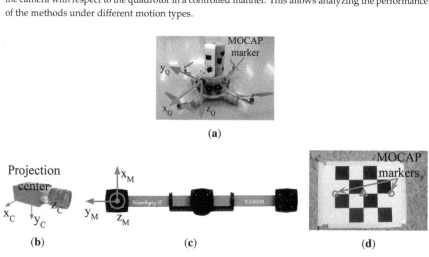

Figure 6. (**a**) The quadrotor used in our study and its body coordinate frame. There are 12 markers mounted roughly 30° apart from each other on the plastic cup of the quadrotor; (**b**) The body coordinate frame of the camera is defined at the projection center; (**c**) The Visualeyez II VZ4000 motion capture system and its body coordinate frame; (**d**) The calibration tool used to obtain 3D-2D correspondence points needed to estimate the transformation matrix, T_M^C, between the motion capture system (MOCAP) and the camera coordinate systems. Circles and the triangle indicate the MOCAP markers and the center of the chess pattern, respectively.

4.1. Ground Truth Extraction

In the indoor experimental setup, the MOCAP captures the motion of active markers mounted on the quadrotor and supplies the ground truth 3D positions of those markers. For our purposes, we need the ground truth bounding box of the quadrotor and the distance between the quadrotor and the camera for each frame.

In order to determine a rectangular ground truth bounding box encapsulating the quadrotor in an image, we need to find a set of 2D pixel points (P'_{Qi}) on the boundaries of the quadrotor in the image (In our derivations, all points in 2D and 3D sets are represented by homogeneous coordinate vectors). These 2D points correspond to a set of 3D points (P_{Qi}) on the quadrotor. In order to find P'_{Qi}, P_{Qi}

should first be transformed from the body coordinate frame of the quadrotor to the MOCAP coordinate frame, followed by a transformation to the camera coordinate frame. These two transformations are represented by the transformation matrices T_Q^M and T_M^C, respectively, and are applied as follows:

$$P_{Mi} = T_Q^M P_{Qi} \text{ for all } i \tag{13}$$

$$P_{Ci} = T_M^C P_{Mi} \text{ for all } i \tag{14}$$

where P_{Mi} and P_{Ci} are the transformed coordinates in the MOCAP and the camera coordinate frames, respectively. After these transformations, we project the points in P_{Ci} to the image plane as:

$$P'_{Qi} = P_c P_{Ci} \text{ for all } i \tag{15}$$

where P_c is the camera matrix and get P'_{Qi}. Then, we can find the bounding box of the quadrotor by calculating the rectangle with the minimum size covering all of the points in P'_{Qi} as follows:

$$x_r = \min(x_i) \tag{16}$$

$$y_r = \min(y_i) \tag{17}$$

$$w_r = \max(x_i) - \min(x_i) \tag{18}$$

$$h_r = \max(y_i) - \min(y_i) \tag{19}$$

where $(x_i, y_i) \in P'_{Qi}$, (x_r, y_r) is the upper left pixel position of the rectangle and w_r and h_r are the width and height of the rectangle, respectively.

It is not possible to place a marker on the quadrotor for every point in P_{Qi}. Therefore, we define a rigid body, a set of 3D points whose relative positions are fixed and remain unchanged under motion, for 12 markers on the quadrotor. The points in P_{Qi} are then defined virtually as additional points of the rigid body.

A rigid body can be defined from the positions of all markers obtained at a particular time instant while the quadrotor is stationary. However, we wanted to obtain a more accurate rigid body and used the method presented in [78,79] with multiple captures of the marker positions. Taking 60 different samples, we performed the following optimization to minimize the spatial distances between the measured points M_i and the points R_i in the rigid body model.

$$\arg\min_{R_i} \sum_i \| M_i - R_i \|^2 \tag{20}$$

where $\| . \|$ denotes the calculation of the Euclidean norm for the given vector.

Once the rigid body is defined for the markers on the quadrotor, if at least four markers are sensed by the MOCAP, T_Q^M can be estimated. Since the MOCAP supplies the 3D position data as labeled and the rigid body is already defined using these labels, there is no correspondence matching problem. Finding such a rigid transformation between two labeled 3D point sets requires the least squares fitting of these two sets and is known as the "absolute orientation problem" [80]. We use the method presented in [78,81] to solve this problem and calculate T_Q^M. Note that the T_Q^M transformation matrix should be calculated whenever the quadrotor and the camera moves with respect to each other.

There is no direct way of calculating T_M^C, since it is not trivial to measure the distances and the angles between the body coordinate frames of the MOCAP and the camera. However, if we know a set of 3D points (P_{Ti}) in the MOCAP coordinate frame and a set of 2D points (P'_{Ti}) which corresponds to the projected pixel coordinates of the points in P_{Ti}, then we can estimate T_M^C as the transformation matrix that minimizes the re-projection error. The re-projection error is given by the sum of squared distances between the pixel points in P'_{Ti} as in the following optimization criterion:

$$\arg\min_{T^C_M} \sum_i \| P'_{Ti} - T^C_M P_{Ti} \|^2 \tag{21}$$

For collecting the data points in P_{Ti} and P'_{Ti}, we prepared a simple calibration tool shown in Figure 6d. In this tool, there is a chess pattern and 2 MOCAP markers mounted on the two edges of the chess pattern. The 3D position of the chess pattern center, shown inside the triangle in Figure 6d, is calculated by finding the geometric center of the marker positions. We obtain the 2D pixel position of the chess pattern center using the camera calibration tools of the Open Source Computer Vision Library (OpenCV) [82]. We collect the data needed for P_{Ti} and P'_{Ti} by moving the tool in front of the camera. Note that, since the MOCAP and the camera are attached to each other rigidly, once T^C_M is estimated, it is valid as long as the MOCAP and the camera assembly remain fixed.

In order to calculate the ground truth distance between the quadrotor and the camera, we use T^M_Q and T^C_M as follows:

$$p'_c = T^C_M T^M_Q p_c \tag{22}$$

where p_c is the 3D position of the quadrotor center in the quadrotor coordinate frame and p'_c is the transformed coordinates of the quadrotor center to the camera coordinate frame. p_c is defined as the geometric center of 4 points where the motor shafts and the corresponding propellers intersect. Once p'_c is calculated, the distance of the quadrotor to the camera (d_Q) is calculated as:

$$d_Q = \| p'_c \| \tag{23}$$

4.2. Data Collection for Training

Indoors: We recorded videos of the quadrotor by moving the MOCAP and the camera assembly around the quadrotor manually while the quadrotor is hanging at different heights from the ground and stationary with its motors running. From these videos, we automatically extracted 8876 image patches, including only the quadrotor using the bounding box extraction method described in Section 4.1 without considering the aspect ratios of the patches. The distribution of the aspect ratios for these images is given in Figure 7 with a median value of 1.8168. Since the training of cascaded classifiers requires image windows with a fixed aspect ratio, we enlarged the bounding boxes of these 8876 images by increasing their width or height only according to the aspect ratio of the originally extracted image window, so that they all have a fixed aspect ratio of approximately 1.8168 (due to floating point rounding, aspect ratios may not be exactly 1.8168). We preferred enlargement to fix the aspect ratios, since this approach keeps all relevant data of the quadrotor inside the bounding box. We also recorded videos of the indoor laboratory environment without the quadrotor in the scene. From these videos, we extracted 5731 frames at a resolution of 1032 × 778 pixels as our background training image set. See Figure 8a,b for sample quadrotor and background images captured indoors.

Outdoors: We used a fixed camera to record the quadrotor while it is flying in front of the camera using remote control. Since the MOCAP is not operable outdoors, the ground truth is collected in a labor-extensive manner: by utilizing the background subtraction method presented in [83], we are able to approximate the bounding box of the quadrotor in these videos as long as there are not any moving objects other than the quadrotor. Nevertheless, it is not always possible to get a motionless background. Therefore, the bounding boxes from background subtraction are inspected manually, and only the ones that bound the quadrotor well are selected. Both the number and aspect ratio of the outdoor training images are the same as the indoor images. For outdoor background training images, we have recorded videos at various places on the university campus. These videos include trees, bushes, grass, sky, roads, buildings, cars and pedestrians without the quadrotor. From these videos, we have extracted frames as the same number of indoor background training images at 1280×720 resolution. See Figure 9a,b for sample images collected outdoors.

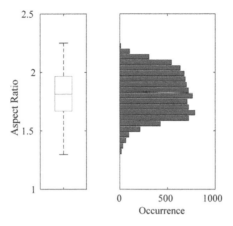

Figure 7. Box-plot (**Left**) and histogram (**Right**) representation for the aspect ratios of 8876 quadrotor images automatically extracted from the training videos. In this figure and the subsequent box-plot figures, the top and bottom edges of the box and the line inside the box represent the first and third quartiles and the median value, respectively. The bottom and top whiskers correspond to the smallest and largest non-outlier data, respectively. The data inside the box lie within the 50% confidence interval, while the confidence interval of the data in between the whiskers is 99.3%. Here, the median value is 1.8168, which defines the aspect ratio of the training images used.

(a)   **(b)**

Figure 8. Example images from indoor (**a**) quadrotor and (**b**) background training image sets. Mostly the challenging examples are provided in the quadrotor images.

(a)   **(b)**

Figure 9. Example images from outdoor (**a**) quadrotor and (**b**) background training image sets. The images are colored; however, their grayscale versions are used in the training. For quadrotor images, mostly the challenging examples are included.

Looking at the training image sets, the following observations can be deduced, which also represent the challenges in our problem: (i) changes in camera pose or quadrotor pose result in very large differences in the quadrotor's visual appearance; (ii) the bounding box encapsulating the quadrotor contains a large amount of background patterns due to the structure of the quadrotor; (iii) vibrations in the camera pose and the agile motions of the quadrotor cause motion blur in the images; (iv) changes in brightness and the illumination direction yield very different images; and (v) motion in the image can also be induced by the motion of the camera or the motion of background objects (e.g., trees swinging due to wind, *etc.*).

4.3. Data Collection for Testing

Indoor and outdoor environments are significantly different from each other, since controlled experiments can only be performed indoors by means of motion capture systems. On the other hand, outdoor environments provide more space, increasing the maneuverability of the quadrotor and causing many challenges that need to be evaluated. These differences directed us to prepare test videos of different characteristics indoors and outdoors.

In order to investigate the performance of the methods (C-HAAR, C-LBP and C-HOG) systematically, we defined 4 different motion types, namely lateral, up-down, yaw and approach-leave, for the indoor test videos. Please note that maneuvers in a free flight are combinations of these motions, and use of these primitive motions is for systematic evaluation purposes. The recording procedure of each motion type is depicted in Figure 10 for two different views, the top view and the camera view. Each motion type has different characteristics in terms of the amount of changes in the scale and appearance of the quadrotor, as well as the background objects, as shown in Table 2. The details of each motion type are as follows:

- Lateral: The camera performs left-to-right or right-to-left maneuvers while the quadrotor is fixed at different positions, as illustrated in Figure 10. As seen in the top view, the perpendicular distance of the quadrotor to the camera motion course is changed by 1 m for each of 5 distances. For each distance, the height of the quadrotor is adjusted to 3 different (top, middle and bottom) levels with 1 m apart, making a total of 15 different position for lateral videos. Left-to-right and right-to-left videos collected in this manner allow us to test the features' resilience against large background changes.

  In each video, the camera is moved along an approximately 5-m path. However, when the perpendicular distance is 1 m and 2 m and, the quadrotor is not fully visible in the videos for the top and bottom levels. Therefore, these videos are excluded from the dataset, resulting in 22 videos with a total of 2543 frames.

- Up-down: The quadrotor performs a vertical motion from the floor to the ceiling for the up motion and *vice versa* for the down motion. The motion of the quadrotor is performed manually with the help of a hanging rope. The change in the height of the quadrotor is approximately 3 m in each video. During the motion of the quadrotor, the camera remains fixed. For each of the 5 different positions shown in Figure 10, one up and one down video are recorded, resulting in 10 videos with a total of 1710 frames. These videos are used for testing the features' resilience against large appearance changes.

- Yaw: The quadrotor turns around itself in a clockwise or counter clockwise direction, while both the camera and the quadrotor are stationary. The quadrotor is positioned at the same 15 different points used in the lateral videos. Since the quadrotor is not fully present in the videos recorded for the top and bottom levels when the perpendicular distance is 1 m and 2 m, these videos are omitted from the dataset. Hence, there are 22 videos with a total of 8107 frames in this group. These videos are used for testing the features' resilience against viewpoint changes causing large appearance changes.

- Approach-leave: In these videos, the camera approaches the quadrotor or leaves from it while the quadrotor is stationary. There are 9 different positions for the quadrotor a with 1-m distance separation, as illustrated in Figure 10. The motion path of the camera is approximately 5 m. Approach and leave videos are recorded separately and we have 18 videos with a total of 3574 frames for this group. These videos are used for testing whether the features are affected by large scale and appearance changes.

We should note that the yaw orientation of the quadrotor is set to random values for each of 50 videos in the lateral, up-down and approach-leave sets, although the quadrotors in Figure 10 are given for a fixed orientation. There are cases where the MOCAP can give the wrong or insufficient data to extract the ground truth for some frames. These frames are not included in the dataset.

For outdoor experiments, we prepared four different videos with distinct characteristics. In all videos, the quadrotor is flown manually in front of a stationary camera. In the first two videos, a stationary background is chosen. These two videos differ in terms of agility, such that in the first video, the quadrotor performs calm maneuvers, whereas in the second one, it is flown in an agile manner. In the third video, the background includes moving objects, like cars, motorcycles, bicycles and pedestrians, while the quadrotor is flown in a calm manner. The fourth video is recorded to test the maximum detection distances of the methods. In this video, the quadrotor first leaves from the camera and then comes back, flying on an approximately straight 110-m path. We will call these videos (i) calm, (ii) agile, (iii) moving background and (iv) distance in the rest of the paper. These videos have 2954, 3823, 3900 and 2468 frames, respectively. The ground truth bounding boxes for each frame of calm, agile and moving background videos are extracted manually. For the distance video, only the ground truth distance of the quadrotor to the camera is calculated by utilizing another video recoded simultaneously by a side view camera. With the help of poles at known locations in the experiment area and by manually extracting the center of the quadrotor from the side view video, we computed the ground truth distance with simple geometrical calculations.

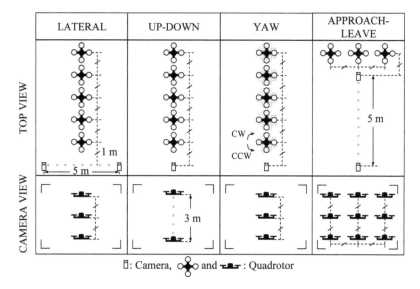

Figure 10. Graphical representation for indoor test videos. There are 4 motion types, namely lateral, up-down, yaw and approach-leave. Each of them is illustrated with the top and camera views. Dashed gray thick lines represent the motion of the camera or the quadrotor along the path with the given length. Dashed black thin lines are used to represent dimensions.

Table 2. Properties of motion types in terms of the amount of changes in the scale and appearance of the quadrotor and the background objects.

	Lateral	Up-Down	Yaw	Approach-Leave
Scale	Moderate	Moderate	Small	Large
Appearance	Moderate	Large	Large	Large
Background	Large	No Change	No Change	Moderate

We should note that the scenes used in testing videos are different from the ones included in the training datasets for both indoors and outdoors.

Our dataset is available at http://www.kovan.ceng.metu.edu.tr/fatih/sensors/.

5. Results

We implemented the cascaded methods introduced in Section 3 using OpenCV [82] and evaluated them on the indoor and outdoor datasets. We trained indoor and outdoor cascade classifiers separately using the corresponding training datasets with the following parameters: The quadrotor image windows were resized to 40×22 pixels. For an image with this window size, C-HAAR extracts $587,408$ features, whereas C-LBP and C-HOG yield $20,020$ and 20 features, respectively. Then, 7900 positive (quadrotor) and $10,000$ negative (background) samples were used for indoors and outdoors. We trained the classifiers with 11, 13, 15, 17 and 19 stages (the upper limit of 19 is due to the enormous time required to train C-HAAR classifiers, as will be presented in Section 5.6.1). During our tests, the classifiers performed multi-scale detections beginning from a minimum window size of 80×44 and enlarging the window size by multiplying it with 1.1 at each scale.

5.1. Performance Metrics

In order to evaluate the detection performance of the classifiers, we use precision-recall (PR) curves, which are drawn by changing the threshold of the classifiers' last stages from -100 to $+100$, as performed by [10,34]. Note that each stage of the cascaded classifiers has its own threshold determined during the training and that decreasing the threshold of a stage S to a low value, such as -100, results in a classifier with $S - 1$ many stages at the default threshold.

Precision is defined as:

$$Precision = \frac{tp}{tp + fp} \tag{24}$$

where tp is the number of true positives (see below) and fp is the number of false positives. Recall is defined as:

$$Recall = \frac{tp}{tp + fn} \tag{25}$$

where fn is the number of false negatives.

A detected bounding box (B_D) is regarded as a true positive if its Jaccard index (J) [84], calculated as follows, is greater than 60%:

$$J(B_D, B_G) = \frac{|B_D \cap B_G|}{|B_D \cup B_G|} \tag{26}$$

where B_G is the ground truth bounding box. Otherwise, B_D is regarded as a false positive. If there are multiple detections in a frame, each B_D is evaluated separately as a tp or fp. If no B_D is found for an image frame by the classifier, then fn is incremented by one.

We use also F-score in our evaluations, calculated as follows:

$$F - Score = 2 \times \frac{Precision \times Recall}{Precision + Recall} \tag{27}$$

A widely-used measure with PR-curves is the normalized area under the curve. If a PR curve, $p(x)$, is defined at the interval $[r_{min}, r_{max}]$, where r_{min} and r_{max} are the minimum and maximum recall values, respectively, the normalized area A_p under curve $p(x)$ is defined as:

$$A_p = \frac{1}{r_{max} - r_{min}} \int_{r_{min}}^{r_{max}} p(x) \tag{28}$$

5.2. Indoor Evaluation

We tested the classifiers trained with the indoor training dataset on indoor test videos having $15,934$ frames in total with four different motion types, namely lateral, up-down, yaw and approach-leave, as presented in Section 4.3. We evaluated the classifiers for five different numbers of stages to understand how they perform while their complexity increases. Figure 11 shows the PR curves, as well as the normalized area under the PR curves for each method and for different numbers of stages. In Table 3, the maximum F-score values and the values at default thresholds are listed.

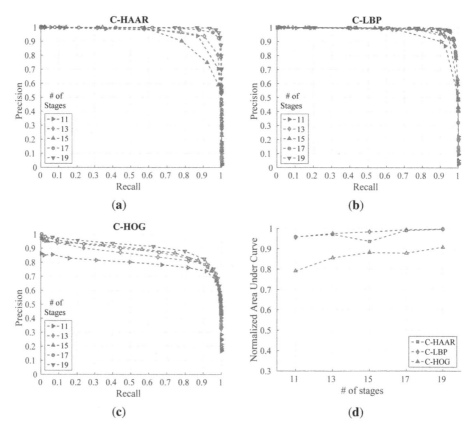

Figure 11. Precision-recall (PR) curves showing the performance of (**a**) C-HAAR, (**b**) C-LBP and (**c**) C-HOG for different numbers of stages on indoor test videos; (**d**) normalized areas under the PR curves in (**a**–**c**).

Table 3. Performance of the methods indoors, reported as F-score values. Bold indicates best performances.

Feature Type	C-HAAR					C-LBP					C-HOG				
Number of Stages	11	13	15	17	19	11	13	15	17	19	11	13	15	17	19
Maximum F-Score	0.903	0.920	0.836	0.958	**0.976**	0.904	0.936	0.940	0.962	**0.964**	0.818	0.848	0.842	0.839	**0.862**
F-Score at Default Threshold	0.058	0.143	0.286	0.570	**0.822**	0.104	0.345	0.774	0.943	**0.954**	0.404	0.550	0.627	0.664	**0.716**

The performances of C-HAAR and C-LBP are close to each other in terms of maximum F-scores (Table 3) and the normalized area under the curve (Figure 11d), except for a decrease at Stage 15 of C-HAAR, and they both perform better than C-HOG in all aspects. The lower performance of C-HOG is due to the low number of features it extracts from a training window. Even with the extension of Zhu *et al.* [37], only 20 features are extracted from a 40×22-pixel^2 training image. For AdaBoost to estimate a better decision boundary, more features are required. The difference between the number of features used by C-HAAR and C-LBP, however, does not result in a considerable performance divergence.

We observe a slight difference between C-HAAR and C-LBP in terms of the lowest points that PR curves (Figure 11) reach. This is related to the performance differences between the methods at their default threshold. As mentioned earlier, decreasing the threshold of a classifier's latest stage, S, to a very low value results in a classifier with a stage number of $S - 1$. Therefore, since the performances of C-LBP classifiers at their default thresholds are greater than the default performances of C-HAAR classifiers, we observe PR curves ending at higher points in the case of C-LBP.

For all methods, training with 19 stages outperforms training with less stages. Therefore, taking 19 as the best stage number for all methods, we present their performances on different motion types in Figure 12 with their overall performances on all motion types. The performance of C-HAAR is slightly better than C-LBP on lateral, up-down and yaw motions, since it has PR curves closer to the rightmost top corner of the figures. C-HOG gives the worst performance in all motion types.

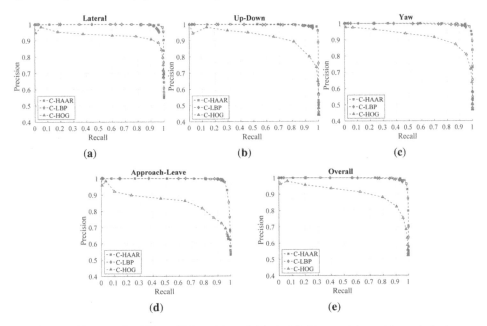

Figure 12. PR curves for: (**a**) lateral left-to-right and right-to-left; (**b**) up and down; (**c**) yaw clockwise and counter-clockwise; (**d**) approach and leave; and (**e**) all motion types.

When we look at the performances of each method individually for each motion type, C-HAAR performs similar on lateral, up-down and yaw motions; however, its performance diminishes on approach-leave, which is the most challenging motion in the indoor dataset. C-LBP has performance degradation on lateral motion, showing that it is slightly affected by the large background changes. Other than this, the performance of C-LBP is almost equal for other motion types. C-HOG performs better on lateral than other motions. Notable performance degradation is observed for the approach-leave motion.

5.3. Outdoor Evaluation

We evaluated the classifiers trained with the outdoor training dataset using all outdoor motion types, namely calm, agile and moving background. For each motion type and for overall performance, we present the resulting PR curves and the normalized area under the curves in Figures 13 and 14, respectively. The F-score performances are listed in Table 4.

We notice that the performances of C-HAAR and C-LBP are remarkably better than C-HOG in all experiments. When comparing C-HAAR and C-LBP, C-HAAR gives slightly better results in terms of all measures. Under the agile maneuvers of the quadrotor, C-LBP and C-HOG display performance degradation, while C-HAAR's performance is hardly affected. This suggests that C-HAAR is more robust against appearance changes due to the rotation of the quadrotor. Slight performance decreases are observed in moving background video for C-HAAR and C-LBP.

When compared to the indoor evaluation, C-HAAR classifiers with low stage numbers perform better outdoors. The performance of C-HOG decreases in outdoor tests. In terms of the F-score, the best performing stage numbers differ for C-HAAR and C-HOG. Unlike indoors, the performances of the C-LBP and C-HAAR classifiers at their default thresholds are close to each other, resulting in PR curves reaching closer end points when compared to indoor results.

In order to determine the maximum distances at which the classifiers can detect the quadrotor successfully, an experiment is conducted with distance test video using the best performing classifiers overall according to the F-scores in Table 4. In this experiment, the minimum detection window size is set to 20×11. The resulting maximum detection distances are 25.71 m, 15.73 m and 24.19 m, respectively, for C-HAAR, C-LBP and C-HOG.

Table 4. Performance of the methods outdoors, reported as F-score values. Bold indicates best performances.

Feature Type	C-HAAR					C-LBP					C-HOG				
Number of Stages	11	13	15	17	19	11	13	15	17	19	11	13	15	17	19
CALM — Maximum F-Score	0.979	0.987	0.991	0.991	**0.997**	0.930	0.951	0.953	0.977	**0.985**	**0.846**	0.822	0.781	0.732	0.842
CALM — F-Score at Default Threshold	0.036	0.112	0.248	0.536	**0.734**	0.040	0.095	0.266	0.670	**0.930**	**0.118**	0.144	0.168	0.189	0.216
AGILE — Maximum F-Score	0.965	0.983	0.988	0.987	**0.989**	0.887	0.902	0.890	**0.947**	0.942	0.719	**0.735**	0.619	0.600	0.713
AGILE — F-Score at Default Threshold	0.034	0.108	0.282	0.727	**0.906**	0.041	0.094	0.260	**0.704**	0.920	0.121	**0.146**	0.168	0.188	0.211
MOVING BACKGROUND — Maximum F-Score	0.955	0.965	**0.969**	0.963	0.967	0.935	0.870	0.940	0.954	**0.964**	0.797	**0.840**	0.785	0.777	0.832
MOVING BACKGROUND — F-Score at Default Threshold	0.030	0.084	**0.169**	0.274	0.441	0.043	0.111	0.269	0.480	**0.747**	0.158	**0.180**	0.199	0.216	0.234
OVERALL — Maximum F-Score	0.955	0.972	**0.977**	0.973	0.975	0.906	0.869	0.915	0.949	**0.957**	0.770	**0.801**	0.707	0.672	0.781
OVERALL — F-Score at Default Threshold	0.033	0.099	**0.221**	0.429	0.627	0.042	0.100	0.265	0.594	**0.850**	0.132	**0.157**	0.178	0.198	0.221

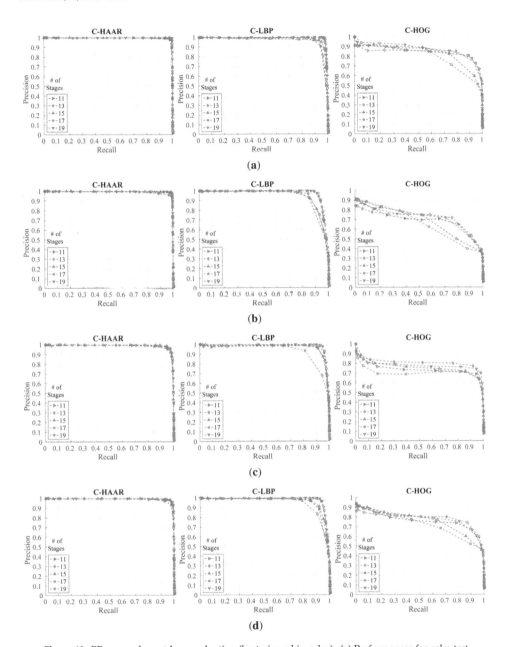

Figure 13. PR curves for outdoor evaluation (best viewed in color). (**a**) Performances for calm test video; (**b**) performances for agile test video; (**c**) performances for moving background test video; (**d**) overall performances.

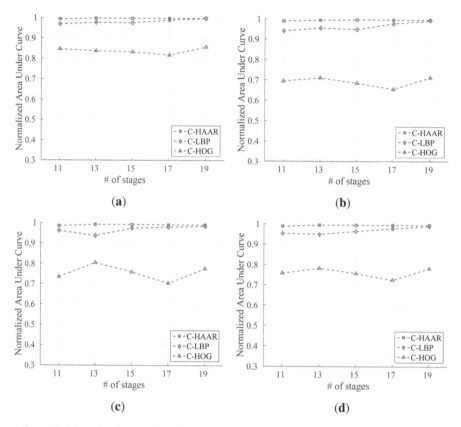

Figure 14. Normalized area under the curves for outdoor evaluation. (**a**) Stationary background calm flight; (**b**) stationary background agile flight; (**c**) moving background calm flight; (**d**) all outdoor flights combined.

5.4. Performance under Motion Blur

We have tested the performance of the methods against motion blur in the images. We utilized a linear motion blur similar to the one used in [85,86]. A motion-blurred version of an image I is generated by convolving it with a filter k (*i.e.*, $\tilde{I} = I*k$), which is defined as:

$$k(x,y) = \begin{cases} 1 \text{ if } y = d/2 \\ 0 \text{ otherwise} \end{cases}$$ (29)

where d is the dimension of the kernel (blur length), determining the amount of motion blur, sampled from a Gaussian distribution $N(\mu = 0, \sigma)$, with μ and σ being the mean and the standard deviation, respectively. We applied this kernel to the video images after a rotation of θ radian (blur angle) chosen from a uniform distribution $U(0, \pi)$. For each frame of a video, a new kernel is generated in this manner, and it is applied to all pixels in that frame. Using this motion blur model, we generated blurred versions of all indoor test videos for five different values of σ, namely, 5, 10, 15, 20 and 25.

We tested the best performing classifiers having 19 stages and giving the maximum F-scores in Table 3 on the blurred and original videos. The tests are performed on the indoor dataset only, for the sake of simplicity, since we do not expect a difference between the effects of motion blur for indoors and outdoors. The results depicting the changes in F-score, precision and recall against the amount of motion blur are given in Figure 15. We see that C-HAAR and C-LBP display a more robust behavior compared to C-HOG, since the decreasing trend in their F-score and recall values is slower than C-HOG. C-LBP performs better than C-HAAR in terms of F-score and recall. However, the precision of C-HAAR and C-HOG increases slightly with the increasing amount of motion blur. The reason for this increase is the decrease in the number of false positives, since they start to be identified as background by C-HAAR and C-HOG when there is more noise. However, this trend has a limit, since, at some point, the noise causes a major decrease in the number of true positives. Here, $\sigma = 25$ is the point where the precision of C-HAAR and C-HOG starts to decrease.

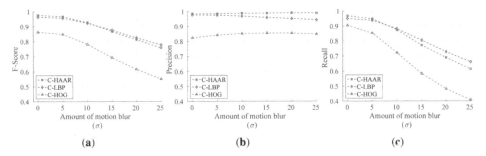

Figure 15. Performance of methods under motion blur. (**a**) F-score; (**b**) precision; and (**c**) recall. $\sigma = 0$ corresponds to original videos without motion blur.

In the case of C-LBP, precision values are continuously decreasing due to an increasing number of false positives. However, this degradation in precision is not so rapid. Moreover, the decreasing trend in the recall of C-LBP is slower than other methods. This slow decline rate in the recall results from a high number of correct detections and a low number of incorrect rejections.

5.5. Distance Estimation

In order to train the distance estimator (Section 3.5), we prepared a training set of 35,570 pairs of $\{(w_i, h_i), d_i\}$, where w_i, h_i are the width and the height of the mUAV bounding box, respectively, and d_i is its known distance, acquired using the motion capture system (see Section 4 for the details).

A support vector regressor (SVR) has been trained on this set with the radial basis function kernel. The values of the parameters are optimized using a grid-search and five-fold cross-validation, yielding the following values: $\nu = 0.09, C = 0.1$ and $\gamma = 0.00225$. With these values, a training error of 6.44 cm as the median is obtained. The distribution of distance estimation errors over the training set is shown in Figure 16a.

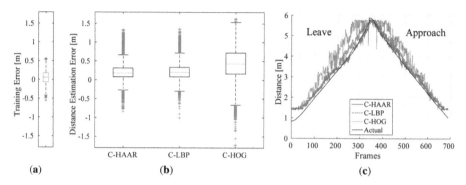

Figure 16. (a) Training error distribution for distance estimation; (b) distribution of distance estimation error for each method; (c) distance estimations during a leaving motion followed by an approach.

Since there is no ground truth distance information at hand for the outdoor dataset, the distance estimation has been evaluated by means of indoor videos only.

As in motion-blur analysis, we tested the best performing classifiers having 19 stages resulting in maximum F-scores tabulated in Table 3. The resulting distance estimation distributions are displayed in Figure 16b.

We see that the performance of C-HAAR is slightly better than C-LBP. The medians of the error for C-HAAR and C-LBP are 18.6 cm and 20.83 cm, respectively. The performance of C-HOG is worse than the other two methods with a median error of 43.89 cm and with errors distributed over a larger span.

In Figure 16c, we plot estimated and actual distances for a leave motion followed by an approach. These plots are consistent with the results provided with Figure 16b, such that the performances of C-HAAR and C-LBP are close to each other and better than C-HOG.

5.5.1. Time to Collision Estimation Analysis

We have analyzed the performance of the methods in the estimation of time to collision (*TTC*). In order to estimate *TTC*, the current speed (v_c) is estimated first:

$$v_c = \frac{d_c - d_p}{\Delta t} \tag{30}$$

where d_c is current distance estimation, d_p is a previous distance estimation and Δt is the time difference between two distance estimations. d_p is arbitrarily selected as the 90th previous distance estimation to ensure a reliable speed estimation. Once v_c is calculated, *TTC* can be estimated as:

$$TTC = \frac{d_c}{v_c} \tag{31}$$

Using this approach, we have evaluated the methods on indoor approach videos. Figure 17a shows the resulting box-plots for errors in estimating *TTC*. Figure 17b illustrates the estimated and actual *TTC*'s for a single approach video. The performances of C-HAAR and C-LBP are close to each other with a smaller median error for C-LBP. C-HOG performs worse than C-HAAR and C-LBP as a result of its low performance in distance estimation.

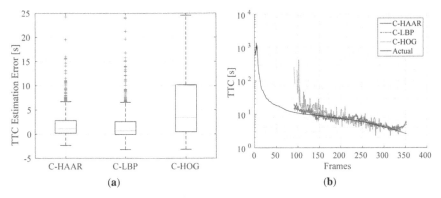

Figure 17. Indoor time to collision estimation performances of the methods for (**a**) all approach motions and (**b**) a single approach motion. In (**a**), there are outliers also outside the limits of the y-axis. However, in order to make differences between the methods observable, y-axis is limited between −5 and 25. In (**b**), the y-axis is in log-scale, and no estimation is available until the 90th frame. The missing points after the 90th frame are due to negative or infinite time to collision estimations.

5.6. Time Analysis

The training and testing time of the methods are analyzed in detail for the indoor and outdoor datasets on a computer with an Intel$^{(r)}$ Core i7-860 processor clocked at 2.80-GHz and 8 GB DDR3-1333MHz memory, running Ubuntu 14.04. Currently, processors with similar computational power are available for mUAVs [87,88].

5.6.1. Training Time Analysis

Figure 18 shows the amount of time required to train each stage of the classifiers, and Table 5 lists the total training times needed for the training of all 19 stages (the upper limit of 19 has been imposed due to the excessive time required for training C-HAAR). We observe that C-HAAR is the most time consuming method, which is succeeded by C-LBP and C-HOG. It is observed that C-HAAR requires on the order of days for training, whereas C-LBP and C-HOG finish in less than an hour.

The main reason behind the differences in the training times of the methods is the number of features extracted by each method from an image window. As mentioned previously (Section 5), the ordering among the methods is C-HAAR, C-LBP and C-HOG, with the decreasing number of associated features with an image window of 40 × 22 pixels. The increase in the number of features amounts to an increase in training the cascaded classifier to select the subset of good features via boosting.

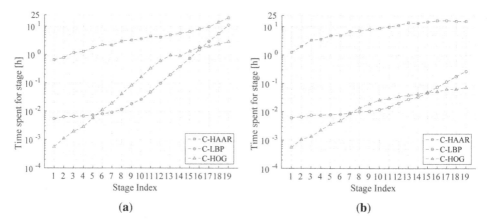

Figure 18. (a) Indoor and (b) outdoor training times consumed for each stage in the cascaded classifier. The y-axes are in log-scale.

Table 5. Time spent for training the cascaded classifiers having 19 stages in hours.

Feature Type	C-HAAR	C-LBP	C-HOG
Indoor	98.31	22.94	13.53
Outdoor	177.59	0.87	0.52

We also observe a significant difference between indoor and outdoor training times for each method. For the outdoor dataset, C-HAAR is twice as slow as for the indoor dataset, where C-LBP and C-HOG are 26-times faster. The reason for this is the fact that the outdoor background images are more distinct, enabling C-LBP and C-HOG to find the best classifier in each stage more quickly. However, this effect is not observed in C-HAAR, since Haar-like features are adversely affected by the illumination changes, which are observed substantially in our outdoor dataset.

5.6.2. Testing Time Analysis

We have measured and analyzed the computation time of each method in two different aspects: (i) on a subset of indoor videos, we measured the computation time by changing the distance of the quadrotor to understand the effect of the distance; and (ii) we analyzed the average running times needed to process indoor and outdoor frames, with respect to the number of stages and the thresholds.

For the first experiment, we have selected five videos from the yaw motion type for 1-, 2-, 3-, 4- and 5-m distances for the middle level height. In total, there were 1938 frames in these videos. We tested the performance of the classifiers having 19 stages at their default thresholds, as shown in Figure 19, with respect to the distance between the quadrotor and the camera. Although there are fluctuations, the time required to process a single frame shows an inverse correlation. This is so because as a quadrotor gets further away, its footprint in the image will decrease, and hence, the bigger scale detectors will reject the candidate windows faster, which will yield a speed up in the overall detection.

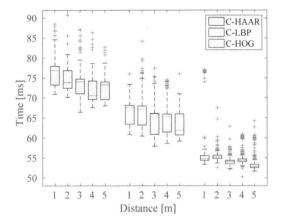

Figure 19. Change of computation time required to process one video frame with respect to the distance of the quadrotor.

In our second experiment, we tested the running time performance of the classifiers with respect to the number of stages. This has been performed both for the classifiers at their default threshold, as well as with thresholds giving the maximum F-score (See Tables 3 and 4).

For indoor experiments, a subset of the indoor dataset consisting of videos from approach, down, lateral left-to-right and yaw-clockwise motion types containing 1366 frames in total was used. For the outdoor experiments, a total of 1500 frames from all motion types, namely calm, agile and moving background, were used. Figure 20 displays the resulting time performance distributions.

When we compare indoor and outdoor results, we observe that all three methods require more time to process outdoor frames. This increase reaches up to three times for C-HAAR and C-LBP. Outdoor frames are bigger than indoor frames by a factor of 1.15. This accounts partially for the increase in the processing time. However, the main reason is the higher complexity of outdoor background patterns, which manage to pass the early simple processing stages of the cascades more; thus, they consume more time before being identified as background.

When the results at the default thresholds and the maximum F-score thresholds are compared, we observe an increase in the time spent on the lower stages of C-HAAR and C-LBP. This is due to the increasing number of candidate bounding boxes that are later merged into the resulting bounding boxes. Both detection and merging of these high number of candidate bounding boxes causes the processing time to increase.

For the maximum F-score thresholds, processing time increases with the number of stages. This is an inherent result due to the increase in the number of stages.

The scatter plots in Figure 21 display the distribution of F-scores with respect to the mean running times both for indoors and outdoors. The classifiers used in these plots are the ones giving maximum F-scores. The F-score values for C-HAAR and C-LBP are close to each other and higher than C-HOG. For C-HAAR, the F-score values are spread over a larger range for indoors, while the deviations in its mean time requirement increase for outdoors. Distributions observed for C-LBP for indoors and outdoors are similar to each other. The F-score values of C-HOG decrease and disperse over a wide range for outdoors, but the spread of its mean time requirement is very similar for indoors and outdoors.

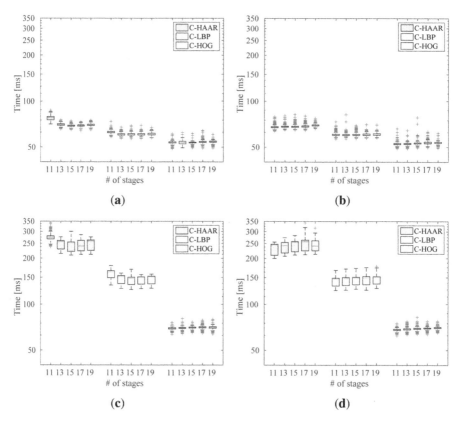

Figure 20. Analysis of time required to process one frame of (**a,b**) indoor and (**c,d**) outdoor videos. In (**a,c**), the classifiers are tested with their default thresholds, whereas in (**b,d**), the thresholds yielding the maximum F-score are used.

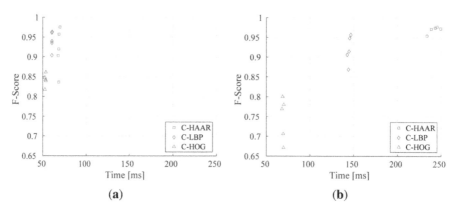

Figure 21. (**a**) Indoor and (**b**) outdoor scatter plots for F-score and mean running times. Each F-score value corresponds to a different classifier with different numbers of stages at the threshold resulting in the maximum F-score.

5.7. Sample Visual Results

In Figure 22, we present samples of successful detection and failure cases. These images are obtained using only the best performing C-LBP classifiers for the sake of space. C-LBP is remarkable among the three methods, since its detection and distance estimation performance is very high and close to that of C-HAAR. Furthermore, it is computationally more efficient than C-HAAR, both in training and testing. Three supplementary videos are also available (at http://www.kovan.ceng.metu. edu.tr/fatih/sensors/) as addendum showing the detection performance of C-LBP on video sequences from the indoor and outdoor test datasets.

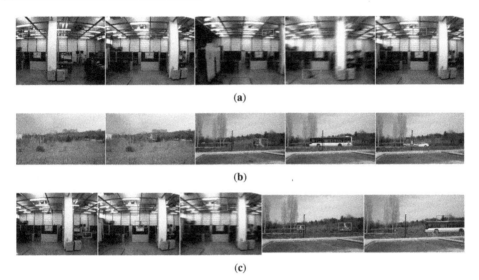

(a)

(b)

(c)

Figure 22. Successful detection and failure examples from indoor and outdoor experiments obtained using the best performing classifiers of C-LBP (only C-LBP results are provided for the sake of space). (a) Successful detections from indoor experiments.; (b) successful detections from outdoor experiments.; (c) failures from indoor and outdoor experiments.

The images in Figure 22a display the performance of the detector in an indoor environment that has extensive T junctions and horizontal patterns. The performance of the detector under motion blur is also displayed. Outdoor images in Figure 22b exemplify the outdoor performance of the detector where there are very complex textures, including also moving background patterns (pedestrians and various types of vehicles). When we look at the failures in Figure 22c, we observe that the regions including T junctions, horizontal patterns and silhouettes very similar to the quadrotor's are the confusing areas for the algorithms.

6. Conclusions

In this article, we have studied whether an mUAV can be detected and its distance can be estimated with a camera through cascaded classifiers using different feature types. In order to demonstrate this in a systematic manner, we performed several experiments indoors and outdoors. For indoor evaluations, a motion platform was built to analyze the performance of the methods in controlled motions, namely in lateral, up-down, rotational and approach-leave motions. For outdoor evaluations, on the other hand, the methods were evaluated for cases where the mUAV was flown in a calm manner, an agile manner or with other moving objects in the background. The maximum detection distances of the methods are also analyzed with an outdoor experiment.

We evaluated the performance of three methods, namely C-HAAR, C-LBP and C-HOG, where, in each method, a different feature extraction approach is combined with the boosted cascaded classifiers and with a distance estimator utilizing SVR. Our experiments showed that near real-time detection and accurate distance estimation of mUAVs are possible. C-LBP becomes prominent among the three methods due to its: (1) high performance in detection and distance and time to collusion estimation; (2) moderate computation time; (3) reasonable training time; and (4) more robustness to the motion blur. When it comes to distance estimation, C-HAAR performs better, since it positions the bounding boxes more accurately compared to the other methods. On the other hand, our time analysis reveals that C-HOG is the fastest, both in training and testing.

We have demonstrated that an mUAV can be detected in about 60 ms indoors and 150 ms outdoors in images with 1032×778 and 1280×720 resolutions, respectively, with a detection rate of 0.96 for the F-score, both indoors and outdoors. Although this cannot be considered real time, a real-time performance with cascaded classifiers is reachable, especially considering that the implementations are not optimized. We also showed that distance estimation of mUAVs is possible using simple geometric cues and the SVR; even the change in the pose of the quadrotor or the camera results in different bounding boxes for the same distance between mUAV and the camera.

The performance of detection can be improved significantly when combined with tracking, e.g., by employing tracking-by-detection methods [89–91]. Such methods limit the search space of the detector in the next frame(s) by using the properties of the current and previous detections. This can improve both the running time and the detection performance substantially.

Cascaded approaches are known to generalize rather well with the increase of the number of objects. By looking at simple, fast, yet effective features at multiple stages to minimize false positives and to maximize detection rates, successful applications on complex and challenging datasets with many exemplars of the same class have been reported [36,37,92]. These indicate that, for mUAV detection, cascaded approaches are very suitable, even if many mUAV variants with appearance characteristics are included.

Acknowledgments: Fatih Gökçe is currently enrolled in the Faculty Development Program (Öğretim Üyesi Yetiştirme Programı - ÖYP) on behalf of Süleyman Demirel University. For the experiments, we acknowledge the use of the facilities provided by the Modeling and Simulation Center of Middle East Technical University (Modelleme ve Simulasyon Merkezi - MODSIMMER). The authors wish to thank Turkish Aerospace Industries, Inc. (TAI) for providing the quadrotor used in this study for a research project called MRC. However, the current study is not a part of the MRC project. The authors would also like to thank Sertaç Olgunsoylu, who installed and wrote software to make sure the indoor camera could capture frames.

Author Contributions: Fatih Gökçe and Göktürk Üçoluk performed the experiments. Göktürk Üçoluk, Sinan Kalkan and Erol Şahin designed the experiments and provided the platforms. Fatih Gökçe, Sinan Kalkan, Erol Şahin and Göktürk Üçoluk wrote the paper.

Conflicts of Interest: The authors declare no conflict of interest.

References

1. Colomina, I.; Molina, P. Unmanned aerial systems for photogrammetry and remote sensing: A review. *ISPRS J. Photogramm. Remote Sens.* **2014**, *92*, 79–97. [CrossRef]
2. Yuan, C.; Zhang, Y.; Liu, Z. A survey on technologies for automatic forest fire monitoring, detection, and fighting using unmanned aerial vehicles and remote sensing techniques. *Can. J. For. Res.* **2015**, *45*, 783–792. [CrossRef]
3. Ackerman, E. When Drone Delivery Makes Sense. *IEEE Spectrum.* 25 September 2014. Available online: http://spectrum.ieee.org/automaton/robotics/aerial-robots/when-drone-delivery-makes-sense (accessed on 19 August 2015).
4. Holmes, K. Man Detained Outside White House for Trying to Fly Drone. *CNN.* 15 May 2015. Available online: http://edition.cnn.com/2015/05/14/politics/white-house-drone-arrest/ (accessed on 19 August 2015).

5.	Martinez, M.; Vercammen, P.; Brumfield, B. Above spectacular wildfire on freeway rises new scourge: Drones. *CNN*. 19 July 2015. Available online: http://edition.cnn.com/2015/07/18/us/california-freeway-fire/ (accessed on 19 August 2015).

6.	Andreopoulos, A.; Tsotsos, J.K. 50 Years of object recognition: Directions forward. *Comput. Vis. Image Underst.* **2013**, *117*, 827–891. [CrossRef]

7.	Campbell, R.J.; Flynn, P.J. A survey of free-form object representation and recognition techniques. *Comput. Vis. Image Underst.* **2001**, *81*, 166–210. [CrossRef]

8.	Lowe, D.G. Object recognition from local scale-invariant features. *Int. Conf. Comput. Vis.* **1999**, *2*, 1150–1157.

9.	Belongie, S.; Malik, J.; Puzicha, J. Shape matching and object recognition using shape contexts. *IEEE Trans. Pattern Anal. Mach. Intell.* **2002**, *24*, 509–522. [CrossRef]

10.	Viola, P.; Jones, M. Rapid object detection using a boosted cascade of simple features. *IEEE Conf. Comput. Vis. Pattern Recognit.* **2001**, *1*, 511–518.

11.	Dalal, N.; Triggs, B. Histograms of oriented gradients for human detection. *IEEE Conf. Comput. Vis. Pattern Recognit.* **2005**, *1*, 886–893.

12.	Serre, T.; Wolf, L.; Bileschi, S.; Riesenhuber, M.; Poggio, T. Robust object recognition with cortex-like mechanisms. *IEEE Trans. Pattern Anal. Mach. Intell.* **2007**, *29*, 411–426. [CrossRef] [PubMed]

13.	Boutell, M.R.; Luo, J.; Shen, X.; Brown, C.M. Learning multi-label scene classification. *Pattern Recog.* **2004**, *37*, 1757–1771. [CrossRef]

14.	Rosten, E.; Drummond, T. Machine Learning for High-Speed Corner Detection. *Eur. Conf. Comput. Vis.* **2006**, *3951*, 430–443.

15.	Trajkovic, M.; Hedley, M. Fast corner detection. *Image Vis. Comput.* **1998**, *16*, 75–87. [CrossRef]

16.	Harris, C.; Stephens, M. A Combined Corner and Edge Detector. In Proceedings of the 4th Alvey Vision Conference, Manchester, UK, 31 August–2 September 1988; pp. 147–151.

17.	Matas, J.; Chum, O.; Urban, M.; Pajdla, T. Robust Wide Baseline Stereo from Maximally Stable Extremal Regions. In Proceedings of the British Machine Vision Conference, Cardiff, UK, 2–5 September 2002; pp. 36.1–36.10.

18.	Shi, J.; Tomasi, C. Good features to track. In Proceedings of the IEEE Conference on Computer Vision and Pattern Recognition (CVPR), Seattle, WA, USA, 21–23 June 1994; pp. 593–600.

19.	Tuytelaars, T.; Mikolajczyk, K. Local invariant feature detectors: A survey. *Found. Trends Comput. Graph. Vis.* **2008**, *3*, 177–280. [CrossRef]

20.	Bay, H.; Ess, A.; Tuytelaars, T.; Van Gool, L. Speeded-Up Robust Features (SURF). *Comput. Vis. Image Underst.* **2008**, *110*, 346–359. [CrossRef]

21.	Lowe, D.G. Distinctive Image Features from Scale-Invariant Keypoints. *Int. J. Comput. Vis.* **2004**, *60*, 91–110. [CrossRef]

22.	Calonder, M.; Lepetit, V.; Strecha, C.; Fua, P. BRIEF: Binary Robust Independent Elementary Features. *Eur. Conf. Comput. Vis.* **2010**, *6314*, 778–792.

23.	Rublee, E.; Rabaud, V.; Konolige, K.; Bradski, G.R. ORB: An efficient alternative to SIFT or SURF. *Int. Conf. Comput. Vis.* **2011**, 2564–2571.

24.	Leutenegger, S.; Chli, M.; Siegwart, R.Y. BRISK: Binary Robust Invariant Scalable Keypoints. In Proceedings of the IEEE International Conference on Computer Vision, Barcelona, Spain, 6–13 November 2011; pp. 2548–2555.

25.	Vandergheynst, P.; Ortiz, R.; Alahi, A. FREAK: Fast Retina Keypoint. In Proceedings of the IEEE Conference on Computer Vision and Pattern Recognition, Providence, RI, USA, 16–21 June 2012; pp. 510–517.

26.	Winn, J.; Criminisi, A.; Minka, T. Object categorization by learned universal visual dictionary. *Int. Conf. Comput. Vis.* **2005**, *2*, 1800–1807.

27.	Murphy, K.; Torralba, A.; Eaton, D.; Freeman, W. Object detection and localization using local and global features. In *Toward Category-Level Object Recognition*; Springer: Berlin/Heidelberg, Germany, 2006; pp. 382–400.

28.	Csurka, G.; Dance, C.R.; Fan, L.; Willamowski, J.; Bray, C. Available online: http://cmp.felk.cvut.cz/eccv2004/files/ECCV-2004-final-programme.png.

29.	Cortes, C.; Vapnik, V. Support-vector networks. *Mach. Learn.* **1995**, *20*, 273–297. [CrossRef]

30. Krizhevsky, A.; Sutskever, I.; Hinton, G.E. ImageNet Classification with Deep Convolutional Neural Networks. In *Advances in Neural Information Processing Systems (NIPS) 25*; Pereira, F., Burges, C., Bottou, L., Weinberger, K., Eds.; Curran Associates, Inc.: New York, NY, USA, 2012; pp. 1097–1105.

31. LeCun, Y.; Bengio, Y.; Hinton, G. Deep learning. *Nature* **2015**, *521*, 436–444. [CrossRef] [PubMed]

32. Dieterich, T.G. Ensemble methods in machine learning. In *Multiple Classifier Systems*; Springer: Berlin/Heidelberg, Germany, 2000; pp. 1–15.

33. Rowley, H.A.; Baluja, S.; Kanade, T. Neural network-based face detection. *IEEE Trans. Pattern Anal. Mach. Intell.* **1998**, *20*, 23–38. [CrossRef]

34. Viola, P.; Jones, M.J. Robust real-time face detection. *Int. J. Comput. Vis.* **2004**, *57*, 137–154. [CrossRef]

35. Freund, Y.; Schapire, R.E. A desicion-theoretic generalization of on-line learning and an application to boosting. In *Computational Learning Theory*; Springer: Berlin/Heidelberg, Germany, 1995; pp. 23–37.

36. Liao, S.; Zhu, X.; Lei, Z.; Zhang, L.; Li, S.Z. Learning multi-scale block local binary patterns for face recognition. In *Advances in Biometrics*; Springer: Berlin/Heidelberg, Germany, 2007; pp. 828–837.

37. Zhu, Q.; Yeh, M.C.; Cheng, K.T.; Avidan, S. Fast human detection using a cascade of histograms of oriented gradients. *IEEE Conf. Comput. Vis. Pattern Recog.* **2006**, *2*, 1491–1498.

38. Heredia, G.; Caballero, F.; Maza, I.; Merino, L.; Viguria, A.; Ollero, A. Multi-Unmanned Aerial Vehicle (UAV) Cooperative Fault Detection Employing Differential Global Positioning (DGPS), Inertial and Vision Sensors. *Sensors* **2009**, *9*, 7566–7579. [CrossRef] [PubMed]

39. Hu, J.; Xie, L.; Xu, J.; Xu, Z. Multi-Agent Cooperative Target Search. *Sensors* **2014**, *14*, 9408–9428. [CrossRef] [PubMed]

40. Rodriguez-Canosa, G.R.; Thomas, S.; del Cerro, J.; Barrientos, A.; MacDonald, B. A Real-Time Method to Detect and Track Moving Objects (DATMO) from Unmanned Aerial Vehicles (UAVs) Using a Single Camera. *Remote Sens.* **2012**, *4*, 1090–1111. [CrossRef]

41. Doitsidis, L.; Weiss, S.; Renzaglia, A.; Achtelik, M.W.; Kosmatopoulos, E.; Siegwart, R.; Scaramuzza, D. Optimal Surveillance Coverage for Teams of Micro Aerial Vehicles in GPS-Denied Environments Using Onboard Vision. *Auton. Robots* **2012**, *33*, 173–188. [CrossRef]

42. Saska, M.; Chudoba, J.; Precil, L.; Thomas, J.; Loianno, G.; Tresnak, A.; Vonasek, V.; Kumar, V. Autonomous deployment of swarms of micro-aerial vehicles in cooperative surveillance. In Proceedings of the 2014 International Conference on Unmanned Aircraft Systems (ICUAS), Orlando, FL, USA, 27–30 May 2014; pp. 584–595.

43. Rosnell, T.; Honkavaara, E. Point Cloud Generation from Aerial Image Data Acquired by a Quadrocopter Type Micro Unmanned Aerial Vehicle and a Digital Still Camera. *Sensors* **2012**, *12*, 453–480. [CrossRef] [PubMed]

44. Shen, S.; Mulgaonkar, Y.; Michael, N.; Kumar, V. Vision-based State Estimation for Autonomous Rotorcraft MAVs in Complex Environments. In Proceedings of the IEEE International Conference on Robotics and Automation (ICRA), Karlsruhe, Germany, 6–10 May 2013.

45. Shen, S.; Mulgaonkar, Y.; Michael, N.; Kumar, V. Vision-Based State Estimation and Trajectory Control Towards Aggressive Flight with a Quadrotor. In Proceedings of the Robotics: Science and Systems (RSS), Berlin, Germany, 24–28 June 2013.

46. Shen, S.; Mulgaonkar, Y.; Michael, N.; Kumar, V. Initialization-Free Monocular Visual-Inertial Estimation with Application to Autonomous MAVs. In Proceedings of the International Symposium on Experimental Robotics, Marrakech, Morocco, 15–18 June 2014.

47. Scaramuzza, D.; Achtelik, M.C.; Doitsidis, L.; Fraundorfer, F.; Kosmatopoulos, E.B.; Martinelli, A.; Achtelik, M.W.; Chli, M.; Chatzichristofis, S.A.; Kneip, L.; *et al.* Vision-Controlled Micro Flying Robots: From System Design to Autonomous Navigation and Mapping in GPS-denied Environments. *IEEE Robot. Autom. Mag.* **2014**, *21*. [CrossRef]

48. Achtelik, M.; Weiss, S.; Chli, M.; Dellaert, F.; Siegwart, R. Collaborative Stereo. In Proceedings of the IEEE/RSJ Conference on Intelligent Robots and Systems (IROS), San Francisco, CA, USA, 25–30 September 2011; pp. 2242–2248.

49. Hesch, J.A.; Kottas, D.G.; Bowman, S.L.; Roumeliotis, S.I. Camera-IMU-based localization: Observability analysis and consistency improvement. *Int. J. Robot. Res.* **2013**, *33*, 182–201. [CrossRef]

50. Krajnik, T.; Nitsche, M.; Faigl, J.; Vanek, P.; Saska, M.; Preucil, L.; Duckett, T.; Mejail, M. A Practical Multirobot Localization System. *J. Intell. Robot. Syst.* **2014**, *76*, 539–562. [CrossRef]

51. Faigl, J.; Krajnik, T.; Chudoba, J.; Preucil, L.; Saska, M. Low-cost embedded system for relative localization in robotic swarms. In Proceedings of the IEEE International Conference on Robotics and Automation (ICRA), Karlsruhe, Germany, 6–10 May 2013; pp. 993–998.
52. Lin, F.; Peng, K.; Dong, X.; Zhao, S.; Chen, B. Vision-based formation for UAVs. In Proceedings of the IEEE International Conference on Control Automation (ICCA), Taichung, Taiwan, 18–20 June 2014; pp. 1375–1380.
53. Zhang, M.; Lin, F.; Chen, B. Vision-based detection and pose estimation for formation of micro aerial vehicles. In Proceedings of the International Conference on Automation Robotics Vision (ICARCV), Singapore, Singapore, 10–12 December 2014; pp. 1473–1478.
54. Lai, J.; Mejias, L.; Ford, J.J. Airborne vision-based collision-detection system. *J. Field Robot.* **2011**, *28*, 137–157. [CrossRef]
55. Petridis, S.; Geyer, C.; Singh, S. Learning to Detect Aircraft at Low Resolutions. In *Computer Vision Systems*; Gasteratos, A., Vincze, M., Tsotsos, J., Eds.; Springer: Berlin/Heidelberg, Germany, 2008; Volume 5008, pp. 474–483.
56. Dey, D.; Geyer, C.; Singh, S.; Digioia, M. Passive, long-range detection of Aircraft: Towards a field deployable Sense and Avoid System. In Proceedings of the Field and Service Robotics, Cambridge, MA, USA, 14–16 July 2009.
57. Dey, D.; Geyer, C.; Singh, S.; Digioia, M. A cascaded method to detect aircraft in video imagery. *Int. J. Robot. Res.* **2011**, *30*, 1527–1540. [CrossRef]
58. Vásárhelyi, G.; Virágh, C.; Somorjai, G.; Tarcai, N.; Szörényi, T.; Nepusz, T.; Vicsek, T. Outdoor flocking and formation flight with autonomous aerial robots. In Proceedings of the IEEE/RSJ International Conference on Intelligent Robots and Systems (IROS), Chicago, IL, USA, 14–18 September 2014; pp. 3866–3873.
59. Brewer, E.; Haentjens, G.; Gavrilets, V.; McGraw, G. A low SWaP implementation of high integrity relative navigation for small UAS. In Proceedings of the Position, Location and Navigation Symposium, Monterey, CA, USA, 5–8 May 2014; pp. 1183–1187.
60. Roberts, J. Enabling Collective Operation of Indoor Flying Robots. Ph.D. Thesis, Ecole Polytechnique Federale de Lausanne (EPFL), Lausanne, Switzerland, April 2011.
61. Roberts, J.; Stirling, T.; Zufferey, J.; Floreano, D. 3-D Relative Positioning Sensor for Indoor Flying Robots. *Auton. Robots* **2012**, *33*, 5–20. [CrossRef]
62. Stirling, T.; Roberts, J.; Zufferey, J.; Floreano, D. Indoor Navigation with a Swarm of Flying Robots. In Proceedings of the IEEE International Conference on Robotics and Automation (ICRA), St. Paul, MN, USA, 14–18 May 2012.
63. Welsby, J.; Melhuish, C.; Lane, C.; Qy, B. Autonomous minimalist following in three dimensions: A study with small-scale dirigibles. In Proceedings of the Towards Intelligent Mobile Robots, Coventry, UK, 6–9 August 2001.
64. Raharijaona, T.; Mignon, P.; Juston, R.; Kerhuel, L.; Viollet, S. HyperCube: A Small Lensless Position Sensing Device for the Tracking of Flickering Infrared LEDs. *Sensors* **2015**, *15*, 16484–16502. [CrossRef] [PubMed]
65. Etter, W.; Martin, P.; Mangharam, R. Cooperative Flight Guidance of Autonomous Unmanned Aerial Vehicles. In Proceedings of the CPS Week Workshop on Networks of Cooperating Objects, Chicago, IL, USA, 11–14 April 2011.
66. Basiri, M.; Schill, F.; Floreano, D.; Lima, P. Audio-based Relative Positioning System for Multiple Micro Air Vehicle Systems. In Proceedings of the Robotics: Science and Systems (RSS), Berlin, Germany, 24–28 June 2013.
67. Tijs, E.; de Croon, G.; Wind, J.; Remes, B.; de Wagter, C.; de Bree, H.E.; Ruijsink, R. Hear-and-Avoid for Micro Air Vehicles. In Proceedings of the International Micro Air Vehicle Conference and Competitions (IMAV), Braunschweig, Germany, 6–9 July 2010.
68. Nishitani, A.; Nishida, Y.; Mizoguch, H. Omnidirectional ultrasonic location sensor. In Proceedings of the IEEE Conference on Sensors, Irvine, CA, USA, 30 October–3 November 2005.
69. Maxim, P.M.; Hettiarachchi, S.; Spears, W.M.; Spears, D.F.; Hamann, J.; Kunkel, T.; Speiser, C. Trilateration localization for multi-robot teams. In Proceedings of the Sixth International Conference on Informatics in Control, Automation and Robotics, Special Session on MultiAgent Robotic Systems (ICINCO), Funchal, Madeira, Portugal, 11–15 May 2008.

70. Rivard, F.; Bisson, J.; Michaud, F.; Letourneau, D. Ultrasonic relative positioning for multi-robot systems. In Proceedings of the IEEE International Conference on Robotics and Automation (ICRA), Pasadena, CA, USA, 19–23 May 2008; pp. 323–328.

71. Moses, A.; Rutherford, M.; Valavanis, K. Radar-based detection and identification for miniature air vehicles. In Proceedings of the IEEE International Conference on Control Applications (CCA), Denver, CO, USA, 28–30 September 2011; pp. 933–940.

72. Moses, A.; Rutherford, M.J.; Kontitsis, M.; Valavanis, K.P. UAV-borne X-band radar for collision avoidance. *Robotica* **2014**, *32*, 97–114. [CrossRef]

73. Lienhart, R.; Maydt, J. An extended set of Haar-like features for rapid object detection. In Proceedings of the International Conference on Image, Rochester, NY, USA, 11–15 May 2002; Volume 1, pp. 900–903.

74. Papageorgiou, C.P.; Oren, M.; Poggio, T. A general framework for object detection. In Proceedings of the International Conference on Computer Vision, Bombay, India, 4–7 January 1998; pp. 555–562.

75. Ojala, T.; Pietikainen, M.; Harwood, D. Performance evaluation of texture measures with classification based on Kullback discrimination of distributions. In Proceedings of the 12th IAPR International Conference on Pattern Recognition, Jerusalem, Israel, 9–13 October 1994; Volume 1, pp. 582–585.

76. Schölkopf, B.; Smola, A.J.; Williamson, R.C.; Bartlett, P.L. New support vector algorithms. *Neural Comput.* **2000**, *12*, 1207–1245. [CrossRef] [PubMed]

77. 3DRobotics. Arducopter: Full-Featured, Open-Source Multicopter UAV Controller. Available online: http://copter.ardupilot.com/ (accessed on 19 August 2015).

78. Gaschler, A. Real-Time Marker-Based Motion Tracking: Application to Kinematic Model Estimation of a Humanoid Robot. Master's Thesis, Technische Universität München, München, Germany, February 2011.

79. Gaschler, A.; Springer, M.; Rickert, M.; Knoll, A. Intuitive Robot Tasks with Augmented Reality and Virtual Obstacles. In Proceedings of the IEEE International Conference on Robotics and Automation (ICRA), Hong Kong, China, 31 May–7 June 2014.

80. Horn, B.K.P.; Hilden, H.; Negahdaripour, S. Closed-Form Solution of Absolute Orientation using Orthonormal Matrices. *J. Opt. Soc. Am.* **1988**, *5*, 1127–1135. [CrossRef]

81. Umeyama, S. Least-squares estimation of transformation parameters between two point patterns. *IEEE Trans. Pattern Anal. Mach. Intell.* **1991**, *13*, 376–380. [CrossRef]

82. Bradski, G. The OpenCV Library. *Dr. Dobb's J. Softw. Tools.* 2000. Available online: http://www.drdobbs.com/open-source/the-opencv-library/184404319?queryText=opencv.

83. Kaewtrakulpong, P.; Bowden, R. An Improved Adaptive Background Mixture Model for Real-time Tracking with Shadow Detection. In *Video-Based Surveillance Systems*; Remagnino, P., Jones, G., Paragios, N., Regazzoni, C., Eds.; Springer: New York, NY, USA, 2002; pp. 135–144.

84. Jaccard, P. The distribution of the flora in the Alpine zone. *New Phytol.* **1912**, *11*, 37–50. [CrossRef]

85. Rekleitis, I.M. Visual Motion Estimation based on Motion Blur Interpretation. Master's Thesis, School of Computer Science, McGill University, Montreal, QC, Canada, 1995.

86. Soe, A.K.; Zhang, X. A simple PSF parameters estimation method for the de-blurring of linear motion blurred images using wiener filter in OpenCV. In Proceedings of the International Conference on Systems and Informatics (ICSAI), Yantai, China, 19–21 May 2012; pp. 1855–1860.

87. Hulens, D.; Verbeke, J.; Goedeme, T. How to Choose the Best Embedded Processing Platform for on-Board UAV Image Processing? In Proceedings of the 10th International Conference on Computer Vision Theory and Applications, Berlin, Germany, 11–14 March 2015; pp. 377–386.

88. AscendingTechnologies. AscTec Mastermind. Available online: http://www.asctec.de/en/asctec-mastermind/ (accessed on 19 August 2015).

89. Leibe, B.; Schindler, K.; van Gool, L. Coupled detection and trajectory estimation for multi-object tracking. In Proceedings of the IEEE International Conference on Computer Vision (ICCV), Rio de Janeiro, Brazil, 14–21 October 2007; pp. 1–8.

90. Huang, C.; Wu, B.; Nevatia, R. Robust object tracking by hierarchical association of detection. 788–801.

91. Stalder, S.; Grabner, H.; van Gool, L. Cascaded confidence filtering for improved tracking-by-detection. In *European Conference on Computer Vision*; Springer: Berlin/Heidelberg, Germany, 2010; pp. 369–382.

92. Dollar, P.; Wojek, C.; Schiele, B.; Perona, P. Pedestrian detection: An evaluation of the state of the art. *IEEE Trans. Pattern Anal. Mach. Intell.* **2012**, *34*, 743–761. [CrossRef] [PubMed]

Article

Unmanned Aerial Vehicles (UAVs) and Artificial Intelligence Revolutionizing Wildlife Monitoring and Conservation

Luis F. Gonzalez [1,*], Glen A. Montes [1], Eduard Puig [1], Sandra Johnson [2], Kerrie Mengersen [2] and Kevin J. Gaston [3]

1   Australian Research Centre for Aerospace Automation (ARCAA),
    Queensland University of Technology (QUT), 2 George St, Brisbane QLD 4000, Australia;
    glen.montes@mail.escuelaing.edu.co (G.A.M.); eduard.puiggarcia@qut.edu.au (E.P.)
2   ARC Centre of Excellence for Mathematical & Statistical Frontiers (ACEMS),
    Queensland University of Technology (QUT), 2 George St, Brisbane QLD 4000, Australia;
    sandra.johnson@qut.edu.au (S.J.); k.mengersen@qut.edu.au (K.M.)
3   Environment and Sustainability Institute, University of Exeter, Penryn, Cornwall TR10 9EZ, UK;
    k.j.gaston@exeter.ac.uk
*   Correspondence: felipe.gonzalez@qut.edu.au; Tel.: +61-41-171-8012

Academic Editor: Vittorio M. N. Passaro
Received: 15 September 2015; Accepted: 5 January 2016; Published: 14 January 2016

Abstract: Surveying threatened and invasive species to obtain accurate population estimates is an important but challenging task that requires a considerable investment in time and resources. Estimates using existing ground-based monitoring techniques, such as camera traps and surveys performed on foot, are known to be resource intensive, potentially inaccurate and imprecise, and difficult to validate. Recent developments in unmanned aerial vehicles (UAV), artificial intelligence and miniaturized thermal imaging systems represent a new opportunity for wildlife experts to inexpensively survey relatively large areas. The system presented in this paper includes thermal image acquisition as well as a video processing pipeline to perform object detection, classification and tracking of wildlife in forest or open areas. The system is tested on thermal video data from ground based and test flight footage, and is found to be able to detect all the target wildlife located in the surveyed area. The system is flexible in that the user can readily define the types of objects to classify and the object characteristics that should be considered during classification.

Keywords: Unmanned Aerial Vehicle (UAV); wildlife monitoring; artificial intelligence; thermal imaging; robotics; conservation; automatic classification; koala; deer; wild pigs; dingo; conservation

1. Introduction

Effective management of populations of threatened and invasive species relies on accurate population estimates [1]. Existing monitoring protocols employing techniques such as remote photography, camera traps, tagging, GPS collaring, scat detection dogs and DNA sampling typically require considerable investment in time and resources [2,3]. Moreover, many of these techniques are limited in their ability to provide accurate and precise population estimates [4]. Some of the challenges in wildlife monitoring include the large size of species' geographic ranges [5], low population densities [3], inaccessible habitat [6,7], elusive behaviour [8] and sensitivity to disturbance [9].

The increase in availability of inexpensive Unmanned Aerial Vehicles (UAVs) provides an opportunity for wildlife experts to use an aerial sensor platform to monitor wildlife and tackle many of these challenges to accurately estimate species abundance [10–12]. In recent years, the use of UAVs that can perform flight paths autonomously and acquire geo-referenced sensor data has increased sharply for agricultural, environmental and wildlife monitoring applications [13–15]. Some issues restricting the wider use of UAVs for wildlife management and research include UAV regulations [9,15–17], operational costs and public perception. One of the most important restrictions, however, is the need to develop or apply advanced automated image detection algorithms designed for this task.

Current examples of the use of UAVs for wildlife management include monitoring sea turtles [18], black bears [8], large land mammals (e.g., elephants [19]), marine mammals (e.g., dugongs [20]) and birds (e.g., flocks of snow geese [21]), wildlife radio collar tracking [22], and supporting anti-poaching operations for rhinos [23]. UAVs with digital and thermal imagery sensors can record high resolution videos and capture images much closer to the animals than manned aerial surveys with fewer disturbances [9,10,18,22]. Jones *et al.* [24] for example conducted a test that involved gathering wildlife video and imagery data from more than 30 missions over two years, and concluded that a UAV could overcome "safety, cost, statistical integrity and logistics" issues associated with manned aircraft for wildlife monitoring. Other advances in this field include autonomous tracking of radio-tagged wildlife [13,25,26].

Overall, UAVs have proven to be effective at carrying out wildlife monitoring surveys however in many cases, the extensive post-processing effort required negates any convenience or time savings afforded by UAVs in the field compared to conventional survey methods. Therefore, for UAVs to become truly efficient wildlife monitoring tools across the entire workflow of data collection through to analysis, improved capabilities to automate animal detection and counting in the imagery collected by UAVs are required. Research into automatic classification of UAV images for wildlife monitoring is emerging. For example, van Gemert *et al.* [14] evaluated the use of UAVs and state-of-the-art automatic object detection techniques for animal detection demonstrating a promising solution for conservation tasks. Although using an elevated structure rather than a UAV, Christiansen *et al.* [4] used thermal imagery and a k-nearest-neighbour classifier to discriminate between animal and non-animal objects, achieving 93.3% accuracy in an altitude range of 3–10m. In this paper, we further address the issue of automated wildlife detection in UAV imagery by describing a system composed of a UAV equipped with thermal image acquisition as well as a video processing pipeline to perform automated detection, classification and tracking of wildlife in a forest setting to obtain a population estimate within the area surveyed.

2. Experimental Design

2.1. System Architecture

The system used in this experiment can be divided into airborne and ground segments as presented in Figure 1a. The airborne system consists of the multirotor UAV, navigation system, thermal camera, gimbal system and video transmitter (Figure 1b). The ground segment consists of the ground station software installed in a laptop, the datalink and video receivers for remote display and recording.

2.1.1. Unmanned Aerial Vehicle (UAV)

The aerial platform weighs approximately 6 kg, including flight and communications systems. It has a recommended maximum take-off weight of 8 kg, thus allowing 2 kg for sensor payload. The UAV has four main sub-systems: the airframe, the power and propulsion subsystem, the navigation subsystem and the communications subsystem. These subsystems are integrated to provide navigation and power during the UAV flight operations.

Airframe

The airframe used in this platform is an S800 EVO Hexacopter [27] weighing 5.4 kg with motors, propellers and ESCs (electronic speed controllers). The frame is fitted with a retractable undercarriage, providing a sensor field of view clear of obstacles.

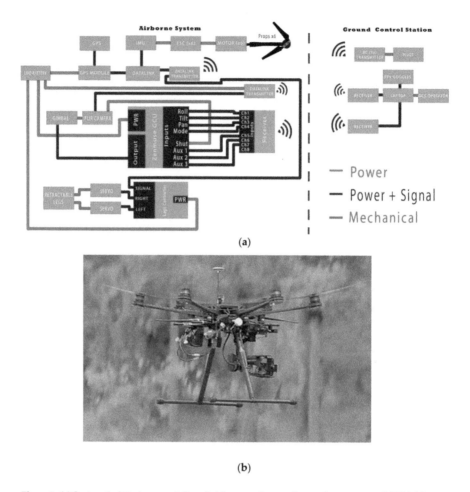

(a)

(b)

Figure 1. (a) System Architecture consisting of airborne and ground control segments and (b) Multirotor UAV, thermal camera, gimbal system and video transmitter.

Power and Propulsion

The UAV uses a 16,000 mAh Lipo 6 cell battery. This provides a maximum hover time of approximately 20 mins with no sensor payload. The maximum motor power consumption of each motor is 500 W operating at 400 rpm/V. These are running in conjunction with 15 × 5.2 inch propellers.

Navigation

The main component of the navigation system is a WooKong-M (WK-M) flight controller autopilot, which comes with a GPS unit with inbuilt compass, stabilization controller, gimbal stabilizer, position and altitude hold, auto go home/landing, with enhanced fail-safe. The system has an IMU located in the centre of the UAV to reduce the vibrations and reduce the risk of damage or failure. The autopilot's role in the aircraft is to navigate towards the desired location by altering the altitude, direction and speed. The autopilot has three main operating modes. The first mode is autonomous which allows the UAV to fly a predefined flight path that is designed using the ground control station (GCS). The second mode is stabilized mode which is designed for pre-flight checks of the control surfaces and autopilot.

This mode allows the aircraft to maintain a level flight when no pilot input is received. The final mode is full manual which is generally used for take-off and landings, as well as any emergency situations. The GPS connects directly to the autopilot multi-rotor controller as seen in Figure 1.

FLIR Camera, Gimbal System and Video Transmission

The FLIR camera used is a Tau 2-640 [28], Figure 2. The camera weighs 100 g and has a 640 × 480 pixels resolution and 25 mm focal lens. The FLIR has a field of view of 25 × 22 degrees. The FLIR video can be sent to the ground via AVL58 5.8 GHz Video Link which comprises a receiver in the laptop and an airborne transmitter. The video received by the laptop is recorded with off-the-shelf video recording software. The sampling frequency of the thermal camera is 9 fps and the sensitivity (NEdT) is <50 mK at f/1.0. The scene range (high gain) is $-25°$ to $+135°$ and the low gain is $-40°$ to $+550°$. The camera is capable of operating at altitudes up to $+40,000$ ft and has an operating temperature of -40 °C to $+80$ °C [28]. A gimbal system is used to hold the camera as shown in Figure 2, stabilize the video signal and dampen the vibrations of the frame. The gimbal control commands are received by the receiver attached to the Zenmuse Gimbal Control Unit (GCU). This allows the pilot or UAV controller to change the roll, pitch, pan and mode of the gimbal, and to control the FLIR camera.

Figure 2. FLIR camera attached to the gimbal system.

Ground Control Station and Datalink

The ground station software together with the datalink enables flight path planning with predefined waypoints and real-time telemetry reception in the ground station. The datalink communication between the UAV and the ground station is performed with a LK900 Datalink that consists of an airborne transmitter and a receiver connected to the ground station laptop. With a frequency of 900 MHz, the range of the datalink is estimated to be up to 1 km outdoors in optimal conditions. This hardware and software combination ensures stable flight performance while providing real-time flight information to the UAV operations crew. Although it is possible to take-off and land autonomously, it is recommended that these flight segments are completed manually via radio transmitter to guarantee safety standards.

Remote Display for Visualization

The ground station is equipped with a First Person View (FPV) system, to visualize the thermal video in real-time. The FPV goggles (Fatshark Dominator) may be used by UAV crew members, ecologists and wildlife experts to monitor aircraft motion in relation to wildlife as the UAV flies over the tree canopy (Figure 3).

Figure 3. Wildlife expert using FPV goggles to observe wildlife while the UAV is flying above the canopy.

2.2. Algorithms for Counting and Tracking

We implemented two algorithms on the ground control station computer that automatically count, track and classify wildlife using a range of characteristics. Different approaches are required depending on the information provided by the image or video. For instance, using a clear image of a koala, deer or a kangaroo, colour, size and position thresholds are applied to determine the object of interest. More complex algorithms are required if the image is less clear, for example if the object is an irregular shape, with no apparent colour, of variable size or in multiple positions. The algorithms were written in the Python programming language using the SimpleCV framework for ease of access to open source computer vision libraries such as OpenCV.

2.2.1. Algorithm 1: Pixel Intensity Threshold (PIT)

This algorithm approaches the problem by using the wildlife's heat signature which creates a good contrast between the background and the target wildlife. This contrast enables an intensity threshold to be applied which in turns eliminates the background and brings the object of interest to the front. Intensity threshold, also known as binarization or segmentation of an image, assigns 0 to all pixels under or equal to the threshold and 255 to all the pixels above the same threshold where 0 represents the black colour and 255 represents the white colour (Figure 4).

Figure 4. Image binarization in algorithm 1.

The algorithm uses the following function:

$$p = image\,(x,y)$$

$$f\left(p\right) = \begin{cases} p \leqslant T, 0 \\ p > T, \ 255 \end{cases}$$

where x and y are the coordinates of a pixel within the image and f (p) is a function that changes in value with respect to the threshold T.

A Graphical User Interface (GUI) was implemented to change this threshold using Tkinter libraries and to assist in finding the most appropriate value for T (Figure 5, slide 1). The second step is to add a morphological operation. Frequently, after applying the intensity threshold, two or more objects may appear as one because they are close to each other, causing a miscount of the number of objects in the frame (Figure 6).

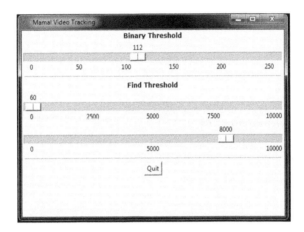

Figure 5. Graphical user interface (GUI) of the intensity threshold algorithm which allows the end user to adjust the thresholds using the sliders.

Figure 6. Two kangaroos appear to be one object.

The morphological operations comprise erosion and dilation steps that clean and separate objects. This process does not differentiate two objects that are on top of one another.

As seen in Figure 7, after applying first the erosion operation and then the dilation, the PIT algorithm is able to separate the kangaroos into two different objects. The third step is to search for the object of interest in the resulting image. Defining a minimum and maximum size threshold (Figure 5, slide 2 and 3) the algorithm searches for clusters of white pixels within the range and groups them to then display and count the clusters. The minimum and maximum size is a function of the number of pixels in the objects of interest. Every threshold value described in this algorithm may be changed during video processing to accommodate changing external conditions such as light, environmental temperature and video quality. The GUI provides slides to adjust the values and fine-tune the final processing.

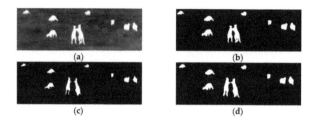

Figure 7. Output from PIT algorithm: (**a**) Original footage (**b**) Image binarized (**c**) Erode operation result (**d**) Dilation operation result.

The complete process can be seen in Figures 8 and 9 for kangaroo footage [29] and deer footage [30], respectively.

Figure 8. (**a**) Result footage for kangaroos (**b**) Image inverted (**c**) Erode operation (**d**) Dilation and group.

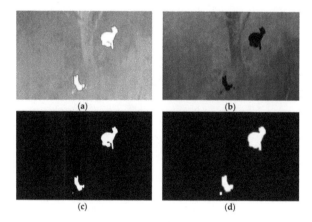

Figure 9. (**a**) Result footage for deer (**b**) Image inverted (**c**) Erode operation (**d**) Dilation and group.

Even though the PIT algorithm works well in many cases, it is dependent on good quality video and good background contrast. Some limitations of the algorithm are the inability to distinguish or classify different species and to track a single object. In order to address these limitations and to improve the processing, a second algorithm was developed.

2.2.2. Algorithm 2: Template Matching Binary Mask (TMBM)

Using a reference image template of the target object, the TMBM algorithm searches for a match in each frame of the video and labels it. The template reflects characteristics of the object of interest. Multiple templates that reflect changes in size, shape and colour of the object can increase the probability of finding the object in the image. The template matching provides the starting point to detect the species of interest. Very clear and distinct differences from the thermal heat signature and the environment at close range are then used to improve the detection.

The algorithm shown in Figure 10 consists of 10 main steps, as follows:

1. *Load templates:* In this first step template images are selected (e.g., koala, kangaroo, deer, pigs or birds). The templates are taken from the original footage or a database of images and then saved as small images of the object of interest. The algorithm is able to search for multiple templates in each frame.
2. *Processes Templates:* The contrast of the template is increased in order to enhance the possibility of finding this template in the footage; white is made lighter and black darker by adding or subtracting a constant, C, from each pixel value, p, depending on a threshold, T.

$$p = image\,(x,y)$$

$$f\,(p) = \begin{cases} p + C,\ p > T \\ p - C,\ p \leqslant T \end{cases}$$

The values of T and C may be changed in the program code, depending on the template quality and animal size, and can be determined by experimenting with different cases.
3. *Search for each template in the video frame:* For a detected animal to be recorded as a match it must be able to pass a scoring threshold. The searching function returns a score from 1 to 10, based on the proximity of the template to the matched object, where 1 indicates the smallest chance of finding the target and 10 indicates a perfect match. A score of 7 was designed to reflect a high quality match that gave an acceptable chance of avoiding false positives. In addition, to avoid false positives, any match found has to be present for at least 10 consecutive frames before it is considered to be a true match.
4. *Assign coordinates:* once a match has been found, the pixel coordinates (x, y) of the location within the frame are stored for later use.
5. *Create a mask using coordinates:* A new image is created with the same dimension as the source footage with a black (pixel with a value of 0) background. Using the coordinates within the image of the match found in the previous step, a white (pixel with a value of 255) bounding box or circle is drawn with a variable area (Figure 11b). The size of this area is defined by calculating the area of the match. The mask image aims to reduce the search area, eliminating what is considered as background.
6. *Logical operation with the mask:* In this step a logical AND is applied using the current frame and the mask. As a result the background is eliminated leaving only the regions of interest at the front.
7. *Pixel intensity threshold:* In this step the function described in Step 2 is used to assign a 0 if the pixel value is less than or equal to the threshold and 255 otherwise.
8. *Tracking:* After obtaining an image comprising only the objects, a function to track is implemented. This function is capable of identifying multiple objects within the same frame and can also distinguish one from another by using their coordinates. The coordinates of the current objects

are compared to the previous mask obtained, therefore making it possible to recognize if an object of interest has moved.

9. *Counting:* This function is able to number, count and display matches in the current frame and throughout the video. This is established by using the object's coordinates within the frame to differentiate multiples objects and to count the number of consecutive frames in which those objects have appeared. If this number increases, it means the object is still in the frame and this is counted as a match. If the object leaves the scene for a specified number of frames after being identified as a match, it is included in the total count.

10. *Last frame Loop:* In this last step the algorithm then checks if the current frame is the last. If not the algorithm restarts at Step 3; otherwise the process ends.

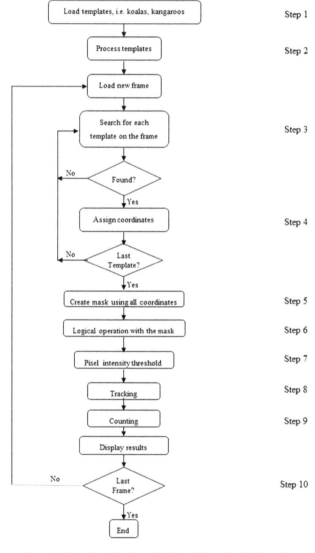

Figure 10. Flowchart Template Matching Binary Mask Algorithm.

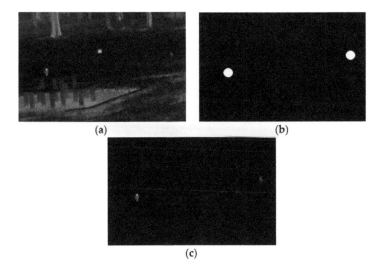

Figure 11. (a) Original footage (b) Mask created using the target's coordinates (birds) (c) Image result eliminating the background.

3. Validation Test

3.1. Focal Species: Koala

The koala (*Phascolarctos cinerus*) was chosen as the focal species for several reasons: it is an iconic native marsupial species of Australia whose status is listed as vulnerable in parts of the region [31] and its sedentary nature makes it ideal to trial the system described in this paper. Northern and eastern koala populations have declined substantially over recent years, mainly due to factors including loss of habitat, disease, road kill and predation by dogs [31]. A Senate inquiry into the status and health of the koala population in Australia was commissioned in 2010 [32]. One of the outcomes from the inquiry recommended a "national monitoring, evaluation and population estimation program for koalas" [31].

3.2. Study Area

The site selected for this experiment was located on the Sunshine Coast, 57 km north of Brisbane, Queensland, Australia. This site is a large rehabilitation enclosure where the number of koalas is determined in advance, which enables the accuracy of our counting algorithms to be assessed (ground-truthing).

The elevation of the treetop canopy varies between 20 m and 30 m, and is more densely populated in the western side than the eastern side (Figure 12). The experiment was conducted on 7th November 2014, with flights taking place between 7:10 a.m. and 8 a.m. Flying early in the morning provides optimal temperature difference for the target species (koalas), which makes it easier to distinguish between koalas, vegetation and soil. On that particular date, and during the specified time window, air temperatures oscillated between 21 and 24 °C. The flight could not start earlier due to fog and light rain, which actually contributed to lower soil and vegetation temperatures than would usually be the case for that time of day. During the flight window koala temperatures remained between 27 and 32 °C, while soil and vegetation remained below 25°.

Figure 12. Location map of the study area, which is on the Sunshine Coast, Queensland, Australia.

3.3. Data Acquisition

Both RGB video and thermal video were obtained simultaneously during the flights over the study area. Survey flights were performed at 60 m and 80 m in autonomous mode following a"lawn mowing" pattern (e.g., Figures 13 and 14), with ground speeds of 2.5 m/s and 4 m/s, respectively. In these flights the gimbal was set to maintain the cameras in down-looking position. Additionally a flight was performed in manual mode at around 20 m flight height, keeping a regular ground speed and with the cameras in lateral-looking position (Figure 15). Camera specifications for both thermal and RGB are given in Table 1.

(a)   (b)

Figure 13. (a) Thermal image (b) digital lateral view at 20 m.

Table 1. Camera specifications for RGB and FLIR Thermal camera.

	Mobius RGB Camera	**FLIR Thermal Camera**
Size	61 mm × 35 mm × 18 mm	44.5 mm × 44.5 mm × 55 mm
Weight	38 g	72 g
Spectrum Wavelength	Visible RGB	7.5 -13.5 µm
Resolution	1080 p	640 × 510
Focal Length	2.1 mm	25 mm
Frame Rate	30 fps	9 fps

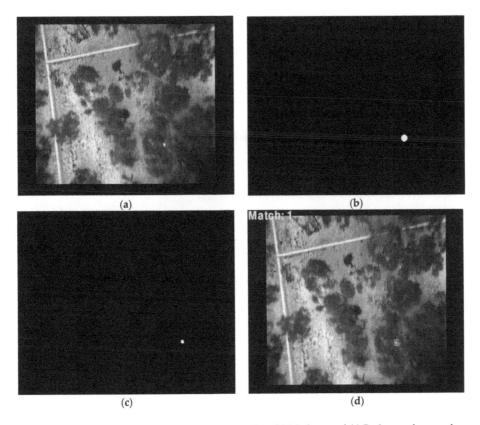

Figure 14. (**a**) Image of Canopy with Detection at 60 m (**b**) Mask created (**c**) Background removal, (**d**) Tracking and displaying.

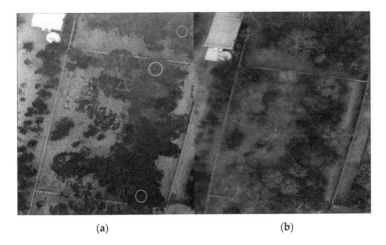

Figure 15. (**a**) Thermal image (**b**) digital RB6 capture at 80 m.

4. Results and Discussion

The detection algorithms described in Sections 2.2.1 and 2.2.2 were applied post-flight on the recorded imagery. The test flights successfully located koalas of different shapes and sizes. Figure 16a,b are examples of results at a height of 60 m. Figure 16a is a greyscale image with thermal imagery tracking. Figure 16b is a digital image of the same view illustrating the difficulty of visually spotting the koala. Similarly, Figure 13a,b show a greyscale thermal image of a koala identified alongside the real world image from a lateral perspective.

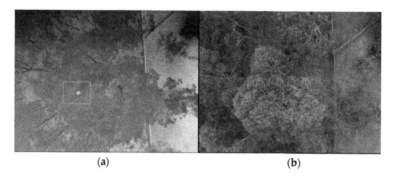

(a)   (b)

Figure 16. (a) Thermal image (b) digital R6B capture at 60 m.

Figure 15 shows an example of an image from footage obtained at 80 m where a thermal signature (e.g., koala) has been detected while searching above the canopy. We found that at this distance, and at heights greater than 60 m, confirming the detection of a particular thermal signature is difficult but this is also highly dependent on the camera characteristics and settings as well as the thermal differential between the target and other objects. Moreover, we found it challenging to distinguish between thermal signatures of different species, due to the lower resolution and difference in target object size relative to the image size. Consequently flying at such heights can feasibly result in false positives. The TMBM algorithm was tested on footage taken while flying at a height lower than 60 m.

The steps of TMBM algorithm were applied as follows. For Step 1, we defined a reference template to search for matching templates within the video. In this case the template was based on a sample of images of koalas seen from above at given altitudes. Processing was then applied to the template to improve the contrast between the shape and the background, as per Step 2. Step 3 involved execution of a search for the templates in the video. Due to the absence of local features to describe the koalas at different altitudes (20 m, 30 m, 60 m and 80 m), it was not possible to apply other types of approaches such as key point descriptors, parameterized shapes or colour matching.

Following Steps 4 and 5, the coordinates of the possible matches were saved as a vector for a follow-up check and a circle or a square box was drawn with the centre of the coordinates creating an image mask (Figure 14b). A logic AND operation was applied to remove what was considered as background and to focus on the region of interest (Figure 14c). Tracking was carried out as described in Step 6 of the TMBM algorithm. This was achieved by assuming the following:

1. If the target wildlife (e.g., koala) is found in at least 10 consecutive frames it is counted as a match
2. The target wildlife (e.g., koala) that has been identified cannot make big jumps in location (coordinates) between consecutive frames.
3. A horizontal displacement of the target wildlife (e.g., koala) is expected to be within the circle surrounding the target in the mask
4. The size of the target wildlife (e.g., koala) cannot suddenly increase drastically (area in pixels).

5. If the target wildlife (e.g., koala) being tracked is not found for 10 consecutive frames it is considered lost or out of the frame.

Figure 14d displays a rectangle containing the identified object of interest based on the above procedure. Results of detection, tracking and counting over multiple frames and at different altitudes (60 m, 30 m and 20 m) are shown in Figures 17–19 respectively.

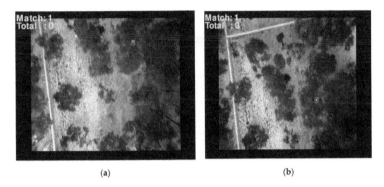

(a) (b)

Figure 17. Koala tracking and detection above the canopy at 60 m.

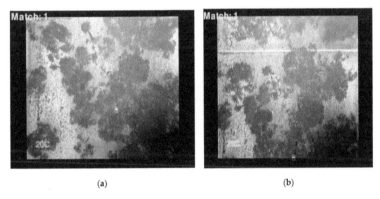

(a) (b)

Figure 18. Koala tracking and detection above the canopy at 30 m.

(a) (b)

Figure 19. (a) Digital image (b) koala tracking and detection above the canopy with image of canopy with side – lateral view detection at 20 m.

In order to evaluate the accuracy of detection, the GPS locations of the detected koalas (Figure 20) were compared with counts obtained by people on the ground. Even though in some cases a single koala was difficult to locate, the vertical traveling of the UAV above the canopy gave the camera a new angle of vision, locating the koala in almost all cases. We filtered all the false positives as described in Section 3, Steps 1 to 5, giving a clean and accurate result. This comparison can be seen in Table 2. Average detection times were defined as the time since the koala was first detected until the algorithm was able to mark it as a "real" match. These were recorded by the algorithm to determine the optimal altitude of detection. As the sampling frequency of the thermal camera is 9 fps, all detections were made in over one second changing as the altitude increases.

	Latitude	Longitude
K1	-26.8341075	152.9599625
K2	-26.8342363	152.9601119
K3	-26.8342885	152.9604243
K4	-26.8342886	152.9604186
K5	-26.8343332	152.9608132
K6	-26.8343332	152.9608298

Figure 20. Orthomosaic thermal image with GPS location of detected koalas.

Table 2. Comparison of number of detections, actual number of koalas, average detection time and average false positives.

Altitude	Number of Detections	Actual Number of Koalas	Average Detection Time (s)	Average False Positives
20 m	6	6	1.3	0
30 m	6	6	1.6	0
60 m	5 to 6	6	2.1	1.5

At altitudes of 20 to 30 m the average detection time was 1.5 s and the false positives were completely filtered by the algorithm. At altitudes above 30 m the algorithm took more time to detect the koala since the UAV had to fly over and then past the koala to detect it. This fractionally increased the number of false positives, although these only appeared in the detection task and was removed in the tracking task.

5. Conclusions

UAVs have proven to be effective at carrying out high-resolution and low-disturbance wildlife aerial surveys in a convenient and timely fashion, especially in habitats that are challenging to access or navigate at ground level [9]. However detecting or counting the species of interest from the large volumes of imagery collected during the flights has often proven to be very time consuming [9]. Other important issues are UAV regulations, operational costs, public perception [9,17], despite increasing effort in this area [15,16].

This paper addresses the challenge of automated wildlife detection in UAV imagery by describing a system that combines UAVs with thermal imaging capabilities and artificial intelligence image processing to locate wildlife in their natural habitats. The TMBM algorithm was able to detect the target koalas at the various locations and from various altitudes and produced an orthomosaic thermal image with GPS location of the species of interest (*i.e.*, koala). The accuracy of detection at altitudes of 20 m, 30 m and 60 m was compared with ground truth detections, showing 100% accuracy at these altitudes. In cases where the species is difficult to locate, vertical traveling of the UAV above the canopy is able to give the camera a new angle of vision and hence improve detection.

The system can be applied to including pest detection (e.g., wild dogs, wild cats, wild pigs, and dingos), locating protected or relocated animals or for search and rescue missions by adjusting the code and templates accordingly. The optimal detection height is likely to vary depending on various factors such as the size, thermal footprint, behaviour and habitat of the target species. This becomes more complex when there are several species of interest. For example, it may be desirable to detect not only the target species, but also its key predators and if applicable, its preferred prey. The GPS orthomosaic map displaying the locations of the desired animals may be used to assist developers and stakeholders to better understand the species' population distribution and abundance before approval is granted for development, or before construction begins in their habitat. Improvements to template accuracy and detection are possible by improving the quality of the original templates. The mask can be updated to adjust to the area being overflown and to distinguish between different animals as well as different sizes or temperatures. Implementing a dynamic template would increase the accuracy of detecting koalas and mammals of different sizes. A dynamic template to account for different possible orientations, positions and angles would mean that the template could be changed and updated during the flight in real time with respect to the object of interest. Radio communication using short ranged antennas could be combined with the system and algorithms to match the thermal imagery with a tagged animal. A copy of the software, algorithms and a User Manual is also available. Please contact felipe.gonzalez@qut.edu.au for more information.

Acknowledgments: The authors wish to thank everyone involved in the trial flight at Australia Zoo and the Australia Zoo Wildlife Hospital, especially the ARCAA Operations team (Dean Gilligan, Dirk Lessner and Gavin Broadbent), Rob Appleby (Griffith University), Amber Gillett and her colleagues at the Australia Zoo Wildlife Hospital. We also like to thank the four anonymous reviewers for their comprehensive reviews and constructive feedback.

Author Contributions: Felipe Gonzalez provided the general concept direction for the project, the selection of the site for the flight trial, advice on algorithm development and contributed to the writing of several sections of the manuscript. Glen Montes developed the algorithms and contributed to the writing of the paper. Eduard Puig developed the UAV flight plan, performed payload management during the mission, provided ideas for the developing the tracking algorithm and wrote several sections of the article. Sandra Johnson collaborated with Felipe Gonzalez regarding new strategies for monitoring and managing urban-wildlife conflict, coordinated the site selection for the trial flight, and contributed to the writing of this paper. Kerrie Mengersen was involved in the planning of the research project and contributed to the paper. Prof Gaston contributed to several sections of the manuscript and on linking the significance of this research to bio-diversity, conservation, environment and sustainability.

Conflicts of Interest: The authors declare no conflict of interest.

References

1. Cristescu, R.H.; Foley, E.; Markula, A.; Jackson, G.; Jones, D.; Frère, C. Accuracy and efficiency of detection dogs: A powerful new tool for koala conservation and management. *Sci. Rep.* **2015**, *1*, 1–5. [CrossRef] [PubMed]
2. Burton, A.C.; Neilson, E.; Moreira, D.; Ladle, A.; Steenweg, R.; Fisher, J.T.; Bayne, E.; Boutin, S. Wildlife camera trapping: A review and recommendations for linking surveys to ecological processes. *J. Appl. Ecol.* **2015**, *52*, 675–685. [CrossRef]
3. Witmer, G.W. Wildlife population monitoring: Some practical considerations. *Wildl. Res.* **2005**, *32*, 259–263. [CrossRef]
4. Christiansen, P.; Steen, K.A.; Jørgensen, R.N.; Karstoft, H. Automated detection and recognition of wildlife using thermal cameras. *Sensors* **2014**, *14*, 13778–13793. [CrossRef] [PubMed]
5. Gaston, K.J.; Fuller, R.A. The sizes of species' geographic ranges. *J. Appl. Ecol.* **2009**, *46*, 1–9. [CrossRef]
6. Murray, J.V.; Low Choy, S.; McAlpine, C.A.; Possingham, H.P.; Goldizen, A.W. The importance of ecological scale for wildlife conservation in naturally fragmented environments: A case study of the brush-tailed rock-wallaby (*petrogale penicillata*). *Biol. Conserv.* **2008**, *141*, 7–22. [CrossRef]

7. Schaub, M.; Gimenez, O.; Sierro, A.; Arlettaz, R. Use of integrated modeling to enhance estimates of population dynamics obtained from limited data. *Conserv. Biol.* **2007**, *21*, 945–955. [CrossRef] [PubMed]
8. Ditmer, M.A.; Vincent, J.B.; Werden, L.K.; Iaizzo, P.A.; Garshelis, D.L.; Fieberg, J.R. Bears show a Physiological but Limited Behavioral Response to Unmanned Aerial Vehicles. *Curr. Biol.* **2015**, *25*, 2278–2283.
9. Chabot, D.; Bird, D.M. Wildlife research and management methods in the 21st century: Where do unmanned aircraft fit in? *J. Unmanned Veh. Syst.* **2015**, *3*, 137–155. [CrossRef]
10. Anderson, K.; Gaston, K.J. Lightweight unmanned aerial vehicles will revolutionize spatial ecology. *Front. Ecol. Environ.* **2013**, *11*, 138–146. [CrossRef]
11. Linchant, J.; Lisein, J.; Semeki, J.; Lejeune, P.; Vermeulen, C. Are unmanned aircraft systems (UAS) the future of wildlife monitoring? A review of accomplishments and challenges. *Mamm. Rev.* **2015**, *45*, 239–252. [CrossRef]
12. Mulero-Pázmány, M.; Barasona, J.Á.; Acevedo, P.; Vicente, J.; Negro, J.J. Unmanned aircraft systems complement biologging in spatial ecology studies. *Ecol. Evol.* **2015**. [CrossRef] [PubMed]
13. Soriano, P.; Caballero, F.; Ollero, A. RF-based particle filter localization for wildlife tracking by using an UAV. In Proceedings of the 40th International Symposium on Robotics, Barcelona, Spain, 10–13 March 2009; pp. 239–244.
14. Van Gemert, J.C.; Verschoor, C.R.; Mettes, P.; Epema, K.; Koh, L.P.; Wich, S.A. Nature conservation drones for automatic localization and counting of animals. In *Computer Vision—ECCV 2014 Workshops, Part I*; Agapito, L., Bronstein, M.M., Rother, C., Eds.; Springer: Cham, Switzerland, 2015; pp. 255–270.
15. Williams, B.P.; Clothier, R.; Fulton, N.; Johnson, S.; Lin, X.; Cox, K. Building the safety case for uas operations in support of natural disaster response. In Proceedings of the 14th Aiaa Aviation Technology, Integration, and Operations Conference, Atlanta, GA, USA, 16–20 June 2014.
16. Cork, L.; Clothier, R.; Gonzalez, L.F.; Walker, R. The future of UAS: Standards, regulations, and operational experiences [workshop report]. *IEEE Aerosp. Electron. Syst. Mag.* **2007**, *22*, 29–44. [CrossRef]
17. Vincent, J.B.; Werden, L.K.; Ditmer, M.A. Barriers to adding UAVs to the ecologist's toolbox. *Front. Ecol. Environ.* **2015**, *13*, 74–75. [CrossRef]
18. Bevan, E.; Wibbels, T.; Najera, B.M.Z.; Martinez, M.A.C.; Martinez, L.A.S.; Martinez, F.I.; Cuevas, J.M.; Anderson, T.; Bonka, A.; Hernandez, M.H.; *et al.* Unmanned aerial vehicles (UAVs) for monitoring sea turtles in near-shore waters. *Mar. Turt. Newsl.* **2015**, 19–22.
19. Vermeulen, C.; Lejeune, P.; Lisein, J.; Sawadogo, P.; Bouché, P. Unmanned aerial survey of elephants. *PLoS ONE* **2013**, *8*. [CrossRef] [PubMed]
20. Hodgson, A.; Kelly, N.; Peel, D. Unmanned aerial vehicles (UAVs) for surveying marine fauna: A dugong case study. *PLoS ONE* **2013**. [CrossRef] [PubMed]
21. Chabot, D.; Bird, D.M. Evaluation of an off-the-shelf unmanned aircraft system for surveying flocks of geese. *Waterbirds* **2012**, *35*, 170–174. [CrossRef]
22. Dos Santos, G.A.M.; Barnes, Z.; Lo, E.; Ritoper, B.; Nishizaki, L.; Tejeda, X.; Ke, A.; Han, L.; Schurgers, C.; Lin, A.; *et al.* Small unmanned aerial vehicle system for wildlife radio collar tracking. In Proceedings of the 2014 IEEE 11th International Conference on Mobile Ad Hoc and Sensor Systems (MASS), Philadelphia, PA, USA, 28–30 October 2014; 2014; pp. 761–766.
23. Mulero-Pázmány, M.; Stolper, R.; van Essen, L.D.; Negro, J.J.; Sassen, T. Remotely piloted aircraft systems as a rhinoceros anti-poaching tool in Africa. *PLoS ONE* **2014**, *9*, e83873. [CrossRef] [PubMed]
24. Jones, G.P.I.V.; Pearlstine, L.G.; Percival, H.F. An assessment of small unmanned aerial vehicles for wildlife research. *Wildl. Soc. Bull.* **2006**, *34*, 750–758. [CrossRef]
25. Korner, F.; Speck, R.; Goktogan, A.; Sukkarieh, S. Autonomous airborne wildlife tracking using radio signal strength. In Proceedings of the 2010 IEEE/RSJ International Conference on Intelligent Robots and Systems (IROS), Taipei, Taiwai, 18–22 October 2010.
26. Leonardo, M.; Jensen, A.; Coopmans, C.; McKee, M.; Chen, Y. A Miniature Wildlife Tracking UAV Payload System Using Acoustic Biotelemetry. In Proceedings of the 2013 ASME/IEEE International Conference on Mechatronic and Embedded Systems and Applications Portland, OR, USA, 4–7 August 2013.
27. DJI. (n.d.). *S800 EVO*, Available online: http://www.dji.com/product/spreading-wings-s800-evo (accessed on 22 March 2015).
28. FLIR. (n.d.). *FLIR*, Available online http://www.FLIR.com/cores/display/?id=54717 (accessed on 23 March 2015).

29. IPI Learning. Kangaroos Boxing in Infrared [Video file]. Available online: https://www.youtube.com/watch?v=aBsvoWfHWXQ (accessed on 9 January 2014).

30. Trail cameras. Imagers for watching wildlife [Video file]. Available online: https://www.youtube.com/watch?v=ZpBgt91Qor8 (accessed on 10 March 2012).

31. Shumway, N.; Lunney, D.; Seabrook, L.; McAlpine, C. Saving our national icon: An ecological analysis of the 2011 Australian Senate inquiry into the status of the koala. *Environ. Sci. Policy* **2015**, *54*, 297–303. [CrossRef]

32. Senate Environment and Communications References Committee. Completed inquiries 2010-2013: The Koala-Saving Our National Icon. Available online: http://www.aph.gov.au/Parliamentary_Business/Committees/Senate/Environment_and_Communications/Completed%20inquiries/2010-13/koalas/index (accessed on 16 November 2015).

Article

UAVs Task and Motion Planning in the Presence of Obstacles and Prioritized Targets

Yoav Gottlieb and Tal Shima *

Technion—Israel Institute of Technology, Technion City, Haifa 3200003, Israel; E-Mail: syoavgo@gmail.com
* E-Mail: tal.shima@technion.ac.il; Tel.: +972-4-829-2705.

Academic Editor: Felipe Gonzalez Toro
Received: 25 June 2015 / Accepted: 12 November 2015 / Published: 24 November 2015

Abstract: The intertwined task assignment and motion planning problem of assigning a team of fixed-winged unmanned aerial vehicles to a set of prioritized targets in an environment with obstacles is addressed. It is assumed that the targets' locations and initial priorities are determined using a network of unattended ground sensors used to detect potential threats at restricted zones. The targets are characterized by a time-varying level of importance, and timing constraints must be fulfilled before a vehicle is allowed to visit a specific target. It is assumed that the vehicles are carrying body-fixed sensors and, thus, are required to approach a designated target while flying straight and level. The fixed-winged aerial vehicles are modeled as Dubins vehicles, *i.e.,* having a constant speed and a minimum turning radius constraint. The investigated integrated problem of task assignment and motion planning is posed in the form of a decision tree, and two search algorithms are proposed: an exhaustive algorithm that improves over run time and provides the minimum cost solution, encoded in the tree, and a greedy algorithm that provides a quick feasible solution. To satisfy the target's visitation timing constraint, a path elongation motion planning algorithm amidst obstacles is provided. Using simulations, the performance of the algorithms is compared, evaluated and exemplified.

Keywords: UAV; task assignment; motion planning; obstacles; prioritized targets; Dubins car

1. Introduction

Unmanned vehicles are currently used in a variety of civil and military missions and are gradually replacing manned vehicles. The need for autonomous capabilities is derived from the fact that the number of unmanned vehicles used in each mission has increased dramatically, and the required collaboration between them for the successful completion of the mission cannot be achieved if each vehicle is operated individually. Furthermore, the complexity of the missions and the number of simultaneous actions to be performed may cause operator overload, which will lead to deterioration in the overall mission performance. In order to maximize performance, unmanned vehicles are expected to work together in coordination as a team. The overall team performance is expected to exceed the sum of the performances of the individual unmanned vehicles.

Two main aspects of the coordination and cooperation of a team of unmanned vehicles are path planning and assignment allocation, usually referred to as motion planning and task assignment problems, respectively. In the task assignment problem, a group of agents needs to be assigned to perform a number of tasks. The tasks can be performed by any of the group's agents while minimizing or maximizing an objective function, depending on the scenario. An assignment task might be presented as a problem in graph theory [1], where the data in the graph are represented by vertices and edges. Such problems are commonly solvable using search algorithms by exploring the data structure level by level (breadth-first search) or reaching the leaf node first and backtracking (depth-first search).

The motion planning problem consists of planning a path for the vehicle while taking into account its kinematic and dynamic constraints, as well as generating feasible paths. The constraints may include a minimum turn radius and/or velocity limits, but there could also be obstacles scattered in the vehicle's environment that need to be taken into consideration. In many cases ([2–5]), for the motion planning, the vehicle is modeled as a Dubins vehicle [6]: a vehicle moving in a plane while having a turn rate constraint. Extending the Dubins model with altitude control, time optimal paths between initial and final configurations are provided in [7]. Considering obstacles, motion planning algorithms for the Dubins vehicle are provided in [8–12]. In [13], a collision-free 3D motion planning algorithm is provided for an aerial vehicle. When using the Dubins model, the resulting trajectory is composed of straight lines and arcs of a minimum turn radius. Discontinuities in the curvature of the trajectory arise at the junctions between the line and arc segments, causing tracking errors when followed by an actual vehicle. To overcome such problems, an algorithm was proposed in works, such as [14,15], for generating a continuous-curvature path between an ordered sequence of waypoints (the junctions between the line and arc segments) produced by the motion planner.

The task assignment problem is usually coupled with that of motion planning, as the assignments allocation process depends on the path length, and the path length depends on the vehicle's assignments. This coupling issue is addressed in the unmanned vehicles cooperative multiple task assignment problem (CMTAP) [16]. The CMTAP includes a scenario in which multiple unmanned vehicles perform multiple tasks on stationary targets. Different approaches based on customized combinatorial optimization methods were employed to solve this problem, including the mixed integer linear programming (MILP) [17,18], the capacitated transhipment network solver [19,20], genetic algorithms [16,21] and tree search methods [22,23]. In [19,21,24], timing and precedence constraints are also considered. In such scenarios, a target can be visited by a vehicle only if a specific task had first been performed on the target and a timing constraint was fulfilled. A method to elongate minimum distance paths for constant speed vehicles to meet the target timing constraints is presented in [25]. The presented works account for the vehicles' constraints, but they simplify the problem by assuming that the environment is obstacle free. Most of the studies that take into account obstacles address only the motion planning subproblem between the initial and final configuration. They include methods such as the rapidly-exploring random trees (RRT) method [26], probabilistic roadmaps [27] and the kinodynamic method [28].

One of the main properties of the problem stated above is the assumption that the targets have the same characteristics and differ only in their position. In many scenarios, each target has unique attributes, which include different importance and priority. The targets' priority may also vary in time depending on the specific scenario. Cases in which targets are assigned with a priority value were studied in [29–31]. The targets' priority was addressed by using an objective function, which includes a constant parameter describing the priority value. In these works, the vehicles' constraints were not taken into account, and the environment was assumed to be free of obstacles, which may lead to infeasible trajectories.

In this paper, the task assignment problem coupled with the problem of motion planning for a team of fixed-winged unmanned aerial vehicles that needs to service (fly over) multiple targets, while taking into account the vehicles' kinematic constraints and the need to avoid obstacles scattered in the environment, is addressed. It is also assumed that vehicles carry downward pointing body-fixed sensors and, thus, are required to approach a target flying straight and level. The main contribution of this paper is incorporating these constraints together with the targets' priority to create a more realistic time-varying priority scenario and by proposing a path elongation algorithm, used to consider the targets' timing constraints dictated by the different scenarios' characteristics. In order to solve this coupled problem, it is represented as a decision tree, and two tree search algorithms are proposed.

The remainder of this paper is organized as follows: In Section 2, a mathematical formulation of the problem is given. Section 3 describes the motion planning subroutine used. In Section 4, a solution to the task assignment problem is proposed. In Section 5, the simulation results of different sample runs are provided, and concluding remarks are offered in Section 6.

2. Problem Formulation

The problem considered in this work includes allocating a group of fixed-winged aerial vehicles to a given set of targets, while taking into account the vehicles' kinematic constraints and avoiding collision with obstacles scattered in the environment. It is assumed that vehicles carry downward pointing body-fixed sensors and, thus, are required to approach a target flying straight and level. Each target is assigned with a time-dependent value (referred to as the target benefit) that represents the target's importance and priority. The objective is to maximize a reward function, which is the sum of all of the benefits gathered by the group of vehicles.

2.1. Example Scenario

The motivation for solving this problem can be explained using the following example: A network of unattended ground sensors (UGS) and a team of unmanned vehicles are used to prevent intruders' access to a restricted zone (base defense) [32]. The UGS network is deployed at critical road junctions, and when a sensor is triggered by an intruder, the location is sent as a target to be visited by the team of unmanned vehicles. If there are multiple intrusions at different times, the group of vehicles must be allocated according to the vehicles' response time (time to target) and the target's priority, which can be based on the order of the UGS triggering time or on the location of the sensor. The target (sensor) priority is time dependent, since, as time passes, the intruder may advance to a different location, and the target relevance decreases. Additionally, the unmanned vehicle may have a timing constraint for visiting the target, e.g., only after it has been classified and cleared from friendly forces by the ground forces. This introduces a timing constraint that needs to be considered when allocating targets to the team of vehicles.

2.2. Vehicles

Let $V = \{V_1, V_2, ..., V_{N_V}\}$ be a set of unmanned aerial vehicles (UAVs) that need to complete the visit requirements of the given set of targets. The vehicles have a minimum turn radius and can move only forward at constant speed. The kinematic constraints need to be accounted for when planning the vehicles' trajectory. The equations of motion are presented below:

Vehicle kinematics:

$$
\begin{aligned}
\dot{x} &= U \cos \psi \\
\dot{y} &= U \sin \psi \\
\dot{\psi} &= \omega
\end{aligned}
\tag{1}
$$

Turn rate constraint (given a minimum turn radius):

$$|\omega| \leqslant U / R_{\min} \tag{2}$$

where (x, y) are the vehicle's Cartesian coordinates, ψ is the vehicle's orientation angle and U and ω are the vehicle's constant speed and turn rate, respectively. A schematic planar view of the vehicle's kinematics is presented in Figure 1. It should be noted that the above kinematics may also represent the motion of other types of vehicles moving in a plane, such as ground vehicles.

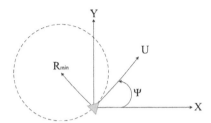

Figure 1. Vehicle kinematics.

The set of initial conditions that represents the vehicles' initial position and orientation is given by $V_{IC} = \{(x_{1_0}, y_{1_0}, \psi_{1_0}), (x_{2_0}, y_{2_0}, \psi_{2_0}), ..., (x_{N_{V_0}}, y_{N_{V_0}}, \psi_{N_{V_0}})\}$.

2.3. Body-Fixed Sensors

UAV sensors can be roughly divided into two categories: gimballed and body fixed. Gimballed sensors are usually more complex and enable pointing the sensor to a desired position, with usually minimal effect of the UAV's state. Body-fixed sensors are usually much simpler and less expensive, but their footprint is determined by the UAV's states, such as pitch and roll angles. Figure ? [33], presents a schematic example of the footprints of gimballed and body-fixed sensors. In the figure, UAV#1 is carrying a gimballed sensor, which can be moved within a larger possible footprint. Target 1 (T1) is enclosed by this larger possible footprint, but the sensor is currently pointing to a different direction. UAV#2 is carrying a body-fixed sensor, and Target 3 (T3) is within its footprint; but, its tracking will not be assured if the UAV rolls or pitches. Target 2 (T2) is outside the footprints of both UAVs.

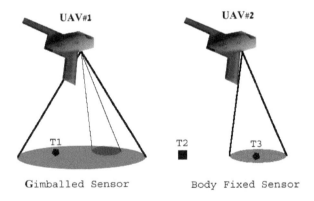

Figure 2. Sensor footprint schematic examples.

In this research, it is assumed that the fixed-winged UAVs carry body-fixed sensors that point directly downwards. Consequently, to ensure that the designated target will be inside the sensor's field-of-view, it is required that the UAVs approach a target flying straight and level. This ensures that the UAVs do not bank before crossing the target.

2.4. Obstacles

Let Ω be the two-dimensional physical environment in which the vehicles move ($(x,y) \in \Omega$) and the targets are located. Let $\mathcal{O} \subset \Omega$ be the set of obstacles that the vehicles need to avoid. $\mathcal{O}$ is considered to be a set of disjoint convex polygons. It is required that:

$$(x(t),y(t)) \cap \text{int}(\mathcal{O}) = \varnothing \tag{3}$$

In this work, it is assumed that a vehicle is allowed to graze the obstacles' boundaries, but it cannot penetrate them. In reality, the obstacles considered by the algorithm would be slightly larger in size than the actual obstacles, so as to ensure that the vehicle is safe, even if the algorithm requires the vehicle to take a path that grazes an obstacle boundary.

2.5. Targets and Benefits

Let $T = \{T_1, T_2, ..., T_{N_T}\}$ be the set of N_T stationary targets, located in Ω, designated to the group of fixed-winged unmanned aerial vehicles. It is assumed that the minimum distance between each pair of targets is larger than $2R_{min}$.

The set of timing constraints assigned to each target is given by $tc = \{tc_1, tc_2, ..., tc_{N_T}\}$. The vehicle is allowed to visit and perform its given task at the target only if the time required for a vehicle to arrive from its initial configuration (x_0, y_0, ψ_0) to the target is greater than or equal to the time constraint specified for the relevant target.

Let $C = \{C_1, C_2, ..., C_{N_T}\}$ be the set of initial benefits assigned to each target, and let $S = \{1, 2, ..., N_T\}$ be the set of stages in which a target is allocated as an assignment to a vehicle. The set of stages S is used to keep track of the vehicles' assignments history, which is of high importance when calculating the vehicles' path length. Furthermore, each stage corresponds to a layer in the tree representation used in Section 3.1.

The target's benefit represents the value granted to a vehicle for visiting the target. Since the benefits are time dependent, a mathematical formulation, which is referred to as the "benefit function", is proposed. This formulation represents the reward granted to a vehicle for visiting a target, depending on the target's priority (represented by its initial benefits) and the time required for a vehicle to arrive at the target from its initial position.

Let $x_{ik}^m \in \{0,1\}$ be a binary decision variable that equals one if vehicle $i \in V$ visits target $k \in T$ at stage $m \in S$ and is zero otherwise, and let $X_m = \{x_{ik}^1, x_{ik}^2, ..., x_{ik}^m\}$ be the set of assignments up to and including stage m. Let:

$$t_{ik}^m = L_{ikm}^{X_{m-1}} / U \tag{4}$$

be the time required for vehicle $i \in V$ to travel to target $k \in T$ at stage $m \in S$. This time is obtained through the division of the vehicle i path length to target k at stage m, notated as $L_{ikm}^{X_{m-1}}$, by its constant speed. Note that $L_{ikm}^{X_{m-1}}$ depends on the position and orientation of vehicle i before stage m, which, in turn, depends on the initial position and orientation of the vehicle and the targets it visited in the subsequent stages until stage m. The assignment history prior to stage m is included in the vehicle path length expression by the notation X_{m-1}.

The benefit function, which represents the reward granted to vehicle i for visiting target j, is formulated as follows:

$$C_j e^{-A \sum_{m=1}^l \sum_{k=1}^{N_T} t_{ik}^m x_{ik}^m} \tag{5}$$

where A is a user-defined coefficient, which defines the benefit function's descent rate, and:

$$\sum_{m=1}^{l} \sum_{k=1}^{N_T} t_{ik}^m x_{ik}^m \tag{6}$$

is the time required for vehicle i to arrive from its initial configuration $(x_{i_0}, y_{i_0}, \psi_{i_0})$ to target j that is visited at stage l.

This formulation helps create a problem in which the vehicle assignments' order depends on the path to each target and not only on the target's initial priority (for example, the highest priority target is not necessarily visited first, and the time to arrive at the target's location is also taken into consideration). In addition, the same formulation can be used to describe the example described in Section 2.1. The example includes a UGS network and a team of unmanned vehicles used for intruder detection and identification. The vehicles' response time is taken into account by calculating the vehicles' path length, and the targets' different initial priority represents the order of the UGS triggering time. Since the time it takes a vehicle to reach a target depends on the vehicle's path length, the latter will be calculated using a motion planning subroutine, described in Section 3. Figure 3a shows the benefit function change over time, each curve beginning with a different initial value (initial benefits three and 10). The benefit function is a monotonically-decreasing function, and as such, the initial value diminishes as time progresses. When the descent rate coefficient is changed (increased by five times), the benefit rapidly diminishes over time, as seen in Figure 3b. The increase of the descent rate may also cause a change in the targets assigned to each vehicle or a different order in which the assigned targets need to be visited.

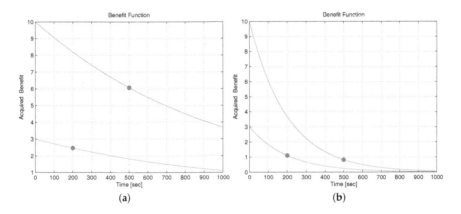

Figure 3. Benefit function. (**a**) Benefit function over time; (**b**) benefit function over time, increased decent rate.

2.6. Cost Function

The objective is to complete the visit requirement (visiting the given set of targets once) so as to maximize a reward function. The reward function considered is the overall benefits acquired by the vehicles:

$$J_1 = \sum_{i=1}^{N_V} \sum_{l=1}^{N_T} \sum_{j=1}^{N_T} [C_j e^{-A\sum_{m=1}^{l} \sum_{k=1}^{N_T} t_{ik}^m x_{ik}^m}] x_{ij}^l \tag{7}$$

where $x_{ij}^l \in \{0,1\}$ is a binary decision variable that equals one if vehicle i visits target j at stage l. Since the benefit function diminishes with time, a "lost" benefit function is formulated and is given by $C_j - C_j e^{-A\sum_{m=1}^{l} \sum_{k=1}^{N_T} t_{ik}^m x_{ik}^m}$. The "lost" benefit function formulation allows the definition of a cost function, which is the equivalent to the reward function defined above, but instead of maximizing the reward function, the objective is to minimize the cost function. The cost function mathematical formulation is given by:

$$J_2 = \sum_{i=1}^{N_V} \sum_{l=1}^{N_T} \sum_{j=1}^{N_T} [C_j - C_j e^{-A\sum_{m=1}^{l} \sum_{k=1}^{N_T} t_{ik}^m x_{ik}^m}] x_{ij}^l \tag{8}$$

The constraints of the problem are given by:

$$\sum_{l=1}^{N_T}\sum_{i=1}^{N_V} x_{ij}^l = 1, \ j = 1,\ldots, N_T \tag{9}$$

$$\sum_{i=1}^{N_V}\sum_{j=1}^{N_T} x_{ij}^l = 1, \ \forall\, l = 1,\ldots, N_T \tag{10}$$

$$\sum_{m=1}^{l}\sum_{k=1}^{N_T} t_{ik}^m x_{ik}^m \geq tc_j, \ \forall\, j = 1,\ldots, N_T, \ tc_j \in tc, \ s.t. \ x_{ij}^l = 1 \tag{11}$$

Equation (9) ensures that each target is visited once. Equation (10) ensures that only a single vehicle is assigned to a target in each stage. In Equation (11), the timing constraint is posed. The time required for a vehicle to arrive at the target location from its initial configuration must be greater than or equal to the time constraint dictated as part of the problem's initial parameters. If the vehicle's arrival time at the target is less than the time constraint tc_j, the path should be elongated, otherwise the third constraint will be violated.

In [16,21,23], somewhat similar problems involving multiple targets and vehicles were solved. The cost function used in the related works is the sum of the path lengths of all of the vehicles and can be formulated as:

$$J_3 = \sum_{i=1}^{N_V}\sum_{l=1}^{N_T}\sum_{j=1}^{N_T} L_{ij}^l x_{ij}^l \tag{12}$$

In these cases, the targets' importance is identical and ignored when solving the problem. In the simulation results' section, this cost function is used to help compare the performance of the proposed algorithms.

The solution process of the problem includes solving two integrated subproblems: task assignment and motion planning problems. To minimize the cost function, the task assignment depends on the underlying motion planning for the path length, while the motion planning depends on the task assignment for the order of the vehicle's targets. This makes the problems coupled.

3. Motion Planning

For the motion planning problem, it is assumed that each vehicle is assigned a list of an ordered set of targets, made by the task assignment algorithm. The goal of the motion planning is to derive a trajectory for each vehicle to visit all of the targets on the list, avoid collision with obstacles and respect the vehicle kinematic constraints (described in Section 2.2) and the timing constraint (described in Equation (11)). Due to the sensor-oriented requirement of having the vehicles fly straight and level when approaching a target (see Section 2.3), the planner needs to issue a trajectory with straight line segments preceding the arrival of a vehicle to a target.

The motion planning problem description and solution presented below is based on the study described in [12]. Since the targets' timing constraints are not included in [12], a path elongation algorithm is proposed in Subsection 3.2.

3.1. Tree Formulation

In order to represent the motion planning problem in the form of a decision tree, it is necessary to generate nodes representing the following: targets position, vehicle's initial configuration and obstacles' vertices (under the assumption of polygonal obstacles). The vehicle's path will either be a direct path (free of obstacles) connecting the initial configuration and the set of targets, or a path that also passes through some of the obstacles' vertices, in case a direct path does not exist. Each branch of the tree represents the described path. The root node (initial configuration) is connected to all of the target nodes, and if a direct path is not feasible, obstacles nodes are also included. The goal is to find the branch that provides the minimum time path. In order to to satisfy the timing constraint described in Equation (11), a path elongation is provided in the following section.

The vehicles in this work are modeled as Dubins vehicles. The Dubins path is a concatenation of arcs of minimum radius turn and straight line segments that connect an initial and final configuration (position and orientation). The optimal path can be achieved by checking six possible path types for the Dubins vehicle [6]. If the orientation angle in the final configuration is removed, the number of possibilities is reduced. This is known as the relaxed Dubins path that include only four possibilities [34].

An important benefit obtained by using the relaxed path is explained using the example given in Figure 4. When calculating the optimal path between an initial (Node 1), final (Node 5) configurations and three additional unordered configurations (for example: obstacle vertices) located between them (Nodes 2–4), the following branches of the tree graph are generated: A branch connecting nodes 1-2-3-4-5 and a branch connecting nodes 1-2-4-3-5. In the relaxed case, the path connecting Nodes 1 and 2 should be calculated only once, as it is independent of the remaining nodes. However, in the non-relaxed case, the arrival angle at Node 2 depends on the order of the following nodes (Node 3 or 4), and the path between Nodes 1 and 2 needs to be calculated separately for each branch. This attribute, where the path between two nodes does not depend on the following nodes, enables us to pose the problem as a tree.

Moreover, as the minimal distance between each of two targets is larger than two times the minimum turn radius of the vehicles, it is guaranteed that the optimal relaxed paths between each pair of targets consist of a terminal straight line segment, satisfying the body-fixed sensor-originated requirement that a UAV approaches a target flying straight and level, discussed in Section 2.3.

In order to find the relaxed optimal path that connects the initial configuration and the targets' set and does not intersect with obstacles, it is necessary to search the tree presented above. The search process includes calculating the relaxed path connecting the different graph nodes: obstacles' vertices or targets' positions. In this search process, the calculation of the relaxed path is repeated multiple times, especially in large-scale scenarios. When real-time scenarios are considered, the use of the relaxed path becomes highly beneficial, as the computational complexity is significantly reduced, compared to the non-relaxed case.

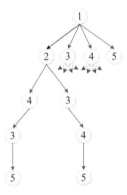

Figure 4. An example of a five-configuration tree.

In this work, an existing motion planning algorithm is used to find an admissible path of bounded curvature through a given ordered set of points among obstacles without timing constraints. The motion planning solution can be achieved by one of two algorithms: (1) an exhaustive algorithm, which explores every branch of the search tree and evaluates every possible visit order sequence in order to find the minimum cost one (See Algorithm B1 in the Appendix); and (2) an A* like heuristic algorithm, which uses Euclidean distances as a heuristic estimation and a greedy approach to find a feasible path (See Algorithm A1 in the Appendix). Since the existing algorithms do not take into account the timing constraints when calculating the vehicles' trajectory, a path elongation algorithm amidst obstacles is proposed in the following section. The combined motion planning algorithm and the path elongation algorithm are used as a subroutine for the developed task assignment algorithm.

3.2. Path Elongation Algorithm

A path elongation algorithm, used to lengthen a vehicle's path that connects an initial configuration and a target position in an environment with obstacles, is now presented. Different methods can be used to elongate the vehicle path. For example, a different (higher cost) branch of the tree can be used instead of the current branch. Another possibility is to increase the vehicle's turn radius to elongate the path. These methods, however, proved to be inefficient in terms of computational run time.

The path elongation method used in this paper is based on appending loitering circles. The technique forces the vehicle to perform a circular flight with a minimum turn radius around a specific point until the timing constraint is fulfilled. One of the motion planning algorithms is used to generate the vehicle's path. If the timing constraint is not fulfilled, the path elongation algorithm is employed.

The algorithm is given in Algorithm 1. The inputs to the algorithm are the vehicle's configuration and speed, obstacle vertices' locations, target position, time constraint and the vehicle's current path, the order and orientation in which nodes are visited. The output is the vehicle's trajectory in terms of the order in which nodes are to be visited, the node in which the loitering circles are performed and the number of cycles.

Algorithm 1 Path elongation algorithm amidst obstacles.

  Input: Initial position and orientation of vehicle; vehicle speed; obstacle information; target position; path nodes position and orientation; path length; time constraint;
  Output: Vehicle trajectory (visit order and elongation circles node if needed)
1: EvList ← generate a list of all elongation vertices in the environment
2: DpList ← Construct a sorted list of direct Dubins path vertices list
3: **if** EvList ∩ pathNodes ≠ ∅ **then**
4:   elongationVertex ← EvList ∩ pathNodes
5:   elongationNode ← elongationVertex(1)
6:   loiterCirclesNumber ← ⌈(timeConstraint-(pathLength/VehicleSpeed))/2πR_{min}⌉
7:   newVisitOrder ← path nodes
8:   **return** newVisitOrder, elongationNode, loiterCirclesNumber
9: **end if**
10: **for** iPathNodes = 1 to nPathNodes **do**
11:   intersectionCounter← checkCirclesIntersection[iPathNodes, obstaclesInfo]
12:   **if** intersectionCounter ≠ ∅ **then**
13:     elongationNode ← pathNodes(iPathNodes)
14:     loiterCirclesNumber ← ⌈(timeConstraint-(pathLength/VehicleSpeed))/2πR_{min}⌉
15:     newVisitOrder ← path nodes
16:     **return** newVisitOrder, elongationNode, loiterCirclesNumber
17:   **end if**
18: **end for**
19: **if** DpList ∩ EvList ≠ ∅ **then**
20:   newDplist=DpList ∩ EvList
21:   **for** j = 1 to newDplistLength **do**
22:     [newVisitOrder, newPathLength, newPathNodesAngles, intersectionCounter] ← motionPlanningAlgo[vehicleConfiguration, newDplist(j), target, obstaclesInfo]
23:     **if** intersectionCounter=∅ **then**
24:       elongationNode ← newDplist(i)
25:       **if** timeConstraint > newPathLength/VehicleSpeed **then**
26:         loiterCirclesNumber ← ⌈(timeConstraint-(newPathLength/VehicleSpeed))/2πR_{min}⌉
27:       **end if**
28:       **return** newVisitOrder, elongationNode, loiterCirclesNumber
29:     **end if**
30:   **end for**
31: **end if**
32: sortedEvList ← euclideanSort[EvList]
33: **for** j = 1 to sortedEvListLength **do**
34:   [newVisitOrder, newPathLength, newPathNodesAngles, intersectionCounter] ← motionPlanningAlgo[vehicleConfiguration, sortedEvList(i), target, obstaclesInfo]
35:   **if** intersectionCounter=∅ **then**
36:     elongationNode ← sortedEvList(i)
37:     **if** timeConstraint > newPathLength **then**
38:       loiterCirclesNumber ← ⌈(timeConstraint-(newPathLength/VehicleSpeed))/2πR_{min}⌉
39:     **end if**
40:     **return** newVisitOrder, elongationNode, loiterCirclesNumber
41:   **end if**
42: **end for**

Since there are obstacles scattered in the environment, the vehicle cannot perform loitering circles around any given point without intersecting with the obstacles. Let us define a set of points that will be referred to as "elongation vertices". This set includes obstacles' vertices in which the vehicle can perform a loiter circle (of minimum turn radius) without intersecting with the obstacles boundaries for any given feasible orientation; infeasible orientations are defined as angles in which the vehicle cannot be positioned without penetrating the obstacle boundary and, therefore, are initially not included in the set of arrival angles, forming the vehicle's path.

It should be noted that, as discussed in Section 2.4, a vehicle is allowed to graze the obstacles' boundaries, but it cannot penetrate them. In reality, the obstacles considered by the algorithm would be slightly larger in size than the actual obstacles, so as to ensure that the vehicle is safe, even if the algorithm requires the vehicle to take a path that grazes an obstacle boundary. Thus, it is safe for the vehicles to perform loitering circles that graze an obstacle edge.

An example of the elongation vertices is given in Figure 5. This example includes a vehicle, a target and two rectangular obstacles. On each of Obstacle 1's vertices, loitering circles are drawn. Each circle corresponds to a specific feasible vehicle orientation (a 30° discretization of the feasible angles' range was used in this example). As can be seen, Vertex 3 can be defined as an elongation vertex, while Vertex 4 is not part of the this set, as several loitering circles in this vertex intersect with Obstacle 2. A zoom-in of Vertex 3 is presented in Figure 6. The discretization of the orientation angles is clearly visible, with each circle corresponding to two feasible angles. All of the loitering circles in each vertex can be enclosed by a polygon, as presented in the figures.

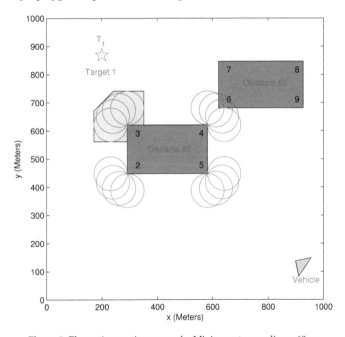

Figure 5. Elongation vertices example: Minimum turn radius = 60 m.

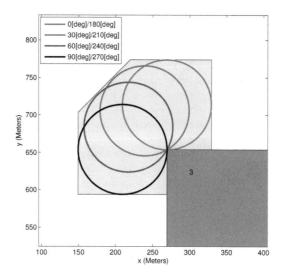

Figure 6. Elongation vertex: 30° discretization.

Figure 7 shows the range of infeasible/feasible orientation angles at a specific obstacle vertex. It is important to notice that each of the rectangle obstacle's vertices has a different distribution of infeasible/feasible orientation.

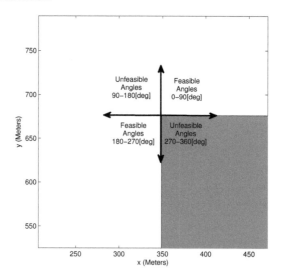

Figure 7. Feasible and unfeasible angles example.

The key idea behind the proposed algorithm is to locate a node of the input path that is also an elongation vertex or to generate a new path that includes at least one elongation vertex. If a new path is generated and the timing constraints are not satisfied, the vehicle can perform the required amount of loitering cycles at the elongation vertex.

4. Task Assignment

The task assignment solutions are now provided. First, the problem is presented as a tree (as can be seen in Figure 8) by generating nodes that describe a vehicle V_i assigned to a target T_j. Figure 8 presents a search tree for a scenario in which three targets need to be assigned to two vehicles. For a concise illustration, only some of the branches of the tree are shown. The branch shown by a dashed line gives an assignment V_1T_1, V_1T_3, V_2T_2. This means that target T_2 is assigned to vehicle V_2, and targets T_1 and T_3 are assigned to vehicle V_1, which must visit the assigned targets in that specific order. Each node of the tree is associated with a cost. For example: node V_iT_j has a cost that equals the "lost" benefit granted to vehicle i for visiting target j. The "lost" benefit value depends on the time that it takes vehicle i to reach target j from its initial position, which depends on the vehicles' path length. The path is obtained using a motion planning subroutine (described in Section 3), which guarantees a feasible path for the vehicles. Since the motion planning subroutine is used in the task assignment process, the problem solution consists of a primary task assignment tree search, which depends on a secondary motion planning tree search.

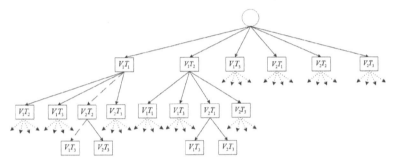

Figure 8. A tree for two vehicles and three targets.

Two algorithms that provide solutions to the task assignment problem are proposed, an exhaustive search algorithm and a greedy algorithm. The greedy algorithm provides a computationally-fast solution that may not be optimal, and the exhaustive algorithm explores all of the assignment possibilities to derive an assignment's allocation with the minimum cost value. The main use of the greedy algorithm may be in scenarios where the assignment cannot be planned beforehand and needs to be planned in real time. Such cases arise when new targets pop-up during the engagement and/or other relevant changes occur in the scenario (like a loss of a vehicle).

4.1. Exhaustive Task Assignment Algorithm

The proposed algorithm that is described in Algorithm 2 explores every branch of the tree to evaluate all of the assignment's possibilities.

The inputs to the algorithm are the initial configuration and constant speed of the vehicles, the locations of the target points, information about the obstacles and the time constraint of each target.

The algorithm's first step includes generating an upper bound on the cost (Line 1), calculated by using the greedy algorithm described in Section 4.2. This upper bound will be useful to bound the branching of the tree, thus preventing unnecessary explorations. Then the following lists are initialized (Lines 2–5). A TargetsList variable is constructed that includes all of the targets that need to be visited by the vehicles, and the corresponding time constraints are entered into a TimeConstraint list. Furthermore, a vehicleTargetList that contains the assigned targets (and order of visit) of each vehicle and an OpenSet that stores all of the nodes to be examined are initialized to an empty set.

Next, the first layer of the tree (as presented in Figure 8) is generated. All possible vehicle-target pair combinations are described as nodes and added to OpenSet for further exploration (Lines 6–26). Each node's vehiclePath and vehicleTargetsList fields are initialized to zero and empty vectors, respectively (Line 8). Then, the current target (T_j) is assigned to the current vehicle (V_i) (Lines 9–10), and the value of the path length and the cost are updated in the corresponding position (*i*-th entry) (Line 11, Line 15).

Algorithm 2 Task assignment exhaustive search algorithm.

Input: Vehicles' initial configuration (V) and constant speed, targets' position and their visit requirements (T), obstacle vertices' positions, targets' time constraints (tc)
Output: Assignment for each vehicle and the order in which the assigned targets needs to be visited

```
1:  UpperBound ← greedy task assignment algorithm solution
2:  TargetsList ← {T_j, j = 1,...,N_T}
3:  TimeConstraint ← {tc_j, j = 1,...,N_T}
4:  vehicleTargetList(i) ← [ ], i = 1,...,N_V
5:  OpenSet ← [ ]
6:  for V_i, i = 1,...,N_V do
7:      for T_j ∈ targetsToVisit do
8:          node.vehiclePath(k) ← 0 , node.vehicleTargetsList(k) ← [ ], k = 1,...,N_V
9:          node.vehicle ← V_i
10:         node.vehicleTargetsList(i) ← T_j
11:         node.vehiclePath(i)=PathLength(V_i, T_j)
12:         if TimeConstraint(T_j) < (node.vehiclePath(i)/VehicleSpeed(V_i)) then
13:             node.vehiclePath(i)=PathElongation(V_i, T_j)
14:         end if
15:         node.cost ← lostBenefitFunction(PathLength(V_i, T_j))
16:         node.targetsList ← targetsList \ T_j
17:         if node.targetsList = ∅ then
18:             if Cost(node) < UpperBound then
19:                 UpperBound ← Cost(node.cost)
20:                 vehicleTargetsList(i) ← node.vehicleTargetsList(i)
21:             end if
22:         else
23:             OpenSet ← OpenSet ∪ node
24:         end if
25:     end for
26: end for
```

Algorithm 2 Task assignment exhaustive search algorithm (continued).

```
27: while OpenSet ≠ ∅ do
28:     parentNode ← OpenSet(last entered node)   – depth-first search
29:     for V_i, i = parentNode.vehicle,...,N_V do
30:         for T_j ∈ targetsToVisit do
31:             childNode.cost(k) ← parentNode.cost(k),
                childNode.vehicleTargetsList(k) ← parentNode.vehicleTargetsList(k),
                childNode.vehiclePath(k) ← parentNode.vehiclePath(k), k = 1,...,N_V
32:             if TimeConstraint(T_j) < (childNode.vehiclePath(i)
                +PathLength(V_i, T_j))/VehicleSpeed(V_i)) then
                PathLength(V_i, T_j)=PathElongation(V_i, T_j)
33:             end if
34:             childNode.vehiclePath(i)=childNode.vehiclePath(i) + PathElongation(V_i, T_j)
35:             childNode.cost(i) ← lostBenefitFunction(childNode.vehiclePath(i))
36:             if Cost(childNode) < UpperBound then
37:                 childNode.vehicle ← V_i
38:                 childNode.vehicleTargetsList(i) ← [childNode.vehicleTargetsList(i) T_j],
39:                 childNode.targetsList ← parentNode.targetsList \ T_j
40:                 if childNode.targetsList = ∅ then
41:                     UpperBound ← Cost(childNode.cost)
42:                     vehicleTargetsList(i) ← childNode.vehicleTargetsList(i)   i = 1,...,N_V
43:                 else
44:                     OpenSet ← OpenSet ∪ childNode
45:                 end if
46:             end if
47:         end for
48:     end for
49:     OpenSet ← OpenSet \ parentNode
50: end while
51: return vehicleTargetsList
```

The path length function returns the length of the relaxed path while avoiding obstacles between the vehicle's current location and the target's position. If the given time constraint is not satisfied, the path elongation algorithm is employed. If the path elongation algorithm is unable to satisfy the time constraint, the vehicle's path length is given a value equal to infinity; this is done in order to ensure that the current branch is pruned, since the problem constraints are not fulfilled (Lines 12–14). The current target can now be removed from the targetsList of the expended branch (Line 16).

If a leaf node is reached (Line 17) and its cost is lower than the current upper bound (Line 18), the upper bound is updated to the cost of the current node (Line 19), and the vehicle's assigned targets are stored in the vehicleTargetList variable (Line 20). Otherwise, the node is added to OpenSet for further exploration (Line 23).

After the initial nodes are generated, the exhaustive search begins. A depth first search is used to expand a branch until a leaf node is generated (Lines 27–52). While OpenSet is not empty, the last node inserted into OpenSet is chosen as the current parent node for further exploration, and the corresponding children nodes are created (Lines 29–49). Each child inherits the cost, vehicleTargetsList and vehiclePath fields from the parent node (Line 31). The child cost is calculated using the lostBenefitFunction, which is based on the vehicle's accumulated path length (Lines 35–36). Additionally, the target's time constraint is again compared to the vehicle's accumulated path length, and if needed, the path elongation algorithm is used (Lines 32–34). If the child cost is lower than the current upper bound (Line 37), the child node's vehicle, targetsList and vehicleTargetsList fields are updated accordingly (Lines 38–40). Otherwise, the branching is bounded.

As before, if the child node has an empty targetList (Line 41), then the UpperBound and the vehicleTargetsList are updated (Lines 42–43). If a leaf node is not reached and the child node targetList is not empty, the child node is added to OpenSet for further exploration (Line 45). Once evaluated, the parent node is removed from OpenSet (Line 50). This process is repeated until all branches have been either bounded or completely explored; OpenSet is empty. The algorithm output is a minimum cost ordered set of targets, assigned to each vehicle (Line 52). Owing to the tree search involved, this algorithm has an exponential time complexity.

4.2. Greedy Task Assignment Algorithm

The proposed algorithm is based on a greedy search method that enables finding an assignment solution quickly and is described in Algorithm 3. Since the algorithm is greedy by nature, the objective function used is the reward function presented in Equation (7).

Algorithm 3 The heuristic greedy algorithm for task assignment.

Input: Vehicles' initial configuration (V) and constant speed, targets' position and their visit requirements (T), obstacle vertices; positions, targets' time constraint (tc)
Output: Vehicle-target list - the targets assigned to each vehicle and the required visitation order.

1: TargetsList $\leftarrow \{T_j, j = 1, \ldots, N_T \mid T_j$ has a visit requirement$\}$
2: TimeConstraint $\leftarrow \{tc_j, j = 1, \ldots, N_T\}$
3: vehicleTargetsList(i) $\leftarrow 0, i = 1, \ldots, N_V$
4: vehicleTotalBenefit(i) $\leftarrow 0, i = 1, \ldots, N_V$
5: accumulatedPathLength(i) $\leftarrow 0, i = 1, \ldots, N_V$
6: **while** TargetsList $\neq \emptyset$ **do**
7:      VehiclePath(V_i, T_j)=PathLength(V_i, T_j), $(i, j) \in \{1, \ldots, N_V\} \times \{1, \ldots, N_T\}$
8:      **if** TimeConstraint $<$ (VehiclePath(V_i, T_j)+accumulatedPathLength(V_i))/VehicleSpeed(V_i) **then**
9:          PathLength(V_i, T_j)=PathElongation(V_i, T_j)
10:      **end if**
11:      VehicleBenefit(V_i, T_j) $\leftarrow$ BenefitFunction(PathLength(V_i, T_j)+accumulatedPathLength(V_i)), $(i, j) \in \{1, \ldots, N_V\} \times \{1, \ldots, N_T\}$
12:      $(i^*, j^*) \leftarrow \arg\max_{(i,j) \in \{1, \ldots, N_V\} \times \{1, \ldots, N_T\}}$ VehicleBenefit(V_i, T_j) subject to: $T_j \in$ targetsList & $T_j \notin$ vehicleTargetsList(i)
13:      vehicleTargetsList(i^*) $\leftarrow$ [vehicleTargetsList(i^*) T_{j^*}]
14:      VehiclePosition(V_{i^*}) $\leftarrow$ VehiclePosition(T_{j^*})
15:      accumulatedPathLength(V_{i^*}) $\leftarrow$ accumulatedPathLength(V_{i^*}) + PathLength(V_{i^*}, T_{j^*})
16:      vehicleTotalBenefit(V_{i^*}) $\leftarrow$ vehicleTotalBenefit(V_{i^*}) + VehicleBenefit(V_{i^*}, T_{j^*})
17:      TargetsList $\leftarrow$ TargetsList$\setminus T_{j^*}$
18: **end while**
19: **return** vehicleTargetsList

The key idea behind the proposed task assignment algorithm is the following: each vehicle is assigned an associated reward, equal to the benefit value acquired by the vehicle until now. A vehicle is assigned to a target only if the benefit value acquired by traveling from its current location to the target is maximum for all vehicle-target pairs. The assigned vehicle is first required to visit the target to which it is assigned. Then, assuming that the vehicle is at the assigned target point and that the target point is already visited, the process is repeated; that is, finding the next target-vehicle pair that has the highest benefit value of all other target-vehicle pairs, and so on.

The inputs to the algorithm are the initial configuration and constant speed of the vehicles, the locations of the target points, information about the obstacles and the time constraint of each target.

In each stage of the proposed algorithm, the motion planning algorithm is used as a subroutine to calculate the vehicle's feasible trajectory to the relevant target point. Thus, the complexity of

Algorithm 3 can be either polynomial or exponential, depending on the motion planning subroutine used (heuristic or exhaustive). If real-time applications are considered, the A*-like motion planning heuristic algorithm should be used as a subroutine in Algorithm 3, leading to polynomial complexity.

5. Simulation Results

In this section, sample runs are used to demonstrate the presented algorithms and to explain the different parameters' (vehicle type, benefit's descent rate, *etc.*) influence on the obtained solution. In the first scenario, the path elongation algorithm is demonstrated. Since only a single target is considered in this scenario, the benefit issue is ignored, and the objective is to obtain a feasible path that satisfies the given time constraint. In all of the remaining scenarios, the task assignment algorithms (exhaustive or greedy) use the motion planning subroutine (exhaustive or heuristic) based on the relaxed Dubins distances; hence, the coupling of the problem is kept. The vehicles' turn radius is set to 60 m (in most cases, except where noted otherwise), and the targets' initial benefit values can be set between 1000 and 10,000 (values are presented in the figures next to each target as a numeral between one and 10).

5.1. Path Elongation Algorithm Demonstration

In the following scenario, an aerial vehicle needs to be assigned to a single target with a time constraint. The vehicle must wait until the time constraint is satisfied before visiting the target. Since the path elongation algorithm is divided into four sequential steps, four figures that present the working of each step are given. In each figure, the scenario is slightly altered to initiate the algorithm's different steps. The vehicle's arrival times at the target point for the different scenarios are summarized in Table 1.

Table 1. Path elongation results summary.

Figure #	Step #	Time Constraint (s)	Vehicle Trajectory Time (s)	Number of Cycles	Solution Time (s)
9a	Step #1	2000	2357	4	3.6×10^{-5}
9b	Step #2	2000	2357	4	9.6×10^{-5}
9c	Step #3	2000	2011	3	0.45
9d	Step #4	2000	2072	0	4

In Figure 9a, the vehicle performs four loitering circles before heading towards the target. According to Step 1, the obstacle's vertex, in which the circles are performed, is part of the elongation vertices list and is also one of the vehicle's original path nodes (without the time constraint). In Figure 9b, an additional obstacle is added to the same scenario. Since in this case, the vertex in which the circles were previously performed is not part of the elongation vertices list, the vehicle performs four loitering circles around the target point. This is done according to Step 2. The scenario in Figure 9c includes additional obstacles, which initiate the third step of the elongation algorithm. Since the vehicle is unable to perform circles around any of the original path nodes, a new path is generated, which includes a direct relaxed path node and the target point. Three cycles are performed around the direct path node to satisfy the given time constraint. In Figure 9d, a new path is generated according to Step 4. In this specific scenario, loitering circles are not performed, even though the new path passes through an elongation vertex. This is due to the fact that the time constraint is already satisfied, as can be seen in Table 1.

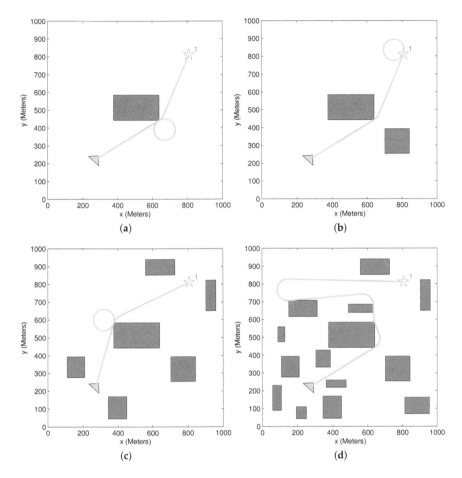

Figure 9. Path elongation demonstration. (**a**) Step #1; (**b**) Step #2; (**c**) Step #3; (**d**) Step #4.

5.2. General Scenario

Figure 10 presents a scenario in which two aerial vehicles need to visit four targets with different initial benefits. The scenario solution is obtained using different algorithm setups in each case. Table 2 summarizes the results of the different sample runs. The highest benefit (lowest lost benefit) and longest running time were gained using an exhaustive algorithm setup (Figure 10a). When heuristic motion planning is used instead (Figure 10c), the cost remains the same, but the running time decreases. In the case of a greedy task assignment and heuristic motion planning algorithms (Figure 10b), the lowest benefit (highest cost) and shortest running time are attained. The solution presented in Figure 10a demonstrates that the vehicles are generally first heading towards targets with high priority while taking into account targets with lower priority. Since the benefit diminishes with time, the vehicle does not head directly towards the high priority targets, but passes through low priority targets, which are closer to its location (upper vehicle on Figure 10a). In Figure 10b, the upper vehicles head directly to Target 6 (initial benefit value = 6) and skips Target 1, since, in this case, the task assignment algorithm is greedy by nature. In the scenarios presented in Sections 5.3 and 5.4, the exhaustive algorithms' setup yields the same results as the exhaustive task assignment algorithm and heuristic motion planning algorithm setup; therefore, only the latter setup is presented. Even though the results presented in these sections are identical, it can be shown that, in certain cases, the exhaustive algorithms' setup provides better results, although the solution running time becomes longer.

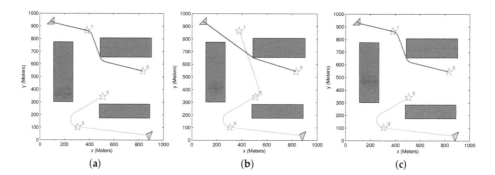

Figure 10. General scenario: two vehicles and four targets among obstacles. (**a**) Exhaustive task assignment algorithm; exhaustive motion planning algorithm; (**b**) greedy task assignment algorithm; heuristic motion planning algorithm; (**c**) exhaustive task assignment algorithm; heuristic motion planning algorithm.

Table 2. Different initial benefits scenarios.

Figure #	Algorithms Used	Initial Benefit	Acquired Benefit	Lost Benefit	Overall Distance	Solution Time (s)
10a	Exhaustive TAExhaustive MP	21,000	9918	11,082	1925	58.1
10b	Greedy TA Heuristic MP	21,000	9537	11,463	2423	1.05
10c	Exhaustive TA Heuristic MP	21,000	9918	11,082	1925	17.3

Figure 11 presents a similar scenario to that of Figure 10, but with different polygon obstacles. Pentagon obstacles are present in Figure 11a, while octagon ones are in Figure 11b. Table 3 summarizes the results of the different sample runs. It can be seen that the type of obstacles has a negligible effect on the acquired benefit. In contrast, it significantly affects the run time of the algorithm, as having more edges enlarges the search space.

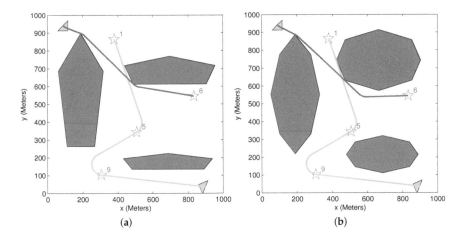

Figure 11. General scenario: two vehicles and four targets among different obstacle types; greedy task assignment algorithm and heuristic motion planning algorithm. (**a**) Pentagon obstacles; (**b**) Octagon obstacles.

Table 3. Different initial benefits scenarios: pentagon and octagon obstacles.

Figure #	Algorithms Used	Initial Benefit	Acquired Benefit	Lost Benefit	Overall Distance	Solution Time (s)
11a	Greedy TA Heuristic MP	21,000	9479	11,521	2446	1.85
11b	Greedy TA Heuristic MP	21,000	9440	11,560	2462	2.73

5.3. Equal Benefit Scenario

The scenario shown in Figure 12 is similar to the scenario shown in Figure 10, where only the targets' initial benefit is equal. Since the targets' priority is identical, it is expected that the results would be similar to the case where the cost function objective is to minimize the overall distance traveled by the vehicles (Equation (12)). In the results summarized in Table 4, the highest benefit (lowest cost) is obtained by the setup of the exhaustive task assignment algorithm (Figure 12b). The overall distance is the same as in the case of using the cost function, which minimizes the sum of the distance traveled (Figure 12c), as expected. As in the previous scenario, the solution running time has the same tendency, when using a greedy and heuristic algorithms' combination to gain the shortest running time; with an exhaustive algorithms' combination, the longest running time is gained. This tendency remains the same through all of the presented scenarios.

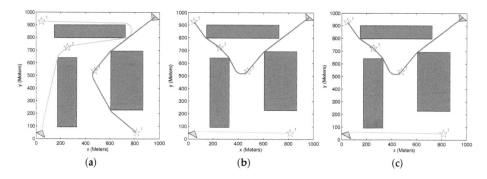

Figure 12. Equal benefits scenario: two vehicles and four targets among obstacles. (**a**) Greedy task assignment algorithm; heuristic motion planning algorithm; (**b**) exhaustive task assignment algorithm; heuristic motion planning algorithm; (**c**) exhaustive task assignment algorithm; exhaustive motion planning algorithm; minimize the sum of the overall distance traveled.

Table 4. Equal initial benefits scenario.

Figure #	Algorithms Used	Initial Benefit	Acquired Benefit	Lost Benefit	Overall Distance	Solution Time (s)
12a	Greedy TA Heuristic MP	4000	1399	2601	3353	1.3
12b	Exhaustive TA Heuristic MP	4000	1623	2377	2075	40.5
12c	Exhaustive TA Heuristic MP (Sum of path length cost function-Equation (12))	4000	1623	2377	2075	40.5

5.4. Comparing Exhaustive and Greedy Task Assignment Algorithms

A scenario involving three vehicles (each having a minimum turn radius of 100 m), five targets and eight obstacles is presented in Figure 13. The results of the scenario are given in Table 5. As expected, the exhaustive algorithm provides better or equal results compared to the greedy algorithm, at the expense of additional run time. This is also evident from Tables 2 and 6.

The main advantage of the greedy algorithm is its low computational time, which makes it suitable for real-time applications. In cases where both the exhaustive task assignment and the motion planning algorithms are used, the best solution coded in the tree is obtained, and the lowest cost assignments allocation and vehicles' paths are provided. However, due to the increased computational burden, such an algorithm may not be applicable for real-time application in a high dimensional problem and can mainly serve as a benchmark to evaluate (off-line) the performance of the greedy algorithm. Figure 14 presents the solution obtained using the greedy algorithm to a high dimensional problem involving seven vehicles, 11 targets and 23 obstacles. The run time of the algorithm was about three minutes, while the exhaustive algorithm did not return a solution within 24 h of run time. These computation times, as all other in this paper, were attained in a MATLAB implementation of the algorithm.

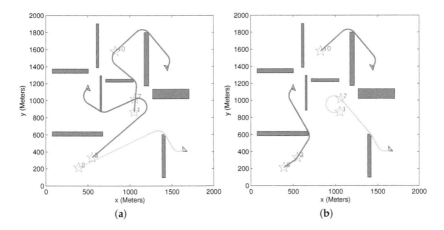

Figure 13. Comparison between exhaustive and greedy task assignment algorithms. (**a**) Greedy task assignment algorithm; heuristic motion planning algorithm; (**b**) exhaustive task assignment algorithm; heuristic motion planning algorithm.

Table 5. Comparing exhaustive and greedy task assignment algorithms.

Figure #	Algorithms Used	Initial Benefit	Acquired Benefit	Lost Benefit	Overall Distance	Solution Time (s)
13a	Greedy TA Heuristic MP	29,000	20,120	8880	5340	5
13b	Exhaustive TA Heuristic MP	29,000	18,050	10,950	3520	150

Table 6. Scenario 1 and Scenario 2.

Figure #	Algorithms Used	Initial Benefit	Acquired Benefit	Lost Benefit	Overall Distance	Solution Time (s)
15	Exhaustive TA Exhaustive MP	15,000	6571	8529	1371	0.06
15	Greedy TA Heuristic MP	15,000	6571	8529	1371	0.016
16	Exhaustive TA Exhaustive MP	15,000	6634	8366	1333	0.06
16	Greedy TA Heuristic MP	15,000	6634	8366	1333	0.015

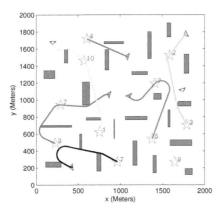

Figure 14. Complex scenario solved only using the greedy TA/MP algorithm.

5.5. Benefit Time Dependency

A simple scenario of one vehicle and two targets (initial benefit of 10 and 5) is used in Figures 15 and 16. The time dependency can be easily explained using these two figures. In Figure 15, the vehicle's path passes through Target 5, even though this causes the vehicle to extend its path toward Target 10. This happens because the time it takes to get to Target 5 is very short compared to Target 10, and it is better to first pass through Target 5 to minimize the "lost benefit" of the two targets. In Figure 16, however, the time it takes to get to Target 5 is still shorter than the time it takes to get to Target 10, but since Target 10 is now located closer to the vehicle, it is better to first pass through Target 10. These two scenarios demonstrate how the arrival time of the vehicle to each target influences the task assignment process. In both of these small-sized simple cases, the greedy algorithm provides an identical result to the exhaustive algorithm's result, and the vehicle's path remains the same.

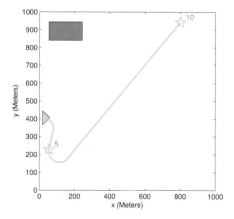

Figure 15. Scenario 1: Exhaustive TA; exhaustive MP and greedy TA; heuristic MP.

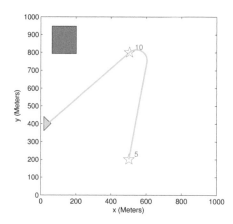

Figure 16. Scenario 2: Exhaustive TA; exhaustive MP and greedy TA; heuristic MP.

5.6. Benefit Descent Rate

The scenario of two targets and one vehicle presented in Figure 17 helps to illustrate the influence of the descent rate on the obtained results. In this scenario, the time it takes the vehicle to get to Target 3 is 200 s from its initial position, and the time it takes the vehicle to get to Target 10 is 500 s from same position. By increasing the value of the descent rate, the benefit rapidly diminishes as time progresses. In Figure 17b, the descent rate is increased to $A = 0.005$, five times compared to Figure 17a, causing a change in the vehicle assignments' order (note that the threshold value for this change of assignment in the examined scenario was $A = 0.0034$). Before the increase of the descent rate, the benefit of Target 10 is significantly higher than that of Target 3 (upper red bullet compared to lower red bullet in Figure 3a), but after the descent rate is increased, the targets have similar benefits (as can be seen by the red bullets' vertical position in Figure 3b). Since the benefit that the vehicle can gather in Target 10 is smaller than the one in Target 3, it is preferable to change the targets' visitation order, as can be seen in Figure 17b. The results summarized in Table 7 emphasize the influence of the descent rate, as the benefit acquired in Figure 17a is much higher than the one in Figure 17b. The benefit decent rate not only influences the benefit gathered, but may also influence the assignments' order, as presented above.

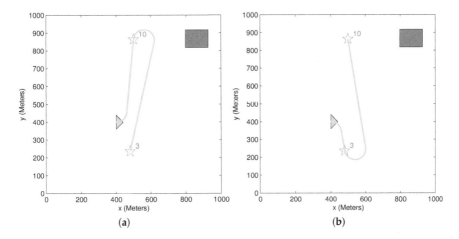

Figure 17. Benefit decent rate. (**a**) Exhaustive task assignment algorithm; exhaustive motion planning algorithm; decent rate: A = 0.001; (**b**) exhaustive task assignment algorithm; exhaustive motion planning algorithm; decent rate: A = 0.005.

Table 7. Benefit decent rate.

Figure #	Algorithms Used	Initial Benefit	Acquired Benefit	Lost Benefit	Overall Distance	Solution Time (s)
17a	Exhaustive TA Exhaustive MP	13,000	6866	6134	1321	0.05
17b	Exhaustive TA Exhaustive MP	13,000	1170	11,830	1004	0.05

6. Conclusions

In this paper, the intertwined problem of assigning and planning paths of UAVs to visit targets (having time-varying priorities) in an environment with obstacles was studied. It was assumed that the vehicles carry body-fixed sensors and, thus, are required to approach a designated target while flying straight and level. In order to address the time dependency of the targets' priority, an objective function incorporating the feasible path length of the vehicles and the targets' initial priority was formulated. Two task assignment algorithms were proposed: an exhaustive search algorithm that provides an optimal (lowest cost) solution and a greedy algorithm that provides a fast feasible solution (also used as an upper bound). A motion planning subroutine based on a tree search of an ordered set of targets and obstacles' vertices is used as part of the task assignment solution and provides feasible vehicle paths. The targets' time constraint was addressed by providing a path elongation algorithm amidst obstacles. Using simulations, the performance of the algorithms was compared, and the influence of the time-varying targets' priority on the task allocation process was demonstrated and investigated. Although the greedy algorithm provides a sub-optimal solution, it is useful in large-scale real-time scenarios, where computational running time is of the essence. The exhaustive algorithm can provide an immediate solution that improves over run time for large-scale scenarios, or it can be used in off-line scenarios. It should be noted that as the Dubins model was used for representing the kinematics of UAVs, similar models may be used for representing the motion of other vehicles moving in a plane (such as ground vehicles), and thus, the developed motion and task assignment algorithms may be used.

Acknowledgments: This work was supported by the U.S. Air Force Office of Scientific Research, Air Force Material Command, under Grant FA8655-12-1-2116.

Author Contributions: Y.G. and T.S. conceived and developed the algorithms; Y.G. performed the simulations; Y.G. and T.S. wrote the paper.

Conflicts of Interest: The authors declare no conflict of interest.

Appendix

The task assignment algorithm uses a motion planning subroutine, which is based on the study described in [12]. The motion planning subroutine includes two types of algorithms. These algorithms are based on a search process that explores the tree described in Section 3. The targets positions, vehicle's initial configuration and obstacles vertices form the tree nodes set $\mathcal{N}$. The two algorithms are described next, an exhaustive search algorithm that finds the minimum cost branch and a heuristic greedy algorithm that finds a feasible solution in a shorter computational time.

Appendix A. Heuristic Motion Planning Algorithm

The algorithm is given in Algorithm A1. Given a set of ordered targets, the vehicle's initial configuration and the obstacle vertices' position, the algorithm performs an A*-like greedy search based on Euclidean distance heuristics. A Euclidean distance between two points exists if and only if the straight line connecting the points does not intersect with the interior of an obstacle. The algorithm steps are as follows:

1. The vehicle is assumed to be located and oriented according to the initial configuration. A visit order list is initialized to null set (Line 1).
2. The following Steps 3–7 are repeated until the entire targets' set is visited (Line 2).
3. The next point of the visit is chosen as n^* until an obstacle-free relaxed path is found from the current point to the next target in the ordered targets' set (Line 4).
4. n^* is the node with the following property: the sum of the relaxed path length (connecting the vehicle's current configuration and n^*) and the shortest Euclidean distance (connecting n^* and the current target) is minimum (Line 5).
5. The node n^* is added to the visit order list (Line 6).
6. The vehicle is assumed to be located at node n^* (Line 7).
7. The current orientation angle becomes the arrival angle at n^* of the relaxed path from the previous point to n^* (Line 8).

The algorithm output is a vehicle trajectory, represented by an ordered set of nodes (including targets and obstacles' vertices) that need to be visited using relaxed paths.

Algorithm A1 Heuristic algorithm for motion planning of a Dubins vehicle amidst obstacles and multiple targets.

Input: Initial position and orientation of vehicle; obstacle information; ordered targets
Output: Vehicle trajectory (visit order)
1: Initialize: currentPosition $\leftarrow$ startPosition; currentOrientation $\leftarrow$ startOrientation; visitOrderList $\leftarrow$ [ ]
2: **for** iTarget = 1 to nTargets **do**
3: $\quad n^* \leftarrow$ currentPosition
4: $\quad$ **while** $n^* \neq$ iTarget **do**
5: $\quad\quad n^* \leftarrow$ arg min$_{n \in \mathcal{N}}$ (RelaxedPath(currentPosition,currentOrientation,n) +
$\quad\quad\quad\quad\quad$ ShortestEuclideanDistance(n,iTarget))
$\quad\quad\quad$ subject to: relaxed path from currentPosition to n is obstacle free
6: $\quad\quad$ visitOrderList $\leftarrow$ [visitOrderList n^*]
7: $\quad\quad$ currentPosition $\leftarrow v^*$
8: $\quad\quad$ currentOrientation $\leftarrow$ final angle of relaxed path from currentPosition to n^*
9: $\quad$ **end while**
10: **end for**

Appendix B. Exhaustive Motion Planning Algorithm

The algorithm explores every branch of the search tree and evaluates every possible visit order sequence to find the minimum cost branch. The algorithm's input is a set of ordered targets, the vehicle's initial configuration and the obstacle vertices' position. The algorithm steps are as follows:

1. Calculate the initial upper bound, using the heuristic algorithm (Line 1).
2. An OPEN list is generated to store the nodes that will be examined as the next node to visit.
3. The initial configuration is entered to OPEN as a node (Line 3).
4. The node with the lowest cost in OPEN (minimum relaxed path connecting the initial configuration and the node) is selected as the current node (Line 5).
5. The neighbors of the current node that can come after it in the visit order are examined (Line 7).

   Their estimated distance is defined as the sum of the following (Line 8):

   (a) Cost of the selected node.
   (b) Relaxed path length connecting the selected node and the neighbor.
   (c) Euclidean distance between the neighbor and the current target to visit.
   (d) The total Euclidean distance that connects the current target and the remaining targets to visit in the targets' set in the required order.

6. The neighbors with an estimated distance that is lower than the current upper bound are added to OPEN as new nodes (Lines 9–10).
7. All of the new nodes added to OPEN are examined.

   (a) In the case a new node is the next target to visit, the target is marked as visited in the current explored branch (Line 12).
   (b) In the case a new node is the last target to visit and the entire targets' set is visited in the required order, a leaf node of the branch is reached, and the entire branch is explored (Line 13).

      i. The upper bound is updated to the relaxed path total length described by the nodes in the branch, and the visit order of the nodes is stored (Lines 14–15).

8. The current node is removed from OPEN, since all of the neighbors have been evaluated (Line 23).
9. This process is repeated until the OPEN list is empty.

Algorithm B1 An exhaustive search algorithm for the motion planning of a Dubins/Reeds–Shepp vehicle amidst obstacles.

Input: Initial position and orientation of vehicle, obstacle information and ordered waypoints
Output: Minimum length vehicle trajectory (visit order) and its length
```
1:  UPPER ← upper bound on the path length obtained from heuristic
2:  initialNode.{position ← initial position, angle ← initial orientation, vertex ← 0,
                targetsVisited ← 0, visitOrder ← [ ], cost ← 0}
3:  OPEN ← initialNode
4:  while notEmpty(OPEN) do
5:      currNode ← arg min_OPEN OPEN.cost
6:      nextTarget ← currNode.targetsVisited+1
7:      for iNode = [nextTarget obstacleVertices] do
8:          pathLengthEstimate ← currNode.cost
                 + relaxedLength(currNode.{position,angle},iNode)
                 + EuclideanDistance(iNode,nextTarget)
                 + EuclideanDistanceCostToGo(nextTarget)
9:          if pathLengthEstimate < UPPER then
10:             newNode.{position ← position(iNode),
                    angle ← arrival angle of relaxed path at iNode,
                    vertex ← iNode, visitOrder ← [currNode.visitOrder iNode],
                    cost ← currNode.cost
                        + relaxedLength(currNode.{position,angle},iNode)}
11:             if iNode = nextTarget then
12:                 newNode.targetsVisited ← nextTarget
13:                 if iNode = lastTarget then
14:                     UPPER ← newNode.cost
15:                     visitOrder ← newNode.visitOrder
16:                 end if
17:             else
18:                 newNode.targetsVisited ← currNode.targetsVisited
19:             end if
20:             OPEN ← OPEN ∪ newNode
21:         end if
22:     end for
23:     OPEN ← OPEN \ currNode
24: end while
```

The algorithm output is identical to the heuristic algorithm output described above.

References

1. Cormen, T.H.; Leiserson, C.E.; Rivest, R.L.; Stein, C. *Introduction to Algorithms*, 2nd ed.; The MIT Press: Cambridge, MA, USA, 2001.

2. LaValle, M.S. *Planning Algorithms*; Cambridge University Press: New York, NY, USA, 2006.

3. Shima, T.; Rasmussen, J. *UAV Cooperative Decision and Control: Challenges and Practical Approaches*; SIAM: Philadelphia, PA, USA, 2009.

4. Enright, J.; Savla, K.; Frazzoli, E.; Bullo, F. Stochastic and Dynamic Routing Problems for Multiple Uninhabited Aerial Vehicles. *AIAA J. Guid. Control Dyn.* **2009**, *32*, 1152–1166.

5. Shanmugavel, M.; Tsourdos, A.; Zbikowski, R.; White, B. Path Planning of Multiple UAVs Using Dubins Sets. In Proceedings of the AIAA Guidance, Navigation, and Control Conference, San Francisco, CA, USA, 15–17 August 2005.

6. Dubins, L.E. On curves of minimal length with a constraint on average curvature, and with prescribed initial and terminal positions and tangents. *Am. J. Math.* **1957**, *79*, 497–516.

7. Chitsaz, H.; LaValle, S.M. Time-optimal paths for a Dubins airplane. In Proceedings of the 2007 46th IEEE Conference on Decision and Control, New Orleans, LA, USA, 12–14 December 2007; pp. 2379–2384.

8. Agarwal, P.K.; Wang, H. Approximation algorithms for curvature-constrained shortest paths. *SIAM J. Comput.* **2001**, *30*, 1739–1772.

9. Backer, J.; Kirkpatrick, D. A Complete approximation algorithm for shortest bounded-curvature paths. In Proceedings of the 19th International Symposium on Algorithms and Computation, Gold Coast, Australia, 15–17 December 2008; pp. 628–643.

10. Jacobs, P.; Canny, J. Planning smooth paths for mobile robots. In Proceedings of the IEEE International Conference on Robotics and Automation, Scottsdale, AZ, USA, 14–19 May 1989; doi:10.1109/ROBOT.1989.99959.

11. Laumond, J.P.; Jacobs, P.; Taix, M.; Murray, R. A motion planner for nonholonomic mobile robots. *IEEE Trans. Robot. Autom.* **1994**, *10*, 577–593.

12. Gottlieb, Y.; Manathara, J.G.; Shima, T. Multi-Target Motion Planning Amidst Obstacles for Aerial and Ground Vehicles. *Robot. Auton. Syst.* **2015**, submitted.

13. Snape, J.; Manocha, D. Navigating multiple simple-airplanes in 3D workspace. In Proceedings of the 2010 IEEE International Conference on Robotics and Automation (ICRA), Anchorage, AK, USA, 3–8 May 2010; pp. 3974–3980.
14. Yang, K.; Sukkarieh, S. Real-time continuous curvature path planning of UAVS in cluttered environments. In Proceedings of the 5th International Symposium on Mechatronics and its Applications, Amman, Jordan, 27–29 May 2008.
15. Yang, K.; Sukkarieh, S. An Analytical Continuous-Curvature Path-Smoothing Algorithm. *IEEE Trans. Robot.* **2010**, *26*, 561–568.
16. Shima, T.; Rasmussen, S.; Sparks, A.; Passino, K. Multiple task assignments for cooperating uninhabited aerial vehicles using genetic algorithms. *Comput. Oper. Res.* **2006**, *33*, 3252–3269.
17. Richards, A.; Bellingham, J.; Tillerson, M.; How, J.P. Coordination and Control of Multiple UAVs. In Proceedings of the AIAA Guidance, Navigation, and Control Conference, AIAA Paper 2002-4588, Monterey, CA, USA, 5–8 August 2002.
18. Schumacher, C.; Chandler, P.R.; Pachter, M.; Pachter, L.S. Optimization of air vehicles operations using mixed-integer linear programming. *J. Oper. Res. Soc.* **2007**, *58*, 516–527.
19. Chandler, P.R.; Pachter, M.; Rasmussen, S.J.; Schumacher, C. Multiple task assignment for a UAV team. In Proceedings of the AIAA Guidance, Navigation, and Control Conference, Monterey, CA, USA, 5–8 August 2002.
20. Schumacher, C.J.; Chandler, P.R.; Rasmussen, S.J. Task allocation for wide area search munitions. In Proceedings of the American Control Conference, Anchorage, AK, USA, 8–10 May 2002; pp. 1917–1922.
21. Edison, E.; Shima, T. Integrated task assignment and path optimization for cooperating uninhabited aerial vehicles using genetic algorithms. *Comput. Oper. Res.* **2011**, *38*, 340–356.
22. Rasmussen, S.J.; Shima, T. Tree search algorithm for assigning cooperating UAVs to multiple tasks. *Int. J. Robust Nonlinear Control* **2008**, *18*, 135–153.
23. Shima, T.; Rasmussen, S.; Gross, D. Assigning micro UAVs to task tours in an urban terrain. *IEEE Trans. Control Syst. Technol.* **2007**, *15*, 601–612.
24. Schumacher, C.; Chandler, P.; Pachter, M.; Pachter, L. *UAV Task Assignment with Timing Constraints*; Defense Technical Information Center: Fort Belvoir, VA, USA, 2003.
25. Schumacher, C.; Chandler, P.R.; Rasmussen, S.J.; Walker, D. *Path Elongation for UAV Task Assignment*; Defense Technical Information Center: Fort Belvoir, VA, USA, 2003.
26. Karaman, S.; Frazzoli, E. Sampling-based algorithms for optimal motion planning. *Int. J. Robot. Res.* **2011**, *30*, 846–894.
27. Kavraki, L.; Svestka, P.; Latombe, J.; Overmars, M. Probabilistic roadmaps for path planning in high-dimensional configuration spaces. *IEEE Trans. Robot. Autom.* **1996**, *12*, 566–580.
28. Donald, B.; Xavier, P.; Canny, J.; Reif, J. Kinodynamic motion planning. *J. ACM (JACM)* **1993**, *40*, 1048–1066.
29. Delle Fave, F.; Rogers, A.; Xu, Z.; Sukkarieh, S.; Jennings, N. Deploying the max-sum algorithm for decentralised coordination and task allocation of unmanned aerial vehicles for live aerial imagery collection. In Proceedings of the 2012 IEEE International Conference on Robotics and Automation (ICRA), Saint Paul, MN, USA, 14–18 May 2012; pp. 469–476.
30. Jiang, L.; Zhang, R. An autonomous task allocation for multi-robot system. *J. Comput. Inf. Syst.* **2011**, *7*, 3747–3753.
31. Shetty, V.; Sudit, M.; Nagi, R. Priority-based assignment and routing of a fleet of unmanned combat aerial vehicles. *Comput. Oper. Res.* **2008**, *35*, 1813–1828.
32. Krishnamoorthy, K.; Casbeer, D.; Chandler, P.; Pachter, M.; Darbha, S. UAV search and capture of a moving ground target under delayed information. In Proceedings of the 2012 IEEE 51st Annual Conference on Decision and Control (CDC), Maui, HI, USA, 10–13 December 2012; pp. 3092–3097.

33. Shaferman, V.; Shima, T. Unmanned aerial vehicles cooperative tracking of moving ground target in urban environments. *AIAA J. Guid. Control Dyn.* **2008**, *31*, 1360–1371.

34. Boissonnat, J.D.; Bui, X.N. *Accessibility Region for a Car That Only Move Forward along Optimal Paths*; Research Report INRIA 2181; INRIA Sophia-Antipolis: Valbonne, France, 1994.

 *sensors*

Article

Towards the Development of a Smart Flying Sensor: Illustration in the Field of Precision Agriculture

Andres Hernandez *, Harold Murcia, Cosmin Copot and Robin De Keyser

Department of Electrical Energy, Systems and Automation (EeSA), Ghent University, 9000 Ghent, Belgium;
E-Mails: harold.murcia@unibague.edu.co (H.M.); cosmin.copot@ugent.be (C.C.);
Robain.DeKeyser@ugent.be (R.D.K.)
* E-Mail: Andres.Hernandez@ugent.be; Tel./Fax: +32-9264-55-84.

Academic Editor: Felipe Gonzalez Toro
Received: 15 May 2015 / Accepted: 3 July 2015 / Published: 10 July 2015

Abstract: Sensing is an important element to quantify productivity, product quality and to make decisions. Applications, such as mapping, surveillance, exploration and precision agriculture, require a reliable platform for remote sensing. This paper presents the first steps towards the development of a smart flying sensor based on an unmanned aerial vehicle (UAV). The concept of smart remote sensing is illustrated and its performance tested for the task of mapping the volume of grain inside a trailer during forage harvesting. Novelty lies in: (1) the development of a position-estimation method with time delay compensation based on inertial measurement unit (IMU) sensors and image processing; (2) a method to build a 3D map using information obtained from a regular camera; and (3) the design and implementation of a path-following control algorithm using model predictive control (MPC). Experimental results on a lab-scale system validate the effectiveness of the proposed methodology.

Keywords: UAV; remote sensor; precision agriculture

---

1. Introduction

In the last few years, there has been great interest from different industries to obtain better product quality at higher production rates, to improve energy efficiency, while decreasing production costs. An essential element in achieving these goals is sensing; without reliable and accurate measurements, it is impossible to quantify productivity and, therefore, unfeasible to make timely corrections.

Depending on the application, different instrumentation structures can be found, but with a clear trend towards the use of multiple sensors, in order to collect all possible information about the system. Therefore, typically, sensor networks are used [1]. This is especially true for applications, such as production machines, automation, mapping, precision agriculture and weather forecasting, where a large area needs to be covered [2].

Sensing of large areas leads to difficulties, such as power supply, calibration procedures, data delays, accessibility issues for installation and maintenance, as well as high costs. Researchers have been working on developing wireless sensor networks to mitigate some of these problems [3]. A possibly more effective solution consists of using a remote sensor to 'freely' move in the space, thus increasing flexibility while diminishing costs, because that solution requires less sensors and is easier to maintain. In this matter, the use of aerial vehicles appears as an interesting option, since they have the ability to maneuver through complex environments. Even more interesting is the use of autonomous unmanned aerial vehicles (UAV) that can execute complex tasks without sustained human supervision, given their capability to perform tasks that are difficult or costly for manned aerial vehicles to accomplish [4]. In order to undertake the challenging task of automated flight and maneuvering, a versatile flight control design is required. One of the aerial vehicles that can accomplish this is a quadrotor due to its relatively small size, ability to hover and mechanical simplicity [5].

A large number of papers have emerged in the literature on quadrotors. Modeling, identification and control of a quadrotor are described by [6] using on-board sensing. Furthermore, simultaneous localization and mapping (SLAM) was implemented to create 3D maps of the environment, as well as to establish the quadrotor position in space [7]. Automatic navigation and object recognition with filtered data from on-board sensors and cameras is reported in [8]. Complex tasks, such as catching a falling object using a single quadrotor, have been accomplished in [5] or for a group of quadrotors in cooperative formation in [9], where high-speed external cameras were used to estimate the position of both the object and UAV. Difficult tasks, such as flying in cities or forests, require further progress on image processing, to achieve reliable detection of obstacles using low-weight hardware and low computational costs [10]. Current implementations in UAVs still require an on-ground operator with visual contact to the aerial vehicle for tasks, like taking off, landing, collision avoidance and adaptive path-planing. There exists a need to design methodologies to cope with these conditions in order to increase the degree of intelligence and therefore autonomy of UAVs.

Regarding applications for remote sensing, precision agriculture, exploration, surveillance and mapping are some of the main activities, because a smart remote sensor can be useful to take information from different angles and/or follow predefined paths in areas that are difficult access for human beings [11,12]. For example, in the field of precision agriculture, the goal is to gather and analyze information about the variability of soil and crop conditions, in order to maximize the efficiency of crop inputs within small areas of the farm field [13]. This requires obtaining reliable information about the crop conditions in real time. A possible solution might be to install a net of ground sensors; unfortunately, this is a costly and difficult to maintain solution. Mapping is also of high interest in activities such as cartography, archeology and architecture [14]. Here, the challenge lies in the accessibility to remote areas for a human being. In the case of exploration and surveillance, a smart remote sensor could autonomously fly above a pipeline of gas or petrol, in order to find leaks or detect corrosion, once the UAV is equipped with the necessary instrumentation.

This paper presents the first steps towards the development of a smart flying sensor, using as a flying platform a low-cost quadrotor. The task of mapping the volume of grain inside a trailer during forage harvesting is used as a test case to demonstrate the concept of smart remote sensing and to assess the performance of the proposed solution. In the context of this research, smart is linked to autonomy or the capacity of the flying sensor to make decisions without human intervention. Thus, automatic detection of the trailer, precise position control to avoid the phantom effect on the pictures and image processing algorithms to automatically generate a 3D profile are some of the features that make this a smart sensor. The main contributions are the development of: (i) a position-estimation method with delay compensation based on sensor fusion; (ii) a method to build a 3D map using information obtained from a regular camera; and (iii) a path-following predictive control algorithm to guarantee the accurate position of the UAV in space.

The paper is structured as follows. A description of the hardware, instrumentation and position estimation using sensor fusion is presented in Section 2. In Section 3, the quadrotor's dynamics and modeling is presented, followed by the path-following control algorithm in Section 4. Next, the effectiveness of the proposed smart remote sensor is tested for the task of mapping in a precision agriculture application, as described in Section 5. The final section summarizes the main outcome of this contribution and presents the next challenges.

2. Quadrotor Characteristics and Sensory Equipment

Quadrotors have four rotating blades, which enable flight similar to that of a helicopter. Movement is attained by varying the speeds of each blade, thereby creating different thrust forces. Today, they are equipped with on-board sensory equipment and the ability to communicate wirelessly with a command station, thus making it possible to implement advanced control algorithms to achieve precise control, even during aggressive aerial maneuvers.

In this work, the commercially available and low-cost AR.Drone 2.0 is used as the flying platform. A description of its main characteristics, sensory equipment and position estimation using sensor fusion is presented in this section.

2.1. Characteristics of the AR.Drone 2.0 Quadrotor

The quadrotor comes with internal in-flight controllers and emergency features, making it stable and safe to fly [15]. The only downside would be that access to the quadrotor's internal controller is restricted. The internal software is a black-box system, and the parameters that refer to control and other calibration procedures are not fully documented. There are four brushless DC motors powered with 14.5 W each, from the three-element 1500-mA/H LiPo rechargeable battery that provides an approximate flight autonomy of 15–20 min. Two video cameras are mounted on the central hull, pointing to the front and to the bottom of the quadrotor.

This on-board black-box system in the AR.Drone 2.0 can be considered the low layer. The high layer is represented by the command station, which defines the references to the internal controllers located in the low layer. A schematic representation is depicted in Figure 1.

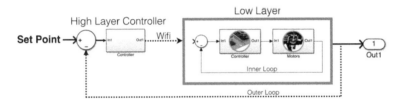

Figure 1. Quadrotor layers: the low layer represents the electronic assistance and the embedded operative system on the AR.Drone; the high layer represents the pilot (natively a smart device *i.e.*, iPhone).

In this application, the high layer consists of a C++ application in Visual Studio, which allows accessing all AR.Drone communication channels, therefore enabling functions to send commands or set configurations, receive and store data from sensors and video-stream. Thus, data can be interpreted off- or on-line for modeling, identification or control purposes. Movement in the quadrotor is achieved by furnishing reference values as input to the internal black-box controllers.

2.2. Technical Specifications and Sensory Equipment

2.2.1. Inertial Measurement Unit Board

The micro-electro-mechanical systems (MEMS)-based sensors are located below the central hull. They consist of:

- a three-axis accelerometer of ±50 mg precision
- a three-axis gyroscope with $2000°/s$ precision
- a three-axis magnetometer of $6°$ precision

which together form the inertial measurement unit (IMU). The IMU provides the software with pitch, roll and yaw angle measurements. The measurements are also used for internal closed-loop stabilizing control (black-box).

2.2.2. Ultrasonic Sensor

The ultrasonic sensor is used for low altitudes (below 3 m); it operates using two different frequencies, 22.22 Hz and 25 Hz, in order to reduce noise or perturbations between quadrotors. It is important to clarify that even if the sound propagates in a cone-like manner, the ultrasonic sensor does not provide a map of what is inside the cone's area. Instead, it outputs a single measured value, which corresponds to the highest object on the ground present somewhere within the cone's area, as illustrated in Figure 2. This effect makes it more difficult to accurately determine the altitude in rough terrain or the presence of ground obstacles. The relation between the altitude and the diameter of the cone is represented in Equation (1).

$$D_{cone}[m] = 2 * tan(0.2182 * altitude[m]) \tag{1}$$

Figure 2. Overview of the ultrasound sensor cone: the green cone indicates the range of the ultrasound sensor [7].

2.2.3. On-Board Cameras and Calibration Procedure

The AR.Drone 2.0 has two cameras. The bottom camera, which uses a CMOS sensor, has an image size of 320 × 240 pixels. It is the fastest camera with a speed of 60 FPS and an angle of vision of approximately 64°. The frontal camera can be used for difference sizes: 1280 × 720 pixels or 640 × 360 pixels; the speed is 30 FPS, and the angle of vision is 80–90°, approximately; this big angle of vision is responsible for the radial distortion in the images, similar to a fish-eye effect.

The cameras represent the main source of information for the system. Therefore, it is important to characterize the camera experimentally, in order to define the relation between pixels and meters. The experiment consists of taking pictures of a reference object of known characteristics (*i.e.*, height, width and color) at different distances. A fitting procedure is performed in order to characterize the cameras using the experimental as depicted in Figure 3, thus obtaining a relation which allows to compute the distance between the quadrotor and a reference object using Equations (2) and (3).

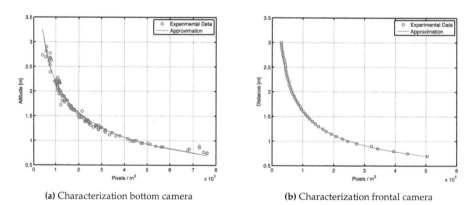

(a) Characterization bottom camera   (b) Characterization frontal camera

Figure 3. Camera characterization. Experimental data and approximation obtained for the (a) bottom and (b) frontal camera.

$$Altitude\ [m] = 148.6 \left(\frac{Pixels}{Area}\right)^{-0.339} - 0.8036 \tag{2}$$

$$Distance\ [m] = 599.3 \left(\frac{Pixels}{Area}\right)^{-0.5138} - 0.006038 \tag{3}$$

Since the quadrotor is equipped with low-cost general-purpose cameras, high distortion is observed when taking images. Fortunately, they can be characterized and corrected using a calibration and mapping procedure. In order to correct the distortion, it is necessary to take into account the radial and tangential factors. The presence of the radial distortion manifests in the form of the "barrel" or "fish-eye" effect, while tangential distortion occurs because the lenses are not perfectly parallel to the image plane.

In this work, we made use of the algorithms available from OpenCV libraries [16]. Currently, OpenCV supports three types of objects for calibration: asymmetrical circle pattern, a symmetrical circle pattern and a classical black-white chessboard. The method used for the elimination of the optical distortion on the images from the frontal camera of the AR.Drone 2.0 was the chessboard method. The procedure consists of taking snapshots of this pattern from different points of view (POV) of the chessboard; the algorithm implemented detects the corners and the intersections on the chessboard and creates an equation. To solve the equation, it is required to have a predetermined number of pattern snapshots to form a well-posed equation system.

In practice, due to the amount of noise present in our input images, good results were obtained using 10 snapshots of the input pattern from different positions. After solving the equation system, the parameters of the correction matrix are obtained and output as XML/YAML files.

This calibration experiment needs to be carried out only once. Then, inside the main application, once the files are loaded, a mapping function from OpenCV libraries is executed to eliminate the camera distortion. Finally, the distortion of the original image (Figure 4a) is eliminated as depicted in Figure 4b. Although a small part of the information is removed during the image processing procedure, the image is distortion free afterward.

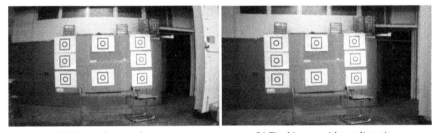

(a) Distorted original image   (b) Final image without distortion

Figure 4. AR.Drone 2.0 pictures from the frontal camera: (**a**) original image with radial distortion; (**b**) image obtained after the remap process with the calculated distortion parameters.

2.2.4. Processing Unit and Communication Channels

Two main circuit boards compose the processing unit of the drone:

- The mother-board holds the 1-GHz 32-bit ARM Cortex A8 processor with 800-MHz video DSP TMS320DMC64X, running a Linux-based real-time operating system.
- The second board uses a 16-bit PIC micro-controller navigation board, which interfaces with the sensors at a frequency of 40 Hz.

Regarding communication, there are four main services to connect with the AR.Drone:

Control and configuration of the drone is realized by sending AT commands on UDP port 5556. The transmission latency of the control commands is critical to the user experience. Those commands are to be sent on a regular basis (usually 30-times per second).

Information about the drone (like its status, its position, speed, engine rotation speed, *etc.*), called navdata, are sent by the drone to its client on UDP port 5554. These navdata also include tag detection information that can be used to create augmented reality games. They are sent approximately 30-times per second.

A video stream is sent by the AR.Drone to the client device on port 5555 with the TCP protocol. Given this protocol has a confirmation step, it presents a video streaming time delay of 360 ms, approximately. Image and video resolution can be selected between 360 p and 720 p. However, changing the video to 720 p creates a very noticeable lag between real time and the video. There was about a two-second delay. Images from this video stream can be decoded using the codec included in this SDK. The embedded system uses a proprietary video stream format, based on a simplified version of the H.263 UVLC (Universal Variable Length Code) format.

A fourth communication channel, called the control port, can be established on TCP port 5559 to transfer critical data, in opposition to the other data that can be lost with no dangerous effect. It is used to retrieve configuration data and to acknowledge important information, such as the configuration.

2.3. Position Estimation Using Sensor Fusion

Depending on the application, there are two possibilities to estimate the position of the quadrotor: (1) from camera images to situate the quadrotor on the X, Y plane; and (2) the AR.Drone provides an estimation of the translational speeds by using its on-board sensors and an optical flow algorithm, making it also possible to estimate the position by integrating the mentioned speeds.

Figure 5 describes the sensor's possibilities to estimate the position on the (X, Y) plane, each having positive and negative characteristics. On the one hand, odometry allows position estimation with almost no delay, but it suffers from drifting, producing an error that increases with time. On the other hand, to compute the position estimation based on optical measurements requires the use of patterns located in known positions, introducing also the problem of the additional time delays and noisy signals due to varying light conditions. An additional obvious difficulty is that once the pattern is out of the image, it is not possible to estimate the position.

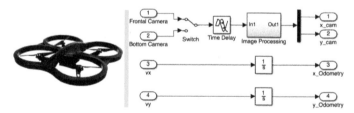

Figure 5. Sensors used for position estimation of the AR.Drone in the x, y plane.

The solution to achieve a reliable position estimation consists of combining the information from the two sensors. In order to reduce the drift effect and noise, odometry is used to read the variations and the optical sensor to find an offset-free measurement. The simplest and functional combination consists of using the optical sensor only when the standard deviation of the last five samples obtained from odometry is bigger than a tolerance value.

A time delay of about 100 ms is present due to latency in the wireless communication. However, the video signal has an additional time delay present in the video channel, which is directly related to the amount of data to be sent. For example, higher camera resolution introduces larger delays (*i.e.*, 330 ms approximately for an image of 640 × 360 pixels).

The position obtained from the camera represents the offset-free position, but "n" samples before, where "n" represents the time delay in samples (*i.e.*, $n = 5$ with $T_s = 66$ (ms)). Next, it is possible to integrate the speed values of the last five samples, in order to obtain the position estimation from odometry up to time "$n-1$". Equation (4) describes the method presented to eliminate the time delay effect on the video signal using a combination with odometry, assuming the dead time is a multiple of the sample time.

$$x_{(k)} = x_{cam}(N_d) + T_s \sum_{k=-(N_d-1)}^{k=0} v_{x(k)} \tag{4}$$

where x is the final position in meters, $N_d = T_d/T_s$, and T_s is the sample time: 66 ms; k represents the samples; v_x is the speed on the x axis; and x_{cam} represents the position obtained from the camera with a constant time delay $T_d = 330$ ms.

Figure 6 presents the performance of the estimation obtained, after using the proposed data fusion to correct the position measurements in the "*Y*" axis.

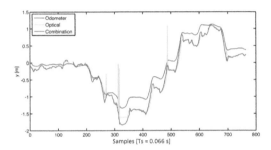

Figure 6. Position values in an open loop obtained from the image processing (green), the odometry (blue) and the fused response (red).

3. Quadrotor Dynamics and Identification

3.1. Coordinates System

The quadrotor's aerial movements are similar to those of a conventional helicopter. The quadrotor has four degrees of freedom (DOF): rotation over pitch, roll and yaw and translational movements over x, y and z, as depicted in Figure 7a. Notice that through rotational movement along the transversal y axis (pitch), translational movement on the x axis is achieved. A similar conclusion can be drawn for rotation over roll and translational movement on y.

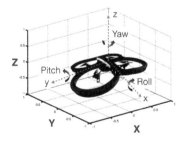

(a) Absolute and relative planes

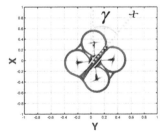

(b) UAV displacement over the yaw angle

Figure 7. UAV axes: (a) difference between absolute axes (X, Y, Z) and relative axes (x, y, z); (b) UAV displacement on the (x, y) plane with respect to the absolute plane.

It is worth noting that the coordinate system described above (x, y, z), represents a relative coordinate system used by the internal controllers (low layer). Using such a coordinate system instead of absolute coordinates (e.g., X, Y, Z) in the high layer will yield errors. For example, notice that by rotating the quadrotor, the relative coordinates (x, y) will change with respect to the absolute coordinates, as depicted in Figure 7b.

Concerning the relationship between the relative and the absolute coordinate systems, four cases were analyzed (Figure 8). The first case represents a null angular displacement on the UAV orientation with respect to the absolute space, thus meaning that the speeds in the relative axes are the same as those of the absolute axes: "$V_X = v_x$" and "$V_Y = v_y$". Cases 2 and 4 represent an angular displacement of $90°$ and $-90°$, respectively. The third case represents an angular displacement of $-180°$, which implies an opposite effect in the "x" and "y" axes.

Clearly, the relationship between the two coordinate systems is defined by the gamma (γ) angle, defined in Figure 7b. Consequently, equations describing the speeds of the UAV in the absolute system are defined as a function of the speed in the relative coordinates and (γ), as follows:

$$V_X = v_x cos(\gamma) - v_y sin(\gamma) \tag{5}$$
$$V_Y = v_x sin(\gamma) + v_y cos(\gamma) \tag{6}$$

After integrating the "V_X" and "V_Y" absolute speeds, it is possible to estimate the position of the UAV in the 3D space; this procedure is known as odometry. It is also important to note that Equation (6) depends on the yaw angle, which suffers from drifting over time, thus producing a "biased" estimation.

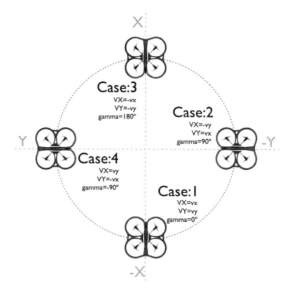

Figure 8. Case studies to describe the relation between the absolute and the relative coordinate systems.

3.2. System Identification

Due to the internal control, the quadrotor behaves as a set of single-input single-output (SISO) systems, therefore making it possible to perform parametric identification on each degree of freedom. This is realized using the prediction error method [17] and a pseudo-random binary signal (PRBS) as the excitation signal. A sampling time of 5 ms for yaw and 66 ms for other degrees of freedom are chosen based on the analysis of dynamics performed in a previous work [18]. The transfer functions obtained are given by:

$$
\begin{aligned}
\frac{v_x(s)}{v_x^*(s)} &= \frac{7.27}{(1.05s + 1)} e^{-0.1s} \\
\frac{v_y(s)}{v_y^*(s)} &= \frac{7.27}{(1.05s + 1)} e^{-0.1s} \\
\frac{v_z(s)}{v_z^*(s)} &= \frac{0.72}{(0.23s + 1)} e^{-0.1s} \\
\frac{\dot{\gamma}(s)}{\dot{\gamma}^*(s)} &= \frac{2.94}{(0.031s + 1)} e^{-0.1s}
\end{aligned}
\tag{7}
$$

where $\dot{\gamma}$ is the angular speed in yaw. The time delay in Equation (7) represents the average time delay present due to the communication in the control channel, *i.e.*, the action on the motors is received approximately 100 milliseconds after it is set on the computer.

Notice that Equation (7) corresponds to the identification of the closed-loop system using information coming from the IMU board. The inputs are the setpoints for speed (*i.e.*, v_x^*, v_y^*, v_z^*, $\dot{\gamma}^*$), and the outputs are the response of the internal control to follow those setpoints (*i.e.*, v_x, v_y, v_z, $\dot{\gamma}$). In other words, what is being identified is the closed-loop dynamics of the quadrotor for each degree of freedom.

4. Path-Following Predictive Control

A robust position controller of the quadrotor is required to follow either a set of way-points or a trajectory and to reject disturbances efficiently. Based on previous work [19,20], it has been found that model predictive control (MPC) fulfills the required specifications for tasks of tracking and positioning in 3D space [21]. In this section, the control structure and the implemented position controller using MPC is introduced.

4.1. EPSAC-MPC Algorithm

MPC refers to a family of control approaches, which makes explicit use of a process model to optimally obtain the control signal by minimizing an objective function [22]. In this contribution, the extended prediction self-adaptive control (EPSAC) approach to MPC is briefly described; for a more detailed description, the reader is referred to [23].

A typical set-up for the MPC optimization problems is as follows:

$$
\Delta U = arg \min_{\Delta U \in \mathbb{R}^{Nu}} \sum_{k=N_1}^{N_2} [r(t + k|t) - y(t + k|t)]^2
\tag{8}
$$

for $k = N_1 \ldots N_2$, where N_1 and N_2 are the minimum and maximum prediction horizons, ΔU is the optimal control action sequence, N_u is the control horizon, $r(t + k|t)$ is a future setpoint or reference sequence and $y(t + k|t)$ is the prediction of the system output.

In EPSAC, the predicted values of the output are given by:

$$
y(t + k|t) = x(t + k|t) + n(t + k|t)
\tag{9}
$$

Then, it follows that $x(t + k|t)$ is obtained by recursion of the process model; using the control input $u(t + k|t)$ and $n(t + k|t)$ represents the prediction of the noise, which includes the effect of the disturbances and modeling errors.

A key element in EPSAC is the use of base and optimizing response concepts. The future response can then be expressed as:

$$y(t+k|t) = y_{base}(t+k|t) + y_{optimize}(t+k|t) \tag{10}$$

The two contributing factors have the following origin:

- $y_{base}(t+k|t)$ is the effect of the past inputs, the *a priori* defined future base control sequence $u_{base}(t+k|t)$ and the predicted disturbance $n(t+k|t)$.
- $y_{optimize}(t+k|t)$ is the effect of the additions $\delta u(t+k|t)$ that are optimized and added to $u_{base}(t+k|t)$, according to $\delta u(t+k|t) = u(t+k|t) - u_{base}(t+k|t)$. The effect of these additions is the discrete time convolution of $\Delta U = \{\delta u(t|t), \dots, \delta u(t+N_u-1|t)\}$ with the impulse response coefficients of the system (G matrix), where N_u is the chosen control horizon.

Once Equation (10) is obtained, then Equation (8) can be solved, thus obtaining the optimal control action that minimizes the cost function. At the next sampling instant, the whole procedure is repeated, taking into account the new measured outputs according to the receding horizon principle, thus introducing feedback into the control law.

4.2. Performance of Position Control

Using the identified model of the quadrotor Equation (7), the EPSAC controller is tuned to achieve the shortest settling time without overshoot. The tuning parameters for the proposed controller are presented in Table 1.

Table 1. Design parameters for the extended prediction self-adaptive control (EPSAC) controllers. SISO, single-input single-output.

SISO System	N_1	N_2	N_u
x, y	1	15	1
z	1	30	1
γ	1	10	1

Furthermore, the position controller is tested for path-following as depicted in Figure 9. It is observed that the controller is able to follow the trajectory with a small tracking error and almost no overshoot.

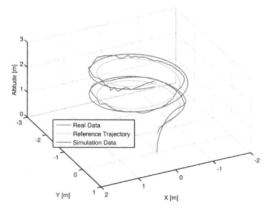

Figure 9. 3D simulated and real response for path following of the AR.Drone 2.0 using EPSAC-MPC on (X, Y and Z) degrees of freedom.

It is also important to highlight that the controllers are expected to work for the absolute coordinate system (X, Y); therefore, a transformation of the control action given by the MPC controllers to the relative coordinate system (x, y) is still necessary. This is performed based on Equations (5) and (6), thus determining the control actions as:

$$v_x^* = V_X cos(\gamma) + V_Y sin(\gamma) \tag{11}$$
$$v_y^* = -V_X sin(\gamma) + V_Y cos(\gamma) \tag{12}$$

where V_X and V_Y represent the absolute control action on the X and Y axis, respectively, at each sample time T_s.

5. Application of Mapping for Precision Agriculture

In this study, we focus on improving the loading process in a forage harvester, by measuring the volume of grain inside a container in real time. This application lies in the field of mapping and precision agriculture.

5.1. Loading Process during Forage Harvesting

Corn silage is a high-quality popular forage for ruminant animals. During this process, the corn on the ground is extracted, chopped and then ensiled. Using combine harvesting machines for forage processes is a very common and demanding operation in agriculture. It requires two operators to operate the combine harvester and the tractor next to it. The forage is discharged from the harvester via a spout, where an orbital motor drives the spouts rotational movement. On the end of the spout, a flipper ensures that the forage is discharged accurately into the trailer. Figure 10 presents a combine harvester machine in a conventional loading process.

Figure 10. Combine harvester machine in a conventional loading process.

The success of the process depends on the two operators and their ability to maintain the correct position of the machines. Good synchrony allows one to reduce the loss of material and to optimize the trailer's space through efficient filling, thus ensuring an ideal flat material profile in the container. However, to achieve this synchronization, the harvester driver must focus on two tasks at once: (1) align the machine with respect to the tractor and the crop; and (2) adjust the actions to manipulate the spout with two degrees of freedom to send the material towards the container in the best way possible.

Automation of this process would enable operators to accurately drive the harvester while the system automatically fills the trailer, disregarding the distance or position of the two vehicles and even when visibility is limited. Consequently, the driver would benefit from improved operating comfort and safety, focusing only on the front of the machine, thus increasing the working rate.

In order to automate this process, the system must perform some special tasks: read arm angles, read the distance between the harvester and the trailer and make a 3D map of the material inside the trailer. In this work, we focus on producing a reliable 3D map of the volume of grain inside the trailer. This information can later be used directly for the operator or by a controller to guide the spout to the interest point of the lower height of the material to get a flat profile. Companies in the field have proposed several alternatives, which are briefly described in the next subsection.

5.2. Commercial Assistant Systems

Several alternatives (academic and commercial) have emerged to automate and improve the performance of the loading process. Here is included a short description for the three most representative solutions.

5.2.1. Asuel

Asuel is prototype intended for position control of the spout of a forage harvester during the loading process [24]. The main objective is to build a mathematical model to estimate the volume and shape of the material inside the trailer. Estimation of the volume is corrected using information from Global Positioning Systems (GPS) and a model of the different actuators. The accuracy of the estimated profile is very low, given that it is difficult to obtain a good mathematical model, with the additional low resolution of GPS.

5.2.2. Auto Fill

Autofill is a stereo camera-based spout control system [25] developed by Claas company. Using a fixed stereo camera has a significant advantage, allowing 3D perception compared to the traditional 2D images. When a trailer approaches the side of the forage harvester, the vision system detects its position. Once the trailer is detected, an overlay is drawn on the picture and shown to the driver. Then, a green line showing the estimated material level is drawn, thus engaging the AutoFill system. The system predicts where the crop jet will hit within the trailer using measurements of the spout and deflector rotations. Due to crop conditions and drift, the precision of the predicted hit point is not sufficient. Thus, the predicted jet trajectory is corrected online by measuring the distance to the jet. For this system, the main problem arises from the disturbance created from the dust created once the material falls inside the trailer.

5.2.3. IntelliFill

Case New Holland (CNH) has recently launched a forage spout guidance called IntelliFill, which is based on a Time-Of-Flight (TOF) camera with 3D images. The optical sensor reads the light reflected back, obtaining a picture of the container filling. This camera allows the system to have functionality in the dark, as well as in bright day light [26]. The camera is mounted on the harvester's spout, and it can measure the distance to the four corners of the container and the filling volume inside the container. Through this information, the spout's turning angles can be automatically controlled.

5.3. Possible Limitation When Using a Fixed Camera

As described above, the goal is to reduce the loss of material and to optimize the trailer's space through an efficient filling, thus ensuring an ideal flat material profile in the container. This is achieved if operators maintain the correct position of the machines and if the operator of the harvester correctly manipulates the spout to send the material towards the container in the best possible way.

Achieving a flat material profile inside the trailer is possible under good visibility conditions and flat terrains, with the additional help of a system to supervise the loading process as depicted in Figure 11a. Nevertheless, some limitations appear when placing the sensor (*i.e.*, camera) in the arm of the harvester machine. For example, noise in the images due to interference coming from dust, the small particles of chopped material and mechanical vibrations (Figure 11b) or a decrease of visibility due to an increase of the distance between the vehicles (Figure 11c). These difficulties can be diminished by using a flying sensor, because the camera can be placed in a better position, thus increasing visibility inside the trailer despite dust or a large distance between the vehicles (Figure 11d).

Additional advantages can be obtained if other information is extracted from images during flight (e.g., density of the crop in front of the harvester to regulate the harvester speed properly) or if other sensors are installed.

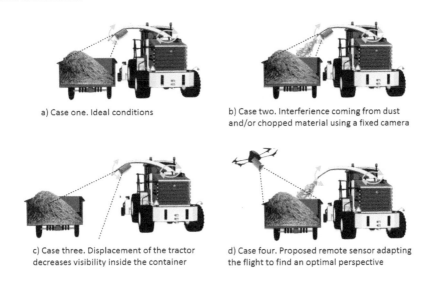

a) Case one. Ideal conditions

b) Case two. Interference coming from dust and/or chopped material using a fixed camera

c) Case three. Displacement of the tractor decreases visibility inside the container

d) Case four. Proposed remote sensor adapting the flight to find an optimal perspective

Figure 11. Possible limitations when using a fixed sensor and advantages of using a smart flying sensor for a loading application during forage harvesting.

5.4. Proposed Alternative by Using a UAV

A solution to the overloading problem could be the use of an UAV acting as a remote sensor, as depicted in Figure 11d. The quadrotor should follow the vehicles, read the profile disposition inside the container and, through image processing, detect the relative distance between the harvester and the trailer, *i.e.*, to minimize forage losses during the discharging process. A simple lab-scale system is utilized as proof-of-concept of the proposed solution. Figure 12 shows the setup platform used to emulate the tractor-trailer with the material. The emulated container has $2.0 \times 1.5 \times 1.0$ m for length, width and height, respectively.

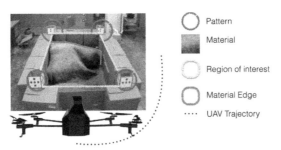

Pattern

Material

Region of interest

Material Edge

UAV Trajectory

Figure 12. Experimental setup description.

The UAV is in charge of collecting information from the process; in other words, it must follow the container and read the profile of the material. The patterns in the corners of the container are placed to provide references to the quadrotor and to delimit the region of interest. Although color patterns can be used, a higher robustness was observed with black and white patterns, given they have less variations for different light conditions.

6. Experimental Validation of the Proposed Smart Flying Sensor

6.1. Structure of the Proposed Solution

The experiment consists of flying with the quadrotor around the emulated container referenced by the patterns on the corners. Once the view is focused on a point of view (POV), the container is segmented by using triangulation with the pattern references. Subsequently, color segmentation can be applied to identify the material profile. Note that at least two pictures are needed to reconstruct a 3D surface representing the relief of the material.

Figure 13. Methodology to build a 3D map of the material inside the trailer during a loading process using a forage harvester.

The complete procedure to build a 3D map of the material inside the container can be represented by the scheme in Figure 13. A description of the steps for the application are hereafter described:

- Take-off and hold on: First of all, the quadrotor must be stabilized in space. Therefore, after the take-off, the quadrotor goes to an altitude setpoint of two meters (*i.e.*, $z^* = 2\,\text{m}$) and tries to hold the position.
- Read the sensors: Once the quadrotor is hovering, the next step is to use the front camera and to take a picture in order to activate the image recognition process. It is important to mention that the camera is at about $-45°$ with respect to the horizontal position. This is done in order to have a better view inside the container.
- Reference the quadrotor: The distance between the UAV and the container is computed using information from the patterns. An image processing algorithm is in charge of computing the area inside the square of the pattern. Since this value is predefined and therefore known, the relative distance between the container and the UAV can be computed using Equation (3) and the time delay correction Equation (4). The controllers are designed to keep a fixed distance between the UAV and the container, centered on the container wall in a similar manner as depicted in Figure 12.
- Read the area of interest: The controllers maintain the quadrotor at a fixed 3D point with respect to the container. When the position error is low and the angles (pitch and roll) are appropriate, the picture is reduced only to the inner container size Figure 14a. Here, color segmentation is used to extract the projection of the material on the walls by using edge detection techniques, as shown in Figure 14b.

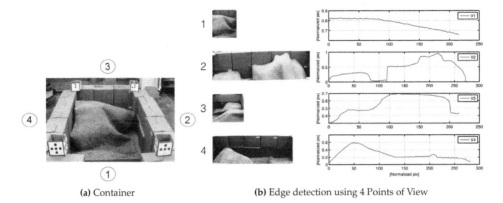

(a) Container   (b) Edge detection using 4 Points of View

Figure 14. Example of (a) material profile in the container and (b) the information extracted using the edge detection algorithm considering four points of view (POV).

- Change the point of view (POV): In computer vision, the methods used to know depth information are based on the combination of images taken from different points of view (e.g., stereo vision). Given the flexibility of the UAVs, since a path-control algorithm using MPC has been implemented, it is possible to fly to other positions in order to get pictures from different points of view. The next step is thus to fly around the container pointing the camera inward, following a trajectory in four degrees of freedom in space, similar to the experiment performed in Figure 9, but using a constant altitude. The trajectory is calculated as a second order polynomial, knowing the actual coordinates and the destination point by using the translation between absolute and relative coordinate systems presented in Equations (11) and (12).
- Extract the information: Once the pictures have been taken, a correction on the pictures was implemented to remove the "fish eye" effect present on the native photos. Then, the material is segmented, and the edge function is applied to calculate the contours of the material for each picture. These contours are in the form of vectors (e.g., V1, V2, V3 and V4) containing the information about the shape of the material on the container's wall. Figure 14 shows the material and its corresponding contour for four different points of view.

6.2. 3D Profile Computation

Image processing is composed of two main parts: the first part corresponds to the pre-processing (*i.e.*, acquisition, segmentation and classification), which is executed in OpenCV at the same sample time as in the UAV; the second part used to compute 3D map is executed in parallel in MATLAB with the process, but using a longer sampling time. Following MATLAB's notation, the surface obtained using four vectors (see Figure 14b) corresponding to the edge of the material inside the container is:

$$S = 0.25 * [V_1(end: -1:1)' * V_2 + V_1(end: -1:1)' * V_4 + V_3' * V_2 + V_3' V_1(end: -1:1)] \qquad (13)$$

A trade-off between accuracy and update time of the 3D map must be considered. Although using four images (four POV) has the advantage that a more accurate profile is obtained, this also implies a longer update time, since the UAV will require more time to take all pictures. A possible solution consists of using only two pictures, for which a good approximation of the 3D map can be computed, as depicted in Figure 15.

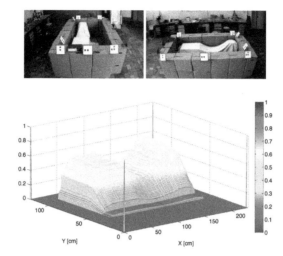

Figure 15. 3D profile obtained experimentally using the smart flying sensor and two pictures.

6.3. Performance of the Path-Following Controller

An important element for this application is the path-following controller, which is required to guarantee an accurate position of the quadrotor in the 3D space and to reject possible wind disturbances. Using the controller designed in Section 4, the quadrotor is able to automatically take-off and follow a pre-defined path around the container while taking the necessary number of pictures to compute the 3D profile.

The performance of the controller for the case of taking two pictures to approximate the material profile is depicted in Figure 16, including the final path followed by the UAV in Figure 16a and the control actions (i.e., setpoints to the low-layer internal controllers) required by the MPC in Figure 16b.

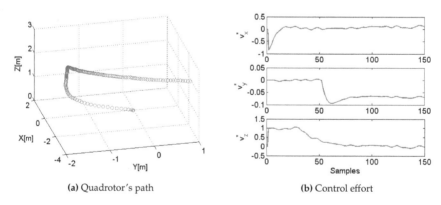

(a) Quadrotor's path   (b) Control effort

Figure 16. Performance of the path-following controller. (**a**) Quadrotor's path around the container in order to take two pictures and (**b**) control effort required by the MPC strategy.

7. Conclusions

In this paper, we have presented the first steps towards the development of a smart flying sensor based on an unmanned aerial vehicle (UAV). The effectiveness of the proposed smart flying sensor is

Sensors **2015**, *15*, 16688–16709

illustrated for the task of mapping the volume of grain inside a trailer during forage harvesting, using a lab-scale system.

The main achievements for this research are: (i) the obtained insight in the dynamics and coordinate systems of the low-cost quadrotor AR.Drone 2.0; (ii) the development of a position-estimation method with time delay compensation based on image processing and the inertial measurement unit (IMU); (iii) a method to build a 3D map using information obtained from a regular camera; and (iv) the design and implementation of a path-following control algorithm using MPC.

Future work includes incorporating a GPS for outdoor flight, together with the development of obstacle avoidance techniques to enhance the autonomy of the quadrotor. An extension to multiple UAVs and/or a combination with ground vehicles is also under investigation.

Acknowledgments: This work has been financially supported by the EeSA department (Electrical Energy, Systems and Automation) at Ghent University (Belgium). This work has been obtained as a result of the bilateral agreement between Ghent University (Belgium) and Universidad de Ibague (Colombia), in the framework of the Masters in Control Engineering. Andres Hernandez acknowledges the financial support provided by the Institute for the Promotion and Innovation by Science and Technology in Flanders (IWT-Vlaanderen), Belgium (grant nr. SBO-110006).

Author Contributions: Andres Hernandez contributed to the implementation of the communication between the quadrotor and the main station, besides performing the system identification and the design of the path-following control algorithm using model predictive control. Harold Murcia developed the algorithm for time delay compensation using sensor fusion and image processing algorithms, contributed to the design and implementation of the control algorithms and performed simulations and experiments. Cosmin Copot and Robin De Keyser contributed to the formulation and analysis of the stated problem, collaborated with advice in control design and image processing and with the analysis of the results. All authors took part in the writing of the manuscript.

Conflicts of Interest: The authors declare no conflict of interest.

References

1. Sahota, H.; Kumar, R.; Kamal, A.; Huang, J. An energy-efficient wireless sensor network for precision agriculture. In Proceedings of the IEEE Symposium on Computers and Communications (ISCC), Washington, DC, USA, 22–25 June 2010; pp. 347–350.
2. Blackmore, S. Precision Farming: An Introduction. *J. Outlook Agric.* **1994**, *23*, 275–280.
3. Siuli Roy, A.; Bandyopadhyay, S. Agro-sense: Precision agriculture using sensor-based wireless mesh networks. In Proceedings of the Innovations in NGN: Future Network and Services, First ITU-T Kaleidoscope Academic Conference, Geneva, Swizerland, 12–13 May 2008; pp. 383–388.
4. Berni, J.; Zarco-Tejada, P.J.; Suárez, L.; Fereres, E. Thermal and narrowband multispectral remote sensing for vegetation monitoring from an unmanned aerial vehicle. *IEEE Trans. Geosci. Remote Sens.* **2009**, *47*, 722–738.
5. Bouffard, P. On-board Model Predictive Control of a Quadrotor Helicopter: Design, Implementation, and Experiments. Master's Thesis, University of California, Berkeley, CA, USA, 2012.
6. Dullerud, G. Modeling, Identification and Control of a Quad-Rotor Drone Using Low-Resolution Sensing. Master's Thesis, University of Illinois at Urbana-Champaign, Champaign, IL, USA, 2012.
7. Dijkshoorn, N.; Visser, A. An elevation map from a micro aerial vehicle for urban search and rescue. In Proceedings of the 16th RoboCup International Symposium, Mexico City, Mexico, 18–24 June 2012.
8. Mogenson, M.N. The AR Drone LabVIEW Toolkit: A Software Framework for the Control of Low-Cost Quadrotor Aerial Robots. Master's Thesis, TUFTS University, Boston, MA, USA, 2012.
9. Ritz, R.; Muller, M.; Hehn, M.; D'Andrea, R. Cooperative quadrocopter ball throwing and catching. In Proceedings of the IEEE/RSJ International Conference on Intelligent Robots and Systems (IROS), Algarve, Portugal, 7–12 October 2012; pp. 4972–4978.
10. Barrows, G. Future visual microsensors for mini/micro-UAV applications. In Proceedings of the IEEE International Workshop on Cellular Neural Networks and Their Applications (CNNA 2002), Washington, DC, USA, 22–24 July 2002; pp. 498–506.
11. Clement, A. Advances in Remote Sensing of Agriculture: Context Description, Existing Operational Monitoring Systems and Major Information Needs. *Remote Sens.* **2013**, *5*, 949–981.
12. Pajares, G. Overview and Current Status of Remote Sensing Applications Based on Unmanned Aerial Vehicles (UAVs). *J. Photogramm. Eng. Remote Sens.* **2015**, *81*, 281–329.

13. Zarco-Tejada, P.; Hubbard, N.; Loudjani, P. Precision Agriculture: An Opportunity for EU Farmers—Potential Support with the CAP 2014-2020; Technical Report, Joint Research Centre (JRC) of the European Commission; Monitoring Agriculture ResourceS (MARS) Unit H04: Brussels, Belgium, 2014.

14. Mesas-Carrascosa, F.; Rumbao, I.; Berrocal, J.; Porras, A. Positional quality assessment of orthophotos obtained from sensors onboard multi-rotor UAV platforms. *Sensors* **2014**, *14*, 22394–22407.

15. Bristeau, P.; Callou, F.; Vissiere, D.; Petit, N. The Navigation and Control Technology inside the AR.Drone Micro UAV. In Proceedings of the 18th IFAC World Congress of Automatic Control, Milano, Italy, 28 August–2 September 2011; pp. 1477–1484.

16. OpenCV-Documentation. Available online: http://docs.opencv.org (accessed on 23 November 2013).

17. Ljung, L. *System Identification: Theory for the User*; Prentice Hall PTR: London, UK, 1999.

18. Vlas, T.E.; Hernandez, A.; Copot, C.; Nascu, I.; de Keyser, R. Identification and Path Following Control of an AR.Drone Quadrotor. In Proceedings of the 17th International Conference on System Theory, Control and Computing, Sinaia, Romania, 11–13 October 2013.

19. Murcia, H.F. A Quadrotor as Remote Sensor for Precision Farming: A Fill-Harvesting Case Study. Master's Thesis, Ghent University, Ghent, Belgium, 2014.

20. Hernandez, A.; Copot, C.; Cerquera, J.; Murcia, H.; de Keyser, R. Formation Control of UGVs Using an UAV as Remote Vision Sensor. In Proceedings of the 19th IFAC World Congress of the International Federation of Automatic Control, Cape Town, South Africa, 24–29 August 2014.

21. Hernandez, A.; Murcia, H.F.; Copot, C.; De Keyser, R. Model Predictive Path-Following Control of an AR.Drone Quadrotor. In Proceedings of the XVI Latin American Control Conference (CLCA'14), Cancun, Quintana Roo, Mexico, 14–17 October 2014; p. 31.

22. Camacho, E.F.; Bordons, C. *Model Predictive Control*, 2nd ed.; Springer-Verlag: London, UK, 2004; p. 405.

23. De Keyser, R. Model Based Predictive Control for Linear Systems. Available online: http://www.eolss.net/sample-chapters/c18/e6-43-16-01.pdf (accessed on 7 July 2015).

24. Happich, G.; Harms, H.H.; Lang, T. Loading of Agricultural Trailers Using a Model-Based Method. *Agric. Eng. Int. CIGR J.* **2009**, *XI*, 1–13.

25. Möller, J. Computer Vision—A Versatile Technology in Automation of Agricultural Machinery. *J. Agric. Eng.* **2010**, *47*, 28–36.

26. Posselius, J.; Foster, C. Autonomous self-propelled units: What is ready today and to come in the near future. In Proceedings of the 23rd Annual Meeting Club of Bologna, Bologna, Italy, 9–10 November 2012.

Article

Flight Test Result for the Ground-Based Radio Navigation System Sensor with an Unmanned Air Vehicle

Jaegyu Jang *, Woo-Guen Ahn, Seungwoo Seo, Jang Yong Lee and Jun-Pyo Park

The 3rd R&D Institute-4, Agency for Defense Development, Yuseong P.O. Box 35, Daejeon 305-600, Korea; wgahn@add.re.kr (W.-G.A.); mcnara82@add.re.kr (S.S.); flukelee@add.re.kr (J.Y.L.); pjp1023@add.re.kr (J.-P.P.)
* Author to whom correspondence should be addressed; jaegyu.jang@gmail.com;
  Tel.: +82-42-821-3643; Fax: +82-42-823-3400 (ext. 16108).

Academic Editor: Felipe Gonzalez Toro
Received: 16 September 2015; Accepted: 2 November 2015; Published: 11 November 2015

Abstract: The Ground-based Radio Navigation System (GRNS) is an alternative/backup navigation system based on time synchronized pseudolites. It has been studied for some years due to the potential vulnerability issue of satellite navigation systems (e.g., GPS or Galileo). In the framework of our study, a periodic pulsed sequence was used instead of the randomized pulse sequence recommended as the RTCM (radio technical commission for maritime services) SC (special committee)-104 pseudolite signal, as a randomized pulse sequence with a long dwell time is not suitable for applications requiring high dynamics. This paper introduces a mathematical model of the post-correlation output in a navigation sensor, showing that the aliasing caused by the additional frequency term of a periodic pulsed signal leads to a false lock (*i.e.*, Doppler frequency bias) during the signal acquisition process or in the carrier tracking loop of the navigation sensor. We suggest algorithms to resolve the frequency false lock issue in this paper, relying on the use of a multi-correlator. A flight test with an unmanned helicopter was conducted to verify the implemented navigation sensor. The results of this analysis show that there were no false locks during the flight test and that outliers stem from bad dilution of precision (DOP) or fluctuations in the received signal quality.

Keywords: pseudolite; pulsed signal; false lock

1. Introduction

GNSS (Global Navigation Satellite System) receivers are currently the most widely-used space-based PNT (positioning, navigation and timing) sensors. From infrastructure, such as the time-keeping systems of wireless telecommunication devices, to mobile devices or military weapons, they are now indispensable. However, the vulnerability of the satellite navigation systems has been an important issue for more than a decade, as GPS sensors must process signals with an extremely low power level. The seriousness of this vulnerability can be found in a statement issued by the Interagency GPS Executive Board, which reads "GPS users must ensure that adequate independent backup systems or procedures can be used when needed" [1]. For the past few years, there have in fact been attacks on the GPS L1-band close to the boarder of Korea, with one article stating that "aircraft had to rely on alternative navigation aids" during the attacks [2].

The ground-based radio navigation system introduced in this paper is an alternative/backup navigation system, which can overcome the vulnerability against intentional interference attacks through its use of a pseudolite network. Before the launching of GPS (Global Positioning System) satellites, pseudolites were used to test transmitters at the desert test ranges [3]. For the last twenty years, many researchers have extended the applications of these systems to various areas, including indoor positioning, the Mars navigation system and regional positioning systems [4–9]. The ground-based radio navigation system has been studied as a regional positioning system of which the main applications are air vehicles. This type of system belongs to the synchronized pseudolite navigation system category, and these systems have a better system survivability than those in the asynchronized system category [4,10].

In a CDMA (code division multiple access)-based system, users share the same frequency channel. This means that users can fail to acquire weak signals if there are strong signals from closely-located transmitters. This is known as the near-far issue. To avoid the near-far problem in the pseudolite sensors, previous researchers suggested a scheme that involved frequency offsetting and pulsing [11,12]. We applied these two methods to avoid the near-far issue in our study. By applying the pulsing, we can allocate time slots to each signal, which means that there are no more collisions between signals. Earlier suggestions, such as RTCM SC-104 and RTCA (radio technical commission for aeronautics) 2000, recommended a randomized sequence, because the pulsed signal may have an aliasing effect on the navigation sensor. Because an additional frequency term caused by a periodic pulse could lead to a possible false lock in the tracking loop, they recommended accumulation of the received signal samples with a randomized pulse position. If the accumulation-and-dump time is long enough to use all of the sequence chips, received signal samples will appear to be continuous in the accumulation-and-dump filter. However, this method requires a long dwell time (e.g., at least 10 ms [11]), and the time is typically too long for some applications with high dynamics.

In this study, we designed pseudolite transmitters with fixed pulse positions, for which the duty ratio is defined as 10%. We also suggested another algorithm that detects possible false locks in the navigation sensor. First, we introduce a mathematical model of the accumulation-and-dump filter output for a periodic pulsed signal, after which we explain why a false lock can occur. Secondly, two algorithms are suggested to detect a false lock in the frequency domain. After detecting a false lock with the multi-correlator characteristics, the algorithm performs a Doppler correction in a signal processing module. The implemented firmware was verified in the flight test with an unmanned helicopter. An analysis of the flight test confirmed that the navigation solutions were reliable and that there were no issues caused by the false lock, although there were a few outliers in the solution caused by bad dilution of precision (DOP) at a low elevation and by channel power fluctuations.

2. Mathematical Model

According to earlier work [13], the sum of the sine and cosine samples of an incoming continuous signal can be approximated by Equation (1) in a complex form.

$$\langle IQ \rangle_{\cos/\sin} \approx \int_0^T e^{j(2\pi \cdot \Delta f \cdot t + \Delta \phi)} dt = \frac{\sin(\pi \cdot \Delta f \cdot T)}{\pi \cdot \Delta f \cdot T} e^{j\Delta\phi} \tag{1}$$

Here, T is the coherent integration time of the accumulation-and-dump filter, $\Delta\Phi$ is the phase error in radian units and Δf is the frequency error in Hz units.

If the coherent integration time is limited to a code sequence period (or a single pulse period), a redefinition of Equation (1) for aperiodic pulsed navigation signals is straightforward, because the integration time is only affected by the duty ratio of the pulse. Therefore, the accumulation-and-dump output model of the incoming signal can be rewritten as Equation (2).

$$\langle IQ \rangle_{aperiodic} = \frac{\sin(\pi \cdot \Delta f \cdot DR \cdot T_{seq})}{\pi \cdot \Delta f \cdot DR \cdot T_{seq}} \cdot \sqrt{2 \frac{C}{N_0} \cdot DR \cdot T_{seq}} \cdot R(\tau) \cdot D \cdot e^{j\Delta\phi} + \eta \tag{2}$$

Here, T_{seq} is the coherent integration time for a code sequence period, DR is the duty ratio of the pulse, C/N_0 is the carrier to noise density ratio, $R(\tau)$ is an auto-correlation function with a code phase error of τ, D is the data bit sign and η denotes the noise in a complex form.

The first term in Equation (2) explains why the null-to-null space of the signal acquisition result in the frequency domain changes when a pulse is applied to an incoming code sequence, as presented in Figure 1. Figure 1a shows the 2D correlation outputs when a coherent integration time of 1 ms is used, as in a GPS navigation signal acquisition, while Figure 1b presents the pulsed signal acquisition result, for which the duty ratio is set to 0.1. Figure 1 shows that the null-to-null frequency space is increased in a manner inversely proportional to the duty ratio DR, which is identical to the first term of Equation (2).

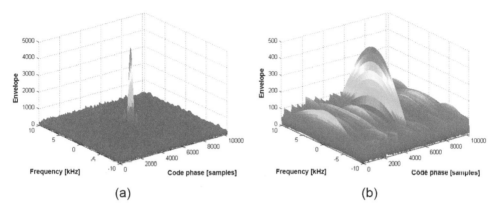

Figure 1. Null-to-null space extension due to a single pulse.

The second term of Equation (2) implies partial correlation effects on the power due to the pulsing scheme acting on a code sequence. Figure 2 shows that the post-correlation power is reduced by a pulse with a duty ratio of 0.1; the second term of Equation (2) explains why in a mathematical expression. Reduced power itself is not critical to a ground-based local navigation system, because it can be resolved by a systematic design, such as transmission power control. The important issue in Figure 2 is the change of the cross-correlation property. As shown in Figure 2, the cross-correlation separation is reduced from 24 to 25 dB to approximately 10 dB by pulsing.

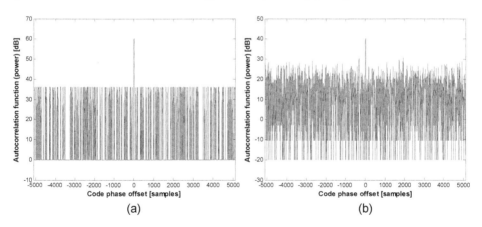

Figure 2. Partial correlation effect on the power: (**a**) duty ratio of 1.0; (**b**) duty ratio of 0.1.

Assuming a Gaussian approximation, the upper bound of the partial correlation (*i.e.*, the cross-correlation separation bound) of pulsed C/A Gold codes can be estimated, as shown in Figure 3.

If the window size is set to 102–103 chips (*DR* = 0.1), the upper bound approaches −10 dB or more, as simulated in Figure 2.

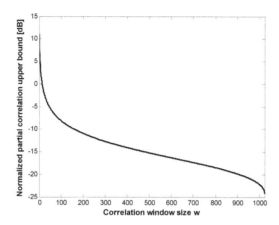

Figure 3. Cross-correlation separation of the pulsed C/A code.

A new spreading code design to obtain good correlation properties is out of the scope of this paper. Therefore, a coherent integration time over a single code sequence period is unavoidable to meet the sensitivity requirements of navigation sensors in general. In this case, we must utilize the periodic pulse train model. Here, we define a rectangular pulse train signal model with Equations (3) and (4):

$$p_0(t) = \begin{cases} 1, & |t| \leq DR \cdot T_{seq}/2 \\ 0, & DR \cdot T_{seq}/2 < |t| \leq T_{seq}/2 \end{cases} \tag{3}$$

$$p(t) = p_0(t \pm n \cdot T_{seq}), \quad n = 0, \pm 1, \pm 2, \dots \tag{4}$$

Equations (3) and (4) correspond to the aperiodic and the periodic model, respectively. The series coefficients of Equation (3) are derived as follows:

$$c_n \approx \frac{1}{T_{seq}} \int_{-T_{seq}/2}^{T_{seq}/2} p(t) \cdot e^{-jn\frac{2\pi}{T_{seq}}t} dt = \frac{1}{n\pi} \sin(DR \cdot n \cdot \pi) = \frac{\sin(DR \cdot n \cdot \pi)}{DR \cdot n\pi} \tag{5}$$

By inserting Equation (5) into the exponential Fourier series definition, the pulse train in Equation (4) can be rewritten as Equation (6).

$$p(t) = DR \cdot \sum_{n=-\infty}^{\infty} \sin c(n \cdot DR) \cdot e^{jn\frac{2\pi}{T_{seq}}t} \tag{6}$$

Equation (1) can be rewritten as Equation (7) by applying Equation (6), as follows:

$$\begin{aligned} \langle IQ \rangle_{\cos/\sin}^{periodic} &= \int_0^{KT_{seq}} e^{j(2\pi \cdot \Delta f \cdot t + \Delta \phi)} \cdot DR \cdot \sum_{n=-\infty}^{\infty} \sin c(n \cdot DR) \cdot e^{jn\frac{2\pi}{T_{seq}}t} dt \\ &= DR \cdot \left[\sum_{n=-\infty}^{\infty} \sin c(n \cdot DR) \int_0^{KT_{seq}} e^{j\{2\pi(\Delta f + nf_{seq})t + \Delta \phi\}} \cdot dt \right] \end{aligned} \tag{7}$$

Here, f_{seq} denotes the code sequence frequency in Hz (=$1/T_{seq}$). The integral term in Equation (7) resembles the form of Equation (1) exactly; thus, we can easily rearrange Equation (7) above into Equation (8):

$$\langle IQ \rangle^{periodic}_{cos/sin} = DR \cdot \left[\sum_{n=-\infty}^{\infty} \sin c(n \cdot DR) \cdot \sin c \cdot ((\Delta f + n f_{seq}) \cdot K \cdot T_{seq}) \cdot e^{j\Delta\phi} \right] \qquad (8)$$

Finally, the accumulation-and-dump output model of the periodic pulsed navigation signal can be defined as Equation (9) by applying Equation (8) as the first term of Equation (2). Because the DR term in Equation (6) will affect the noise density, as well, we normalized Equation (8) by DR to define Equation (9).

$$\langle IQ \rangle_{periodic} = \left[\sum_{n=-\infty}^{\infty} \sin c(n \cdot DR) \cdot \sin c((\Delta f + n \cdot f_{seq}) \cdot K \cdot T_{seq}) \right] \cdot \sqrt{2\frac{C}{N_0} \cdot DR \cdot T_{seq}} \cdot R(\tau) \cdot D \cdot e^{j\Delta\phi} + \eta \qquad (9)$$

Equation (9) indicates that the sinc function train repeats every f_{seq} Hz and that the overall shape on the frequency axis is limited by the sinc function presented in Equation (2). This is a type of aliasing phenomenon caused by additional periodic pulse samples, which may lead to a false lock during the signal acquisition process. In Figure 4, we showed the two-dimensional signal acquisition results as simulated using the FFT technique. A GPS C/A code-like sequence was used with a duty ratio of 0.1, and the length of the coherent integration time was two sequences. As derived by Equation (9), we note that there is a sinc function train under the wide sinc function in Figure 4. The false lock issue caused by the sinc function train (or the side lobes in the frequency domain) is a major concern in the pulsed signal processing. The various algorithms utilized in earlier work and the algorithm suggested in this paper to dissolve the false lock issue will be described in the next section.

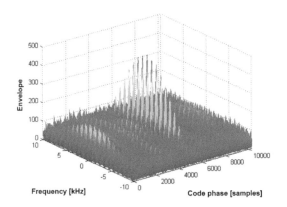

Figure 4. Side lobes due to periodic pulsed signals.

In Figure 5, we present the signal acquisition results of pulsed signal samples with a duty ratio of 0.1 while varying the number of coherent integrations K. For the simulation, we used a Gold code, for which the code sequence length is 1023 chips.

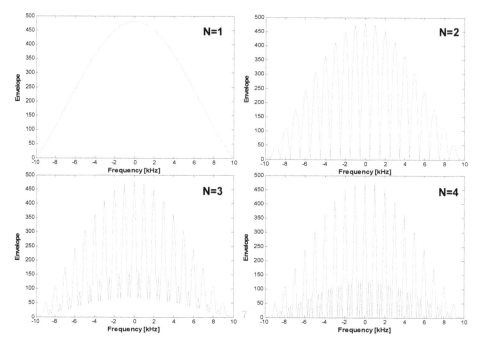

Figure 5. Auto-correlation function (*N* = the number of coherent integrations in sequence units).

The upper left figure shows the signal acquisition results when the number of coherent integrations *K* is set to one (1 ms, 1023 chips). As redefined in Equation (2), the null-to-null space is extended by the duty ratio. However, in the case of multiple coherent integrations (*K* > 1), we note that the repetition pattern of the sinc function train with f_{seq} (*i.e.*, 1 kHz) appears, as shown in both Figure 4 and Equation (9). As expected from Equation (9), the null-to-null space of the sinc train becomes narrower if *K* becomes larger. Therefore, we must use smaller frequency bins during the signal acquisition process. It is important to note that a long integration time always causes a long signal acquisition time or increased hardware complexity.

3. The False Lock

The pulsing scheme is known as the most effective solution to avoid interference between pseudolite signals and GNSS signals or between pseudolite signals. However, the pulsing scheme causes an additional frequency in the original signal, which leads to the false lock effect during the signal acquisition and tracking processes. As shown in Figure 5 in the previous section, the signal acquisition process in the FPGA (field programmable gate array) module may be locked onto a side lobe in the frequency domain, as the BOC (binary offset carrier) signal acquisition process experiences a false lock due to multiple side lobes in the code chip domain.

To resolve the false lock effect of the pulsed signal, previous research results, such as RTCM SC-104 or the RTCA special committee SC-159, suggested a randomized pulse sequence [11,12]. In the case of the RTCM suggestion, the pulse duration is 93 code chips, corresponding to 1/11 of a code sequence. The average duty ratio is defined as 0.1, and the pulse position is altered for every code sequence, indicating that a complete code sequence (1023 chips) can be used after a coherent integration time of 10 ms to prevent a false lock.

In Figure 6, we described how the randomized pulse sequence works to resolve the aliasing issue due to the additional frequency. In the case of the RTCM suggestion, a single pulse has 93 code chips, as explained in the above lines. The sum of 11 time slots creates a complete code sequence (1023 chips) in the correlation process, indicating that the signal acquisition and tracking process are independent of pulsing, because there are no periodic pulsed samples. The RTCA 2000 scheme uses a similar concept, but with more pseudo-randomly-distributed pulse positions. However, recently, this appears to be removed from the RTCA document due to the difficulties related to standard GPS sensors [11].

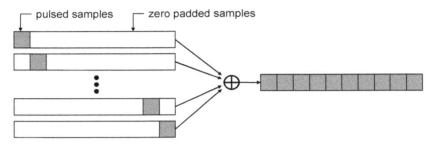

Figure 6. Principle of the randomized pulse sequence scheme (e.g., the RTCM SC-104 suggestion).

The randomized pulse position is a very useful scheme, because it can avoid the false lock effect and reduce the possibility of pulse collisions. However, a long integration time of the randomized pulse scheme (e.g., 10 or 20 ms) is not suitable for high dynamic applications, which require a limited coherent integration time. Within the framework of the ground-based radio navigation system, a deterministic pulse position scheme is assumed for signal transmissions for a similar reason. In fact, this method enables a zero pulse collision within a specific area assuming a sophisticated system design, which is another advantage for certain applications. Due to the deterministic pulse position, it was necessary to overcome the false lock issue on the navigation sensor side instead of the signal transmission side.

In the tracking process, a false lock arises despite the fact that a single code sequence is used for the accumulation-and-dump process. This is clearly explained in Figure 7, which presents both the tracking loop discriminator and the accumulation-and-dump filter output in the frequency domain. In this paper, we used a second order FLL (frequency-locked loop) and the atan2 discriminator, as defined in Equation (10) [14].

$$D_{FLL} = \frac{\alpha \tan 2(cross, dot)}{(t_2 - t_1) \cdot 2\pi} \tag{10}$$

Here,

$$dot = I(t_1) \cdot I(t_2) + Q(t_1) \cdot Q(t_2)$$

$$cross = I(t_1) \cdot Q(t_2) - I(t_2) \cdot Q(t_1)$$

If a coherent integration time of 1 ms is used for signal tracking, from Equation (10), we can determine that the distance between lock points is 1 kHz, as presented in Figure 7. Doppler values presented in Figure 8 show what occurs when the false lock problem arises.

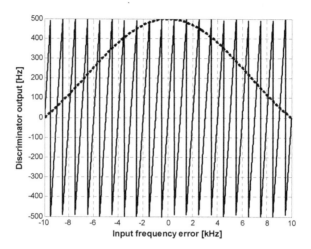

Figure 7. Tracking loop discriminator *vs.* correlator output.

Figure 8a shows that the tracking loops were locked at the correct point. The same Doppler is estimated regardless of the duty ratio value. Figure 8b shows that the tracking loop was locked in the incorrect position, in this case 1 kHz away from the true point. The false lock in the frequency domain causes two symptoms with regard to the tracking loop. First, it degrades the power (*i.e.*, a relatively lower signal-to-noise ratio). Second, it gives biased velocity information to the code-tracking loop (*e.g.*, the carrier-aided first-order DLL) and to the navigation filter.

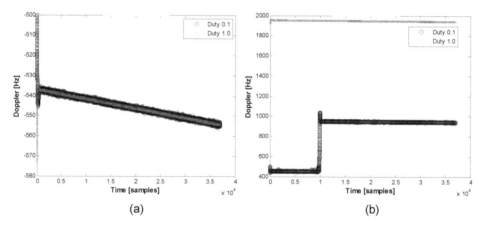

Figure 8. Estimated Doppler: (**a**) correctly locked; (**b**) falsely locked.

The scale factors for the carrier-aided codes are 1/1, 540 and 1/154 for the C/A code and the P(Y) code, respectively. If there is a Doppler error of 1 kHz, the velocity biases in the code NCO are 0.65 Hz and 6.5 Hz for the C/A and the P(Y), respectively. This will likely have minor effects on the loop. However, the velocity bias is converted to approximately 190 m/s assuming the L1-band range domain. Because our system did not use the L1-band to avoid possible jamming attacks, the velocity bias will have different values with the above example; however, it is unquestionably a critical error source for the navigation filter. If the number of coherent integrations K is larger than one, the power degradation becomes critical, and the tracking loops may experience a frequent loss of lock. In the frequency-locked loop, the distance between the zero crossing points of the discriminator decreases; consequently, the locked point can be located in the middle of the sinc main lobe or on small side lobes.

To prevent a false lock in the frequency domain, we used additional correlator arms in the fast signal-acquisition module implemented in the FPGA. This concept is similar to the concept of the bump-and-jump technique, which was proposed to find a possible false lock in the code chip domain during the GNSS BOC signal acquisition and tracking processes. Figure 9 shows the proposed channel block diagram, which serves to detect a false lock in the frequency domain. The additional correlator arms located at adjacent peaks of the side frequency lobe are compared to the prompt frequency correlator arm to find the maximum peak position or they can be combined to compose more complex comparators to use the slopes between the lobes. In this paper, we refer to the former as the maximum peak finder and the latter as the slope comparator. Figure 10 presents two schemes graphically to assist with the understanding of this. The values of n and the m for frequency separation are assumed to be one for simplicity.

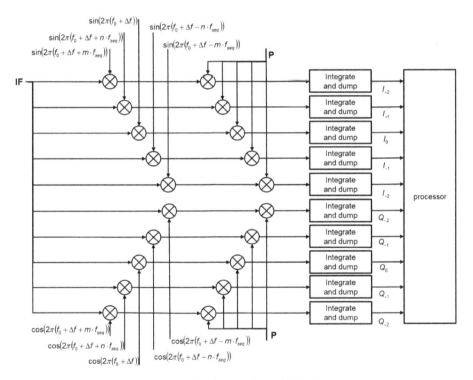

Figure 9. Channel block diagram for the false lock detection.

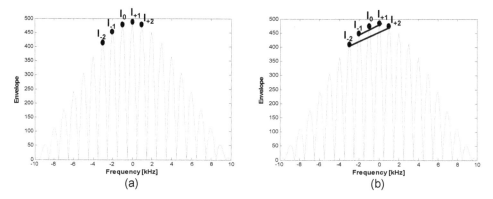

Figure 10. The false lock detection schemes: (**a**) The max peak finder; (**b**) The slope comparator.

The maximum peak finder uses the same logic as the bump-and-jump (BJ) method suggested earlier [15], while the slope comparator is similar to the heuristic detection metrics of the SQM (signal quality monitoring) scheme with multi-correlator functionality, also proposed in an earlier work [16]. In the BJ method, early and late correlators are used in the code-tracking loop, and very late (VL) and very early (VE) correlators are used to monitor the power of other possible peaks close to the puncture correlator. Here, two additional correlator arms (VE and VL) are used only for the false lock, with a simple up/down counter mechanism applied for the decision. Counting is performed for all integrate and dump periods during the code-tracking process in the BJ. However, in the false lock detection scheme used here, the decision is designed to be made in the signal-acquisition module to reduce the computational burden. We used a hard decision rule instead of a simple up/down counting mechanism and fully utilized all correlator arms to ensure the maximum peak finding performance (e.g., five arms for I-channels, five for Q-channels). Several linear combinations to compose the heuristic test metrics have been suggested [16]. These can be categorized as follows:

(1) Delta (Δ) metrics: differential between two symmetric slopes (normalized by a puncture).
(2) Average ratio metrics: ratio between a symmetric slope and the puncture correlator.
(3) Single sided ratio metrics: ratio between a puncture and another correlator.
(4) Asymmetric ratio metrics: ratio between asymmetric correlators or ratio between an asymmetric slope and the puncture correlator.

It is challenging to find appropriate correlator positions and statistical thresholds to apply the above test metrics directly into the frequency domain. For our application, the first two metrics are appropriate, because our purpose is not to monitor signal deformation, such as an evil waveform. For high-SNR situations (e.g., during flights), the max peak finder showed generally good performance when used to detect an incorrect frequency lock.

4. Flight Experiment Results

Real-time flight tests were conducted to evaluate an implemented user sensor using a ground-based radio navigation system, in this case the synchronized pseudolite navigation system shown in Figure 11. In this system, the transmission times of the pseudolites are synchronized to a specific pseudolite time known as the master pseudolite time. Detailed information about time synchronization can be found in earlier work [10].

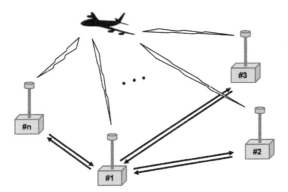

Figure 11. The ground-based radio navigation system using the synchronized pseudolites.

Because the master pseudolite time is a local system time, we created a log file that included both a GPS NMEA (National Marine Electronics Association) time tag and the local system time for the purpose of analyzing the navigation accuracy performance. Assuming 10 Hz data output from a sensor, the ideal log time synchronization error from the GPS time is 0.05 s, which corresponds to one meter considering the 20 m/s dynamics of the test vehicle. We did not synchronize the local system time to the GPS time, because the basic idea of this local navigation system is to assume the vulnerability of the GNSS signal. Log synchronization using the GPS NMEA was only done for a performance analysis during the developmental stage. We did not use an IMU (inertial measurement unit) sensor to compensate for the lever arm between the local navigation system antenna and the GPS antenna. We can expect that a horizontal bias of less than one meter would be added to the navigation results due to the lever arm. In the case of the vertical direction, it can be corrected mostly with body-frame information, because the UAV helicopter is supposed to keep its pitch constant. As described earlier [10], pseudolites installed on the ground use dual-frequency bands as one of the near-far problem resolutions. During the flight test, the master pseudolite and the slave pseudolites uses different bands. Possible near-far problems between the slave pseudolite signals are assumed to be mitigated by the pulsing scheme.

To avoid cross-correlations in the satellite navigation system, pseudolites can use the frequency offset. For pseudolite sensors using both GPS and pseudolite transmitted signals, in-band offsets at satellite-signal spectral nulls are used with a single RF frontend. Because our system is an alternative navigation system, it uses out-of-band offsets to prevent hostile energy sources from interfering with the GPS signal.

The unmanned helicopter shown in Figure 12 was used to carry the ground-based navigation sensor during the flight test. A radio navigation sensor using a pseudolite signal from the ground does not have good vertical performance characteristics, because it has a relatively large vertical DOP (dilution of precision). Therefore, the cruise height of an air vehicle should be kept as high as possible so as to prevent poor condition numbers during the positioning calculations. During the flight test, we planned the height of the unmanned helicopter to be several hundred meters, because the safety of the vehicle should be guaranteed against strong wind. While the planned height was high enough to obtain the line of sight from all transmitters during the cruise mission, the expected vertical DOP was not good.

Figure 12. The unmanned helicopter and navigation sensor used for the flight tests.

The pseudolites were installed in a coastal area; the horizontal distances between them were set to dozens of kilometers. The multipath signal reflected from the sea surface was one of the critical issues to resolve before the flight test could be conducted. We attempted to install ground stations as far as possible from the coast line to avoid possible direct reflections from the sea surface, and we used a spatial diversity combining technique with the time synchronization sensors installed at the same site with the transmitters, as described in previous research [10,17]. The time synchronization performance is an important measure to operate a ground-based radio navigation system within a specific accuracy level in real time. In this study, all of the ground sensors used multiple correlators in the code-tracking loop to reduce the multipath errors from inland areas. However, a simple narrow correlator was implemented in the user sensor installed in the unmanned helicopter to reduce the possibility of a loss of lock during the flight.

Figure 13 depicts a computed version of the horizontal DOP of the flight test area. The master station is centered on four slave stations optimally placed in the test area. The DOP has a minimum value close to the master station, which is located at the center of Figure 13, and this value increases exponentially if a user vehicle moves away from the center. The trajectory of the unmanned vehicle equipped with the DUT (device under test) was scheduled to cross the center area to ascertain the condition that led to the best performance. In this paper, we assumed a horizontal DOP (HDOP) of two or less as the operational criteria to define the effective mission area of the vehicle.

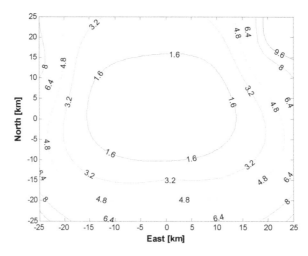

Figure 13. Horizontal dilution of precision (HDOP) of the test area.

The real-time flight trajectories using the GRNS signals, as well as the post-processed CDGPS results using GNSS signals are shown in Figure 14 together. Position calculation in the developed sensor was done, and the results were stored at 10 Hz. In the unmanned helicopter, a NovAtel GNSS receiver was equipped to store the log files. The precise positions of the vehicle were computed to mm to cm accuracy by the GrafNav software. To provide correction information to the program, a DGPS reference station was installed very close to the master station with a NovAtel OEM6 receiver. We had conducted flight tests several times with UAV helicopters; however, from Figure 14 to Figure 16, we focused on the first flight result to explain the flight analysis clearly.

In Figure 14, we present a 2D trajectory of the first flight in ENU (east-north-up) coordinates. Figure 14a shows a full trajectory, and Figure 14b presents a zoomed-in trajectory of the right upper corner of Figure 14a to show the positioning error in detail. Very large outliers were noted, as marked with the blue dots in Figure 14a. Poor condition numbers in the least squares estimator caused the outliers, which occurred during a low-elevation stage. When the altitude of the flight vehicle is low, the navigation sensor attached to the vehicle experiences low visibility due to the geometrical environment, unstable signal power conditions and the poor DOP situation. The last factor is the most critical error source related to the outliers. The trajectory shown with the black circle only close to the upper left corner does not have a three-dimensional navigation solution due to a visibility problem (*i.e.*, the low elevation). The upper right corner of Figure 14a corresponds to the best DOP area, which is above the area of the master station. At every corner of the trajectory, the unmanned helicopter stayed for some minutes in hovering mode and changed its heading to an appropriate direction. Just after the rotation of the head, the navigation sensor experienced positioning errors distributed in a specific direction, as can be seen in the corner area of the zoomed-in Figure 14b. We suspect that blockages by the leg structures of the unmanned helicopter and the gain patterns of the equipped antennas likely caused these errors during the rotation. Because we applied the spatial diversity combining technique to the stations installed on the ground and to the navigation sensor equipped on the vehicle, the performance degradation due to rotation must be minimized.

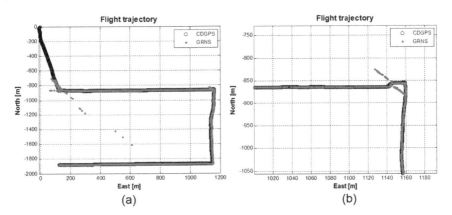

Figure 14. 2D trajectory: (**a**) Full trajectory; (**b**) Zoom-in figure of the upper right corner of (**a**).

Figures 15 and 16 show SNR graphs created after the analysis of the received signal power variations during the hovering, as described in Figure 14. RF1 and RF2 in the Figure 15 denote the SNR from the first and second RF antennas used for the diversity combining process. Figure 16 presents the SNRs from a single antenna, as there was a cable connection problem at a specific RF band during the flight. As shown in the two figures, at the initial stage of hovering, the SNR values from most of the channels fluctuated severely due to the rotation of the heading. A few epochs in the RF1 of the master station and dozens of epochs from the slave Stations 2 and 3 dropped below a threshold and affect the geometry matrix for the positioning computation. Except for the rotation time (~about 40 s), the signal powers maintain relatively stable values.

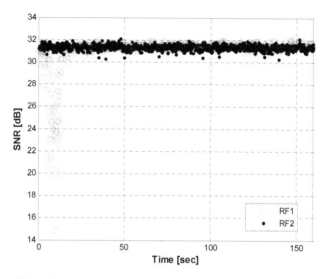

Figure 15. Received signal power transmitted from the master station while hovering.

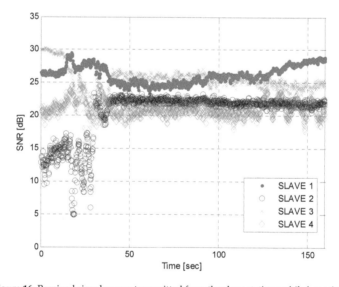

Figure 16. Received signal power transmitted from the slave stations while hovering.

From Figures 14–16, we can conclude that the horizontal navigation performance and the received signal power showed stable results, except at the low-elevation area with a very high DOP (e.g., < several thousand PDOP) and during rotation while hovering. There were no position jumps or lock losses, which can stem from a false lock. An additional and more direct measure to check if there were false locks is to analyze the estimated velocity values, as a false lock represents the Doppler frequency bias. The velocity errors are plotted in Figure 17, for which the time scope corresponds to a cruise mode with a nearly constant altitude. Outliers caused by the rotation of the heading are visible, caused by the hovering time, as explained in Figures 15 and 16. Except for the velocity outliers, the figure shows that the estimated unsmoothed velocities are stable while in cruise mode.

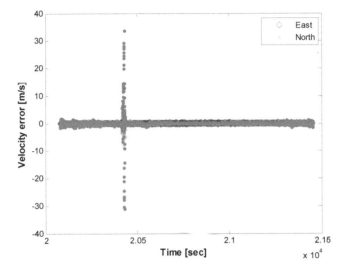

Figure 17. Velocity error while cruising, horizontal components only.

During flight trials, we had conducted three sorties, and performance analysis results are plotted in Figure 18. Solutions for which horizontal DOP is less than 2.0 are included for the analysis. Due to the lever arm, each trial has a slightly different RMS error, as can be seen in Figure 18a. Mean horizontal RSS (root sum square) error was approximately 2.4 m for all flight trials. Clock synchronization error, residual tropospheric delay, tracking jitter and the lever arm can be considered as dominant error sources. In the case of velocity performance, horizontal velocity RSS error was approximately 0.14 m, and all trials show similar performances. In this paper, vertical performance was not considered for the analysis, because vertical DOP is not good in tested altitudes, as explained previously.

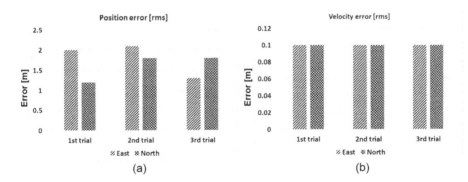

Figure 18. Performance analysis. (**a**) Horizontal error; (**b**) Horizontal velocity error.

5. Conclusions

The ground-based radio navigation system is a local navigation system that uses synchronized pseudolites. It has been studied as an alternative to satellite navigation systems, such as GPS or Galileo. One of the major problems to be resolved in relation to a pseudolite-based navigation sensor is the near-far problem. The transmission of pulsed signals is known to be the most effective means of mitigating the interference caused by the near-far problem. Previous studies suggested pseudolite transmitters using a randomized pulsed sequence to overcome the aliasing issue caused by the additional frequency term of the periodic pulse. However, a randomized pulse sequence may not be adequate in some applications, due to the dwell time required to use all of the sequence chips.

This paper introduced a mathematical model for the post-correlation of the periodic pulsed signal and proposed algorithms to resolve the false lock issue due to the aliasing process. A flight test using an unmanned helicopter was conducted to verify the implemented algorithm. The results showed that there were a few outliers caused by an extremely poor DOP condition at a low elevation, as well as power fluctuations and signal blockages due to the rotation of the heading while hovering. However, the overall navigation performance was quite good, *i.e.*, nearly equivalent to that of a satellite navigation system. Most importantly, there were no side effects caused by a false lock, such as a velocity bias or a frequent loss of the lock.

Author Contributions: Jaegyu Jang contributed to the design of the navigation sensors for a pseudolite time synchronization system and a user vehicle. Woo-Guen Ahn collaborated in the ground segment design. Seungwoo Seo worked on designing a time synchronization algorithm. Jang Yong Lee contributed to developing a control station for the ground-based radio navigation system. Jun-Pyo Park led this study as the project leader of the ground-based radio navigation system.

Conflicts of Interest: The authors declare no conflict of interest.

Disclaimer: The contents of this paper do not reflect the official opinions or policies of the Agency of Defense Development. The views and opinions expressed in this paper are solely those of the authors.

References

1. Thomas, M. Evolution of GPS Systems Architecture and Its Impacts. *Commun. IIMA* **2010**, *10*, 27–40.
2. The Washington Times. Available online: http://washingtontimes.com/news/2012/aug/23/north-korean-jamming-gps-shows-systems-weakness (accessed on 2 November 2015).
3. Cobb, H. GPS Pseudolites: Theory, Design, and Applications. Ph.D. Thesis, Stanford University, Palo Alto, CA, USA, 1997.
4. Lee, T. A Study on the Smart Pseudolite Navigation System Using Two-way Measuring Technique. Ph.D. Thesis, Seoul National University, Seoul, Korea, 2008.
5. Kee, C.; Jun, H.; Yun, D. Indoor Navigation System using Asynchronous Pseudolites. *J. Navig.* **2003**, *56*, 443–455. [CrossRef]
6. Lemaster, E.; Rock, S. A Local-Area GPS Pseudolite-Based Navigation System for Mars Rovers. *Auton. Robot.* **2003**, *14*, 209–224. [CrossRef]
7. Park, B.; Kim, D.; Lee, T.; Kee, C.; Paik, B.; Lee, K. A Feasibility Study on a Regional Navigation Transceiver System. *J. Navig.* **2008**, *61*, 177–194. [CrossRef]
8. Kim, C.; So, H.; Lee, T.; Kee, C. A Pseudolite-Based Positioning System for Legacy GNSS Receivers. *Sensors* **2014**, *14*, 6104–6123. [CrossRef] [PubMed]
9. Lee, K.; Noh, H.; Lim, H. Airborne Relay-Based Regional Positioning System. *Sensors* **2015**, *15*, 12682–12699. [CrossRef] [PubMed]
10. Seo, S.; Park, J.; Suk, J.; Song, K. A Design of Dual Frequency Bands Time Synchronization System for Synchronized-Pseudolite Navigation System. *J. Position. Navig. Timing* **2014**, *3*, 71–81. [CrossRef]
11. Abt, T; Soualle, F.; Martin, S. Optimal Pulsing Schemes for Galileo Pseudolite Signals. *J. Glob. Position. Syst.* **2007**, *6*, 133–141.
12. *GNSS-Based Precision Approach Local Area Augmentation System (LAAS) Signal-In-Space Interface Control Document (ICD)*; RTCA Inc.: Washington, DC, USA, 2001.
13. Parkinson, B.; Spilker, J. *Global Positioning System: Theory and Applications*; AIAA: Washington DC, USA, 1996.
14. Kaplan, E. *Understanding GPS, Principles and Applications*; Artech House: Boston, MA, USA, 1996.
15. Fine, P.; Wilson, W. Tracking Algorithm for GPS Offset Carrier Signals. In Proceedings of the 1999 National Technical Meeting of The Institute of Navigation, San Diego, CA, USA, 25–27 January 1999; pp. 671–676.

Sensors **2015**, *15*, 28472–28489

16. Phelts, R.; Walter, T.; Enge, P. Toward Real-Time SQM for WAAS: Improved Detection Techniques. In Proceedings of the 16th International Technical Meeting of the Satellite Division of the Institute of Navigation, Portland, OR, USA, 23 September 2003; pp. 2739–2749.

17. Jang, J.; Ahn, W.; Park, J.; Song, K. Optimum Diversity Combining for Pseudolite Navigation System. In Proceedings of the ISGNSS 2014 in conjunction with KGS Conference, Jeju Island, Korea, 22 October 2014; pp. 35–38.

Article

Multisensor Super Resolution Using Directionally-Adaptive Regularization for UAV Images

Wonseok Kang, Soohwan Yu, Seungyong Ko and Joonki Paik *

Department of Image, Chung-Ang University, 84 Heukseok-ro, Dongjak-gu, Seoul 156-756, Korea;
E-Mails: kandws12@cau.ac.kr (W.K.); shyu@cau.ac.kr (S.Y.); fantaaltanix@cau.ac.kr (S.K.)
* E-Mail: paikj@cau.ac.kr; Tel.: +82-2-820-5300; Fax: +82-2-814-9110. Acacemic Editors: Felipe Gonzalez Toro and Antonios Tsourdos

Received: 27 March 2015 / Accepted: 20 May 2015 / Published: 22 May 2015

Abstract: In various unmanned aerial vehicle (UAV) imaging applications, the multisensor super-resolution (SR) technique has become a chronic problem and attracted increasing attention. Multisensor SR algorithms utilize multispectral low-resolution (LR) images to make a higher resolution (HR) image to improve the performance of the UAV imaging system. The primary objective of the paper is to develop a multisensor SR method based on the existing multispectral imaging framework instead of using additional sensors. In order to restore image details without noise amplification or unnatural post-processing artifacts, this paper presents an improved regularized SR algorithm by combining the directionally-adaptive constraints and multiscale non-local means (NLM) filter. As a result, the proposed method can overcome the physical limitation of multispectral sensors by estimating the color HR image from a set of multispectral LR images using intensity-hue-saturation (IHS) image fusion. Experimental results show that the proposed method provides better SR results than existing state-of-the-art SR methods in the sense of objective measures.

Keywords: multisensor super-resolution (SR); UAV image enhancement; regularized image restoration; image fusion

---

1. Introduction

Multispectral images contain complete spectrum information at every pixel in the image plane and are currently applied to various unmanned aerial vehicle (UAV) imaging applications, such as environmental monitoring, weather forecasting, military intelligence, target tracking, *etc.* However, it is not easy to acquire a high-resolution (HR) image using a multispectral sensor, because of the physical limitation of the sensor. A simple way to enhance the spectral resolution of a multispectral image is to increase the number of photo-detectors at the cost of sensitivity and signal-to-noise ratio due to the reduced size of pixels. In order to overcome such physical limitations of a multispectral imaging sensor, an image fusion-based resolution enhancement method is needed [1,2].

Various enlargement and super-resolution (SR) methods have been developed in many application areas over the past few decades. The goal of these methods is to estimate an HR image from one or more low-resolution (LR) images. They can be classified into two groups: (i) single image based; and (ii) multiple image based. The latter requires a set of LR images to reconstruct an HR image. It performs the warping process to align multiple LR images with a sub-pixel precision. If LR images are degraded by motion blur and additive noise, the registration process becomes more difficult. To solve this problem, single image-based SR methods became popular, including: an interpolation-based SR [3–6], patch-based SR [7–12], image fusion-based SR [13–17], and others [18].

In order to solve the problem of simple interpolation-based methods, such as linear and cubic-spline interpolation [3], a number of improved and/or modified versions of image interpolation methods have been proposed in the literature. Li *et al.* used the geometric duality between the LR and HR images using local variance in the LR image [4]. Zhang *et al.* proposed an edge-guided non-linear interpolation algorithm using directionally-adaptive filters and data fusion [5]. Giachetti *et al.* proposed a curvature-based iterative interpolation using a two-step grip filling and an iterative correction of the estimated pixels [6]. Although the modified versions of interpolation methods can improve the image quality in the sense of enhancing the edge sharpness and visual improvement to a certain degree, fundamental interpolation artifacts, such as blurring and jagging, cannot be completely removed, due to the nature of the interpolation framework.

Patch-based SR methods estimate an HR image from the LR image, which is considered as a noisy, blurred and down-sampled version of the HR image. Freeman *et al.* proposed the example-based SR algorithm using the hidden Markov model that estimates the optimal HR patch corresponding to the input LR patch from the external training dataset [7]. Glasner *et al.* used a unified SR framework using patch similarity between in- and cross-scale images in the scale space [8]. Yang *et al.* used patch similarity from the learning dataset of HR and LR patch pairs in the sparse representation model [9]. Kim *et al.* proposed a sparse kernel regression-based SR method using kernel matching pursuit and gradient descent optimization to map the pairs of trained example patches from the input LR image to the output HR image [10]. Freedman *et al.* used non-dyadic filter banks to preserve the property of an input LR image and searched a similar patch using local self-similarity in the locally-limited region [11]. He *et al.* proposed a Gaussian regression-based SR method using soft clustering based on the local structure of pixels [12]. Existing patch-based SR methods can better reduce the blurring and jagging artifacts than interpolation-based SR methods. However, non-optimal patches make the restored image look unnatural, because of the inaccurate estimation of the high-frequency components.

On the other hand, image fusion-based SR methods have been proposed in the remote sensing fields. The goal of these methods is to improve the spatial resolution of LR multispectral images using the detail of the corresponding HR panchromatic image. Principal component analysis (PCA)-based methods used the projection of the image into the differently-transformed space [13]. Intensity-hue-saturation (IHS) [14,15] and Brovery [17] methods considered the HR panchromatic image as a linear combination of the LR multispectral images. Ballester *et al.* proposed an improved variational-based method [16]. These methods assume that an HR panchromatic image is a linear combination of LR multispectral images. Therefore, conventional image fusion-based SR methods have the problem of using an additional HR panchromatic imaging sensor.

In order to improve the performance of the fusion-based SR methods, this paper presents a directionally-adaptive regularized SR algorithm. Assuming that the HR monochromatic image is a linear combination of multispectral images, the proposed method consists of three steps: (i) acquisition of monochromatic LR images from the set of multispectral images; (ii) restoration of the monochromatic HR image using the proposed directionally-adaptive regularization; and (iii) reconstruction of the color HR image using image fusion. The proposed SR algorithm is an extended version of the regularized restoration algorithms proposed in [19,20] for optimal adaptation to directional edges and uses interpolation algorithms in [21–23] for resizing the interim images at each iteration.

The major contribution of this work is two-fold: (i) the proposed method can estimate the monochromatic HR image using directionally-adaptive regularization that provides the optimal adaptation to directional edges in the image; and (ii) it uses an improved version of the image fusion method proposed in [14,15] to reconstruct the color HR image. Therefore, the proposed method can generate a color HR image without additional high-cost imaging sensors using image fusion and the proposed regularization-based SR method. In experimental results, the proposed SR method is compared with seven existing image enlargement methods, including interpolation-based, example-based SR and patch similarity-based SR methods in the sense of objective assessments.

The rest of this paper is organized as follows. Section 2 summarizes the theoretical background of regularized image restoration and image fusion. Section 3 presents the proposed directional adaptive regularized SR algorithm and image fusion. Section 4 summarizes experimental results on multi- and hyper-spectral images, and Section 5 concludes the paper.

2. Theoretical Background

The proposed multispectral SR framework is based on regularized image restoration and multispectral image fusion. This section presents the theoretical background of multispectral image representation, regularized image restoration and image fusion in the following subsection.

2.1. Multispectral Image Representation

A multispectral imaging sensor measures the radiance of multiple spectral bands whose range is divided into a series of contiguous and narrow spectral bands. On the other hand, a monochromatic or single-band imaging sensor measures the radiance of the entire spectrum of the wavelength. The relationship between multispectral and monochromatic images is assumed to be modeled as the gray-level images between wavelength ω_1 and ω_2 as [24]:

$$I = \int_{\omega_1}^{\omega_2} R(\omega)Kq(\omega)d\omega + \eta_{(\omega_1 \sim \omega_2)} \tag{1}$$

where $R(\omega)$ represents the spectral radiance through the sensor's entrance pupil and K a constant that is determined by the sensor characteristics, including the electronic gain, detector saturation, quantization levels and the area of the aperture. $q(\omega)$ is the spectral response function of the sensor in the wavelength range between ω_1 and ω_2. $\eta_{(\omega_1 \sim \omega_2)}$ is the noise generated by the dark signal.

Since the spectral radiance $R(\omega)$ does not change by the sensor, the initial monochromatic HR image is generated by Equation (1), and it is used to reconstruct the monochromatic HR image from multispectral LR images using image fusion [14,15]. Figure 1 shows the multispectral imaging process, where a panchromatic image is acquired by integrating the entire spectral band, and the corresponding RGB image is also acquired using, for example, 33 bands. The high-resolution (HR) panchromatic and low-resolution (LR) RGB image are fused to generate an HR color image.

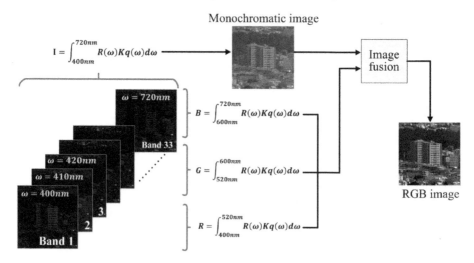

Figure 1. The multispectral imaging process.

2.2. Multispectral Image Fusion

In order to improve the spatial resolution of multispectral images, the intensity-hue-saturation (IHS) image fusion method is widely used in remote sensing fields [14,15]. More specifically, this method converts a color image into the IHS color space, where only the intensity band is replaced by the monochromatic HR image. The resulting HR image is obtained by converting the replaced intensity and the original hue and saturation back to the RGB color space.

2.3. Regularized Image Restoration

Regularization-based image restoration or enlargement algorithms regard the noisy, LR images as the output of a general image degradation process and incorporate *a priori* constraints into the restoration process to make the inverse problem better posed [19–23].

The image degradation model for a single LR image can be expressed as:

$$g = Hf + \eta \tag{2}$$

where g represents the observed LR image, H the combined low-pass filtering and down-sampling operator, f the original HR image and η the noise term.

The restoration problem is to estimate the HR image f from the observed LR image g. Therefore, the regularization approach minimizes the cost function as:

$$\begin{aligned} J(f) &= \tfrac{1}{2}\left\{\|g - Hf\|^2\right\} + \tfrac{1}{2}\lambda\|Cf\|^2 \\ &= \tfrac{1}{2}(g - Hf)^T(g - Hf) + \tfrac{1}{2}\lambda f^T C^T C f \end{aligned} \tag{3}$$

where C represents a two-dimensional (2D) high-pass filter, λ is the regularization parameter and $\|Cf\|^2$ the energy of the high-pass filtered image representing the amount of noise amplification in the restoration process.

The derivative of Equation (3) with respect to f is computed as:

$$\nabla J(f) = (-H^T g + H^T H f) + \lambda C^T C f \tag{4}$$

which becomes zero if:

$$f = \left(H^T H + \lambda C^T C\right)^{-1} H^T g \tag{5}$$

Thus, Equation (5) can be solved using the well-known regularized iteration process as:

$$f^{k+1} = f^k + \beta\left\{H^T g - \left(H^T H + \lambda C^T C\right) f^k\right\} \tag{6}$$

where $H^T H + \lambda C^T C$ represents the better-posed system matrix, and the step length β should be sufficiently small for the convergence.

3. Directionally-Adaptive Regularization-Based Super-Resolution with Multiscale Non-Local Means Filter

Since the estimation of the original HR image from the image degradation process given in (2) is almost always an ill-posed problem, there is no unique solution, and a simple inversion process, such as inverse filtering, results in significant amplification of noise and numerical errors [7–12]. To solve this problem, regularized image restoration incorporates *a priori* constraints on the original image to make the inverse problem better posed.

In this section, we describe a modified version of the regularized SR algorithm using a non-local means (NLM) filter [25] and the directionally-adaptive constraint as a regularization term to preserve edge sharpness and to suppress noise amplification. The reconstructed monochromatic HR image is used to generate a color HR image together with given LR multispectral images using IHS image fusion [14,15]. The block-diagram of the proposed method is shown in Figure 2.

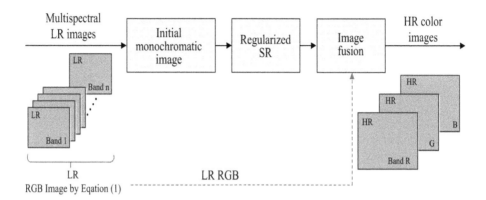

Figure 2. Block-diagram of the proposed super-resolution method.

3.1. Multispectral Low-Resolution Image Degradation Model

Assuming that the original monochromatic image is a linear combination of multispectral images [24], the observed LR multispectral images are generated by low-pass filtering and down-sampling from the differently translated version of the original HR monochromatic image. More specifically, the observed LR image in the i-th multispectral band is defined as:

$$g_i = Hf'_{(x_i,y_i)} + \eta_i = H_i f + \eta, \text{for}, \quad i = 1, ..., L \tag{7}$$

where $f'_{(x_i,y_i)}$ represents the translated version of the original HR monochromatic image f by (x_i, y_i), H the image degradation operator, including both low-pass filtering and down-sampling, and η the additive noise. In this paper, we assume that there is no warping operation in the image degradation model, because LR images are acquired by the multispectral sensor for the same scene.

3.2. Multiscale Non-Local Means Filter

If noise is present in the image degradation model given in Equation (2), the estimated HR image using the simple inverse filter yields:

$$\hat{f} = H^{-1}g = H^{-1}(Hf - \eta) = f^* + \Delta f \tag{8}$$

where the term $\Delta f = H^{-1}\eta$ amplifies the noise in an uncontrollable manner. This process can be considered to solve $g = Hf$ with the observation noise or perturbation η, which result in the amplified error $\Delta f = H^{-1}\eta$ in the solution.

If Δf is unbounded, the corresponding image restoration of the SR problem is ill posed. To solve the ill-posed restoration problem, we present an improved multiscale non-local means (NLM) filter to minimize the noise before the main restoration problem. The estimated noise-removed monochromatic image can be obtained using the least-squares optimization as:

$$\hat{f}_m = \arg\min_{f_m} \sum_{n \in \Omega_s} \left[g_{s,n}^P - f_m \right]^2 w_{m,n}^P, \text{ for}, \quad m, n = 1, ..., M, N \tag{9}$$

where f_m represents the m-th underlying pixel, Ω_s the local region in the 1.25^S-times down-scaled image, for $s = \{-1, -2, -3\}$, using the cubic-spline interpolation kernel [3], $g_{s,n}^P$ the local patches of g_m corresponding to Ω_s, $w_{m,n}^P$ the similarity weighting value between the local patches $g_{s,n}^P$ in the down-scaled image and the corresponding patch in g_m and the superscript **P** a patch.

Since the cubic-spline kernel performs low-pass filtering, it decreases the noise variance and guarantees searching sufficiently similar patches [8]. The similarity weight value is computed in the down-scaled image as:

$$w_{m,n}^P = \exp\left(-\frac{\|g_{s,n}^P - g_m^P\|_G^2}{1.25^s \sigma^2}\right) \tag{10}$$

where $g_{s,n}^P$ represents pixels in the patch centered at the location of $g_{s,m}$ and g_m^P pixel in the patch centered at the location of g_m in the original scale image. The parameter **G** is a Gaussian kernel that controls the exponential decay in the weighting computation.

The solution of the least-squares estimation in Equation (9) is given as:

$$\hat{f}_m = \left(\sum_{n\in\Omega_s} w_{m,n}^P\right)^{-1}\left(\sum_{n\in\Omega_s} w_{m,n}^P g_{s,n}^P\right) \tag{11}$$

3.3. Directionally-Adaptive Constraints

In minimizing the cost function in Equation (3), minimization of $\|g - Hf\|^2$ results in noise amplification, while minimization of $\|Cf\|^2$ results in a non-edge region. In this context, conventional regularized image restoration or SR algorithms [21–23] tried to estimate the original image by minimizing the cost function that is a linear combination of the two energies as $\|g - Hf\|^2 + \lambda\|Cf\|^2$. In this paper, we incorporate directionally-adaptive smoothness constraints into regularization process to preserve directional edge sharpness and to suppress noise amplification as:

$$\lambda\|C_D f\|^2, \text{ for } D = 1, ..., 5 \tag{12}$$

where the directionally-adaptive constraints C_D, for $D = 1, ..., 5$, suppress the noise amplification along the corresponding edge direction. In this work, we use the edge orientation classification filter [23]. The proposed directionally-adaptive constraints can be implemented using four 2D different high-pass filters as:

$$C_1^{0°} = \begin{pmatrix} 0 & 0 & 0 \\ 0 & 1 & 0 \\ 0 & 0 & 0 \end{pmatrix} - \frac{1}{6}\begin{pmatrix} 0 & 0 & 0 \\ 1 & 1 & 1 \\ 1 & 1 & 1 \end{pmatrix} = \begin{pmatrix} 0 & 0 & 0 \\ -0.1677 & 0.8333 & -0.1677 \\ -0.1677 & 0.8333 & -0.1677 \end{pmatrix} \tag{13}$$

$$C_2^{45°} = \begin{pmatrix} 0 & 0 & 0 \\ 0 & 1 & 0 \\ 0 & 0 & 0 \end{pmatrix} - \frac{1}{6}\begin{pmatrix} 1 & 0 & 0 \\ 1 & 1 & 0 \\ 1 & 1 & 1 \end{pmatrix} = \begin{pmatrix} -0.1677 & 0 & 0 \\ -0.1677 & 0.8333 & 0 \\ -0.1677 & 0.8333 & -0.1677 \end{pmatrix} \tag{14}$$

$$C_3^{90°} = \begin{pmatrix} 0 & 0 & 0 \\ 0 & 1 & 0 \\ 0 & 0 & 0 \end{pmatrix} - \frac{1}{6}\begin{pmatrix} 0 & 1 & 1 \\ 0 & 1 & 1 \\ 0 & 1 & 1 \end{pmatrix} = \begin{pmatrix} 0 & -0.1677 & -0.1677 \\ 0 & 0.8333 & -0.1677 \\ 0 & -0.1677 & -0.1677 \end{pmatrix} \tag{15}$$

and:

$$C_4^{135°} = \begin{pmatrix} 0 & 0 & 0 \\ 0 & 1 & 0 \\ 0 & 0 & 0 \end{pmatrix} - \frac{1}{6}\begin{pmatrix} 1 & 1 & 1 \\ 0 & 1 & 1 \\ 0 & 0 & 1 \end{pmatrix} = \begin{pmatrix} -0.1677 & -0.1677 & -0.1677 \\ 0 & 0.8333 & -0.1677 \\ 0 & 0 & -0.1677 \end{pmatrix} \tag{16}$$

By applying the directionally-adaptive constraints, an HR image can be restored from the input LR image. In the restored HR image, four directional edges are well preserved. In order to suppress noise amplification in the non-edge (NE) regions, the following constraint is used.

$$C_5^{NE} = \begin{pmatrix} 0 & 0 & 0 \\ 0 & 1 & 0 \\ 0 & 0 & 0 \end{pmatrix} - \frac{1}{9}\begin{pmatrix} 1 & 1 & 1 \\ 1 & 1 & 1 \\ 1 & 1 & 1 \end{pmatrix} = \begin{pmatrix} -0.1111 & -0.1111 & -0.1111 \\ -0.1111 & 0.8333 & -0.1111 \\ -0.1111 & -0.1111 & -0.1111 \end{pmatrix} \tag{17}$$

3.4. Combined Directionally-Adaptive Regularization and Modified Non-Local Means Filter

Given the multispectral LR images g_i, for $i = 1, ..., L$, the estimated monochromatic HR image $\hat{f}$ is given by the following optimization:

$$\hat{f} = \arg\min_{f} J(f) \tag{18}$$

where the multispectral extended version of Equation (3) is given as:

$$J(f) = \frac{1}{2}\sum_{i=1}^{L}\|g_i - H_i f\|^2 + \frac{\lambda}{2}\|C_D f\|^2$$
$$= \frac{1}{2}\sum_{i=1}^{L}(g_i - H_i f)^T(g_i - H_i f) + \frac{\lambda}{2}f^T C_D^T C_D f \tag{19}$$

The derivative of Equation (19) with respect to f is computed as:

$$\nabla J(f) = \sum_{i=1}^{L}(-H_i^T g_i + H_i^T H_i f) + \lambda C_D^T C_D f$$
$$= \left\{\sum_{i=1}^{L}H_i^T H_i + \lambda C_D^T C_D\right\}f - \sum_{i=1}^{L}H_i^T g_i \tag{20}$$

which becomes zero if:

$$f = \left(\sum_{i=1}^{L}H_i^T H_i + \lambda C_D^T C_D\right)^{-1}\sum_{i=1}^{L}H_i^T g_i \tag{21}$$

Finally, Equation (21) can be solved using the well-known iterative optimization with the proposed multiscale NLM filter as:

$$f^{k+1} = f_{NLM}^k + \beta\left\{\sum_{i=1}^{L}H_i^T g_i - \left(\sum_{i=1}^{L}H_i^T H_i + \lambda C_D^T C_D\right)f_{NLM}^k\right\} \tag{22}$$

where the matrix $\sum_{i=1}^{L}H_i^T H_i + \lambda C_D^T C_D$ is better-conditioned, and the step length β should be small enough to guarantee the convergence. f_{NLM}^k represents the multiscale NLM filtered version that can be expressed as:

$$f_{NLM}^k = \sum_{m=1}^{M}\left[\left(\sum_{n\in\Omega_s}w_{m,n}^P\right)^{-1}\left\{\sum_{n\in\Omega_s}w_{m,n}^P\left(\sum_{i=1}^{L}H_i^T g_{s,i,n}^P\right)\right\}\right] \tag{23}$$

For the implementation of Equation (22), the term $H_i^T g_i = S_i^T H^T g_i$ implies that the i-th multispectral LR images are first enlarged by simple interpolation as:

$$H^T = \tilde{H}^T \otimes \tilde{H}^T \tag{24}$$

where $\otimes$ represents the Kronecker product of matrices and $\tilde{H}$ represents the one-dimensional (1D) low-pass filtering and subsampling process with a specific magnification ratio.

In order to represent the geometric misalignment among different spectral bands, pixel shifting by $(-x_i, -y_i)$ is expressed as $S_i = \tilde{S}_{xi} \otimes \tilde{S}_{yi}$, where $\tilde{S}_p$ is the 1D translating matrix that shifts a 1D vector by

p samples. The term $H_i^T H_i f^k = S_i^T H^T H S_i f^k$ implies that the k-th iterative solution is shifted by (x_i, y_i), down-sampled by H, enlarged by interpolation H^T and then shifted by $(-x_i, -y_i)$, respectively.

3.5. Image Fusion-Based HR Color Image Reconstruction

A multispectral imaging sensor measures the radiance of multiple spectral bands whose ranges are divided into a series of contiguous and narrow spectral bands. In this paper, we adopt the IHS fusion method mentioned to estimate the HR color image from multispectral LR images as [14,15]:

$$\begin{bmatrix} I \\ H \\ S \end{bmatrix} = \begin{bmatrix} \frac{1}{3} & \frac{1}{3} & \frac{1}{3} \\ \frac{-\sqrt{2}}{6} & \frac{-\sqrt{2}}{6} & \frac{2\sqrt{2}}{6} \\ \frac{1}{\sqrt{2}} & \frac{-1}{\sqrt{2}} & 0 \end{bmatrix} \begin{bmatrix} R \\ G \\ B \end{bmatrix} \tag{25}$$

where R, G and B bands are computed as:

$$R = \int_{400\text{nm}}^{520\text{nm}} R(\omega) K q(\omega) d\omega \tag{26}$$

$$G = \int_{520\text{nm}}^{600\text{nm}} R(\omega) K q(\omega) d\omega \tag{27}$$

and:

$$B = \int_{600\text{nm}}^{720\text{nm}} R(\omega) K q(\omega) d\omega \tag{28}$$

where $R(\omega)$ represents the spectral radiance, K a constant gain and $q(\omega)$ is the spectral response function of the multispectral sensor as defined in Equation (1). The intensity component I is replaced with the estimated monochromatic HR image, and then, the IHS color space is converted back to the RGB color space as:

$$\begin{bmatrix} R_H \\ G_H \\ B_H \end{bmatrix} = \begin{bmatrix} 1 & \frac{-1}{\sqrt{2}} & \frac{1}{\sqrt{2}} \\ 1 & \frac{-1}{\sqrt{2}} & \frac{-1}{\sqrt{2}} \\ 1 & \sqrt{2} & 0 \end{bmatrix} \begin{bmatrix} \hat{f} \\ H \\ S \end{bmatrix} \tag{29}$$

where $\hat{f}$ represents the estimated monochromatic HR image and C_H, for $C \in \{R, G, B\}$, is the fused color HR image.

In this paper, we used the cubic-spline interpolation method to enlarge the hue (H) and saturation (S) channels by the given magnification factor. Figure 3 shows the image fusion-based HR color image reconstruction process.

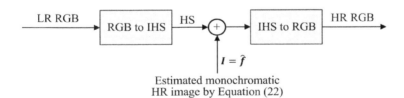

Figure 3. Block-diagram of the proposed fusion-based high-resolution (HR) color image reconstruction process.

279

Sensors **2015**, *15*, 12053–12079

4. Experimental Results

In this section, the proposed method is tested on various simulated, multispectral, real UAV and remote sensing images to evaluate the SR performance. In the following experiments, parameters were selected to produce the visually best results. In order to provide comparative experimental results, various existing interpolation and state-of-the-art SR methods were tested, such as cubic-spline interpolation [3], advanced interpolation-based SR [4–6], example-based SR [7] and patch-based SR [9–12].

| (a) | (b) | (c) | (d) | (e) |

Figure 4. Five multispectral test images.

To compare the performance of several SR methods, we used a set of full-reference metrics of image quality, including peak-to-peak signal-to-noise ratio (PSNR), structural similarity index measure (SSIM) [26], multiscale-SSIM (MS-SSIM) [27] and feature similarity index (FSIM) [28]. On the other hand, for the evaluation of the magnified image quality without the reference HR image, we adopted the completely blind image quality assessment methods, including the blind/referenceless image spatial quality evaluator (BRISQUE) [29] and natural image quality evaluator (NIQE) [30]. The higher image quality results in lower BRISQUE and NIQE values, but higher PSNR, SSIM, MS-SSIM and FSIM values.

The BRISQUE quantifies the amount of naturalness using the locally-normalized luminance values on *a priori* knowledge of both natural and artificially-distorted images. The NIQE takes into account the amount of deviations from the statistical regularities observed in the undistorted natural image contents using statistical features in natural scenes. Since BRISQUE and NIQE are referenceless metrics, they may not give the same ranking to the well-known metrics with reference, such as PSNR and SSIM.

4.1. Experiment Using Simulated LR Images

In order to evaluate the qualitative performance of various SR algorithms, we used five multispectral test images, each of which consists of 33 spectral bands in the wavelength range from 400 to 720 nanometers (nm), as shown Figure 4. In order to compare the objective image quality measures, such as PSNR, SSIM, MS-SSIM, FSIM and NIQE, the original multispectral HR image is first down-sampled by a factor of four to simulate the input LR images. Next, the input LR images are magnified four times using the nine existing methods and the proposed multispectral SR method.

In simulating LR images, the discrete approximation of Equation (1) is used, and the RGB color image is degraded by Equation (2). Given the simulated LR images, existing SR algorithms enlarge the RGB channels, whereas the proposed method generates the monochromatic HR image, including all spectral wavelengths, using the directionally-adaptive SR algorithm, and the IHS image fusion finally generates the color HR image using the monochromatic HR and RGB LR images [14,15].

Figures 5–9 show the results of enhancing the resolution of multispectral images using nine existing SR methods and the proposed multispectral SR method. Interpolation-based SR methods proposed in [3–6] commonly generate the blurring and jagging artifacts and cannot successfully recover the edge and texture details. The example-based method [7] and patch-based SR methods proposed in [9–12] can reconstruct clearer HR images than interpolation-based methods, but they cannot avoid unnatural artifacts in the neighborhood of the edge.

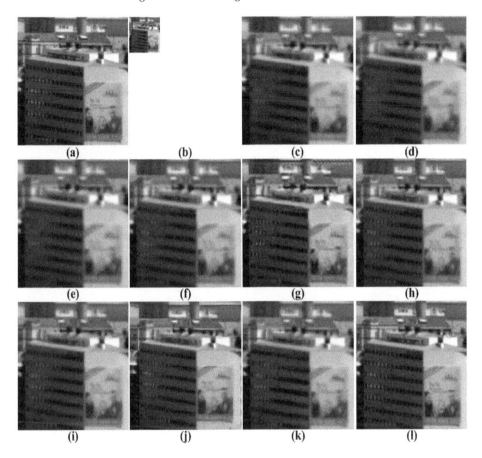

Figure 5. Results of resolution enhancement by enlarging a simulated low-resolution (LR) multispectral image: (**a**) cropped original HR image in Figure 4a; (**b**) the four-times down-sampled LR image; results of: (**c**) cubic-spline interpolation [3]; (**d**) interpolation-based SR [4]; (**e**) interpolation-based SR [5]; (**f**) interpolation-based SR [6]; (**g**) example-based SR [7]; (**h**) patch-based SR [9]; (**i**) patch-based SR [10]; (**j**) patch-based SR [11]; (**k**) patch-based SR [12] and (**l**) the proposed method.

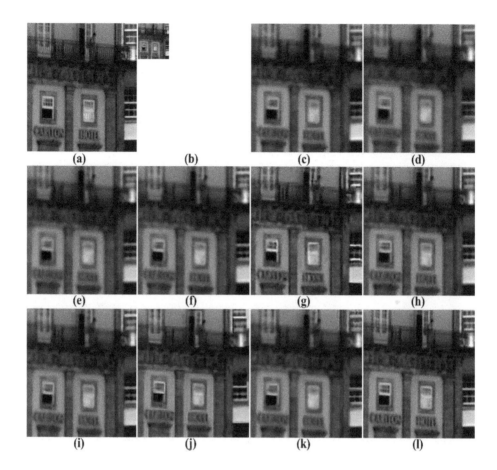

Figure 6. Results of resolution enhancement by enlarging a simulated LR multispectral image: (a) cropped original HR image in Figure 4b; (b) the four-times down-sampled LR image; results of: (c) cubic-spline interpolation [3], (d) interpolation-based SR [4], (e) interpolation-based SR [5], (f) interpolation-based SR [6], (g) example-based SR [7], (h) patch-based SR [9], (i) patch-based SR [10], (j) patch-based SR [11], (k) patch-based SR [12] and (l) the proposed method.

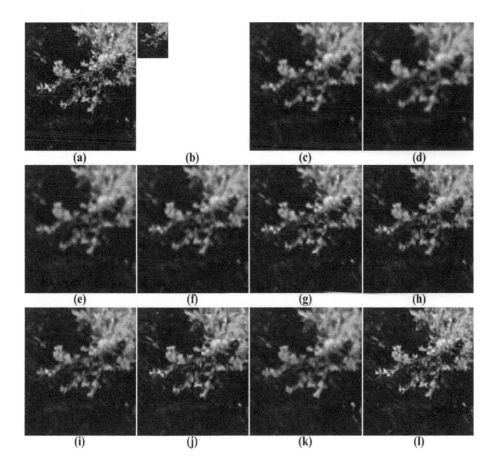

Figure 7. Results of resolution enhancement by enlarging a simulated LR multispectral image: (**a**) cropped original HR image in Figure 4c; (**b**) the four-times down-sampled LR image; results of: (**c**) cubic-spline interpolation [3], (**d**) interpolation-based SR [4], (**e**) interpolation-based SR [5], (**f**) interpolation-based SR [6], (**g**) example-based SR [7], (**h**) patch-based SR [9], (**i**) patch-based SR [10], (**j**) patch-based SR [11], (**k**) patch-based SR [12] and (**l**) the proposed method.

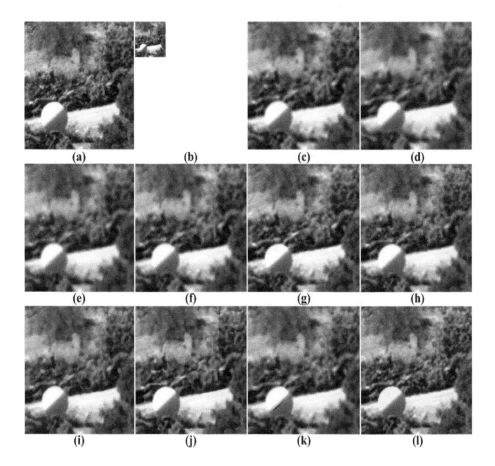

Figure 8. Results of resolution enhancement by enlarging a simulated LR multispectral image: (a) cropped original HR image in Figure 4d; (b) the four-times down-sampled LR image; results of: (c) cubic-spline interpolation [3], (d) interpolation-based SR [4], (e) interpolation-based SR [5], (f) interpolation-based SR [6], (g) example-based SR [7], (h) patch-based SR [9], (i) patch-based SR [10], (j) patch-based SR [11], (k) patch-based SR [12] and (l) the proposed method.

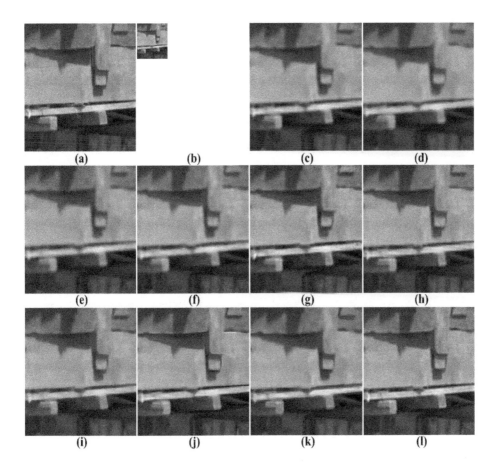

Figure 9. Results of resolution enhancement by enlarging a simulated LR multispectral image: (**a**) cropped original HR image in Figure 4e; (**b**) the four-times down-sampled LR image; results of: (**c**) cubic-spline interpolation [3], (**d**) interpolation-based SR [4], (**e**) interpolation-based SR [5], (**f**) interpolation-based SR [6], (**g**) example-based SR [7], (**h**) patch-based SR [9], (**i**) patch-based SR [10], (**j**) patch-based SR [11], (**k**) patch-based SR [12] and (**l**) the proposed method.

On the other hand, the proposed method shows a significantly improved SR result by successfully reconstructing the original high-frequency details and sharpens edges without unnatural artifacts. The PSNR, SSIM, MS-SSIM, FSIM, BRISQUE and NIQE values of the simulated multispectral test images shown in Figure 4 are computed for nine different methods, as summarized in Table 1.

Table 1. Comparison of peak-to-peak signal-to-noise ratio (PSNR), structural similarity index measure (SSIM), multiscale-SSIM (MS-SSIM), feature similarity index (FSIM), blind/referenceless image spatial quality evaluator (BRISQUE) and natural image quality evaluator (NIQE) values of the resulting images shown in Figure 4 using nine existing and the proposed SR methods.

Images	Methods	[3]	[4]	[5]	[6]	[7]	[9]	[10]	[11]	[12]	Proposed
	PSNR	28.13	24.82	25.03	28.50	25.84	30.14	30.21	-	27.49	**36.64**
	SSIM [26]	0.844	0.751	0.763	0.844	0.784	0.878	0.877	-	0.827	**0.964**
Figure 4a	MS-SSIM [27]	0.945	0.885	0.891	0.955	0.920	0.974	0.971	-	0.930	**0.993**
	FSIM [28]	0.875	0.825	0.830	0.877	0.854	0.903	0.905	-	0.863	**0.969**
	BRISQUE [29]	65.30	64.29	72.66	63.22	51.98	55.51	53.84	57.78	58.68	**48.30**
	NIQE [30]	9.65	11.18	12.45	8.47	7.50	9.07	11.35	9.29	8.46	**7.10**
	PSNR	27.76	24.94	25.09	26.12	25.21	28.97	29.28	-	27.27	**34.37**
	SSIM [26]	0.794	0.695	0.707	0.743	0.738	0.816	0.818	-	0.777	**0.931**
Figure 4b	MS-SSIM [27]	0.939	0.870	0.877	0.909	0.919	0.961	0.962	-	0.935	**0.990**
	FSIM [28]	0.855	0.795	0.804	0.823	0.843	0.869	0.871	-	0.848	**0.951**
	BRISQUE [29]	65.83	70.40	69.35	58.66	48.66	62.82	53.32	53.97	56.96	**48.33**
	NIQE [30]	9.65	16.13	11.37	7.97	7.15	9.91	8.47	8.12	8.41	**7.05**
	PSNR	25.04	24.69	24.91	25.03	26.54	28.68	25.51	-	26.29	**33.21**
	SSIM [26]	0.712	0.680	0.694	0.708	0.766	0.824	0.819	-	0.772	**0.932**
Figure 4c	MS-SSIM [27]	0.882	0.853	0.866	0.876	0.926	0.965	0.962	-	0.914	**0.987**
	FSIM [28]	0.817	0.782	0.793	0.810	0.850	0.870	0.858	-	0.834	**0.948**
	BRISQUE [29]	58.51	66.21	70.53	61.71	51.29	51.78	49.35	51.19	57.65	**49.06**
	NIQE [30]	8.59	13.35	13.00	8.12	7.07	**6.61**	7.15	7.17	7.91	6.79
	PSNR	23.22	23.08	23.22	23.24	24.77	29.73	29.86	-	27.34	**32.38**
	SSIM [26]	0.679	0.649	0.665	0.677	0.736	0.843	0.837	-	0.796	**0.939**
Figure 4d	MS-SSIM [27]	0.874	0.858	0.865	0.871	0.917	0.974	0.971	-	0.950	**0.988**
	FSIM [28]	0.824	0.793	0.804	0.818	0.853	0.895	0.886	-	0.867	**0.959**
	BRISQUE [29]	64.41	68.59	74.00	67.63	61.84	**56.83**	57.22	60.81	64.12	60.83
	NIQE [30]	8.17	13.85	11.49	8.22	7.12	7.14	7.43	**7.07**	8.32	7.85
	PSNR	26.43	26.45	26.37	26.49	26.80	32.45	33.25	-	29.26	**37.50**
	SSIM [26]	0.852	0.846	0.850	0.853	0.861	0.925	0.930	-	0.896	**0.969**
Figure 4e	MS-SSIM [27]	0.918	0.915	0.915	0.918	0.935	0.985	0.985	-	0.967	**0.996**
	FSIM [28]	0.872	0.856	0.869	0.873	0.883	0.927	0.930	-	0.895	**0.967**
	BRISQUE [29]	68.90	74.64	74.39	66.52	**51.99**	56.45	57.68	61.63	61.92	56.55
	NIQE [30]	9.93	13.78	10.64	9.91	9.86	**7.72**	9.07	8.97	8.94	8.37
	PSNR	26.12	24.79	24.93	25.88	25.83	29.99	29.62	-	27.53	**34.82**
	SSIM [26]	0.776	0.724	0.736	0.765	0.777	0.857	0.856	-	0.814	**0.947**
Average	MS-SSIM [27]	0.912	0.876	0.883	0.906	0.923	0.972	0.970	-	0.939	**0.991**
	FSIM [28]	0.849	0.810	0.820	0.840	0.857	0.893	0.890	-	0.861	**0.959**
	BRISQUE [29]	64.59	68.83	72.19	63.55	53.15	56.68	54.28	57.08	59.87	**52.61**
	NIQE [30]	9.20	13.66	11.79	8.54	7.74	8.09	8.69	8.12	8.41	**7.43**

Based on Table 1, the proposed method gives better results than existing SR methods in the sense of PSNR, SSIM, MS-SSIM and FSIM. Although the proposed method did not always provide the best results in the sense of NIQR and BRISQUE, the averaged performance using the extended set of test images shows that the proposed SR method performs the best.

In the additional experiment, an original monochromatic HR image is down-sampled and added by zero-mean white Gaussian noise with standard deviation $\sigma = 10$ to obtain a simulated version of the noisy LR image. The simulated LR image is enlarged by three existing SR [7,9,12] and the proposed methods, as shown in Figure 10. As shown in Figure 10, existing SR methods can neither remove the noise, nor recover the details in the image, whereas the proposed method can successfully reduce the noise and successfully reconstruct the original details. Table 2 shows PSNR and SSIM values of three existing SR methods for the same test image shown in Figure 10.

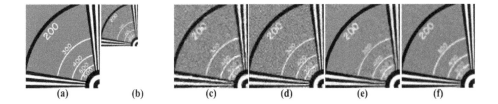

    (a)   (b)   (c)   (d)   (e)   (f)

Figure 10. Results of resolution enhancement by enlarging a simulated noisy LR monochromatic image: (**a**) original HR image; (**b**) the two-times down-sampled LR image with additive white Gaussian noise ($\sigma = 10$); results of: (**c**) example-based SR [7], (**d**) patch-based SR [9]; (**e**) patch-based SR [12] and (**g**) the proposed method.

Table 2. Comparison of PSNR and SSIM values of the resulting images using three existing SR and the proposed SR methods.

Methods	[7]	[9]	[12]	Proposed
PSNR	20.15	21.83	18.95	**24.45**
SSIM	0.566	0.523	0.802	**0.869**

The original version of the example-based SR method was not designed for real-time processing, since it requires a patch dictionary before starting the SR process [7]. The performance and processing time of the patch searching process also depend on the size of the dictionary. The sparse representation-based SR method needs iterative optimization for the ℓ_1 minimization process [9], which results in indefinite processing time. Although the proposed SR method also needs iterative optimization for the directionally-adaptive regularization, the regularized optimization can be replaced by an approximated finite-support spatial filter at the cost of the quality of the resulting images [31]. The non-local means filtering is another time-consuming process in the proposed work. However, a finite processing time can be guaranteed by restricting the search range of patches.

4.2. Experiment Using Real UAV Images

The proposed method is tested to enhance real UAV images, as shown Figure 11. More specifically, the remote sensing image is acquired by QuickBird equipped with a push broom-type image sensor to obtain a 0.65-m ground sample distance (GSD) panchromatic image.

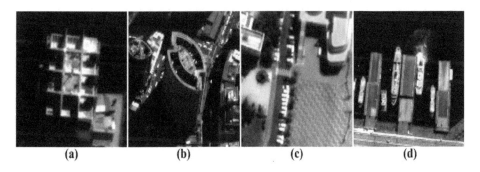

    (a)   (b)   (c)   (d)

Figure 11. Four real UAV test images.

Figures 12 to 15 show the results of enhanced versions using nine different SR and the proposed methods. In order to obtain no-reference measures, such as NIQE and BRISQUE values, Figure 11a–d are four-times magnified. In addition, the original UAV images are four-times down-sampled to generate simulated LR images and compared by the full-reference image quality measures, as summarized in Table 3.

As shown in Figures 12–15, the interpolation-based SR methods cannot successfully recover the details in the image. Since they are not sufficiently close to the unknown HR image, their NIQE and BRISQUE values are high, whereas example-based SR methods generate unnatural artifacts near the edge because of the inappropriate training dataset. Patch-based and the proposed SR methods provide better SR results.

The PSNR, SSIM, MS-SSIM, FSIM, NIQE and BRISQUE values are computed using nine different SR methods, as summarized in Table 3. Based on Table 3, the proposed method gives better results than existing SR methods in the sense of PSNR, SSIM, MS-SSIM and FSIM. Although the proposed method did not always provide the best results in the sense of NIQE and BRISQUE, the averaged performance using the extended set of test images shows that the proposed SR method performs the best.

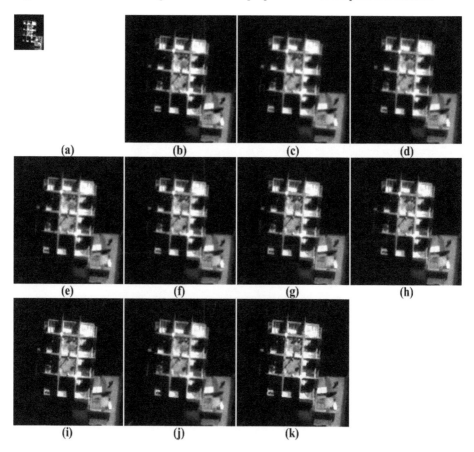

Figure 12. Results of resolution enhancement by enlarging a real UAV image: (a) original HR image in Figure 11a; result of: (b) cubic-spline interpolation [3]; (c) interpolation-based SR [4]; (d) interpolation-based SR [5]; (e) interpolation-based SR [6]; (f) example-based SR [7]; (g) patch-based SR [9]; (h) patch-based SR [10]; (i) patch-based SR [11]; (j) patch-based SR [12] and (k) the proposed method.

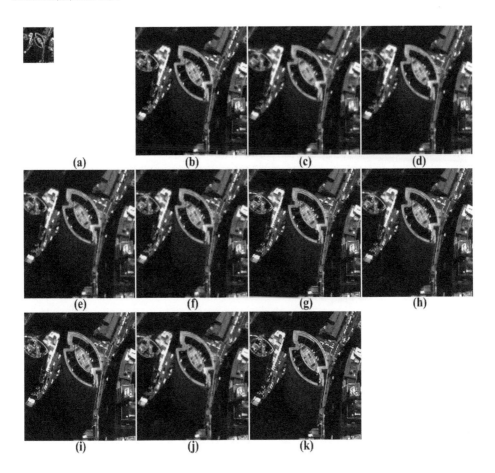

Figure 13. Results of resolution enhancement by enlarging a real UAV image: (**a**) original HR image in Figure 11b; result of: (**b**) cubic-spline interpolation [3]; (**c**) interpolation-based SR [4]; (**d**) interpolation-based SR [5]; (**e**) interpolation-based SR [6]; (**f**) example-based SR [7]; (**g**) patch-based SR [9]; (**h**) patch-based SR [10]; (**i**) patch-based SR [11]; (**j**) patch-based SR [12] and (**k**) the proposed method.

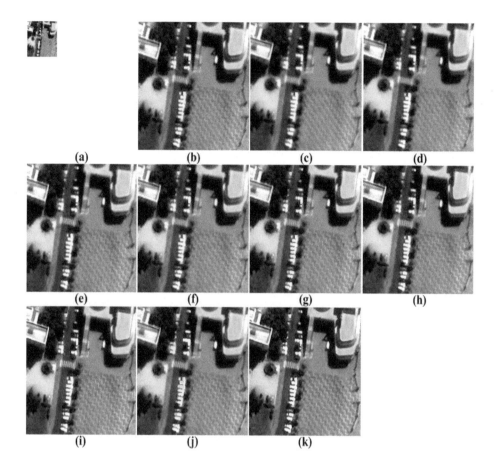

Figure 14. Results of resolution enhancement by enlarging a real UAV image: (**a**) original HR image in Figure 11c; result of: (**b**) cubic-spline interpolation [3]; (**c**) interpolation-based SR [4]; (**d**) interpolation-based SR [5]; (**e**) interpolation-based SR [6]; (**f**) example-based SR [7]; (**g**) patch-based SR [9]; (**h**) patch-based SR [10]; (**i**) patch-based SR [11]; (**j**) patch-based SR [12] and (**k**) the proposed method.

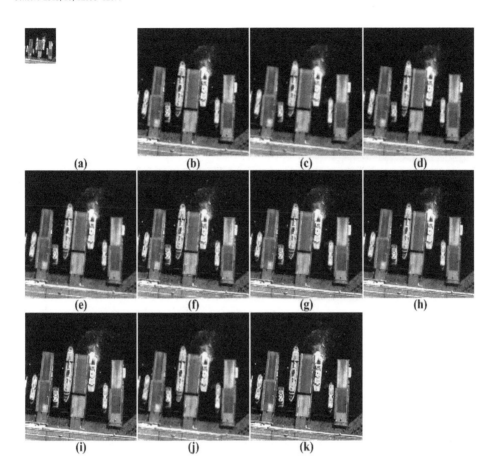

Figure 15. Results of resolution enhancement by enlarging a real UAV image: (**a**) original HR image in Figure 11d; result of: (**b**) cubic-spline interpolation [3]; (**c**) interpolation-based SR [4]; (**d**) interpolation-based SR [5]; (**e**) interpolation-based SR [6]; (**f**) example-based SR [7]; (**g**) patch-based SR [9]; (**h**) patch-based SR [10]; (**i**) patch-based SR [11]; (**j**) patch-based SR [12] and (**k**) the proposed method.

Table 3. Comparison of PSNR, SSIM, MS-SSIM, FSIM, BRISQUE and NIQE values of the resulting images shown in Figure 11 using nine existing and the proposed SR methods.

Images	Methods	[3]	[4]	[5]	[6]	[7]	[9]	[10]	[11]	[12]	Proposed
	PSNR	17.85	17.56	17.75	17.82	18.61	21.94	21.97	-	19.12	**26.86**
	SSIM [26]	0.710	0.681	0.698	0.709	0.723	0.824	0.830	-	0.766	**0.930**
Figure 11a	MS-SSIM [27]	0.837	0.844	0.843	0.840	0.882	0.948	0.955	-	0.907	**0.990**
	FSIM [28]	0.795	0.784	0.791	0.795	0.812	0.855	0.589	-	0.823	**0.929**
	BRISQUE [29]	63.35	38.87	71.87	63.29	63.91	52.78	55.73	58.04	64.69	**50.90**
	NIQE [30]	8.31	10.24	10.10	8.50	8.92	7.30	7.83	7.24	8.51	**5.97**
	PSNR	18.14	18.82	18.86	18.21	18.89	19.60	19.82	-	18.19	**22.53**
	SSIM [26]	0.641	0.654	0.662	0.666	0.618	0.670	0.680	-	0.590	**0.824**
Figure 11b	MS-SSIM [27]	0.821	0.840	0.825	0.825	0.886	0.915	0.922	-	0.853	**0.971**
	FSIM [28]	0.772	0.778	0.780	0.790	0.740	0.769	0.768	-	0.721	**0.865**
	BRISQUE [29]	52.93	46.00	67.31	50.33	46.24	51.71	39.59	41.50	39.03	**37.12**
	NIQE [30]	6.81	8.08	9.80	6.21	6.63	6.53	6.77	4.76	**4.71**	5.36
	PSNR	19.89	19.81	20.29	19.98	20.62	21.64	21.74	-	20.14	**24.94**
	SSIM [26]	0.634	0.609	0.651	0.643	0.615	0.660	0.661	-	0.600	**0.856**
Figure 11c	MS-SSIM [27]	0.805	0.802	0.822	0.807	0.876	0.903	0.897	-	0.847	**0.973**
	FSIM [28]	0.815	0.796	0.813	0.826	0.780	0.812	0.811	-	0.776	**0.901**
	BRISQUE [29]	61.30	63.46	71.18	63.32	64.93	53.94	**53.28**	54.58	62.75	55.88
	NIQE [30]	7.75	9.83	10.10	7.94	8.66	6.58	7.18	6.63	7.63	**5.85**
	PSNR	18.68	19.00	19.07	18.52	19.59	20.36	20.47	-	18.99	**23.98**
	SSIM [26]	0.633	0.646	0.656	0.646	0.635	0.679	0.686	-	0.635	**0.841**
Figure 11d	MS-SSIM [27]	0.818	0.828	0.835	0.816	0.882	0.912	0.910	-	0.872	**0.973**
	FSIM [28]	0.776	0.776	0.780	0.778	0.746	0.778	0.776	-	0.753	**0.873**
	BRISQUE [29]	55.20	52.13	68.70	54.06	51.35	53.70	**44.93**	50.01	48.33	47.51
	NIQE [30]	7.11	8.69	9.48	6.66	5.06	5.65	5.96	5.11	5.33	**5.56**
	PSNR	18.64	18.80	18.99	18.63	19.43	20.88	21.00	-	19.11	**24.58**
	SSIM [26]	0.654	0.648	0.667	0.666	0.648	0.708	0.714	-	0.648	**0.863**
Average	MS-SSIM [27]	0.820	0.828	0.831	0.822	0.881	0.919	0.921	-	0.870	**0.977**
	FSIM [28]	0.789	0.783	0.791	0.797	0.770	0.803	0.736	-	0.768	**0.892**
	BRISQUE [29]	58.19	50.12	69.77	57.75	56.61	53.03	48.38	51.03	53.70	**47.85**
	NIQE [30]	7.49	9.21	9.87	7.33	7.32	6.52	6.94	5.93	6.55	**5.68**

5. Conclusions

In this paper, we presented a multisensor super-resolution (SR) method using directionally-adaptive regularization and multispectral image fusion. The proposed method can overcome the physical limitation of a multispectral image sensor by estimating the color HR image from a set of multispectral LR images. More specifically, the proposed method combines the directionally-adaptive regularized image reconstruction and a modified multiscale non-local means (NLM) filter. As a result, the proposed SR method can restore the detail near the edge regions without noise amplification or unnatural SR artifacts. Experimental results show that the proposed method provided a better SR result than existing state-of-the-art methods in the sense of objective measures. The proposed method can be applied to all types of images, including a gray-scale or single-image, RGB color and multispectral images.

Acknowledgments: This work was supported by Institute for Information & communications Technology Promotion (IITP) grant funded by the Korea government (MSIP) (B0101-15-0525, Development of global multi-target tracking and event prediction techniques based on real-time large-scale video analysis), and by the Technology Innovation Program (Development of Smart Video/Audio Surveillance SoC & Core Component for Onsite Decision Security System) under Grant 10047788.

Acknowledgments: Wonseok Kang initiated the research and designed the experiments. Soohwan Yu performed experiments. Seungyong Ko analyzed the data. Joonki Paik wrote the paper.

Conflicts of Interest: The authors declare no conflict of interest.

References

1. Li, X.; Hu, Y.; Gao, X.; Tao, D.; Ning, B. A multi-frame image super-resolution method. *Signal Process.* **2010**, *90*, 405–414.
2. Zhang, Y. Understanding image fusion. *Photogramm. Eng. Remote Sens.* **2004**, *70*, 657–661.
3. Wick, D.; Martinez, T.; Adaptive optical zoom. *Opt. Eng.* **2004**, *43*, 8–9.
4. Li, X.; Orchard, M. New edge-directed interpolation. *IEEE Trans. Image Process.* **2001**, *10*, 1521–1527.
5. Zhang, L.; Wu, X. An edge-guided image interpolation algorithm via directional filtering and data fusion. *IEEE Trans. Image Process.* **2006**, *15*, 2226–2238.
6. Giachetti, A.; Asuni, N. Real-time artifact-free image upscaling. *IEEE Trans. Image Process.* **2011**, *20*, 2760–2768.
7. Freeman, W.; Jones, T.; Pasztor, E.C. Example-based super-resolution. *IEEE Comput. Graph. Appl. Mag.* **2002**, *22*, 56–65.
8. Glasner, D.; Bagon, S.; Irani, M. Super-resolution from a single image. In Proceedings of the IEEE International Conference on Computer Vision, Kyoto, Japan, 29 September 2009; pp. 349–356
9. Yang, J.; Wright, J.; Huang, S.; Ma, Y. Image super resolution via sparse representation. *IEEE Trans. Image Process.* **2010**, *19*, 2861–2873.
10. Kim, K.; Kwon, Y. Single-image super-resolution using sparse regression and natural image prior. *IEEE Trans. Pattern Anal. Mach. Intell.* **2010**, *32*, 1127–1133.
11. Freedman, G.; Fattal, R. Image and video upscaling from local self-examples. *ACM Trans. Graph.* **2011**, *30*, 1–10.
12. He, H.; Siu, W. Single image super-resolution using gaussian process regression. In Proceedings of the IEEE Computer Society Conference on Computer Vision, Pattern Recognition, Providence, RI, USA, 20–25 June 2011; pp. 449–456.
13. Shettigara, V.K. A generalized component substitution technique for spatial enhancement of multispectral images using a higher resolution data set. *Photogramm. Remote Sens.* **1992**, *58*, 561–567.
14. Tu, T.-M.; Huang, P.S.; Hung, C.-L.; Chang, C.-P. A fast intensity hue-saturation fusion technique with spectral adjustment for IKONOS imagery. *IEEE Geosci. Remote Sens. Lett.* **2004**, *1*, 309–312.
15. Choi, M. A new intensity-hue-saturation fusion approach to image fusion with a tradeoff parameter. *IEEE Geosci. Remote Sens. Lett.* **2006**, *44*, 1672–1682.
16. Ballester, C.; Caselles, V.; Igual, L.; Verdera, J. A Variational Model for P+XS Image Fusion. *Int. J. Comput. Vis.* **2006**, *69*, 43–58.
17. Du, Q.; Younan, N.; King, R.; Shah, V. On the performance evaluation of pan-sharpening techniques. *IEEE Geosci. Remote Sens. Lett.* **2007**, *4*, 518–522.
18. Nasrollahi, K.; Moeslund, T.B. Super-resolution: A comprehensive survey. *Mach Vis. Appl.* **2014**, *25*, 1423–1468.
19. Katsaggelos, A.K. Iterative image restoration algorithms. *Opt. Eng.* **1989**, *28*, 735–748.
20. Katsaggelos, A.K.; Biemond, J.; Schafer, R.W.; Mersereau, R.M. A regularized iterative image restoration algorithms. *IEEE Trans. Signal Process.* **1991**, *39*, 914–929.
21. Shin, J.; Jung, J.; Paik. J. Regularized iterative image interpolation and its application to spatially scalable coding. *IEEE Trans. Consum. Electron.* **1998**, *44*, 1042–1047.
22. Shin, J.; Choung, Y.; Paik, J. Regularized iterative image sequence interpolation with spatially adaptive contraints. In Proceedings of the IEEE International Conference on Image Processing, Chicago, IL, USA, 4–7 October 1998; pp. 470–473
23. Shin, J.; Paik, J.; Price, J.; Abidi, M. Adaptive regularized image interpolation using data fusion and steerable contraints. *SPIE Vis. Commun. Image Process.* **2001**, *4310*, 798–808.
24. Zhao, Y.; Yang, J.; Zhang, Q.; Song, L.; Cheng, Y.; Pan, Q. Hyperspectral imagery super-resolution by sparse representation and spectral regularization. *EURASIP J. Adv. Signal Process.* **2011**, *2011*, 1–10.
25. Buades, A.; Coll, B.; Morel, J.M. A non-local algorithm for image denoising. In Proceedings of the IEEE Computer Society Conference on Computer Vision, Pattern Recognition, San Diego, CA, USA, 20–25 June 2005; pp. 60–65.
26. Wang, Z.; Bovik, A.; Sheikh, H.; Simoncelli E. Image quality assessment: From error visibility to structural similarity. *IEEE Trans. Image Process.* **2004**, *13*, 600–612.

27. Wang, Z.; Simoncelli, E. P.; Bovik, A. C. Multi-scale structural similarity for image quality assessment. In Proceedings of the IEEE Asilomar Conference on Signals, Systems and Computers, Pacific Grove, CA, USA, 9–12 November 2003; pp. 1398–1402.

28. Zhang, L.; Zhang, L.; Mou, X.; Zhang, D. FSIM: A feature similarity index for image quality assessment. *IEEE Trans. Image Process.* **2011**, *20*, 2378–2386.

29. Mittal, A.; Moorthy, A. K.; Bovik, A.C. No-reference image quality assessment in the spatial domain. *IEEE Trans. Image Process.* **2012**, *21*, 4695–4708.

30. Mittal, A.; Soundararajan, R.; Bovik, A.C. Making a completely blind image quality analyzer. *IEEE Signal Process. Lett.* **2013**, *22*, 209–212.

31. Kim, S.; Jun, S.; Lee, E.; Shin, J.; Paik, J. Real-time bayer-domain image restoration for an extended depth of field (EDoF) camera. *IEEE Trans. Consum. Electron.* **2009**, *55*, 1756–1764.

Article

UAV Control on the Basis of 3D Landmark Bearing-Only Observations

Simon Karpenko, Ivan Konovalenko, Alexander Miller, Boris Miller * and Dmitry Nikolaev

Institute for Information Transmission Problems RAS, Bolshoy Karetny per. 19, Build. 1, GSP-4, Moscow 127051, Russia; simon.karpenko@gmail.com (S.K.); konovalenko@iitp.ru (I.K.); amiller@iitp.ru (A.M.); dimonstr@iitp.ru (D.N.)

* Correspondence: bmiller@iitp.ru; Tel.: +7-495-650-4781; Fax: +7-495-650-0579

Academic Editors: Felipe Gonzalez Toro and Antonios Tsourdos
Received: 15 August 2015; Accepted: 10 November 2015; Published: 27 November 2015

Abstract: The article presents an approach to the control of a UAV on the basis of 3D landmark observations. The novelty of the work is the usage of the 3D RANSAC algorithm developed on the basis of the landmarks' position prediction with the aid of a modified Kalman-type filter. Modification of the filter based on the pseudo-measurements approach permits obtaining unbiased UAV position estimation with quadratic error characteristics. Modeling of UAV flight on the basis of the suggested algorithm shows good performance, even under significant external perturbations.

Keywords: UAV; visual odometry; projective geometry; video stream; feature points; modified Kalman filter; control

---

1. Introduction

Modern UAV's navigation systems use the standard elements of INS (inertial navigation systems) along with GPS, which permits correcting the bias and improving the UAV localization, which are necessary for resolving mapping issues, targeting and reconnaissance tasks [1]. The use of computer vision as a secondary or a primary method for autonomous navigation of UAVs has been discussed frequently in recent years, since the classical combination of GPS and INS systems cannot sustain autonomous flight in many situations [2]. UAV autonomous missions usually need so-called data fusion, which is a difficult task, especially for standard INS and vision equipment. It is clear that cameras provide visual information in a different form, inapplicable to UAV direct control, and therefore, one needs an additional on-board memory and special recognition algorithms.

1.1. Visual-Based Navigation Approaches

Several studies have demonstrated the effectiveness of approaches based on motion field estimation and feature tracking for visual odometry [3]. Vision-based methods have been proposed even in the context of autonomous landing management [2]. In [4], visual odometry based on geometric homography was proposed. However, the homography analysis uses only 2D reference points coordinates, though for the evaluation of the current UAV altitude, the 3D coordinates are necessary. All such approaches presume the presence of some recognition system in order to detect the objects nominated in advance. Examples of such objects can be special buildings, crossroads, tops of mountains, and so on. The principal difficulties are the different scale and aspect angles of observed and stored images, which leads to the necessity of huge template libraries in the memory of the UAV control system. Here, one can avoid this difficulty, because of the usage of another approach based on the observation of so-called feature points [5] that are scale and aspect angle invariant. For this purpose, the technology of feature points [6] is used. In [7], the approach based on the coordinate correspondence of the reference points observed by the on-board camera and the reference points on the map loaded into the UAV's memory before the mission start had been suggested. During the flight, these maps are compared to the frame of the land, directly observed with the help of an on-board video camera. As a result, one can detect the current location and orientation without time-error accumulation. These methods are invariant to some transformations, and they are noise-stable, so that predetermined maps can be different in scale, aspect angle, season, luminosity, weather conditions, *etc*. This technology appeared in [8]. The contribution of this work is the usage of a modified unbiased pseudo-measurements filter for bearing-only observations of some reference points with known terrain coordinates.

1.2. Kalman Filter

In order to obtain metric data from visual observations, one needs first to make observations from different positions *i.e.*, *triangulation* and then use nonlinear filtering. However, all nonlinear filters either have unknown bias [9] or are very difficult for on-board implementation, like the Bayesian-type estimation [10,11]. Approaches for position estimation based on bearing-only observations had been analyzed long ago, especially for submarine applications [12] and nowadays for UAV applications [1].

A comparison of different nonlinear filters for bearing-only observations in the issue of ground-based object localization [13] shows that the EKF (extended Kalman filter), the unscented Kalman filter, the particle filter and the pseudo-measurement filter give almost the same level of accuracy, while the pseudo-measurement filter is usually more stable and simple for on-board implementation. This observation is in accordance with older results [12], where all of these filters were compared in the issue of moving object localization. It has been mentioned that all of these filters have bias, which makes their use in data fusion issues rather problematic [14]. The principle requirement for such filters in data fusion is the non-biased estimate with the known mean square characterization of the error. Among the variety of possible filters, the pseudo-measurement filter can be easily modified to satisfy the data fusion demands. The idea of such nonlinear filtering was developed by V.S. Pugachev and I. Sinitsyn in the form of so-called conditionally-optimal filtering [15], which provides the non-biased estimation within the class of linear filters with the minimum mean squared error. In this paper, we develop such a filter (the so-called pseudo-measurement Kalman filter (PKF)) for the UAV position estimation and give the algorithm for path planning along with the reference trajectory under external perturbations and noisy measurements.

1.3. Optical Absolute Positioning

Some known aerospace maps of a terrain in a flight zone are loaded into the aircraft memory before the start of a flight. During the flight, these maps are compared to the frame of the land, directly observed with the help of an on-board video camera. For this purpose, the technology of feature points [6] is used. As a result, one can detect the current location and orientation without time-error accumulation. These methods are invariant to some transformations and are also noise-stable, so that predetermined maps can vary in height, season, luminosity, weather conditions, *etc*. Furthermore, from the moment of the previous plane surveying, the picture of this landscape can be changed due to human and natural activity. All approaches based on the capturing of the objects assigned in advance presume the presence of some on-board recognition system in order to detect and recognize such objects. Here, we avoid this difficulty by using the observation of feature points [5] that are scale and aspect angle invariant. In addition, the modified pseudo-measurements Kalman filtering (PKF) is used for the estimation of UAV positions and the control algorithm.

1.4. Outline of the Approach and the Article Structure

One of the principal parts of this research is an approach to the estimation of the UAV position on the basis of the bearing-only observations. The original filter that uses the idea of pseudo-measurements had been suggested in reference [16] for the case of the azimuth bearing of the terrain objects nominated in advance. In reference [17], this approach had been extended to the case of two angle measurements, namely azimuth and elevation. However, the usage of this approach as a real navigation tool needs huge on-board memory and a sophisticated recognition algorithm, since the template and in-flight observed images, even of the same object, are rather different due to the changes of illuminance, the altitude of flight and the aspect angles. That is why the method based on the observation of feature points looks more attractive for in-flight implementation. In reference [7], an algorithm joining together the feature points approach and modified PKF had been suggested, though for 2D feature point localization, while the more advanced 3D localization had been suggested in references [18,19], which are the shortened versions of the methodology presented in this article.

In this work, we use a computer simulation of a UAV flight and on-board video camera imaging. The simulation program is written in MATLAB. The type of feature points is ASIFT, realized in OpenCV (Python) [20]. Feature points in this model work as in a real flight, because the images for the camera model and for the template images were transformed by projective mapping and created by observations from different satellites.

The next section presents the original RANSAC algorithm for 3D feature point localization. Sections 3 and 4 give the description of PKF, providing the unbiased estimation of the UAV position with the estimate of quadratic errors. Section 5 describes the locally optimal control algorithm for tracking the reference trajectory on the basis of PKF estimation of the UAV position. In Section 6, we give a new approach to the RANSAC robustness with the use of the UAV motion model. Section 7 presents the modeling results, and Section 8 is the conclusions.

2. Random Sample Consensus for Isometry

At every step, the algorithm deals with two images of a 3D landscape. An example of the landscape used for modeling is shown in Figure 1.

a)   b)

Figure 1. (a) Image I_c of the on-board camera of the UAV; (b) template image loaded in the UAV memory in advance.

The first image I_c is obtained from the on-board UAV camera, the position of which is unknown and should be estimated. The second image I_m was taken previously from a known position and uploaded into the UAV memory. The ASIFT method is used for both images to detect feature points, which are specified in pixels:

$$c_i = \begin{vmatrix} c_{xi} \\ c_{yi} \end{vmatrix}$$

and:

$$m_i = \begin{vmatrix} m_{xi} \\ m_{yi} \end{vmatrix}$$

and calculates their descriptors. The correspondence between images is constructed by using these descriptors, and thereby, the feature points are combined in pairs. However, many of these pairs are wrong, and therefore, these pairs are considered as outliers or they are not. The result of ASIFT correspondence is shown in Figure 2.

The Earth coordinate system is the Cartesian coordinate system, which is rigidly connected with the Earth. Therefore, the algorithm uses a 3D terrain map of the area from which the image I_m was taken and over which the UAV flies. Therefore, one can determine the coordinates of the points:

$$r_i = \begin{vmatrix} x_i \\ y_i \\ z_i \end{vmatrix}$$

which generated m_i points in the Earth coordinate system. However, if i corresponds to the pair of points that is not an outlier, then the point r_i also generates a point c_i in the UAV camera.

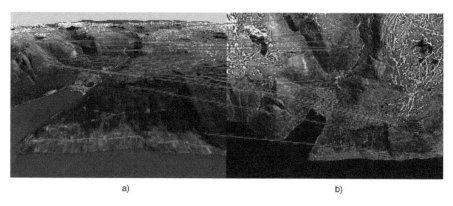

a)   b)

Figure 2. (a) Image I_c of the on-board camera of the UAV; (b) template image loaded in the UAV memory in advance.

Another Cartesian coordinate system is rigidly connected with the UAV camera. The axis of the camera is parallel to the axis z. The transformation from the Earth coordinate system to the UAV coordinate system has the form:

$$r' = A(r - b)$$

where b represents the coordinates of the camera in the Earth coordinate system and A is the orthogonal ($AA^T = I$) rotation matrix defining the orientation of the UAV camera. Then, the points r_i in the camera coordinate system are:

$$r'_i = A(r_i - b)$$

To define the relation between r_i and the feature points c_i, one can use the model of the camera obscura. This model gives a central projection on the plane:

$$c_i = \begin{vmatrix} \frac{\rho_{xi}}{\rho_{zi}} \\ \frac{\rho_{yi}}{\rho_{zi}} \end{vmatrix}$$

where:

$$\begin{vmatrix} \rho_{xi} \\ \rho_{yi} \\ \rho_{zi} \end{vmatrix} = \rho_i = Kr'_i = KA(r_i - b)$$

where K is a known calibration matrix of the camera.

Thus, the task is to estimate A and b on the basis of known c_i, r_i, K. The minimum number of feature point pairs needed to solve this task is three.

One can give the solution of the problem under the assumption that there are just three pairs:

$$i = \{1,2,3\}$$

Points r_i' form a triangle in the space with the following sides:

$$\rho_1 = ||r_2' - r_3'||_2, \rho_2 = ||r_3' - r_1'||_2, \rho_3 = ||r_1' - r_2'||_2$$

Meanwhile, due to the rectilinear propagation of light, each point r_i' lies on the beam $r' = a_i t$, where t is a scalar parameter, and:

$$a_i = K^{-1} \begin{vmatrix} c_{xi} \\ c_{yi} \\ 1 \end{vmatrix}$$

In order to find r_i', we have to determine parameters t_i, $i = \{1,2,3\}$ that satisfy the system of quadratic equations:

$$\begin{cases} (a_2 t_2 - a_3 t_3)^T (a_2 t_2 - a_3 t_3) = \rho_1^2 \\ (a_3 t_3 - a_1 t_1)^T (a_3 t_3 - a_1 t_1) = \rho_2^2 \\ (a_1 t_1 - a_2 t_2)^T (a_1 t_1 - a_2 t_2) = \rho_3^2 \end{cases}$$

For the given t_1, this system may be either solved analytically or has no solution. A determination of t_1 can be done numerically, for example by the bisection method.

Finally, one can obtain the coordinates of three points on the Earth's surface in the camera coordinate system r_i' and, at the same time, in the Earth coordinate system r_i. The connection between them is: $r_i' = A(r_i - b)$. Since A is the orthogonal matrix, then $y = Ax$ implies $||y||_2 = ||x||_2$; thereby:

$$r_i'^T r_i' = (r_i - b)^T (r_i - b)$$

Therefore, we have eliminated A and obtained the problem of finding the intersection of three spheres, which can be solved analytically. This problem may have two solutions; one of them will be rejected later. When b has been found, solutions for A are as follows:

$$A = \begin{vmatrix} r_1' & r_2' & r_3' \end{vmatrix} \begin{vmatrix} r_1 - b & r_2 - b & r_3 - b \end{vmatrix}^{-1}$$

Therefore, there are two options, and only one of them is correct:

$$\begin{vmatrix} a_{11} \\ a_{21} \\ a_{31} \end{vmatrix} = \begin{vmatrix} a_{12} \\ a_{22} \\ a_{32} \end{vmatrix} \times \begin{vmatrix} a_{13} \\ a_{23} \\ a_{33} \end{vmatrix} \text{ or } \begin{vmatrix} a_{11} \\ a_{21} \\ a_{31} \end{vmatrix} = - \begin{vmatrix} a_{12} \\ a_{22} \\ a_{32} \end{vmatrix} \times \begin{vmatrix} a_{13} \\ a_{23} \\ a_{33} \end{vmatrix}$$

If the first one is correct, then the second one corresponds exactly to the turnover.

As a result, A and b have been found by using three pairs of feature points and the height map. However, this approach alone is not suitable as the final solution, due to the following problems:

1. The method gives either knowingly false solution or no solution at all if among the three points there are outliers.
2. There is a strong dependence on the noise in the feature points' location.

Both problems may be solved with random sample consensus (RANSAC) [21,22]. From the general selection of points, one needs to select N times a subsample of size three. For each subsample

$j = \{1, 2, \cdots, N\}$, one can calculate A_j and b_j, which allows one to simulate the generation of all feature points on the UAV camera:

$$c_{ji} = \begin{vmatrix} \rho_{jxi} \\ \rho_{jzi} \\ \rho_{jyi} \\ \rho_{jzi} \end{vmatrix}, \rho_j = Kr'_{ji} = KA_j(r_{ji} - b_j)$$

Then, one can evaluate which points are the outliers by the threshold $s_{ji} = \mathbf{1}(||c_i - c_{ji}||_2 < d)$, where d is the threshold. Here, $s_{ji} = 0$ means that projection of the i-th point on the j-th point is counted as an outlier, otherwise $s_{ji} = 1$. The answer will be the following:

$$A = A_J, b = b_J, J = \arg\max_j \sum_i s_{ji}$$

Therefore, we really solve the problem of outliers. Next, we find the required number of N subsamples of a size of three, such that among them, there will be at least one subsample without outliers with probability p. Let the proportion of outliers be $1 - w$. It is easy to see that [21]:

$$N(p) = \frac{\log(1 - p)}{\log(1 - w^3)}$$

In the case when $w = \frac{1}{2}$: $N(0.9999) \approx 69$, which shows the high efficiency of algorithm. After that, the points marked as outliers are removed from consideration. The clarification of the response is made by the numerical solution of the following optimization problem on the set of remaining points:

$$\{A^*, b^*\} = \arg\min_{A,b} \sum_i ||c_i - c_i(A, b)||_2^2$$

Thereby, the second problem of noise reduction may be solved. However, one can use a more advanced procedure, which takes into account the motion model. A more stable solution may be obtained with the aid of so-called robust RANSAC [23]; the idea is to use predicted values of (A, b) for the preliminary rejection of outliers from the pairs of observed feature points. Therefore, if on the k-th step of the filtering procedure, the values (A_k, b_k) have been obtained, one can use the following values on the $(k + 1)$-th step:

$$(A_{k+1}, b_{k+1}) = (A_k, \hat{b}_{k+1}) \tag{1}$$

where $\hat{b}_{k+1}$ is the predicted estimate of the UAV attitude obtained on the basis of the PKF estimate. The filtering approach is described in Section 5.

3. Filtering Problem Statement

The problem of bearing-only filtering is considered to determine the coordinates of the UAV, which can observe some objects with known coordinates. These objects can be either the well recognizable objects or a network of radio-beacon stations with a well-specified frequency and known coordinates. In this work, the function of beacons is performed by the feature points determined with the aid of the RANSAC algorithm. The UAV has the standard set of INS devices, which enables it to perform the flight with some degree of accuracy, which, however, is not sufficient for mission completion.

3.1. Model of the UAV's Motion

We assume that a UAV motion is described by three coordinates $(X(t_k), Y(t_k), Z(t_k))$ and velocities $(V_x(t_k), V_y(t_k), V_z(t_k))$. At times $t_k = k\Delta t, k = 1, 2, ...$, these coordinates satisfy the following equations:

$$X(t_{k+1}) = FX(t_k) + Ba(t_k) + W(t_k) \tag{2}$$

where:

$$\mathbf{X}(t_k) = \left(X(t_k), Y(t_k), Z(t_k), V_x(t_k), V_y(t_k), V_z(t_k)\right)^T$$

is the vector of state-velocities,

$$\mathbf{a}(t_k) = \left(a_x(t_k), a_y(t_k), a_z(t_k)\right)^T$$

is the vector of accelerations, which we consider as controls,

$$\mathbf{W}(t_k) = \left(0, 0, 0, W_x(t_k), W_y(t_k), W_z(t_k)\right)^T$$

is the vector of current perturbations, modeling the turbulence components of the wind and the autopilot errors, as well. The matrices A and B are equal:

$$F = \begin{pmatrix} 1 & 0 & 0 & \Delta t & 0 & 0 \\ 0 & 1 & 0 & 0 & \Delta t & 0 \\ 0 & 0 & 1 & 0 & 0 & \Delta t \\ 0 & 0 & 0 & 1 & 0 & 0 \\ 0 & 0 & 0 & 0 & 1 & 0 \\ 0 & 0 & 0 & 0 & 0 & 1 \end{pmatrix}$$

$$B = \begin{pmatrix} \frac{\Delta t^2}{2} & 0 & 0 \\ 0 & \frac{\Delta t^2}{2} & 0 \\ 0 & 0 & \frac{\Delta t^2}{2} \\ \Delta t & 0 & 0 \\ 0 & \Delta t & 0 \\ 0 & 0 & \Delta t \end{pmatrix}$$

and stochastic Equation (2) describes a controlled and perturbed UAV motion.

3.2. Measurements

Assume that (X_i, Y_i, Z_i) are the coordinates of the i-th reference point and $\phi_i(t_k), \lambda_i(t_k)$ are the bearing angles on that point. The measuring scheme is shown in Figure 3.

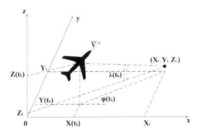

Figure 3. Scheme of the measurements of bearing angles. X_i, Y_i, Z_i are the coordinates of the i-th feature point; $\lambda(t_k), \phi(t_k)$ are the elevation and azimuth bearing angles, measured at the moment t_k.

At moment t_k, these angles satisfy the relations:

$$\frac{Y_i + \varepsilon_k^y - Y(t_k)}{X_i + \varepsilon_k^x - X(t_k)} I_i(t_k) = \tan \phi_i(t_k) + \varepsilon_k^\phi$$

$$\frac{Z_i + \varepsilon_k^z - Z(t_k)}{\sqrt{\left(X_i + \varepsilon_k^x - X(t_k)\right)^2 + \left(Y_i + \varepsilon_k^y - Y(t_k)\right)^2}} I_i(t_k) = \tan \lambda_i(t_k) + \varepsilon_k^\lambda$$

(3)

where $\varepsilon_k^x \sim \mathcal{WN}(0, \sigma_x^2), \varepsilon_k^y \sim \mathcal{WN}\left(0, \sigma_y^2\right), \varepsilon_k^z \sim \mathcal{WN}(0, \sigma_z^2), \varepsilon_k^\phi \sim \mathcal{WN}\left(0, \sigma_\phi^2\right), \varepsilon_k^\lambda \sim \mathcal{WN}(0, \sigma_\lambda^2)$ are uncorrelated random variables with zero means and variances $\sigma_x^2, \sigma_y^2, \sigma_z^2, \sigma_\phi^2, \sigma_\lambda^2$, defined as errors in the measurement of the coordinates of the *i*-th reference point and of the tangents of angles $\phi_i(t_k), \lambda_i(t_k)$ and forming the white noise sequences.

Remark 1. *In the majority of works based on the method of pseudo-measurements, another model is used. It assumes the measurements of the angles with Gaussian errors (see [12,14] and most of the successive works). However, in the real definition of the object position in the image or in the matrix of sensors, the system measures the distance between the object image and the center of the sensor, that is the tangent of the bearing angle. This simple observation allows one to find the unbiased estimate of the UAV coordinates.*

One can rewrite Equation (3) for angle $\lambda_i(t_k)$ as follows:

$$\frac{Z_i + \varepsilon_k^z - Z(t_k)}{Y_i + \varepsilon_k^y - Y(t_k)} \sin \phi_i(t_k) I_i(t_k) = \frac{\sin \lambda_i(t_k)}{\cos \lambda_i(t_k)} + \varepsilon_k^\lambda \qquad (4)$$

Remark 2. *The indicator function $I_i(t_k) = 1$ if at t_k the bearing of the i-th reference point occurs, and $I_i(t_k) = 0$ otherwise. For convenience, we assume that $I_i(t_k) = 1$.*

Therefore, at the moment t_k, the UAV control system determines the angles $\phi_i(t_k)$ and $\lambda_i(t_k)$, related to the coordinates of the UAV $(X(t_k), Y(t_k), Z(t_k))$ as follows:

$$\begin{aligned}
\left(Y_i + \varepsilon_k^y - Y(t_k)\right) \cos \phi_i(t_k) - \left(X_i + \varepsilon_k^x - X(t_k)\right) \sin \phi_i(t_k) \\
= \varepsilon_k^\phi \left(X_i + \varepsilon_k^x - X(t_k)\right) \cos \phi_i(t_k)
\end{aligned}$$

$$\begin{aligned}
\left(Z_i + \varepsilon_k^z - Z(t_k)\right) \sin \phi_i(t_k) \cos \lambda_i(t_k) - \left(Y_i + \varepsilon_k^y - Y(t_k)\right) \sin \lambda_i(t_k) \\
= \varepsilon_k^\lambda \left(Y_i + \varepsilon_k^y - Y(t_k)\right) \cos \lambda_i(t_k)
\end{aligned} \qquad (5)$$

4. Modified Kalman Filtering on the Basis of Pseudo-Measurements

4.1. Linear Measurements Model

The idea of the pseudo-measurement method is to separate in Equation (5) the observable and non-observable values, which gives the following observation equations:

$$\begin{aligned}
m_k^\phi &= Y_i \cos \phi_i(t_k) - X_i \sin \phi_i(t_k) = Y(t_k) \cos \phi_i(t_k) - X(t_k) \sin \phi_i(t_k) \\
&\quad -\varepsilon_k^y \cos \phi_i(t_k) + \varepsilon_k^x \sin \phi_i(t_k) + \varepsilon_k^\phi \left(X_i + \varepsilon_k^x - X(t_k)\right) \cos \phi_i(t_k)
\end{aligned}$$

$$\begin{aligned}
m_k^\lambda &= Z_i \sin \phi_i(t_k) \cos \lambda_i(t_k) - Y_i \sin \lambda_i(t_k) = Z(t_k) \sin \phi_i(t_k) \cos \lambda_i(t_k) - Y(t_k) \sin \lambda_i(t_k) \\
&\quad -\varepsilon_k^z \sin \phi_i(t_k) \cos \lambda_i(t_k) + \varepsilon_k^y \sin \lambda_i(t_k) + \varepsilon_k^\lambda \left(Y_i + \varepsilon_k^y - Y(t_k)\right) \cos \lambda_i(t_k)
\end{aligned} \qquad (6)$$

where X_i, Y_i, Z_i represent the coordinates of the *i*-th feature point determined with the aid of the RANSAC algorithm and ϕ_i, λ_i are the corresponding observable bearing angles measured by the system. Thus, the left-hand side of Equation (6), that is (m_k^ϕ, m_k^λ), corresponds to the observable values, whereas the right-hand side containing the coordinates of the UAV corresponds to the non-observable ones. The aim is to estimate the coordinates and velocities of the UAV on the basis of linear observation Equation (6) and the motion model Equation (2). Therefore, the measurement vector has the following form:

$$m_k = \begin{pmatrix} m_k^\phi \\ m_k^\lambda \end{pmatrix} = \begin{pmatrix} Y(t_k)\cos\phi_i(t_k) - X(t_k)\sin\phi_i(t_k) - \varepsilon_k^y\cos\phi_i(t_k) + \varepsilon_k^x\sin\phi_i(t_k) \\ +\varepsilon_k^\phi(X_i + \varepsilon_k^x - X(t_k))\cos\phi_i(t_k) \\ \\ Z(t_k)\sin\phi_i(t_k)\cos\lambda_i(t_k) - Y(t_k)\sin\lambda_i(t_k) - \varepsilon_k^z\sin\phi_i(t_k)\cos\lambda_i(t_k) \\ +\varepsilon_k^y\sin\lambda_i(t_k) + \varepsilon_k^\lambda\left(Y_i + \varepsilon_k^y - Y(t_k)\right)\cos\lambda_i(t_k) \end{pmatrix} \quad (7)$$

Thereby, we obtain the system Equation (7) of linear measurement equations, though the noise variance depends on non-observable coordinates. By using V.S. Pugachev's method [15], one can obtain the unbiased estimate and the variance with the aid of a prediction-correction filter [24]. Moreover, we do not need to assume the Gaussian distribution of errors that is not valid in bearing observations with optical-electronic cameras with discrete image sensors.

4.2. Prediction-Correction Estimation

Assume that at the moment t_k, we have unbiased estimates $\hat{X}(t_k)$, such that:

$$E(\hat{X}(t_k)) = X(t_k) \quad (8)$$

with the following matrix of the mean-square errors:

$$\hat{P}(t_k) = E\left\{ (\hat{X}(t_k) - X(t_k))(\hat{X}(t_k) - X(t_k))^T \right\}$$

$$= \begin{pmatrix}
\hat{p}^{xx}(t_k) & \hat{p}^{xy}(t_k) & \cdots & \cdots & \cdots & \hat{p}^{xV_z}(t_k) \\
\hat{p}^{xy}(t_k) & \hat{p}^{yy}(t_k) & \cdots & \cdots & \cdots & \hat{p}^{yV_z}(t_k) \\
\hat{p}^{xz}(t_k) & \hat{p}^{yz}(t_k) & \cdots & \cdots & \cdots & \hat{p}^{zV_z}(t_k) \\
\hat{p}^{xV_x}(t_k) & \hat{p}^{yV_x}(t_k) & \cdots & \cdots & \cdots & \hat{p}^{V_xV_z}(t_k) \\
\hat{p}^{xV_y}(t_k) & \hat{p}^{yV_y}(t_k) & \cdots & \cdots & \cdots & \hat{p}^{V_yV_z}(t_k) \\
\hat{p}^{xV_z}(t_k) & \hat{p}^{yV_z}(t_k) & \cdots & \cdots & \cdots & \hat{p}^{V_zV_z}(t_k)
\end{pmatrix} \quad (9)$$

Problem 1. *Find the unbiased estimates $\hat{X}(t_{k+1})$ and matrix $\hat{P}(t_{k+1})$ on the basis of estimates at the moment t_k, m_k, the known position of the i-th observable object (X_i, Y_i, Z_i) and the UAV's motion parameter Equation (2). These estimates must satisfy Equation (8) and give the matrix Equation (9) for the moment t_{k+1}.*

4.2.1. Prediction

The prediction is obtained by assuming that at the moment t_{k+1}, the values of $\phi(t_{k+1})$, $\lambda(t_{k+1})$ will be known:

$$\widetilde{X}(t_{k+1}) = F\hat{X}(t_k) + Ba(t_k)$$

$$\widetilde{m}_{k+1} = \begin{pmatrix} \widetilde{m}_{k+1}^\phi \\ \widetilde{m}_{k+1}^\lambda \end{pmatrix} = \begin{pmatrix} I(t_{k+1})(\widetilde{Y}(t_{k+1})\cos\phi(t_{k+1}) - \widetilde{X}(t_{k+1})\sin\phi(t_{k+1})) \\ I(t_{k+1})(\widetilde{Z}(t_{k+1})\sin\phi(t_{k+1})\cos\lambda(t_{k+1}) - \widetilde{Y}(t_{k+1})\sin\lambda(t_{k+1})) \end{pmatrix} \quad (10)$$

Assuming that the motion perturbations and the UAV position are uncorrelated, we obtain:

$$\widetilde{P}(t_{k+1}) = \begin{pmatrix} \widetilde{P}^{xx}(t_{k+1}) & \widetilde{P}^{xy}(t_{k+1}) & \cdots & \cdots & \cdots & \widetilde{P}^{xV_z}(t_{k+1}) \\ \widetilde{P}^{xy}(t_{k+1}) & \widetilde{P}^{yy}(t_{k+1}) & \cdots & \cdots & \cdots & \widetilde{P}^{yV_z}(t_{k+1}) \\ \widetilde{P}^{xz}(t_{k+1}) & \widetilde{P}^{yz}(t_{k+1}) & \cdots & \cdots & \cdots & \widetilde{P}^{zV_z}(t_{k+1}) \\ \widetilde{P}^{xV_x}(t_{k+1}) & \widetilde{P}^{yV_x}(t_{k+1}) & \cdots & \cdots & \cdots & \widetilde{P}^{V_xV_z}(t_{k+1}) \\ \widetilde{P}^{xV_y}(t_{k+1}) & \widetilde{P}^{yV_y}(t_{k+1}) & \cdots & \cdots & \cdots & \widetilde{P}^{V_yV_z}(t_{k+1}) \\ \widetilde{P}^{xV_z}(t_{k+1}) & \widetilde{P}^{yV_z}(t_{k+1}) & \cdots & \cdots & \cdots & \widetilde{P}^{V_zV_z}(t_{k+1}) \end{pmatrix} \tag{11}$$

where the elements of this matrix are:

$$\widetilde{P}^{xx}(t_{k+1}) = \hat{P}^{xx}(t_k) + 2\hat{P}^{xV_x}(t_k)\Delta t + \hat{P}^{V_xV_x}(t_k)\Delta t^2$$
$$\widetilde{P}^{yy}(t_{k+1}) = \hat{P}^{yy}(t_k) + 2\hat{P}^{yV_y}(t_k)\Delta t + \hat{P}^{V_yV_y}(t_k)\Delta t^2$$
$$\widetilde{P}^{zz}(t_{k+1}) = \hat{P}^{zz}(t_k) + 2\hat{P}^{zV_z}(t_k)\Delta t + \hat{P}^{V_zV_z}(t_k)\Delta t^2$$
$$\widetilde{P}^{xy}(t_{k+1}) = \hat{P}^{xy}(t_k) + \hat{P}^{xV_y}(t_k)\Delta t + \hat{P}^{yV_x}(t_k)\Delta t + \hat{P}^{V_xV_y}(t_k)\Delta t^2$$
$$\widetilde{P}^{xz}(t_{k+1}) = \hat{P}^{xz}(t_k) + \hat{P}^{xV_z}(t_k)\Delta t + \hat{P}^{zV_x}(t_k)\Delta t + \hat{P}^{V_xV_z}(t_k)\Delta t^2$$
$$\widetilde{P}^{yz}(t_{k+1}) = \hat{P}^{yz}(t_k) + \hat{P}^{yV_z}(t_k)\Delta t + \hat{P}^{zV_y}(t_k)\Delta t + \hat{P}^{V_yV_z}(t_k)\Delta t^2$$
$$\widetilde{P}^{xV_x}(t_{k+1}) = \hat{P}^{xV_x}(t_k) + \hat{P}^{V_xV_x}(t_k)\Delta t$$
$$\widetilde{P}^{xV_y}(t_{k+1}) = \hat{P}^{xV_y}(t_k) + \hat{P}^{V_xV_y}(t_k)\Delta t$$
$$\widetilde{P}^{xV_z}(t_{k+1}) = \hat{P}^{xV_z}(t_k) + \hat{P}^{V_xV_z}(t_k)\Delta t$$
$$\widetilde{P}^{yV_x}(t_{k+1}) = \hat{P}^{yV_x}(t_k) + \hat{P}^{V_xV_y}(t_k)\Delta t$$
$$\widetilde{P}^{yV_y}(t_{k+1}) = \hat{P}^{yV_y}(t_k) + \hat{P}^{V_yV_y}(t_k)\Delta t \tag{12}$$
$$\widetilde{P}^{yV_z}(t_{k+1}) = \hat{P}^{yV_z}(t_k) + \hat{P}^{V_yV_z}(t_k)\Delta t$$
$$\widetilde{P}^{zV_x}(t_{k+1}) = \hat{P}^{zV_x}(t_k) + \hat{P}^{V_xV_z}(t_k)\Delta t$$
$$\widetilde{P}^{zV_y}(t_{k+1}) = \hat{P}^{zV_y}(t_k) + \hat{P}^{V_yV_z}(t_k)\Delta t$$
$$\widetilde{P}^{zV_z}(t_{k+1}) = \hat{P}^{zV_z}(t_k) + \hat{P}^{V_zV_z}(t_k)\Delta t$$
$$\widetilde{P}^{V_xV_x}(t_{k+1}) = \hat{P}^{V_xV_x}(t_k) + \sigma_{\mathbf{X}}^2$$
$$\widetilde{P}^{V_yV_y}(t_{k+1}) = \hat{P}^{V_yV_y}(t_k) + \sigma_{\mathbf{Y}}^2$$
$$\widetilde{P}^{V_zV_z}(t_{k+1}) = \hat{P}^{V_zV_z}(t_k) + \sigma_{\mathbf{Z}}^2$$
$$\widetilde{P}^{V_xV_y}(t_{k+1}) = \hat{P}^{V_xV_y}(t_k)$$
$$\widetilde{P}^{V_xV_z}(t_{k+1}) = \hat{P}^{V_xV_z}(t_k)$$
$$\widetilde{P}^{V_yV_z}(t_{k+1}) = \hat{P}^{V_yV_z}(t_k)$$

Note that $\sigma_{\mathbf{X}}^2$ is not the same as σ_x^2 and similarly for the other indices.

Then, the following values $\widetilde{P}^{xm}(t_{k+1})$, $\widetilde{P}^{ym}(t_{k+1})$, $\widetilde{P}^{zm}(t_{k+1})$, $\widetilde{P}^{V_xm}(t_{k+1})$, $\widetilde{P}^{V_ym}(t_{k+1})$, $\widetilde{P}^{V_zm}(t_{k+1})$, $\widetilde{P}^{mm}(t_{k+1})$ should be calculated using the following relations:

$$\begin{pmatrix} m_{k+1}^{\phi} - \widetilde{m}_{k+1}^{\phi} \\ m_{k+1}^{\lambda} - \widetilde{m}_{k+1}^{\lambda} \end{pmatrix}$$

$$= \begin{pmatrix} (Y(t_{k+1}) - \widetilde{Y}(t_{k+1}))\cos\phi_i(t_{k+1}) - (X(t_{k+1}) - \widetilde{X}(t_{k+1}))\sin\phi_i(t_{k+1}) \\ -\varepsilon_{k+1}^{y}\cos\phi_i(t_{k+1}) + \varepsilon_{k+1}^{x}\sin\phi_i(t_{k+1}) + \varepsilon_{k+1}^{\phi}\left(X_i + \varepsilon_{k+1}^{x} - X(t_{k+1})\right)\cos\phi_i(t_{k+1}) \\ (Z(t_{k+1}) - \widetilde{Z}(t_{k+1}))\sin\phi_i(t_{k+1})\cos\lambda_i(t_{k+1}) - (Y(t_{k+1}) - \widetilde{Y}(t_{k+1}))\sin\lambda_i(t_{k+1}) \\ -\varepsilon_{k+1}^{z}\sin\phi_i(t_{k+1})\cos\lambda_i(t_{k+1}) + \varepsilon_{k+1}^{y}\sin\lambda_i(t_{k+1}) + \varepsilon_{k+1}^{\lambda}\left(Y_i + \varepsilon_{k+1}^{y} - Y(t_{k+1})\right)\cos\lambda_i(t_{k+1}) \end{pmatrix}$$

and the identities:

$$X_i - X(t_{k+1}) = X_i - \widetilde{X}(t_{k+1}) - (X(t_{k+1}) - \widetilde{X}(t_{k+1}))$$
$$Y_i - Y(t_{k+1}) = Y_i - \widetilde{Y}(t_{k+1}) - (Y(t_{k+1}) - \widetilde{Y}(t_{k+1}))$$
$$Z_i - Z(t_{k+1}) = Z_i - \widetilde{Z}(t_{k+1}) - (Z(t_{k+1}) - \widetilde{Z}(t_{k+1}))$$

where we consider that the position of the *i*-th object is known and use the non-correlatedness of $\varepsilon_{k+1}^x, \varepsilon_{k+1}^y, \varepsilon_{k+1}^z, \varepsilon_{k+1}^\phi, \varepsilon_{k+1}^\lambda$ and differences $(X(t_{k+1}) - \widetilde{X}(t_{k+1}))$, $(Y(t_{k+1}) - \widetilde{Y}(t_{k+1}))$ and $(Z(t_{k+1}) - \widetilde{Z}(t_{k+1}))$.

Finally, we get:

$$
\left[\widetilde{P}^{xm}(t_{k+1})\right]^T = E\left[(X(t_{k+1}) - \widetilde{X}(t_{k+1}))\begin{pmatrix} m_{k+1}^\phi - \widetilde{m}_{k+1}^\phi \\ m_{k+1}^\lambda - \widetilde{m}_{k+1}^\lambda \end{pmatrix}\right]
$$
$$
= \begin{pmatrix} \widetilde{P}^{xy}(t_{k+1})\cos\phi_i(t_{k+1}) - \widetilde{P}^{xx}(t_{k+1})\sin\phi_i(t_{k+1}) \\ \widetilde{P}^{xz}(t_{k+1})\sin\phi_i(t_{k+1})\cos\lambda_i(t_{k+1}) - \widetilde{P}^{xy}(t_{k+1})\sin\lambda_i(t_{k+1}) \end{pmatrix},
\tag{13}
$$

$$
\left[\widetilde{P}^{ym}(t_{k+1})\right]^T = E\left[(Y(t_{k+1}) - \widetilde{Y}(t_{k+1}))\begin{pmatrix} m_{k+1}^\phi - \widetilde{m}_{k+1}^\phi \\ m_{k+1}^\lambda - \widetilde{m}_{k+1}^\lambda \end{pmatrix}\right]
$$
$$
= \begin{pmatrix} \widetilde{P}^{yy}(t_{k+1})\cos\phi_i(t_{k+1}) - \widetilde{P}^{xy}(t_{k+1})\sin\phi_i(t_{k+1}) \\ \widetilde{P}^{yz}(t_{k+1})\sin\phi_i(t_{k+1})\cos\lambda_i(t_{k+1}) - \widetilde{P}^{yy}(t_{k+1})\sin\lambda_i(t_{k+1}) \end{pmatrix},
\tag{14}
$$

$$
\left[\widetilde{P}^{zm}(t_{k+1})\right]^T = E\left[(Z(t_{k+1}) - \widetilde{Z}(t_{k+1}))\begin{pmatrix} m_{k+1}^\phi - \widetilde{m}_{k+1}^\phi \\ m_{k+1}^\lambda - \widetilde{m}_{k+1}^\lambda \end{pmatrix}\right]
$$
$$
= \begin{pmatrix} \widetilde{P}^{yz}(t_{k+1})\cos\phi_i(t_{k+1}) - \widetilde{P}^{xz}(t_{k+1})\sin\phi_i(t_{k+1}) \\ \widetilde{P}^{zz}(t_{k+1})\sin\phi_i(t_{k+1})\cos\lambda_i(t_{k+1}) - \widetilde{P}^{yz}(t_{k+1})\sin\lambda_i(t_{k+1}) \end{pmatrix},
\tag{15}
$$

$$
\left[\widetilde{P}^{V_x m}(t_{k+1})\right]^T = E\left[(V_x(t_{k+1}) - \widetilde{V}_x(t_{k+1}))\begin{pmatrix} m_{k+1}^\phi - \widetilde{m}_{k+1}^\phi \\ m_{k+1}^\lambda - \widetilde{m}_{k+1}^\lambda \end{pmatrix}\right]
$$
$$
= \begin{pmatrix} \widetilde{P}^{y V_x}(t_{k+1})\cos\phi_i(t_{k+1}) - \widetilde{P}^{x V_x}(t_{k+1})\sin\phi_i(t_{k+1}) \\ \widetilde{P}^{z V_x}(t_{k+1})\sin\phi_i(t_{k+1})\cos\lambda_i(t_{k+1}) - \widetilde{P}^{y V_x}(t_{k+1})\sin\lambda_i(t_{k+1}) \end{pmatrix},
\tag{16}
$$

$$
\left[\widetilde{P}^{V_y m}(t_{k+1})\right]^T = E\left[(V_y(t_{k+1}) - \widetilde{V}_y(t_{k+1}))\begin{pmatrix} m_{k+1}^\phi - \widetilde{m}_{k+1}^\phi \\ m_{k+1}^\lambda - \widetilde{m}_{k+1}^\lambda \end{pmatrix}\right]
$$
$$
= \begin{pmatrix} \widetilde{P}^{y V_y}(t_{k+1})\cos\phi_i(t_{k+1}) - \widetilde{P}^{x V_y}(t_{k+1})\sin\phi_i(t_{k+1}) \\ \widetilde{P}^{z V_y}(t_{k+1})\sin\phi_i(t_{k+1})\cos\lambda_i(t_{k+1}) - \widetilde{P}^{y V_y}(t_{k+1})\sin\lambda_i(t_{k+1}) \end{pmatrix},
\tag{17}
$$

$$
\left[\widetilde{P}^{V_z m}(t_{k+1})\right]^T = E\left[(V_z(t_{k+1}) - \widetilde{V}_z(t_{k+1}))\begin{pmatrix} m_{k+1}^\phi - \widetilde{m}_{k+1}^\phi \\ m_{k+1}^\lambda - \widetilde{m}_{k+1}^\lambda \end{pmatrix}\right]
$$
$$
= \begin{pmatrix} \widetilde{P}^{y V_z}(t_{k+1})\cos\phi_i(t_{k+1}) - \widetilde{P}^{x V_z}(t_{k+1})\sin\phi_i(t_{k+1}) \\ \widetilde{P}^{z V_z}(t_{k+1})\sin\phi_i(t_{k+1})\cos\lambda_i(t_{k+1}) - \widetilde{P}^{y V_z}(t_{k+1})\sin\lambda_i(t_{k+1}) \end{pmatrix}.
\tag{18}
$$

In a similar way, we calculate:

$$
(\widetilde{P}^{mm}(t_{k+1}))^{-1}
$$
$$
= \left[E\begin{pmatrix} (m_{k+1}^\phi - \widetilde{m}_{k+1}^\phi)^2 & (m_{k+1}^\phi - \widetilde{m}_{k+1}^\phi)(m_{k+1}^\lambda - \widetilde{m}_{k+1}^\lambda) \\ (m_{k+1}^\phi - \widetilde{m}_{k+1}^\phi)(m_{k+1}^\lambda - \widetilde{m}_{k+1}^\lambda) & (m_{k+1}^\lambda - \widetilde{m}_{k+1}^\lambda)^2 \end{pmatrix}\right]^{-1}
\tag{19}
$$

Therefore:

$$E\left[\left(m_{k+1}^{\phi} - \tilde{m}_{k+1}^{\phi}\right)^2\right]$$
$$= \tilde{P}^{yy}(t_{k+1}) \cos^2 \phi_i(t_{k+1}) - \tilde{P}^{xy}(t_{k+1}) \sin 2\phi_i(t_{k+1})$$
$$+ \tilde{P}^{xx}(t_{k+1}) \sin^2 \phi_i(t_{k+1}) + \sigma_y^2 \cos^2 \phi_i(t_{k+1}) + \sigma_x^2 \sin^2 \phi_i(t_{k+1}) + \sigma_\phi^2 ((X_i - \hat{X}(t_k))^2$$
$$+ \tilde{P}^{xx}(t_{k+1}) + \sigma_x^2) \cos^2 \phi_i(t_{k+1})$$

$$E\left[\left(m_{k+1}^{\phi} - \tilde{m}_{k+1}^{\phi}\right)\left(m_{k+1}^{\lambda} - \tilde{m}_{k+1}^{\lambda}\right)\right]$$
$$= \tilde{P}^{yz}(t_{k+1}) \sin \phi_i(t_{k+1}) \cos \phi_i(t_{k+1}) \cos \lambda_i(t_{k+1})$$
$$- \tilde{P}^{yy}(t_{k+1}) \cos \phi_i(t_{k+1}) \sin \lambda_i(t_{k+1}) - \tilde{P}^{xz}(t_{k+1}) \sin^2 \phi_i(t_{k+1}) \cos \lambda_i(t_{k+1})$$
$$+ \tilde{P}^{xy}(t_{k+1}) \sin \phi_i(t_{k+1}) \sin \lambda_i(t_{k+1})$$

$$E\left[\left(m_{k+1}^{\lambda} - \tilde{m}_{k+1}^{\lambda}\right)^2\right]$$
$$= \tilde{P}^{zz}(t_{k+1}) \sin^2 \phi_i(t_{k+1}) \cos^2 \lambda_i(t_{k+1}) - \tilde{P}^{yz}(t_{k+1}) \sin \phi_i(t_{k+1}) \sin 2\lambda_i(t_{k+1})$$
$$+ \tilde{P}^{yy}(t_{k+1}) \sin^2 \lambda_i(t_{k+1}) + \sigma_z^2 \sin^2 \phi_i(t_{k+1}) \cos^2 \lambda_i(t_{k+1}) + \sigma_y^2 \sin^2 \lambda_i(t_{k+1}) + \sigma_\lambda^2((Y_i - \hat{Y}(t_k))^2$$
$$+ \tilde{P}^{yy}(t_{k+1}) + \sigma_y^2) \cos^2 \lambda_i(t_{k+1})$$

4.2.2. Correction

After getting the measurements at the moment t_{k+1}, one can obtain the estimate of the UAV position at this moment. Therefore, the solution of Problem 1 has the form:

$$\hat{\mathbf{X}}(t_{k+1}) = \tilde{\mathbf{X}}(t_{k+1}) + \tilde{\mathbf{P}}(t_{k+1})\left(\tilde{P}^{mm}(t_{k+1})\right)^{-1}(m_{k+1} - \tilde{m}_{k+1}) \tag{20}$$

and the matrix of the mean square errors is equal to:

$$\hat{\mathbf{P}}(t_{k+1}) = \tilde{P}(t_{k+1}) - \tilde{\mathbf{P}}(t_{k+1})\left(\tilde{P}^{mm}(t_{k+1})\right)^{-1}\tilde{\mathbf{P}}(t_{k+1})^T \tag{21}$$

where:

$$\tilde{\mathbf{P}}(t_{k+1}) = \left(\tilde{P}^{xm}(t_{k+1}), \tilde{P}^{ym}(t_{k+1}), ..., \tilde{P}^{V_zm}(t_{k+1})\right)^T$$

5. Robust Filtering on the Basis of the UAV Motion Model

The RANSAC method calculates the rotation matrix and the coordinates of the camera $\{A^*, b^*\}$ in the Earth coordinate system with some minor error. However, the RANSAC method can give quite the wrong answer, called an outlier. It could happen, for example, if the frames I_c and I_m do not depict common objects. We provide further a method that makes a decision about whether $\{A^*, b^*\}$ is an outlier or not. This problem has been considered in relation to the exclusion of outliers in the RANSAC-type procedures [25,26]. Here, we use the modification of the robust RANSAC [23] based on PKF for bearing-only observations [17].

After the prediction step of the Kalman filter, one can estimate the UAV (camera) position and the matrix of the mean square errors:

$$\mathbf{X} = \tilde{\mathbf{X}}(\mathbf{t_{k+1}}), \tilde{\mathbf{P}} = \tilde{\mathbf{P}}(\mathbf{t_{k+1}}).$$

Like in [26], one can suppose that the corresponding probability density is Gaussian. The reason is that the PKF gives the best linear estimates obtained until the current time t. This estimate is the sum of uncorrelated random variables with almost the same variations, at least on the short intervals preceding the current time. It gives the opportunity to approximate the probability density distribution by the Gaussian one. Further extension of the robust RANSAC technique is based on the prior distribution of the UAV attitude. The approach has been developed in [23,27] on the basis of the extended Kalman filter (EKF). However, the estimate given by the EKF has an unknown bias and, of course, does not give the posterior Gaussian distribution. Therefore, the PKF, which gives an unbiased estimate, looks more preferable under the hypothesis of the posterior Gaussian distribution. Therefore, at the $(k+1)$-th step, the posterior distribution of r_i' corresponding to an inlier is assumed to be Gaussian, that is according to Equation (1):

$$\mathcal{N}(A_k(r_i - \widetilde{\mathbf{X}}(t_{k+1})), A_k(\widetilde{\mathbf{P}}(t_{k+1}) + P_{rr})A_k^T)$$

where P_{rr} is the covariance matrix of the landmarks localization. Further, in the estimation algorithm, the pair $\{A_k, \widetilde{\mathbf{X}}(t_{k+1})\}$ is considered as an outlier at the confidence level 95% if:

$$\| r_i' - A_k(r_i - \widetilde{\mathbf{X}}(t_{k+1}))\| \geq 2 * Sp[A_k(\widetilde{\mathbf{P}}(t_{k\,|\,1}) + P_{rr})A_k^T]$$

Otherwise, the correction step is based on $\{A_k, \widetilde{\mathbf{X}}(t_{k+1})\}$. Of course, all such nonlinear conjectures need confirmation on the basis of statistical modeling, which is one of the results of this article. One can observe the performance of robust filtering in the following figures. Figure 4 shows the correspondence between feature points obtained without a UAV motion model. Next, Figure 5 shows the correspondence established on the basis of the UAV motion model. The number of outliers reduces substantially.

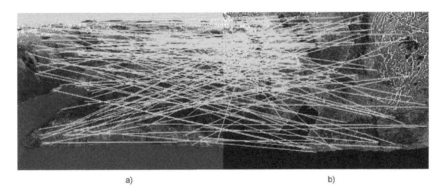

a)   b)

Figure 4. Correspondence between (**a**) (image I_c of the on-board camera of the UAV) and (**b**) (template image loaded in the UAV memory in advance) feature points found without the UAV motion model. One can observe chaotic correspondence, which gives a huge number of outliers.

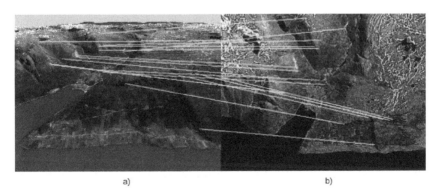

a)   b)

Figure 5. Correspondence between (**a**) (image I_c of the on-board camera of the UAV) and (**b**) (template image loaded in the UAV memory in advance) feature points found with the aid of RANSAC based on the UAV motion model. The number of outliers reduces substantially.

6. Control of the UAV

Control of a UAV that ensures its motion along the reference trajectory may be determined on the basis of the standard deterministic linear-quadratic approach [28]. However, the problem of control on the basis of bearing-only observation is different from the standard one. It should be underlined that this problem is a non-linear one and cannot be solved by the standard way. The problem of the optimal control for the system Equation (2) is the stochastic one with incomplete information and does not have an explicit solution. However, for practical reasons, one can simplify it by considering the locally optimal control. Here, we discuss the following problem:

Problem 2. *Find the locally optimal controls $a_x(t_k)$, $a_y(t_k)$ and $a_z(t_k)$ aimed to keep the motion of the UAV along the reference trajectory.*

Assume that we have some reference trajectory $X_{nom}(t_k)$.

Therefore, at the moment t_{k+1}, we have to minimize the following expressions:

$$E_1 = E\{(X(t_{k+1}) - X_{nom}(t_{k+1}))^2 + (V_x(t_{k+1}) - V_{x_{nom}}(t_{k+1}))\Delta t)^2\} \to \min_{a_x(t_k)}$$

$$E_2 = E\{(Y(t_{k+1}) - Y_{nom}(t_{k+1}))^2 + (V_z(t_{k+1}) - V_{z_{nom}}(t_{k+1}))\Delta t)^2\} \to \min_{a_y(t_k)}$$

$$E_3 = E\{(Z(t_{k+1}) - Z_{nom}(t_{k+1}))^2 + (V_z(t_{k+1}) - V_{z_{nom}}(t_{k+1}))\Delta t)^2\} \to \min_{a_z(t_k)}$$

Let us consider the components of the E_1 expression:

$$X(t_{k+1}) - X_{nom}(t_{k+1}) = X(t_k) - \hat{X}(t_k) - (X_{nom}(t_k) - \hat{X}(t_k))$$
$$+(V_x(t_k) - \hat{V}_x(t_k))\Delta t - (V_{x_{nom}}(t_k) - \hat{V}_x(t_k))\Delta t + (a_x(t_k) - a_{x_{nom}}(t_k))\frac{\Delta t^2}{2}$$

$$V_x(t_{k+1}) - V_{x_{nom}}(t_{k+1}) = V_x(t_k) - \hat{V}_x(t_k) - (V_{x_{nom}}(t_k) - \hat{V}_x(t_k)) + (a_x(t_k) - a_{x_{nom}}(t_k))\Delta t$$
$$+W_x(t_k)$$

Then, we square these components and take the derivative of the sum with respect to $a_x(t_k)$ given that some components are uncorrelated:

$$E\{(X(t_k) - \hat{X}(t_k))(X_{nom}(t_k) - \hat{X}(t_k))\} = 0$$
$$E\{(V_x(t_k) - \hat{V}_x(t_k))(V_{x_{nom}}(t_k) - \hat{V}_x(t_k))\} = 0$$

Finally, we get:

$$a_x(t_k) = a_{x_{nom}}(t_k) + \frac{2(X_{nom}(t_k) - \hat{X}(t_k))}{5\Delta t^2} + \frac{6(V_{x_{nom}}(t_k) - \hat{V}_x(t_k))}{5\Delta t} \tag{22}$$

We take into account that the acceleration of the UAV has limitations $[a_{x_{min}}, a_{x_{max}}]$, so the control acceleration has the form:

$$a_x^c(t_k) = \begin{cases} a_{x_{min}} \, if \, a_x(t_k) < a_{x_{min}}, \\ a_x(t_k) \, if \, a_{x_{min}} \le a_x(t_k) \le a_{x_{max}}, \\ a_{x_{max}} \, if \, a_x(t_k) > a_{x_{max}}. \end{cases}$$

Similarly, we obtain the expressions for $a_y^c(t_k)$ and $a_z^c(t_k)$. Thus, we get the following solution of Problem 2 [19]:

$$\hat{a}(t_k) = \hat{a}_{nom}(t_k) + \frac{2}{5\Delta t^2}(\hat{X}_{nom}(t_k) - \hat{X}(t_k)) + \frac{6}{5\Delta t}(\hat{V}_{nom}(t_k) - \hat{V}(t_k)) \tag{23}$$

where:

$$\hat{a}(t_k) = \left(a_x(t_k), a_y(t_k), a_z(t_k)\right)^T$$
$$\hat{X}(t_k) = \left(X(t_k), Y(t_k), Z(t_k)\right)^T$$
$$\hat{V}(t_k) = \left(V_x(t_k), V_y(t_k), V_z(t_k)\right)^T$$

and similarly for the nominal trajectory components.

7. Experimental Results

In this section, we give the results of the algorithm's modeling. The UAV is virtually flying over the landscape shown in Figure 1. This image has been obtained from Google Maps, and for verification of the algorithm, the image of the same region obtained from another Bing satellite was used. Therefore, these two images modeled the preliminary downloaded template and the image obtained by the on-board camera. The result of virtual flight experiment is shown in Figure 6.

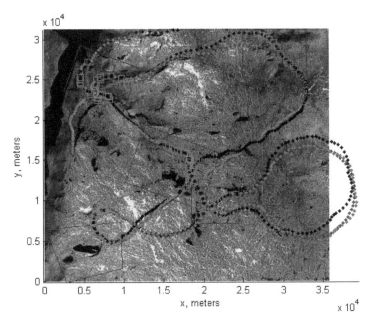

Figure 6. Blue dots corresponds to the reference trajectory and black dots to the real path. Blue squares show the localization of the terrain areas corresponding to the template images, and red squares show the moments where the estimates of the UAV positions have been obtained and assumed to be reliable according to the robust RANSAC algorithm described in Section 5.

8. Results and Discussion

In the modeling of the control algorithm, we use the UAV moving approximately with a velocity of 50 m/s, though the change of the altitude is assumed to be rather substantial. The control algorithm takes into account the constraints imposed on linear acceleration and angular velocities. The software developed for modeling may be used in a real on-board navigation system, as well. Moreover, the filtering algorithm based on unbiased estimation may be easily incorporated with the INS, since it gives also the unbiased square error estimates, which opens the way to correct data fusion. The quality of tracking for x, y, z components is shown below in Figures 7–9, respectively. In all of these figures, blue dots correspond to the reference trajectory, black dots to the real path and red squares to the moments, where the estimates of the UAV positions have been obtained and assumed to be reliable according to the robust RANSAC algorithm. One can observe that in the "measurement" areas, the algorithm estimates the coordinates with high accuracy, and the control provides the tracking with high accuracy, as well.

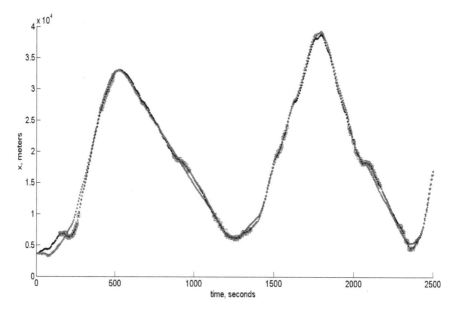

Figure 7. Tracking of the *x*-coordinate.

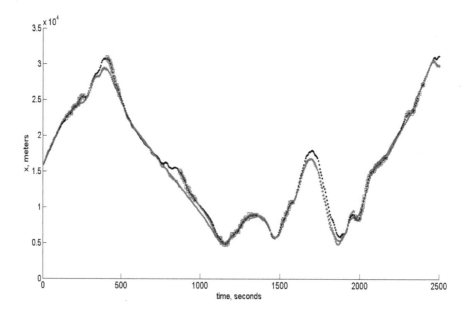

Figure 8. Tracking of the *y*-coordinate.

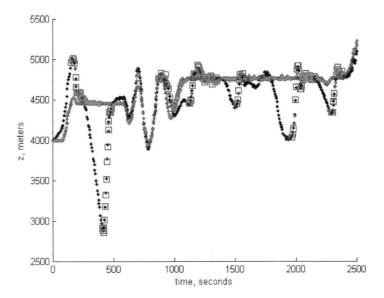

Figure 9. Tracking of the z-coordinate.

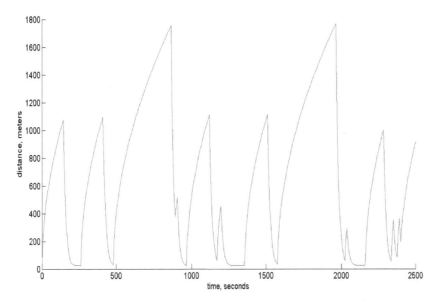

Figure 10. Averaged standard deviation of the position estimation error. The limit value in the observation areas is close to 24.5 m. One can see that in the areas of no observations, the SD monotonically increases.

The high accuracy is in accordance with the standard deviation (SD) theoretically calculated from the PKF. The value of the averaged standard deviation, which is the square root of $P_{xx}(t) + P_{yy}(t) + P_{zz}(t)$, is shown in Figure 10 below.

9. Conclusions

The main result of the work is the new algorithm of the UAV control based on the observation of the landmarks in a 3D environment. The new RANSAC based on the UAV motion model permits one to exclude the huge number of outliers and, by that, to provide the reliable set of data for the estimation of the UAV position on the basis of the novel non-biased PKF algorithm. This work is just the beginning of the implementation of this approach in the navigation of UAVs during long-term autonomous missions.

Acknowledgments: This research was supported by Russian Science Foundation Grant 14-50-00150.

Author Contributions: The work presented here was carried out in collaboration among all authors. All authors have contributed to, seen and approved the manuscript.

Conflicts of Interest: The authors declare no conflict of interest.

References

1. Osborn, R.W.; Bar-Shalom, Y. Statistical Efficiency of Composite Position Measurements from Passive Sensors. *IEEE Trans. Aerosp. Electron. Syst.* **2013**, *49*, 2799–2806. [CrossRef]
2. Cesetti, A.; Frontoni, E.; Mancini, A.; Zingaretti, P.; Longhi, S. A Vision-Based Guidance System for UAV Navigation and Safe Landing Using Natural Landmarks. *J. Intell. Robot. Syst.* **2010**, *57*, 233–257. [CrossRef]
3. Caballero, F.; Merino, L.; Ferruz, J.; Ollero, A. Vision-Based Odometry and SLAM for Medium and High Altitude Flying UAVs. *J. Intell. Robot. Syst.* **2009**, *54*, 137–161. [CrossRef]
4. Wang, C.L.; Wang, T.M.; Liang, J.H.; Zhang, Y.C.; Zhou, Y. Bearing-only Visual SLAM for Small Unmanned Aerial Vehicles in GPS-denied Environments. *Int. J. Autom. Comput.* **2013**, *10*, 387–396. [CrossRef]
5. Konovalenko, I.; Kuznetsova, E. Experimental comparison of methods for estimation the observed velocity of the vehicle in video stream. In Proceedings of the SPIE 9445, Seventh International Conference on Machine Vision (ICMV 2014), Milan, Italy, 19–21 November 2014; Volume 9445.
6. Lowe, D.G. Object recognition from local scale-invariant features. In Proceedings of the International Conference on Computer Vision, Kerkyra, Greece, 20–27 September 1999; Volume 2, pp. 1150–1157.
7. Konovalenko, I.; Miller, A.; Miller, B.; Nikolaev, D. UAV navigation on the basis of the feature points detection on underlying surface. In Proceedings of the 29th European Conference on Modeling and Simulation (ECMS 2015), Albena (Varna), Bulgaria, 26–29 May 2015; pp. 499–505.
8. Guan, X.; Bai, H. A GPU accelerated real-time self-contained visual navigation system for UAVs. In Proceedings of the IEEE International Conference on Information and Automation, Shenyang, China, 6–8 June 2012; pp. 578–581.
9. Belfadel, D.; Osborne, R.W.; Bar-Shalom, Y. Bias Estimation for Optical Sensor Measurements with Targets of Opportunity. In Proceedings of the 16th International Conference on Information Fusion, Istanbul, Turkey, 9–12 July, 2013; pp. 1805–1812.
10. Bishop, A.N.; Fidan, B.; Anderson, B.D.O.; Dogancay, K.; Pathirana, P.N. Optimality analysis of sensor-target localization geometries. *Automatica* **2010**, *46*, 479–492. [CrossRef]
11. Jauffet, C.; Pillon, D.; Pignoll, A.C. Leg-by-Leg Bearings-only Target Motion Analysis without Observer Maneuver. *J. Adv. Inf. Fusion* **2011**, *6*, 24–38.
12. Lin, X.; Kirubarajan, T.; Bar-Shalom, Y.; Maskell, S. Comparison of EKF, Pseudomeasurement and Particle Filters for a Bearing-only Target Tracking Problem. *Proc. SPIE Int. Soc. Opt. Eng.* **2002**, *4728*, 240–250.
13. Miller, B.M.; Stepanyan, K.V.; Miller, A.B.; Andreev, K.V.; Khoroshenkikh, S.N. Optimal filter selection for UAV trajectory control problems. In Proceeedings of the 37th Conference on Information Technology and Systems, Kaliningrad, Russia, 1–6 September 2013; pp. 327–333.
14. Aidala, V.J.; Nardone, S.C. Biased Estimation Properties of the Pseudolinear Tracking Filter. *IEEE Trans. Aerosp. Electron. Syst.* **1982**, *18*, 432–441. [CrossRef]
15. Pugachev, V.S.; Sinitsyn, I.N. *Stochastic Differential Systems—Analysis and Filtering*; Wiley: Chichester, UK, 1987.
16. Amelin, K.S.; Miller, A.B. An Algorithm for Refinement of the Position of a Light UAV on the Basis of Kalman Filtering of Bearing Measurements. *J. Commun. Technol. Electron.* **2014**, *59*, 622–631. [CrossRef]

17. Miller, A.B. Development of the motion control on the basis of Kalman filtering of bearing-only measurements. *Autom. Remote Control* **2015**, *76*, 1018–1035. [CrossRef]

18. Karpenko, S.; Konovalenko, I.; Miller, A.; Miller, B.; Nikolaev, D. Stochastic control of UAV on the basis of 3D natural landmarks. In Proceedings of the 17th International Conference on Machine Vision (ICMV), Barselona, Spain, 19–20 November 2015.

19. Miller, A.B.; Miller, B.M. Stochastic control of light UAV at landing with the aid of bearing-only observations. In Proceedings of the 17th International Conference on Machine Vision (ICMV), Barselona, Spain, 19–20 November 2015.

20. GitHub. Available online: https://github.com/Itseez/opencv/blob/master/samples/python2/ asift.py (accessed on 23 September 2014).

21. Fischler, M.A.; Bolles, R.C. Random Sample Consensus: A Paradigm for Model Fitting with Applications to Image Analysis and Automated Cartography. *Commun. ACM* **1981**, *24*, 381–395. [CrossRef]

22. Zuliani, M.; Kenney, C.S.; Manjunath, B.S. The MultiRANSAC Algorithm and Its Application to Detect Planar Homographies. In Proceedings of the 12th IEEE International Conference on Image Processing (ICIP 2005), Genova, Italy, 11–14 September 2005; Volume 3, pp. III-153–156.

23. Civera, J.; Grasa, O.G.; Davison, A.J.; Montiel, J.M.M. 1-Point RANSAC for Extended Kalman Filtering: Application to Real-Time Structure from Motion and Visual Odometry. *J. Field Robot.* **2010**, *27*, 609–631. [CrossRef]

24. Miller, B.M.; Pankov, A.R. *Theory of Random Processes (in Russian)*; Phizmatlit: Moscow, Russia, 2007.

25. Torr, P.H.S. Geometric motion segmentation and model selection. *Phil. Trans. R. Soc. A* **1998**, *356*, 1321–1340. [CrossRef]

26. Torr, P.H.S.; Zisserman, A. MLESAC: A New Robust Estimator with Application to Estimating Image Geometry. *Comput. Vis. Image Underst.* **2000**, *78*, 138–156. [CrossRef]

27. Civera, J.; Grasa, O.G.; Davison, A.J.; Montiel, J.M.M. 1-Point RANSAC for EKF-Based Structure from Motion. In Proceedings of the 2009 IEEE/RSJ International Conference on Intelligent Robots and Systems, St. Louis, MO, USA, 11–15 October 2009; pp. 3498–3504.

28. Sujit, P.B.; Saripalli, S.; Borges Sousa, J. Unmanned aerial vehicle path following: A survey and analysis of algorithms for Fixed-Wing unmanned aerial vehicless. *IEEE Contol Syst.* **2014**, *34*, 42–59. [CrossRef]

Article

Cooperative Surveillance and Pursuit Using Unmanned Aerial Vehicles and Unattended Ground Sensors

Jonathan Las Fargeas *, Pierre Kabamba and Anouck Girard

Department of Aerospace Engineering, University of Michigan, Ann Arbor, MI 48105, USA;
E-Mails: daninman@umich.edu (P.K.); anouck@umich.edu (A.G.)
*   E-Mail: jfargeas@umich.edu; Tel.: +1-734-763-1305; Fax: +1-734-763-0578.

Academic Editor: Felipe Gonzalez Toro
Received: 17 October 2014 / Accepted: 4 January 2015 / Published: 13 January 2015

Abstract: This paper considers the problem of path planning for a team of unmanned aerial vehicles performing surveillance near a friendly base. The unmanned aerial vehicles do not possess sensors with automated target recognition capability and, thus, rely on communicating with unattended ground sensors placed on roads to detect and image potential intruders. The problem is motivated by persistent intelligence, surveillance, reconnaissance and base defense missions. The problem is formulated and shown to be intractable. A heuristic algorithm to coordinate the unmanned aerial vehicles during surveillance and pursuit is presented. Revisit deadlines are used to schedule the vehicles' paths nominally. The algorithm uses detections from the sensors to predict intruders' locations and selects the vehicles' paths by minimizing a linear combination of missed deadlines and the probability of not intercepting intruders. An analysis of the algorithm's completeness and complexity is then provided. The effectiveness of the heuristic is illustrated through simulations in a variety of scenarios.

Keywords: sensor networks; target tracking; path planning for multiple UAVs

1. Introduction

A team of small unmanned aerial vehicles (UAVs) is tasked with patrolling a network of roads near a friendly base. Ground intruders (e.g., trucks) use the road network to reach the base and do not know of the presence of the UAVs. The UAVs patrol the roads to detect and take pictures of any intruders present on the roads.

However, small UAVs possess limited onboard processing resources, and the current detection capability of small aircraft using electro-optical or infrared sensors is not sufficient to ascertain whether an intruder is present or not [1]; thus, the UAVs in this problem are assumed to not possess automated target recognition capabilities. Instead, intruder detection is performed by unattended ground sensors (UGSs) placed on the roads. These sensors measure a given property (e.g., weight of a vehicle driving by), perform classification on the measurement to decide whether it corresponds to an intruder or not and register the time of the detection if the measurement was classified as an intruder. The use of UGSs in conjunction with UAVs enables the pursuit of an intruder along the road network, which is otherwise not possible solely using UAVs.

To maximize the coverage of the road network, the UGSs are placed far apart. The UGSs possess short-range communication devices, but have limited long-range communication capabilities; they require a line of sight to a dedicated communication device with a permanent power source. Placing communication devices to meet these requirements can be difficult depending on the terrain and conditions (e.g., contested environment). However, this problem can be circumvented by using mobile UAVs instead of immobile communication devices. The UAVs do not require the difficult placement of communication devices and power sources in the area, such that the line of sight is maintained between devices; instead, they act as mobile communication devices by querying UGSs directly below using a short-range communication device. Thus, it is assumed that the UGSs cannot communicate with one another, but only with UAVs directly overhead. In contrast, the UAVs are capable of communicating between one another and a central authority (e.g., the base) via a low bandwidth, long-range communication link. The low bandwidth link allows for the transmission of small amounts of information, such as UGS detections or waypoints, but prohibits the transmission of large datasets, such as images. The link is assumed to cover the entire operating area. The short-range and long-range communication devices are assumed to transmit in the electromagnetic spectrum, and as such, communication times are small compared to the other time constants involved in the problem.

This work is motivated by base defense scenarios within the Talisman Saber biennial U.S./Australian military exercise [2], where UAVs are tasked with obtaining intelligence (e.g., location and imagery) about intruders. In these base defense scenarios, the UAVs have limited onboard processing capabilities and, thus, cannot autonomously detect intruders; the UAVs thus rely on UGSs for intruder detection, pursuit and interception [3].

Figure 1 shows a visualization of such a scenario with two UAVs and a single intruder attempting to reach the base. The UAV on the bottom of the figure is communicating with a UGS directly below. Through the communication, the UAV learns of a recent detection from the UGS. The detection is shared with the central authority and used to predict possible future locations of the intruder. The central authority then selects destinations for the UAVs where they are likely to image the intruder.

A mission designer is assumed to have set the UGSs locations and patrol parameters before the mission begins; optimal UGSs placement and patrol parameter selection for this base defense scenario is treated in [4].

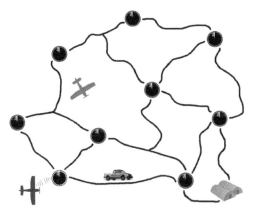

Figure 1. Cooperative surveillance and pursuit scenario.

The UAVs are forced to revisit UGSs to maintain up-to-date and accurate information on the intruder; this is enforced through the use of revisit deadlines, *i.e.*, the maximum time that can elapse between two consecutive visits to a UGS. Forced revisits also help mitigate the impact of false alarms by UGSs. The UAVs cannot detect a target autonomously; hence, they loiter above a UGS and capture an image when the UGS below detects an intruder passing.

The objective is to generate paths for the UAVs (*i.e.*, select waypoints in real time) that satisfy the revisit constraints of UGSs and capture images of intruders before the latter reach the base (*i.e.*, maximize the likelihood of a UAV and intruder being at the same location). A heuristic is provided to compute such paths, and its completeness and complexity are assessed. While the heuristic directs the paths taken by the UAVs, it also indirectly interacts with the UGSs used for monitoring the roads. The heuristic selects when information from the UGSs is obtained, and the heuristic uses the information acquired from the UGSs to predict possible intruder locations in the future. In addition, the capture of an image of the intruder is achieved when a UGS that a UAV is loitering (as directed by the heuristic) detects an intruder.

1.1. Literature Review

The UAVs' task consists of patrolling the road network and pursuing the intruders. A broad amount of research exists on patrolling problems and pursuit-evasion games; as such, literature pertaining to these topics is reviewed.

The problem of persistently monitoring a given area with one or more vehicles has been studied extensively. The problem of finding cycles, such that all points in the patrol area are covered by the vehicle's sensor footprint, is treated in [5]. In [6,7], the authors investigate a patrolling problem for multiple UAVs and maximize the minimum frequency of visitation for different partitions in the area. In [8,9], the authors find paths with a minimized uniform frequency of visitation for all partitions. In [10], the authors investigate the problem of finding paths for mobile agents that minimize the accumulation of an uncertainty metric in the mission area. In [11,12], the authors investigate heuristics, such that the time between visits to the same object of interest is minimized. The problem of patrolling multiple targets cooperatively is also treated in [13], where the authors provide algorithms to compute trajectories for vehicles that minimize the weighted refresh time of the targets being visited. The dual objective of satisfying patrolling constraints (revisit deadlines) and intercepting the intruder separates the problem studied in the current paper from many patrolling problems investigated in the literature.

This problem also shares similarities to the traveling salesman problem (TSP), where an agent is tasked with visiting a certain number of locations while minimizing the distance traveled. TSP is a non-deterministic polynomial-time (NP)-complete problem, and thus, heuristics are used to find solutions; [14] contains a survey on existing heuristics for TSP. In [15], the authors study stochastic and dynamic variations of TSP, where the target locations are generated randomly; the authors then provide algorithms to compute paths for the vehicles that optimize criteria, such as the length of the path or the time between the generation of the target and its observation by the vehicle. The problem treated in the current paper differs from vehicle routing problems and TSPs due to the added goal of capturing images of mobile intruders.

The problem considered in this paper has several components in common with pursuit-evasion games, where defenders are to capture an intruder. Pursuit-evasion games often occur on graphs; the defenders win the game if the intruder is caught; otherwise, the intruder wins. The defenders and intruder move in turns and can only travel to adjacent nodes. The pursuit-evasion problem was studied, and conditions on the number of defenders necessary to guarantee capture were derived in [16–18]. Many variations of the problem exist, such as the addition of constraints on the topology of the graph, the velocities of the defenders or intruders and the amount of information held by one team about the graph or the other team; a survey of pursuit-evasion research relevant to mobile robotics is presented in [19]. Security games introduce targets that the defenders must protect from intruder attacks; in addition, the intruder can observe the defenders. In [20,21], the authors study and provide algorithms to solve security games where the intruder is attempting to infiltrate nodes in a graph, and defenders must visit the various nodes in the graph, such that the intruder never has enough time to infiltrate. The current paper differs from the pursuit-evasion literature, because the defenders and intruders do not move in turns.

In [22], the problem of monitoring an area with intruders using UGSs and UAVs is discussed, and approaches to finding paths for the UAVs that maximize the number of interceptions are presented. The related problem of designing a sensor network for surveillance of a moving target has been studied in [23], where the authors provide methods for multi-point surveillance and demonstrate their effectiveness with regards to tracking probability. The problem of estimating a vehicle's position in a graph given its velocity distribution and a previous detection at a known location and time is treated in [24]. The authors use a histogram filter to predict the vehicle's potential locations in a graph. The approach used in this paper coordinates multiple UAVs (using a different method for prediction) to intercept an intruder on a graph in a centralized fashion while handling uncertainty in the target motion.

While the current literature examines many variations of patrolling problems and pursuit problems, no method treats a framework where the mobile agents fully rely on static sensors placed on an arbitrary road network to track and intercept intruders and provides a path planning algorithm for the mobile agents. The current paper concentrates on this subject. Addressing the reliance on static sensors is important, since the proper tracking of a target by a pursuer cannot be guaranteed in a large number of scenarios and, in some cases, may not be possible at all (as was indicated earlier with small UAVs currently being incapable of autonomous target recognition). In addition, not making assumptions about the topology of the network enables the handling of cases where the environment does not permit favorable sensor placement.

1.2. Original Contributions

The original contributions of this work are as follows: the problem of cooperative surveillance and pursuit is formulated and shown to be NP-hard; a heuristic to solve the problem is given; and an analysis of the heuristic's completeness and complexity is provided.

In previous work, the persistent visitation problem was introduced in which a single vehicle persistently visits a set of nodes, each with a revisit deadline [25]. The goal was to find a path for the vehicle such that no revisit deadline is missed. The existence of periodic solutions was proven; the complexity class of the problem was derived; and heuristics that could solve the problem were presented and characterized.

A version of this problem that included fuel constraints and refueling costs was also studied, and an algorithm that found the minimal cost path satisfying the revisit deadlines and fuel constraints was provided [26].

The current paper differs from the authors' previous work, which focused solely on patrolling, in the following ways: multiple vehicles are considered; an adversary is present in the area; sensors that provide information from which decisions need to be made are included; and the vehicles pursue the adversary while patrolling a number of locations with revisit deadlines.

1.3. Paper Outline

The remainder of the paper is as follows. In Section 2, the model for the defenders and intruders is presented, including the kinematics of the mobile agents and the properties of the UGS. The mathematics for the defenders' pursuit of the intruder are also presented in Section 2. The problem is formulated in Section 3. The heuristic that generates the defenders' actions to solve the problem is presented in Section 4. Results from simulations are shown in Section 5 and discussed in Section 6. Conclusions and future work are discussed in Section 7.

2. Modeling

In this section the model for the defenders (UAVs and UGSs) is presented, followed by a description of the intruder model.

2.1. Defenders

There are n UGSs placed in a planar area along a road network with Cartesian coordinates (ξ_i, ζ_i), $1 \leq i \leq n$. The UGSs and the roads that the intruder can use to travel between them are modeled as a graph $G(N, E)$, where $e_{i,j} \in E$, if there exists a road connecting UGSs i and j. Let $d_{i,j}$ be the length of road $e_{i,j}$. The road network and UGSs' placement is assumed to remain the same over the duration of the mission. The UGSs can measure a continuous property in their proximity and make a classification decision based on the measurement; this classification results in a detection if the measurement is above a certain threshold, e.g., in [22], the UGSs use a small Doppler radar to detect intruders nearby. The UGSs are not equipped with perfect sensors or classifiers and, thus, may emit false alarms.

In addition, m UAVs are patrolling the area each with constant velocity v, finite fuel capacity F and fuel consumption rate $\dot{f}_c$. There is a single base that is capable of refueling the UAVs located at (ξ_{n+1}, ζ_{n+1}). The UGS are assumed to be distant from one another; thus, the UAVs' limited turn radius is not taken into account, and the UAVs are modeled as point masses moving in straight lines [27]. If straight line travel cannot be assumed, then curvature must be accounted for; this can be done using the results from [28], where the authors describe methods to convert paths on a graph with straight line edges to paths on a graph with edges that satisfy the turning constraints of the UAVs.

The UAVs are equipped with a long-range communication device, which enables communication with a central authority, and a short-range communication device, which enables the UAVs to query the status of a UGS directly below (in [22], short-range communication is performed via WiFi radios). The UAVs, UGSs and central authority are assumed to possess synchronized clocks; this can be achieved by calibration before the mission and periodic UGS clock synchronization during UAV visits (it can also be accomplished by equipping each UAV and each UGS with a clock synchronization device, such as a GPS).

Once a UAV obtains information from a UGS, it is immediately relayed to the central authority. The central authority decides which nodes the UAVs are to visit next once their respective destinations are reached. There is no benefit from allowing multiple UAVs to visit the same UGSs simultaneously, since a detection is shared immediately and only a single UAV is required to capture an image; hence, a UGS can only be visited by a single UAV at a time.

2.1.1. Revisit Deadlines

Each UGS has a revisit deadline, $r_i > 0, 1 \leq i \leq n$, set by the mission designer, which under ideal conditions, is the maximum time that can elapse between two visits to the UGS by the defenders. The revisit deadlines are used as a method to keep the defenders' knowledge of intrusions up-to-date. The revisit deadlines are also used to indicate the relative importance of the various UGSs, such that the UAVs can prioritize their actions accordingly.

2.1.2. State Space Model

UGS queries and intruder interceptions only occur above UGSs, and as such, the arrival of a UAV at a UGS is the event that advances the system for the defenders. Thus, a number of states evolve in discrete time corresponding to the arrival of a UAV at a UGS, while other states (such as the amount of fuel onboard a UAV) evolve in continuous time.

The increments (or steps) in discrete time are delimited by the arrivals of UAVs at UGSs. $\tau(k) \in \mathbb{R}$ is the total time elapsed since the beginning of the mission upon completion of step k. $\Omega(k)$ is the set of upcoming UAV arrival times at step k. The current destination of UAV j at step k is indicated by $p_j(k) \in 1, 2, ..., (n+1), 1 \leq j \leq m$. The Cartesian coordinates of UAV j at step k are $(\xi_j'(k), \zeta_j'(k)) \in \mathbb{R}^2, 1 \leq j \leq m$. Let $y(k)$ denote the discrete time states at step k, i.e., $y(k) = [\, \tau(k)\ \Omega(k)\ p_j(k)\ (\xi_j(k), \zeta_j(k)) \,]$.

The continuous time states are linked to the discrete time states by applying impulses in their dynamics if certain conditions are met when a step is completed. The amount of fuel UAV j is carrying at time t is indicated by $f_j(t) \in \mathbb{R}, 1 \leq j \leq m$. Let $x_i(t) \in \mathbb{R}, 1 \leq i \leq n$ be the slack time of UGS i at time t, which indicates how much time remains before a visit to UGS i is overdue. Let the input $u_j(k)$ be the destination of UAV j at step k.

2.1.3. Initial Conditions

The initial time is set to zero. The initial set of arrival times is initialized to the empty set. Without loss of generality, the UAVs are assumed to start at the base with full fuel capacity, and the slack time for each UGS is initialized to its respective revisit deadline, *i.e.*,

$$\tau(0) = 0,$$
$$\Omega(0) = \emptyset,$$
$$p_1(0) = n + 1,$$
$$(\xi_1{}'(0), \zeta_1{}'(0)) = (\xi_{n+1}, \zeta_{n+1}),$$
$$p_2(0) = n + 1,$$
$$(\xi_2{}'(0), \zeta_2{}'(0)) = (\xi_{n+1}, \zeta_{n+1}),$$
$$\vdots$$
$$p_m(0) = n + 1,$$
$$(\xi_m{}'(0), \zeta_m{}'(0)) = (\xi_{n+1}, \zeta_{n+1}), \tag{1}$$
$$f_1(0) = F,$$
$$f_2(0) = F,$$
$$\vdots$$
$$f_m(0) = F,$$
$$x_1(0) = r_1,$$
$$x_2(0) = r_2,$$
$$\vdots$$
$$x_n(0) = r_n$$

2.1.4. Dynamics

Let $q_j(i, k)$ be the Euclidean distance between UAV j and UGS i at step k. When a UAV arrives at its destination, its next destination is set by the input, and the travel time to that destination is added to the set of arrival times:

$$\text{if } q_j(p_j(k), k) = 0, \begin{cases} p_j(k+1) & := u_j(k) \\ \Omega(k) & := \{\frac{q_j(u_j(k),k)}{v}\} \cup \Omega(k) \end{cases}$$

$$\text{else,} \qquad\qquad p_j(k+1) \quad := p_j(k) \tag{2}$$

The time of the next step is set by the minimum UAV arrival time in $\Omega(k)$; that time is then removed from $\Omega(k)$:

$$\tau(k+1) := minimum(\Omega(k)) \tag{3}$$

$$\Omega(k+1) := \Omega(k) \setminus \{\tau(k+1)\} \tag{4}$$

The amount of fuel onboard a UAV decreases as dictated by the fuel consumption rate and is reset to full capacity whenever a visit to the base occurs:

$$\dot{f}_j(t) = -\dot{f}_c + \sum_{l=1}^{k-1} \dot{f}_c \times (1 - \delta_{p_j(l)p_j(l+1)} \times \delta_{(n+1)p_j(l)}) \times H(t - \tau(l)) \times H(\tau(l+1) - t)$$

$$+ \sum_{l=1}^{k} \delta_{(n+1)p_j(l)} \times (F - f_j(\tau(l))) \times \delta(t - \tau(l)), \, t \leq \tau(k) \, , \, 1 \leq j \leq m \quad (5)$$

where δ_{ij} is the discrete Kronecker delta function, $H(t - \tau(l))$ is the Heaviside step function, and $\delta(t - \tau(j))$ is the continuous Dirac delta function. The slack time of a given UGS decreases linearly in time and is reset to its revisit deadline whenever it is visited by a UAV:

$$\dot{x}_i(t) = -1 + \sum_{l=1}^{k-1}\sum_{j=1}^{m}(1 - \delta_{p_j(l)p_j(l+1)} \times \delta_{ip_j(l)}) \times H(t - \tau(l)) \times H(\tau(l+1) - t)$$

$$+ \sum_{l=1}^{k}\sum_{j=1}^{m} \delta_{ip_j(l)} \times (r_i - x_i(\tau(l))) \times \delta(t - \tau(l)), \, t \leq \tau(k) \, , \, 1 \leq j \leq m \quad (6)$$

2.2. Intruder

A single intruder is traveling on the road network at a time; but there may be multiple intruders over the course of the mission. The intruders move inside their adversary's territory and suspect that they are under observation. Thus, the intruders move stochastically to reduce the predictability of their actions. However, the intruders do not know how they are being observed and cannot perceive the UAVs flying above [3]. If the intruders can detect the UAVs in the area, then a different intruder model is needed, e.g., using results from the pursuit-evasion literature or Stackelberg games.

The intruder moves from one node to the next in the graph according to a first order Markov process, *i.e.*, the next location that it visits only depends on its current location. The central authority is assumed to possess the Markov model for the intruder's movement; this model could have been obtained through intelligence or deduced from prior observation. The intruder enters the graph at a random node according to the initial distribution of the Markov model. Since the intruder's movement is memoryless, the UAVs and central authority only use the most recent intruder detection for their computations and do not keep track of the trail of UGSs detections left by the intruder.

The distance traveled by the intruder at time t is modeled as the process X_t:

$$X_0 = 0,$$
$$X_t - X_\rho = \mathcal{N}(\mu_v \times (t - \rho), \sigma_v^2 \times (t - \rho)^2), t > \rho > 0 \quad (7)$$

where $\mu_v > 0, \sigma_v > 0$. The fuel consumption of the intruder's vehicle is assumed to be negligible.

2.3. Intruder Interception

To direct the UAVs, such that interception is likely to occur, the probability of the intruder passing by a UGS that a UAV is loitering above must be calculated.

2.3.1. Probability of Intruder Passing a Node in a Given Time Window

Given a path s of UGSs, where $s(i)$ indicates the $i-th$ UGS visited and the fact that the intruder passed $s(1)$ at time zero, the probability of the intruder passing $s(2)$ between times t_i and t_f (where $t_i > 0$ and $t_f > t_i$) is:

$$P_{int}(s(2), s, t_i, t_f) = \int_{t_i}^{t_f} P[X_\rho = d_{s(1),s(2)}]d\rho \quad (8)$$

The probability of the intruder passing $s(3)$ between times t_i and t_f is:

$$P_{int}(s(3), s, t_i, t_f) = \int_{t_i}^{t_f} P[X_\rho = d_{s(1),s(2)} + d_{s(2),s(3)}] d\rho \tag{9}$$

The probability of the intruder passing $s(b+1)$ between times t_i and t_f is:

$$P_{int}(s(b+1), s, t_i, t_f) = \int_{t_i}^{t_f} P[X_\rho = \sum_{f=1}^{b} d_{s(f),s(f+1)}] d\rho \tag{10}$$

Generalizing, the probability of the intruder passing node j (while traveling along the path s) between times t_i and t_f is:

$$P_{int}(j, s, t_i, t_f) = \sum_{b=1}^{|s|-1} \delta_{s(b+1),j} \int_{t_i}^{t_f} P[X_\rho = \sum_{f=1}^{b} d_{s(f),s(f+1)}] d\rho \tag{11}$$

The probability that the intruder will pass node j between times t_i and t_f before reaching the base is:

$$P_{int}(j, s, t_i, t_f) = \sum_{b=1}^{|s|-1} (1 - 1_{s(1:b)}(n+1)) \delta_{s(b+1),j} \int_{t_i}^{t_f} P[X_\rho = \sum_{f=1}^{b} d_{s(f),s(f+1)}] d\rho \tag{12}$$

where $1_{s(1:b)}(n+1)$ is the indicator function.

A set of paths S is introduced, where $S(a)$ indicates the ath path within the set and $S(a, f)$ indicates the fth node visited in the ath path within the set. $P(S(a))$ is the probability of the ath path occurring in the set where $\sum_{a=1}^{|S|} P(S(a)) = 1$; $P(S(a))$ is computed using the Markov model for the intruder's motion along the road network. Thus, the probability that the intruder will pass node j between times t_i and t_f before reaching the base given a set of paths S is:

$$P_{int}(j, S, t_i, t_f) = \sum_{a=1}^{|S|} P(S(a)) \sum_{b=1}^{|S(a)|-1} (1 - 1_{S(a,1:b)}(n+1)) \delta_{S(a,b+1),j} \int_{t_i}^{t_f} P[X_\rho = \sum_{f=1}^{b} d_{S(a,f),S(a,f+1)}] d\rho \tag{13}$$

2.3.2. Probability of Intruder Interception

Let $g_i(j, k)$ be the Euclidean distance between the positions of UAV i at steps j and k. Given the probability of an intruder passing a UGS during a certain time window, the probability of intruder interception can be calculated by accounting for the UAV locations:

$$P_{capture}(S, y(l), y(l+1)) = \sum_{j=1}^{m} \alpha_j \times P_{int}(p_j(l+1), S, \tau(l), \tau(l+1)) \tag{14}$$

$$\text{where: } \alpha_j = \begin{cases} 1 & \text{if } g_j(l, l+1)=0 \\ 0 & \text{otherwise} \end{cases}.$$

This equation consists of a sum over all of the UAVs, where j indicates the UAV index. α_j is only one when UAV j is loitering over a UGS during the step, in which case, the probability of an intruder passing the node where the UAV is located during the time window of the step is added.

3. Problem Formulation

Using the models for the defenders and intruders presented in the previous section, the problem is now formulated. Given the UAV states' at the current step, the UGSs' revisit deadlines and the results from the UGSs' queries, the UAVs are to find paths that satisfy the revisit deadlines and maximize the probability of intercepting the intruder. This revisit deadline satisfaction version of the problem does not allow for missing any revisit deadline and, hence, limits the UAVs' ability to pursue the intruder.

Thus, a revisit deadline optimization version problem is formulated to give the UAVs flexibility in meeting revisit deadlines and pursuing the intruder. Instead of satisfying the revisit deadlines, the amount by which revisit deadlines are missed is minimized. Let $t_{mission}$ be the duration of the UAVs' base defense mission. Let S_l be the set of possible intruder paths given the most recent intruder detection at step l. Let $\lfloor h \rfloor (u, t)$ indicate the smallest step in sequence u of UAV actions where $\tau(\lfloor h \rfloor (u, t)) \geq t$. Let $\tilde{x}_i(t)$ and $\underline{x}_i(t)$, respectively, be the slack time left and the slack time overdue at time t,

$$\tilde{x}_i(t) = \frac{(|x_i(t)| + x_i(t))}{2} \tag{15}$$

$$\underline{x}_i(t) = \frac{(|x_i(t)| - x_i(t))}{2} \tag{16}$$

Let β be the cost of not capturing the intruder, and let γ be the cost associated with missing a deadline, where $\beta > 0$ and $\gamma > 0$. Let $C(l)$ be the cost of the UAV actions taken at step l,

$$C(l) = \beta \times (\tau(l+1) - \tau(l)) \times (1 - P_{capture}(S_l, y(l), y(l+1)))$$
$$+ \frac{\gamma}{n} \times \sum_{i=1}^{n} (\tau(l+1) - \tau(l) - \tilde{x}_i(\tau(l))) \times \frac{x_i(\tau(l+1)) + x_i(\tau(l))}{2} \tag{17}$$

The first component of the cost penalizes the UAVs for not intercepting the intruder over the course of the step. The second component of the cost penalizes revisit deadlines that are overdue by adding the integral of the slack time missed over the course of the step.

Based on the cost function above, the revisit deadline optimization version of the problem is formulated as follows: the central authority is to find a sequence $u_j(k)$, $1 \leq j \leq m, k \in \mathbb{N}$, such that, under Equations (1)–(6), $\sum_{k=1}^{\lfloor h \rfloor (u, t_{mission})} C(k)$ is minimized and $0 \leq t \leq t_{mission}, 1 \leq j \leq m, f_j(t) \geq 0$.

3.1. Problem Complexity

Proposition 1. *The revisit deadline satisfaction version problem of cooperative surveillance and pursuit is NP-hard.*

Proof. If no intruders are present in the road network, then an optimal solution is one where the slack times are kept positive. The problem of keeping slack times positive for a single UAV is the persistent visitation problem and is proven to be NP-complete in [25]. The persistent visitation problem can be reduced to the problem of cooperative surveillance and pursuit by selecting one UAV to patrol the same graph with the same revisit deadlines without intruders present. Thus, the problem of cooperative surveillance and pursuit is NP-hard. □

Corollary 1. *The revisit deadline optimization version problem of cooperative surveillance and pursuit is NP-hard.*

4. UAV Path Selection

The problem is NP-hard, hence a heuristic algorithm is used to select the paths of the UAVs. This algorithm searches ahead for a sequence of UAV actions that minimizes the cost function within a certain time window and executes the first set of UAV actions in the sequence.

4.1. System Structure

The algorithm used to simulate the defenders' system for a cooperative surveillance and pursuit problem is provided in Algorithm 1. The states are first initialized (Lines 1–3); then at each step, the possible intruder paths within t_{search} are computed (Line 5), followed by the computation of the minimal cost action for the UAVs within the same time horizon using the possible intruder paths (Line 6). In addition, at each step k, new detections are added from the UGS queries (Lines 9–11), stale queries rejected (Lines 12–16) and UGS queries without detections added (Lines 17–18). A UGS query is characterized by the status of the UGS, the time of the detection if one occurred (otherwise, the time of the query) and the index of the queried UGS. The resulting data from querying the UGSs is used in the intruder path generation process in the next time step. This process of finding intruder paths, selecting UAV actions and gathering UGSs queries is repeated until the mission completion time is reached (Line 4).

Algorithm 1: Defenders' system for the cooperative surveillance and pursuit problem.

    Data: $G(N,E),r,d,m,v,F,\dot{f_c},t_{mission},t_{search},t_{stale}$

1 $k \leftarrow 0; t \leftarrow 0$

2 $(y(k),f(t),x(t)) \leftarrow$ (Equation (1))

3 $t_D \leftarrow \varnothing; n_D \leftarrow \varnothing; T_U \leftarrow \varnothing; N_U \leftarrow \varnothing$

4 **while** $\tau(k) < t_{mission}$ **do**

5      $(S_k, P(S_k)) \leftarrow paths(n_D, 1, t_D, t_D, T_U, N_U, \tau(k) + t_{search}, \varnothing, \varnothing)$

6      $u(k) \leftarrow uavsAction(y(k), f(\cdot), x(\cdot), t_{search}, S_k, P(S_k))$

7      $(y(k+1), f(\cdot), x(\cdot)) \leftarrow$ (Equations $(2) - (6)$), $y(k), u(k))$

8      $k \leftarrow k + 1; t \leftarrow \tau(k+1)$

9   **for** $(detection) \in queries(k)$ **do**

10   **if** $(detection).t > t_D$ **then**

11            $(t_D, n_D) \leftarrow (detection)$

12   **if** $t_D + t_{stale} < \tau(k)$ **then**

13        $t_D \leftarrow \varnothing; n_D \leftarrow \varnothing; T_U \leftarrow \varnothing; N_U \leftarrow \varnothing$

14   **for** $(t_u, n_u) \in (T_U, N_U)$ **do**

15   **if** $t_u \leq (\tau(k) - t_{stale})$ **then**

16             $(T_U, N_U) \leftarrow (T_U, N_U) \setminus (t_u, n_u)$

17   **for** $(\neg detection) \in queries(k)$ **do**

18        $(T_U, N_U) \leftarrow (T_U, N_U) \cup (\neg detection)$

    Result: $u_j(\cdot)$

4.2. Intruder Path Generation

Possible intruder paths are generated using the information from UGS queries. The central authority stores the time and UGS index of the most recent intruder detection. It also stores the times and UGS indices of recent UGS queries without detections. This information is used to generate the potential intruder paths at each step using a recursive breadth first search methodology. If a detection is too stale, then it is ignored. In simulations, the threshold for a detection to be considered stale was selected to be half the mean intruder travel time between the two UGSs most distant from one another.

A recursive algorithm is used to compute the possible intruder paths (Algorithm 2); this algorithm is used by the heuristic to assist in its decision making process. This algorithm to find intruder paths is the method by which the heuristic uses the information obtained from the UGSs. The input of the algorithm is the current candidate path for the intruder (which at the start of the algorithm's execution, is the most recent detection). At each iteration in the search, the nodes adjacent to the last node of the current working path, s, are obtained (Line 1). Possible travel times to these adjacent nodes are then computed using the intruder's velocity probability distribution. If any of the visits to the adjacent nodes violate the constraints set by recent UGS queries without detections, then they are removed (Lines 2–10). Valid adjacent nodes are then appended to the current working path (Lines 11–12). If the minimum travel time of the new path is larger than the search depth, then the path is admitted to the set of possible paths (Lines 14–15); otherwise, the search continues for that path (Lines 16–17). When all of the candidate paths reach the search depth, the algorithm terminates and returns the possible intruder paths. These possible intruder paths are then used to select the actions of the UAVs.

Algorithm 2: The paths()algorithm to find possible intruder paths.

Data: $s,p,t_{min},t_{max},T_U,N_U,t_f,S,P(S)$

1   $\Gamma \leftarrow adjacentNodes(s(|s|),G)); l \leftarrow |\Gamma|$

2   **while** $l > 0$ **do**

3      $T'_{max}(l) \leftarrow t_{max} + \frac{d_{s(|s|),\Gamma(l)}}{min(v_{int})}$

4   **for** $(t_U,n_U) \in (T_U,N_U)$ **do**

5   **if** $\Gamma(l) = n_U$ **then**

6   **if** $T'_{max}(l) \leq t_U$ **then**

7            $\Gamma \leftarrow \Gamma \setminus \{\Gamma(l)\}$

8            $T'_{max} \leftarrow T'_{max} \setminus \{T'_{max}(l)\}$

9   **break**

10     $l \leftarrow (l-1)$

11   **for** $l \leftarrow 1$ **to** $|\Gamma|$ **do**

12     $s' \leftarrow [s\,\Gamma(l)]; p' \leftarrow p \times \frac{1}{|\Gamma|}$

13     $t'_{min} \leftarrow t_{min} + \frac{d_{s(|s|),\Gamma(l)}}{max(v_{int})}$

14   **if** $t'_{min} \geq t_f$ **then**

15       $S \leftarrow S \cup \{s'\}; P(S = s') \leftarrow p'$

16   **else**

17       $(S,P(S)) \leftarrow paths(s',p',t'_{min},T'_{max}(l),T_U,N_U,t_f,S,P(S))$

Result: $S, P(S)$

4.3. Selection of UAV Actions

The following algorithm is used by the heuristic to make its decisions. The algorithm starts by searching for all possible sequences of actions for the team of UAVs within the search horizon t_{search}. For each sequence of actions available to the team of UAVs, $\tilde{u}$, it then assesses the cost using the following equation:

$$\sum_{l=k}^{\lfloor h \rfloor(\tilde{u},t_{search})} C(l) \tag{18}$$

where k is the current step. The cost calculations use the intruder paths generated earlier.

The algorithm starts by using the current state of the UAVs to compute feasible actions. The action is then applied, and the corresponding UAV states and costs are computed. These actions, states and costs are then added to sets of candidate sequences of actions, candidate states of the UAVs after the application of the corresponding sequence of actions and candidate costs after the application of the corresponding sequence of actions. The algorithm then proceeds to iterate over the set of candidate sequences of actions for the UAVs.

The procedure for computing potential sequences of actions and their costs is described in Algorithm 3. Several intermediate variables are used: $\tilde{U}$ is the working set of sequences of UAV actions; W is the corresponding set of states after the sequences of actions in $\tilde{U}$ have been applied; and Θ is the set of costs for the sequences of actions. u is the current minimal cost sequence of actions, and ϕ is the current minimal cost. These variables are initialized (Lines 1–2). The algorithm then proceeds to loop over the working set of sequences of UAV actions until none remain (Line 3). At each iteration, the first elements in the working sets of states, sequences of actions and costs are obtained and removed from their parent sets (Lines 4–5). The set of possible actions, Z, given the current working state $w \in W$

is then computed (Lines 6–14). For each possible action for the team of UAVs, $z \in Z$, the next state, λ, is computed (Line 15). If λ results in positive fuel for all of the UAVs and has reached the search depth, then its cost is computed (Lines 16–17); if that cost is smaller than the current minimal cost, then the current minimal cost sequence of actions is set to the sequence of actions that resulted in λ (Lines 18–19). If λ results in positive fuel for all of the UAVs without reaching the search depth, then the working state, sequence of actions and cost are added to the corresponding parent working sets (Lines 20–24). The algorithm terminates when the set of candidate sequences of UAV actions is empty (*i.e.*, all feasible actions in the search horizon have been considered) and returns the minimal cost sequence of actions. The heuristic directs the UAVs to take the first step in this minimal cost sequence of actions; this step results in the UAVs visiting or loitering above certain UGSs, thereby obtaining new information from the UGSs and possibly capturing an image of the intruder. The possibility of imaging the intruder exists when a UAV is loitering above a UGS; the capture of an image is triggered when the UGS which the UAV is loitering detects the intruder.

4.4. Algorithm Completeness

The search depth for the UAV actions affects the existence of solutions; if the search depth is less than the endurance of the UAVs, $t_{search} < (F/\dot{f}_c)$, then there is no guarantee that the generated paths will lead to the satisfaction of the UAVs' fuel constraints. If $t_{search} \geq t_{mission}$, then the heuristic is complete, *i.e.*, it will find a solution if one exists.

Algorithm 3: The uavsAction() heuristic to find the minimal cost action for the UAVs within a given search horizon.

Data: $y(k), f(\cdot), x(\cdot), S, P(S)$

1 $W \leftarrow \{(y(k), f(\cdot), x(\cdot))\}$
2 $\tilde{U} \leftarrow \varnothing; \Theta \leftarrow \varnothing; u \leftarrow \varnothing; \phi \leftarrow \infty$
3 **while** $|W| > 0$ **do**
4 $w \leftarrow W(1); \tilde{u} \leftarrow \tilde{U}(1); \theta \leftarrow \Theta(1)$
5 $W \leftarrow W \setminus \{w\}; \tilde{U} \leftarrow \tilde{U} \setminus \{\tilde{u}\}; \Theta \leftarrow \Theta \setminus \{\theta\}$
6 $Z \leftarrow \varnothing$
7 **if** $q_1(w.p_1) = 0$ **then**
8   $Z \leftarrow N \cup \{n+1\}$
9 **for** $l \leftarrow 2$ **to** m **do**
10   **if** $q_l(w.p_l) = 0$ **then**
11    $Z \leftarrow Z \times \{N \cup \{n+1\}\}$
12   **else**
13    $Z \leftarrow Z \times \varnothing$

14 **for** $z \in Z$ **do**
15   $\lambda \leftarrow$ (Equations $(2) - (6)$), w, z
16   **if** $(\forall j, \lambda.f_j(\lambda.\tau) \geq 0) \wedge (\lambda.\tau \geq \tau(k) + t_{search})$ **then**
17    $\theta' \leftarrow \theta + $ (Equation(17)), $S, P(S), w, \lambda$
18   **if** $\theta' < \phi$ **then**
19     $u \leftarrow [\tilde{u}\, z]; \phi \leftarrow \theta'$

20   **else if** $\forall j, \lambda.f_j(\kappa.\tau) \geq 0$ **then**
21    $\theta' \leftarrow \theta + $ (Equation (17)), $S, P(S), w, \lambda$
22   **if** $\theta' < \phi$ **then**
23     $W \leftarrow W \cup \{\lambda\}; \tilde{U} \leftarrow \tilde{U} \cup [\tilde{u}\, z]$
24     $\Theta \leftarrow \Theta \cup \{\theta'\}$

Result: u

4.5. Algorithm Complexity

The topology of the road network, the number of UAVs, the UAVs' velocity, the mean intruder velocity and t_{search} affect the algorithm complexity. Let $\bar{d}$ be the mean distance between any two UGSs. By inspection, the time complexity of the heuristic per step is:

$$O\left(|E|^{\frac{t_{search}\times\mu v}{d}} + \left(\frac{n!}{m!(n-m)!}\right)^{\frac{t_{search}\times v}{d}}\right) \tag{19}$$

The complexity increases polynomially with respect to the number of UGS and the number of roads, increases exponentially with respect to the vehicle velocities and the search depth and decreases exponentially with respect to the mean distance between any two UGSs.

5. Simulations

To illustrate the performance of the *uavsAction*() heuristic, several simulations with varying configurations are shown. A local search heuristic is used as a baseline comparison to the *uavsAction*() heuristic developed in the previous sections. The local search heuristic is detailed in Appendix A.

Four scenarios are considered: Scenario A2 occurs in Area A with a UAV-to-intruder velocity ratio of two; Scenario A3 occurs in Area A with a UAV-to-intruder velocity ratio of three; Scenario B1 occurs in Area B with a UAV-to-intruder velocity ratio of one; and Scenario B2 occurs in Area B with a UAV-to-intruder velocity ratio of two. Two UAVs are operating in Area A, while three UAVs are operating in Area B; UAV fuel consumption is not accounted for in these simulations. Visualizations of Areas A and B are shown in Figure 2. Detailed parameters for the scenarios are given in Tables 1 and 2; these tables contain the locations of the base and UGSs, revisit deadlines for the UGSs and velocity for the UAVs and intruders.

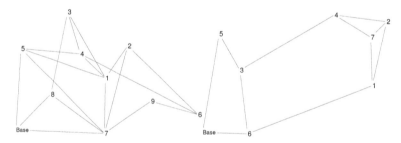

Figure 2. Visualization with base, UGSs and roads of Scenario A (**left**) and B (**right**).

Table 1. Base and unattended ground sensor (UGS) parameters for Scenarios A and B.

Destination	Scenarios A			Scenarios B		
	$\tilde{\varsigma}_i$	ς_i	r_i	$\tilde{\varsigma}_i$	ς_i	r_i
Base	6	6	N/A	6	6	N/A
UGS 1	94.66	70.33	11.4	150.03	56.21	8.1
UGS 2	117.05	109.94	12.4	162.18	124.23	10.7
UGS 3	57.17	151.44	10.5	37.41	72.06	3.0
UGS 4	70.00	100.00	2.6	117.54	131.08	8.4
UGS 5	10.79	106.16	10.8	19.65	110.68	9.5
UGS 6	186.80	25.98	8.3	44.29	4.66	7.1
UGS 7	93.88	2.38	5.6	148.70	107.66	12.6
UGS 8	40.00	50.00	7.0	N/A	N/A	N/A
UGS 9	140.00	42.00	11.0	N/A	N/A	N/A

Table 2. Vehicle parameters for Scenarios A and B.

Vehicle	Scenario A2		Scenario A3		Scenario B1		Scenario B2	
	μ_v	σ_v	μ_v	σ_v	μ_v	σ_v	μ_v	σ_v
Intruder	37.5	1.57	25	1.57	25	2.33	25	2.33
UAVs	75	0	75	0	25	0	50	0

The metric used to assess the performance of the approaches is the intruder capture index. The intruder capture index is the number of intruders whose image was captured subtracted by the number of intruders that reached the base divided by the total number of intruders over the course of the mission. Three cost configurations, indicated by (β, γ), are considered: $(1, t_{mission})$, $(t_{mission}, 1)$, $(t_{mission}^2, 1)$, where the mission time, $t_{mission}$, is 30 in the simulations. Missing revisit deadlines is weighted heavily in the first cost configuration, while not capturing an image of the intruder is weighted heavily in the last cost configuration. Eighty simulations were run per search depth per scenario per cost configuration; Figures 3–6 show the average intruder capture index (represented on the ordinate) for these 80 simulations for the three different cost configurations (represented by the three different line styles) as a function of search depth for Scenarios A and B (represented on the abscissa).

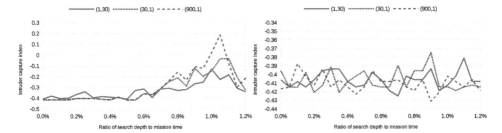

Figure 3. Intruder capture index for Scenario A2 using *uavsAction*() (**left**) and *localSearch*() (**right**).

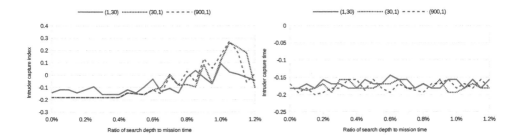

Figure 4. Intruder capture index for Scenario A3 using *uavsAction*() (**left**) and *localSearch*() (**right**).

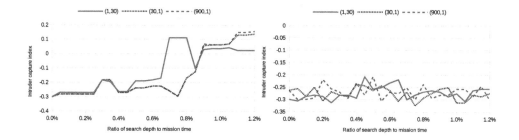

Figure 5. Intruder capture index for Scenario B1 using *uavsAction*() (**left**) and *localSearch*() (**right**).

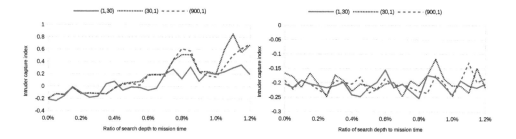

Figure 6. Intruder capture index for Scenario B2 using *uavsAction*() (**left**) and *localSearch*() (**right**).

In the left subfigures of Figures 3–6, where *uavsAction*() is used, the intruder capture index increases as a function of search depth for all cost functions. The intruder capture index for *localSearch*() (right subfigures of Figures 3–6) does not increase significantly with a larger search depth. While *uavsAction*() and *localSearch*() perform similarly for short search depths, *uavsAction*() performs better than *localSearch*() for larger search depths.

For Scenarios A2, A3 and B2 (Figures 3, 4 and 6), weighting intruder capture heavily when using *uavsAction*() resulted in more captures for large search depths, while only a marginal difference between cost configurations is seen for small search depths. In Scenario B1 (Figure 5), weighting missed revisit deadlines more heavily for *uavsAction*() resulted in more captures for search depths less than 0.9% of the mission completion time. The performance of *localSearch*() is not significantly affected by different cost configurations.

In Scenario A, an optimal search depth is seen for both velocity advantages at 1.05% of the mission completion time for *uavsAction*(). An optimal search depth is not seen in Scenario B; however, it may occur at a search depth greater than performed in the simulations. As expected, the intruder capture indices in Scenario A3 are larger than in Scenario A2, *i.e.*, the UAVs perform better when they are faster. This trait is also seen when comparing Scenario B2 to Scenario B1. The UAVs do not perform well when they do not have a velocity advantage (Figure 5), even when they are more numerous. When the UAVs are faster than the intruder and the topography is advantageous, as is the case in Scenario B, they can perform very well (Figure 6).

Videos of simulations for a variety of other scenarios can be found at [29].

6. Discussion

The performance of the *uavsAction*() heuristic largely depends on the problem instance, as can be seen from the differences in the results from the scenarios presented. In the simulations, the UAVs perform better with a larger velocity advantage compared to the intruder; this behavior is expected in all scenarios. Problem instances with bottlenecks in the road network topology lead to better performance, since the UAVs' pursuit and interception of the intruder are simplified. Topologies with highly connected nodes close together lead to poorer performance, since the UAVs cannot adequately pursue the intruder. The effect of the topology on the performance of the heuristic highlights the importance of the selection of UGSs' locations.

The simulations demonstrate that the heuristic can be used to plan the paths of UAVs in cooperative surveillance and pursuit tasks. Cooperative surveillance and pursuit systems, such as the motivating Talisman Saber exercise, are realistic and can be constructed with current technology. Details on a UAV and UGS system to monitor and pursue intruders for the Talisman Saber exercise is provided in [22]. This system uses small, commercially available UAVs and small Doppler radars as UGSs. UAVs and UGSs communicate using WiFi radios and the UAVs capture images of the intruder using a gimballed video camera. The Talisman Saber area is approximately 2450 km^2, and the road network within the area is sparse. The velocity of the UAVs ranges between 50 km/h and 100 km/h, and typical cruising altitude ranges between 60 m and 900 m, which is within the 1-km communication range of the UGSs' WiFi radios. With these parameters, the simulations shown in Section 5 can be viewed as taking place in an area nine-times larger than the Talisman Saber area or with UAVs that travel at a third of the velocity of the commercially available UAVs discussed. Simulating faster UAVs in a smaller area with more UGSs, similar to Talisman Saber, takes much longer to simulate, because of the complexity of the heuristic: in Equation (19), faster UAVs and smaller areas increase the computation time exponentially, and increasing the network size increases the computation time polynomially.

While this work is motivated by the Talisman Saber exercise, the heuristic can be used for other monitoring and pursuit tasks occurring on graphs using vehicles and stationary sensors with the same assumptions and dynamics that are described above; examples of such tasks include a pollutant monitoring task in a body of water using autonomous boats in conjunction with stationary buoys or the monitoring of a forest fire or flood using appropriate sensors on the ground and autonomous aircraft to visit these sensors.

The assumptions made in this work allow for paths to be computed while still capturing the essence of the problem; however, they do not allow for the modeling of certain features that do occur in real scenarios. For example, the UAV constant velocity and constant fuel consumption rate assumption does not handle wind disturbances. Wind disturbances affect the speed of the UAVs and alter the arrival times of UAVs at UGSs. If the wind is steady, the decision making heuristic can account for wind in its computations. If the wind is stochastic, *i.e.*, turbulence, the constant fuel consumption rate assumption needs to be relaxed, so that more fuel can be consumed during turbulence to allow the vehicles to remain on their nominal paths. The assumption that the intruder moves according to a Markov chain may be valid in some scenarios, but does not allow for intelligent intruder behavior, such as reacting to seeing a UAV flying overhead or attempting to deceive the UAVs. The assumptions made for communication reflect the capabilities of the hardware used for the Talisman Saber exercise; however, they do not handle disturbances, such as communication jamming.

7. Conclusions

In this paper, a path planning problem for a team of UAVs patrolling a network of roads and pursuing intruders using UGSs is formulated. The problem is shown to be NP-hard. A heuristic algorithm to solve the problem is presented, and its completeness and complexity are assessed. The heuristic primarily plans the paths for the UAVs; however, it also interacts with the UGSs by acquiring information from the UGSs through the UAVs, using the information to predict possible intruder paths and relying on the UGSs to trigger the capture of an image of the intruder by a loitering UAV.

The heuristic demonstrates that intercepting the intruder is possible using a sensing scheme relying entirely on UGSs. Given a good topology for the road network and well selected revisit deadlines, such as Scenario B in Section 5, the heuristic performs well and can intercept a majority of the intruders. The heuristic also exhibits intuitive behavior at times, such as orbiting several UGSs nearby and trapping the intruder, thereby forcing an interception.

In future work, other heuristics to solve the problem and decentralized approaches will be investigated. The ability to handle multiple intruders simultaneously will also be included. This could be achieved by the UGSs tagging their measurements to certain intruders. Variable levels of communication between the UGSs and UAVs will also be considered.

Acknowledgments: The authors would like to thank Corey Schumacher, David Casbeer and Derek Kingston of the Air Force Research Laboratory for their feedback on this problem. The research was supported in part by the United States Air Force grant, FA 8650-07-2-3744.

Author Contributions: Anouck Girard, Pierre Kabamba and Jonathan Las Fargeas contributed equally in the formulation and analysis of the stated problem as well as in the writing of the manuscript. Jonathan Las Fargeas developed the algorithms and ran the simulations.

Conflicts of Interest: The authors declare no conflict of interest.

Appendix

Appendix A. Local Search Heuristic

Algorithm A1: The localSearch() heuristic to find a local minimal cost action for the UAVs.

Data: $y(k), f(\cdot), x(\cdot), S, P(S), search_{init}, search_{iter}$

1   $\tilde{U} \leftarrow \varnothing; u \leftarrow \varnothing; \phi \leftarrow \infty$

2   **for** $l \leftarrow 1$ **to** $search_{init}$ **do**

3      $\kappa \leftarrow$ Generate a feasible sequence of actions of a given depth

4      $S, P(S) \leftarrow$ Compute possible intruder paths for the duration of the sequence of actions

5      $i \leftarrow -1$

6      $\tilde{U} \leftarrow$ Compute the k-neighborhood of the sequence of actions

7      $S, P(S) \leftarrow$ Extend possible intruder paths to the maximum end time of the k-neighborhood

8      $\tilde{U}.cost \leftarrow$ Compute costs for all actions

9      $\kappa_2 \leftarrow \min_{\tilde{u} \in \tilde{U}}(\tilde{u}.cost)$

10   **while** $\kappa_2 \neq \kappa \wedge i < search_{iter}$ **do**

11         $\kappa \leftarrow \kappa_2$

12         $\tilde{U} \leftarrow$ Compute the k-neighborhood of the sequence of actions

13         $S, P(S) \leftarrow$ Extend possible intruder paths to the maximum end time of the k-neighborhood

14         $\tilde{U}.cost \leftarrow$ Compute costs for all actions

15         $\kappa_2 \leftarrow \min_{\tilde{u} \in \tilde{U}}(\tilde{u}.cost)$

16   **if** $search_{iter} > 0 \wedge i < 0$ **then**

17            $i \leftarrow i + 2$

18   **else if** $search_{iter} > 0$ **then**

19            $i \leftarrow i + 1$

20   **if** $\kappa_2.cost < \phi$ **then**

21         $u \leftarrow \kappa_2$

Result: u

For comparison, a local search heuristic is also implemented to solve this problem (Algorithm A1). At each step, the algorithm randomly generates a feasible sequence of actions of a given depth (Line 3) and computes possible intruder paths that can occur in the same time window (Line 4). The k-neighborhood of the initial sequence of actions is then computed (Line 6) for k = 2, and the possible intruder paths are extended due to the increased time window (Line 7). For each sequence of actions, the corresponding cost is computed using the intruder paths (Line 8), and the minimum cost sequence of actions is extracted (Line 9). This procedure of generating a k-neighborhood and selecting the optimal sequence is repeated until convergence of the minimal cost sequence of actions or a number of iterations, $search_{iter}$, is reached (Lines 10–21). The overall procedure can also be repeated for multiple initial random feasible sequences of actions to further increase the search space. $search_{init}$ indicates the number of these initial guesses. For the search to continue until convergence, $search_{iter}$ is indicated as zero. The simulations shown in this paper use $search_{init} = 5$ and $search_{iter} = 0$.

References

1. Ratches, J. Review of Current Aided/Automatic Target Acquisition Technology for Military Target Acquisition Tasks. *Opt. Eng.* **2011**, *50*, 072001.
2. Kingston, D.B.; Rasmussen, S.J.; Mears, M.J. Base defense using a task assignment framework. In Proceedings of the AIAA Guidance, Navigation, and Control Conference, Chicago, IL, USA, 10–13 August 2009.
3. Krishnamoorthy, K.; Casbeer, D.; Chandler, P.; Pachter, M.; Darbha, S. UAV search and capture of a moving ground target under delayed information. In Proceedings of the 51st IEEE Conference on Decision and Control, Maui, HI, USA, 10–13 December 2012; pp. 3092–3097.
4. Las Fargeas, J.; Kabamba, P.; Girard, A. Optimal Configuration of Alarm Sensors for Monitoring Mobile Ergodic Markov Phenomena on Arbitrary Graphs. *IEEE Sens. J.* **2015**, accepted.
5. Hokayem, P.; Stipanovic, D.; Spong, M. On Persistent Coverage Control. In Proceedings of the 46th IEEE Conference on Decision and Control, New Orleans, LA, USA, 12–14 December 2007; pp. 6130–6135.
6. Nigam, N.; Kroo, I. Persistent Surveillance Using Multiple Unmanned Air Vehicles. In Proceedings of the IEEE Aerospace Conference, Big Sky, MT, USA, 1–8 March 2008; pp. 1–14.
7. Nigam, N.; Bieniawski, S.; Kroo, I.; Vian, J. Control of Multiple UAVs for Persistent Surveillance: Algorithm and Flight Test Results. *IEEE Trans. Control Syst. Technol.* **2011**, *20*, 1–17.
8. Elmaliach, Y.; Agmon, N.; Kaminka, G. Multi-Robot Area Patrol under Frequency Constraints. In Proceedings of the IEEE International Conference on Robotics and Automation, Roma, Italy, 10–14 April 2007; pp. 385–390.
9. Elmaliach, Y.; Shiloni, A.; Kaminka, G. A Realistic Model of Frequency-Based Multi-Robot Fence Patrolling. In Proceedings of the 7th International Joint Conference on Autonomous Agents and Multi-Agent Systems, Estoril, Portugal, 12–16 May 2008; Volume 1, pp. 63–70.
10. Cassandras, C.; Ding, X.; Lin, X. An Optimal Control Approach for the Persistent Monitoring Problem. In Proceedings of the 50th IEEE Conference on Decision and Control and European Control Conference, Orlando, FL, USA, 12–15 December 2011; pp. 2907–2912.
11. Chevaleyre, Y. Theoretical Analysis of the Multi-agent Patrolling Problem. In Proceedings of the IEEE/WIC/ACM International Conference on Intelligent Agent Technology, Beijing, China, 20–24 September 2004; pp. 302–308.
12. Wolfler Calvo, R.; Cordone, R. A Heuristic Approach to the Overnight Security Service Problem. *Comput. Oper. Res.* **2003**, *30*, 1269–1287.
13. Pasqualetti, F.; Durham, J.; Bullo, F. Cooperative Patrolling via Weighted Tours: Performance Analysis and Distributed Algorithms. *IEEE Trans. Robot.* **2012**, *28*, 1181–1188.
14. Oberlin, P.; Rathinam, S.; Darbha, S. Today's Traveling Salesman Problem. *IEEE Robot. Autom. Mag.* **2010**, *17*, 70–77.
15. Savla, K.; Bullo, F.; Frazzoli, E. On Traveling Salesperson Problem for Dubins' Vehicle: Stochastic and Dynamics Environments. In Proceedings of the 44th IEEE Conference on Decision and Control and European Control Conference, Seville, Spain, 15 December 2005; pp. 4530–4535.
16. Quilliot, A. A Short Note About Pursuit Games Played on a Graph with a given Genus. *J. Comb. Theory Ser. B* **1985**, *38*, 89–92.
17. Nowakowski, R.; Winkler, P. Vertex-to-vertex Pursuit in a Graph. *Discrete Math.* **1983**, *43*, 235–239.
18. Aigner, M.; Fromme, M. A Game of Cops and Robbers. *Discrete Appl. Math.* **1984**, *8*, 1–12.
19. Chung, T.H.; Hollinger, G.A.; Isler, V. Search and Pursuit-evasion in Mobile Robotics. *Auton. Robot.* **2011**, *31*, 299–316.

20. Basilico, N.; Gatti, N.; Amigoni, F. Developing a Deterministic Patrolling Strategy for Security Agents. In Proceedings of the IEEE/WIC/ACM International Joint Conference on Web Intelligence and Intelligent Agent Technologies, Milan, Italy, 15–18 September 2009; Volume 2, pp. 565–572.

21. Basilico, N.; Gatti, N.; Villa, F. Asynchronous Multi-Robot Patrolling against Intrusions in Arbitrary Topologies. In Proceedings of the 24th AAAI Conference on Artificial Intelligence, Atlanta, GA, USA, 11–15 July 2010; pp. 1224–1229.

22. Kingston, D. Intruder Tracking using UAV Teams and Ground Sensor Networks. In Proceedings of the German Aerospace Congress, Berlin, Germany, 11 September 2012; German Society for Aeronautics and Astronautics: Bonn, Germany, 2012.

23. Tsukamoto, K.; Ueda, H.; Tamura, H.; Kawahara, K.; Oie, Y. Deployment design of wireless sensor network for simple multi-point surveillance of a moving target. *Sensors* **2009**, *9*, 3563–3585.

24. Niedfeldt, P.; Kingston, D.; Beard, R. Vehicle State Estimation within a Road Network using a Bayesian Filter. In Proceedings of the American Control Conference, San Francisco, CA, USA, 29 June–1 July 2011; pp. 4910–4915.

25. Las Fargeas, J.; Hyun, B.; Kabamba, P.; Girard, A. Persistent Visitation under Revisit Constraints. In Proceedings of the 2013 International Conference on Unmanned Aircraft Systems, Atlanta, GA, USA, 28–31 May 2013; pp. 952–957.

26. Las Fargeas, J.; Hyun, B.; Kabamba, P.; Girard, A. Persistent Visitation with Fuel Constraints. *Proc. Soc. Behav. Sci.* **2012**, *54*, 1037–1046.

27. Klesh, A.; Kabamba, P.; Girard, A. Path Planning for Cooperative Time-optimal Information Collection. In Proceedings of the American Control Conference, Seattle, WA, USA, 11–13 June 2008; pp. 1991–1996.

28. Dai, R.; Cochran, J. Path Planning and State Estimation for Unmanned Aerial Vehicles in Hostile Environments. *J. Guid. Control Dyn.* **2010**, *33*, 595–601.

29. Las Fargeas, J. Videos of Cooperative Surveillance and Pursuit Simulations. Available online: http://arclab.engin.umich.edu/?page_id=450 (accessed on 7 January 2015).

Article

A Multispectral Image Creating Method for a New Airborne Four-Camera System with Different Bandpass Filters

Hanlun Li [1], Aiwu Zhang [1,*] and Shaoxing Hu [2]

[1]   Key Laboratory of 3D Information Acquisition and Application of Ministry, Capital Normal University, Beijing 100048, China; lihanlun@126.com
[2]   School of Mechanical Engineering & Automation, Beihang University, Beijing 100083, China; husx@buaa.edu.cn
*   Author to whom correspondence should be addressed; zhangaw98@163.com; Tel.: +86-10-6890-3003.

Academic Editor: Felipe Gonzalez Toro
Received: 28 April 2015; Accepted: 6 July 2015; Published: 20 July 2015

Abstract: This paper describes an airborne high resolution four-camera multispectral system which mainly consists of four identical monochrome cameras equipped with four interchangeable bandpass filters. For this multispectral system, an automatic multispectral data composing method was proposed. The homography registration model was chosen, and the scale-invariant feature transform (SIFT) and random sample consensus (RANSAC) were used to generate matching points. For the difficult registration problem between visible band images and near-infrared band images in cases lacking manmade objects, we presented an effective method based on the structural characteristics of the system. Experiments show that our method can acquire high quality multispectral images and the band-to-band alignment error of the composed multiple spectral images is less than 2.5 pixels.

Keywords: panchromatic camera; filters; SIFT; RANSAC; registration; multispectral

1. Introduction

With the rapid development of unmanned aerial vehicle (UAV) technology, we urgently need a low cost multispectral system which can acquire multispectral images at the wavelengths based on actual requirements. Our research group developed an airborne high resolution multispectral system (Figure 1) which is mainly composed of a set of digital video recorders (DVR), a ruggedized Getac B300 PC, four identical Hitachi KPF120CL monochrome cameras (2/3 inch Interline type, Progressive Scan CCD), and four bandpass filters. The four identical monochrome cameras are sensitive in the 400 to 1000 nm spectral range, have the capability of obtaining 8-bit images with 1392 × 1040 pixels, and are respectively equipped with near-infrared (800 nm), red (650 nm), green (550 nm) and blue (450 nm) bandpass filters. As a result, it has the flexibility to change filters to acquire other band images in the 400 to 1000 nm spectral range for specific requirements. Because the four cameras are independent, it has the advantage that each camera can be individually adjusted for optimum focus and aperture setting. However, for the multiple optical systems, it is nearly impossible to align different band images taken by the cameras at one exposal optically or mechanically [1], so a registration method is needed.

Figure 1. The four-camera multispectral mapping system.

In recent years, many multispectral mapping systems were developed. For example, a multispectral image system called MS4100 (Geospatial Systems, Inc., West Henrietta, NY, USA) used a beam splitting prism and three charge coupled device (CCD) sensors to acquire images in three spectral bands within the 400–1100 nm. Although the alignment issue may become easier, it is difficult to change the band-pass filters once integrated. Oppelt [2] introduced an imaging system. As with the MS4100, once it is integrated, the band images at other wavelengths cannot be acquired. A complex imaging system equipped with a lens, a cold mirror, a beamsplitter, three rear lens units, three filters and three monochrome cameras was introduced by Kise [3]. It is not easy to change filters and hard to extend to get other band images at different wavelengths. Everitt [4] proposed a multispectral digital video system which is comprised of three video CCD analog video cameras and three filters. Its hardware design and video data synchronization acquisition were introduced in detail. Gorsevski [5] designed an airborne mapping system that can provide valuable experiential learning opportunities for students. It is a very complex system, and the multispectral system is only one of its subsystems. Most of these four papers focus on hardware design and data synchronization acquisition but fail to introduce alignment methods in detail. Honkavaara [6] introduced a complex weight UAV spectral camera. This spectral camera acquires different band images by changing its air gap. The dislocation between different bands depends on flight speed and flying height. Due to different structures, its band matching methods are not very suitable for our multispectral system. Yang [7,8] used first-order and second-order polynomial transformation models to register band images and successfully obtained multispectral images. However, the polynomial model is just a generic registration model; the structural characteristics of the system itself were not fully considered, so the transformation may not be modeled properly. In addition, the method to generate matching points has not been introduced.

Due to great altitude and limited undulating ground, this paper considers the land surface as a plane and has proved that the homography registration model is very suitable. Currently, many papers [9–13] use SIFT [14,15] and RANSAC [16] to get the parameters of transformation models. The RANSAC algorithm is a learning technique to estimate parameters of a specific model by random sampling of observed data and uses a voting scheme to find the optimal fitting result; the voting scheme is based on an assumption that there are enough inliers (correct matches in this paper) to satisfy the specific model. When the number of inliers is less than 50% [17], it usually performs badly. However, the different band images acquired by our system differ from each other in the image intensity, and the correct rate of initial matches declines rapidly. Especially for pairs including an infrared band image and a visible band image, the correct rate is even less than 10% if the images comprise a considerable part of vegetation and very few manmade objects. So, the RANSAC cannot be used directly. Very few papers focused on registering pairs of infrared images and visible images. In light of this, according to the structural characteristics of the four-camera multispectral system, this paper proposed an effective method to remove most of the false matches and greatly increase the correct rate, and then uses the RANSAC to eliminate the remaining false matches. Finally, parameters of the registration model are calculated using the least squares method. Experiments show that this method does not only improve the registration performance, but also solves the matching problem between near-infrared images and visible images in the case of lack of manmade objects.

This article developed an airborne high resolution four-camera multispectral system, proposed an excellent registration method and introduces this method in detail. The second part describes the derivation of the matching model and its parameters calculation method; the third part introduces the method of eliminating false matches; the fourth part shows the different band image registration experiments; the fifth part shows the data acquisition and accuracy assessment; and the final part gives a conclusion.

2. Registration Model and Parameters Calculation

Because the four cameras are independent, transformation relationships among different band images must be known before the process of multispectral image composition. The derivation of the transformation model and the parameters calculation method are as follows.

2.1. The Derivation of the Transformation Model

The multispectral system consists of four cameras, and their IDs respectively are 0, 1, 2, and 3. For any point W in the three dimensional space, its coordinate is $\mathbf{W} = [X_W \ Y_W \ Z_W]$ in matrix form, $\begin{bmatrix} \mathbf{W} & 1 \end{bmatrix}^T$ in homogeneous coordinates. $\mathbf{v} = [x \ y \ -f]^T$ denotes its corresponding image/space point. According to the pinhole imaging principle [1], the relationship between $\mathbf{W}$ and $\mathbf{v}$ can be described as Equation (1):

$$\lambda \mathbf{v} = \begin{bmatrix} \mathbf{K} & \mathbf{0} \end{bmatrix} \begin{bmatrix} \mathbf{R} & \mathbf{t} \\ \mathbf{0}^T & 1 \end{bmatrix} \begin{bmatrix} \mathbf{W} & 1 \end{bmatrix}^T = \mathbf{KRW}^T + \mathbf{Kt} \tag{1}$$

$\mathbf{K}$, $\mathbf{R}$ and $\mathbf{t}$ are a 3×3 camera calibration matrix, 3×3 rotation matrix, and 3×1 translation vector respectively. $\mathbf{v}_i$ is the corresponding image/space point in the i-th camera. When World Coordinate System (WCS) coincides with the Camera Coordinate System (CCS) of the 0-th camera, we obtain:

$$\lambda_0 \mathbf{v}_0 = \mathbf{K}_0 \mathbf{IW} + \mathbf{K0} = \mathbf{K}_0 \mathbf{W} \tag{2}$$

$$\lambda_i \mathbf{v}_i = \mathbf{K}_i \mathbf{R}_i \mathbf{W} + \mathbf{K}_i \mathbf{t}_i \tag{3}$$

Due to the great altitude, small field angle, limited undulating ground, and negligible offsets, the land surface is considered as a plane. Assuming that the normal vector of the plane is $\mathbf{m}$, $\mathbf{m}^T \mathbf{W} = N$, and N is a constant, Equation (3) can be written as Equation (4):

$$\lambda_i \mathbf{v}_i = \mathbf{K}_i \mathbf{R}_i \mathbf{W} + \mathbf{K}_i \mathbf{t}_i \frac{\mathbf{m}^T \mathbf{W}}{N} \tag{4}$$

According to Equations (2) and (4), we obtain:

$$\mathbf{v}_i = \frac{\lambda_0}{\lambda_i} \mathbf{K}_i \left(\mathbf{R}_i + \frac{\mathbf{t}_i \mathbf{m}^T}{N} \right) \mathbf{K}_0^{-1} \mathbf{v}_0 = \mathbf{H}_i \mathbf{v}_0 \tag{5}$$

$\mathbf{H}_i$, the so-called homography, is a 3×3 matrix determined by the scale factors λ_0, λ_i, calibration matrix $\mathbf{K}_0$, $\mathbf{K}_i$, normal vector $\mathbf{m}$, rotation matrix $\mathbf{R}_i$, and translation vector $\mathbf{t}_i$; so it can model the transformation relationship between the images properly.

335

2.2. Transformation Model Parameters Calculation

As a 3×3 matrix, $\mathbf{H}_i$ has nine unknown variables, eight of which are independent. In total, it has eight degrees of freedom; every pair of matched points can build two equations, so we need four matches to solve the $\mathbf{H}_i$. However, the noise and the other errors lessen credibility for the lack of the rest observations when we only have four pairs of matching points. To calculate the $\mathbf{H}_i$, we need more than four matches for adjustment so as to promote credibility and accuracy. Equation (5) can be written as Equation (6):

$$\begin{bmatrix} x_i \\ y_i \\ 1 \end{bmatrix} = \begin{bmatrix} h_{11}^i & h_{12}^i & h_{13}^i \\ h_{21}^i & h_{22}^i & h_{23}^i \\ h_{31}^i & h_{32}^i & h_{33}^i \end{bmatrix} \begin{bmatrix} x_0 \\ y_0 \\ 1 \end{bmatrix} \tag{6}$$

For any feature point (x_0, y_0) in the image taken by the 0-th camera, where (x_i, y_i) is its matching point taken by the i-th camera, the error equation can be described as Equation (7):

$$\varepsilon_{xi} = x_i - \frac{h_{11}^i x_0 + h_{12}^i y_0 + h_{13}^i}{h_{31}^i x_0 + h_{32}^i y_0 + h_{33}^i}, \ \varepsilon_{yi} = y_i - \frac{h_{21}^i x_0 + h_{22}^i y_0 + h_{23}^i}{h_{31}^i x_0 + h_{32}^i y_0 + h_{33}^i} \tag{7}$$

In order to get the $\mathbf{H}_i$, we just need to use the least square method to minimize the ε_i. The process of multispectral image composition is as follows: three empty output matrixes are created at the same pixel size as the reference image. For any one of the three, using the transformation model $\mathbf{H}_i$, the coordinates of each pixel in the output image matrix are transformed to determine their corresponding location in the original input image. However, because of small differences in the CCD sensors and orientations of the cameras in the aluminum frame, this transformed cell location will not directly overlay a pixel in the input matrix. Interpolation is used to assign a Digital Number (DN) value to the output matrix pixel determined on the basis of the pixel values that surround its transformed position in the input image matrix. So far, there are some common interpolation methods can be chosen, including nearest neighbor algorithm, bilinear interpolation, cubic convolution, and Inverse Distance Weighted (IDW). The nearest neighbor algorithm is adopted because it simply assigns the closest pixel value in the input image to the new pixel and does not change the original data values. Of course, we may use other complex interpolation methods if necessary.

3. Rejecting False Matches

Image feature points are extracted by the SIFT detector. The SIFT feature is a kind of local feature of digital images using 128-dimensional vectors to describe feature points. It maintains invariance of scaling and rotation and also keeps a certain degree of stability with change of brightness, the viewing change, affine transformation and noise. It consists of four major stages: (1) scale-space peak selection; (2) keypoint localization; (3) orientation assignment; and (4) keypoint descriptor. There are a lot of false matches in the initial matches, so the SIFT initial matches cannot be used to calculate the parameters of the transformation model directly. The conventional approach is to apply RANSAC to eliminate false matches. The RANSAC method is based on the random sampling theory and requires the correct rate of initial matches higher than 50%. In general, the correct rate of initial matches can satisfy the requirement of the RANSAC, such as a pair of adjacent aerial images. However, owing to the four monochrome cameras equipped with different bandpass filters lead to the significant difference in the image intensity among different band images. Thus the number of incorrect matches will increase rapidly and the correct rate will decline sharply, especially for pairs of infrared and visible images. If the image is full of vegetation, the variation of the intensity between pairs of infrared and visible images is more significant. In this case, the correct rate is very low, and the RANSAC is not reliable.

In light of this situation, based on the structural characteristics of the multispectral system, the paper presents an effective method to eliminate most of the false matches and to improve the correct rate for meeting the requirement of the RANSAC. The principle of the method is as follows.

Figure 2 shows two cameras of the four-camera system, the flying altitude H, the focal length f, and the length of CCD d. The field of a camera can be described as Equation (8):

$$DX = H\frac{d}{f} \tag{8}$$

If the CCD has M cells, the ground resolution can be calculated as Equation (9):

$$R = \frac{DX}{M} = \frac{Hd}{Mf} \tag{9}$$

dx indicates the distance between the optical centers of the two cameras. If these two cameras are the same and parallel, the displacement between the fields of the two cameras also is dx. The displacement between the two images taken by the two cameras in X direction is $xcol$, as Equation (10):

$$xcol = \frac{dx}{R} = \frac{dxMf}{Hd} \tag{10}$$

dx is 0.136 m, and M is 1040 in X direction; we could know that f/d is 2.376 after calibrating the camera intrinsic parameters using the Camera Calibration Toolbox [18]. To ensure the safety of the aircraft, we need the flying height to be large enough, at least 50 m, then $xrow$ is 5.74 pixels. Assuming that the height change of ground is less than 10 m, the stereo parallax will be less than 1.16 pixels. Therefore, the stereo effect can be ignored. Using the same method, we can also calculate the displacement in Y direction $yrow$.

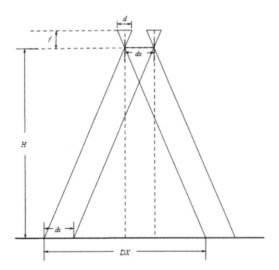

Figure 2. Schematic diagram of two camera imaging.

If the flying height is fixed, there are no changes in flight attitude and no ups and downs on the ground, the camera plane parallels to the ground surface, the cameras are arrangement in parallel, there is no camera lenses distortion and CCD distortion, then the displacement ($xrow$, $ycol$) between the two images is fixed. If all of the dx, M, d, H and f are known, the ($xrow$, $ycol$) can be calculated by using Equation (10) directly. However, if not all of these parameters are known, we need other methods. An effective histogram method is proposed, and we will describe it in Section 4. For any feature point p_0 with coordinate (x, y) in reference image I_0, the coordinate of its matching point p_i is ($x + xrow$, $y + ycol$) in input image I_i. Although these ideal situations do not occur in actuality, the influence of all these factors is limited for our multispectral system because the arrangement of the four cameras is almost parallel, as Figure 1. So, it can be estimated that p_i is near $(x + xrow, y + ycol)$ in the input image I_i. We just need to set a threshold to check whether p_i is near $(x + xrow, y + ycol)$. If the threshold is too low, some correct matches can be rejected. And, if it is too high, the removed false matches will decrease. In this paper, the threshold is set to one tenth of the image size, 104 pixels. If the Euclidean distance between p_i and $(x + xrow, y + ycol)$ is lower than this threshold, the match will be retained in this step, described as the solid line in Figure 3; otherwise, it will be taken as a false matching point and be eliminated, described as the dotted line in Figure 3. Due to this, the correct rate of initial matches will increase significantly and the RANSAC approach will become more reliable.

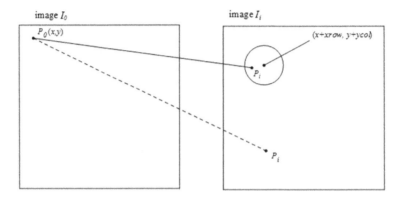

Figure 3. The dotted line shows a false match because p_i is too far away from $(x + xrow, y + ycol)$. And, the match showed by the solid line will be remained.

4. Different Band Image Registration

4.1. Experiment 1

We select four synchronous images acquired by our multispectral system, which contain many artificial objects, as shown in Figure 4. The number of SIFT feature points of the four images respectively is 5097, 6481, 5911, and 6618. Because green vegetation has higher reflection levels in the infrared and green bands, and lower reflection levels in the red and blue bands, green bands are more similar to infrared bands in vegetation areas, and the registration between the green band and the other two visible bands is easier, so the green band is chosen as the reference band image, and the other three will be registered to it. The total initial matches respectively are 1446, 1139, and 485. For all initial matches, the coordinates of the key points in the reference band image are respectively (x_{01}, y_{01}), (x_{02}, y_{02}), ... , (x_{0n}, y_{0n}) where the coordinates of the corresponding points in the input band image are respectively (x_{i1}, y_{i1}), (x_{i2}, y_{i2}), ... , (x_{in}, y_{in}), and $i = 1, 2, 3$ represents the points in the other three bands. Letting $d_{xi} = x_0 - x_i$, $d_{yi} = y_0 - y_i$, d_{x3} and d_{y3} respectively indicates the matches' coordinate displacements between the green band image and the infrared band image in X direction and Y direction. To represent distribution of d_{x3} and d_{y3}, two histograms have been created, as shown in Figure 5. We can see that the number of matches reaches the top when d_{x3} is at -6.17; the number of matches reaches a maximum when d_{y3} is at -8.84. It can be estimated that $(xrow, ycol)$ is approximately equal to $(-6.17, -8.84)$ for the green band and infrared band pair. We can use the same method to calculate the displacements between the green band and other two bands. If flying height and flying attitude vary a little, the evaluated $(-6.17, -8.84)$ can be used to register other images taken at other times. The correct matches respectively are 1278, 973, and 257, with the correct rate of 88%, 85% and 53%. The correct rate of initial matches between the green band and the infrared band is significantly lower than the correct rates between the green band and the other two bands. However, because of many manmade objects, all the correct rates are higher than 50%, so we can still use the RANSAC method to eliminate the error matches directly.

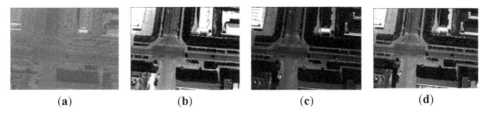

(a)   **(b)**   **(c)**   **(d)**

Figure 4. The band images; (**a**) the infrared band; (**b**) the red band; (**c**) the green band; (**d**) the blue band.

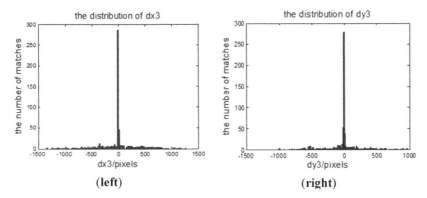

(left)   **(right)**

Figure 5. The matches' coordinate displacements; (**left**) the distribution of dx_3; (**right**) the distribution of dy_3.

4.2. Experiment 2

As shown in Figure 6, there are four different band images acquired by the multispectral system at one exposure. Compared with Figure 4, the four pictures shown in Figure 6 contain fewer artificial objects. The number of SIFT feature points in the four images respectively is 4739, 7964, 7497 and 8999. For the same reason as in Experiment 1, the green band is chosen to be the reference band image, and the other three band images are registered to it. As shown in Table 1, the initial matches (IM) respectively are 1446, 1139, 537, and the correct matches (CM) respectively are 1038, 926, 36; the correct rate (CR) respectively is 72%, 85%, 7%. The correct rate (CR) between the green band and the infrared band is significantly lower than 50%. It is unable to get the correct result using the RANSAC directly, as shown in Figure 7 (left). Because Figures 4 and 6 are from a same flight strip, there is little change in flying height and flying attitude. So, the evaluated (*xrow*, *ycol*) in Experiment 1 can be used to eliminate false matches. 398, 185, and 491 false matches have respectively been removed, as shown in Table 1. After eliminating, the matches respectively become 1048, 954, 46, and the correct rates respectively become 99%, 97%, 78% , so the registration performance is significantly improved. All of these are much higher than 50%, making RANSAC more reliable, and solving the registration problem between the green band and infrared band, as shown in Figure 7 (right).

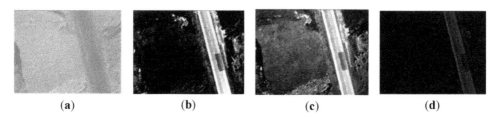

| (a) | (b) | (c) | (d) |

Figure 6. The band images with less artificial objects; (**a**) the infrared band; (**b**) the red band; (**c**) the green band; (**d**) the blue band.

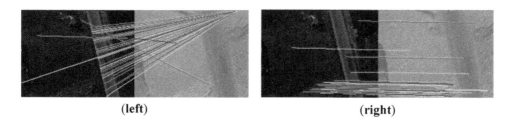

| (left) | (right) |

Figure 7. Image registration; (**left**) the result obtained by using the RANSAC directly; (**right**) the result obtained by using our method.

Table 1. The number of matches and correct rate.

Image Pairs	Experiment 2					Experiment 3				
	IM	CM	CR (%)	RFM	CR (%)	IM	CM	CR (%)	RFM	CR (%)
G-B	1446	1038	72	398	99	555	342	61	210	99
G-R	1139	926	85	185	97	792	525	66	255	98
G-IR	537	36	7	491	78	303	19	6	277	73

4.3. Experiment 3

As shown in Figure 8, four different band images acquired by the multispectral system at one other exposal are selected. Compared with Figures 4 and 6, Figure 8 contains no artificial objects. The number of SIFT feature points in the four images are 4137, 5578, 5469, and 7385, respectively. For the same reason as in the above two experiments, the green band is chosen as the reference band, and the other three band images are registered to it. As shown in Table 1, the numbers of initial matches (IM) are 555, 792 and 303, and the correct matches (CM) respectively are 342, 525 and 19, thus the correct rate (CR) respectively is 61%, 66% and 6%. The correct rate between the green band and the infrared band is evidently lower than 50%. Therefore the RANSAC cannot be used directly, as shown in Figure 9 (left). Because Figures 4, 6 and 8 are from a same flight strip, they have little change in flying height and attitude. So, the evaluated ($xrow$, $ycol$) in Experiment 1 can also be used to eliminate error matches in this experiment. As shown in Table 1, 210, 255 and 277 false matches (RFM) have been eliminated respectively. After that, the number of matches respectively become 345, 537 and 26, and the correct rates (CR) respectively becomes 99%, 97%, 73%; the registration performance is significantly improved. All of these percentages are much higher than 50%, and the RANSAC is more reliable and can be used directly to obtain correct matching results, as shown in Figure 9 (right).

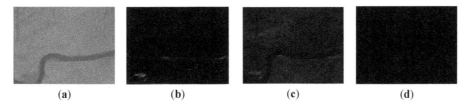

Figure 8. The band images with no artificial objects; (**a**) the infrared band; (**b**) the red band; (**c**) the green band; (**d**) the blue band.

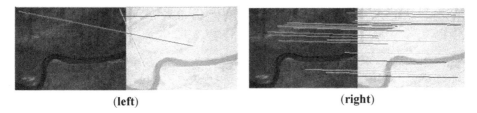

(**left**)   (**right**)

Figure 9. Image registration; (**left**) the result obtained by using the RANSAC directly; (**right**) the result obtained by using our method.

5. Four-Band Multispectral Data Acquisition and Accuracy Assessment

The multispectral mapping system was mounted on a metal protective box installed on an airship, named ASQ-HAA380, which was developed by our research group. On 18 August 2014, the research group carried a flight experiment in Haibei Tibetan Autonomous Prefecture, Qinghai Province, China, and the experimental scenario is shown in Figure 10. Qinghai TV and many other media sources reported this experiment [19]. The experimental data will be used mainly for pasture biomass assessment and the survey of urban green space in high altitude area. In order to guarantee the image quality of each multispectral camera, according to flight altitude and weather conditions, each camera was individually adjusted for optimum focus and aperture setting. Four different band images acquired by the multispectral system at one exposal are shown in Figure 4 in the Section 3, and their histograms are shown in Figure 11. The histograms of these images from diverse target areas spread well within the dynamic range without saturation and indicate that the system is able to capture high quality multispectral data.

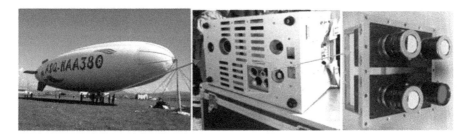

Figure 10. Experimental scene and equipment installation.

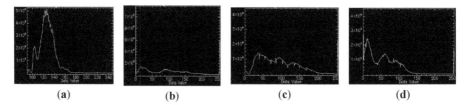

Figure 11. Histograms of the four band images; (a–d) respectively the histograms of the infrared band, red band, green band and blue band.

5.1. Experiment 1

In this experiment, four images containing a lot of man-made objects are used, as shown in Figure 4. Because it contains a lot of man-made objects, the SIFT operator can extract enough effective feature points, and the correct rate is higher than 50%, so the RANSAC can be used directly. Using the homography transformation model and the nearest neighbor interpolation method mapping the blue band, the red band, and the near infrared band to the green band, a four-band multispectral image is obtained. Figures 12 and 13 depict the true-color composite and the CIR composite of the four-band image respectively. Figures 12a and 13a depict the unregistered multispectral image, and Figures 12b and 13b, respectively, display their enlarged partial regions. Figures 12c and 13c depict the registered multispectral images, and Figures 12d and 13d displays their enlarged partial regions. There are severe dislocations between different bands of the unregistered multispectral image. In contrast, these dislocations disappear in the registered multispectral image.

Figure 12. The true-color composite (red, green, blue); (a) the unregistered multispectral image; (b) the enlarged partial region of (a); (c) the registered multispectral image; (d) the enlarged partial region of (c).

The difference among the results of the four methods is in subpixel level, unable to be recognized by the naked eye. In order to have a quantitative measure of it objectively, an inverse transformation is performed using the inverse matrix of H_i, which can transform the coordinates of the input image to the reference coordinates. For any input point (x_i, y_i), its retransformed point (x_r, y_r) in the reference image can be derived from Equation (6) by replacing (x_0, y_0) with (x_r, y_r), shown in Equation (11):

$$\begin{bmatrix} x_r \\ y_r \\ 1 \end{bmatrix} = \begin{bmatrix} h_{11}^i & h_{12}^i & h_{13}^i \\ h_{21}^i & h_{22}^i & h_{23}^i \\ h_{31}^i & h_{32}^i & h_{33}^i \end{bmatrix}^{-1} \begin{bmatrix} x_i \\ y_i \\ 1 \end{bmatrix} \tag{11}$$

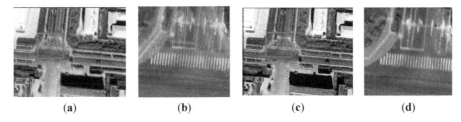

(a) (b) (c) (d)

Figure 13. The CIR composite (infrared, red, green); (**a**) the unregistered multispectral image; (**b**) the enlarged partial region of (**a**); (**c**) the registered multispectral image; (**d**) the enlarged partial region of (**c**).

The residual, XR, is the difference between x_0 and x_r, and YR is the difference between y_0 and y_r. The root mean square error, R, is the distance between the reference point and the retransformed point in the reference image coordinate system. XR, YR and R, for any match are calculated with distance formulas:

$$XR = \sqrt{(x_r - x_0)^2}, \; YR = \sqrt{(y_r - y_0)^2}, \; R = \sqrt{XR^2 + YR^2} \qquad (12)$$

with n, as the amount of the point pairs, j, as the serial number, R_x, the root mean square error in x direction, R_y, the root mean square error in y direction, R_t, the total root mean square error, can be calculated as the following formulas:

$$R_y = \sqrt{\frac{1}{n}\sum_{j=1}^{n} YR_j^2}, \; R_x = \sqrt{\frac{1}{n}\sum_{j=1}^{n} XR_j^2}, \; R_t = \sqrt{\frac{1}{n}\sum_{j=1}^{n} XR_j^2 + \frac{1}{n}\sum_{j=1}^{n} YR_j^2} \qquad (13)$$

These errors indicate how good the registration is between the input band image and the reference band image. The smaller these errors are, the higher the quality of the four-band multispectral data is. Table 2 indicates that although all of them have a high accuracy, the methods using homography model have a higher precision than those using the polynomial model.

Table 2. The errors for registering near infrared, red and green band to the blue band/pixels.

	G-B			G-R			G-IR		
	R_x	R_y	R_t	R_x	R_y	R_t	R_x	R_y	R_t
R1	1.17	1.12	1.62	0.98	0.84	1.3	1.7	1.14	2.05
R2	1.03	1.21	1.59	1	0.82	1.3	1.52	1.17	1.92
R3	1.02	1.14	1.57	0.98	0.8	1.27	1.46	1.09	1.82
R4	1.02	1.14	1.57	0.98	0.8	1.27	1.46	1.09	1.82

R1, R2 respectively represents the two methods using the first-order and second-order polynomial model. R3 represents the methods using the homography model. Compared to R3, R4 uses the false matches rejecting method to remove most of the wrong matches first.

5.2. Experiment 2

Compared with Figure 4, nearly all of the ground objects shown in Figure 6 are the grass except a road. The four methods in the previous section are used to compose one four-band multispectral image. Figure 14 depicts the true-color composite and CIR composite. Figure 14a shows the true-color composite of the unregistered multispectral image, and Figure 14b is its enlarged partial region. An obvious dislocation can be seen in Figure 14a,b. Figure 14c shows the registered true-color composite, and Figure 14d is its enlarged partial region. The dislocation is missing. Because of the lack of artificial objects in these images, there is a great difference between visible SIFT features and infrared SIFT features, and the correct rate of initial matching point pairs is significantly lower than 50%. The first three methods use the RANSAC directly; therefore they cannot get a correct infrared band in the four-band multispectral image, as shown in the first three severely distorted images of Figure 15. However, the fourth method uses the rejecting false matches method mentioned in Section 3 to remove most false matches first for promoting the correct rate, and then uses the RANSAC, so it can get a correct infrared band, as shown in the fourth picture of Figure 15. So, the first three methods cannot get a correct CIR composite, as shown in the first three images of Figure 16, but the fourth can, as shown in the fourth image of Figure 16. Table 3, quantitative evaluation of registration error, shows that the errors of the third method and the fourth method are the same, and a little better than the first and second method at column G-B and G-R. At the G-IR column, the first three methods cannot obtain correct results, but the fourth method can get correct result; and its total error, about 2.4 pixels, is still very low.

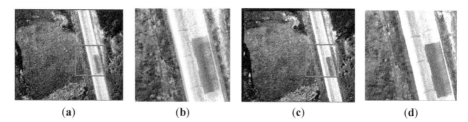

(a)   (b)   (c)   (d)

Figure 14. The true-color composite (red, green, blue); (**a**) the unregistered multispectral image; (**b**) the enlarged partial region of (**a**); (**c**) the registered multispectral image; (**d**) the enlarged partial region of (**c**).

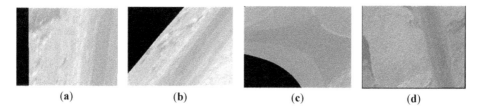

(a)   (b)   (c)   (d)

Figure 15. The single infrared bands of the multispectral images; (**a**–**c**) respectively the infrared bands obtained by using the methods using the first-order polynomial, second-order polynomial and homography mode and using the RANSAC directly; (**d**) the infrared band obtained by using our method.

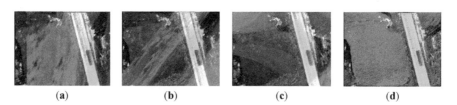

(a)   (b)   (c)   (d)

Figure 16. The CIR composite of the four-band images; (**a**–**c**) respectively the multispectral images obtained by using the methods using the first-order polynomial, second-order polynomial and homography model and using the RANSAC directly; (**d**) the multispectral images obtained by using our method.

Table 3. The errors for registering near infrared, red and green band to the blue band.

	G-B			G-R			G-IR		
	R_x	R_y	R_t	R_x	R_y	R_t	R_x	R_y	R_t
R1	1.03	1.26	1.63	0.83	0.93	1.25	146	309	342
R2	1.02	1.27	1.63	0.82	0.93	1.24	328	377	500
R3	0.92	1.14	1.48	0.78	0.84	1.16	195	249	295
R4	0.92	1.14	1.48	0.78	0.84	1.16	1.3	2.06	2.43

5.3. Experiment 3

Compared with Figures 4 and 6, Figure 8 contains no man-made objects. The four methods are used respectively to compose one four-band image. Figure 17a shows the true-color composite of the unregistered multispectral image, and Figure 17b is its enlarged partial region. The dislocation is obvious. Figure 17c,d show the true-color composite of the registered multispectral image. There is no dislocation between different bands. Because there are no artificial objects in these images, there is a great difference between the visible band SIFT features and infrared band SIFT features, and the correct rate of initial matches is lower than 50%, so the first three methods cannot get a correct infrared band in the four-band image, as shown in the first three severely distorted images of Figure 18. Compared with the other three methods, the fourth method uses the rejecting false matches method mentioned in the Section 3 to remove most false matches firstly, and then uses the RANSAC; therefore it can get a correct infrared band, as shown in the fourth image of Figure 18. So, the first three methods cannot get the CIR composite correctly, as shown in the first three images of Figure 19, but the fourth method can, as shown in the fourth image of Figure 19. Table 4 shows that the errors of the third method and the fourth method are the same at the column G-B and G-R, and a little better than the first and second method. At the G-IR column, the first three methods cannot get correct results, but the fourth method can get correct results with a very low error, about 2.5 pixels.

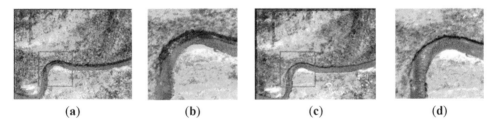

$\qquad$ (a) $\qquad\qquad$ (b) $\qquad\qquad$ (c) $\qquad\qquad$ (d)

Figure 17. The true-color composite (red, green, blue); (**a**) the unregistered multispectral image; (**b**) the enlarged partial region of (**a**); (**c**) the registered multispectral image; (**d**) the enlarged partial region of (**c**).

$\qquad$ (a) $\qquad\qquad$ (b) $\qquad\qquad$ (c) $\qquad\qquad$ (d)

Figure 18. The single infrared band of the multispectral images; (**a–c**) respectively the infrared bands obtained by using the methods using the first-order polynomial, second-order polynomial and homography model and using the RANSAC directly; (**d**) the infrared band obtained by using our method.

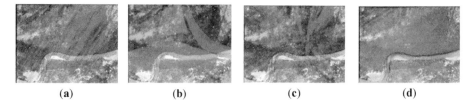

(a)　　　　　　(b)　　　　　　(c)　　　　　　(d)

Figure 19. The CIR composite of the four-band images; (**a–c**) respectively the multispectral images obtained by using the methods using the first-order polynomial, second-order polynomial and homography model and using the RANSAC directly; (**d**) the multispectral images obtained by using our method.

Table 4. The errors for registering near infrared, red and green band to the blue band.

	G-B			G-R			G-IR		
	R_x	R_y	R_t	R_x	R_y	R_t	R_x	R_y	R_t
R1	1.07	1.31	1.69	0.85	0.97	1.29	227	308	383
R2	1.05	1.28	1.66	0.84	0.93	1.25	125	249	279
R3	0.95	1.12	1.47	0.83	0.92	1.24	194	237	306
R4	0.95	1.12	1.47	0.83	0.92	1.24	1.4	2.08	2.5

6. Conclusions

Using four changeable bandpass filters, four identical monochrome cameras, a set of digital video recorders (DVR) and a ruggedized PC, we developed an airborne high resolution multispectral mapping system. The homography registration model was chosen, and we provided a calculation method for its parameters. In the case of fewer or no artificial objects, the conventional SIFT registration approach cannot solve the matching problem between visible images and infrared images. Toward this situation, based on the structural characteristics of the four-camera multispectral system, an effective method to reject most false matches was proposed, and then the RANSAC and the homography model were used to remove the remaining false matches and calculate registration parameters. Experimental results show that for this four-camera multispectral imaging system, the multispectral image creating method can obtain a high quality multispectral image, and band-to-band alignment error is less than 2.5 pixels.

Acknowledgments: This work is supported by program for National Natural Science Foundation of China (NSFC 41071255), the National Key Technologies R&D Program (2012BAH31B01), and the Key Project B of Beijing Natural Science Foundation (KZ201310028035). This work is funded by Specialized Research Fund for the Doctoral Program of Higher Education (SRFDP20131108110005) and the Importation and Development of High-Caliber Talents Project of Beijing Municipal Institutions (CIT&TCD20150323).

Author Contributions: Hanlun Li proposed the multispectral image composition method. Aiwu Zhang and Hanlun Li performed and analyzed the experiments. Shaoxing Hu designed the multispectral system.

Conflicts of Interest: The authors declare no conflict of interest.

References

1. Hartley, R.; Zisserman, A. *Multiple View Geometry in Computer Vision*; Cambridge University Press: Cambridge, UK, 2003.
2. Oppelt, N.; Mauser, W. Airborne Visible/Infrared Imaging Spectrometer Avis: Design, Characterization and Calibration. *Sensors* **2007**, *7*, 1934–1953. [CrossRef]
3. Kise, M.; Park, B.; Heitschmidt, G.W.; Lawrence, K.C.; Windham, W.R. Multispectral imaging system with interchangeable filter design. *Comput. Electron. Agric.* **2010**, *72*, 61–68. [CrossRef]

Sensors **2015**, *15*, 17453–17469

4. Everitt, J.H.; Escobar, D.E.; Cavazos, I.; Noriega, J.R.; Davis, M.R. A three-camera multispectral digital video imaging system. *Remote Sens. Environ.* **1995**, *54*, 333–337. [CrossRef]

5. Gorsevski, P.V.; Gessler, P.E. The design and the development of a hyperspectral and multispectral airborne mapping system. *ISPRS J. Photogramm. Remote Sens.* **2009**, *64*, 184–192. [CrossRef]

6. Honkavaara, E.; Saari, H.; Kaivosoja, J.; Pölönen, I.; Hakala, T.; Litkey, P.; Mäkynen, J.; Pesonen, L. Processing and assessment of spectrometric, stereoscopic imagery collected using a lightweight UAV spectral camera for precision agriculture. *Remote Sens.* **2013**, *5*, 5006–5039. [CrossRef]

7. Yang, C.; Everitt, J.H.; Davis, M.R.; Mao, C. A CCD camera-based hyperspectral imaging system for stationary and airborne applications. *Geocarto Int.* **2003**, *18*, 71–80. [CrossRef]

8. Yang, C. A high-resolution airborne four-camera imaging system for agricultural remote sensing. *Comput. Electron. Agric.* **2012**, *88*, 13–24. [CrossRef]

9. Schwind, P.; Suri, S.; Reinartz, P.; Siebert, A. Applicability of the SIFT operator to geometric SAR image registration. *Int. J. Remote Sens.* **2010**, *31*, 1959–1980. [CrossRef]

10. Ke, Y.; Sukthankar, R. PCA-SIFT: A more distinctive representation for local image descriptors. In Proceedings of the 2004 IEEE Computer Society Conference on Computer Vision and Pattern Recognition, CVPR 2004, Washington, DC, USA, 27 June–2 July 2004; Volume 2, pp. II-506–II-513.

11. He, J.; Li, Y.; Lu, H.; Ren, Z. Research of UAV aerial image mosaic based on SIFT. *Opto-Electron. Eng.* **2011**, *2*, 122–126.

12. Wang, X.; Fu, W. Optimized SIFT image matching algorithm. In Proceedings of the IEEE International Conference on Automation and Logistics (ICAL 2008), Qingdao, China, 1–3 September 2008; pp. 843–847.

13. Hasan, M.; Jia, X.; Robles-Kelly, A.; Zhou, J.; Hasan, M. Multi-spectral remote sensing image registration via spatial relationship analysis on sift keypoints. In Proceedings of the 2010 IEEE International Geoscience and Remote Sensing Symposium (IGARSS), Honolulu, HI, USA, 25–30 July 2010; pp. 1011–1014.

14. Lowe, D.G. Distinctive image features from scale-invariant keypoints. *Int. J. Comput. Vis.* **2004**, *60*, 91–110. [CrossRef]

15. Lowe, D.G. Object recognition from local scale-invariant features. In Proceedings of the Seventh IEEE International Conference on Computer Vision, Kerkyra, Greece, 20–27 September 1999; Volume 2, pp. 1150–1157.

16. Fischler, M.A.; Bolles, R.C. Random Sample Consensus: A Paradigm for Model Fitting with Applications to Image Analysis and Automated Cartography. *Commun. ACM* **1981**, *24*, 381–395. [CrossRef]

17. Hast, A.; Nysjö, J.; Marchetti, A. Optimal RANSAC—Towards a Repeatable Algorithm for Finding the Optimal Set. *J. WSCG* **2013**, *21*, 21–30.

18. Bouguet, J.-Y. Camera Calibration Toolbox for Matlab. Available online: http://www.vision.caltech.edu/bouguetj/calib_doc/ (accessed on 14 July 2015).

19. Qinghai TV. Test Flight of a New High Altitude Airship ASQ-HAA380. Available online: http://news.cntv.cn/2014/08/18/VIDE1408368121196123.shtml?ptag=vsogou (accessed on 14 July 2015).

Review

Vision and Control for UAVs: A Survey of General Methods and of Inexpensive Platforms for Infrastructure Inspection

Koppány Máthé * and Lucian Buşoniu

Automation Department, Technical University of Cluj-Napoca, Memorandumului Street no. 28,
400114 Cluj-Napoca, Romania; E-Mail: lucian@busoniu.net
* E-Mail: koppany.mathe@aut.utcluj.ro; Tel.: +40-264-401-587.

Academic Editor: Felipe Gonzalez Toro
Received: 27 December 2014 / Accepted: 26 May 2015 / Published: 25 June 2015

Abstract: Unmanned aerial vehicles (UAVs) have gained significant attention in recent years. Low-cost platforms using inexpensive sensor payloads have been shown to provide satisfactory flight and navigation capabilities. In this report, we survey vision and control methods that can be applied to low-cost UAVs, and we list some popular inexpensive platforms and application fields where they are useful. We also highlight the sensor suites used where this information is available. We overview, among others, feature detection and tracking, optical flow and visual servoing, low-level stabilization and high-level planning methods. We then list popular low-cost UAVs, selecting mainly quadrotors. We discuss applications, restricting our focus to the field of infrastructure inspection. Finally, as an example, we formulate two use-cases for railway inspection, a less explored application field, and illustrate the usage of the vision and control techniques reviewed by selecting appropriate ones to tackle these use-cases. To select vision methods, we run a thorough set of experimental evaluations.

Keywords: unmanned aerial vehicle; control; planning; camera-based sensing; infrastructure inspection

---

1. Introduction

Unmanned vehicles, including UAVs, offer new perspectives for transportation and services. Although the legal requirements are still quite restrictive [1], UAV applications are becoming widespread, from military usage to civil applications, such as aerial imaging [2] or various inspection tasks [3,4].

We focus here on low-cost (under $1500), small-scale (diameter under 1 m) and lightweight (under 4 kg) UAVs that can be reliably used outdoors. Examples for UAVs that fit these criteria are the Parrot AR.Drone [5], the Arducopter platforms [6] or others presented in Section 4. While the specific limits on size, weight and cost are, of course, arbitrary to an extent, we are also motivated to select them by the railway inspection application we discuss in Section 6: since the UAVs are small they are unlikely to damage a train in the event of an unavoidable collision, and their low cost makes them easily replaceable. More generally, inexpensive UAVs are accessible to civilian users and commercially attractive for companies, making them more likely to become widespread. Among small-scale UAVs, higher-cost platforms, such as the AscTec, Draganflyer and MikroKopter products, offer improved flight stability [7,8] and advanced sensing units, such as laser rangefinders [9,10] or thermal infrared cameras [11,12]. While such platforms are sometimes also referred to as low-cost UAVs [12,13], we consider here significantly less expensive UAVs.

UAVs under $1500 use less expensive hardware, especially for sensing and processing units. They still possess basic navigation units, such as inertial measurement units (IMU) and possibly Global Positioning System (GPS) modules, but the measurement accuracy is usually reduced. Color cameras are used, which are useful only in daytime and do not provide depth and scale information for the captured environment. Nevertheless, the cameras are the richest data sources, so computer vision usually plays a central role in UAV automation. Building on vision, advanced sensing and control methods are used to compensate for the performance and capability limitations.

Therefore, in the first part of the paper, we survey general techniques for vision and control. We describe methods that work in any application, but are specifically motivated by infrastructure inspection, so we point out the connections of vision and control with this area. We begin by presenting at a high level existing vision methodologies and highlight those shown to be successful in UAV applications. The discussion is structured along several classes of techniques: feature detectors and descriptors to identify objects in the image, optical flow for motion detection and visual servoing and mapping techniques. The latter two techniques blur the line between vision and control, e.g., visual servoing tightly integrates visual feedback on the position relative to an object and the control actions taken to maintain the position. We continue by detailing UAV control methods, on two separate levels. For low-level stabilization and path following, we briefly introduce a simplified quadrotor model and methods for attitude and position control. We overview platform-independent high-level planning tasks and methods, focusing on methods that can integrate obstacle avoidance and other constraints.

Already for low-level control, and also for the remainder of our paper, we select quadrotors as our preferred class of UAVs. Among the two main types of UAVs, fixed-wing and rotorcrafts, rotorcrafts have the important capability of hovering. Subtypes are helicopters and multirotors, where multirotors are preferred for the robustness and modularity of the fuselage, being less subject to damage and easier to repair. Furthermore, quadrotors are the most widespread and least costly multirotors. Since we are interested in automating broadly-used UAVs, in this paper, we will focus on quadrotor platforms.

We start the second part of the paper by overviewing several low-cost quadrotor platforms. We then introduce the main focus of this second part, infrastructure inspection, and review UAV applications in this area. Selecting a less explored subfield, namely railway inspection, we develop two use-cases in this field to illustrate semi-autonomous short-range and fully-autonomous long-range inspection. We start with a detailed comparison of the performance of various feature detectors, based on real data from the use-cases. This also serves as a detailed illustration of the practical performance of many of the vision techniques we review. Based on the test results, we select appropriate vision algorithms. In addition, we select control methods based on our literature review, for each scenario, adapted to the facilities of the Parrot AR.Drone quadrotor.

In related work, a number of surveys address specific sensing, vision and control topics for UAVs. For example, the recent survey of Whitehead *et al.* [14] evaluates sensor types used for remote sensing applications. In the context of vision methods, [15] presents extensive datasets and benchmarks for optical flow techniques, and [16] discusses in detail edge detection methods. On the control side, [17] presents techniques for low-level stabilization and control, ranging from simple linear to more accurate nonlinear controllers, while [18] discusses high-level planning. Another survey [19] overviews planning methods from the perspective of uncertainties. The definitive book on robotic planning [20] also addresses low-level dynamics as constraints on planning. We do not intend to duplicate these efforts here. Instead, we provide an overall, high-level overview of both vision and control, focusing on methods relevant to recent low-cost quadrotors and infrastructure inspection applications. Indeed, we refer to these existing works, drawing on their comparisons between available methods and synthesizing their results, so in a sense, our paper is a meta-review of the area. Of course, due to the wide research fields discussed, the overview provided by us is not exhaustive. In particular, we do not include state estimation methods, used for example to infer unmeasurable variables from measured data or to perform sensor fusion. Our major goal is to help the practitioner in the areas of inspection with UAVs understand what methods are available and how they organize in a coherent landscape, so as to select an array of techniques for their specific application; and we provide the relevant references needed to implement the chosen techniques.

The next two sections present our survey on vision (Section 2) and control methods (Section 3). Then, Section 4 lists popular low-cost UAVs; Section 5 discusses common UAV monitoring and inspection applications; Section 6 evaluates vision techniques and selects control methods from those discussed for two illustrative use-cases; and Section 7 concludes our survey. Throughout, we pay special attention to sensor suites and present them whenever this information appears in the cited papers. Specifically, in Section 2.5, we discuss sensors used in vision-based applications. We highlight in Section 4 sensors found in low-cost platforms, and in Section 5.4, we present the sensor suites considered in infrastructure applications.

2. Vision: Camera-Based Sensing and Image Processing

Automated navigation of UAVs inherently requires sensing. Usually, ultrasonic sensors, color, thermal or infrared cameras or laser rangefinders are used to acquire information about the surrounding environment. From these sensor types, low-cost UAVs often possess color cameras. Information is then extracted using computer vision techniques: the acquired images are processed for stabilization, navigation and further information collection.

In UAV navigation, feature detectors and extractors are often used for object detection; optical flow techniques are used to distinguish motion in a scene; visual servoing is employed to translate image frame motion into UAV displacement; whereas 3D reconstruction methods are exploited for navigation and mapping. The literature on computer vision is rich (see surveys [15,21–23]), and instead of reproducing these efforts, here, we briefly introduce the aforementioned classes of vision techniques, highlighting popular methods and providing a list of relevant references for further reading. We also exemplify the use of specific methods in inspection applications. Later on, in Section 6, we will evaluate the performance of existing implementations of several vision methods on real data taken from our railway inspection use-cases.

An important remark regarding most vision methods is that they are well known as being difficult to rank by performance. Each method is suitable for specific types of environments and target objects. Evaluation methodologies, like the one presented by Rockett [24], exist, but the performance of vision methods remains subject to the fine-tuning of their parameters according to the problem at hand. Nevertheless, certain methods are preferred either for their robustness, flexibility in parameter selection or lower computational demands, as highlighted in the sequel.

2.1. Feature Detection and Description Methods

Feature detection and description algorithms are basic tools for object detection and tracking. These methods are used, for example, to extract UAV position and motion information. Methods differ from each other in the preprocessing used (grayscaling, blurring, masking), in the way the features are interpreted and selected, and in the mathematical operations used in the processing steps.

Features determine regions of interest in images, which are classified roughly as edges, corners and blobs. Detection methods are responsible for identifying them, whereas descriptors are used to match a feature in two images (e.g., images from a different perspective or subsequent frames from a video stream). Detectors in combination with descriptors and matching methods form complete tools for motion tracking. They can be also used for object detection given a reference image of an object, and in this context, additional tools, like the model fitting methods, can be considered. In what follows, we first discuss edge, corner and region detectors, then descriptor methods. We present then feature matching, homography-based detection and model fitting methods for object detection.

Edge detection is usually employed to identify lines and planes in images. Some of the classic methods are Canny, Sobel, Laplacian and Scharr edge detectors [25]. Several surveys exist that compare the performances of these and other algorithms [21,26,27]. A survey performed by Oskoei *et al.* [16] highlights the good performance of step edge models used for feature lookup and Gaussian filtering considered for further image processing. A classic example of such a method is the Canny edge detector. However, it is known to produce false edges for noisy images, and therefore, methods like the Haar wavelet transform [26] can be considered when the performance of the former is not satisfactory.

Corner and region detectors are mainly used for finding and tracking objects. The Harris–Stephens [28] and Shi–Tomasi [29] methods are often used. A recent study performed by Tulpan *et al.* [30] compares four corner detectors in their performance of identifying distant objects for sense-and-avoid applications. They compare the Harris–Stephens, smallest uni-value segment assimilating nucleus (SUSAN) [31], features from accelerated segment test (FAST) [32] and Shi–Tomasi methods on real video streams. Their results show that the Shi–Tomasi and Harris–Stephens methods outperform the others when looking at the execution time, while the Shi–Tomasi method has the best results concerning the detection range and the ratio of frames containing the detected target.

Feature descriptors are used for matching features in image pairs, either for detecting motion or for finding objects. Well-known methods are speeded up robust features (SURF) [33] and scale-invariant feature transform (SIFT) [34], whereas from recent years, we mention binary robust independent elementary (BRIEF) [35] and oriented FAST and rotated BRIEF (ORB) [36]. SIFT and SURF are older methods and are superseded by BRIEF and ORB in execution time while keeping comparable accuracy. On the other hand, ORB overcomes the limitations of BRIEF in processing rotations in images.

Given the feature descriptors, features can be matched in pairs of images. This can be achieved by simply comparing the descriptors from the two processed images and marking the closest descriptor pairs. This approach is called brute force feature matching. Other solutions consider search trees for comparing the descriptors. A popular matcher uses approximate nearest-neighbor (ANN) search [37] that offers a more efficient way to match features than the brute force approach. A well-known implementation of ANN is the Fast Library for ANN (FLANN) [38]. Using matched features, methods like the homography transform [39] determine the transformation of the image compared to the reference image, and from there, they infer displacement and rotation. A recent comparison of camera pose estimation algorithms using stereo cameras shows that the homography method provides an acceptable level of detection and is useful for applications with computational constraints [40].

Model fitting methods are often used in object detection. They categorize image features to find inliers and outliers according to a model (a curve or a shape described mathematically). A well-known example is the random sample consensus (RANSAC) method. A comprehensive performance evaluation of the RANSAC family is performed by Choi *et al.* [23], showing an accuracy and robustness improvement for maximum likelihood estimation SAC (MLESAC). RANSAC is used in several UAV applications, e.g., for wall plane detection [41] or for identifying the ground plane [42,43]. Although RANSAC methods aid the identification of objects having various shapes, they require high processing power and, thus, are less preferred in applications where computation is limited. Another classical model fitting method class uses the Hough transform; see surveys [44,45]. Hough transforms are most often used as straight line detectors and, thus, are preferred for linear structure detection. From the numerous variants [45], we point out the progressive probabilistic Hough transform for its faster computational performance and for the available implementation in OpenCV [46].

In UAV inspection applications, feature detectors are useful for detecting targets (buildings, objects) or references that have to be followed (like linear structures or the horizon). For linear structure detection, edge detectors can be combined, e.g., with line extractors. To achieve target detection, feature detectors can be coupled with descriptors, matchers and reference images. When combined with descriptors and matchers, feature detectors can also be used to track moving objects or to keep a reference position relative to a detected object.

2.2. Optical Flow Techniques

Optical flow is a family of techniques that focuses on determining motion from images. More precisely, optical flow can be defined as the apparent motion of feature points or patterns in a 2D image of the 3D environment [47]. Often, optical flow detection is performed, e.g., with an optical mouse sensor, as it works on a similar principle and is a popular, well-tested device [47]. A comprehensive analysis of existing optical flow techniques is performed by Baker *et al.* in [15]. As stated in their concluding remarks, at the time of their publication, the most promising optical flow detection approach was the one presented by Sun *et al.* [48]. In a more recent study by Sun *et al.* [49], they show, among others, the good performance of classic optical flow formulations.

Although providing useful navigation information, optical flow algorithms are usually time consuming. Chao *et al.* [47] remark for instance that usually, optical flow algorithms perform well with image sequences presenting much slower motions compared to UAV flights. These methods can nevertheless still be considered, for example, in near-hovering operation of UAVs.

2.3. Visual Servoing

Building on the techniques discussed before, visual servoing deals with controlling the camera (and, therefore, vehicle) motion using visual data. In the context of UAV navigation, visual servoing methods offer solutions for translating image frame motions into real-world displacement. Examples of applications of visual servoing in inspection are flight around traffic signals, following gas pipeline or scanning building façades.

In its classical formulation, visual servoing builds upon feature detectors. Novel approaches consider other parameters (mainly global histogram parameters) of the images as features and perform visual servoing using this information [50–52]. We concentrate next on feature-based visual servoing, which has two major directions: position-based visual servoing (PBVS) and image-based visual servoing (IBVS). PBVS finds a position estimate of the camera by calculating the pose of a known object in the image and uses this position to correct the motion path [53]. IBVS simply selects and tracks features in the images, using feature descriptors shown in Section 2.1, and corrects the position of the camera so as to keep the desired position of the features in the image frame [54]. Although simpler, IBVS may suffer from drift, and thus, it is often used only in combination with other navigation methods (e.g., GPS/IMU-based control) or as a fallback solution. The pros and cons are thus complementary for the two approaches: PBVS can ensure accurate positioning, but at high computational costs, while IBVS is faster, but may result in following undesired trajectories. Implementation examples and further discussions can be found, e.g., in papers [55,56].

2.4. 3D Reconstruction Methods: Mapping

Beyond finding the pose and the motion of the camera, simultaneous localization and mapping (SLAM) constructs a map of the environment while localizing the vehicle on this map. Localization then allows for autonomous navigation in unknown environments. Of course, SLAM is useful for mapping applications. In the case of camera-based SLAM, depth information, needed for finding distances and scaling factors of objects, can be collected by means of extra sensors or by using reference tags. For low-cost solutions, the latter technique is preferred, which calculates the distance from the size of known objects or tags, captured by the camera.

SLAM requires high processing performance, and its implementation is time consuming. A common implementation of SLAM is parallel tracking and mapping (PTAM). PTAM addresses the processing issue by executing UAV tracking (localization) and mapping in two parallel threads: the first thread tracks the camera motion (localization step), while the second one adds new information to the map (map refinement step) [42]. Despite parallel execution, PTAM remains a computationally-consuming algorithm, especially due to the map that requires more processing as it grows.

Yang *et al.* [42] consider PTAM to find an indoor landing site autonomously, using a single down-looking camera. They propose to obtain a constant computation time of PTAM by avoiding refinement of the entire map and performing it only locally, around the current position of the quadrotor. Similarly, Schauwecker *et al.* [43] consider refining maps only locally and clearing old frames in order to boost PTAM. They find that using down-looking cameras, the algorithm can have problems in tracking the quadrotor's motion in the case of yaw rotation (around the z axis). Using stereo cameras, they can correct for these errors in the case of slow rotations.

Although research in the field of computer vision is extensive, there remain several open challenges. Most of the vision algorithms are subject to fine-tuning and are limited in use to certain conditions, specifically due to illumination conditions, pattern types and the motion of captured objects. Another issue is the increased processing time of these methods, which needs to be further reduced for online applications on devices with limited processing power. Additionally, the lack of scaling and distance information in the case of 2D images leads to the need for further tools (sensors or techniques) for acquiring this information. Though the existing solutions are promising, further development is needed for having more robust techniques with wider applicability.

2.5. Sensors in Image Processing Applications

From the above listed works, several provide details on sensor platforms. We have found, e.g., that Tulpan *et al.* [30] use a Prosilica GC2450 5 MP monochrome camera for image processing, operating at 14 fps. Flores *et al.* [41] consider Kinect, a commercially available camera system, consisting of a low-cost RGB-D and infrared camera, providing 640 × 480 pixel RGB images and 320 × 240 pixel depth images, both at 30 fps. Yang *et al.* [42] work with a Firefly MV monochrome camera that provides 640 × 480 pixel images at 60 fps and a 90-degree viewing angle. Schauwecker *et al.* [43] consider a dual stereo-camera solution, mounting two pairs of 640 × 480 grayscale cameras on their UAV, one pair facing downwards, recording at 15 fps, and another facing ahead, recording at 30 fps.

One may conclude that, in most cases, images not larger than 0.3 MP are used for image processing, at frame rates up to 30 fps. This resolution is far lower than those offered by the currently available cameras, though it is preferred for keeping computation low (higher resolution images would require more processing time) and also for requiring only low-cost devices. Most research results, including the above ones, show that this resolution is good enough for proper detection. Furthermore, in terms of control, the acquisition frame rate of 30 fps is high enough, e.g., for controlling small-scale UAVs, and exceeds, in many cases, the processing rate that vision methods can offer.

3. Flight Control and Planning

Relying on information provided by sensing, effective control can help in overcoming the limitations of inexpensive sensors. Indeed, several works report good results using low-cost platforms with advanced controllers [57–59]. In general, the task of an unmanned aircraft is to safely navigate on a desired path and/or visit points of interest in order to perform certain missions. Control tasks behind these terms were grouped by Amelink [60] in the following levels of abstraction, from top to bottom: mission, navigation, aviation, flight control and flight. In this manner, he clearly classifies control problems and offers a modular approach to UAV control.

Instead of such a detailed decomposition of UAV control problems, we prefer to discuss control tasks at two levels: low-level flight control and high-level flight planning. With this grouping, the lower level covers control tasks that are often already implemented in UAVs: hovering ability and disturbance rejection, achieved mainly by attitude control, and trajectory following, the result of position control. By an abuse of terminology, we will refer to all of these tasks as UAV control in this section. On the other hand, high-level flight planning is then made responsible for mission and path planning, including obstacle avoidance. We will call these problems together UAV planning.

Many of the methods presented, especially the higher level planning techniques, are platform independent. Low-level control will be discussed from the perspective of quadrotors, the platform type we will use in our illustrative use-cases and also in our future work. Therefore, first, we introduce the dynamic model of a quadrotor, followed by detailing low-level control and high-level planning methods.

3.1. Quadrotor Dynamics

A basic scheme of a quadrotor is shown in Figure 1, where $\mathcal{E}$ denotes the Earth frame, also called the inertial frame; $\mathcal{B}$ denotes the body frame, attached to the center of mass of the quadrotor; x, y and z mark the coordinates of the center of mass of the quadrotor, in the Earth frame $\mathcal{E}$; ϕ, θ and ψ correspond to the conventional roll, pitch and yaw angles; and ω_i marks the angular velocity of each rotor separately.

In the "plus" configuration, where the vehicle axes correspond to the x and y axes of frame $\mathcal{B}$, displacement along the x axis can be obtained by pitch rotation, which results from keeping $\omega_1 = \omega_3$ and setting $\omega_2 \neq \omega_4$. Similarly, flight on the y axis results from yaw rotation, *i.e.*, $\omega_2 = \omega_4$ and $\omega_1 \neq \omega_3$. Hovering, lift and landing can be achieved by having velocities of the same magnitude on all propellers, whereas rotation around the z axis is the result of $\omega_2 = \omega_4 \neq \omega_1 = \omega_3$. Then, control commands can be defined as:

$$
\begin{aligned}
U_{coll} &= b\left(\omega_1^2 + \omega_2^2 + \omega_3^2 + \omega_4^2\right) \\
U_\phi &= b\left(\omega_1^2 - \omega_3^2\right) \\
U_\theta &= b\left(\omega_4^2 - \omega_2^2\right) \\
U_\psi &= d\left(\omega_2^2 + \omega_4^2 - \omega_1^2 - \omega_3^2\right)
\end{aligned}
\tag{1}
$$

where U_{coll} denotes the collective input (responsible for vertical displacement), U_ϕ the roll (y axis movement), U_θ the pitch (x axis displacement) and U_ψ the yaw forces, b is the thrust coefficient and d is the drag coefficient. With these four inputs, the quadrotor can be operated simply in non-acrobatic flight mode (non-acrobatic flight maneuvers mean that the quadrotor's velocity is changed slowly and the vehicle is used to fly most of the time parallel to the Earth, up to some tilt being necessary for horizontal displacement). Note that all of these motions are with respect to frame $\mathcal{B}$, which then have to be transformed into Earth frame $\mathcal{E}$.

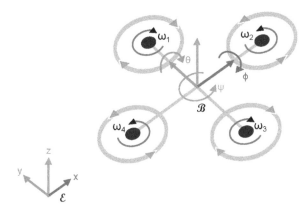

Figure 1. Quadrotor model.

Now, the reaction to the control inputs, *i.e.*, the flight dynamics of the quadrotor, can be modeled using the Euler–Lagrange approach. A simplified form of the dynamics in near-hovering mode can be written as [8]:

$$\ddot{x} = \theta g \quad \ddot{y} = -\phi g \quad \ddot{z} = \frac{\Delta U_{coll}}{m}$$
$$\ddot{\phi} = \frac{1}{I_x} U_\phi \quad \ddot{\theta} = \frac{1}{I_y} U_\theta \quad \ddot{\psi} = \frac{1}{I_z} U_\psi \tag{2}$$

Often, more general forms of this model are considered [61,62]. However, even those works build on the assumptions of near-hovering operation mode, low flight speeds and that the quadrotor can be modeled as a rigid body and has a symmetric structure. As these are realistic considerations for most low-cost small-sized quadrotors, controllers based on the principle of Equation (2) can be easily transferred from one platform to another, where only some parameters have to be adjusted to match the new vehicle.

3.2. Low-Level Flight Control

The main low-level control tasks are: achieving flight, stabilizing the UAV and following a flight path. These tasks are addressed by attitude and position control, which, in the case of quadrotors, are commonly coupled as a nested control loop, shown in Figure 2. Attitude control is then responsible for flight stabilization and tracking the desired heading, while position control serves for trajectory following.

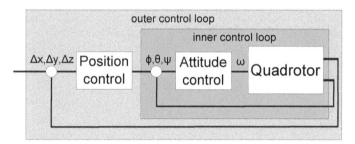

Figure 2. Nested low-level control loop: attitude and position control.

Attitude control is often addressed by using proportional integral derivative (PID) controllers. Often, the PID controllers are set up by experimental tuning. They have the advantage of requiring no complex model of the system dynamics. Based on the results from papers [57,63], this control method, although simple, provides good results for the attitude control. However, the attitude controller is usually enhanced with robust features for obtaining improved stability [57,64].

Concerning position control, Raffo *et al.* [64], for example, propose to use an error-model based model-predictive control that simulates a virtual quadrotor following a desired trajectory. The role of their controller is to minimize the position error between a real and a virtual quadrotor. They compare this solution in a simulation to a backstepping approach. Both methods show robust tracking performance, though the former solution leads to smaller errors and smoother control.

Attitude and position control are often discussed together. Cui *et al.* [63] perform, for instance, trajectory tracking, using PID controllers for position control and several controller types for setting the attitude of the simulated vehicle. In simulations, they show that the PID controller provides the smallest tracking error and lowest settling time for attitude control. Salazar *et al.* [62] combine PID with robust sliding mode control (SMC) for attitude and position control in performing trajectory tracking. They conclude that, despite the good control performance of SMC, chattering of the control input may lead to quick wearing of the actuators, and thus, the use of SMC might be a less preferred solution.

Despite the good performance of current low-level stabilization systems and controllers, several further challenges remain to be addressed. Among these are the various types of uncertainties appearing in outdoor operation, the integration of saturation limits in the control schemes and the underactuated nature of the systems. In recent works, such challenges were addressed by more advanced control methods, such as adaptive controllers [8] or model-predictive control [65].

3.3. High-Level Flight Planning

The higher level problems of UAV automation relate mainly to defining and planning missions, as well as to planning flight paths that fulfil these missions. The goal is to make UAVs fly autonomously based on a mission plan and to make the flight paths feasible and optimal.

Path planning provides flight paths for the lower-level trajectory tracker. It results in optimization problems where certain costs (e.g., energy consumption, execution time) have to be minimized to find an optimal path. To achieve online execution, often, receding horizon techniques are considered. In the sequel, we detail the principle of receding horizon planning and present works that use it for UAV path planning. Afterwards, we discuss planning constraints that ensure flight path feasibility and methods to address them.

Exhaustive overviews of path planning methods can be found in [18–20,66]; see also the references therein. Here, we provide some reference works in the field of UAV path planning.

The idea of receding horizon planning is to reevaluate solutions while closing the control loop. In control, this approach is called model predictive control (MPC). At each call, MPC evaluates possible control sequences for a period of time called a horizon. Each sequence has an associated cost. After the simulation stops, MPC follows the control sequence with the best cost for a time period called the execution horizon. The algorithm is then repeated from the new state. MPC can cover nonlinear dynamics problems, too, such as UAV path planning and path following under uncertainties. However, especially in that case, MPC becomes time consuming, mainly due to "repeated" calculation of the cost values that usually requires simulating the model of the system.

For reducing computation, a common planning method used for UAV planning is the rapidly-exploring random tree (RRT) algorithm. RRT propagates random samples in the search space of control sequences, rapidly covering in this manner the possible solutions. Lin *et al.* [66] use RRT in a receding horizon fashion and couple it with Dubins curves in order to find feasible paths online, paths that avoid moving obstacles. Dubins curves are curves that connect two points while respecting constraints on the curvature. Although the RRT method cannot ensure global optimality [66], shows the practical success of the method both in simulations and in real flights.

Bellingham *et al.* [67] use MPC for path planning among obstacles. They work with linearized models and simplify the cost value calculation by using a cost estimator in order to reduce the execution time. They manage to obtain near-optimal solutions in around a minute [67]. More recent works report on using MPC for more complex situations, such as path planning for multi-UAV formation flights among obstacles [68] and with communication constraints [69]. However, the computational demand of these solutions is not transparent and is likely to exceed the processing performance of low-cost platforms.

An important challenge comes from the feasibility constraints considered in the planning. These integrate kinematic constraints (coming, e.g., from obstacles) and dynamic limitations (mainly due to the velocity and acceleration limits of the vehicle) [70]. Feasibility constraints can be treated in several ways. Commonly, they can be implemented as equality or inequality constraints in the optimizer [67,68,71] or can be simulated as curves [66,72,73]. For example, mixed-integer linear programming (MILP) is an optimizer that can integrate both linear and binary constraints. Bellingham *et al.* [67] use, for instance, MILP for solving path planning, while considering both the continuous constraints resulting from the dynamics of the vehicle and binary constraints coming from obstacle avoidance rules. In their RRT approach discussed before, Lin *et al.* [66] use Dubins curves for their planning method in order to cover the dynamic constraints of the vehicle.

Alternatively, kinematic constraints (e.g., obstacles) in particular can be addressed by means of control methods from computer vision; see Section 2. Navigation methods, such as visual servoing or control based on optical-flow, can be considered. Furthermore, to reduce the problem complexity, obstacle avoidance can be addressed with position control, where the avoidance maneuver represents a temporary deviation from the planned trajectory [74]. Similarly, dynamic constraints can be applied to adjust the flight path after it has been planned. Methods, such as the artificial potential field, can be used to smoothen the trajectory [75]. Such solutions are often less time consuming and may be preferred when computation power is limited. However, they do not offer performance guarantees and might be applicable only in certain scenarios, where waypoints are not mandatory to be reached, but deviations from the planned trajectory that are small in a certain sense are acceptable.

On top of flight planning sits mission planning, which covers the descriptive part of UAV automation. Mission planning includes all formal and informal requirements and specifications regarding usage of the vehicle. Furthermore, it is responsible for translating the missions into a well-defined sequence of subtasks that can be interpreted by path planners. Such subtasks are: take-off, fly to a coordinate, track an object, *etc.* Mission planning can be formalized, and the translation of missions into subtasks can be then performed automatically, based on a set of subtasks and rules. Looking at the literature, only sparse work has been done in this direction, mainly related to airspace integration efforts [1,76,77].

Several open challenges exist in UAV high-level planning. The computational demand of the advanced path planning methods is often high compared to the processing capabilities of low-cost UAVs, meaning that adaptations of these techniques are required in order to implement them onboard and online. Computational limitations also require approximate models, linearization or discretization of the system. Regarding the automation tasks, obstacle avoidance remains the most demanding problem, mainly due to the big variety of avoidance cases, which are hard to address all at once. On the other hand, mission planning challenges mainly relate to airspace integration issues, for which it is important to have a proper formalization of tasks in order to clearly define the expected and possible flight phases and events.

3.4. Flight Control and Planning Methods for Inspection Applications

Low-level stabilization is a basic flight requirement, and thus, attitude and position control are mandatory. However, the required flight planning techniques will vary. If one can assume obstacle-free flight when keeping a reference distance to the inspected target, visual servoing or preplanned trajectories can be considered. For online obstacle avoidance or for power consumption optimization online replanning, methods like the RRT should be considered. MPC-based techniques are able to explicitly deal with more complex constraints on the inspection (e.g., covering both flight time and obstacle avoidance) and are thus preferred in environments with more obstacles (e.g., urban roads, train stations).

4. Low-Cost Quadrotor Platforms

Having overviewed vision and control techniques, we move on to available low-cost platforms and UAV applications. The market of low-cost UAVs has exploded in the past two years. New brands and models of UAVs continuously appear, mainly quadrotors and helicopters, and several communities [78,79] and websites [80–82] focus on the evolution of these products. Instead of looking at the newest platforms resulting from startups and other research projects, we focus on popular brands. Working with a widespread platform usually ensures that the hardware is well tested and that the platform has long-term support, compared to more custom solutions. We therefore list several representative UAV platforms, available until April 2015, that also satisfy our criteria defined in Section 1.

We focus mainly on ready-to-fly (RTF) vehicles, having the advantage of less time needed for setup and calibration. However, RTF vehicles are usually limited in programmability and, therefore, are less customizable. This aspect does not limit the manually-teleoperated inspection applications, but is relevant for automated flights. Thus, we discuss with each presented platform the level of customization, as well. We also compare these platforms with a category of modular solutions that overcome these limitations. For each platform, we highlight the sensors used. In general, all of the quadrotors listed have onboard stabilization, some type of camera (embedded or as a separate device) and wireless connection for data transmission, and most of them include a GPS module, as well. Details are presented below.

Parrot released its popular AR.Drone 2.0 in 2012 citeweb:parrot. Edition 2.0 costs $400 and weighs 420 g. It has two onboard, built-in cameras, a bottom quarter video graphics array (QVGA) camera (320 × 240 px resolution) with 60 fps for ground speed measurement and an HD front camera (1280 × 720 px resolution) with 30 fps for video recording. The flight stabilization system consists of a set of three-axis gyroscope, accelerometer and magnetometer, enhanced with a pressure sensor that, together with the previous, makes the quadrotor able to withstand winds of up to 24 km/h [83]. A pair of ultrasonic sensors help in altitude keeping in close-to-terrain flights (up to 6 m). The onboard flight controller runs in Linux and is closed-source; however, using the Parrot SDK or other APIs, one may remotely access functionalities, such as: receive online camera image frames, navigation and other sensor data and issue high-level velocity commands on all axes, as well as take-off and land commands. These operations are limited to the range of the WiFi signal used for communication with the quadrotor. Various projects target extending the control range or allow for onboard programmability, though not as part of the official product. An official extension by Parrot is a GPS module that, for an additional $120, among others, allows for flight path scheduling by specifying waypoints, even outside the WiFi range. However, the quadrotor is not controllable outside the WiFi range, which limits it to short-range usage. Besides this popular model, Parrot released the Bebop drone at the end of 2014. For $500, among others, Bebop provides improved processing image capturing capabilities (14 Mpx image and 30 fps video recording at 1920 × 1080 px), has a built-in geo-location system (GNSS, including GPS and GLONASS) and an operation range of 250 m [84]. The additional Skycontroller, costing around $400, allows for extending the operation range to 2 km.

Similar products are the Xaircraft X650 Pro [85], having a SuperX onboard flight controller [86], and the Mikrokopter Quadrokopter XL [87], using the FlightCtrl ME flight controller [88]. The price of these UAVs is around $1000, and they weigh around 1000–1500 g. Based on the flight controller specifications, both controllers use for stabilization a set of sensors like those presented with the AR.Drone: pressure sensors, three-axis gyroscopes, magnetometers and accelerometers. A special feature of the controller of the Mikrokopter UAV is that its altitude sensing works up to 5000 m. Both UAVs are meant for mounting external cameras, where the X650 Pro flight controller has built-in functionalities for the camera gimbal stabilization and control. Theoretically, any recording device can be attached to these UAVs, up to the payload limit (around 1 kg for both UAVs).

DJI produces quadrotors for aerial imaging [2]. For the price of around $1000 and weights of about 1200 g, these vehicles are known to be stable. However, the flight controllers used with them allow only for path scheduling based on GPS waypoints, using a graphical interface. The publicly available specifications provide less technical information about the flight controllers [89]. Some of their proprietary flight controllers support the attachment of a GPS module. The newer platforms have proprietary cameras attached to the UAV through a two-axis gimbal, providing HD 1080 p recording at 30 fps, and taking 14 MP (4384 × 3288 px) photos. However, certain platforms support the use of other cameras than the proprietary ones. Furthermore, DJI specifies an extended range of operation for the newer products, up to 2 km.

In contrast with the above-listed RTF vehicles, which are readily assembled UAVs with the possibility of limited customization, 3D Robotics and the Arducopter come with modular solutions, based on the Arduino platform [90]. These UAVs use Ardupilot Mega (APM) or Pixhawk flight controllers that are known to be custom-programmable. Furthermore, these controllers support the attachment of various external devices, such as sensors, controllers and communication units. The

Arducopter UAVs are usually custom-built, but there exist complete kits, as well, offering RTF or almost-RTF solutions. An RTF quadrotor is the 3DR Iris+, which costs $750 and weighs 1300 g [91], whereas an almost-RTF solution is the Arducopter Quad [6], costing $620 and having a similar weight. Compared to the platforms from the other producers, the Arducopter UAVs are highly customizable, though requiring more knowledge of UAV programming and operation. A more recent platform is the 3DR Solo, a user-friendly RTF UAV [92] with enhanced capabilities (among others, increased flight time up to 20 min with a mounted GoPro camera and an improved onboard controller).

Regarding sensors in Arducopters, these platforms come with the ability of customization. The newer Pixhawk controller comes with built-in gyroscopes, accelerometers and pressure sensors [93]. On the vision part, the Iris+ has, for instance, the possibility of mounting an external camera using a gimbal system. Pixhawk offers an optical flow module called PX4FLOW [94], which can be used for navigation purposes. It is not meant for video recording, though, due to its reduced performance (it has a resolution of 752×480 px).

The presented UAVs are meant mainly for aerial imaging and gaming applications. Despite the different price ranges, the types of sensors used for stabilization are similar. Obviously, the more costly solutions offer better stabilization. On the vision side, platforms supporting the mount of external cameras offer improved recording experience. However, working with integrated vision units eases the usage of the platform, as in the case of the AR.Drone or with the PX4FLOW module. The quality and frame rate offered, for example, by the AR.Drone 2.0 are already good enough for use for vision-based navigation and environment capturing.

5. UAVs for Infrastructure Inspection Applications

A growing interest is shown for using UAVs for inspection and other remote sensing applications, starting from public area monitoring, to infrastructure inspection, intelligent farming or aerial mapping. Reaching satisfactory performance with low-cost UAVs can offer new perspectives for industrial applications and public services.

As we discuss in the sequel, several projects already focus on inspection use-cases, although they use more costly UAVs. We present specific applications by grouping them into two common classes, namely power line inspection and building monitoring. We also dedicate a subsection to railway inspection applications, a less explored field. Furthermore, we highlight the level of autonomy of the UAVs used in the works discussed. Finally, we list and discuss briefly the sensors used in the presented applications.

5.1. Power Line Inspection

Power line inspection applications roughly cover the tasks of following the lines and stopping at interesting points, to scan in detail certain parts of the infrastructure. This procedure is similar to the case of inspecting pipelines or any other linear structures, such as roads, walls, coasts, *etc.* Although projects can be found that target gas pipeline monitoring [95] or structure inspection in general [96], the topic of power line inspection appears to be more popular in recent years.

In their survey in 2010, Montambault *et al.* [97] already point out several research projects all around the world that use quadrotors for power line inspection. As stated in the dissertation of Ellis, in August 2013, in Australia alone, there were already 16 companies listed as licensed UAV operators for power line inspection [98]. We do not have details on the level of automation in these applications. However, the following projects clearly focus on automating power line inspection.

Li *et al.* [4] already perform fully-automated power line inspection. They use a helicopter of 31 kg, which also carries enough fuel for up to one hour of flight. Furthermore, their helicopter is equipped with a more advanced sensor suite, which eases flight automation. Although promising, such platforms do not fit into the inexpensive category.

Several projects strongly focus on the image processing part of inspection, from various perspectives. Zhang *et al.* [99], for instance, compare some line detection algorithms for identifying and tracking power lines from videos captured by quadrotors. Luque-Vega *et al.* Luque-Vega:14 combine a color camera with a thermal infrared camera to inspect infrastructure components. Larrauri *et al.* citeLarrauri:13 deal with calculating distance to vegetation, trees and buildings, based on video frames. Martinez *et al.* citeMartinez:14 perform power line tower inspection. Their tracking approach steers the camera so as to keep an inspected tower in focus while flying along lines.

5.2. Building Monitoring

Another popular application field is the inspection of building façades and other surfaces, with the aim of examining their integrity. In such use-cases, the primary goal is to design flight plans that allow for proper data acquisition.

Baiocchi *et al.* [3] use quadrotors in a post-seismic environment for inspecting historic buildings for cracks and other damages. The GPS-based path planner developed by them optimizes flight paths, reducing redundant acquisition. Furthermore, their processing software allows for the 3D reconstruction of building façades from pairs of images. Another project, led by Eschmann *et al.* [103], presents a similar application, though with manual flight control.

Nikolic *et al.* [104] deal with power plant boiler inspection. They design an automated trajectory-following system and a personalized sensor suite for visual navigation. Using these, they examine the interior walls of boilers, expanding the GPS navigation functionality of the quadrotor with visual navigation, in order to be able to operate in GPS-denied regions, as well.

5.3. Railway Infrastructure Inspection

Railway inspection comprises the tasks of structure inspection and linear structure following. In this context, railway inspection is related to the previous two fields. It is an application area not yet considered in public research projects, to the best of our knowledge. As presented below, newsletters report on several companies that intend to or already use UAVs in railways, based mainly on manual teleoperation or automated waypoint navigation. However, we have limited technical information on the work performed by these groups.

In the spring of 2013, news appeared about German Railways (DB – Deutsche Bahn) regarding their intention of using UAVs for catching graffiti sprayers [105]. French National Railways (SNCF – Société nationale des chemins de fer français) announced in the autumn of 2013 a project of railway bridge and viaduct inspection using UAVs [106]. The international company Elimco is also offering inspection of various infrastructures, including railways [107]. Similar applications can be found at Microdrones, where automated waypoint navigation is already implemented [108]. An article from Smartrail World reports on further companies that use or plan to work with UAVs: NetworkRail from the UK, ProRail from The Netherlands, Union Pacific from the USA and the Jerusalem light rail network [109]. Although few technical details are publicly available about these projects, they mainly seem to be based on manual teleoperation.

5.4. Sensors in Infrastructure Inspection Applications

Most papers discussed above provide generic or little information about the sensors used. In general, some kind of color cameras are used to perform data acquisition. From the works where more details are provided, Hausamann *et al.* [95] use a combination of optical and infrared sensors, discussing sensor types with different spectral bands; and synthetic aperture radars that have higher availability than color cameras due to the fact that radars are independent of weather and light conditions. Luque-Vega *et al.* [100] combine thermal-infrared cameras with color cameras for improving background subtraction and, thus, object detection. Larrauri *et al.* [101] use an HD camera in their project, though for processing data at only 4 fps. Eschmann *et al.* [103] perform navigation based on GPS data and use an external 12-MP commercial camera for offline post-processing only. Nikolic *et al.* [104] consider a CMOS image sensor and low-cost inertial sensors on a custom-built integrated circuit that aids UAV navigation and data collection in GPS-denied environments.

From these various details, one can conclude that color cameras are usually considered for data acquisition. Often, the performance of these cameras is reduced, but still good enough, while aiming at keeping the device low cost or at respecting payload constraints. Among the applications discussed, in certain cases, infrared cameras are additionally used to improve detection.

6. Illustration in Railway Inspection Use-Cases

Finally, we illustrate in this section the use of the camera-based sensing and control methods presented in Sections 2 and 3 for automated railway inspection. We formulate two use-cases: one where the UAV performs infrastructure inspection in close, but difficult-to-access areas (such as long bridges or tracks separated from the road by, e.g., a river). A second use-case is meant for railway track following for the sake of recording the infrastructure, such as tracks, sleepers, points or cabling. In the first use-case, the automated control runs on a remote station, while in the second one, full control is onboard. The chosen UAV platform is the Parrot AR.Drone 2.0 quadrotor, presented in Section 4. In certain cases, the implementation ideas come from the specifics of this particular platform, but they are usually valid for other low-cost quadrotors, as well.

We continue with evaluating the vision methods presented in Section 2, namely edge detectors for rail track detection in Section 6.1 and feature detectors for target (signal light) detection in Section 6.2. Then, we break down the use-cases into subtasks. We select detection solutions based on the results from our experiments detailed in Sections 6.1 and 6.2. Furthermore, we indicate other vision and control methods needed for our setup, based on the discussions from Sections 2 and 3. In this manner, our method selection from Section 6.3 considers, to an extent, our own experimental results and further solutions from the conclusions of our survey.

6.1. Evaluating Feature Detectors for Target Detection

In the first use-case, the quadrotor must find the target object before inspecting it. Considering that the target is close enough to the quadrotor (below 10 m) and is in its field of view, one can apply feature detection, description and matching methods that, given a reference image of the target, can identify it in a new image, called the scene image.

Numerous detection, description and matching methods are readily implemented. We consider the OpenCV 2.4.9 library [110] that, from the methods discussed in Section 2.1, comes with implementations for the SURF, SIFT, FAST and Shi–Tomasi feature detectors and the SURF, SIFT, BRIEF and ORB descriptors. Each of these methods has a number of tuning parameters, as presented on the OpenCV website [110]. Based on prior tuning tests, we select a grid of meaningful values for these parameters (values marked with bold are the best according to the optimization criterion we describe below).

- For SURF, the Hessian threshold is taken in $\{300, 310, ..., \mathbf{420}\}$, the number of octaves in $\{2, 3, ..., 8\}$ and the number of octave layers in $\{1, 2, 3, \mathbf{4}, 5, 6\}$, and we will use up-right features and extended descriptors.
- For SIFT, we allow all features to be retained, and we set the number of octaves in $\{1, \mathbf{2}, 3\}$, the contrast threshold in $\{0.01, \mathbf{0.02}, ..., 0.16\}$, the edge threshold in $\{5, \mathbf{7}, ..., 15\}$ and the standard deviation of the Gaussian in $\{1.2, 1.3, \mathbf{1.4}, ..., 2.1\}$.
- For FAST, we use threshold values between $\{0, 1, ..., \mathbf{31}, ..., 80\}$ and test both **enabling** and disabling non-max suppression.
- For Shi–Tomasi, we allow all features to be retained, set a quality level from $\{0.001, 0.006, ..., \mathbf{0.021}, ..., 0.041\}$, a minimum feature distance of $\{0, \mathbf{1}, ..., 5\}$, an evaluation block size of $\{3, 4, 5, \mathbf{6}\}$, allow for enabling or **disabling** the Harris detector and set parameter k to $\{0.02, \mathbf{0.04}, ..., 0.1\}$.

With each setting, we combine the detectors with all of the descriptors and a FLANN-based matcher [38] in order to evaluate the performance of the methods. The descriptors and the matcher require no parameter tuning.

As the data source, we consider a 46×188 px reference image and 50 scene images, all 640×360 px in size and rectified. The reference image was taken with the same camera as the scene images. Each scene image contains the target object at different distances. Figure 3 shows a sample matching result.

Figure 3. Target detection. Matching result with the reference and scene image.

After testing the various combinations of detector and descriptor methods, we obtain meaningful results only when using the SIFT descriptor. The other descriptors, in general, fail to detect the target object. Then, testing all of the detectors on the grid of parameters, we select the best parameter set for each detection algorithm that maximizes the detection rates during the simulations. Table 1 summarizes the performance of the four detectors in the case of these parameter sets.

Table 1. Performance evaluation of the feature detectors for target detection.

Method	Detect. Rate (%)	Execution Time (ms)	Position Error (px)	Scaling Error (%)
SURF	8	71.5	41.7	5
SIFT	54	17.4	2.7	45
FAST	98	8.3	2.2	37
Shi–Tomasi	96	7.4	2.2	41

As shown in Table 1, we evaluate the detection success rate, the average execution time per frame and average errors on the horizontal position and the scaling of the detected object. The detection rate tells in how many frames the algorithm detected the reference object. The average execution time considers the total time required for detection, description and matching, for a single frame, *i.e.*, the total image processing time. The average position error tells the horizontal distance in pixels between the real center of the object in the scene image and the center of the detected area. Finally, the scaling error indicates the average difference between the size of the reference image and the size of the detected object, expressed in percentages, *i.e.*, it is a size detection error. Note that the latter two parameters indicate the accuracy of the detection, while the first two inform about the usefulness of the algorithm in the case of online use.

According to the results from Table 1, the FAST and Shi–Tomasi methods outperform the other two, both in detection rate and execution time. The detection rate of SIFT is also acceptable. Looking at the average position error, all of the methods perform well, except SURF. However, the scaling error indicates the opposite. SURF, although it has a far lower detection rate, detects the object size more precisely. Furthermore, higher scaling errors in the case of the other methods appear since the object detection algorithm was implicitly allowed to have these errors in favor of higher detection rates by the optimization procedure. Additionally, the small position errors indicate that, with the last three methods, the target was identified in the correct place.

Based on these results, we select the FAST algorithm for target detection in the given use-case, in combination with the SIFT descriptor. However, we remark that the parameter sets were fine-tuned for the given dataset, and other combinations might turn out to have better results for different scenarios where other reference/target objects or other image parameters are considered. Nevertheless, we highlight the good performance of the FAST and Shi–Tomasi detectors, which, with an average execution time below 10 ms, allow for at least 100-Hz control, good enough for target detection during flight with a quadrotor. With respect to the discussion in Section 2.1, the selection of the Shi–Tomasi detector confirms the conclusions from there. However, the results of the FAST detector are even better in our particular setup, while the Harris detector seems to reduce the performance when applied in the Shi–Tomasi algorithm.

6.2. Evaluating Edge Detectors for Track Detection

In the second use-case, the quadrotor has to follow track lines. This can be accomplished by vanishing point-based control, as introduced briefly below in Section 6.3.1 and presented in [111]. To achieve this aim, edge detector methods can be used in combination with line extractors that help with finding the track lines, which are then used to find their vanishing point.

From the edge detectors mentioned in Section 2.1, OpenCV comes with implementations for Canny, Sobel, Laplacian and Scharr algorithms. We will test these methods on a grid of meaningful parameter sets and combine them with a probabilistic Hough transform (PHT) for line extraction and a custom filtering method for line selection. This method removes all of the lines up to a vertical angle, after which it progressively filters out the lines that do not point to an average vanishing point. An example of the processed image and obtained lines and vanishing point is shown in Figure 4.

(a)   (b)

Figure 4. (**a**) Track line detection; (**b**) Canny edge detection result and vanishing point detection.

The parameters of the detectors are taken according to the following grid (best values marked with bold or stated separately).

- For all of the methods, we worked with default image color depth and edge modeling border type and set the x and y derivative orders in $\{0, 1, 2\}$, kernel sizes in $\{1, 3, 5, 7\}$, derivative scaling factors in $\{0.001, 0.002, ..., 0.01, 0.015, ..., 0.14\}$ and a delta value, added to the pixels of the output image during the convolution, in $\{0, 0.01, ..., 0.40\}$. Note that some parameters appear with only some of the methods.
- In the case of Canny, the two hysteresis thresholds are taken in the interval $\{0, 10, ..., 90, 91, ..., \mathbf{106}, ..., 110, 120, ..., 300\}$; the best kernel size was 3, and we allowed for **enabling** or disabling the use of the L_2 gradient in image gradient magnitude calculation.
- For Sobel, the best values of the parameters are: derivative orders $x = 1$ and $y = 0$, kernel size three, scaling factor 0.0125 and delta 0.037.
- For Laplacian, the best values are: kernel size five, scaling factor 0.002 and delta 0.095.
- For Scharr, the best values are: derivative orders $x = 1$ and $y = 0$, scaling factor 0.001 and delta 0.23.

From the common parameters, the kernel size is the size of the matrix used in calculating the transformed image. In our experiments, we observe that keeping this value low provides better results. The scaling factor determines the image dimming, and better results are obtained when keeping this value low, *i.e.*, having almost completely dimmed images. The delta value has no visible influence on the image processing, although it turns out that the lower its value, the better the detection.

We applied these detectors for a set of 165 scene images of size 640 × 360 px, all containing a pair of track lines with different orientations, after which the line extraction and selection algorithms were executed. We evaluated the average execution time and the average and maximum position errors. The execution time is calculated per frame, for the detection algorithms only, as these are the subject of our evaluation. The position errors determine the difference in pixels between the real (ground truth) and measured horizontal position of the vanishing point of the tracks. Based on these indicators, Table 2 summarizes the performance of the detection methods for the parameter sets for which the average position error was the lowest.

Table 2. Performance evaluation of edge detectors for line detection.

Method	Exec Time (ms)	Average Position Error (px)	Max Positioning Error (px)
Canny	1.60	17.5	126
Sobel	2.27	16.4	193
Laplacian	2.74	13.8	79
Scharr	2.19	16.9	193

From Table 2, one can see that all of the methods have an execution time below 3 ms. Recall that we considered only the duration of the detection, which together with the line extraction and selection results in times up to 20–25 ms. Still, this offers a 40–50-Hz control rate in the case of any detection algorithm, high enough for quadrotor control. The maximum position error indicates some false detections, which is the least severe in the case of the Laplacian algorithm. However, given the 640 px image width, all of the methods have an average position error below 2.5% that indicates an overall correct detection. From all of these parameters, we prefer to consider the position error the most important and use, therefore, the Laplacian method in our use-case of track following. Nevertheless, the test results confirm the discussion from Section 2.1 on the good performance of the Canny detector and its weakness of generating false positives, when comparing with the performance of the Laplacian method.

6.3. Use-Case Subtasks

We select solutions for the two use-cases, the short-range inspection in difficult-to-access areas and the long-range, track following-based infrastructure recording. This selection is just an illustration of how the presented methods can be used for UAV navigation. We provide no details on the settings of the considered techniques. However, flight tests were also already performed that demonstrate the suitability of several selected methods.

First, we need to break down the use-cases into subtasks. These are mainly: take-off, fly on a path, find and inspect targets, fly home and land. Next, we detail these subtasks and propose solutions to the related control problems. The take-off and landing tasks are, in the case of the AR.Drone and with many other RTF quadrotors, already solved by built-in functions. Furthermore, the newest products come with a fly-home function that, based on GPS coordinates, makes the quadrotor return autonomously to its take-off location and land there. For automation of the other tasks, additional processing and control methods are required.

6.3.1. Flying on a Path

Flying on a path poses different challenges in the two use-cases. In the case of remotely-controlled local inspection, it can be solved by GPS waypoint navigation with obstacle avoidance. Here, three subtasks can be identified: planning the waypoint sequence, navigating using GPS data and obstacle detection and avoidance. The waypoints have to be planned, e.g., to optimize the flight time or to avoid obstacles. For online planning, we suggest the use of the RRT and MPC-based algorithms. Regarding GPS-based waypoint navigation, quadrotors with a GPS module like the AR.Drone usually have implementations for this task. We will consider the software from [112]. Finally, obstacle avoidance is one of the most challenging tasks for quadrotors. Here, based on the experiments from Section 6.1, we suggest using Laplacian filtering for detection, combined with optical flow techniques that can determine motion. Then, we recommend the previously mentioned online planning methods for the avoidance maneuver.

In the second use-case, railway following, the path planning and following problem boils down to line following. With an onboard camera looking ahead, this can be achieved, for example, by finding and tracking the vanishing point of the lines formed by the pair of rails [111]. The vanishing point detection consists of image preprocessing for edge detection, line detection and line filtering in order to identify the tracks and their vanishing point. Based on [111] and on our experiments from Section 6.2, we propose the use of the Laplacian operator for edge detection. Then, the probabilistic Hough transform (PHT) can be applied to find lines from the resulting contours. These lines can then be filtered simply based on their lengths and slopes: we select long enough lines (e.g., longer than a quarter of the image height) that are almost vertical. These lines have a vanishing point that matches the vanishing point of the tracks. The tracking subtask can be solved by simple visual servoing: the vanishing point can be kept in the middle of the camera image through a proportional derivative controller. More precisely, we propose to perform yaw (z axis) angle rotations, while additionally correcting with lateral displacement and forward velocity reduction if the vanishing point is outside the desired range.

6.3.2. Finding and Inspecting Targets

For both use-cases, a method to find the target and a navigation strategy are needed, whereas in the first use-case (local inspection), a further navigation solution is needed for inspection. We consider a database of images of the possible targets. Then, based on the conclusions from Section 6.1, we propose to use FAST feature detectors with SIFT descriptors and the FLANN-based matcher. Together, these methods can track an object in subsequent frames. Above a threshold for the detection rate over time, we consider the target being found. We also point out the solution presented in [102], where machine learning is used in combination with edge detectors to find the target using a database of reference images. Yet another idea is to test RANSAC model fitting, introduced in Section 2.1.

To navigate around a detected target, we consider visual servoing techniques. An alternative would be GPS-based navigation, although for the AR.Drone, the GPS accuracy is of 2 m [113], not enough for safe navigation close to objects. We propose therefore the use of PBVS methods, introduced in Section 2.3. Furthermore, based on our experiments from Section 6.1, a basic visual servoing solution is to use the target detection described above together with homography-based identification. The obtained homography can then be used to determine the scaling and image-frame position of the target. Knowing the distance to the target in the reference image, the scaling and image-frame position can be then transformed into longitudinal and lateral distances to the object, which indicate the relative position to the target. Based on this, simple controllers can be applied to correct the distances so as to track the desired inspection trajectories.

7. Summary and Outlook

In the first major part of this paper, we reviewed vision and control methods for UAVs, with the final goal of using them in railway inspection applications. In the second part, we presented several popular low-cost quadrotor platforms, overviewed research concerning UAV inspection applications and formulated two use-cases. The use-cases address the novel application field of railway inspection and focus on short-range inspection in difficult-to-access areas and long-range track following. We performed an exhaustive evaluation of feature detectors for track following and target detection. Finally, we devised a strategy to accomplish the use-case task using results from our experiments and from the conclusions of the survey.

The survey of vision and control techniques revealed several open challenges, like the difficult problem of fine-tuning in the case of vision methods, the high computational demands of both vision and flight planning tools compared to the onboard processing capacity of the low-cost UAVs or the limitations appearing due to the lack of adequate mission formalization and due to the restrictive regulations. Further open issues are the lack of a general obstacle avoidance solution, which is crucial for fully-automated UAV navigation, and limitations derived from the short battery life of low-cost UAVs. Our future work is motivated by the railway inspection use-cases, and we are currently continuing our research by evaluating additional vision techniques for object classification and for obstacle avoidance, by developing trajectory and path planning techniques for automated flight of low-cost quadrotors.

Acknowledgments: This paper is supported by a grant from Siemens, Reference No. 7472/3202246859; by the Sectoral Operational Programme Human Resources Development (SOP HRD), ID 137516 (POSDRU/159/1.5/S/137516), financed by the European Social Fund and by the Romanian Government; and by a grant from the Romanian National Authority for Scientific Research, Project Number PNII-RU-TE-2012-3-0040. We are grateful to Péter Virág from University of Miskolc, Hungary, for his contribution regarding the testing of image processing techniques; and to Cristea-Ioan Iuga from Technical University of Cluj-Napoca, Romania, for his help in performing the flight tests and data collection.

Author Contributions: Koppány Máthé is the main author, having conducted the survey and written the content. Lucian Buşoniu has contributed with numerous comments, as well as content in the Introduction and Abstract.

Conflicts of Interest: The authors declare no conflict of interest.

References

1. Bakx, G.; Nyce, J. UAS in the (Inter)national airspace: Approaching the debate from an ethnicity perspective. In Proceedings of the International Conference on Unmanned Aircraft Systems (ICUAS), Atlanta, GA, USA, 28–31 May 2013; pp. 189–192.

2. DJI Products. Available online: http://www.dji.com/products (accessed on 22 December 2014).

3. Baiocchi, V.; Dominici, D.; Mormile, M. UAV application in post-seismic environment. *Int. Arch. Photogramm. Remote Sens. Spatial Inf. Sci. XL-1 W* **2013**, *2*, 21–25.

4. Li, H.; Wang, B.; Liu, L.; Tian, G.; Zheng, T.; Zhang, J. The design and application of SmartCopter: An unmanned helicopter based robot for transmission line inspection. In Proceedings of the IEEE Chinese Automation Congress (CAC), Changsha, China, 7–8 November 2013; pp. 697–702.

5. Parrot AR.Drone. Available online: http://ardrone2.parrot.com/ar-drone-2/specifications/ (accessed on 22 December 2014).

6. ArduCopter Quadrocopter. Available online: http://www.uav-store.de/diy-kits/3dr-quadrocopter (accessed on 22 December 2014).

7. Grabe, V.; Bulthoff, H.; Giordano, P. On-board Velocity Estimation and Closed-loop Control of a Quadrotor UAV based on Optical Flow. In Proceedings of the IEEE International Conference on Robotics and Automation (ICRA), St Paul, MN, USA, 14–18 May 2012; pp. 491–497.

8. Dydek, Z.; Annaswamy, A.; Lavretsky, E. Adaptive Control of Quadrotor UAVs: A Design Trade Study with Flight Evaluations. *IEEE Trans. Control Syst. Technol.* **2013**, *21*, 1400–1406.

9. Pestana, J.; Mellado-Bataller, I.; Sanchez-Lopez, J.L.; Fu, C.; Mondragón, I.F.; Campoy, P. A General Purpose Configurable Controller for Indoors and Outdoors GPS-Denied Navigation for Multirotor Unmanned Aerial Vehicles. *J. Intell. Robotic Syst.* **2014**, *73*, 387–400.

10. Fossel, J.; Hennes, D.; Claes, D.; Alers, S.; Tuyls, K. OctoSLAM: A 3D mapping approach to situational awareness of unmanned aerial vehicles. In Proceedings of the International Conference on Unmanned Aircraft Systems (ICUAS), Atlanta, GA, USA, 28–31 May 2013; pp. 179–188.

11. Stark, B.; Smith, B.; Chen, Y. Survey of thermal infrared remote sensing for Unmanned Aerial Systems. In Proceedings of the IEEE International Conference on Unmanned Aircraft Systems (ICUAS), Orlando, FL, USA, 27–30 May 2014; pp. 1294–1299.

12. Bendig, J.; Bolten, A.; Bareth, G. Introducing a low-cost mini-UAV for thermal-and multispectral-imaging. *Int. Arch. Photogramm. Remote Sens. Spat. Inf. Sci.* **2012**, *39*, 345–349.
13. Wenzel, K.E.; Masselli, A.; Zell, A. Automatic take off, tracking and landing of a miniature UAV on a moving carrier vehicle. *J. Intell. Robot. Syst.* **2011**, *61*, 221–238.
14. Whitehead, K.; Hugenholtz, C.H. Remote sensing of the environment with small unmanned aircraft systems (UASs), part 1: A review of progress and challenges 1. *J. Unmanned Veh. Syst.* **2014**, *2*, 69–85.
15. Baker, S.; Scharstein, D.; Lewis, J.; Roth, S.; Black, M.J.; Szeliski, R. A database and evaluation methodology for optical flow. *Int. J. Comput. Vis.* **2011**, *92*, 1–31.
16. Oskoei, M.A.; Hu, H. *A Survey on Edge Detection Methods*; University of Essex: Essex, UK, 2010.
17. Hua, M.D.; Hamel, T.; Morin, P.; Samson, C. Introduction to Feedback Control of Underactuated VTOL Vehicles. *IEEE Control Syst. Mag.* **2013**, *33*, 61–75.
18. Goerzen, C.; Kong, Z.; Mettler, B. A survey of motion planning algorithms from the perspective of autonomous UAV guidance. *J. Intell. Robot. Syst.* **2010**, *57*, 65–100.
19. Dadkhah, N.; Mettler, B. Survey of motion planning literature in the presence of uncertainty: Considerations for UAV guidance. *J. Intell. Robot. Syst.* **2012**, *65*, 233–246.
20. LaValle, S.M. *Planning Algorithms*; Cambridge University Press: Cambridge, UK, 2006.
21. Roushdy, M. Comparative study of edge detection algorithms applying on the grayscale noisy image using morphological filter. *GVIP J.* **2006**, *6*, 17–23.
22. Tuytelaars, T.; Mikolajczyk, K. Local invariant feature detectors: A survey. *Found. Trends Comput. Graph. Vis.* **2008**, *3*, 177–280.
23. Choi, S.; Kim, T.; Yu, W. Performance Evaluation of RANSAC Family. In Proceedings of the British Machine Vision Conference (BMVC 2009), British Machine Vision Association (BMVA), London, UK, 7–10 September 2009; pp. 81.1–81.12.
24. Rockett, P. Performance assessment of feature detection algorithms: A methodology and case study on corner detectors. *IEEE Trans. Image Proc.* **2003**, *12*, 1668–1676.
25. Jain, R.; Kasturi, R.; Schunck, B.G. *Machine Vision*; McGraw-Hill: New York, NY, USA, 1995.
26. Rufeil, E.; Gimenez, J.; Flesia, A. Comparison of edge detection algorithms on the undecimated wavelet transform. In Proceedings of the IV CLAM, Latin American Congress in Mathematics, Córdoba, Argentina, 6–10 August 2012.
27. Senthilkumaran, N.; Rajesh, R. Edge detection techniques for image segmentation—A survey of soft computing approaches. *Int. J. Recent Trends Eng.* **2009**, *1*, 250-254.
28. Harris, C.; Stephens, M. A combined corner and edge detector. In Proceedings of the Alvey Vision Conference, Manchester, UK, 31 August–2 September 1988; Volume 15, p. 50.
29. Shi, J.; Tomasi, C. Good features to track. In Proceedings of the IEEE Computer Society Conference on Computer Vision and Pattern Recognition (CVPR'94), Seattle, WA, USA, 21–23 June 1994; pp. 593–600.
30. Tulpan, D.; Belacel, N.; Famili, F.; Ellis, K. Experimental evaluation of four feature detection methods for close range and distant airborne targets for Unmanned Aircraft Systems applications. In Proceedings of the IEEE International Conference on Unmanned Aircraft Systems (ICUAS), Orlando, FL, USA, 27–30 May 2014; pp. 1267–1273.
31. Smith, S.M.; Brady, J.M. SUSAN—A new approach to low level image processing. *Int. J. Comput. Vis.* **1997**, *23*, 45–78.
32. Rosten, E.; Drummond, T. Rapid rendering of apparent contours of implicit surfaces for real-time tracking. In Proceedings of the 14th British Machine Vision Conference (BMVC 2003), British Machine Vision Association (BMVA), Norwich, UK, 9–11 September 2003; pp. 719–728.
33. Bay, H.; Tuytelaars, T.; van Gool, L. Surf: Speeded up robust features. In *Computer Vision—ECCV 2006*; Springer: Berlin/Heidelberg, Germany, 2006; Volume 3951, pp. 404–417.
34. Lowe, D.G. Distinctive image features from scale-invariant keypoints. *Int. J. Comput. Vis.* **2004**, *60*, 91–110.
35. Calonder, M.; Lepetit, V.; Strecha, C.; Fua, P. Brief: Binary robust independent elementary features. In *Computer Vision —ECCV 2010*; Springer: Berlin/Heidelberg, Germany, 2010; Volume 6314, pp. 778–792.
36. Rublee, E.; Rabaud, V.; Konolige, K.; Bradski, G. ORB: An efficient alternative to SIFT or SURF. In Proceedings of the IEEE International Conference on Computer Vision (ICCV), Barcelona, Spain, 6–13 November 2011; pp. 2564–2571.

37. Arya, S.; Mount, D.M.; Netanyahu, N.S.; Silverman, R.; Wu, A.Y. An optimal algorithm for approximate nearest neighbor searching fixed dimensions. *J. ACM* **1998**, *45*, 891–923.

38. Muja, M.; Lowe, D.G. Fast matching of binary features. In Proceedings of the IEEE Ninth Conference on Computer and Robot Vision (CRV), Toronto, ON, Canada, 28–30 May 2012; pp. 404–410.

39. Dubrofsky, E. Homography Estimation. Ph.D Thesis, University of British Columbia (Vancouver), Kelowna, BC, Canada, March 2009.

40. Zsedrovits, T.; Bauer, P.; Zarándy, A.; Vanek, B.; Bokor, J.; Roska, T. Error Analysis of Algorithms for Camera Rotation Calculation in GPS/IMU/Camera Fusion for UAV Sense and Avoid Systems. In Proceedings of the IEEE International Conference on Unmanned Aircraft Systems (ICUAS), Orlando, FL, USA, 27–30 May 2014.

41. Flores, G.; Zhou, S.; Lozano, R.; Castillo, P. A Vision and GPS-Based Real-Time Trajectory Planning for a MAV in Unknown and Low-Sunlight Environments. *J. Intell. Robot. Syst.* **2014**, *74*, 59–67.

42. Yang, S.; Scherer, S.A.; Schauwecker, K.; Zell, A. Autonomous Landing of MAVs on an Arbitrarily Textured Landing Site Using Onboard Monocular Vision. *J. Intell. Robot. Syst.* **2014**, *74*, 27–43.

43. Schauwecker, K.; Zell, A. On-Board Dual-Stereo-Vision for the Navigation of an Autonomous MAV. *J. Intell. Robot. Syst.* **2014**, *74*, 1–16.

44. Herout, A.; Dubská, M.; Havel, J. Review of Hough Transform for Line Detection. In *Real-Time Detection of Lines and Grids*; Springer: Berlin/Heidelberg, Germany, 2013; Volume 1, pp. 3–16.

45. Mukhopadhyay, P.; Chaudhuri, B.B. A survey of Hough Transform. *Pattern Recog.* **2015**, *48*, 993–1010.

46. OpenCV Probabilistic Hough Transform implementation. Available online: http://docs.opencv.org/modules/imgproc/doc/feature_detection.html (accessed on 8 May 2015).

47. Chao, H.; Gu, Y.; Napolitano, M. A Survey of Optical Flow Techniques for Robotics Navigation Applications. *J. Intell. Robot. Syst.* **2014**, *73*, 361–372.

48. Sun, D.; Roth, S.; Black, M.J. Secrets of optical flow estimation and their principles. In Proceedings of the IEEE Conference on Computer Vision and Pattern Recognition (CVPR), San Francisco, CA, USA, 13–18 June 2010; pp. 2432–2439.

49. Sun, D.; Roth, S.; Black, M.J. A quantitative analysis of current practices in optical flow estimation and the principles behind them. *Int. J. Comput. Vis.* **2014**, *106*, 115–137.

50. Dame, A.; Marchand, E. Mutual information-based visual servoing. *IEEE Trans. Robot.* **2011**, *27*, 958–969.

51. Caron, G.; Dame, A.; Marchand, E. Direct model based visual tracking and pose estimation using mutual information. *Image Vis. Comput.* **2014**, *32*, 54–63.

52. Collewet, C.; Marchand, E. Photometric visual servoing. *IEEE Trans. Robot.* **2011**, *27*, 828–834.

53. Thuilot, B.; Martinet, P.; Cordesses, L.; Gallice, J. Position based visual servoing: Keeping the object in the field of vision. In Proceedings of the IEEE International Conference on Robotics and Automation (ICRA '02), Washington, DC, USA, 11–15 May 2002; Volume 2, pp. 1624–1629.

54. Chaumette, F.; Malis, E. 2 1/2 D visual servoing: A possible solution to improve image-based and position-based visual servoings. In Proceedings of the IEEE International Conference on Robotics and Automation (ICRA'00), San Francisco, CA, USA, 24–28 April 2000; Volume 1, pp. 630–635.

55. Bourquardez, O.; Mahony, R.; Guenard, N.; Chaumette, F.; Hamel, T.; Eck, L. Image-Based Visual Servo Control of the Translation Kinematics of a Quadrotor Aerial Vehicle. *IEEE Trans. Robot.* **2009**, *25*, 743–749.

56. Fahimi, F.; Thakur, K. An alternative closed-loop vision-based control approach for Unmanned Aircraft Systems with application to a quadrotor. In Proceedings of the IEEE International Conference on Unmanned Aircraft Systems (ICUAS), Atlanta, GA, USA, 28–31 May 2013; pp. 353–358.

57. Bai, Y.; Liu, H.; Shi, Z.; Zhong, Y. Robust control of quadrotor unmanned air vehicles. In Proceedings of the IEEE 31st Chinese Control Conference (CCC), Hefei, China, 25–27 July 2012; pp. 4462–4467.

58. Krajnik, T.; Nitsche, M.; Pedre, S.; Preucil, L.M.A. A simple visual navigation system for an UAV. In Proceedings of the 9th International Multi-Conference on Systems, Signals and Devices (SSD), Chemnitz, Germany, 20–23 March 2012; pp. 1–6.

59. Ross, S.; Melik-Barkhudarov, N.; Shankar, K.S.; Wendel, A.; Dey, D.; Bagnell, J.A.; Hebert, M. Learning monocular reactive uav control in cluttered natural environments. In Proceedings of the IEEE International Conference on Robotics and Automation (ICRA), Karlsruhe, Germany, 6–10 May 2013; pp. 1765–1772.

60. Amelink, M.H.J. Ecological Automation Design, Extending Work Domain Analysis. Ph.D Thesis, Delft University of Technology, Delft, Holland, October 2010.

61. García Carrillo, L.R.; Dzul López, A.E.; Lozano, R.; Pégard, C. Modeling the Quad-Rotor Mini-Rotorcraft. In *Quad Rotorcraft Control*; Springer: London, UK, 2013; Volume 1, pp. 23–34.

62. Salazar, S.; Gonzalez-Hernandez, I.; López, J.R.; Lozano, R.; Romero, H. Simulation and Robust Trajectory-Tracking for a Quadrotor UAV. In Proceedings of the IEEE International Conference on Unmanned Aircraft Systems (ICUAS), Orlando, FL, USA, 27–30 May 2014.

63. Cui, Y.; Inanc, T. Controller design for small air vehicles—An overview and comparison. In Proceedings of the IEEE International Conference on Unmanned Aircraft Systems (ICUAS), Atlanta, GA, USA, 28–31 May 2013; pp. 621–627.

64. Raffo, G.V.; Ortega, M.G.; Rubio, F.R. An integral predictive/nonlinear H∞ control structure for a quadrotor helicopter. *Automatica* **2010**, *46*, 29–39.

65. Alexis, K.; Nikolakopoulos, G.; Tzes, A. On trajectory tracking model predictive control of an unmanned quadrotor helicopter subject to aerodynamic disturbances. *Asian J. Control* **2014**, *16*, 209–224.

66. Lin, Y.; Saripalli, S. Path planning using 3D Dubins Curve for Unmanned Aerial Vehicles. In Proceedings of the IEEE International Conference on Unmanned Aircraft Systems (ICUAS), Orlando, FL, USA, 27–30 May 2014; pp. 296–304.

67. Bellingham, J.; Richards, A.; How, J.P. Receding horizon control of autonomous aerial vehicles. In Proceedings of the IEEE 2002 American Control Conference, Anchorage, AK, USA, 8–10 May 2002; Volume 5, pp. 3741–3746.

68. Chao, Z.; Zhou, S.L.; Ming, L.; Zhang, W.G. Uav formation flight based on nonlinear model predictive control. *Math. Probl. Eng.* **2012**, *2012*, doi:10.1155/2012/261367

69. Grancharova, A.; Grøtli, E.I.; Ho, D.T.; Johansen, T.A. UAVs Trajectory Planning by Distributed MPC under Radio Communication Path Loss Constraints. *J. Intell. Robot. Syst.* **2014**, 1–20, doi:10.1007/s10846-014-0090-1.

70. Donald, B.; Xavier, P.; Canny, J.; Reif, J. Kinodynamic motion planning. *J. ACM* **1993**, *40*, 1048–1066.

71. Franze, G.; Mattei, M.; Ollio, L.; Scordamaglia, V. A Receding Horizon Control scheme with partial state measurements: Control Augmentation of a flexible UAV. In Proceedings of the IEEE Conference on Control and Fault-Tolerant Systems (SysTol), Nice, France, 9–11 October 2013; pp. 158–163.

72. Sahingoz, O.K. Generation of Bezier Curve-Based Flyable Trajectories for Multi-UAV Systems with Parallel Genetic Algorithm. *J. Intell. Robot. Syst.* **2014**, *74*, 499–511.

73. Schopferer, S.; Adolf, F.M. Rapid trajectory time reduction for unmanned rotorcraft navigating in unknown terrain. In Proceedings of the IEEE International Conference on Unmanned Aircraft Systems (ICUAS), Orlando, FL, USA, 27–30 May 2014; pp. 305–316.

74. Zufferey, J.C.; Beyeler, A.; Floreano, D. Autonomous flight at low altitude with vision-based collision avoidance and GPS-based path following. In Proceedings of the IEEE International Conference on Robotics and Automation (ICRA), Anchorage, AK, USA, 3–8 May 2010; pp. 3329–3334.

75. Qu, Y.; Zhang, Y.; Zhang, Y. Optimal Flight Path Planning for UAVs in 3-D Threat Environment. In Proceedings of the IEEE International Conference on Unmanned Aircraft Systems (ICUAS), Orlando, FL, USA, 27–30 May 2014.

76. Gimenes, R.A.; Vismari, L.F.; Avelino, V.F.; Camargo, J.B., Jr.; de Almeida, J.R., Jr.; Cugnasca, P.S. Guidelines for the Integration of Autonomous UAS into the Global ATM. *J. Intell. Robot. Syst.* **2014**, *74*, 465–478.

77. Faughnan, M.; Hourican, B.; MacDonald, G.; Srivastava, M.; Wright, J.; Haimes, Y.; Andrijcic, E.; Guo, Z.; White, J. Risk analysis of Unmanned Aerial Vehicle hijacking and methods of its detection. In Proceedings of the IEEE Systems and Information Engineering Design Symposium (SIEDS), Charlottesville, VA, USA, 26 April 2013; pp. 145–150.

78. RCgroups Community. Available online: http://www.rcgroups.com/forums/index.php (accessed on 11 May 2015).

79. Diydrones Community. Available online: http://diydrones.com/, (accessed on 11 May 2015).

80. My First Drone Website. Available online: http://myfirstdrone.com/ (accessed on 11 May 2015).

81. Drones for Good Website. Available online: https://www.dronesforgood.ae/en (accessed on 11 May 2015).

82. Best Quadcopters Website. Available online: http://www.bestquadcoptersreviews.com/ (accessed on 11 May 2015).

83. Parrot AR.Drone Pressure Sensor. Available online: http://ardrone2.parrot.com/ardrone-2/altitude (accessed on 22 December 2014).

84. Parrot Bebop Drone. Available online: http://www.parrot.com/products/bebop-drone/ (accessed on 11 May 2015).

85. Xaircraft Products. Available online: http://xaircraftamerica.com/collections/xaircraft-products/products/x650-pro-kit-1 (accessed on 22 December 2014).

86. Xaircraft SuperX Flight Controller. Available online: http://www.xaircraft.com/products/superx/ (accessed on 22 December 2014).

87. MikroKopter Products. Available online: http://www.quadrocopter.com/MK-Basicset-QuadroKopter-XL_p_283.html (accessed on 22 December 2014).

88. MikroKopter Flight Controller. Available online: http://wiki.mikrokopter.de/en/FlightCtrl_ME_2_5 (accessed on 22 December 2014).

89. DJI Flight Controllers. Available online: http://www.dji.com/info/spotlight/whats-difference-of-naza-m-litenaza-m-v1naza-m-v2 (accessed on 22 December 2014).

90. ArduCopter Programming. Available online: http://copter.ardupilot.com/wiki/programming-arducopter-with-arduino/ (accessed on 22 December 2014).

91. ArduCopter Iris+. Available online: http://store.3drobotics.com/products/iris (accessed on 22 December 2014).

92. 3DR Solo. Available online: http://3drobotics.com/solo/ (accessed on 11 May 2015).

93. PIXHAWK Autopilot. Available online: http://pixhawk.org/modules/pixhawk (accessed on 22 December 2014).

94. PIXHAWK PX4FLOW. Available online: https://pixhawk.org/modules/px4flow (accessed on 22 December 2014).

95. Hausamann, D.; Zirnig, W.; Schreier, G.; Strobl, P. Monitoring of gas pipelines—A civil UAV application. *Aircraft Eng. Aerosp. Technol.* **2005**, *77*, 352–360.

96. Rathinam, S.; Kim, Z.W.; Sengupta, R. Vision-Based Monitoring of Locally Linear Structures Using an Unmanned Aerial Vehicle 1. *J. Infrastruct. Syst.* **2008**, *14*, 52–63.

97. Montambault, S.; Beaudry, J.; Toussaint, K.; Pouliot, N. On the application of VTOL UAVs to the inspection of power utility assets. In Proceedings of the IEEE 1st International Conference on Applied Robotics for the Power Industry (CARPI), Montreal, QC, Canada, 5–7 October 2010; pp. 1–7.

98. Ellis, N. Inspection of Power Transmission Lines Using UAVs. Master's Thesis, University of Southern Queensland, Toowoomba, Queensland, Australia, October 2013.

99. Zhang, J.; Liu, L.; Wang, B.; Chen, X.; Wang, Q.; Zheng, T. High speed automatic power line detection and tracking for a UAV-based inspection. In Proceedings of the IEEE International Conference on Industrial Control and Miscs Engineering (ICICEE), Xi'an, China, 23–25 April 2012; pp. 266–269.

100. Luque-Vega, L.F.; Castillo-Toledo, B.; Loukianov, A.; Gonzalez-Jimenez, L.E. Power line inspection via an unmanned aerial system based on the quadrotor helicopter. In Proceedings of the IEEE 17th Mediterranean Electrotechnical Conference (MELECON), Beirut, Lebanon, 13–16 April 2014; pp. 393–397.

101. Larrauri, J.; Sorrosal, G.; González, M. Automatic system for overhead power line inspection using an Unmanned Aerial Vehicle - RELIFO project. In Proceedings of the IEEE International Conference on Unmanned Aircraft Systems (ICUAS), Atlanta, GA, USA, 28–31 May 2013; pp. 244–252.

102. Martinez, C.; Sampedro, C.; Chauhan, A.; Campoy, P. Towards autonomous detection and tracking of electric towers for aerial power line inspection. In Proceedings of the IEEE International Conference on Unmanned Aircraft Systems (ICUAS), Orlando, FL, USA, 27–30 May 2014; pp. 284–295.

103. Eschmann, C.; Kuo, C.; Kuo, C.; Boller, C. Unmanned aircraft systems for remote building inspection and monitoring. In Proceedings of the Sixth European Workshop on Structural Health Monitoring, Dresden, Germany, 3–6 July 2012.

104. Nikolic, J.; Burri, M.; Rehder, J.; Leutenegger, S.; Huerzeler, C.; Siegwart, R. A UAV system for inspection of industrial facilities. In Proceedings of the IEEE Aerospace Conference, Big Sky, MT, USA, 2–9 March 2013; pp. 1–8.

105. DB about Using Drones against Graffiti Sprayers. Available online: http://www.bbc.com/news/world-europe-22678580 (accessed on 22 December 2014).

106. SNCF Using Drones for Railway Viaduct Inspection. Available online: http://www.personal-drones.net/french-trains-company-sncf-is-starting-to-experiment-with--drones-for-safety-and-inspection-purposes/ (accessed on 22 December 2014).

107. ELIMCO Industrial Aerial Inspection Solutions. Available online: http://www.elimco.com/eng/l_ Infrastructure-Monitoring-and-Inspection_46.html (accessed on 22 December 2014).

108. Microdrones UAVs. Available online: http://www.microdrones.com/en/applications/areas-of-application/inspection/ (accessed on 22 December 2014).

109. Drones in railway maintenance. Available online: http://www.smartrailworld.com/how_drones_are_ already-being-used-by-railways-around-the-world (accessed on 22 December 2014).

110. OpenCV Feature Detectors. Available online: http://docs.opencv.org/modules/features2d/doc/common_ interfaces_of_feature_detectors.html (accessed on 30 March 2015).

111. Páll, E. Vision-Based Quadcopter Navigation for Following Indoor Corridors and Outdoor Railways. Master's Thesis, The Technical University of Cluj-Napoca, Cluj-Napoca, Romania, June 2014.

112. AR.Drone Autonomy GPS Waypoints Package. Available online: https://github.com/AutonomyLab/ ardrone_autonomy/tree/gps-waypoint (accessed on 8 May 2015).

113. Parrot AR.Drone Flight Recorder. Available online: http://ardrone2.parrot.com/ apps/flight-recorder/ (accessed on 8 May 2015).

Article

Feasibility of Using Synthetic Aperture Radar to Aid UAV Navigation

Davide O. Nitti [1], Fabio Bovenga [2,*], Maria T. Chiaradia [3], Mario Greco [4] and Gianpaolo Pinelli [4]

[1] Geophysical Applications Processing s.r.l., Via Amendola 173, 70126 Bari, Italy; davide.nitti@gapsrl.eu
[2] National Research Council of Italy, ISSIA institute, Via Amendola 173, 70126 Bari, Italy
[3] Dipartimento di Fisica "M. Merlin", Politecnico di Bari, Via Amendola 173, 70126 Bari, Italy;
    chiaradia@ba.infn.it
[4] IDS—Ingegneria Dei Sistemi S.p.A., Via Enrica Calabresi 24, 56121 Pisa, Italy;
    m.greco@idscorporation.com (M.G.); g.pinelli@idscorporation.com (G.P.)
*   Author to whom correspondence should be addressed; bovenga@ba.issia.cnr.it;
    Tel.: +39-80-592-9425; Fax: +39-80-592-9460.

Academic Editor: Felipe Gonzalez Toro
Received: 4 May 2015; Accepted: 17 July 2015; Published: 28 July 2015

Abstract: This study explores the potential of Synthetic Aperture Radar (SAR) to aid Unmanned Aerial Vehicle (UAV) navigation when Inertial Navigation System (INS) measurements are not accurate enough to eliminate drifts from a planned trajectory. This problem can affect medium-altitude long-endurance (MALE) UAV class, which permits heavy and wide payloads (as required by SAR) and flights for thousands of kilometres accumulating large drifts. The basic idea is to infer position and attitude of an aerial platform by inspecting both amplitude and phase of SAR images acquired onboard. For the amplitude-based approach, the system navigation corrections are obtained by matching the actual coordinates of ground landmarks with those automatically extracted from the SAR image. When the use of SAR amplitude is unfeasible, the phase content can be exploited through SAR interferometry by using a reference Digital Terrain Model (DTM). A feasibility analysis was carried out to derive system requirements by exploring both radiometric and geometric parameters of the acquisition setting. We showed that MALE UAV, specific commercial navigation sensors and SAR systems, typical landmark position accuracy and classes, and available DTMs lead to estimate UAV coordinates with errors bounded within ± 12 m, thus making feasible the proposed SAR-based backup system.

Keywords: UAV; navigation; Geo-referencing; SAR; interferometry; ATR; feasibility

1. Introduction

This study is devoted to explore the potentials of synthetic aperture radar (SAR) and Interferometric SAR (InSAR) to aid unmanned aerial vehicle (UAV) navigation. Over the past decades, UAVs have been increasingly used for a wide range of both civilian and military applications, such as reconnaissance, surveillance and security, terrain mapping and geophysical exploration. The feasible use of a UAV for a certain application relies on the accuracy and robustness of the navigation system. The navigation of a UAV is controlled by the inertial navigation system (INS), which exploits different sensors such as inertial measurement unit (IMU), radar altimeter (RALT) and global positioning system (GPS) receiver. These sensors allow the UAV to measure the status vectors of the aircraft (position, velocity, acceleration, Euler attitude angle and rates) needed to infer the actual trajectory with enhanced accuracy. Thus, the INS defines the commands needed to change the status vectors in order to guide the aircraft along the mission reference trajectory, or can be used to provide geo-referenced products acquired onboard for ground control point free applications [1].

Sensors **2015**, *15*, 18334–18359

The INS performance and (consequently) the navigation accuracy depend on the performance of such sensors and, in particular, on the IMU. It is made of three gyroscopes and three accelerometers, which are used to calculate attitude, absolute acceleration and velocity of the aircraft [2]. One of the main drawbacks of IMU is the rapid growth of systematic errors as bias and drift, which have to be compensated by using the absolute positions measured by the GPS [3]. However, due to the low power of the ranging signals, the received GPS signal is easily corrupted or completely obscured by either intentional (e.g., jamming) or unintentional interferences. Moreover, along the UAV trajectory the GPS signal can be also absent. When GPS data are not reliable or absent, the correction of the aircraft trajectory through the IMU is consequently unfeasible.

This problem can affect in particular medium-altitude long-endurance (MALE) UAV, which has to fly for thousands of kilometers before reaching its target. Cumulative drift (up to hundreds of meters) during a long endurance flight can lead the UAV to miss the planned target with catastrophic consequences on the mission [4]. Thus, future guidance systems for UAV autonomous missions have challenging requirements for high reliability and integrity. To fully meet these new requirements, the following capacities need to be radically improved: enhanced tolerance to GPS denial/jamming, higher IMU performance, and reduced INS drift.

The aim of this work is to assess the feasibility of using SAR to aid a standard navigation system when IMU measurements are not accurate enough to eliminate drifts from a planned trajectory. Examples of SAR systems mounted onboard UAVs already exist (e.g., [5,6]) but not devoted to aid the platform navigation. The basic idea is to infer position and attitude of an aerial platform by inspecting both amplitude and phase of SAR images provided by a SAR system onboard the platform. SAR data provide information on the electromagnetic microwave backscatter characteristics of the Earth surface with day/night and all-weather capability, thanks to the active nature of radar sensors [7]. Moreover, the SAR imaging is able to illuminate a wide area on the ground (up to several square kilometers) from a long distance (up to tens of kilometers) by ensuring high spatial resolution (meters or less depending on the bandwidth). These characteristics are advantageous with respect to other sensors (e.g., optical) and make suitable the proposed backup navigation system.

In case of the amplitude-based approach, the system navigation correction can be based on a comparison between processed SAR images and a terrain landmark Data Base (DB), which contains the geographic coordinates of expected (during the UAV mission) ground landmarks (*i.e.*, conspicuous objects on land that unequivocally mark a locality). Let us assume that the scene acquired by the SAR onboard the platform is populated by terrain landmarks (e.g., roads, buildings), which is quite likely in several operating scenarios (e.g., rural, industrial, suburban). The image coordinates (range/cross-range) of the expected landmarks can be automatically extracted by an Automatic Target detection and Recognition (ATR) algorithm for SAR images [8]. Then, the coordinates of an image landmark have to be correlated with the geographic coordinates of the corresponding landmark in the mission DB. Once a match is found, a mathematical relation between range/cross-range landmark coordinates and landmark geographic coordinates can be found and exploited to retrieve aircraft position. The whole ATR and SAR based geo-referencing block is depicted in the processing flow in Figure 1. The onboard SAR acquires an image under a certain viewing geometry; terrain landmarks in the focused image are extracted by an ATR processing chain that is fed by a landmark DB; mission planned landmarks are recognized by the ATR chain; the ATR block provides the coordinates of each landmark both in the SAR coordinate system and in the DB inertial coordinate system; local and inertial coordinates of each landmark, as well the initial aircraft attitude/position measurements (from the INS) are exploited by the geo-referencing algorithm; the final output is the position of the aircraft in the inertial coordinate system.

When the use of SAR amplitude is unfeasible, the phase content of the SAR image is exploited through a real time InSAR system mounted onboard the platform. By synthetically coupling SAR images acquired by different positions [9], the InSAR system provides information about the radiation path delay between two acquisitions, which includes the topography, and any difference occurred between the two passes on the ground surface as well as in the atmospheric refractivity profile. The InSAR phase derived by using a single-pass interferometry system avoids contributions from both ground deformation and atmosphere, and is related only to the aircraft position and to the topography. Therefore, by using both approximated position and attitude values of the platform, and a reference DTM, it is possible to generate a synthetic InSAR phase model to be compared with respect to (w.r.t.) that derived by processing the InSAR images. The geometrical transformation needed to match these two terrain models depends on the difference between the actual values of position and attitude, and those derived by the instruments available onboard. Hence, this matching provides a feedback to be used for adjusting position and attitude when a lack of GPS signal leads to unreliable IMU data.

The goal of the paper is to propose an advance in the integration of radar imaging into UAV navigation systems and to prove the technological feasibility of the proposed integrated system. Nowadays, medium-altitude long-endurance (MALE) UAV class [10] permits heavy (in the order of hundreds of kilograms) and wide (in the order of few square meters) payloads to be carried onboard, during a long mission (in the order of few thousands of kilometers), thus making feasible the use of SAR systems, which are more demanding than other sensors in terms of weight and space (the InSAR configuration in particular). Some examples of MALE UAVs are Predator B, Global Hawk and Gray Eagle, which are already equipped with SAR systems and can be also equipped with InSAR systems. The paper provides a feasibility analysis performed by simulating realistic scenarios according to the characteristics of commercial navigation sensors and SAR systems, typical landmark position accuracy and classes, and available DTMs.

Section 2 describes the amplitude-based approach, and presents a feasibility analysis that provides requirements for aerial platform class, onboard navigations sensors, the SAR system and acquisition geometry, landmark DB and mission planning. In Section 3, we first introduce the InSAR configuration with preliminary consideration concerning the expected InSAR phase quality. Then, we evaluate the InSAR sensitivity to changes on aircraft position and attitude. Requirements for the DTM are also provided. Finally, in the conclusions section, we resume the limits of applicability and provide indications concerning the system parameters.

2. SAR Amplitude Exploitation

2.1. Proposed Approach

A concept scheme of the proposed SAR amplitude-based approach is depicted in Figure 1. Novel and reliable system navigation correction is necessary and can be based on the best matching between planned (latitude/longitude/altitude) and automatically extracted (azimuth/slant-range) landmark (*i.e.*, a conspicuous object on land that unequivocally marks a locality) coordinates as a function of aircraft state variables. The automatic extraction step is performed by an ATR chain, which is able to automatically extract from the SAR image the landmarks (e.g., buildings, crossroads) in the mission DB, which contains several features (e.g., landmark position, size). Finally, both extracted radar coordinates (those obtained under the actual SAR viewing geometry) and the world coordinates (those in the mission DB) of the same landmark are used for retrieving aircraft state variables. For simplicity of representation, a landmark can be modeled as a set of Ground Control Points (GCPs), e.g., a cross-road landmark can be represented by the coordinates of four GCPs.

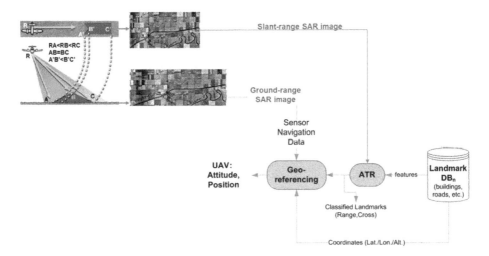

Figure 1. Whole ATR and SAR based geo-referencing concept: processing-flow.

As a GPS denial occurs, the backup system triggers the SAR sensor for the next Way Point (WP) and controls the radar beam steering onto the expected terrain landmark previously stored in the onboard DB. The beam steering must ensure enough time on the landmark to gather the echo returns needed to focus the raw data image. It is worth noting that the proposed approach and the feasibility analysis here presented are more focused on the UAV platform, SAR system and navigation sensors than on the image processing chain (SAR data autofocusing [5,6], ATR chain), which is beyond the scope of this paper.

2.2. Basic Concepts and Geo-Referencing Procedure

The whole (SAR amplitude-based) geo-referencing procedure is depicted in the processing-flow in Figure 1. It can be noted that viewing geometry and SAR geometrical distortions have to be considered in order to derive a meaningful geo-referencing algorithm based on the recognition of landmarks in SAR images. A basic viewing geometry of an SAR is shown in Figure 2a [8]: a platform flying with a given velocity at altitude h carries a *side-looking* radar antenna that illuminates the Earth's surface with pulses of electromagnetic radiation. The direction of travel of the platform is known as the *azimuth* direction, while the distance from the radar track is measured in the *slant range* direction. The *ground range* and its dependence on the angle θ is also depicted. Note that in Figure 2b we assumed the flat Earth approximation, which can be considered valid for the airborne case, even for long-range systems [11]. Before presenting the proposed procedure, we introduce some quantities that will be used in the following. A complete list can be found in Table 1.

Inertial system of coordinates: classically, it is the Cartesian system of coordinates. In the geodetic literature, an inertial system is an Earth-Centred-Earth-Fixed (ECEF) frame where the origin is at the Earth's centre of mass, the (x_{ecef}, y_{ecef}) plane coincides with the equatorial plane, and the (x_{ecef}, z_{ecef}) plane contains the first meridian) [12].

Local flat Earth: position of a point $\mathbf{p}_H = [x_H, y_H, z_H]^t$ can be computed from a geodetic latitude-longitude-altitude frame or an ECEF frame by assuming that the flat Earth z_H-axis is normal to the Earth only at the initial geodetic coordinates [13]. For our application, the flat Earth model can be assumed if $h \ll R_E = 6370$ km, where R_E is the Earth radius and h is the SAR (or aircraft) altitude [11]. As a consequence, the coordinates for a specific ellipsoid planet can be easily computed resorting to commercial software tool, such as in [14]. A pictorial view of ECEF and flat Earth (H) frame can be found in Figure 2b.

Euler angles: angles that are used to define the orientation of a rigid body within an inertial system of coordinates [15] (see Figure 3). Such angles hereinafter will be also referred to as attitude (ψ, θ, φ).

Now, let us define a coordinate transformation of a point $\mathbf{p}_H$ from an inertial (H) frame to a new coordinate system (C) [15,16]:

$$\mathbf{p}_C = \mathbf{t}_C + \gamma_H^C \mathbf{R}_H^C(\psi, \theta, \varphi)\mathbf{p}_H \tag{1}$$

where every quantity is defined in Table 1. This transformation has some interesting properties [15]: if the ground was planar ($z_H = 0$), the transformation would be *affine*; $\mathbf{R}_H^C(\psi, \theta, \phi)$ can be factorized into the product of three orthonormal-rotation-matrices; a parallel pair in the H domain remains parallel in C; the transformation either stretches or shrinks a vector; typical SAR distortions (*i.e.*, *foreshortening*, *layover*, *shadow* in Figure 1 and in Figure 2a) are correctly modelled [16,17].

Equation (1) can be inverted as follows:

$$\mathbf{p}_H = \mathbf{t}_H + \gamma_C^H \mathbf{R}_C^H(\psi, \theta, \varphi)\mathbf{p}_C \tag{2}$$

where every quantity is defined in Table 1. The main geometrical properties considered here are: invariance to rotation of an SAR image with respect to its corresponding geographical map; invariance to scale factor (stretch or shrink); invariance to translation; invariance to slant range, foreshortening and layover distortions for $\theta \neq 0$ (Figure 2b). To perform the feasibility analysis of the SAR amplitude-based geo-referencing approach, we propose a procedure relying on a well-established and simple coordinate transformation of a point from an inertial frame (H) to a new coordinate system (C), hereinafter the radar frame [16]. Equation (1) allows us to transform three-coordinates of the local flat Earth (H) frame into three-coordinates of the system (C). However, our goal is to transform a 3-coordinate system into a new two-coordinate system, where every point is a pixel of the scene imaged by the SAR. As a consequence, for the third coordinate the following condition holds:

$$z_C = 0 \tag{3}$$

Note that, z_C can be still computed as a function of the focal length for an optical image, but it cannot be generally derived for a SAR image if (ψ, θ, φ), $\mathbf{t}_C$ and γ_H^C are unknown.

By exploiting Equations (1) and (2), the following relation for the aircraft position (or equivalently the SAR position) estimation can be written:

$$\mathbf{t}_H = \mathbf{p}_H - \mathbf{R}_C^H(\psi, \theta, \varphi)\mathbf{p}_C \tag{4}$$

where it γ_C^H was set equal to one. The previous equation clearly states that SAR position ($\mathbf{t}_H = \mathbf{p}_H^{(SAR)}$) can be found if and only if: SAR attitude (ψ, θ, φ) is "correctly" measured or estimated; coordinates of a point landmark ($\mathbf{p}_H = \mathbf{p}_H^{(GCP)}$) are known (from a DB) in the inertial coordinate system H; the same landmark point is imaged by the SAR and its coordinates in the non-inertial coordinate system C, *i.e.*, $\mathbf{p}_C = \mathbf{p}_C^{(GCP)}$, are correctly extracted by the ATR processing chain.

Thus, SAR position can be computed by re-writing Equation (4) as follows:

$$\mathbf{p}_H^{(SAR)} = \mathbf{p}_H^{(GCP)} - \mathbf{R}_C^H(\psi, \theta, \varphi)\mathbf{p}_C^{(GCP)} \tag{5}$$

An SAR sensor essentially measures the slant-range between the sensor position $\mathbf{p}_H^{(SAR)}$ and a GCP, e.g.,: $\mathbf{p}_H^{(n)} = [0, R_n \cos\theta_n, 0]^t$, point A in Figure 2a located at the near range R_n and corresponding to a SAR attitude ($\psi = 0°$, $\theta = \theta_n$, $\varphi = 0°$). Thus, if the GCP is in the landmark DB (*i.e.*, $\mathbf{p}_H^{(n)}$ is known) and is correctly extracted from the SAR image by the ATR chain (*i.e.*, $\mathbf{p}_C^{(n)}$ is known), then the SAR sensor position $\mathbf{p}_H^{(SAR)}$ can be retrieved by using Equation (5). Note also that $(\psi, \theta_n, \varphi)$ have to be estimated or measured in order to exploit Equation (5).

It is worth noting that this section does not aim to provide an SAR-amplitude based geo-referencing procedure, but only a simplified mathematical approach to perform a feasibility analysis.

Table 1. Parameter classification, definition and measurement unit.

	Parameter	Definition	Measurement Unit
Coordinate Systems	H	Local flat Earth coordinate system in Figure 2b	–
	C	Local coordinate system, e.g., radar coordinates	–
	h	Altitude respect to frame H in Figure 2a	(km)
	$\mathbf{p}_H = [x_H, y_H, z_H]^t$	Position of a point in the H frame in Figure 2b, e.g., SAR or GCP ($\mathbf{p}_H^{(GCP)}$)	(m)
	$\mathbf{t}_H$	Translation vector (3×1) in the H frame	(m)
	γ_H^C	Scale parameter of the transformation from H to C	–
	$\mathbf{R}_H^C(\psi, \theta, \phi)$	Rotation matrix (3×3) from H to C frame	–
	$\mathbf{p}_C = [x_C, y_C, z_C]^t$	Position of a point in the C frame (e.g., GCP position $\mathbf{p}_C^{(GCP)}$)	(m)
	$\mathbf{t}_C$	Translation vector (3×1) in the C frame	(m)
	γ_C^H	Scale parameter of the transformation from C to H	–
	$\mathbf{R}_C^H(\psi, \theta, \phi)$	Rotation matrix (3×3) from C to H frame	–
	$\mathbf{p}_H^{(SAR)}, \mathbf{p}_H^{(n)}, \mathbf{p}_H^{(c)}, \mathbf{p}_H^{(f)}$	H frame coordinates of SAR/aircraft, point A, C and B in Figure 2a	(m)
	(ψ, θ, φ)	Euler angles in Figure 3	(°)
Radar	θ_0	Depression angle of radar beam-centre in Figure 2a	(°)
	$\theta_{inc} = 90° - \theta_0$	Incidence angle of radar beam-centre on a locally flat surface in Figure 2a	(°)
	d	Swath-width in Figure 2a	(km)
	$\Delta\theta_0$	Beam-width in the elevation plane in Figure 2a	(°)
	Δ_r, Δ_{cr}	Image resolution along range and cross-range	(m)
	R_n, R_0, R_f	Near, center of the beam and far range in Figure 2a	(km)
	R_{max} or R_f	Maximum detection range (or far range)	(km)
	$\theta_n, \theta_0, \theta_f$	Near, center of the beam and far depression angle in Figure 2a	(°)
Navigation	Δh	RALT accuracy in [%] (root mean square error—rmse)	–
	$\sigma_\psi, \sigma_\theta, \sigma_\phi$	IMU attitude accuracy on each component (rmse)	(°)
Processing	Δp	Inaccuracy on landmark position extraction from a SAR image	(pixel)

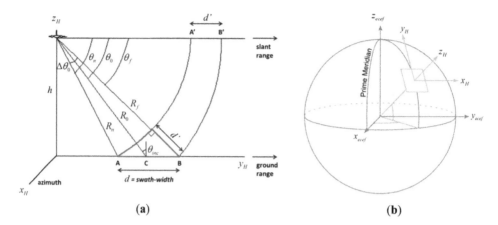

(a)   (b)

Figure 2. (a) Airborne side looking SAR geometry; (b) Basic flat-Earth geometry.

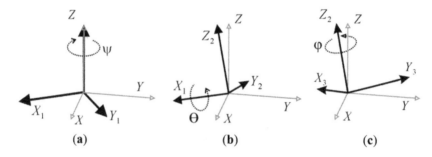

(a)   (b)   (c)

Figure 3. (a) Euler angles (ψ, θ, φ): (a) ψ defines the first rotation about the z-axis (note that the angle ψ in the figure is negative); (b) θ defines the second rotation about the x_1-axis (note that θ is negative); (c) φ defines the third rotation about the z_2-axis (note that φ is positive).

2.3. Feasibility Analysis

As already stated in Section 1, the feasibility analysis refers in particular to MALE UAV class [10], which permits heavy and wide payloads to be carried onboard, and can be affected by a dramatic cumulative drift during long mission when GPS data are not reliable. Some examples of MALE UAVs are Predator B, Global Hawk and Gray Eagle, which are actually equipped with SAR system and can be also equipped with InSAR system (more details in Section 3). Several competing system parameters have to be considered in the feasibility analysis as detailed in the following.

X band choice is mainly driven by both the limited payload offered by X-band SAR/InSAR systems and by the wavelength robustness to "rain fading" [17].

Concerning the polarisation, a single channel leads to a small and light SAR system that can fit onboard a UAV. Moreover, in the case of single polarisation, well-suited statistical distribution for terrain modelling can be employed by a fast and high-performance ATR chain. Finally, backscattering in VV polarization is higher than in HH polarisation for incidence angle greater than 50° and X band [17].

SAR image resolution has to be chosen as a trade-off among competing requirements. Range/cross-range resolution of about 1 m is suitable for recognizing terrain landmarks, *i.e.*, large targets such as buildings, cross-roads, whose shorter linear dimension is at least 10–20 times the suggested resolution [18]. On the contrary, pixel resolution higher than 1m would be unnecessary and increase ATR computational load.

Stripmap imaging mode allows us a shorter observation/integration time (as opposed to spotlight mode), lower computational complexity, and easier autofocusing [19]; moreover, it also requires a mechanical antenna steering mechanism simpler and lighter than other imaging modes [18].

Requirements on SAR acquisition geometry (*i.e.*, altitude h and attitude (ψ, θ, φ)), IMU attitude measurement, SAR swath-width, landmark characteristics, ATR accuracy, and operative scenarios, were derived through Montecarlo simulations according to the procedure described in Section 2.1.

In particular, the simulated scenarios assume good position accuracy on landmark coordinates derived by ATR chain, while exploring the various parameters in Section 2.2 to achieve the "best" aircraft position estimates through the SAR-amplitude based approach. Table 2 reports a résumé of the specific case studies presented in the following.

The swath-width d is defined by the near-range R_n and the far-range R_f reported in Table 2 for all the explored configurations. Position estimation is based on a single GCP and computed in three cases: point A (*i.e.*, $\mathbf{p}_H^{(n)}$), C (*i.e.*, $\mathbf{p}_H^{(c)}$) and B (*i.e.*, $\mathbf{p}_H^{(f)}$) in Figure 2a and in Table 2.

Table 2. Synthetic aperture radar (SAR) amplitudes-based approach: common parameter settings, case studies and results.

	SAR	Coordinates (SAR, GCPs) and Route	Source of Inaccuracy
Common Parameters of Case Studies	VV polarization X band Stripmap mode $\Delta_r = \Delta_{cr} = 1$ m	$\mathbf{p}_H^{(SAR)} = \mathbf{t}_H^{(SAR)} = [0,0,h]^t,$ $\mathbf{p}_H^{(n)} = [0, R_n \cos\theta_n, 0]^t,$ $\mathbf{p}_H^{(c)} = [0, R_0 \cos\theta_0, 0]^t,$ $\mathbf{p}_H^{(f)} = [0, R_f \cos\theta_f, 0]^t,$ $\psi = 0°, \phi = 0°; d = 2,\ldots,10$ km	$\sigma_\psi = \sigma_\theta = \sigma_\phi \in$ $[0.05°, 1°]$ $\Delta h = 0.5\%, 1\%, 2\%$ $\Delta p = \pm2, \pm4$

Case Study	SAR/Platform Position (m)	Other Settings (km)	SAR Position Estimates
CS#1	$\mathbf{p}_H^{(SAR)} = [0,0,8000]^t$	$R_n \in [45.6, 41.7], R_f \in [46.6, 51.5]$	Figure 4: accuracy lower than in CS#2–5, $\theta_0 = 10°$
CS#2	$\mathbf{p}_H^{(SAR)} = [0,0,6000]^t$	$R_n \in [11.6, 9.2], R_f \in [12.5, 18.0]$	Figure 5: $\theta_0 = 30°$
CS#3	$\mathbf{p}_H^{(SAR)} = [0,0,6000]^t$		Figure 6: accuracy lower than in CS#2, $\theta_0 = 40°$
CS#4	$\mathbf{p}_H^{(SAR)} = [0,0,4000]^t$	$R_n \in [7.6, 5.7], R_f \in [8.5, 14.7]$	Figure 7: $\theta_0 = 30°$
CS#5	$\mathbf{p}_H^{(SAR)} = [0,0,4000]^t$		Figure 8: accuracy lower than in CS#4, $\theta_0 = 40°$

We assumed that: SAR attitude (ψ, θ, φ) is measured by the IMU with a rmse ranging from 0.05° to 1° (without GPS correction), which is allowed by high accurate (*i.e.*, navigation grade class) commercial IMU such as LN-100G IMU [20]; h is measured by a RALT with accuracy equal to $\Delta h = 0.5\%, 1\%, 2\%$, which is allowed by commercial systems compliant with the regulation in [21]; SAR image resolution is $\Delta_r = \Delta_{cr} = 1$ m; inaccuracy on landmark position extraction in SAR image is $\Delta p = \pm2, \pm4$ pixels, which is compatible with the performance of the ATR algorithms [5,8]. Finally, without loss of generality, we refer to the simplified geometry in Figure 2a defined by the following relations: $\mathbf{p}_H^{(SAR)} = \mathbf{t}_H^{(SAR)} = \left[x_H^{(SAR)}, y_H^{(SAR)}, z_H^{(SAR)}\right]^t = [0,0,h]^t, \psi = 0°, \varphi = 0°$.

The first case study (CS#1 in Table 2) corresponds to a high altitude flight (*i.e.*, $h = 8$ km), which is allowed by a MALE UAV class [10]. Figure 4 depicts UAV/SAR position estimates under the "best" configuration (*i.e.*, $\theta_0 = 10°, \sigma_\psi = \sigma_\theta = \sigma_\phi = 0.05°, \Delta h = 0.5\%, \Delta p = \pm2$), with error bars proportional to the error standard deviation (std): no significant bias can be noted, but the std, which is very stable w.r.t. the swath-width, is too large on the 1st and 3rd component ($x_H^{(SAR)}, y_H^{(SAR)}$). Even by increasing θ_0 (from 10° to 40°), suitable results cannot be achieved. Note that if the IMU rmse was $\sigma_\psi = \sigma_\theta = \sigma_\phi = 0.5°$, the std of each component of $\mathbf{p}_H^{(SAR)}$ would be about 10 times greater than in Figure 4 (where IMU rmse = 0.05°). Analogously, if RALT accuracy was $\Delta h = 1\%$ and 2%, the std of both the 2nd and 3rd component of the estimated SAR position would be about 1.25 and 2.5 times greater than in Figure 4 (where $\Delta h = 0.5\%$). On the contrary, inaccuracy on landmark position in the image (derived by the ATR chain) has no appreciable impact on SAR position estimates (e.g., $\Delta p = \pm 2$, ± 4).

Thus, to reach suitable estimates, we decreased the SAR altitude to $h = 6000$ m (case study CS#2 in Table 2). The results from the "best" configuration are shown in Figure 5. It can be seen that all the estimated components of $\mathbf{p}_H^{(SAR)}$ have no bias and their variability is always bounded within ± 20 m. Fairly worse results are achieved by exploiting $\mathbf{p}_H^{(c)}$ as landmark, because of the higher uncertainty on the corresponding depression angle θ_c. Moreover, the estimates of $\mathbf{p}_H^{(SAR)}$ based on $\mathbf{p}_H^{(n)}$ are generally better than those based on $\mathbf{p}_H^{(f)}$, because the impact of IMU inaccuracy is stronger on θ_f than on θ_n ($\theta_f < \theta_n$).

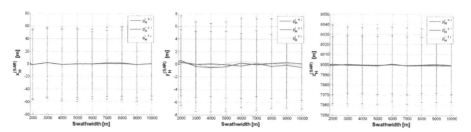

Figure 4. SAR position estimates (CS#1 in Table 2) based on a single GCP ($\mathbf{p}_H^{(n)}$, $\mathbf{p}_H^{(c)}$, $\mathbf{p}_H^{(f)}$): $\mathbf{p}_H^{(SAR)} = [0, 0, 8000]^t$ m, $\theta_0 = 10°$, $\sigma_\psi = \sigma_\theta = \sigma_\phi = 0.05°$, $\Delta h = 0.5\%$, $\Delta p = \pm 2$. Error (on each SAR position component) mean value curve as a function of swath-width (d); error bars show the error standard deviation along the curve.

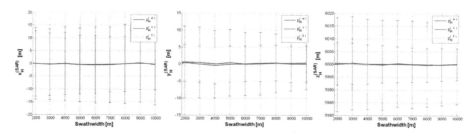

Figure 5. SAR position estimates (CS#2 in Table 2) based on a single GCP ($\mathbf{p}_H^{(n)}$, $\mathbf{p}_H^{(c)}$, $\mathbf{p}_H^{(f)}$): $\mathbf{p}_H^{(SAR)} = [0, 0, 6000]^t$ m, $\theta_0 = 30°$, $\sigma_\psi = \sigma_\theta = \sigma_\phi = 0.05°$, $\Delta h = 0.5\%$, $\Delta p = \pm 2$.

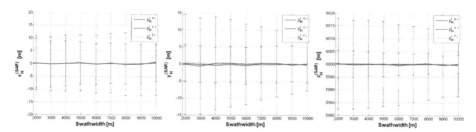

Figure 6. SAR position estimates (CS#3 in Table 2) based on a single GCP ($\mathbf{p}_H^{(n)}$, $\mathbf{p}_H^{(c)}$, $\mathbf{p}_H^{(f)}$): $\mathbf{p}_H^{(SAR)} = [0, 0, 6000]^t$ m, $\theta_0 = 40°$, $\sigma_\psi = \sigma_\theta = \sigma_\phi = 0.05°$, $\Delta h = 0.5\%$, $\Delta p = \pm 2$.

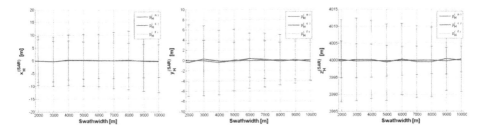

Figure 7. SAR position estimates (CS#4 in Table 2) based on a single GCP ($\mathbf{p}_H^{(n)}, \mathbf{p}_H^{(c)}, \mathbf{p}_H^{(f)}$): $\mathbf{p}_H^{(SAR)} = [0, 0, 4000]^t$ m, $\theta_0 = 30°$, $\sigma_\psi = \sigma_\theta = \sigma_\phi = 0.05°$, $\Delta h = 0.5\%$, $\Delta p = \pm 2$.

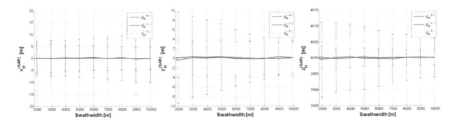

Figure 8. SAR position estimates (CS#5 in Table 2) based on a single GCP ($\mathbf{p}_H^{(n)}, \mathbf{p}_H^{(c)}, \mathbf{p}_H^{(f)}$): $\mathbf{p}_H^{(SAR)} = [0, 0, 4000]^t$ m, $\theta_0 = 40°$, $\sigma_\psi = \sigma_\theta = \sigma_\phi = 0.05°$, $\Delta h = 0.5\%$, $\Delta p = \pm 2$.

We also considered a case study (CS#3 in Table 2) with greater SAR depression angle ($\theta_0 = 40°$), while keeping platform h as in CS#2. Figure 6 show CS#3 results, which keep on satisfying the constraint on position accuracy. Note that the accuracy on $x_H^{(SAR)}$ is improved compared with CS#2, but it is worse on $y_H^{(SAR)}$, because of the wider $\Delta \theta_0$ for $\theta_0 = 40°$ than for $\theta_0 = 30°$ (*i.e.*, CS#2) needed to illuminate the same swath-width. For depression angles higher than 40°, the accuracy on the estimates further worsens: such a trend can be also observed by comparing Figures 5 and 6.

As a limiting case, we reduced the UAV altitude down to $h = 4000$ m (CS#4 in Table 2). Figure 7 shows position estimates under the "best" configuration ($\theta_0 = 30°$): the estimated $\mathbf{p}_H^{(SAR)}$ components show negligible bias and their variability is even lower than in the previous case studies, *i.e.*, ± 15 m. Note that, for swath-width greater than 7000 m, only $x_H^{(SAR)}$ exceeds the previous boundary.

We also considered a further case study (CS#5 in Table 2) with a depression angle $\theta_0 = 40°$ greater than in CS#4. All the estimates shown in Figure 8 are rigorously bounded within ± 15 m and SAR position estimates are generally very similar to those achieved in CS#4. Only the inaccuracy on $y_H^{(SAR)}$ slightly increases, because of the wider $\Delta \theta_0$ for $\theta_0 = 40°$ than for $\theta_0 = 30°$. Again, for depression angles higher than 40°, the accuracy on the estimates further worsens.

A UAV altitude lower than 4000 m cannot be taken into account because it would have severe consequences on both SAR system requirements and SAR data processing. In fact, in order to keep constant the values of both θ_0 and swath-width, the lower the altitude, the wider $\Delta \theta_0$, thus leading to considerably increasing the required transmitted power. Moreover, a wide $\Delta \theta_0$ means large resolution changes across the ground swath [8], which negatively impact the performance of the ATR algorithm: features that are clearly distinguishable at far range can become nearly invisible at near range.

In conclusion, feasible parameter settings are those relative to CS#2 and CS#4 configurations, which provide estimated UAV coordinates with errors bounded within ± 18 m and ± 12 m, respectively. In Table 3, a résumé of the suggested requirements, *i.e.*, a first trade-off, is listed. Note that the suggested range of SAR swath-width d (*i.e.*, few kilometres) allows us to be confident about the presence of the desired landmarks within the illuminated area, even if the SAR system (due to uncertainty on attitude and position) points at the wrong direction. It is worth noting that SAR requirements in Table 3 can be easily fulfilled by commercial systems, e.g., Pico-SAR radar produced by Selex-ES [22]; MALE UAV [10] such as Predator B, Global Hawk and Gray Eagle, which easily carry a Pico-SAR radar; navigation grade class IMU such as LN-100G IMU [20]; any RALT compliant with the regulation in [21].

Table 3. Platform, viewing geometry, navigation sensor requirements and corresponding reference system which allows a feasible SAR position retrieval: *first trade-off.*

	First Setting (CS#2)	Second Setting (CS#4)
Aircraft altitude (km)	6	4
Depression angle θ_0 (°)	30	30
Swath-width d (km)	few units	few units
Elevation beam-width $\Delta\theta_0$ (°)	≥11.5	≥16.6
Image resolution (m)	~1	
SAR band	X	
SAR polarization	VV	
R_n (km); R_0 (km); R_f (km)	10.3−9.2; 12; 14.6−18.0	6.5−5.7; 8; 10.8−14.7
Maximum detection range R_f (km)	≥14.6	≥10.8
Δh (%)	0.5 (or fairly worse)	0.5 (or fairly worse)
$\sigma_\psi, \sigma_\theta, \sigma_\phi$ (°)	0.05	0.05
Estimated UAV based on $\mathbf{p}_H^{(c)}$ (bias ± std) (m)	0 ± 15, 0 ± 11, 0 ± 18 (Figure 5)	0 ± 12.5, 0 ± 7.5, 0 ± 12 (Figure 7)
Reference systems	MALE UAV, e.g.,: Predator B, Global Hawk, Gray Eagle	
	Pico-SAR	
	Any RALT compliant with the regulation in [21]	
	Navigation grade class IMU, e.g.,: LN-100G IMU	

2.4. Landmark DB and Mission Planning

In order to retrieve SAR position, the proposed algorithm has to exploit landmark points. In the previous section, it is assumed that landmark coordinates in the mission DB are ideal, *i.e.*, without errors. Such an assumption is quite unlikely, and, consequently, we also evaluated the impact of landmark coordinate inaccuracy on the estimated UAV position. According to the previous analysis, preliminary requirements were derived for the landmark DB reference domain, typology, and accuracy. Concerning the reference domain, an inertial system of coordinates (*H*) can be adopted, e.g., Earth-centered frame or local map coordinate system. Concerning the landmark typology, planar landmarks (e.g., crossroad, roundabout, railway crossing) are strongly suggested because they are very recurrent in several operating scenarios, can be precisely extracted, and do not introduce any vertical distortion due to elevation with respect to the ground surface [8]. Small 3D-landmarks can be also exploited but in the limits of the visibility problems due to the shadow areas occurring close to high structures. According to this, a mission planning should avoid scenarios densely populated by buildings, while preferring suburban or rural areas.

Concerning landmark DB accuracy, we exploited the procedure in Section 2.2 and the most promising two settings in Section 2.3. Table 4 presents the platform geo-referencing accuracy derived under two different error settings: a "moderate" DB accuracy with RMSE equal to $[3, 3, 3]^t$ in (m) on landmark coordinates of $\mathbf{p}_H^{(c)}$, and a "low" DB accuracy with RMSE equal to $[3, 3, 10]^t$ in (m). Results show a limited increase in the final std values with respect to the ideal results in Section 2.2.

The DB accuracy assumed above is quite reasonable, because the coordinates of interest refer to centroid and corners of small landmarks, and can be derived by exploiting a Geographic Information System (GIS) archive (or similar) [23]. The mapping standards employed by the United States Geological Survey (USGS) specifies that [24]: 90% of all measurable horizontal (or vertical) points must be within ±1.01 m at a scale of 1:1200, within ±2.02 m at a scale of 1:2400, and so on. Finally, the errors related to the geo-referencing could be modelled as a rigid translation of a landmark with respect to its ideal position, with RMSE smaller than 10 m.

Table 4. Platform position retrieval (bias ± standard deviation of each coordinate) as a function of the settings in Table 1 and Data Base (DB) accuracy on landmark coordinates of $\mathbf{p}_H^{(c)}$.

Landmark DB Accuracy	CS#2	CS#4
"Moderate" (rmse [3, 3, 3]^t in (m))	0 ± 14.8 m, 0 ± 10.2 m, 0 ± 17.8 m	0 ± 12.4 m, 0 ± 05.8 m, 0 ± 12.1 m
"Low" (rmse [3, 3, 10]^t in (m))	0 ± 14.6 m, 0 ± 10.2 m, 0 ± 20.2 m	0 ± 12.4 m, 0 ± 05.8 m, 0 ± 15.3 m

3. InSAR Phase Exploitation

When the use of SAR amplitude is unfeasible because of unreliable surveyed scenario, the InSAR phase can be exploited as derived by a single pass interferometer mounted onboard the platform. The proposed approach is sketched in Figure 9 and basically consists in comparing the real InSAR phase with a synthetic one derived by using both approximated position and attitude values and a reference DTM. The matching block provides a feedback to be used for adjusting position and attitude values derived by the instruments available onboard.

3.1. InSAR Setting

According to the acquisition geometry sketched in Figure 10, the phase difference (Φ) between two slightly displaced SAR observations S1 and S2 is related to the difference in time delay from a given pixel to each antenna of the SAR Interferometer. In the hypothesis of bistatic acquisitions (real interferometer), the InSAR phase is not affected by neither atmosphere, nor ground deformation. It depends only on the geometrical distance and can be expressed in terms of the reference phase (Φ_{ref}), which accounts for the difference between the slant range geometry and the reference elevation model (ellipsoidal or flat), while the topographic phase (Φ_h) is related to the elevation h w.r.t. the reference elevation model [25]:

$$\Phi(p) = \Phi^{ref}(p) + \Phi^h(p) \approx \frac{2\pi}{\lambda} \cdot \left[L_{//,p} - \frac{h(p)}{h_a} \right] \tag{6}$$

where λ is the SAR wavelength, $L_{//,p}$ is the component parallel to the master slant direction of the geometrical distance between the two acquisitions (or baseline, L), and h_a is the so-called *height of ambiguity*. This last is defined as:

$$h_a = \frac{\lambda R_{S1,p} sen\left(\theta_p - \alpha_p\right)}{L_{\perp,p}} \tag{7}$$

where $R_{S1,p}$ and $\theta_p{}^{ref}$ are, respectively, the range distance and the look angle relative to the master (S1) view computed w.r.t. the reference elevation at the pixel p; $L_{\perp,p}$ is the component of the baseline orthogonal to the slant direction of the master computed at pixel p; α_p is the terrain slope. h_a defines the InSAR height sensitivity, depends on the SAR system specifications, and drives the performance of the InSAR processing for height computation. In particular, the smaller the value of the height of ambiguity, the smaller the detectable height difference.

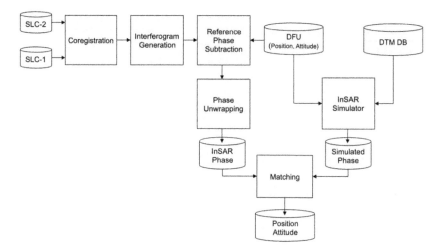

Figure 9. Flow chart of the InSAR based navigation system.

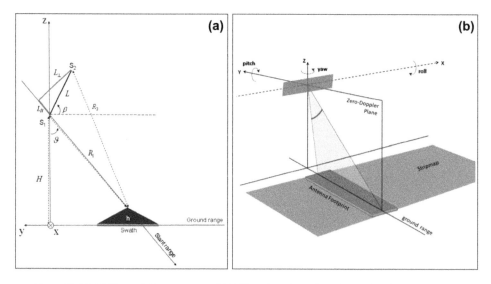

Figure 10. (a) InSAR acquisition geometry; (b) UAV position (X, Y, Z) and attitude (roll, pitch, yaw) in the basic SAR geometry.

Since h_a is inversely proportional to the effective baseline $L_\perp$, high baseline values are recommended for accurate topography mapping. Basically, the aircraft size and shape limit the baseline values and consequently h_a. The baseline component $L_\perp$ depends also on both the look angle and the orientation angle β (see Figure 10), which can be optimized in order to maximize $L_\perp$. In the following, we assume a vertical baseline ($\beta = 90°$), which guarantees a symmetric weight distribution across the aircraft fuselage. An inclination angle β equal to the mean look angle ($\beta = \theta$ in Figure 10) maximizes the orthogonal component of the baseline $L_\perp$ leading to increase the height sensitivity up to 5% for low incident angle.

The following analysis was carried out assuming, as in the case of amplitude-based procedure, a MALE UAV class [10], which can carry on a maximum payload of 150×130 cm^2 and reach a maximum flight altitude of 8000 m.

3.2. Analysis with Respect to InSAR Phase Integrity

The reliability of the InSAR processing depends on the quality of the InSAR phase, which is related to the correlation between the two acquisitions. Changes of the terrain conditions between the two acquisitions, system noise, and approximations in the processing lead to correlation loss. The image correlation can be assessed in terms of the InSAR coherence γ [9], which is a normalized correlation coefficient varying from 0 (full decorrelation) to 1 (full correlation). The coherence is directly related to the signal to noise ratio of the interferogram and can be partitioned according to different sources of decorrelation. In particular, the geometric decorrelation, δ_{geo}, is due to the difference between the incidence angles of the two acquisitions. This geometrical difference causes a spectral shift between the signal spectral bands, which is proportional to the effective geometrical baseline [26]. When the spectral shift equals the signal bandwidth the SAR images are totally uncorrelated and the baseline has a value known as critical baseline.

The volume decorrelation, δ_{vol}, is due to the penetration of the radar waves into the soil or vegetation and depends on both radar wavelength and scattering medium (bare soil, urban structures, vegetation or forest) [27]. The volume scattering can be neglected in case of low penetration (bare soil) or high extinction rate (forested area) as for X band or higher. In the following we assume to refer to an X band system thus allowing us to neglect the volume decorrelation. This choice is also able to guarantee better height sensitivity, as from Equation (7).

A further important source of errors comes from the fact that the InSAR phase field is obtained by extracting the phase term of the complex interferogram leading to an ambiguity of 2π in the real phase measurement. The Phase Unwrapping (PU) consists in deriving the correct map of the multiples of 2π to be added to the "wrapped" phase in order to infer the absolute phase field correlated to the ground topography images [28]. This processing step is critical in terms of both computational needs and reliability. SAR systems at high spatial resolution, smooth terrains, and proper baselines can help in performing a reliable PU. In particular, the height of ambiguity, which depends on the baseline value according to Equation (7), provides also the sensitivity to the wrapping of the InSAR phase. Thus, large baseline values improve the accuracy on height estimations, but also lead to a wide range of terrain slopes affected by aliasing. According to the geometric and radiometric system parameters, the critical slope angle α_C can be defined so that if the terrain slope α is in the interval $[\theta - \alpha_C, \theta + \alpha_C]$, then the InSAR phase is affected by aliasing and PU fails.

Therefore, height of ambiguity, geometrical decorrelation and critical angle are useful indicators of the expected performances of the interferometer. The first one provides the InSAR sensitivity to the height variation. The second impacts on the InSAR phase noise (usually estimated through the coherence) which causes artifacts in the phase unwrapping, thus decreasing the accuracy of the final measurement. The latter provides a direct indication on the areas where phase unwrapping could fail.

We explored these three figures to derive reliable radiometric and geometric parameters for the InSAR configuration. In particular, high baselines are required for increasing the InSAR sensitivity to the height; however, at the same time, they increase the geometrical decorrelation as well as the probability of phase aliasing to occur, thus making PU more problematic. High carrier frequencies, or equivalently short wavelengths, increase the geometrical decorrelation but in general ensure less penetration, thus limiting the effect of volumetric decorrelation.

Moreover, by increasing the bandwidth, or equivalently the spatial resolution, the geometrical decorrelation (and consequently the critical angle α_C) decreases as well, thus leading to improved performances of the interferometer.

Figure 11 shows the behavior of δ_{geo}, h_a and α_C with respect to different values of vertical baseline and incident angle. We simulated a bistatic InSAR system with: baseline $L = [0.3, 0.4, 0.6, 0.9, 1.3]$ m, orientation angle $\beta = 90°$ (*i.e.*, vertical orientation—see Figure 10a), aircraft elevation of 8 km, look angles ranging from 10° up to 80°. Moreover, we assumed a SAR sensor with high spatial resolution, B = 150 MHz, working at X band. Figure 11a shows the trends for δ_{geo}. In this case, the performance improves (δ_{geo} decreases) by increasing the look angle and by decreasing the baseline L. In general, the geometrical decorrelation is very limited (<0.006) for every configuration. In case of bistatic acquisitions (no temporal decorrelation) and short wavelengths (no volume decorrelation), the InSAR coherence is expected to be quite high except for the noise related to both processing (autofocusing and co-registration) and electronic devices. Figure 11b shows the trends for h_a. For a fixed incidence angle, the InSAR height sensitivity increases (or equivalently h_a decreases) as the baseline L increases. Moreover, the sensitivity decreases by increasing the look angle. In general, the sensitivity can be quite limited and, for some configurations, h_a is really too high (>1000 m) to guarantee reliable performances (<1/100 rad). Figure 11c shows the trends for α_C. In this case, the performance improves (α_C decreases) by decreasing the baseline L and it is not monotonic with the look angle. However, the values of α_C are always less than 1°, leading to very limited range of terrain slopes forbidden due to phase aliasing. Thus, the major problem in conventional SAR interferometry (*i.e.*, the reconstruction of absolute phase values starting from the InSAR wrapped principal phase fields, alias PU) is strongly reduced.

From this preliminary investigation we can conclude that the explored configurations are not problematic in terms of both geometrical decorrelation and phase unwrapping. Thus, the height sensitivity should drive the selection of requirements for both radiometric and geometric parameters. In particular, high baseline values should be preferred. In the present case, the baseline value is limited by the aircraft size and shape (also considering possible external aerodynamics appendices).

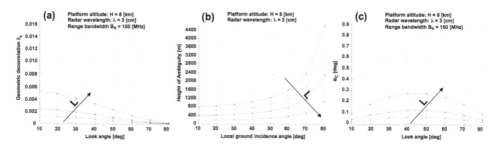

Figure 11. (a) Height of ambiguity h_a, (b) geometric decorrelation δ_{geo}, and (c) critical angle α_C *vs.* look angle for B = 150 MHz, λ = 3 cm , platform altitude H = 8 km, vertical baseline L = [0.3, 0.4, 0.6, 0.9, 1.3] m (from blue to purple), baseline orientation angle β = 90° (vertical orientation see Figure 2).

3.3. Analysis with Respect to Altitude and Position Changes

In the previous section, we assessed the feasibility of the proposed approach (sketched in Figure 9) with respect to the quality of the InSAR phase field. The same assessment is required with respect to the InSAR sensitivity to changes on aircraft position and attitude, which are the parameters to be estimated. To this aim, we refer to the SAR looking geometry sketched in Figure 10b (body frame, or b-frame), where: the azimuth direction defines the X coordinate, the Y axis is opposite to the ground range direction (assuming right looking systems), the Z axis is coincident with the local zenith direction, the roll error angle is measured in the YZ plane from the Z axis towards the Y axis, the yaw error angle is measured counter-clockwise in the XY plane from the X axis, the pitch error angle is defined in the XZ plane from the nadir direction towards the X axis.

We developed a geometrical model that, referring to a point target on the ground with topographic height h, relates changes on aircraft attitude/position to changes on both InSAR phase and target SAR coordinates (slant range, azimuth). According to this model, we investigated the effects that errors in the UAV attitude/position have on both the InSAR phase and the slant range/azimuth misalignments.

The same parameters of the previous analysis were assumed (H = 8 km; λ = 3 cm). Incidence angle values between 15° and 60° were explored in order to limit the height of ambiguity. As we will see, the influence of the topographic height on the sensitivity analysis is very poor and it is almost negligible for high incidence angles ($\theta \approx 60°$).

The sensitivity of the InSAR phase to roll errors does not vary appreciably with the platform height, while it increases with the incidence angle, and, as expected, is heavily affected by the baseline length. Plots (a) and (b) in Figure 12 show, respectively for incident angles at near (θ = 15°) and far range (θ = 60°), the residual InSAR phase cycles evaluated for roll errors between −3° to 3°, by assuming baseline L = 1.3 m and topographic heights ranging from −100 m to 1 km. With regard to the sensitivity on the position of the selected target, the analysis is restricted to the slant range, since a roll error cannot introduce a misalignment along the azimuth direction. Roll errors even up to 3° (in absolute value) lead to slant range variations less than 1 m. Hence, the slant range sensitivity is very poor in this case.

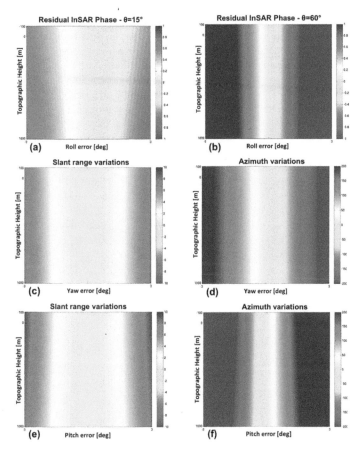

Figure 12. Subplots (a,b) show residual InSAR phase cycles evaluated for roll errors between −3° and 3°, and incident angles at near range (a) and far range (b). Subplots (c–f) show the misalignments in slant range and azimuth evaluated for yaw (c,d) and pitch errors (e,f) between −3° and 3°, assuming topographic heights ranging from −100 m to 1 km, and an incident angle at mid-range (θ_{inc} = 30°). For all the plots, the topographic height ranges from −100 m to 1 km.

Concerning errors on the yaw angle, the analytical model predicts a poor sensitivity of the InSAR phase and a high sensitivity of the azimuth misalignment, which increases with both incident angle and aircraft altitude. In particular, a yaw error of just 1° may lead to azimuth shifts close to 200 m (see Figure 12d), with incidence angles close to 60°. The slant range sensitivity is almost independent from the aircraft altitude and it slightly increases with the incidence angle, while a yaw error of 1° never produces a slant range variation exceeding 2 m (see Figure 12c) for all the explored configurations. Moreover, an intrinsic uncertainty on the sign of the yaw error affects the estimation based on slant range misalignment.

As for the yaw angle, sensitivity of InSAR phase to errors on pitch angle is generally very poor. Also the slant range misalignment is very low in particular for pitch errors not exceeding 1° (see Figure 12e). On the contrary, azimuth misalignments are important, regardless the platform height and the incidence angle. 1° pitch error may lead to azimuth misalignments higher than 200 m, as shown in Figure 12f. Errors on the along-track coordinate (*i.e.*, X coordinate) do not affect the InSAR phase or the Zero-Doppler distance between the target and the sensor after SAR focusing. They only lead to a corresponding azimuth misalignment in the focused image.

Let us consider now an error only on the UAV position, *i.e.*, on the Y coordinate (ground range in Figure 10a). The residual InSAR phase can be computed analytically through geometrical modeling. The sensitivity of the InSAR phase is high for low orbits and at mid-range, and it increases significantly with the baseline length (Equation (6)). In general for a platform altitude of 8 km the InSAR phase sensitivity is quite limited (see Figure 13a). Concerning sensitivity of the target position to Y errors, the analysis is restricted to the slant range, since an error on the Y position cannot introduce a misalignment along the azimuth direction. Slant range sensitivity significantly increases with the incidence angle (see plots (c) and (d) in Figure 13 derived at near and fare range, respectively), while the effect of the platform altitude is negligible. Errors in slant range up to 200 m can occur.

The last parameter to investigate is the Z position of the UAV platform. In this case, the sensitivity of the InSAR phase is high for low orbits and at far range (see Figure 13b), and again it increases significantly with the baseline length. In this case, the analysis of the sensitivity on the position of a selected target to Z errors is also restricted to the slant range, since an error on the Z position cannot introduce a misalignment along the azimuth direction. Slant range sensitivity is in general considerable and increases at near range (see plots (e) and (f) in Figure 13), while the effect of the platform altitude is negligible, as for the Y parameter.

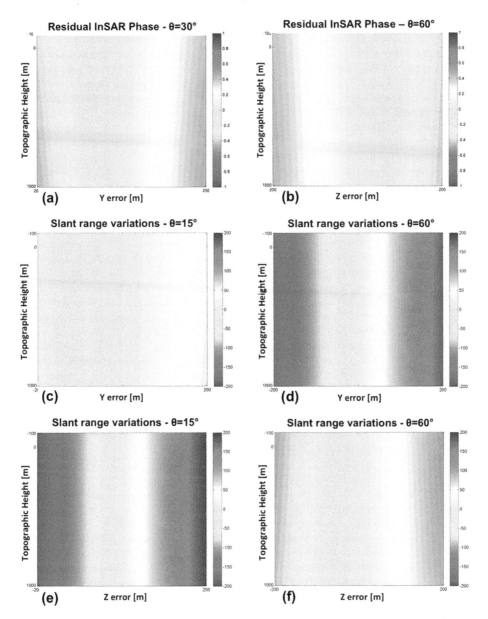

Figure 13. Subplots (**a**) and (**b**) show residual InSAR phase cycles evaluated for Y errors (**a**) and Z errors (**b**) ranging from −200 to 200 m, and incident angles at mid-range and far range respectively. Subplots (**c–f**) show the variation of the target position along slant rage (meters) evaluated for Y errors (**c,d**) and Z errors (**e,f**) ranging from −200 to 200 m, and for incident angle at near range (**c,e**) and far range (**d,f**). For all the plots, the topographic height ranges from −100 m to 1km.

3.4. Analysis with Respect to DTM Characteristics

The performances of the InSAR-based geo-referencing procedure depend on the Matching algorithm, which, in turn, depends on the InSAR height sensitivity, the spatial variability of the terrain profile and the specifications of the reference DTM available onboard. The implementation of a specific matching algorithm is beyond the scope of this paper, but a possible processing scheme can designed, based on SIFT technique [29] and, in general, strong spatial variability is required in order to improve the correlation performance. Thus, it is expected that the InSAR-based geo-referencing procedure would be reliable on hilly or mountainous areas, and when using the configuration with the highest baseline value (1.3 m in our simulated scenario), which ensures the best InSAR sensitivity to the terrain elevation.

Both spacing of the grid points and accuracy of the elevation values determine the quality of a DTM. For our purposes, a reliable DTM spacing should be comparable to the spatial resolution of the InSAR products. This, in general, is worse than the resolution of the original SAR images because of the multi-viewing required to decrease the phase noise. The number of looks adopted in this smoothing procedure can be set to the minimum value, which provides a multi-looked resolution cell comparable to the horizontal geolocation accuracy. Assuming the most favorable InSAR configuration of L = 1.3 m, and an InSAR coherence of 0.9, a reliable value for the number of looks is about four, both in range and azimuth.

The impact of the DTM spacing on the InSAR-based geo-referenced procedure was evaluated by using the difference between the simulated InSAR phase fields derived by using DTM with different postings. A statistical analysis was performed by computing the std of the InSAR phase errors. Assuming the most favorable configuration of L = 1.3 m, and a multi-look value of 4×4, the DTM spacing does not exceed 10 m. This result meets the specifications of the Level 3 DTM standard defined by the National Imagery and Mapping Agency (NIMA) [30] as: spatial resolution of 12 m, absolute vertical accuracy less than 10m, relative vertical accuracy less than 2 m.

The requirement on the DTM height accuracy can be derived by considering as reliable a value at least comparable to the InSAR height sensitivity. Assuming the most favorable configuration of L = 1.3 m, and an InSAR coherence of 0.9, the height sensitivity is in general lower than $10 \div 13$ m, depending on the look angle. Therefore, a DTM of Level 3 for the NIMA standard again meets the requirements for the height accuracy.

Finally, the TanDEM-X mission is generating a DEM fulfilling these requirements [31] and covering the whole Earth surface, so that reliable input data to the InSAR-based geo-referencing procedure will be soon available globally.

4. Conclusions

The goal of the paper is to propose an advanced integration of SAR imaging into UAV navigation systems and to prove the technological feasibility of such integrated system. We address the problem of drifts in planned trajectory due to inaccurate INS measurements occurring when a GPS signal is absent or corrupted by either intentional or unintentional interferences. This problem can affect in particular the MALE UAV class during a long endurance flight. However, the same UAV class permits to carry out onboard heavy and wide payloads, thus making feasible the use of SAR/InSAR technology. SAR sensors are valuable for a UAV navigation backup system thanks to the day/night and all-weather imaging, and to the wide area illuminated from a long distance by ensuring high spatial resolution.

A feasibility analysis was aimed at deriving requirements for the geo-referencing procedure performed on a UAV platform and based on the SAR amplitude (if terrain landmarks are available) and on the SAR Interferometry (if terrain landmarks are not available).

Simple and well-established approaches were presented to derive a feasibility analysis concerning both the amplitude-based and the InSAR-based procedures, and considering different competitors: SAR system, image acquisition geometry, navigation instruments, UAV platform, landmarks and DTMs.

Results detailed in Section 2.4, 2.5 and 3 finally show that both SAR amplitude- and InSAR-based approaches here proposed are technologically feasible by resorting to a certain class of UAV platform, IMU, RALT, both SAR and InSAR systems, typical landmark position accuracy and class (e.g., buildings and road network in a suburban/rural scenario), and DTM accuracy. The requirements derived from the feasibility analysis can be fulfilled by commercial systems e.g., Pico-SAR radar produced by Selex-ES [22]; MALE UAV [10] such as Predator B, Global Hawk and Gray Eagle, which easily carry a Pico-SAR radar; navigation grade class IMU such as LN-100G IMU [20], and any RALT compliant with the regulation in [21]. Furthermore, the size of these UAVs permits hosting SAR interferometric system with baseline values able to provide reliable performances in terms of both phase noise and sensitivity. We also proved that a DTM of Level 3 for the NIMA standard meets the requirements for both DTM spacing and height accuracy. Moreover, a global DTM of such quality will be soon available thanks to the TanDEM-X mission.

Concerning the amplitude-based approach, feasible parameter settings are those relative to CS#2 and CS#4 configurations in Table 3, which provide estimated aircraft coordinates with errors bounded within about ±15 m. It can be clearly stated that the parameter settings or requirements (here derived) are more focused on aerial platform, SAR systems and navigation sensors than on the image processing chain (SAR data Autofocusing, ATR chain), which is out of the scope of this paper but already exploited also for UAVs (e.g., [5,6]).

Concerning the InSAR-based approach, the analysis showed that the phase variation does not seem to be a useful figure to correct position and attitude. On the contrary, the exploration of SAR coordinates variation due to changes in position and attitude of the aircraft appears feasible. This result suggests that, for the matching algorithm, it seems promising to explore the difference in range and azimuth location between the real and simulated InSAR phase profiles, instead of looking at the phase differences. A possible approach can be based on SIFT technique [29].

Future work will be devoted to develop an *ad hoc* and performing ATR chain (for the amplitude-based approach) and a matching algorithm (for the InSAR-based approach), as well as to process real data.

Acknowledgments: The authors acknowledge the support of the SARINA project A-0932-RT-GC, which is coordinated by the European Defence Agency (EDA) and partially funded by 10 contributing Members (Cyprus, France, Germany, Greece, Hungary, Italy, Norway, Poland, Slovakia, Slovenia and Spain) in the framework of the Joint Investment Programme on Innovative Concepts and Emerging Technologies (JIP-ICET).

Author Contributions: All authors contributed to the literature survey and result analysis. Nitti, Bovenga and Chiaradia worked mainly on the SAR interferometry based approach, while Greco and Pinelli on the landmark based approach.

Conflicts of Interest: The authors declare no conflict of interest.

References

1. Chiang, K.; Tsai, M.; Chu, C. The Development of an UAV Borne Direct Georeferenced Photogrammetric Platform for Ground Control Point Free Applications. *Sensors* **2012**, *12*, 9161–9180. [CrossRef] [PubMed]
2. McLean, D. *Automatic Flight Control Systems*; Prentice Hall: New York, NY, USA, 1990.
3. Hasan, A.M.; Samsudin, K.; Ramli, A.R.; Azmir, R.S.; Ismaeel, S.A. A Review of Navigation Systems (Integration and Algorithms). *Aust. J. Basic Appl. Sci.* **2009**, *3*, 943–959.
4. Greco, M.; Querry, S.; Pinelli, G.; Kulpa, K.; Samczynski, P.; Gromek, D.; Gromek, A.; Malanowski, M.; Querry, B.; Bonsignore, A. SAR-based Augmented Integrity Navigation Architecture: SARINA project results presentation. In Proceedings of 13th IEEE International Radar Symposium (IRS), Warsaw, Poland, 23–25 May 2012; pp. 225–229.
5. González-Partida, J.; Almorox-González, P.; Burgos-Garcia, M.; Dorta-Naranjo, B. SAR System for UAV Operation with Motion Error Compensation beyond the Resolution Cell. *Sensors* **2008**, *8*, 3384–3405. [CrossRef]
6. Aguasca, A.; Acevo-Herrera, R.; Broquetas, A.; Mallorqui, J.; Fabregas, X. ARBRES: Light-Weight CW/FM SAR Sensors for Small UAVs. *Sensors* **2013**, *13*, 3204–3216. [CrossRef] [PubMed]
7. Curlander, J.C.; McDonough, R.N. *Synthetic Aperture Radar: Systems and Signal Processing*; J. Wiley & Sons, Inc.: New York, NY, USA, 1991.

8.   Oliver, C.; Quegan, S. *Understanding Synthetic Aperture Radar Images*; SciTech Publishing, Inc.: Raleigh, NC, USA, 2004.

9.   Rosen, P.A.; Hensley, S.; Joughin, I.; Li, F.K.; Madsen, S.N.; Rodriguez, E.; Goldstein, R.M. Synthetic Aperture Radar Interferometry. *IEEE Proc.* **2000**, *88*, 333–382. [CrossRef]

10. Weibel, R.E.; Hansman, R.J. Safety Considerations for Operation of Different Classes of UAVs in the NAS. In Proceedings of the AIAA 3rd "Unmanned Unlimited" Technical Conference, Workshop and Exhibit, Chicago, IL, USA, 20–23 September 2004.

11. Barton, D.K. Radar Equations for Clutter and Jamming. In *Radar Equations for Modern Radar*; Artech House, Inc.: Norwood, MA, USA, 2013; pp. 55–107.

12. *World Geodetic System 1984: Its Definition and Relationships with Local Geodetic Systems*; Technical Report TR8350.2; National Imagery and Mapping Agency (NIMA): St. Louis, MO, USA, 2000.

13. Etkin, B.; Reid, L.D. The Stability Derivatives. In *Dynamics of Atmospheric Flight, Stability and Control*, 3rd ed.; John Wiley & Sons: New York, NY, USA, 1972; pp. 129–160.

14. Matlab. Available online: http://it.mathworks.com/help/aerotbx/ug/lla2flat.html (accessed on 27 March 2015).

15. Siciliano, B.; Sciavicco, L.; Villani, L.; Oriolo, G. Kinematics. In *Robotics—Modelling, Planning and Control*, 2nd ed.; Springer-Verlag: London, UK, 2009; pp. 39–103.

16. Glasbey, C.A.; Mardia, K.V. A penalized approach to image warping. *J. R. Stat. Soc. Ser. B Stat. Methodol.* **2001**, *63*, 465–492. [CrossRef]

17. Richards, J.A. *Remote Sensing with Imaging Radar*; Springer-Verlag GmbH: Berlin/Heidelberg, Germany, 2009.

18. Hopper, G.S. Forward-looking Infrared Systems. In *The Infrared & Electro-Optical Systems Handbook, Passive Electro-Optical Systems*; Campana, S.B., Ed.; SPIE—International Society for Optical Engine: Bellingham, WA, USA, 1993; Volume 5, pp. 105–158.

19. Carrara, W.G.; Goodman, R.S.; Majewski, R.M. *Spotlight Synthetic Aperture Radar*; Artech House Inc.: Norwood, MA, USA, 1995.

20. Northrop Grumman LN-100G Inertial Measurement Unit datasheet. Available online: http://www.northropgrumman.com/Capabilities/LN100GInertialNavigationSystem/Documents/ln100g.png (accessed on 20 July 2015).

21. Air Data Computer. Technical Report TSO-C106. Department of Transportation, Federal Aviation Administration, January 1988; Washington DC, USA. Available online: http://rgl.faa.gov/Regulatory_and_Guidance_Library/rgTSO.nsf/0/fdc3133eed60bdb986256dc600696543/\protect\T1\textdollarFILE/C106.png (accessed on 20 July 2015).

22. PicoSAR. Available online: http://www.selex-es.com/it/-/picosar-1 (accessed on 27 March 2015).

23. ESGS GIS. Available online: http://egsc.usgs.gov/isb//pubs/gis_poster/ (accessed on 27 March 2015).

24. Digital Terrain Elevation Data (DTED): Performance Specification. Technical Report MIL-PRF-89020B; National Imagery and Mapping Agency (NIMA), May 2000; St. Louis, USA. Available online: https://dds.cr.usgs.gov/srtm/version2_1/Documentation/MIL-PDF-89020B.png (accessed on 20 July 2015).

25. Hanssen, R.F. *Radar Interferometry: Data Interpretation and Error Analysis*; Kluwer Academic Publishers: Dordrecht, The Netherlands, 2001.

26. Gatelli, F.; Monti Guarnieri, A.; Parizzi, F.; Pasquali, P.; Prati, C.; Rocca, F. The wavenumber shift in SAR interferometry. *IEEE Trans. Geosci. Remote Sens.* **1994**, *32*, 855–865. [CrossRef]

27. Dall, J. InSAR Elevation Bias Caused by Penetration into Uniform Volumes. *IEEE Trans. Geosci. Remote Sens.* **2007**, *45*, 2319–2324. [CrossRef]

28. Ghiglia, D.C.; Pritt, M.D. *Two Differential Phase Unwrapping: Theory, Algorithms and Software*; J. Wiley & Sons, Inc.: New York, NY, USA, 1998.

29. Lingua, A.; Marenchino, D.; Nex, F. Performance Analysis of the SIFT Operator for Automatic Feature Extraction and Matching in Photogrammetric Applications. *Sensors* **2009**, *9*, 3745–3766. [CrossRef] [PubMed]

Sensors **2015**, *15*, 18334–18359

30. Heady, B.; Kroenung, G.; Rodarmel, C. High resolution elevation data (HRE) specification overview. In Proceedings of ASPRS/MAPPS Conference, San Antonio, TX, USA, 16–19 November 2009.
31. Rizzoli, P.; Bräutigam, B.; Kraus, T.; Martone, M.; Krieger, G. Relative height error analysis of TanDEM-X elevation data. *ISPRS J. Photogramm. Remote Sens.* **2012**, *73*, 30–38. [CrossRef]

Article

Towards an Autonomous Vision-Based Unmanned Aerial System against Wildlife Poachers

Miguel A. Olivares-Mendez [1,*], Changhong Fu [2], Philippe Ludivig [1], Tegawendé F. Bissyandé [1], Somasundar Kannan [1], Maciej Zurad [1], Arun Annaiyan [1], Holger Voos [1] and Pascual Campoy [2]

[1] Interdisciplinary Centre for Security, Reliability and Trust, SnT - University of Luxembourg, 4 Rue Alphonse Weicker, L-2721 Luxembourg, Luxembourg; ludivig@hotmail.com (P.L.); tegawende.bissyande@uni.lu (T.F.B.); somasundar.kannan@uni.lu (S.K.); maciej.zurad@gmail.com (M.Z.); arun.annaiyan@uni.lu (A.A.); holger.voos@uni.lu (H.V.)
[2] Centre for Automation and Robotics (CAR), Universidad Politécnica de Madrid (UPM-CSIC), Calle de José Gutiérrez Abascal 2, 28006 Madrid, Spain; fu.changhong@upm.es (C.F.); pascual.campoy@upm.es (P.C.)
* Corresponce: miguel.olivaresmendez@uni.lu; Tel.: +352-46-66-44-5478; Fax: +352-46-66-44-35478

Academic Editor: Felipe Gonzalez Toro
Received: 28 September 2015; Accepted: 2 December 2015; Published: 12 December 2015

Abstract: Poaching is an illegal activity that remains out of control in many countries. Based on the 2014 report of the United Nations and Interpol, the illegal trade of global wildlife and natural resources amounts to nearly $213 billion every year, which is even helping to fund armed conflicts. Poaching activities around the world are further pushing many animal species on the brink of extinction. Unfortunately, the traditional methods to fight against poachers are not enough, hence the new demands for more efficient approaches. In this context, the use of new technologies on sensors and algorithms, as well as aerial platforms is crucial to face the high increase of poaching activities in the last few years. Our work is focused on the use of vision sensors on UAVs for the detection and tracking of animals and poachers, as well as the use of such sensors to control quadrotors during autonomous vehicle following and autonomous landing.

Keywords: unmanned aerial vehicles; computer vision; animal tracking; face detection; vision-based control; object following; autonomous navigation; autonomous landing; anti-poaching

1. Introduction

The use of unmanned aerial vehicles (UAVs) has increased rapidly in the last few decades and is now common in a variety of domains, ranging from leisure to rescue missions. UAVs have indeed become accessible to common consumers thanks to: the miniaturization of electronic components, including sensors, driven by other technologies, such as smart-phones; the increase in computational power for onboard CPUs; and the reduction in costs for this type of platform for robots. Thus, nowadays, UAVs are no longer solely reserved for military purposes. Several civilian applications (e.g., in agriculture, filming, *etc.*) have been developed recently. To accomplish the final take-off of UAV technology, many legal issues still need to be addressed for regulating the use of remotely-piloted or fully-autonomous UAVs. Nonetheless, there are specific scenarios where legal issues are irrelevant. These include areas of natural disasters or animal reserves where UAV technology can be essential in saving lives, either human or animal. Our current work focuses on the latter for supporting anti-poaching missions, which play an important role in protecting different species of animals around the world [1]. In Africa, animal poaching has reached critical levels due to the lack of resources on security and protection of the wildlife. The large size of national parks make it almost impossible to control different areas with the traditional surveillance methods with the limited number of security guards. In the necessary fight against poachers, reaction time is crucial.

For example, it is estimated that a 10 minute delay is sufficient for killing and de-horning a rhino. Currently, the use of UAVs for anti-poaching activities is only limited to remotely-piloted systems with continuously-supervised images streamed from the onboard cameras [2]. In such approaches, however, expert UAV piloting knowledge is required. To maximize the usefulness of the UAVs, some autonomous capabilities must be included in the onboard systems. These capabilities must be designed to provide support in the surveillance of groups of animals, the recognition of poachers, as well as the tracking and following of their vehicles. In this work, we have focused on the development of these capabilities to define completely autonomous surveillance missions in the fight against poachers. In this work, an autonomous surveillance mission definition comprises the autonomous taking-off and following of a predefined position list (already available in almost all commercial UAVs systems), tracking animals, detecting poachers faces, tracking and following poachers vehicles and return, via an autonomously-landing scenario, on specific stations in order to recharge the batteries and prepare for the next surveillance flight.

Taking into account the large amount of security forces needed to cover the huge area (e.g. the Kruger National Park covers an area of $19,480$ km^2) of natural parks, we propose to equip all park security patrol vehicles with a UAV. These vehicles will have a landing/recharging platform on their roof (referred to as moving landing/charging stations). A set of static landing/charging stations should also be installed throughout the natural park area. These stations are needed to recover UAVs in cases where the closest security patrol vehicle is out of the range of a UAV's battery endurance. Once the UAV takes off from one of these moving or static stations, it will follow a predefined patrolling trajectory. This trajectory could be modified autonomously in cases where an animal or a group of animals is detected. We assume that the onboard cameras are streaming the images to a security base or patrol vehicle in which a security guard can identify if an animal is hurt or under attack by poachers. The predefined trajectory may also be autonomously modified in case a group of poachers is detected. In this case, the system must detect the faces of the poachers to add them into a database of poachers. This database should be shared with international authorities in order to identify these individuals and have them prosecuted. Another case in which the UAV system should modify its trajectory is when a vehicle is detected. In this case, the system must be able to follow the vehicle in order to help the security authorities to pursue and catch potential poachers. Once the mission is completed or when the batteries of an UAV are about to be depleted, it will return to the closest moving or static landing/charging station and perform an autonomous landing. Based on the previously-established mission scenarios, this paper presents a number of vision-based algorithms. Some of these algorithms use aerial images for the tracking of animals, as well as for face detection. Other included algorithms use vision-based control systems in order to follow vehicles or to land UAVs autonomously on both moving and static platforms.

The remainder of this paper is organized as follows. Section 2 presents the related works regarding the computer vision techniques for visual tracking and face detection, the vision-based control approaches for UAVs and the anti-poaching activities with UAVs. Section 3 presents the related actions against poachers using UAVs. Section 4 shows the adaptive visual animal tracking approach. Section 6 presents the face detection approach for the identification of the poachers. Section 7 presents the vision-based control approach to control a quadrotor to follow vehicles and to accomplish autonomous landings on mobile platforms. Finally, Section 8 presents the conclusions and future works.

2. Related Works

2.1. Computer Vision Using Aerial Images

Similarly to other robotics platforms, depending on the application, UAVs may be set up with different sensor configurations. However, in the case of UAVs, the selection of the sensor configuration is more critical than in robotics platforms, such as ground vehicles, because of its limited payload.

For this reason, the onboard sensors are required to present a wide working range, in order to be useful in different scenarios and for different purposes. Such requirements make the vision sensor the most auspicious sensor to be mounted onboard because of its low cost, high efficiency and its similarities to human vision. Numerous vision algorithms have already been developed to allow vision sensors to be used in a variety of applications. A vision sensor is commonly used for the detection and tracking of objects in images. It has been used this way for decades now with static cameras and on ground vehicles. More recently, such sensors started to be used on UAVs, as well. However, UAVs constitute a specific case where the complexity of vision tasks is substantially increased due to the rapid movements that such a platform can experience. These movements are not limited to lateral-forward, but also involve movements along the vertical axis, which affect the visioned size (*i.e.*, scale) of the tracked object. Therefore, well-tested vision algorithms which have been used for a long time on other platforms (e.g., on ground robots) are not as usable in UAV scenarios. They usually must be adapted to fit a particular use-case scenario, for example by including background subtraction capabilities. In previous research, one can find many object tracking surveys that show the alternative techniques and methodologies to follow [3–5]. Object tracking with a vision-based sensor relies on features present in the images. Those features can be detected based on color, edges, textures and optical flow. All of these tracking methods require an object detection mechanisms, as explained in detail by Yilmaz *et al.* [3].

Our work presents two different approaches of vision algorithms for specific purposes: the first one is related to object tracking, specifically animal tracking, while the second one is focused on face detection in order to identify and create a database of poachers' identities. In the remainder of this sub-section, we will discuss research related to these specific computer vision approaches.

2.1.1. Visual Tracking

Recently, visual tracking has been researched and developed fruitfully in the robot community. However, real-time robust visual tracking for arbitrary 3D animals (also referred to as visual animal model-free tracking), especially in UAV control and navigation applications, remains a challenging task due to significant animal appearance changes, variant illumination, partial animal occlusion, blur motion, rapid pose variation, cluttered background environments and onboard mechanical vibration, among others.

The typical visual tracking system should take into account three main requirements: (1) Adaptivity: this requires a reliable and sustained online adaptation mechanism to learn the real appearance of 3D animals; (2) Robustness: the tracking algorithm should be capable of following an animal accurately, even under challenging conditions; (3) Real time: this requires the tracking algorithm to process live images at a high frame rate and with an acceptable tracking performance, in order to generate consecutive and fast feedback vision estimations.

In the literature, visual tracking algorithms, based on the Lucas–Kanade optical flow, have been frequently utilized to track objects (e.g., [6] for UAVs). The 3D position of a UAV is estimated using a pre-defined reference object selected on the first image frame. However, this type of tracker cannot learn the animal appearance during tracking, and RANSAC [7] requires a large number of iterations (which implies heavy time consumption) to reach optimal estimation. Similarly, SIFT [8] and SURF [9] features have been used in visual tracking algorithms for object tracking. In summary, all such methods are known as feature-based visual tracking approaches.

The direct tracking method (*i.e.*, directly represent the object using the intensity information of all pixels in the image) was used to track objects from a UAV (*cf.* [10]). This type of tracker has been shown to perform better than the previously-mentioned and well-known feature-based algorithms. However, the direct tracking method also employs a fixed object template for the whole UAV tracking process. Although this type of tracker has been improved in [11] by manually adding many other templates, it still does not provide online self-taught learning. Moreover, the gradient descent method often falls into local minimum values and is relatively slow at achieving the global minimum.

An off-line learning algorithm for recognizing specified objects in UAV applications has been applied in [12] where a large amount of image training data is used to train off-line using a multi-layer perceptron artificial neural network (MLP-ANN). However, the object recognition is fixed/predefined instead of freewill objects, which are selected online. Besides the collection of the training image data, it is difficult to cover all of the challenging conditions that the UAV could encounter during an actual UAV flight. Furthermore, it is time consuming to empirically compute the optimal parameters for these kinds of off-line learning methods.

The adaptively-discriminative tracking method (referred to as the model-free tracking approach, where the tracked object is separated from its dynamic surrounding background using an adaptive binary classifier, which is updated with some positive and negative image samples) was applied in our previous work [13] and allowed for obtaining the accurate location of objects in an object tracking scenario from a UAV. Nevertheless, this tracker cannot provide the estimations of other motion model parameters, such as the rotation or scale information of the object. Even though incorporating these new parameter estimations into the tracker is straightforward, as declared by B. Babenko *et al.* [14] and tested in our different tracking experiments [15], the three performances mentioned above will dramatically decrease.

In our work, to handle the problems of drift, rapid pose variation and variant surrounding illumination, motivated by several works [16–19], the low-dimensional subspace representation scheme is applied as the practicable method to represent/model the 3D animal. The online incremental learning approach is utilized as the effective technique for learning/updating the appearance of a 3D animal. Moreover, the particle filter (PF) [20] and hierarchical tracking strategy are also employed to estimate the motion model of the 3D animal for UAV anti-poaching.

2.1.2. Face Detection

A comprehensive survey on face detection algorithms has been published by Yang, Kriegman and Ahuja [21], where they list multiple approaches from the literature. We only discuss a few in the following:

One approach to face detection consists of looking for structural features that can be found in faces even when expression, viewpoint or lighting conditions change. An example of such an approach has been presented by Leung, Burl and Perona [22].

Another way to perform face detection is to make use of machine learning algorithms. As shown by Lanitis, Taylor and Cootes [23], it is possible to build machine learning classifiers to detect faces by training them with images of faces or specific features of faces. The resulting detection systems can than compare input images against learned patterns.

The third approach also makes use of machine learning algorithms, but in this case, the patterns that are used for the detection are not selected by the user, but by the machine learning algorithm itself. Examples of this approach are making use of a variety of machine learning algorithms, such as neural networks [24], hidden Markov models [25], support vector machines [26] or boosting [27]. This approach appears to be popular within the community due to the difficulty of reducing a face down to just a handful of features, especially when considering different lighting conditions and multiple viewing angles. Instead, machine learning algorithms manage to process a large number of training images in order to select a number of reliable features with a limited amount of human supervision.

For the purpose of our research, however, most of the presented methods are not suitable because of the real-time requirement of the face detection in our scenario. One commonly-used algorithm for real-time applications is the Viola and Jones boosting cascade algorithm [27] that we leverage in our work.

2.2. Vision as the Sensor for Control Applications in Robotics

Vision is a useful robotic sensor, since it mimics the human vision sense and allows one to extract non-contact measurements from the environment. The use of vision with robots has a long history, starting with the work of Shirai and Inoue [28], who describe how a visual feedback loop can be used to correct the position of a robot to increase the task accuracy. Today, vision systems are available from major vendors, and they are widely integrated in robotic systems. Typically, visual sensing and manipulation are combined in an open-loop fashion, "looking" and then "moving". The accuracy of the results is directly dependent on the accuracy of the visual sensor, the robot end-effector and the controller. An alternative for increasing the overall accuracy of the system is to use a visual feedback control loop.

Visual servoing is no more than the use of vision at the lowest level, with simple image processing to provide reactive or reflexive behavior to servo position a robotic system. A classical visual servo control was developed for serial link robotic manipulators with the camera typically mounted on the end-effector, also called eye-in-hand. Even tough the first visual servoing systems had been presented by Sanderson [29] back in the 1980s, the development of visual control systems for robots has been fairly slow. However, many applications have appeared in the last two decades, due to the increase in computing power, which enables the analysis of images at a sufficient rate to "servo" a robotic manipulator.

Vision-based robot control using an eye-in-hand system can be classified into two groups: position-based and image-based visual servoing, PBVS and IBVS, respectively. PBVS involves the reconstruction of the target pose with respect to the robot and leads to a Cartesian motion planning problem. This kind of control is based on the three-dimensional information from the scene, so the geometric model of the object to track and a calibrated model of the camera are needed. Then, the estimation of the position and orientation of the object is obtained. The PBVS design is sensitive to the camera calibration, which is particularly challenging when using a low quality camera. In contrast, for IBVS, the control task is defined in terms of image features. A controller is designed to maneuver the image features to a desired configuration. The original Cartesian motion planning problem is solved. The approach is inherently robust to camera calibration and target modeling errors which in turn reduces the computational cost. However, this configuration implies an extra complexity for the control design problem.

2.3. Vision-Based UAV Control

There are many visual servoing applications for UAVs present in the literature. Different vision-based algorithms have been used to follow a car from a UAV [30–32]. Visual terrain following (TF) methods have been developed for a vertical take-off and landing (VTOL) UAVs [33]. In [34], a description of a vision-based algorithm to follow and land on a moving platform and other related tasks are proposed. A cooperative strategy has been presented in [35] for multiple UAVs to pursue a moving target in an adversarial environment. The low-altitude road-following problem for UAVs using computer vision technology was addressed in [36]. The people-following method with the parallel tracking and mapping (PTAM) algorithm has been developed in [37]. Contrary to the above discussed research, the autonomous target following and landing approach presented in this work is based on the control of the lateral, longitudinal, vertical and heading velocities of the quadrotor to modify its position to follow and land on a predefined platform.

Related to the autonomous landing, there exists previous work that is focused on the theoretical control part of this problem, which has been examined in simulated environments, such as [38]. This presents a classical PID control using the SIFT vision algorithm, proving the feasibility of this algorithm for this specific task and testing the controllers in a simulated environment. In [39], the authors have evaluated the use of visual information at different stages of a UAV control system, including a visual controller and a pose estimation for autonomous landing using a chessboard pattern. In [40], a visual system is used to detect, identify a landing zone (helipad) and confirm the landing direction of the vehicle. The work in [41,42] proposed an experimental method of autonomous landing on a moving target, by tracking a known helipad and using it to complement the controller IMU + GPS state estimation. Other research has also been able to demonstrate autonomous landing with a VTOL aircraft [43]. This research makes use of a fusion sensor control system using GPS to localize the landmark, vision to track it and sonar for the last three meters of the autonomous landing task.

The work in [44,45] used a method to fuse visual and inertial information in order to control an autonomous helicopter landing on known landmarks. In [46], the authors presented the results of a fusion sensor of GPS, compass and vision with a PID controller to track and follow the landing location and land on a landmark. Overall, all of the aforementioned works are related to fixed wing aircraft or helicopters.

Nowadays, the increasing popularity of multi-copters (commonly quadcopters) calls the attention of the research to this topic. Some examples include work presented by Lange in [47] where a visual system is used to estimate a vehicle position relative to a landing place. In [48,49], a decomposition of a quadrotor control system to an outer-loop velocity control and an inner-loop attitude control system is proposed. In this work, the landing controller consists of a linear altitude controller and a nonlinear 2D-tracking controller. The work in [50] shows a deep theoretical work of a non-linear controller of a quadrotor that is built on homography-based techniques and Lyapunov design methods. Recently, [51,52] have shown two different methods to be used with micro-UAVs for autonomous takeoff, tracking and landing on a moving platform. This work is based on optical flow and IR landmarks to estimate the aircraft's position. The work in [53] displays the experiments of the autonomous landing of an AR.Drone on a landing pad mounted on top of a kayak. A deep review of the different control techniques for autonomous navigation, guidance and control for UAVs is presented in [54–57].

3. Related Actions against Poachers

To the best of our knowledge, the only working approach that relies on UAVs for fighting against poaching is the initiative conducted by SPOTS-Air Rangers [2]. As is mentioned on their web page, they are a "registered Section 21 Conservation Company focused purely on the conservation and the protection of any and all threatened species". The company is using UAVs in Africa, specifically South Africa, for the specific task of poacher detection. They use fixed-wing aircraft equipped with a high quality thermal camera, among other sensors needed to fly. The aircraft is provided by Shadowview, which is also a non-profit organization providing multiple UAS solutions for conservation and civilian projects [58]. The main goal of Air Rangers is to participate in the fight against poaching, through the detection of potential poachers. They also use UAVs to track animals and thus improve the wildlife census system. Furthermore, this company leveraged their aircraft for burn assessment and biomass management. Based on what we could extract from SPOTS's website, they are not using any software for autonomous detection of poaching situations, by, e.g., detecting poachers, vehicles, camps and strange behaviors. No specific algorithms to increase the effectiveness of the patrol trajectories are shown in their tasks descriptions either.

We claim that the work of this company and its project in Africa could be improved with the automation techniques that are discussed in our work [1] and whose design and implementation are presented in this paper. Our objective is to increase the number of detected poachers and to improve the effectiveness and efficiency of autonomous flights. It is noteworthy that similar initiatives are currently in the process of being applied. Those include the Air Sheppard project by the Lindberg Foundation and the wildlife conservation UAV challenge organized by Princess Aliyah Pandolfi, as well as the Kashmir World Foundation [59], in which more than 50 teams are involved.

4. Adaptive Visual Animal Tracking Using Aerial Images

In this section, the details of the presented adaptive visual animal tracking onboard a UAV for anti-poaching are introduced. This visual algorithm can overcome the problems generated by the various challenging situations, such as significant appearance change, variant surrounding illumination, partial animal occlusion, rapid pose variation and onboard mechanical vibration.

4.1. Adaptive Visual Animal Tracking Algorithm

Online incremental subspace learning methods, e.g., by G. Li *et al.* [60], T. Wang *et al.* [61], D. Wang *et al.* [62] and W. Hu *et al.* [63], have obtained promising tracking performances. Recently, D. Ross *et al.* [64] have presented an online incremental learning approach for effectively modeling and updating the tracking of objects with a low dimensional principal component analysis (PCA) subspace representation method, which demonstrated that PCA subspace representation with online incremental updating is robust to appearance changes caused by rapid pose variation, variant surrounding illumination and partial target occlusion, as expressed by Equation (1) and shown in Figure 1. In addition, PCA has also been demonstrated in [19,65] to have those advantages in tracking applications. A. Levey *et al.* [66] and P. Hall *et al.* [67] have done works similar to those in [64], although [66] did not consider the changing of the subspace mean when new data arrive, while the forgetting factor is not integrated in [67], which generates a higher computational cost during the tracking process.

$$\mathbf{O} = \mathbf{U}\mathbf{c} + \mathbf{e} \tag{1}$$

In Equation (1), $\mathbf{O}$ represents an observation vector, $\mathbf{c}$ indicates the target coding coefficient vector, $\mathbf{U}$ denotes the matrix of column basis vectors and $\mathbf{e}$ is the error term, which is the Gaussian distribution with small variances.

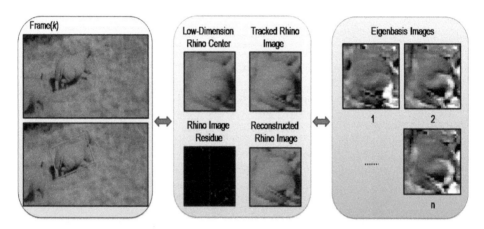

Figure 1. The PCA subspace-based tracking of a 3D rhino in our work, where each rhino image is re-sized to 32 × 32 pixels, and the reconstructed rhino image is constructed using the eigenbasis. Moreover, the eigenbasis images are sorted based on their according eigenvalues.

The main procedures of the online incremental PCA subspace learning algorithm with subspace mean updating [64] are as follows: Given a set of training images $\mathcal{S}_a = \{\mathbf{S}_1, \mathbf{S}_2, ..., \mathbf{S}_n\} \in \mathbb{R}^{d \times n}$, the appearance model of the 3D animal can be computed by the singular value decomposition (SVD) of the centered data matrix $[(\mathbf{S}_1 - \bar{\mathbf{S}}_a) \cdots (\mathbf{S}_n - \bar{\mathbf{S}}_a)]$, denoted by $(\mathcal{S}_1 - \bar{\mathbf{S}}_a)$, *i.e.*, $(\mathcal{S}_a - \bar{\mathbf{S}}_a) = U\Sigma V^\top$, where $\bar{\mathbf{S}}_a = \frac{1}{n} \sum_{i=1}^{n} \mathbf{S}_i$ is the sample mean of the training images.

If a new set of images $\mathcal{S}_b = \{\mathbf{S}_{n+1}, \mathbf{S}_{n+2}, ..., \mathbf{S}_{n+m}\} \in \mathbb{R}^{d \times m}$ arrives, then the mean vectors of $\mathcal{S}_b$ and $\mathcal{S}_c = [\mathcal{S}_a \, \mathcal{S}_b]$ are computed, *i.e.*, $\bar{\mathbf{S}}_b = \frac{1}{m} \sum_{i=n+1}^{n+m} \mathbf{S}_i$, $\bar{\mathbf{S}}_c = \frac{n}{n+m} \bar{\mathbf{S}}_a + \frac{m}{n+m} \bar{\mathbf{S}}_b$. Because the SVD of $(\mathcal{S}_c - \bar{\mathbf{S}}_c)$ is equal to the SVD of concatenation of $(\mathcal{S}_a - \bar{\mathbf{S}}_a)$, $(\mathcal{S}_b - \bar{\mathbf{S}}_b)$ and $\sqrt{\frac{nm}{n+m}}(\bar{\mathbf{S}}_a - \bar{\mathbf{S}}_b)$, which is denoted as $(\mathcal{S}_c - \bar{\mathbf{S}}_c) = U'\Sigma'V'^\top$, this can be done efficiently by the R-SVD algorithm, *i.e.*:

$$U' = [U \, \tilde{E}]\tilde{U}, \qquad \Sigma' = \tilde{\Sigma} \tag{2}$$

where, $\tilde{U}$ and $\tilde{\Sigma}$ are calculated from the SVD of R: $\begin{bmatrix} \Sigma & U^\top E \\ 0 & \tilde{E}(E - UU^\top E) \end{bmatrix}$, E is the concatenation of $(S_b - \bar{S}_b)$ and $\sqrt{\frac{nm}{n+m}}(\bar{S}_a - \bar{S}_b)$, $\tilde{E}$ represents the orthogonalization of $E - UU^\top E$ and U and Σ are the SVD of $(S_a - \bar{S}_a)$.

Taking the forgetting factor, *i.e.*, $\theta \in (0,1]$, into account for balancing between previous and current observations to reduce the storage and computation requirements, the R and $\bar{S}_c$ are modified as below:

$$R = \begin{bmatrix} \eta\Sigma & U^\top E \\ 0 & \tilde{E}(E - UU^\top E) \end{bmatrix} \tag{3}$$

$$\bar{S}_c = \frac{\eta n}{\eta n + m}\bar{S}_a + \frac{m}{\eta n + m}\bar{S}_b \tag{4}$$

where $\theta = 1$ means that all previous data are included to adapt to the changing appearance of the 3D animal.

For the visual animal tracking task of the UAV, it can be formulated as an inference problem with a Markov model and hidden state variables. Given a set of observed images $\mathcal{O}_k = \{\mathbf{O}_1, \mathbf{O}_2, ..., \mathbf{O}_k\}$ at the k-th frame, the hidden state variable $\mathbf{X}_k$ can be estimated as below:

$$p(\mathbf{X}_k|\mathcal{O}_k) \propto p(\mathbf{O}_k|\mathbf{X}_k) \cdot$$
$$\int p(\mathbf{X}_k|\mathbf{X}_{k-1})p(\mathbf{X}_{k-1}|\mathcal{O}_{k-1})d\mathbf{X}_{k-1} \tag{5}$$

where $p(\mathbf{X}_k|\mathbf{X}_{k-1})$ is the dynamic (motion) model between two consecutive states and $p(\mathbf{O}_k|\mathbf{X}_k)$ represents the observation model that estimates the likelihood of observing $\mathbf{O}_k$ at the state $\mathbf{X}_k$. The optimal state of the tracking animal given all of the observations up to the k-th frame is obtained by the maximum *a posteriori* estimation over N samples at time k by:

$$\hat{\mathbf{X}}_k = \arg\max_{\mathbf{X}_k^i} p(\mathbf{O}_k^i|\mathbf{X}_k^i)p(\mathbf{X}_k^i|\mathbf{X}_{k-1}), i = 1, 2, ..., N \tag{6}$$

where $\mathbf{X}_k^i$ is the i-th sample of the state $\mathbf{X}_k$ and $\mathbf{O}_k^i$ denotes the image patch predicted by $\mathbf{X}_k^i$.

4.1.1. Dynamic Model

In this application, we aim to use four parameters for constructing the motion model $\mathbf{X}_k$ of the 3D animal to close the vision control loop: (I) location x and y; (II) scale factor s; (III) rotation angle θ of the 3D animal in the image plane, *i.e.*, $\mathbf{X}_k = (x_k, y_k, s_k, \theta_k)$, which can be modeled between two consecutive frames; it is called similarity transformation in [68], as shown in the Figure 2. The state transition is formulated by a random walk:

$$p(\mathbf{X}_k|\mathbf{X}_{k-1}) = \mathcal{N}(\mathbf{X}_k; \mathbf{X}_{k-1}, \mathbf{\Psi}) \tag{7}$$

In Equation (7), $\mathbf{\Psi}$ is the diagonal covariance matrix, *i.e.*, $\mathbf{\Psi} = (\sigma_x^2, \sigma_y^2, \sigma_s^2, \sigma_\theta^2)$. However, the efficiency (*i.e.*, how many particles should be generated) and effectiveness (*i.e.*, how well particle filter should approximate the *a posteriori* distribution, which depends on the values in $\mathbf{\Psi}$) of the PF should be a trade off. Larger values in $\mathbf{\Psi}$ and more particles will obtain higher accuracy, but at the cost of more storage and computation expenses. We solved this problem in Section 4.2.

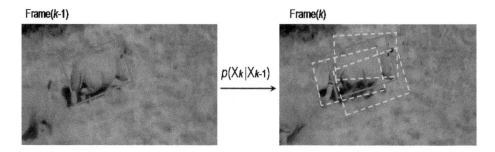

Figure 2. The dynamic model of visual rhino tracking.

4.1.2. Observation Model

In this work, we apply the low-dimensional PCA subspace representation to describe the tracked 3D animal. Thus, a probabilistic interpretation of PCA should be modeled for the image observations. The probability is inversely proportional to the distance from the sample to the reference point (*i.e.*, center) of the subspace, which includes two types of distances: (i) the distance-to-subspace, d_{to}; (ii) the distance-within-subspace, d_{within}.

The probability of d_{to} is defined as:

$$p_{d_{to}}(\mathbf{O}_k|\mathbf{X}_k) = \mathcal{N}(\mathbf{O}_k; \mu, UU^\top + \varepsilon I) \tag{8}$$

where μ is the center of the subspace, I represents the identity matrix and εI denotes the Gaussian noise.

$$p_{d_{within}}(\mathbf{O}_k|\mathbf{X}_k) = \mathcal{N}(\mathbf{O}_k; \mu, U\Sigma^{-2}U^\top) \tag{9}$$

where Σ represents the matrix of singular values corresponding to the columns of U.

Hence, the probability of the observation model is as follows:

$$\begin{aligned} p(\mathbf{O}_k|\mathbf{X}_k) &= p_{d_t}(\mathbf{O}_k|\mathbf{X}_k)p_{d_w}(\mathbf{O}_k|\mathbf{X}_k) \\ &= \mathcal{N}(\mathbf{O}_k; \mu, U\Sigma^{-2}U^\top)\mathcal{N}(\mathbf{O}_k; \mu, U\Sigma^{-2}U^\top) \end{aligned} \tag{10}$$

Moreover, the robust error norm, *i.e.*, $\rho(x,y) = \frac{x^2}{x^2+y^2}$, rather than the quadratic error norm, has been applied to reduce the noise effects.

4.2. Hierarchy Tracking Strategy

In the visual animal tracking application of UAVs, we find that an incremental PCA subspace learning-based (IPSL) visual tracker is sensitive to large displacements or strong motions. Although the value in Ψ (in Equation 7) can be set to be larger and more particles can be generated to get more tolerance for these problems, more noises will be incorporated from those particles, however, and the requirements of storage and the computation cost will be higher, which influences the real-time and accuracy performances. Therefore, the hierarchical tracking strategy, based on the multi-resolution structure, has been proposed in our work to deal with these problems, as shown in Figure 3. Nevertheless, there must be a compromise between the number of levels required to overcome the large inter-frame motion and the amount of visual information required to update the appearance of a 3D animal for estimating the motions. The main configurations for hierarchical visual animal tracking are as follows.

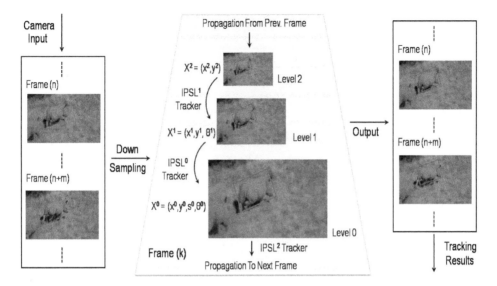

Figure 3. Our adaptive visual tracker for 3D animal tracking. The *k*-th frame is downsampled to create the multi-resolution structure (middle). In the motion model propagation, lower resolution textures are also initially used to reject the majority of samples at relatively low cost, leaving a relatively small number of samples to be processed at higher resolutions. The IPSL^p represents the incremental PCA subspace learning-based (IPSL) tracker in the *p*-th level of the pyramid.

4.3. Hierarchical Structure

Considering that the image frames are down-sampled by a ratio factor of two, the number of pyramid levels (N_{PL}) of the multi-resolution structure is defined as a function below:

$$N_{PL} = \lfloor log_2 \frac{min\{\mathbf{T}_W, \mathbf{T}_H\}}{minSizes} \rfloor \tag{11}$$

where $\lfloor * \rfloor$ is the largest integer not greater than value $*$, $\mathbf{T}_W$ and $\mathbf{T}_H$ represent the width and height of animal $\mathbf{T}$ in the highest resolution image (*i.e.*, the highest level of pyramid: Level 0), respectively. Additionally, *minSizes* is the minimum size of the animal in the lowest resolution image (*i.e.*, the lowest-level of pyramid: p_{min} level, $p_{min} = N_{PL}$-1), in order to have enough information to estimate the motion model at that level. Thus, if the *minSizes* is set in advance, the N_{PL} directly depends on the width/height of tracking animal $\mathbf{T}$. In this application, the number of pyramid levels is $N_{PL} = 3$; then, *p* is initialized as $p = \{2, 1, 0\}$.

4.4. Particle Filter Setup

Since the multi-resolution structure provides the computational advantage to analyze textures and update the appearance model in low resolution images and the lowest resolution image is good for estimating the location of tracking the animal, with the increase of resolution, more details from visual information can be used to estimate more parameters in the motion model. In this work, the motion models estimated in different resolution frames are defined as follows based on [68]:

Level 2:
$$\mathbf{X}_k^2 = (x_k^2, y_k^2), \text{ i.e., translation}$$
Level 1:
$$\mathbf{X}_k^1 = (x_k^1, y_k^1, \theta_k^1), \text{ i.e., translation + rotation}$$
Level 0:
$$\mathbf{X}_k^0 = (x_k^0, y_k^0, s_k^0, \theta_k^0), \text{ i.e., similarity}$$

4.5. Motion Model Propagation

Taking into account the fact that the motion model estimated at each level is used as the initial estimation of motion for the next higher resolution image, the motion model propagation is defined as follows:

$$x_k^{p-1} = 2x_k^p, y_k^{p-1} = 2y_k^p$$

$$\theta_k^{p-1} = \theta_k^p \tag{12}$$

$$s_k^{p-1} = s_k^p$$

where p represents the p-th level of the pyramid, $p = \{p_{min}, p_{min} - 1, ..., 0\} = \{N_{PL} - 1, N_{PL} - 2, ..., 0\}$, and k is the k-th frame.

After finding the motion model in the k-th frame, this motion model is sent as the initial estimation to the highest pyramid level of the $(k+1)$-th frame, as shown in Figure 3:

$$x_{k+1}^{p_{min}} = \frac{x_k^0}{2^{p_{min}}}, y_{k+1}^{p_{min}} = \frac{y_k^0}{2^{p_{min}}}$$

$$\theta_{k+1}^{p_{min}} = \theta_k^0 \tag{13}$$

$$s_{k+1}^{p_{min}} = s_k^0$$

Besides the propagation of motion models, the majority of particles will be rejected based on their particle weights in the lower resolution image. In other words, it is not necessary to generate a larger number of samples to estimate the same parameters in the higher resolution image, leaving a higher tracking speed and better accuracy than a single full-resolution-based voting process. The reject particle number is defined as:

$$N_R^p = \alpha^p N_P^p \tag{14}$$

where α^p is the reject ratio $(0 < \alpha^p < 1)$ at the p-th level in the pyramid and N_P^p is the number of generated particles.

In the rejected particles, the particle with maximum weight is called the critical particle (C_k^p). Taking the x position for example, the distance between x of C_k^p and x_k^p is denoted as the heuristic distance (H_k^p). Therefore, for the searching range propagation, *i.e.*, σ_x, it is defined as:

$$\sigma_{(k,x)}^{p-1} = 2H_k^p \tag{15}$$

where $\sigma_{(k,x)}^{p-1}$ is the variance of x translation at the $(p-1)$-th level of the pyramid of the k-th frame. Additionally, the other motion model parameters have similar propagations during the animal tracking process.

4.6. Block Size Recursion

The multi-block size adapting method has been used to update the 3D animal with different frequencies, *i.e.*, a smaller block size means more frequent updates, making it faster for modeling appearance changes. Because the image at the lowest level of the pyramid has less texture information, thus the recursion of block size (N_B) is given as below:

$$N_B^{p-1} = \lfloor \frac{N_B^p}{log_2(1+p)} \rfloor \qquad (16)$$

where $\lfloor * \rfloor$ is the largest integer not greater than value $*$, p represents the p-th level in the pyramid and k is the k-th frame.

All of the approaches introduced in this section are integrated to ensure higher accuracy and the real-time performance of the 3D animal tracking from UAVs. The real tests are discussed in the section below.

5. Visual Animal Tracking Evaluation and Discussion

This section discusses the 3D animal tracking performance of our presented algorithm, which is evaluated with the ground truth datasets generated from real UAV flight tests in Africa. In the different experiments, we have used the Robot Operating System (ROS) framework to manage and process image data.

5.1. Ground Truth Generation

Manually-generated ground truth databases have been leveraged to analyze the performance of our visual animal tracker. Figure 4 shows a reference rectangle for generating the ground truth. The center location, rotation, width and height of the tracked animal can be obtained frame-to-frame based on the reference rectangle.

Figure 4. Reference rectangle of the ground truth. The reference rectangle has included all of the pixels of the tracked animal, and the pink points are key pixels for locating the reference rectangle.

5.2. Real Test of Visual Animal Tracking

In the first experimental settings, we have selected the rhino and elephant as typical threatened animal species in Africa to test our proposed visual tracking algorithm. Some tracking results are shown in Figures 5–8.

5.2.1. Test 1: Rhino Tracking

In this test, the visual rhino tracking estimation contains three main challenging factors: (I) 3D appearance change (e.g., different views from the onboard camera and random running of the rhino); (II) partial rhino occlusion (e.g., occluded by bushes); (III) protective coloration (*i.e.*, the color of the rhino body is similar to the background). The tracking results, as illustrated by some randomly-selected examples in Figure 5, show that our presented visual tracker can locate the running rhino (*i.e.*, the front (right) rhino with the red rectangle) accurately from a flying UAV, even in in face of varying running speeds. Although those two running rhinos have extremely similar appearances, our visual tracker has not been confused to then track the front (right) rhino stably. The center location error is defined as the Euclidean distance from the tracked animal center to the ground truth center at each image frame. The average errors of the center location, rotation angle, width and height are three pixels, two degrees, three pixels and two pixels, respectively. The performance of tracking the back (left) rhino also displays similar results.

Figure 5. Visual rhino tracking. The red rectangle shows the estimated location of the running rhino.

5.2.2. Test 2: Rhino Tracking

Compared to Test 1, the flying UAV is closer to the running rhinos, as shown in Figure 6. The vision-based rhino tracking estimation now contains three main challenging factors: (I) 3D appearance change; (II) rapid pose variation; (III) partial rhino occlusion. In this test, the real-time and adaptive performances of our presented visual tracker have guaranteed the accuracy of the location for this visual animal tracking application. The average errors of the center location, rotation angle, width and height are 11 pixels, four degrees, seven pixels and four pixels, respectively.

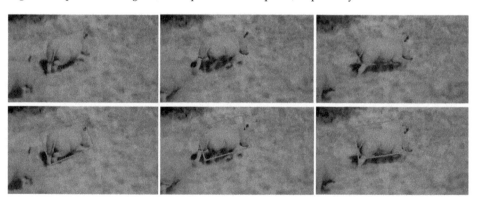

Figure 6. Visual rhino tracking.

5.2.3. Test 3: Elephant Tracking

In this test, our presented visual tracker has been used to locate an elephant, by trying to track one moving elephant in a group of moving elephants, as shown in Figure 7. The visual tracking estimation contains three main challenging factors: (I) 3D appearance change; (II) clustered tracking background; (III) partial elephant occlusion. The average errors of the center location, rotation angle, width and height are three pixels, three degrees, two pixels and four pixels, respectively.

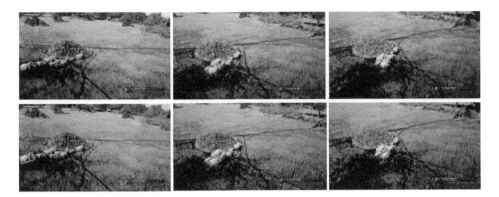

Figure 7. Visual elephant tracking.

5.2.4. Test 4: Elephant Tracking

In this test, the tracking object is the same as the one in Test 3, *i.e.*, moving elephant, although with different challenges. The vision-based tracking estimation contains three main challenging factors: (I) 3D appearance change; (II) clustered tracking background; (III) illumination variation; as the shadow areas shown in the right-top corner of the images in Figure 8. The average error of the center location, rotation angle, width and height are three pixels, two degrees, four pixels and two pixels, respectively.

Figure 8. Visual elephant tracking.

In general, the vision estimations are well-matched with the real location (ground truth) of animals in all of the visual animal tracking tests, without many salient outliers.

The videos related to some of the experiments presented in this section are available on a website [69].

6. Face Detection Using Aerial Images for Poachers' Detection and Identification

In this section, we detail our work on face detection for identifying poachers. The information acquired may be used for a number of purposes, including future UAV control applications to follow poachers, as well as to create and maintain a database of poachers, which can be distributed to the competent authorities. Our work leverages existing face detections systems and implements them for UAV platforms. While similar research involving face recognition on UAVs [70] and face detection for UAVs [71] or for robotic platforms [72] has been investigated previously, these works cover very specific scenarios where the subjects are relatively close to the drone while flying in controlled indoor environments. There is still a lack of research regarding real-time face detection for more real-world outdoor scenarios.

6.1. Face Detection Approach

To satisfy real-time requirements in face detection, we are using a commonly-applied face detection algorithm, which was presented by Viola and Jones [27]. The boosting cascade algorithm has three main characteristics:

- It is easy to calculate integral features
- It uses machine learning using AdaBoost
- It leverages the cascade system for speed optimization

6.1.1. Feature Detection

For faster feature detection, the system avoids using pixel-based features. Instead, it makes use of Haar-like features [73], where the system checks the sums of rectangular areas against the sums of other rectangular areas. There are four types of integral image features, which can be seen in Figure 9. For the detection, the pixels within the white rectangles are added together and then subtracted by the sum of all of the pixels within the grey area. Since the algorithm only works on grey-scale images, there is only one value for each pixel. This type of feature can be calculated with a single pass over the image. In addition to this, the implementation that we used applies an additional extension [74] for the Haar-Like features, as seen in Figure 10. This extension allows for features that are rotated at a 45°angle, as well as center-surround features, which define a point feature rather than an edge or a line.

In Figure 11, one can see two features selected by Adaboost. Feature (a) capitalizes on the fact that the nose-bridge tends to be brighter than the eyes. Feature (b) works with the forehead being brighter than the eye region below.

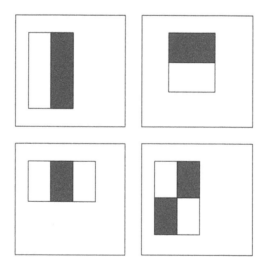

Figure 9. Integral image features used in boosting cascade face detections.

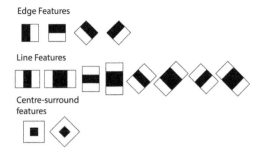

Edge Features

Line Features

Centre-surround
features

Figure 10. Total set of features used by the OpenCV detection.

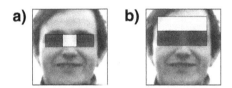

a) b)

Figure 11. Example of features. (**a**) the nose-bridge tends to be brighter than the eyes; (**b**) the forehead being brighter than the eye region below.

6.1.2. AdaBoost

The detection of the features is performed by a machine learning algorithm known as AdaBoost [75] (adaptive boosting). In principle, many weak classifiers are needed to produce one much more accurate strong classifier. In our case, this system is ideal for the integral features, since the initial weak classifiers only have to be better than random. Adaboost requires an initial learning period with labeled and carefully cropped training data.

6.1.3. Cascade System

To reduce the processing time of a large number of features, we use a cascade system where the most defining features are tested first. This aims to solve the main problem of the AdaBoost system where a large number of features are required, due to the very basic nature of the Haar-Like features. The cascade approach allows the algorithm to discard many false detections without having to check all of the features in the system. The order in which the features are being tested may significantly speed up the face detection. To put this into perspective, the test system from [27] had 6000 features, which were split into 38 cascade stages. The use of the cascade system resulted in a 15-fold speed increase. The downside to using this method is an additional training period, which is used to determine in which order the features are being applied. The boosting cascade system may have long training phases by an "order of weeks" ([76], p. 583).

6.2. Implementation

For the purpose of our research, the existing OpenCV cascade classifier implementation has been applied. The version used is OpenCV 2.4.8 since it comes with ROS Indigo and because there is an existing CV_bridge which translates the ROS images topic into OpenCV Mat images. It is important to know that the classifiers need to be trained beforehand. Fortunately, OpenCV comes with a set of pre-trained models, which are perfectly suited for face detection. The next step of the detection is to prepare the images for the detection. The classifier requires the color images to be transformed into a grey-scale color space. For improved detection, depending on the lighting conditions, either additional contrast or a histogram equalization is applied. Optionally, the may also be scaled at this stage. With respect to larger resolution images, this is the best way to improve performance. In this case, however, this is not ideal, since it reduces the detection of people that are further away from the camera. This is partially due to the wide angle lens of the camera, which makes people appear smaller in images. One should also keep in mind that people tend to keep their distance from UAVs.

At the detection level, there are a number of settings that may be used to improve the number of correct detections and reduce the number of false detections:

- *Cascade*: OpenCV comes with a number of different face detection cascade systems. For this research, haarcascade_frontalface_alt_tree.xml has been selected.
- *scaleFactor*= 1.2: The algorithm scans the input image in multiple iterations, each time increasing the detection window by a scaling factor. A smaller scaling factor increases the number of detections, but also increases the processing time.
- *minNeighbors* = 3: This refers to the number of times a face needs to be detected before it is accepted. Higher values reduce the number of false detections.
- *Flags*: There a number of additional flags in existence. In this particular case, the flag CV_HAAR_DO_CANNY_PRUNING is used because it helps to reduce the number of false detections and it also improves the speed of the detection.
- *minSize* and *maxSize*: These parameters limit the size of the search window. With fixed cameras, this is beneficial to improve speed, but in this case, these values are set very broadly because the distance from the subjects is constantly changing.

6.3. Drone Setup

The UAV featured is an AscTec Firefly (as seen in Figure 12), with a processing computer on board. It runs ROS Indigo on Ubuntu 14.04. As a camera, a UEye UI-1240ML-C-HQ with a wide-angle lens has been mounted. The camera produces a 1280 × 1024 uncompressed eight-bit color image. It also has a global shutter to avoid rolling shutter issues, which are particularly problematic for UAV-mounted cameras. While it is possible to run the face detection on the UAV, this has been avoided to be able to run multiple tests on the same film material. For the recording process, the ROS rosbag package has been used. The footage is played back on an off-line computer with the same rosbag package, where the images are then processed by the face detection system. The processing computer also runs ROS Indigo in an Ubuntu 14.04 environment. The machine has an Intel Xeon(R) CPU E565 running 12 cores at 2.40 GHz with 5.8 GB of RAM.

Figure 12. AscTec Firefly with the mounted uEye camera.

6.4. Experiments

To ensure a realistic test environment, the drone was flown outside. The UAV was piloted with manual controls at a height of about 2.5–3.5 m. The direct sunlight provided additional difficulty, as there is much more contrast in the images. Multiple test situations where set up in order to prove the robustness of the system:

- Standing in the shadow: Faces are well exposed, and there are no harsh shadows. The face detection works well under these conditions, even during direction changes of the drone. One frame of this test is shown in Figure 13.
- Direct sunlight: When filming people standing in direct sunlight, the harsh shadows make the detection more difficult (Figure 14). In this case, the detection is not as consistent as in the previous

test, but it still manages to detect all of the faces at some point. In Figure 14, it also shows how the system is able to detect a person who is standing in the shadow (left), even though the camera was not exposed for those lighting conditions, and it is difficult for humans eyes to even detect the body of this person.

- Fly-over: For the last experiment, the UAV was set to fly over a group of moving people. An example frame of this footage can be seen in Figure 15. Due to the close proximity to the subjects, this tests required the detection of a lot of faces of different sizes. The proximity also makes the motion blur on the subjects stronger. Because of the wide angle of the lens, lens distortion can also cause problems with closer subjects. In this case, the detection also works well, mainly because of the large size of the faces.

Figure 13. Shadow example: The detection is stable even during faster movement of the drone.

Figure 14. Direct sunlight example: note the detection of the person standing in the shadow.

Figure 15. Fly-over example.

6.5. Calculation Speed

The calculation speed can fluctuate depending on the input images. On average, the detection speed has been consistent with a difference of only 0.03 frames per second over the three different test cases. While running on the previously-mentioned off-line machine, the system has been able to process an average of 4.258 frames per second.

6.6. Limitations

The current OpenCV implementation is very robust, even in difficult lighting conditions. One notable weakness of the face detection is people that are further away from the camera. Below a size of 10×10 pixels, the algorithm struggles to produce a good detection. This problem could, however, easily be fixed by using a higher resolution image sensor or by applying a lens with a narrower field of view. The detection also has difficulties with the orientation of the faces. Faces that are tilted sideways usually fall outside the detection range of the training model. Faces that are not front facing cannot be detected by the front-facing cascade model. OpenCV does, however, include a trained model for faces in profile view. In this case, only the frontal face model has been applied, since using a single model already requires much processing power. One should note here that people usually look at drones once they can hear them, which thereby leads to an easier detection. Lastly, partially-occluded faces are not always detected.

The videos related to some of the experiments presented in this section are available at [69].

7. Vision-Based Control Approach for Vehicle Following and Autonomous Landing

In previous sections, an animal tracking approach and a face detection technique using aerial images have been presented. However, for these to be promising in the fight against poachers, an autonomous control of the UAV is mandatory. This section presents a vision-based control approach to close the control loop using the result of an image processing algorithm as the input of the control. Our work includes vision-based control to follow suspicious vehicles (potential poachers) and to accomplish autonomous landing to recover the UAV after the end of a surveillance mission to recharge the batteries. This way, we are presenting a potential full surveillance mission in which the UAV takes off from a specific place and then follows GPS waypoints (these tasks are already integrated in most of the commercial available UAVs) to patrol a specific area of natural parks. If there is any animal or group of them detected during patrolling, it should be tracked to get information about the animal status to determine a potential poacher attack. In case people are found, the system has to detect the faces and store these data for security authorities (these two image processing algorithms were presented previously in this paper). Any potential vehicle should also be tracked. In this case, the

UAV is able to follow the suspicious vehicle in a specific trajectory relatively to it. During the vehicle following task, the GPS position is shared with security authorities. Finally, the UAV has to come back to the closest (moving or static) base station and accomplish the autonomous landing to recharge its batteries and/or to be prepared for the next mission. In this section, we present the control approach to follow vehicles and to autonomously land on both static and moving bases.

7.1. Vision-Based Fuzzy Control System Approach

The presented control approach for UAVs is designed to use an onboard downwards looking camera and an inertial measurement unit (IMU). The information extracted from these two sensors is used to estimate the pose of the UAV in order to control it to follow vehicles and to land on the top part of ground vehicles or moving targets. The computer vision algorithm used is 3D estimations based on homographies. The homography detection is done with regards to a known target, which is an augmented reality (AR) code. The detection of this type of code is done with an ROS-implementation of the ArUco library [77]. The result of this algorithm is the pose estimation of the multi-copter with respect to the AR code. Multi-copters, as well as rotary wing platforms have a singularity with respect to fixed wings, which is the relation of the movement and the tilt of the thrust vector. It is not possible to move longitudinally (forward/backward) or laterally (left/right) without modifying the thrust vector. Because the camera is attached to the frame of the UAV, this movement significantly affects the estimations calculated by the vision algorithm. The estimation of the rotation of the UAV is retrieved from the gyroscope of the IMU. The subtraction of the roll and pitch rotations of the UAV from the image estimation is called de-rotation [78,79]. The relevant formulas are presented in Equation (17).

$$
\begin{aligned}
x' &= x - tan(\phi) \times z \\
y' &= y + tan(\theta) \times z \\
z' &= \sqrt{z^2 - y'^2}
\end{aligned}
\tag{17}
$$

where x, y, z are the translation estimation on the x, y, z UAV axis, respectively, ϕ, θ are the roll and pitch rotations of the UAV, respectively, and x', y', z' are the resulting de-rotated translations of the x, y, z of the UAV, respectively.

The de-rotated values of the pose estimation of the UAV are given as input to the control system. The control approach presented in this work consists of four controllers that are working in parallel, commanding the longitudinal, lateral and vertical velocities, as well as the orientation of the UAV. These four controllers are implemented as fuzzy logic PID-like controllers. The main reason for using this specific technique for the control loop is the way that this technique manages the uncertainty that is derived from the noisy data received from the vision-based detection algorithms and the IMU, as well as how it manages the high complexity of this type of non-linear robotics platform. Furthermore, the use of linguistic values by the fuzzy logic controllers simplifies the tuning process of the control system. The four fuzzy controllers were implemented using an in-house software called MOFS [80]. An initial configuration of the controllers was done based on heuristic information and was then subsequently tuned by using the Virtual Robotics Experimental Platform (V-REP) [81] and self-developed ROS modules. Detailed information about the tuning process can be found in [82]. In the present work, we use the same controller definition for the longitudinal, lateral and vertical velocity controllers, which is shown in Figure 16. The inputs are given in meters, and the output is calculated in meters × seconds. The orientation velocity controller gives the inputs in degrees, and the output is calculated in degrees × seconds. The control system has two different working states: In the first state, the UAV is set to follow the vehicles of a poacher while the height is predefined. The second state is used to recover the UAV for the next mission by landing it autonomously on top of the security entity's vehicles.

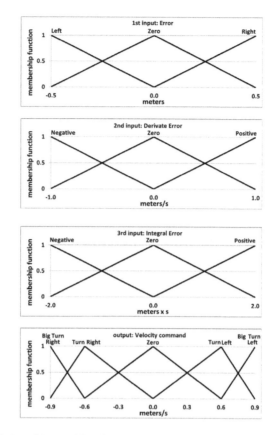

Figure 16. Final design of the variables of the fuzzy controller after the manual tuning process in the virtual environment (V-REP).

In this work, an additional tuning phase has been included. It has been developed during experiments with a quadrotor tracking the target. Comparing to previous work [82], in this work, a weight value was assigned to each of the three inputs of each control variable, as well as to the output of each controller. The tuning process of these weights was done with the real quadrotor in real experiments. Table 1 shows the final values of the weight for all of the controllers after the tuning process.

Table 1. Tuned weight values for the four controllers.

Controller Weight	Lateral	Longitudinal	Vertical	Heading
Error	0.3	0.3	1.0	1.0
Derivative of the error	0.5	0.5	1.0	1.0
Integral of the error	0.1	0.1	1.0	1.0
Output	0.4	0.4	0.4	0.16

7.2. Experiments

The experiments have been done in the laboratory of the Automation & Robotics Research Group at the University of Luxembourg. The flight arena has a size of 6 × 5 m and a height of 5 m. The UAV used in the experiments is an AR.Drone v2.0 [83]. This platform is not equipped with an onboard computer to process the images. Therefore, the image processing and the control process are calculated remotely on a ground station. The delay of the WiFi communication affects the system by increasing the complexity of the non-linear system of the vision-based UAV control. A WiFi-router is used to reduce the maximum variations of the image rate. A Youbot mobile robot from KUKA [84] has been used as the target vehicle to follow, as well as for the autonomous landing task on a moving target. The top of this robot was equipped with a landing platform. The platform was covered with an ArUco code in order for it to be detected by the vision algorithm, as is shown in Figure 17. This ground platform was controlled randomly via ROS with a remote computer. The omnidirectional wheels of this ground robot allowed for the position of the tracked target to be modified in all directions. This freedom of movement and the height limitation increase the complexity of the following and landing tasks. This type of movement cannot be performed by normal vehicles.

Figure 17. Youbot platform with the ArUco target.

Two different kinds of experiments were preformed. In the first experiment, the UAV had to follow the moving target from a fixed altitude of 3.5 m. In the second experiment, the UAV had to land on the moving ground platform. Table 2 shows the results of seven different experiments. The results are expressed with the root mean squared error (RMSE) for the evolution of the error for each controller. Depending on the controller, the RMSE is in meters (lateral, longitudinal and vertical velocity controller) or in degrees (heading velocity controller). The speed of the target platform was set to 0.5 m/s for the following Test #1, following Test #4 and the landing Test #3. For all other tests, the speed of the target platform was set to 0.3 m/s.

Table 2. Root mean square error for the lateral, longitudinal, vertical and heading velocity controllers for the autonomous following and landing on a moving target.

Controller Experiment	Lateral (RMSE, m)	Longitudinal (RMSE, m)	Vertical (RMSE, m)	Heading (RMSE, Degrees)	time (s)
Following #1	0.1702	0.1449	0.1254	10.3930	300
Following #2	0.0974	0.1071	0.1077	8.6512	146
Following #3	0.1301	0.1073	0.1248	5.2134	135
Following #4	0.1564	0.1101	0.0989	12.3173	144
Landing #1	0.1023	0.0.096	1.1634	4.5843	12
Landing #2	0.0751	0.0494	1.1776	3.5163	11
Landing #3	0.0969	0.0765	0.9145	4.6865	31

The most important experiments are shown in the next graph. Figure 18 shows the behavior of the system in the first target-following experiment. While this was also the experiment with the longest duration, the RMSE error of the lateral and longitudinal is under 15 cm. An error in the estimation of the orientation of the target can be seen between the 50th and the 80th second of the test. This error, which was produced by the computer vision algorithm, did not affect the estimations for the lateral, longitudinal and vertical controllers.

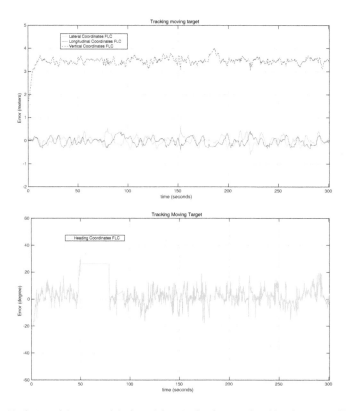

Figure 18. Evolution of the error of the lateral, longitudinal, vertical and heading controllers on the first moving target-following experiment.

Figure 19 shows the evolution of the error rate for all of the controllers in the second target-following experiment presented in Table 2. In this case, several orientation movements were applied to the target in order to evaluate the behavior of the heading controller in detail. This controller performs quickly, as can be seen in the two big changes at the first 50 s of the experiment. In this section of the test, the error reaches up to 35°, but the controller manages to reduce it in just a few seconds. During the other tests, more changes have been applied to the orientation of the target platform with similar performances of the heading controller. In this case, the RMSE of the heading controller was 8.6°.

Figure 20 shows the behavior of the controller for the second autonomous landing experiment. In this test, a gradual reduction of the altitude was performed by the UAV, reducing it from 3.5 m to 1 m in 8 s with an almost zero error for the lateral and longitudinal controllers. It has to be taken into account that the control system was set to reduce the vertical error up to 1 m, and then, a predefined landing command was sent to the UAV, which reduces the speed of the motors gradually.

Figure 21 shows the behavior of the control system for the third autonomous landing experiment. In this case, one can observe how the vertical controller pauses a couple of times for a few seconds in between the 25th and the 30th second of the test. This is because the vertical control system only sends commands when the errors of the heading, lateral and longitudinal controllers are smaller than some predefined values. This predefined behavior stabilized the landing process, reducing the potential loss of the target during the landing.

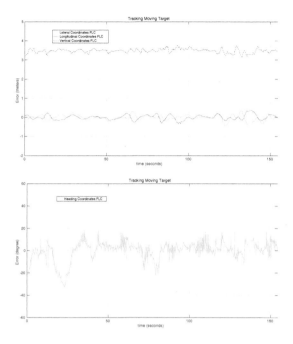

Figure 19. Evolution of the error of the lateral, longitudinal, vertical and heading controllers on the second moving target-following experiment.

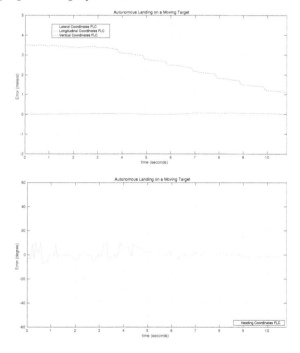

Figure 20. Evolution of the error of the lateral, longitudinal, vertical and heading controllers on the second autonomous landing on a moving target experiment.

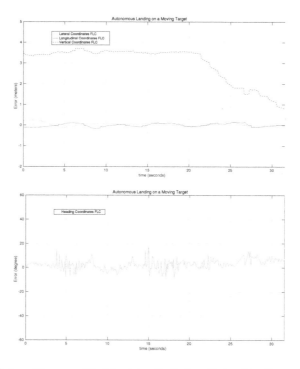

Figure 21. Evolution of the error of the lateral, longitudinal, vertical and heading controllers on the third autonomous landing on a moving target experiment.

The videos related to some of the experiments presented in this section are available online [69].

8. Conclusions and Future Works

In this paper, the main engineering challenge and contribution is the integration of different techniques and algorithms to have a potential solution for setting up an autonomous surveillance system against anti-poaching activities. A surveillance mission consists of an autonomous take-off, detection and tracking of animals, detection and storage of poachers' face data, an autonomous navigation of the UAV to follow suspicious vehicles and an autonomous landing for specific moving and/or static landing/recharging platforms.

An adaptive visual tracking algorithm was presented to track wildlife animals in their natural environment. The adaptive algorithm was tested with different aerial videos taken by quadcopters in Africa. In these videos, different animals in different natural environments were successfully tracked. A vision-based face detection algorithm is also presented to help with the detection, identification and creation of a database of poachers. The algorithm was successfully tested with aerial images captured with a quadrotor in different light conditions. The limitations of the camera configuration, which does not provide zoom optics, and the low resolution of the camera sensor reduced the distance at which people can be detected. An improvement of the camera configuration should improve the detecting distance for this algorithm. One of the remaining issues is that the detection is not always continuous. We are currently working on improving the detection with the help of image tracking algorithms. These algorithms are used to bridge the gap between successful detections. The mean shift algorithm was selected for this task, because it is not computationally intensive. Promising results in laboratory tests were achieved with the help of this technique, which still have to be tested in outdoor environments.

A vision-based control system approach was also presented in this paper in order to control a quadrotor to follow vehicles, as well as to land on a moving platform in order to achieve the recovery of the aircraft. This part of the work did not focus on developing a visual tracking system for the detection of the vehicles for the estimation of the relative position of the UAV. Instead, this issue was solved using augmented reality codes. In this case, we focused on the development of a vision-based control approach to command the UAV for these specific tasks. The control approach was done using fuzzy logic techniques, and the controllers were initially tuned in a simulated environment. A final tuning phase was conducted with the real quadrotor during real experiments in an indoor environment. A series of tests for the following and landing on a moving target was presented in this paper. The use of vision tracking algorithms is the next step to improve this vision-based control approach.

Acknowledgments: The authors would like to acknowledge Raphael Hinger from the University of Luxembourg for the technical support.

Author Contributions: Miguel A. Olivares-Mendez and Tegawendé F. Bissyandé are the project idea initiators and supervisors of the full paper. Miguel A. Olivares-Mendez developed the fuzzy control system and carried out the vision-based control approach and experiments in collaboration with Maciej Zurad and Arun Annaiyan. Changhong Fu developed the animal tracker algorithm and carried out the experiments in collaboration with Miguel A. Olivares-Mendez. Philippe Ludivig developed the face detection algorithm and carried out the experiments in collaboration with Miguel A. Olivares-Mendez and Maciej Zurad. Macied Zurad helped with the overall software integration. Somasundar Kannan and Tegawendé F. Bissyandé contributed to the writing of the manuscript and provided significant suggestions on the development. Holger Voos and Pascual Campoy provided additional guidance for the whole study.

Conflicts of Interest: The authors declare no conflict of interest.

References

1. Olivares-Mendez, M.A.; Bissyandé, T.F.; Somasundar, K.; Klein, J.; Voos, H.; le Traon, Y. The NOAH project: Giving a chance to threatened species in Africa with UAVs. In Proceedings of the Fifth International EAI Conference on e-Infrastructure and e-Services for Developing Countries, Blantyre, Malawi, 25–27 November 2013; pp. 198–208.

2. SPOTS-Air Rangers. Available online: http://www.spots.org.za/ (accessed on 15 October 2013).

3. Yilmaz, A.; Javed, O.; Shah, M. Object tracking: A survey. *ACM Comput. Surv.* **2006**, *38*, doi:10.1145/1177352.1177355.

4. Smeulder, A.; Chu, D.; Cucchiara, R.; Calderara, S.; Deghan, A.; Shah, M. Visual Tracking: An Experimental Survey. *IEEE Trans. Pattern Anal. Mach. Intell.* **2013**, *36*, 1442–1468.

5. Fotiadis, E.P.; Garzon, M.; Barrientos, A. Human Detection from a Mobile Robot Using Fusion of Laser and Vision Information. *Sensors* **2013**, *13*, 11603–11635.

6. Mondragon, I.; Campoy, P.; Martinez, C.; Olivares-Mendez, M. 3D pose estimation based on planar object tracking for UAVs control. In Proceedings of the 2010 IEEE International Conference on Robotics and Automation (ICRA), Anchorage, Alaska, 3–7 May 2010; pp. 35–41.

7. Fischler, M.A.; Bolles, R.C. Random Sample Consensus: A Paradigm for Model Fitting with Applications to Image Analysis and Automated Cartography. *Commun. ACM* **1981**, *24*, 381–395.

8. Lowe, D. Distinctive Image Features from Scale-Invariant Keypoints. *Int. J. Comput. Vis.* **2004**, *60*, 91–110.

9. Bay, H.; Ess, A.; Tuytelaars, T.; van Gool, L. Speeded-Up Robust Features (SURF). *Comput. Vis. Image Underst.* **2008**, *110*, 346–359.

10. Martinez, C.; Mondragon, I.; Campoy, P.; Sanchez-Lopez, J.; Olivares-Mendez, M. A Hierarchical Tracking Strategy for Vision-Based Applications On-Board UAVs. *J. Intell. Robot. Syst.* **2013**, *72*, 517–539.

11. Martinez, C.; Richardson, T.; Thomas, P.; du Bois, J.L.; Campoy, P. A vision-based strategy for autonomous aerial refueling tasks. *Robot. Auton. Syst.* **2013**, *61*, 876–895.

12. Sanchez-Lopez, J.; Saripalli, S.; Campoy, P.; Pestana, J.; Fu, C. Toward visual autonomous ship board landing of a VTOL UAV. In Proceedings of the 2013 International Conference on Unmanned Aircraft Systems (ICUAS), Atlanta, GA, USA, 28–31 May 2013; pp. 779–788.

13. Fu, C.; Suarez-Fernandez, R.; Olivares-Mendez, M.; Campoy, P. Real-time Adaptive Multi-Classifier Multi-Resolution Visual Tracking Framework for Unmanned Aerial Vehicles. In Proceedings of the 2013 2nd International Workshop on Research, Education and Development of Unmanned Aerial Systems (RED-UAS), Compiegne, France, 20–22 November 2013; pp. 532–541.

14. Babenko, B.; Yang, M.H.; Belongie, S. Robust Object Tracking with Online Multiple Instance Learning. *IEEE Trans. Pattern Anal. Mach. Intell.* **2011**, *33*, 1619–1632.

15. Fu, C.; Carrio, A.; Olivares-Mendez, M.; Suarez-Fernandez, R.; Campoy, P. Robust Real-time Vision-based Aircraft Tracking from Unmanned Aerial Vehicles. In Proceedings of the 2014 IEEE International Conference on Robotics and Automation (ICRA), Hong Kong, China, 31 May–7 June 2014; pp. 5441–5446.

16. Black, M.; Jepson, A. EigenTracking: Robust Matching and Tracking of Articulated Objects Using a View-Based Representation. *Int. J. Comput. Vis.* **1998**, *26*, 63–84.

17. Murase, H.; Nayar, S. Visual learning and recognition of 3-d objects from appearance. *Int. J. Comput. Vis.* **1995**, *14*, 5–24.

18. Belhumeur, P.; Kriegman, D. What is the set of images of an object under all possible lighting conditions? Computer Vision and Pattern Recognition, 1996. In Proceedings of the 1996 IEEE Computer Society Conference on CVPR '96, San Francisco, CA, USA, 18–20 June 1996; pp. 270–277.

19. Ke, Y.; Sukthankar, R. PCA-SIFT: A more distinctive representation for local image descriptors. In Proceedings of the 2004 IEEE Computer Society Conference on Computer Vision and Pattern Recognition, Washington, DC, USA, 27 June–2 July 2004; Volume 2, pp. 506–513.

20. Arulampalam, M.; Maskell, S.; Gordon, N.; Clapp, T. A tutorial on particle filters for online nonlinear/non-Gaussian Bayesian tracking. *IEEE Trans. Sign. Process.* **2002**, *50*, 174–188.

21. Yang, M.H.; Kriegman, D.J.; Ahuja, N. Detecting faces in images: A survey. *IEEE Trans. Pattern Anal. Mach. Intell.* **2002**, *24*, 34–58.

22. Leung, T.K.; Burl, M.C.; Perona, P. Finding faces in cluttered scenes using random labeled graph matching. In Proceedings of the Fifth International Conference on Computer Vision, Cambridge, MA, USA, 20–23 June 1995; pp. 637–644.

23. Lanitis, A.; Taylor, C.J.; Cootes, T.F. Automatic face identification system using flexible appearance models. *Image Vis. Comput.* **1995**, *13*, 393–401.

24. Rowley, H.; Baluja, S.; Kanade, T. Neural network-based face detection. *IEEE Trans. Pattern Anal. Mach. Intell.* **1998**, *20*, 23–38.

25. Rajagopalan, A.N.; Kumar, K.S.; Karlekar, J.; Manivasakan, R.; Patil, M.M.; Desai, U.B.; Poonacha, P.; Chaudhuri, S. Finding faces in photographs. In Proceedings of the Sixth International Conference on Computer Vision, 1998, Bombay, India, 4–7 January 1998; pp. 640–645.

26. Osuna, E.; Freund, R.; Girosi, F. Training support vector machines: An application to face detection. In Proceedings of the 1997 IEEE Computer Society Conference on Computer Vision and Pattern Recognition, San Juan, Puerto Rico, 17–19 Jun 1997; pp. 130–136.

27. Viola, P.; Jones, M. Rapid object detection using a boosted cascade of simple features. In Proceedings of the 2001 IEEE Computer Society Conference on Computer Vision and Pattern Recognition, Kauai, HI, USA, 8–14 December 2001; Volume 1, pp. I511–I518.

28. Shirai, Y.; Inoue, H. Guiding a robot by visual feedback in assembling tasks. *Pattern Recognit.* **1973**, *5*, 99–108.

29. Sanderson, A.C.; Weiss, L.E. Adaptative visual servo control of robots. In *Robot Vision*; Pugh, A., Ed.; Springer: Berlin, Germany; Heidelberg, Germany, 1983; pp. 107–116.

30. Campoy, P.; Correa, J.; Mondragón, I.; Martínez, C.; Olivares, M.; Mejías, L.; Artieda, J. Computer Vision Onboard UAVs for Civilian Tasks. *J. Intell. Robot. Syst.* **2009**, *54*, 105–135.

31. Ding, W.; Gong, Z.; Xie, S.; Zou, H. Real-time vision-based object tracking from a moving platform in the air. In Proceedings of the 2006 IEEE/RSJ International Conference on Intelligent Robots and Systems, Beijing, China, 9–15 October 2006; pp. 681–685.

32. Teuliere, C.; Eck, L.; Marchand, E. Chasing a moving target from a flying UAV. In Proceedings of the 2011 IEEE/RSJ International Conference on Intelligent Robots and Systems, San Francisco, CA, USA, 25–30 September 2011; pp. 4929–4934.

33. Ruffier, F.; Franceschini, N. Visually guided micro-aerial vehicle: Automatic take off, terrain following, landing and wind reaction. In Proceedings of the 2004 IEEE International Conference on Robotics and Automation (ICRA'04), New Orleans, LA, USA, 26 April–1 May 2004; pp. 2339–2346.

34. Lee, D.; Ryan, T.; Kim, H. Autonomous landing of a VTOL UAV on a moving platform using image-based visual servoing. In Proceedings of the 2012 IEEE International Conference onRobotics and Automation (ICRA), St Paul, MN, USA, 14–18 May 2012; pp. 971–976.

35. Zengin, U.; Dogan, A. Cooperative target pursuit by multiple UAVs in an adversarial environment. *Robot. Auton. Syst.* **2011**, *59*, 1049–1059.

36. Egbert, J.; Beard, R.W. Low-altitude road following using strap-down cameras on miniature air vehicles. In Proceedings of the 2007 American Control Conference, New York, NY, USA, 9–13 July 2011; pp. 831–843.

37. Rodríguez-Canosa, G.R.; Thomas, S.; del Cerro, J.; Barrientos, A.; MacDonald, B. A Real-Time Method to Detect and Track Moving Objects (DATMO) from Unmanned Aerial Vehicles (UAVs) Using a Single Camera. *Remote Sens.* **2012**, *4*, 1090–1111.

38. Cesetti, A.; Frontoni, E.; Mancini, A.; Zingaretti, P.; Longhi, S. A Vision-Based Guidance System for UAV Navigation and Safe Landing using Natural Landmarks. *J. Intell. Robot. Syst.* **2010**, *57*, 233–257.

39. De Wagter, C.; Mulder, J. Towards Vision-Based UAV Situation Awareness. In Proceedings of the AIAA Guidance, Navigation, and Control Conference and Exhibit, San Francisco, CA, USA, 17 August 2005.

40. Fucen, Z.; Haiqing, S.; Hong, W. The object recognition and adaptive threshold selection in the vision system for landing an Unmanned Aerial Vehicle. In Proceedings of the 2009 International Conference on Information and Automation (ICIA '09), Zhuhai/Macau, China, 22–25 June 2009; pp. 117–122.

41. Saripalli, S.; Sukhatme, G.S. Landing a helicopter on a moving target. In Proceedings of the IEEE International Conference on Robotics and Automation, Roma, Italy, 10–14 April 2007; pp. 2030–2035.

42. Saripalli, S.; Montgomery, J.F.; Sukhatme, G.S. Visually-Guided Landing of an Unmanned Aerial Vehicle. *IEEE Trans. Robot. Autom.* **2003**, *19*, 371–381.

43. Saripalli, S.; Montgomery, J.; Sukhatme, G. Vision-based autonomous landing of an unmanned aerial vehicle. In Proceedings of the 2002 IEEE International Conference on Robotics and Automation (ICRA'02), Washinton, DC, USA, 11–15 May 2002; Volume 3, pp. 2799–2804.

44. Merz, T.; Duranti, S.; Conte, G. Autonomous Landing of an Unmanned helicopter Based on Vision and Inbertial Sensing. In Proceedings of the The 9th International Symposium on Experimental Robotics (ISER 2004), Singapore, 18–21 June 2004.

45. Merz, T.; Duranti, S.; Conte, G. Autonomous landing of an unmanned helicopter based on vision and inertial sensing. In *Experimental Robotics IX*; Ang, M., Khatib, O., Eds.; Springer Berlin: Heidelberg, Germany, 2006; Volume 21, pp. 343–352.

46. Hermansson, J. Vision and GPS Based Autonomous Landing of an Unmanned Aerial Vehicle. Ph.D. Thesis, Linkoping University, Department of Electrical Engineering, Automatic Control, Linkoping, Sweeden, 2010.

47. Lange, S.; Sünderhauf, N.; Protzel, P. Autonomous landing for a multirotor UAV using vision. In Proceedings of the SIMPAR International Conference on Simulation, Modeling and Programming for Autonomous Robots, Venice, Italy, 3–7 November 2008; pp. 482–491.

48. Voos, H. Nonlinear landing control for quadrotor UAVs. In *Autonome Mobile Systeme 2009*; Dillmann, R., Beyerer, J., Stiller, C., Zöllner, J., Gindele, T., Eds.; Springer: Berlin, Germany; Heidelberg, Germany, 2009; pp. 113–120.

49. Voos, H.; Bou-Ammar, H. Nonlinear Tracking and Landing Controller for Quadrotor Aerial Robots. In Proceedings of the 2010 IEEE International Conference on Control Applications (CCA), Yokohama, Japan, 8–10 September 2010; pp. 2136–2141.

50. Chitrakaran, V.; Dawson, D.; Chen, J.; Feemster, M. Vision Assisted Autonomous Landing of an Unmanned Aerial Vehicle. In Proceedings of the 2005 44th IEEE Conference on Decision and Control European Control Conference (CDC-ECC'05), Seville, Spain, 12–15 December 2005; pp. 1465–1470.

51. Nonami, K.; Kendoul, F.; Suzuki, S.; Wang, W.; Nakazawa, D.; Nonami, K.; Kendoul, F.; Suzuki, S.; Wang, W.; Nakazawa, D. Guidance and navigation systems for small aerial robots. In *Autonomous Flying Robots*; Springer: Tokyo, Japan, 2010; pp. 219–250.

52. Wenzel, K.; Masselli, A.; Zell, A. Automatic Take Off, Tracking and Landing of a Miniature UAV on a Moving Carrier Vehicle. *J. Intell. Robot. Syst.* **2011**, *61*, 221–238.

53. Venugopalan, T.; Taher, T.; Barbastathis, G. Autonomous landing of an Unmanned Aerial Vehicle on an autonomous marine vehicle. *Oceans 2012* **2012**, 1–9.

54. Kendoul, F. Survey of Advances in Guidance, Navigation, and Control of Unmanned Rotorcraft Systems. *J. Field Robot.* **2012**, *29*, 315–378.

55. Chao, H.; Cao, Y.; Chen, Y. Autopilots for small unmanned aerial vehicles: A survey. *Int. J. Control Autom. Syst.* **2010**, *8*, 36–44.

56. Zulu, A.; John, S. A Review of Control Algorithms for Autonomous Quadrotors. *Open J. Appl. Sci.* **2014**, *4*, 547–556.

57. Martinez, S.E.; Tomas-Rodriguez, M. Three-dimensional trajectory tracking of a quadrotor through PVA control. *Rev. Iberoam. Autom. Inform. Ind. RIAI* **2014**, *11*, 54–67.

58. Shadowview. Available online: http://www.shadowview.org (accessed on 15 October 2013).

59. Wildlife Conservation UAV Challenge. Available online: http://www.wcuavc.com (accessed on 23 August 2015).

60. Li, G.; Liang, D.; Huang, Q.; Jiang, S.; Gao, W. Object tracking using incremental 2D-LDA learning and Bayes inference. In Proceedings of the 2008 15th IEEE International Conference on Image Processing (ICIP 2008), San Diego, CA, USA, 12–15 October 2008; pp. 1568–1571.

61. Wang, T.; Gu, I.H.; Shi, P. Object Tracking using Incremental 2D-PCA Learning and ML Estimation. In Proceedings of the 2007 IEEE International Conference on Acoustics, Speech and Signal Processing (ICASSP 2007), Honolulu, HI, USA, 15–20 April 2007; Volume 1, pp. 933–936.

62. Wang, D.; Lu, H.; wei Chen, Y. Incremental MPCA for color object tracking. In Proceedings of the 2010 20th International Conference on Pattern Recognition (ICPR), Istanbul, Turkey, 23–26 August 2010; pp. 1751–1754.

63. Hu, W.; Li, X.; Zhang, X.; Shi, X.; Maybank, S.; Zhang, Z. Incremental Tensor Subspace Learning and Its Applications to Foreground Segmentation and Tracking. *Int. J. Comput. Vis.* **2011**, *91*, 303–327.

64. Ross, D.A.; Lim, J.; Lin, R.S.; Yang, M.H. Incremental Learning for Robust Visual Tracking. *Int. J. Comput. Vis.* **2008**, *77*, 125–141.

65. Juan, L.; Gwon, O. A Comparison of SIFT, PCA-SIFT and SURF. *Int. J. Image Process. IJIP* **2009**, *3*, 143–152.

66. Levey, A.; Lindenbaum, M. Sequential Karhunen-Loeve basis extraction and its application to images. *IEEE Trans. Image Process.* **2000**, *9*, 1371–1374.

67. Hall, P.; Marshall, D.; Martin, R. Adding and subtracting eigenspaces with eigenvalue decomposition and singular value decomposition. *Image Vis. Comput.* **2002**, *20*, 1009–1016.

68. Hartley, R.I.; Zisserman, A. *Multiple View Geometry in Computer Vision*, 2nd ed.; Cambridge University Press: Cambridge, United Kingdom, 2004.

69. Automation & Robotics Research Group at SnT—University of Luxembourg. Available online: http://wwwen.uni.lu/snt/research/automation_research_group/projects#MoRo (accessed on 28 November 2015).

70. Davis, N.; Pittaluga, F.; Panetta, K. Facial recognition using human visual system algorithms for robotic and UAV platforms. In Proceedings of the 2013 IEEE International Conference on Technologies for Practical Robot Applications (TePRA), Woburn, MA, USA, 22–23 April 2013; pp. 1–5.

71. Gemici, M.; Zhuang, Y. *Autonomous Face Detection and Human Tracking Using ar Drone Quadrotor*. Cornell University: Ithaca, NY, USA, 2011.

72. Pan, Y.; Ge, S.; He, H.; Chen, L. Real-time face detection for human robot interaction. In Proceedings of the 18th IEEE International Symposium on Robot and Human Interactive Communication (RO-MAN 2009), Toyama, Japan, 27 September–2 October 2009; pp. 1016–1021.

73. Papageorgiou, C.P.; Oren, M.; Poggio, T. A general framework for object detection. In Proceedings of the 1998 Sixth International Conference on Computer vision, Bombay, India, 4–7 January 1998; pp. 555–562.

74. Lienhart, R.; Kuranov, A.; Pisarevsky, V. Empirical analysis of detection cascades of boosted classifiers for rapid object detection. In *Pattern Recognition*; Springer: Berlin, Germany, 2003; pp. 297–304.

75. Freund, Y.; Schapire, R.E. A decision-theoretic generalization of on-line learning and an application to boosting. *J. Comput. Syst. Sci.* **1997**, *55*, 119–139.

76. Szeliski, R. *Computer Vision: Algorithms and Applications*, 1st ed.; Springer-Verlag New York, Inc.: New York, NY, USA, 2010.

77. Garrido-Jurado, S.; Munoz-Salinas, R.; Madrid-Cuevas, F.; Marin-Jimenez, M. Automatic generation and detection of highly reliable fiducial markers under occlusion. *Pattern Recognit.* **2014**, *47*, 2280–2292.

78. Briod, A.; Zufferey, J.C.; Floreano, D. Automatically calibrating the viewing direction of optic-flow sensors. In Proceedings of the 2012 IEEE International Conference on Robotics and Automation (ICRA), St Paul, MN, USA, 14–18 May 2012; pp. 3956–3961.

79. Sa, I.; Hrabar, S.; Corke, P. Inspection of Pole-Like Structures Using a Visual-Inertial Aided VTOL Platform with Shared Autonomy. *Sensors* **2015**, *15*, 22003–22048.

80. Mondragón, I.; Olivares-Méndez, M.; Campoy, P.; Martínez, C.; Mejias, L. Unmanned aerial vehicles UAVs attitude, height, motion estimation and control using visual systems. *Auton. Robot.* **2010**, *29*, 17–34.

81. Rohmer, E.; Singh, S.; Freese, M. V-REP: A versatile and scalable robot simulation framework. In Proceedings of the 2013 IEEE/RSJ International Conference on Intelligent Robots and Systems (IROS), Tokyo, Japan, 3–7 November 2013; pp. 1321–1326.

82. Olivares-Mendez, M.; Kannan, S.; Voos, H. Vision based fuzzy control autonomous landing with UAVs: From V-REP to real experiments. In Proceedings of the 2015 23th Mediterranean Conference on Control and Automation (MED), Torremolinos, Spain, 16–19 June 2015; pp. 14–21.

83. Ar.Drone Parrot. Available online: http://ardrone.parrot.com (accessed on 1 March 2012).

84. KUKA Youbot. Available online: http://www.kuka-robotics.com/en/products/education/youbot/ (accessed on 17 July 2015).

Article

Formation Flight of Multiple UAVs via Onboard Sensor Information Sharing

Chulwoo Park, Namhoon Cho, Kyunghyun Lee and Youdan Kim *

Department of Mechanical & Aerospace Engineering, Seoul National University, Daehak-dong, Gwanak-gu, Seoul 151-744, Korea; bakgk@snu.ac.kr (C.P.); nhcho91@snu.ac.kr (N.C.); damul731@gmail.com (K.L.)
* Author to whom correspondence should be addressed; ydkim@snu.ac.kr; Tel.: +82-2-880-7398; Fax: +82-2-888-0321.

Academic Editors: Felipe Gonzalez Toro and Antonios Tsourdos
Received: 5 June 2015; Accepted: 14 July 2015; Published: 17 July 2015

Abstract: To monitor large areas or simultaneously measure multiple points, multiple unmanned aerial vehicles (UAVs) must be flown in formation. To perform such flights, sensor information generated by each UAV should be shared via communications. Although a variety of studies have focused on the algorithms for formation flight, these studies have mainly demonstrated the performance of formation flight using numerical simulations or ground robots, which do not reflect the dynamic characteristics of UAVs. In this study, an onboard sensor information sharing system and formation flight algorithms for multiple UAVs are proposed. The communication delays of radiofrequency (RF) telemetry are analyzed to enable the implementation of the onboard sensor information sharing system. Using the sensor information sharing, the formation guidance law for multiple UAVs, which includes both a circular and close formation, is designed. The hardware system, which includes avionics and an airframe, is constructed for the proposed multi-UAV platform. A numerical simulation is performed to demonstrate the performance of the formation flight guidance and control system for multiple UAVs. Finally, a flight test is conducted to verify the proposed algorithm for the multi-UAV system.

Keywords: multiple UAV operation; onboard sensor information sharing; monitoring multiple environment; close UAV formation; circular formation; triangular formation

---

1. Introduction

Many studies on multiple UAVs performing various missions have recently been conducted to address the increasing demands for unmanned aerial vehicle (UAV) applications [1–8]. Multiple UAVs can monitor multiple targets simultaneously, and multiple agents can complement each other in response to failures. A formation flight guidance law must be implemented to operate multiple UAVs. The formation flight guidance law enables each UAV to maintain relative positions in the formation, which allows the UAVs to be efficiently and safely controlled while satisfactorily performing a given mission. In [9], multiple quadrotor UAVs performed a boundary tracking formation flight while maintaining a phase angle of a geometrical boundary. In this manner, a boundary monitoring mission, such as a mission concerning an oil spill in an ocean, can be conducted rapidly and efficiently. The concept of miniature UAVs used to perform atmospheric measurements was proposed in [10]. The UAVs operated as low-cost aerial probes to measure temperature or wind profiles in the atmosphere. Therefore, multiple UAVs can effectively perform environmental monitoring missions based on formation flight guidance. The formation flight of multiple UAVs is an area wherein the robustness of the formation when subject to stochastic perturbations is crucial. Therefore, a robust analysis of the formation with respect to external disturbances, including wind, should be performed, although this is beyond the scope of this study. Additionally, the failure of a single UAV can lead to the failure of the entire system; therefore, it is important to increase the robustness of multi-UAV systems.

In the formation flight guidance law, information about other UAVs is utilized to generate formation guidance commands. Sensor information of each UAV, including position, velocity, and attitude, is measured by sensors such as instrument measurement units (IMUs), a global positioning system (GPS), and air data sensors; then, the sensor data obtained following signal processing are transferred to the other UAVs using a communication device. The sensor information of each UAV should be carefully and reliably treated and shared to successfully obtain formation flight.

Depending on the flight geometry, formation flight can be classified as a circular formation or close formation flight. Multiple UAVs typically perform a circular formation flight around an area in large-area monitoring missions. However, when the UAVs move to the next mission area, they should perform flight in close formation to reduce the total aerodynamic drag and increase survivability. Each formation guidance law has been studied and verified by numerical simulations or flight tests. A standoff tracking guidance for multiple UAVs was introduced in [11]. A nonlinear model predictive method for UAVs was proposed to track a moving target. A coordinated guidance using the second-order sliding mode was introduced in [12]. Three UAVs performed a coordinated circular formation using a standoff flight guidance law. However, the algorithms in [11,12] were developed based on the assumption that each UAV was a 3-degree-of-freedom (DOF) point-mass object. Therefore, the performance of the proposed guidance logics may be degraded when applied to a fixed-wing UAV because the longitudinal and lateral dynamics are different. A simulation environment based on a 6-DOF UAV model was introduced in [13]. The formation reconfiguration of six UAVs was performed to validate the developed environment. However, the results of the simulation may differ from the actual flight because there was an assumption of perfect information sharing between the UAVs. In contrast to the simulation environment, a communication limitation exists in the flight test. Therefore, the formation flight guidance law should be developed while considering the UAV dynamics and should be verified in the simulation while considering the communications limitations. A non-linear dynamic inversion (NLDI) guidance law was introduced in [14] to enable triangular formations. A guidance command was generated from the dynamic inversion of a simple point-mass UAV model. A simulation and a flight test of three UAVs were conducted. The cooperative formation flight of two fixed-wing UAVs was studied in [15]. A hardware-in-the-loop simulation was conducted, and a flight test that utilized the leader-follower formation pattern was performed. A cooperative controller for three rotary wing UAVs was proposed in [16] for heavy payload transport. Three engine-powered rotary UAVs formed a triangle formation, considering an unbalanced force generated by the payload. In [14–16], a ground control station (GCS) was used to relay information between UAVs; thus, the GCS could be a single point of formation failure. In addition, the mission range was limited within the surrounding GCS. To avoid a centralized GCS, UAVs should be interconnected using a full duplex and multipoint-to-multipoint topology to exchange information. A full duplex topology is required to enable a bidirectional connection, and a multipoint-to-multipoint topology is required for a decentralized connection between the UAVs. However, a communication system supporting this communication topology requires an expensive and heavy radiofrequency (RF) system. The system also consumes a large amount of power, which makes it unsuitable for small-scale UAV applications. Therefore, to perform a formation flight using a small UAV system, the communication system should be carefully designed by considering multiple communication topologies, low weight, low power, and low latency. This issue is why small UAVs cannot easily perform precise formation flight.

In this study, formation flight guidance algorithms are introduced and verified via both numerical simulation and a flight test using three fixed-wing UAVs. The contributions of this paper are as follows. First, an onboard sensor information-sharing system for small UAVs is developed. A wireless device is proposed considering the decentralized communication topology. Delays in the communication path are analyzed step by step, and the performance of the sensor information sharing between UAVs is verified. Second, a precise formation guidance law is developed considering the longitudinal and lateral dynamics of a fixed-wing UAV. A 6-DOF numerical simulation that includes a communication model is utilized to validate the developed guidance law. Third, an integrated formation flight test composed of various formation shapes is performed in a single flight test. Continuous circular formation flight and continuous close formation flight are conducted for a complete mission, including area monitoring and formation movement. Intermediate formation reconfiguration is also performed.

The remainder of this paper is organized as follows. In Section 2, the sensor and developed multi-UAV system are presented, as well as the system identification results. In Section 3, the onboard sensor information sharing system is explained in detail. In Section 4, a circular formation flight algorithm and close formation flight algorithm are proposed. Utilizing the algorithms, an integrated formation flight scenario is designed, and the proposed guidance algorithms are verified via a numerical simulation and flight experiment in Section 5. The conclusions of this research are presented in Section 6.

2. Sensor and UAV System

To perform formation flight, each UAV should be precisely controlled using sensors, and the same UAV hardware system should be used in each of the UAVs in the system to ensure uniform performance. In this study, three UAVs equipped with wind sensors were developed based on a commercial radio control (RC) aircraft. A linear dynamic model is also developed using a system identification scheme to ensure the accuracy of the numerical simulation.

2.1. Sensors

The developed UAV is equipped with multiple sensors to achieve precise flight control [17]. Figure 1 shows the sensors and avionics in the UAV, and detailed specifications of sensors and actuators are summarized in Table 1. A Microstrain 3DM-GX3-45 inertial navigation sensor is used to measure the vehicle attitude, position, and velocity. The external high-gain GPS antenna of the 3DM-GX3-45 is located at the tail of the UAV. A US Digital MA3 miniature absolute magnetic encoder is used to measure the angle of attack (AOA) and angle of sideslip (AOS), which are relative to the direction of the wind.

Figure 1. Sensors and avionics of an Unmanned Aerial Vehicle.

This miniature encoder sensor outputs an absolute rotation angle based on a non-contact method with low friction; therefore, it is appropriate for measuring wind direction in a small-scale UAV. A pitot tube is connected to a MPX7002 differential pressure sensor to measure the relative wind speed. The measured dynamic pressure is converted to air speed by a conversion formula. All of the sensors are connected to the developed ARM Cortex-M3-based embedded flight control computer (FCC), which is shown in Figure 2a. The embedded FCC sends control commands to the control surfaces, and an onboard 900 MHz ZigBee modem communicates with the other UAVs.

Table 1. Sensors and avionics specifications.

Component	Model Name	Manufacturer	Data Rate	Specification
Inertial navigation sensor	3DM-GX3-45	Microstrain	50 Hz	Typ. Attitude accuracy ±0.35° Typ. Velocity accuracy ± 0.1 m/s Typ. Position accuracy ± 2.5 m RMS
AOA, AOS sensor	3MA-A10-125-B	US Digital	50 Hz	12-bit resolution, 0.08° accuracy
Airspeed sensor	MPX7002DP	Freescale	50 Hz	Typ. Pressure accuracy ± 1.6 Pa
RF telemetry	XBP09-DMUIT-156	Digi	10 Hz	3 Km LOS range, 900 MHz
FCC	ARM FCC-M3	Self-developed	400 Hz	ARM Cortex-M3, 72 MHz Clock, 6 Ch PWM In & Out

2.2. UAV Airframe

For reliable formation flight, identical UAVs, each with an identical avionics system, are chosen as the platform for the multi-UAV system. An off-the-shelf RC airplane is chosen to ensure uniform and reliable flight characteristics of the UAVs. The RC airplane should be selected considering the structural strength necessary to tolerate frequent takeoffs and landings as well as its portability to enable easy transportation of the UAVs. For these reasons, a pusher-type airplane (Hitec Skyscout) with a 1.4-m wingspan, as shown in Figure 2b, is selected as the UAV. A Hitec Optima 6 RC receiver and Hitec Aurora 9 RC controller are used for manual control. Using a 2200 mAh Li-Polymer battery, each UAV can fly approximately 20 min under level flight conditions.

(a)   (b)

Figure 2. (a) Developed ARM Cortex-M3-based FCC; (b) UAV system.

2.3. UAV System Identification

An accurate dynamic model of the UAV is required to design a guidance and control algorithm for multiple UAVs. A system identification flight has been conducted to obtain a linear 6-DOF dynamic model that reflects the characteristics of the fixed-wing airplane [18,19]. During the system identification, predefined control surface inputs are used to excite the airplane dynamics, and the FCC stores the measurement data from the sensors. Lateral and longitudinal dynamic models are obtained by analyzing the recorded data. The lateral and longitudinal dynamics are identified separately. For aileron and rudder input, multistep 3-2-1-1 inputs are used to identify the lateral dynamics. For throttle and elevator control, multistep 3-2-1-1 inputs of the elevator and doublet input of the throttle are used to identify the longitudinal dynamics. The lateral and longitudinal system models are obtained as follows:

$$
\begin{bmatrix} \dot{\beta} \\ \dot{\phi} \\ \dot{p} \\ \dot{r} \end{bmatrix} = \begin{bmatrix} 0 & 0.7231 & -0.1718 & -1.6319 \\ 0 & 0 & 0.9957 & -0.0942 \\ -53.0053 & 0 & -10.5207 & 10.4250 \\ 7.3627 & 0 & -3.3925 & -4.8898 \end{bmatrix} \begin{bmatrix} \beta \\ \phi \\ p \\ r \end{bmatrix} + \begin{bmatrix} -1.4129 & -1.7054 \\ 0 & 0 \\ -51.1925 & 8.2723 \\ -9.0546 & -13.6097 \end{bmatrix} \begin{bmatrix} \delta_{ail} \\ \delta_{rud} \end{bmatrix} \quad (1)
$$

$$
\begin{bmatrix} \dot{V} \\ \dot{\alpha} \\ \dot{\theta} \\ \dot{q} \end{bmatrix} = \begin{bmatrix} 0 & 25.0414 & -9.7545 & 0 \\ -0.4130 & -12.0086 & 0.9435 & 2.2203 \\ 0 & 0 & 0 & 1 \\ -0.1204 & -23.5155 & 0 & -0.6636 \end{bmatrix} \begin{bmatrix} V \\ \alpha \\ \theta \\ q \end{bmatrix} + \begin{bmatrix} 0.0080 & 10.0711 \\ 0.0003 & 3.2113 \\ 0 & 0 \\ -0.0039 & -15.3344 \end{bmatrix} \begin{bmatrix} \delta_{thr} \\ \delta_{ele} \end{bmatrix} \quad (2)
$$

Figure 3 shows the lateral and longitudinal system identification results with the corresponding control inputs. System responses of the acquired model (solid line in Figure 3) are well matched with the recorded flight results (dashed line in Figure 3). The estimated lateral and longitudinal dynamic models are implemented in a MATLAB/Simulink environment to perform the 6-DOF simulation.

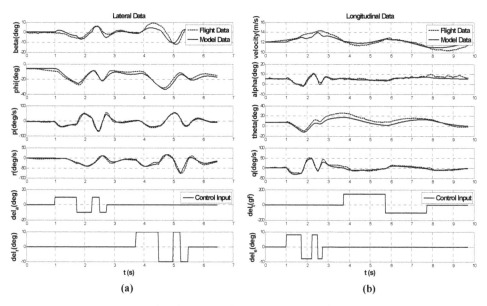

Figure 3. (a) Lateral; and (b) longitudinal system ID results for target airplane.

3. Onboard Sensor Information Sharing via Wireless Communication

Developing a method of sensor information sharing is a challenging issue in multi-UAV operation because unsynchronized information may cause incorrect decisions to be made when incorrect sensor information is shared among the UAVs. If all UAV communication is relayed via a leader aircraft or ground control system, then the central system may become a single point of failure, which would result in the UAVs having a limited mission range. Therefore, decentralized communication should be realized. In the actual flight environment, communication is conducted via a wireless device; as a result, the device should be selected considering both synchronization and decentralization.

Typical wireless communication devices are summarized in Table 2. In general, Wi-Fi is widely used for wireless communication. However, Wi-Fi requires a router to connect individual devices; therefore, decentralized communication cannot be realized. Bluetooth has a relatively short communication range and only provides a connection to a single device. In contrast, the ZigBee modem can communicate with multiple devices without a router and can cover a large distance. Therefore, the ZigBee modem (Digi XBP09) is selected as the wireless device for the system. In particular, the ZigBee modem acts as a transparent transceiver without requiring additional encoding/decoding, thereby enabling real-time communication.

Table 2. Comparison of wireless technology standards.

	Wi-Fi	**Bluetooth**	**ZigBee**
Range	50–100 m	10–100 m	100 m–1 km
Network Topology	Point to Hub Ad-hoc	Ad-hoc	Ad-hoc, peer to peer, star, mesh
Frequency	2.4 GHz and 5 GHz	2.4 GHz	868 MHz, 900 MHz, 2.4 GHz
Complexity	High	High	Low
Power Consumption	High	Middle	Very low
Security	WEP	64-bit or 128-bit encryption	128 AES or AL
Applications	-Wireless LAN	-Wireless device connection	-Industrial monitoring-Sensor network

Using the selected wireless device, *i.e.*, the ZigBee modem, UAVs share their sensor information with each other. The sensor information to be shared includes the UAV's status, such as attitude, position, velocity, and other essential parameters. Table 3 presents the communication packet used in the developed multi-UAV system. In Table 1, Vel denotes a velocity, Air denotes a barometric output, and Cmd denotes a command generated from the guidance algorithm.

Table 3. Communication packets of the multi-UAV system.

Data	Header	Status	Latitude	Longitude	Height	GPS Time	Roll	Pitch	Yaw	P
Byte	1–2	3–6	7–10	11–14	15–18	19–22	23–26	27–30	31–34	35–38
Q	R	Vel N	Vel E	Vel D	Acc X	Acc Y	Acc Z	Air Height	Air speed	Gamma Cmd
39–42	43–46	47–50	51–54	55–58	59–62	63–66	67–70	71–74	75–78	79–82
Roll Cmd	Battery	Elevator	Rudder	Throttle	Aileron	Alpha	Beta	Stage	Reserved	Putter
83–86	87–90	91–94	95–98	99–102	103–106	107–110	111–114	115–118	119–122	12–124

UAV 1 Data Transmit   **UAV 2 Data Transmit**   **UAV 3 Data Transmit**

Figure 4. Description of sequential cyclic communication.

To achieve robust onboard sensor information sharing between UAVs, a sequential cyclic communication method is used, as shown in Figure 4. The trigger UAV (UAV1) starts broadcasting its sensor information. Next, the neighboring UAV (UAV2) transmits its sensor information. This procedure continues until the last UAV (UAV3) performs a transmission. This procedure can prevent data loss from occurring due to simultaneous data transmission. The concept of sequential cyclic communication is realized at a low information sharing rate (<1 Hz). However, this approach requires accurate timing control to achieve a high information sharing rate (>10 Hz) because physical delays may exist during data transmission. Figure 5 describes a communication flow in the FCC. Three major physical delays can be found in the communication flow. The first delay is from microprocessor 1 (MCU1) to Zigbee1, $t_{MCU1toZigbee1}$. If UAV1 data consist of n bytes ($8n$ bits) and the baud rate is P bps (bits per second), then the delay from MCU1 to Zigbee1 can be calculated as follows:

$$t_{MCU1toZigbee1} = \frac{8n}{p} \tag{3}$$

The second delay is from Zigbee1 to Zigbee2, $t_{Zigbee1\ to\ Zigbee2}$. This delay depends on the air rate specification, that is, the RF transmission speed in air, which is inversely proportional to the carrier frequency of the modem. The data processing time inside the ZigBee modem is also included in that delay.

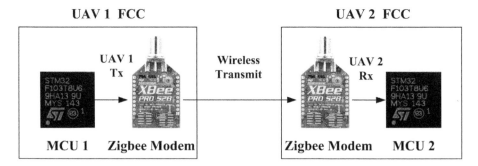

Figure 5. Communication flow for multiple UAVs.

If the ZigBee modem has a baud rate of P bps, then the delay from Zigbee1 to Zigbee2 can be calculated as:

$$t_{Zigbee1\ to\ Zigbee2} = \frac{8n}{q} + t_{Zigbee\ processing} \tag{4}$$

The third delay is from Zigbee2 to MCU2, $t_{Zigbee2toMCU2}$, which is equal to $t_{MCU1toZigbee1}$.

$$t_{Zigbee\ 2\ to\ MCU2} = \frac{8n}{p} \tag{5}$$

Therefore, the total communication delay can be expressed as:

$$t_{total} = t_{MCU1toZigbee1} + t_{Zigbee1\ to\ Zigbee2} + t_{Zigbee2toMCU2} \tag{6}$$

The considered ZigBee modem, Xbee-Pro DigiMesh900, has a baud rate of 230,400 bps and an air rate of 156,000 bps. In this case, nearly 24 ms passes between the first transmission to the second transmission for the $n = 124$-byte case. The total communication delay can be measured using a signal analyzer; as shown in Figure 6, 10 Hz of onboard sensor information sharing is possible because one cycle of four UAVs takes 24 ms × 4 = 96 ms. Accurate 10 Hz cyclic communication can be achieved if the FCC has less than 4 ms of additional processing delay. Because the typical minimum time interval of Microsoft Windows OS is in the range of 10 ms–20 ms, the Windows OS is not suitable for handling the cyclic communication. In this study, a real-time embedded FCC is used because it can control the communication timing with a 1 ms resolution. The developed embedded FCC and cyclic communication sequence is shown in Figure 7.

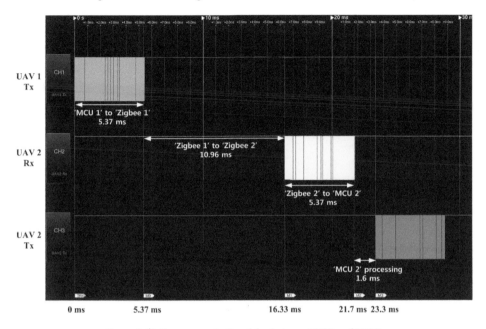

Figure 6. ZigBee communication delay between UAV1 and UAV2.

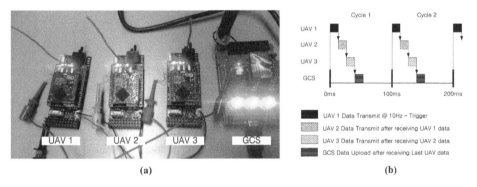

Figure 7. (**a**) Embedded FCCs and GCS; (**b**) sequential cyclic communication.

When UAV1 transmits its own data to other UAVs using the ZigBee modem, other UAVs receive the data. A periodic 1 ms watching process checks the received data, and the next UAV transmits its own data according to the given transmission order. The GCS also acts as a virtual UAV for monitoring and command uploading purposes, which does not affect the formation flight or communication structure. To verify the performance of the sequential cyclic communication, the transmission timing of the FCCs is measured, as shown in Figure 8. The sequential cyclic communication is found to function well, with 10 Hz onboard sensor information sharing properly conducted among the UAVs.

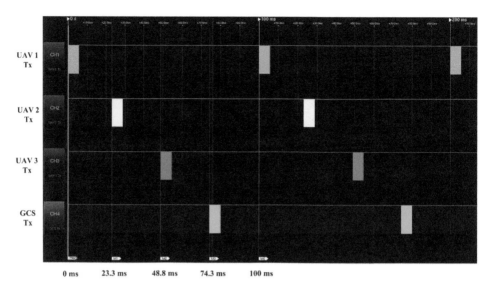

Figure 8. Timing measurement result of sequential cyclic communication.

4. Guidance Algorithms for Formation Flight

The formation flight of multiple UAVs requires various formation flight algorithms for large-area monitoring and formation movement. Circular formation flight guidance is suitable for the omni-directional monitoring of large areas because such guidance can control the phase angles among UAVs on the circular path. However, close formation flight guidance is proper for the polygonal shape formation of fixed-wing UAVs because such guidance can decrease the aerodynamic drag when flying to the next mission area. These algorithms are explained in Sections 4.1 and 4.2, respectively.

4.1. Circular Formation Flight Guidance

When a fixed-wing UAV monitors an area, a loitering flight is typically performed. If multiple UAVs are monitoring the area, the phase angles between UAVs should be controlled to ensure efficient area surveillance on the circular path. Nonlinear path-following guidance [20–22] was proposed to make UAVs fly on a circular path, assuming that all UAVs are moving on a two-dimensional surface, *i.e.*, flying at the same altitude. The stability of nonlinear path-following guidance was proven using the Lyapunov stability theorem [20], and therefore, all UAVs asymptotically converge to the predefined path. The reference altitude is set sufficiently high to cover the ground slopes and hills. The lateral guidance geometry of the nonlinear path following is shown in Figure 9, where V is the airspeed, L is the constant guidance distance, and η is the angle between V and L.

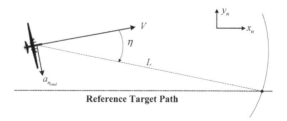

Figure 9. Nonlinear path-following guidance.

The UAV can be guided to the reference path by the following lateral acceleration command $a_{n_{cmd}}$:

$$a_{n_{cmd}} = \frac{2V^2}{L} \sin \eta \tag{7}$$

The following roll command can be used to make the UAV follow the lateral acceleration command $a_{n_{cmd}}$:

$$\phi_{cmd} = \tan^{-1}\left(\frac{a_{n_{cmd}}}{g}\right) \tag{8}$$

where g is the acceleration of gravity. Wind and sideslip motion are not considered in this guidance law.

A circular path can be generated from the target point p_c^n and the loitering radius R, as shown in Figure 10. The target point p_c^n is assumed to be the origin (0,0) without loss of generality. Once the UAVs are flying on the circular path using the nonlinear path following guidance, a phase angle can be maintained by controlling the airspeed of the UAV.

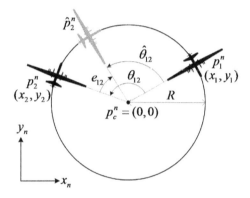

Figure 10. Phase angle on a circular path.

In this study, the relative phase angles of three UAVs are calculated in counter-clockwise order, starting from the first UAV. The phase angle between the UAVs is calculated using the positions p_1^n and p_2^n of the UAVs as:

$$\theta_{12} = atan2(y_2 - y_1, x_2 - x_1) \tag{9}$$

Because the phase angle starts from the 1st UAV, it has a positive value in the range of $[0, 2\pi)$. The target phase angle $\hat{\theta}_{12}$ is determined by considering the number of UAVs. Using the phase angle θ_{12} and target phase angle $\hat{\theta}_{12}$, the phase angle error e_{12} can be calculated as:

$$e_{12} = \hat{\theta}_{12} - \theta_{12} \tag{10}$$

A speed command for controlling the phase angle can be generated using the phase angle error as:

$$V_{cmd} = V_{cruise} + sat(K_v e_{12}), K_v > 0 \tag{11}$$

where V_{cruise} is the cruise velocity during level flight and K_v is a proportional velocity guidance gain. To prevent each UAV from entering a stall speed, a saturation value ΔV is set to $K_v e_{12}$ in Equation (11). In addition, the flight path angle command, γ_{cmd}, is generated to track the target height h_{ref} as follows:

$$h_{err} = h_{ref} - h$$
$$\gamma_{cmd} = K_p h_{err} + K_i \int h_{err} dt + K_d \dot{h}_{err} \tag{12}$$

where K_p, K_i, K_d are the proportional, integral, and derivative gains of the height controller, respectively.

4.2. Close Formation Flight Guidance

For close formation flight of the fixed-wing UAVs, the use of the separated design of the longitudinal and lateral guidance laws increases the performance of the formation flight because fixed-wing airplanes have different flight characteristics in the longitudinal and lateral axes. The lateral guidance law of the close formation flight is designed based on the nonlinear path-following guidance, as shown in Figure 11. In this study, the UAVs in the close formation are classified as a leader or as followers. The leader UAV follows the prescribed target path, and it shares all of the sensor information with the followers. The follower UAVs estimate the leader's path based on the most recent position data of the leader UAV. Using the estimated path of the leader UAV, the follower UAVs generate their own formation paths using the formation guidance law. Figure 11 shows a geometric description of the path estimation process.

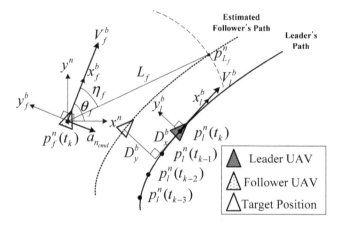

Figure 11. Formation path generation of a follower UAV.

In Figure 11, (x^n, y^n) denotes an inertial navigation frame, (x_l^b, y_l^b) denotes the body frame of the leader UAV, (x_f^b, y_f^b) denotes the body-fixed frame of the follower UAV, and (D_x^b, D_y^b) is the distance of the desired longitudinal and lateral formation between the leader and follower. The subsequent position data of the leader UAV, $p_l^n(t_k)$, are recorded by the follower UAV to estimate the leader's future trajectory.

$$p_l^n(t_k) = \begin{bmatrix} x_l^n(t_k) \\ y_l^n(t_k) \end{bmatrix}, t_k = t_0 + kT, k = 0, 1, 2, 3 \tag{13}$$

The recoded positions depend on a communication interval T. To estimate the leader's path based on the follower's position, the follower's position p_f^n is subtracted from $p_l^n(t_k)$ as

$$\Delta p_f^n = p_l^n - p_f^n, \begin{cases} \Delta p^n(t_k) = p_l^n(t_k) - p_f^n(t_k) \\ \Delta p^n(t_{k-1}) = p_l^n(t_{k-1}) - p_f^n(t_k) \\ \Delta p^n(t_{k-2}) = p_l^n(t_{k-2}) - p_f^n(t_k) \\ \Delta p^n(t_{k-3}) = p_l^n(t_{k-3}) - p_f^n(t_k) \end{cases} \tag{14}$$

The relative position Δp_f^n is transformed into the follower's body frame axis x_f^b using the follower's heading angle θ_f as:

$$\Delta p_f^b = \Omega_n^b \times \Delta p_f^n, \Omega_n^b = \begin{bmatrix} \cos(\theta_f) & -\sin(\theta_f) \\ \sin(\theta_f) & \cos(\theta_f) \end{bmatrix} \tag{15}$$

Using the relative position $\Delta p_f^b(t_k) = \left[\Delta x_f^b(t_k) \Delta y_f^b(t_k) \right]^T$, the coefficients of the cubic polynomial function $[c_{11}c_{21}c_{31}c_{41}]^T$ can be calculated using the least squares method.

$$\begin{bmatrix} c_{11} \\ c_{21} \\ c_{31} \\ c_{41} \end{bmatrix} = (A^TA)^{-1}A^TY, A = \begin{bmatrix} \left(\Delta x_f^b(t_k)\right)^3 & \left(\Delta x_f^b(t_k)\right)^2 & \Delta x_f^b(t_k) & 1 \\ \left(\Delta x_f^b(t_{k-1})\right)^3 & \left(\Delta x_f^b(t_{k-1})\right)^2 & \Delta x_f^b(t_{k-1}) & 1 \\ \left(\Delta x_f^b(t_{k-2})\right)^3 & \left(\Delta x_f^b(t_{k-2})\right)^2 & \Delta x_f^b(t_{k-2}) & 1 \\ \left(\Delta x_f^b(t_{k-3})\right)^3 & \left(\Delta x_f^b(t_{k-3})\right)^2 & \Delta x_f^b(t_{k-3}) & 1 \end{bmatrix}, Y = \begin{bmatrix} \Delta y_f^b(t_k) \\ \Delta y_f^b(t_{k-1}) \\ \Delta y_f^b(t_{k-2}) \\ \Delta y_f^b(t_{k-3}) \end{bmatrix} \tag{16}$$

The polynomial path function of the leader UAV can be estimated as:

$$\hat{l}_l^b(x) = c_{11}\left(x_f^b\right)^3 + c_{21}\left(x_f^b\right)^2 + c_{31}x_f^b + c_{41} \tag{17}$$

Based on the estimated polynomial path function $\hat{l}_l^b(x)$, the follower UAV generates a formation path $l_f^b(x)$ with the lateral distance D_y^b aligned with the y_f^b-axis as

$$l_f^b(x) = \hat{l}_l^b(x) + D_y^b \tag{18}$$

The follower UAV can follow the generated formation path by applying the nonlinear path-following guidance law. A roll angle ϕ_{cmd} command is calculated using the lateral acceleration command $a_{n_{cmd}}$ as

$$\phi_{cmd} = \tan^{-1}\left(\frac{a_{n_{cmd}}}{g}\right) where \ a_{n_{cmd}} = \frac{2V_f^2}{L_{L_f}}\sin\eta_f \tag{19}$$

Finally, the attitude controller makes the UAV follow the roll angle command ϕ_{cmd}.

Once The follower UAV is on the formation path, a longitudinal velocity command V_{ref} is generated to control the longitudinal distance D_x^b aligned to the x_f^b-axis as

$$V_{cmd} = V_{cruise} - K_d\left(D_x^b - (x_l^b - x_f^b)\right) \tag{20}$$

5. Simulation and Experimental Results

5.1. Procedure for Autonomous Formation Flight

A formation flight scenario that consists of area monitoring and formation movement is conducted to demonstrate the performance of the formation flight guidance laws. In the area monitoring scenario, the UAVs circle the area based on the circular formation flight guidance. In the formation movement scenario, the UAVs fly closely in a triangular formation as they move to the next target point area.

An integrated formation flight is introduced with five subscenarios, which are conducted in sequence. Table 4 summarizes the formation flight scenarios considered in this study. Each scenario has a specified mission and stage number depending on the longitudinal and lateral guidance mode. The mission variable specifies the observation or movement scenario, and the stage variable addresses the guidance mode transition. During the integrated formation flight, the stage variable is automatically changed based on the consensus of the UAVs. Figure 12 shows the path description of the integrated formation flight scenarios. Five subscenarios are described below.

Table 4. Integrated formation flight scenarios.

Scenario	Maneuver	Mission	Stage	Longitudinal Guidance	Lateral Guidance
-	Sequential takeoff	-	-	-	-
1	Circular formation/Separated altitude	1	1,2	$360/N_1$ deg phase separation/$\delta h = \pm 10$ m	Circular path following approximately 1st target
2	Circular formation/ Same altitude	1	3	$360/N_1$ deg phase separation/$\delta h = 0$ m	Circular path following approximately 1st target
3	Separation and Reconfiguration ofCircular formation	1	4,5	$360/N_2$ deg phase separation/$\delta h = 0$ m	Transition from 1st target to 2nd target
4	Close circular formation	1	6	30deg phase separation/$\delta h = 0$ m	Circular path following approximately 2nd target
5	Close triangular formation	2	-	-10 m rear position of leader UAV/$\delta h = 0$ m	± 10 m left/right position of leader UAV
-	Sequential landing	-	-	Longitudinal guidance	-

5.1.1. Circular Formation Flight with Separated Altitude

In the flight experiment, takeoff and landing are manually conducted by human pilots. Three UAVs takeoff in sequence, and the manual mode of each UAV is immediately switched to autonomous formation flight mode when it reaches the reference height. To prevent collisions between the UAVs during takeoff, each UAV follows a circular path with an altitude difference δh. The reference altitude is UAV01 = 60 m, and the altitude difference between UAV2 and UAV3 is set as $\delta h = \pm 10$ m; therefore, UAV2 is at a target altitude of 50 m, and UAV3 is at 70 m (Stage 1). During the separated altitude flight, the relative phase angles among UAVs are regulated to $360/N_1$ deg, where N_1 is a number of UAVs. The relative phase angles are controlled by the circular formation flight guidance law. If all three UAVs are on the circular path with 120 ± 10 deg phase angles (Stage 2), then UAV2 and UAV3 move to the reference altitude 60 m by increasing/decreasing their altitude via a flight path angle control.

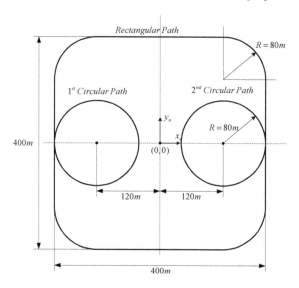

Figure 12. Path description of the integrated formation flight scenarios.

5.1.2. Circular Formation Flight at the Same Altitude

During the altitude transitions, the phase angles among the UAVs may change because of the acceleration of UAV3 and the deceleration of UAV2. Phase angle formation flight guidance regulates the phase angles of the UAVs to be 120 ± 10 deg (Stage 3). In this scenario, omni-directional surveillance of the target can be performed, wherein the target is at the center of the circular path.

5.1.3. Separation and Reconfiguration Formation Flight of the Circular Formation

To monitor multiple areas, the UAVs on the circular path should be separated and move to the second target. In the separation stage (Stage 4), the UAVs fly from the 1st circular path to the 2nd circular path one by one. During the separation, the phase angles of the remaining UAVs are modified to the phase angle of $360/N_1$deg, where N_1 is the number of UAVs remaining. In the reconfiguration stage (Stage 5), UAVs on the 2nd circular path are reconfigured to maintain a phase angle of $360/N_2$deg, where N_2 is the number of the UAVs in the 2nd circular path. In this stage, UAVs can monitor multiple areas while the circular formation flight is performed.

5.1.4. Close Circular Formation Flight

After the UAVs complete their monitoring of the mission area, they should fly together to the next mission area while maintaining the close formation flight. To perform this task, the UAVs on the circular path should converge to establish a close formation. To accomplish this maneuver, the phase angles between the UAVs are adjusted to 30 ± 10 deg in this stage (Stage 6). Once the specified phase angle has been reached, the circular formation flight mode is switched to the close formation flight mode.

5.1.5. Close Triangular Formation Flight

The close formation flight guidance law makes three UAVs follow a prescribed path while keeping them in a triangular formation. In the triangular configuration, UAV1, UAV2, and UAV3 are located at (0 m,0 m) $(-10$ m,-10 m), and $(-10$ m,10 m), respectively. UAV1 becomes the leader UAV, which tracks the predefined path, and the other UAVs become follower UAVs. The formation path of each of the followers can be calculated based on the leader's path. To ensure collision avoidance, a safety radius of 2 m is considered for each of the following UAVs. If one of the follower UAV's relative distance becomes less than 2 m, then an additional lateral command is activated to make the follower UAV fly at a distance from the formation.

5.2. Simulation Results

The proposed integrated formation flight scenario is composed of switching logics and multiple formation flights. The algorithms are thoroughly examined using a 6-DOF numerical simulation in the MATLAB/Simulink environment. The onboard sensor information sharing method and communication delay are considered in the simulation to emulate a real multi-UAV environment. The simulation blocks of the MATLAB/Simulink are shown in Figure 13. The identical integrated formation flight guidance block is used in all UAVs with the assigned UAV number. The sensor information sharing block emulates data transfer among the UAVs. Different execution rates of the guidance loop and communication loop are also implemented in the simulation block.

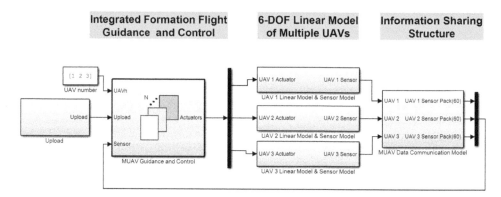

Figure 13. Simulation configuration in MATLAB/Simulink.

The numerical simulation results are shown in Figure 14 and indicate that the formation flight of multiple UAVs is performed well using both the close formation flight algorithm and close formation flight algorithm.

5.3. Experimental Results

After demonstrating the performance of the proposed algorithm via numerical simulation, a flight test is performed using the developed multi-UAV system. The integrated formation guidance algorithm block is directly converted into embedded C code via the MATLAB/Embedded Coder®.

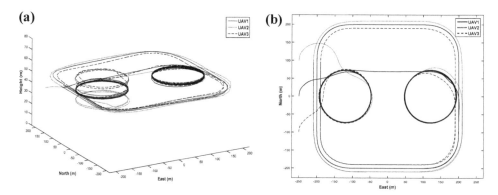

Figure 14. (a) 3D; and (b) 2D simulation results of integrated formation flight.

The integrated formation flight scenario in Table 4 is conducted in sequence. The Mission and State variables are sequentially and automatically changed. The fight test is conducted in an area that measures 400 m × 400 m, and the results are shown in Figure 15. The integrated formation flight was conducted successfully. In Figure 15, solid, dotted, and dashed lines correspond to the trajectories of UAV1, UAV2, and UAV3, respectively. During the sequential takeoff stage (Figure 15a,b), the UAVs form a 120-degree circular formation at different altitudes to ensure safety (Figure 15c). Once the phase angle of the circular formation is stabilized, UAV2 and UAV3 move to the same altitude, 60 m, and execute circular formation flight (Figure 15d). After performing a circular formation flight along the circular path, UAV1 is separated from the formation and moves to the next circular path, and UAV2 and UAV3 continue the circular formation of 180° along the first circular path (Figure 15e). UAV2 moves to the second circular path and reconfigures the 180-degree circular formation with UAV1

(Figure 15f) at the second circular path. Next, UAV3 moves to the second circular path (Figure 15g), and finally, the UAVs reconfigure the circular formation of 120° (Figure 15h). The UAVs reduce the phase angles to 30° to prepare for close formation flight (Figure 15i); then, they perform the close triangular formation flight (Figure 15j).

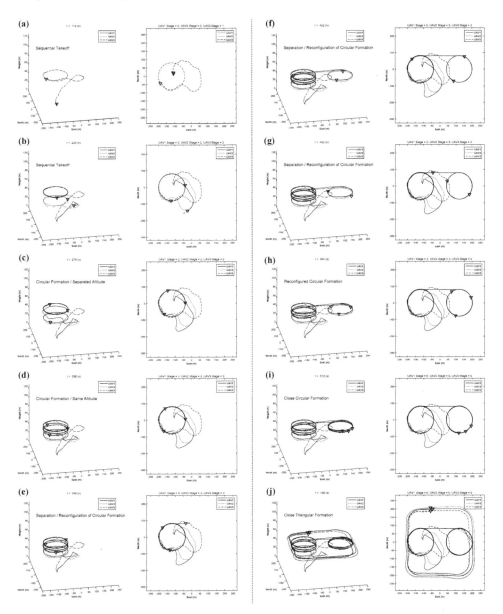

Figure 15. Flight results of integrated formation flight: (**a**)–(**b**) sequential takeoff; (**c**) circular formation at different altitudes; (**d**) circular formation at same altitude; (**e**)–(**h**) separation and reconfiguration of circular formation; (**i**) close circular formation; (**j**) close triangular formation.

The position histories of the integrated formation flight are shown in Figure 16. The mode variable indicates a flight mode controlled by an RC controller, where Mode 0 is the manual flight mode, Mode 1 is the stabilized co-pilot flight mode, and Mode 2 is the automatic mission flight mode. Depending on the status of the UAVs, the stage of each UAV may be different. The circular formation flight starts at 230 s, and the triangular formation flight starts at 591 s. The lateral and longitudinal control histories are shown in Figure 17. As shown in Figure 17, the inner-loop controllers are found to perform well at following the guidance commands. A detailed triangular formation flight result is shown in Figure 18. As shown in Figure 18, UAV1 follows the rounded rectangular path, and the follower UAVs generate their own formation path based on the estimated path of UAV1. Due to the east wind effect, the formation paths of UAV2 and UAV3 are slightly shifted to the west; nevertheless, the triangular shape is maintained well during the formation flight.

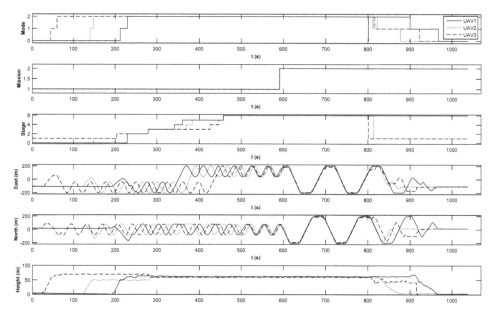

Figure 16. Position histories of integrated formation flight.

Figure 19a shows the site of the flight test with the UAVs and the GCS, and Figure 19b shows the photo of the close formation flight using three UAVs in a triangular formation captured by a ground camera. The triangular formation is shown to be maintained well during the flight test [23].

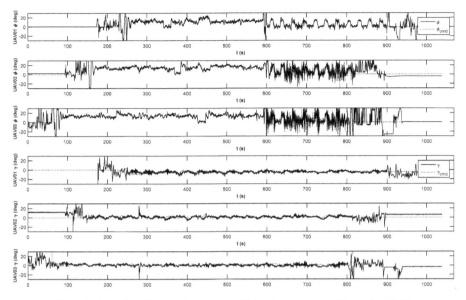

Figure 17. Lateral and longitudinal command histories of integrated formation flight.

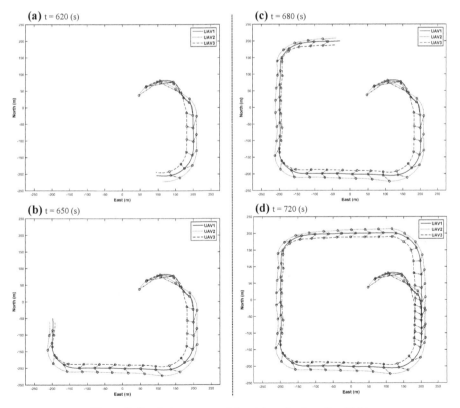

Figure 18. Detailed sequences of close triangular formation flight.

Figure 19. (**a**) Flight test setup; (**b**) Multiple UAVs in close triangular formation

6. Conclusions

The formation flight of three fixed-wing UAVs was performed based on circular formation flight guidance and a close formation flight algorithm. A multi-UAV system was developed, and the developed UAV dynamics were identified using a system identification scheme. A decentralized onboard sensor information sharing system for a miniature UAV system was developed using ZigBee modems, and the performance of the system was verified by analyzing the communication status between UAVs. Formation flight guidance laws of circular formation flight and close formation flight were proposed. The phase angle control scheme was used for circular formation flight, and the leader-follower guidance law was used for the close triangular formation flight during a formation movement. To verify the proposed algorithms for an entire formation flight, the integrated formation flight scenario was designed, and various formation flights were performed in sequence. The proposed guidance laws were first examined using a 6-DOF numerical simulation, and an actual flight test was conducted using the developed multi-UAV system.

Acknowledgments: This work was supported by the National Research Foundation of Korea (NRF) grant funded by the Korean government (MSIP) (2014R1A2A1A12067588).

Author Contributions: Chulwoo Park developed the sensor information sharing algorithm and close formation flight algorithm for the multi-UAV formation flight. Namhoon Cho developed the circular formation flight algorithm. Kyunghyun Lee designed the embedded FCC for the integrated formation flight experiment. Youdan Kim was the principal investigator of the project, and all of the algorithms were developed under the guidance of Kim.

Conflicts of Interest: The authors declare no conflict of interest.

References

1. Giulietti, F.; Pollini, L.; Innocenti, M. Formation Flight Control: A Behavioral Approach. In Proceedings of the AIAA Guidance, Navigation, and Control Conference, Montreal, QC, Canada, 6–9 August 2001.
2. Price, C.R. The Virtual UAV Leader. In Proceedings of the AIAA Infotech, Rohnert Park, CA, USA, 7–10 May 2007.
3. Li, N.H.M.; Liu, H.H.T. Multiple UAVs Formation Flight Experiments Using Virtual Structure and Motion Synchronization. In Proceedings of the AIAA Guidance, Navigation, and Control Conference, Chicago, IL, USA, 10–13 August 2009.
4. Teo, R.; Jang, J.S.; Tomlin, C.J. Automated Multiple UAV Flight—The Stanford Dragon Fly UAV Program. In Proceedings of the 43rd IEEE Conference on Decision and Control, Atlantis, Paradise Island, Bahamas, 14–17 December 2004.
5. Verma, A.; Wu, C.; Castelli, V. UAV Formation Command and Control Management. In Proceedings of the 2nd AIAA Unmanned Unlimited System, Technologies, and Operations, San Diego, CA, USA, 15–18 September 2003.

6. Schmitt, L.; Fichter, W. Collision-Avoidance Framework for Small Fixed-Wing Unmanned Aerial Vehicles. *J. Guid. Control. Dyn.* **2014**, *37*, 1323–1328. [CrossRef]

7. Beard, R.W.; McLain, T.W.; Nelson, D.B.; Kingston, D.; Johanson, D. Decentralized Cooperative Aerial Surveillance Using Fixed-Wing Miniature UAVs. *IEEE Proc.* **2006**, *94*, 1306–1324. [CrossRef]

8. Mahboubi, Z.; Kolter, Z.; Wang, T.; Bower, G. Camera Based Localization for Autonomous UAV Formation Flight. In Proceedings of the AIAA Infotech, St. Louis, MO, USA, 29–31 March 2011.

9. Lee, H.B.; Moon, S.W.; Kim, W.J.; Kim, H.J. Cooperative Surveillance and Boundary Tracking with Multiple Quadrotor UAVs. *J. Inst. Control. Robot. Syst.* **2013**, *19*, 423–428. [CrossRef]

10. Geoffrey, B.; Ted, M.; Michael, L.; Mark, M.; Joseph, B. "Mini UAVs" for Atmospheric Measurements. In Proceedings of the AIAA Infotech, Rohnert Park, CA, USA, 7–10 May 2007; pp. 461–470.

11. Kim, S.; Oh, H.; Tsourdos, A. Nonlinear Model Predictive Coordinated Standoff Tracking of Moving Ground Vehicle. In Proceedings of the AIAA Guidance, Navigation, and Control Conference, Portland, OR, USA, 8–11 August 2011.

12. Yamasaki, T.; Balakrishnan, S.N.; Takano, H.; Yamaguchi, I. Coordinated Standoff Flights for Multiple UAVs via Second-Order Sliding Modes. In Proceedings of the AIAA Guidance, Navigation, and Control Conference, Kissimmee, FL, USA, 5–9 January 2015.

13. Venkataramanan, S.; Dogan, A. A Multi-UAV Simulation for Formation Reconfiguration. In Proceedings of the AIAA Modeling and Simulation Technologies Conference, Providence, RI, USA, 16–19 August 2004.

14. Gu, Y.; Seanor, B.; Campa, G.; Napolitano, M.R.; Rowe, L.; Gururajan, S.; Wan, S. Design and Flight Testing Evaluation of Formation Control Laws. *IEEE Trans. Control. Syst. Technol.* **2006**, *14*, 1105–1112. [CrossRef]

15. Bayraktar, S.; Fainekos, G.E.; Pappas, G.J. Experimental Cooperative Control of Fixed-Wing Unmanned Aerial Vehicles. In Proceedings of the Conference on Decision and Control, Los Angeles, CA, USA, 15–17 December 2004.

16. Maza, I.; Kondak, K.; Bernard, M.; Ollero, A. Multi-UAV Cooperation and Control for Load Transportation and Deployment. *J. Intell. Robot. Syst.* **2010**, *57*, 417–449. [CrossRef]

17. Park, C.; Kim, H.J.; Kim, Y. Real-Time Leader-Follower UAV Formation Flight Based on Modified Nonlinear Guidance. In Proceedings of the 29th Congress of the International Council of the Aeronautical Sciences, St. Petersburg, Russia, 7–12 September 2014.

18. Oh, G.; Park, C.; Kim, M.; Park, J.; Kim, Y. Small UAV System Identification in Time Domain. In Proceedings of the Spring Conference of KSAS, High One Resort, Gangwon-Do, Korea, 11–13 April 2012.

19. Kim, H.J.; Kim, M.; Lim, H.; Park, C.; Yoon, S.; Lee, D.; Choi, H.; Oh, G.; Park, J.; Kim, Y. Fully Autonomous Vision-Based Net-Recovery Landing System for a Fixed-Wing UAV. *IEEE/ASME Trans. Mechatron.* **2013**, *18*, 1320–1332. [CrossRef]

20. Park, S.; Deyst, J.; How, J.P. Performance and Lyapunov Stability of a Nonlinear Path-Following Guidance Method. *J. Guid. Control. Dyn.* **2007**, *30*, 1718–1728. [CrossRef]

21. Kim, D.; Park, S.; Nam, S.; Suk, J. A Modified Nonlinear Guidance Logic for a Leader-Follower Formation Flight of Two UAVs. In Proceedings of the International Conference on Control, Automation Systems-SICE, Fukuoka, Japan, 18–21 August 2009.

22. Lee, D.; Lee, J.; Kim, S.; Suk, J. Design of a Track Guidance Algorithm for Formation Flight of UAVs. *J. Inst. Control. Robot. Syst.* **2014**, *20*, 1217–1224. [CrossRef]

23. Formation Flight of Multiple UAV. Available online: https://youtu.be/6NVlgST9agQ (accessed on 16 July 2015).

 *sensors*

Article

Mini-UAV Based Sensory System for Measuring Environmental Variables in Greenhouses

Juan Jesús Roldán [1], Guillaume Joossen [2], David Sanz [1], Jaime del Cerro [1] and Antonio Barrientos [1,*]

1   Centre for Automation and Robotics (UPM-CSIC), José Gutiérrez Abascal 2, 28006 Madrid, Spain;
    jj.roldan@upm.es (J.J.R.); d.sanz@upm.es (D.S.); j.cerro@upm.es (J.C.)
2   ENSTA ParisTech, Boulevard des Maréchaux, 828, 91120 Palaiseau, France;
    guillaume.joossen@ensta-paristech.fr
*   Author to whom correspondence should be addressed; antonio.barrientos@upm.es;
    Tel.: +34-913-363-061; Fax: +34-913-363-010.

Received: 26 September 2014; Accepted: 26 January 2015; Published: 2 February 2015

Abstract: This paper describes the design, construction and validation of a mobile sensory platform for greenhouse monitoring. The complete system consists of a sensory system on board a small quadrotor (*i.e.*, a four rotor mini-UAV). The goals of this system include taking measures of temperature, humidity, luminosity and CO_2 concentration and plotting maps of these variables. These features could potentially allow for climate control, crop monitoring or failure detection (e.g., a break in a plastic cover). The sensors have been selected by considering the climate and plant growth models and the requirements for their integration onboard the quadrotor. The sensors layout and placement have been determined through a study of quadrotor aerodynamics and the influence of the airflows from its rotors. All components of the system have been developed, integrated and tested through a set of field experiments in a real greenhouse. The primary contributions of this paper are the validation of the quadrotor as a platform for measuring environmental variables and the determination of the optimal location of sensors on a quadrotor.

Keywords: greenhouse; UAVs; sensory system; environmental monitoring; agriculture; robotics

1. Introduction

Greenhouse farming is one of the most suitable areas for employing automation, robotic and computing technologies. Many greenhouses have climate control systems, which are usually composed of temperature and humidity sensors as well as irrigation, ventilation, and heating systems. These technologies offer a wide range of possibilities including climate control, production monitoring or detection of infestations or weeds. However, they suffer from several limitations, primarily due to their cost and reliability issues, which can make their implementation unprofitable and complex.

The emergence of Wireless Sensor Networks (WSNs) has initiated a revolution in these types of projects: WSNs provide flexibility (*i.e.*, the network can be constructed without a fixed architecture), modularity (*i.e.*, the network can incorporate new devices) and fault tolerance (*i.e.*, the network can work with failures in some motes) with low power consumption (*i.e.*, the motes usually have a sleep mode) to facilities [1].

Thus, these networks have been used in many applications in fields related to agriculture and food [2,3]: environmental monitoring [4] (e.g., climate monitoring and fire detection), precision agriculture [5] (e.g., rationalization of chemical products and optimization of irrigation) or the food industry (e.g., quality control and product traceability).

However, in the context of greenhouse farming, WSNs have been implemented more experimentally than productively. The previous literature contains several proposals concerning WSN deployment in greenhouses, but they are restricted to small fields [6]. For example, [7,9] deployed WSNs with nodes that measured temperature, humidity or luminosity in small greenhouses.

Some of the limitations of WSNs in greenhouse farming are the fixed locations of the motes and the corresponding costs, particularly for large greenhouses. Two possible alternatives to WSNs, which solve the problems of movement and costs, are mobile ground or aerial robots, which have been partially tested in previous studies [10,12].

Unmanned Aerial Vehicles (UAVs) are used in diverse fields related to environmental monitoring, such as in the acquisition of meteorological information [13,14]; the measurement of greenhouse gases in the atmosphere, which primarily includes carbon dioxide, methane and water vapor [15]; the surveillance of clouds of contaminant gases produced by human activities [16]; and the mapping of distribution of different gases [17].

In the context of agriculture, UAVs are typically used in some precision agriculture (PA) tasks: the measurement of vegetation density [18], the determination of irrigation needs [19], the construction of mosaics of fields for weeds detection [20] and the support of WSNs in crop monitoring [21]. This last article can be considered as a previous step of this paper.

Although the use of UAVs is growing in outdoor farming, it is still limited in indoor farming. There are several tasks in greenhouse agriculture that could be performed using mini-UAVs: the measurement of climate variables, the monitoring of plants and the surveillance of the perimeter. Thus, despite of their current limitations (*i.e.*, autonomy, payload capacity and safety), their wide range of applications, low cost, versatility and precision augur a promising future for UAVs in indoor farming [22].

This paper presents a quadrotor-based sensory system for measuring environmental variables in a greenhouse. Aspects related to the navigation of the quadrotor in a restricted and irregular place are reflected in the bibliography [23]. Two challenges of this work, namely the quadrotor's limited payload and the possible influence of rotors on the sensors' measurements are successfully overcome.

2. System Overview

The proposal of a mini-UAV-based sensory system is expected to be integrated in a greenhouse farming management system (Figure 1). The first one performs the acquisition of the environmental variables that can be measured in the air, while the second one encompasses not only the sensing (*i.e.*, acquisition of all environmental variables) but also the actuation (*i.e.*, climate control, crop monitoring and failure detection).

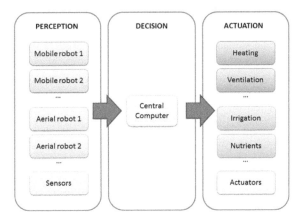

Figure 1. Architecture of the complete system.

The complete system has a centralized architecture. A central computer receives data from sensing devices (e.g., ground or aerial robots and static sensors), compiles the data, makes decisions and sends commands to the actuation devices. The centralized architecture has advantages (e.g., all information is collected, managed and saved on a single computer) and weaknesses (e.g., the system cannot recover from a failure in the central computer).

In addition, the system has flexibility and modularity; it may be composed of different aerial and ground robots with different purposes (e.g., acquisition of air variables, determination of ground properties and supply of water, nutrients, fertilizers or protection products). The flexible and modular character of this system allows adding or removing robots to adapt to the needs of different greenhouses. All modules (*i.e.*, sensing, processing and effecting modules) and components (*i.e.*, aerial and ground robots, central computer, heating and ventilation systems, and other actuators) communicate via a wireless local area network.

However, one must remember that the aim of this work is the description of the quadrotor-based sensory system; this complete system is only the framework of this study. In the next subsections, a platform analysis together with the selection and integration of sensors are described.

2.1. Platform Analysis

The primary alternatives to the proposed mini-UAV sensory system are Wireless Sensor Networks (WSNs) and Unmanned Ground Vehicles (UGVs). Both of them are well-known solutions that have been applied in the context of greenhouse agriculture. Despite this fact, the WSNs and UGVs are hindered by limitations that UAVs can overcome.

The primary advantage of WSNs is simultaneous measurements at various points, which a UGV or UAV cannot perform and may be desirable for this application. WSNs are a robust solution due to their simplicity, which reduces the probability of failure, and their modularity, which allows working with damaged motes. Conversely, in contrast to UGVs and UAVs, WSNs are not able to move within the workspace to take measures at points of interest. In addition, the costs of WSNs strongly depend on the number of motes, which may reach hundreds in a medium size greenhouse. This multiplication of motes (e.g., sensors, controllers, batteries and communication modules) makes their costs higher than the costs of UGVs or UAVs.

UGVs tend to have lower costs than WSNs and are competitive against UAVs. The simplicity of their mechanic elements and control systems makes their costs lower than the costs of UAVs. In addition, UGVs can move to the points of interest; however, these movements are restricted to the ground, preventing them from reaching certain points of interest due to obstacles such as plants and covers. Conversely, UAVs can obtain measurements at nearly any point in a three dimensional space including at different altitudes. This fact is interesting not only for reducing the number of sensors and therefore the total cost of such a system but also for obtaining local data for production monitoring, problem detection (e.g., a break in a plastic cover) and local climate control. In summary, the characteristics of UAVs make them a competitive option for measuring the environmental variables of greenhouses and justify this research.

2.2. Selection of Sensors

Sensors have been selected based on the needs of climate control and crop monitoring activities. Multiple models of climate [24,25], temperature [26] and humidity [27] in greenhouses can be found in the literature. Additionally, a model of the growth and maturation of plants is available in [28]. The study of these models has determined the variables that should be measured; these include air temperature, air humidity, carbon dioxide concentration, ethylene concentration, ground temperature, ground humidity, nutrient concentration and solar radiation.

Among these variables, the ground temperature, ground humidity and nutrient concentration can be measured by a UGV with less risk and cost than by a UAV. Nevertheless, current ethylene sensors are too heavy to be placed on-board a quadrotor; thus, the incorporation of an ethylene sensor should be investigated in future works. This study will focus on demonstrating the capability of measuring gases using a mini-UAV this can be accomplished by testing and validating the use of a mini-UAV with a carbon dioxide concentration sensor.

Table 1 lists sensors for air temperature and humidity, carbon dioxide concentration and solar radiation measurement that are commercially available and their features. The final selection has been performed according to the criteria of weight, size, range, resolution and cost. Specifically, the RHT03 temperature and humidity sensor, and the MG811 carbon dioxide concentration sensor have been selected. The primary features of these sensors are shown in Table 2.

Table 1. Analysis of environmental variables.

Variable	Is It Better to Measure It in the Air or on the Ground?	Is There a Sensor That Can be Attached to a Mini-UAV?	Result
Air temperature	√	√	√
Air humidity	√	√	√
CO_2 concentration	√	√	√
Ethylene concentration	√	×	×
Ground temperature	×	√	×
Ground humidity	×	√	×
Nutrient concentration	)(	×	×
Solar radiation	×	√	√

Table 2. Sensor features. Source: Sensor datasheets.

Features	Sensors		
	Temperature/Humidity: RHT03	Luminosity: TSL2561	CO_2 Concentration: MG811
Power supply	3.3–6.0 V	2.7–3.3 V	6.0 V
Measurement range	T: [−40; 80] °C H: [0; 100]%	[0; 40,000] lux	[350; 10,000] ppm
Sensitivity	T: 0.1 °C H: 0.1%	1 lux	Variable
Accuracy	T: 0.5 °C H: 2%	Not available	Not available
Preparation time	0–5 s	0–1 s	30–60 s
Response time	0–5 s	0–1 s	15–30 s
Communications	Digital	I2C	Analog

2.3. Integration of Sensors

The sensors have been integrated to satisfy two needs: the collection and storage of measurements including space and time references; and the communication between the mini-UAV sensory system and the greenhouse management system.

Several alternatives for the integration of sensors have been studied, and two prototypes have been developed: one with an Arduino UNO [29] (Figure 2) and another with a Raspberry Pi [30] (Figure 2). Both prototypes have been compared with multiple criteria including size, weight, performance and connectivity.

447

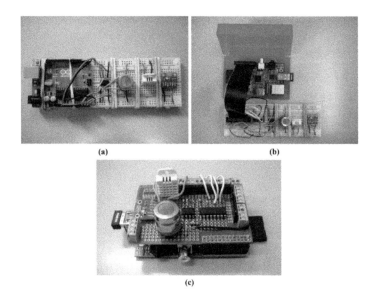

Figure 2. (a) Arduino UNO preliminary prototype; (b) Raspberry Pi preliminary prototype; (c) Raspberry Pi final prototype.

The Raspberry Pi has ultimately been chosen instead of the Arduino UNO due to its performance, connectivity and programming (Figure 2). The Raspberry Pi has better performance than the Arduino UNO, both in hardware (e.g., processor speed and memory) and software (*i.e.*, operating system), which allows it to preprocess data while measuring. Additionally, the Raspberry Pi typically has better performance when connected to Wi-Fi networks and exchanging data with other devices. Finally, the Raspberry Pi provides additional programming capabilities including full integration with Robot Operating System (ROS) [31].

3. Location of Sensors

The location of sensors on the quadrotor is not a trivial issue and requires the study of some conditions. Both the air-flows produced by the rotors and the light and shadow conditions can affect the sensor measurements. Additionally, the weights of the sensors influence the weight and inertia of the quadrotor, which can in turn affect navigation.

Specifically, the temperature and humidity sensor can be affected by solar radiation and the airflows of the rotors, the luminosity sensor is obviously conditioned by solar radiation and the carbon dioxide sensor can be affected by the air-flows of the rotors.

Two previous studies have addressed quadrotor aerodynamics with similar results [32,33]. Their conclusions stated that when considering an isolated rotor, the airspeed is maximum at its perimeter and minimum in the center and the exterior of the quadrotor; moreover, considering all rotors, the airspeed is maximum near the rotors and minimum in the center and outside the quadrotor.

Based on the quadrotor aerodynamics and considering the effects of solar radiation, there are two possible sensor locations to consider: the center part of the top side of the quadrotor and outside the quadrotor at some distance. Considering both options, the first does not require a complex assembly that could modify the center of gravity of the quadrotor (e.g., an extension for the sensors) and therefore is selected for the location of the sensors. Unfortunately, the conclusions of both publications were focused on quadrotor design and modeling instead of sensor allocation. Therefore, a complementary exhaustive study of quadrotor aerodynamics oriented to sensor allocation was necessary.

Simulations of computational fluid dynamics (CFD) and real experiments for determining quadrotor airflows were performed to determine the relevant aerodynamics and validate the location of sensors. These simulations and experiments are described in Sections 3.1 and 3.2.

3.1. CFD Simulation of a Quadrotor

A set of CFD simulations has been performed to determine the aerodynamics of a quadrotor and the evolution of the airflows of the rotors. These simulations have been performed using Autodesk Simulation CFD 2014.

The simulation model of a quadrotor has been designed with maximum detail near the propellers to increase the precision of the results and lower complexity in the other parts of quadrotor to reduce the computational cost of simulations. The quadrotor designed has a wingspan of 400 mm (*i.e.*, the distance between opposite rotors) and its center is a square with sides 125 mm long.

These simulations have been performed in a transient regimen (*i.e.*, initially, the rotors were stopped but later accelerated up to 3000 rpm) instead of at steady state for two reasons: the range of speeds is wider, and these conditions are more unfavorable (*i.e.*, worst case scenario). In addition, the hypotheses of incompressibility and turbulence of airflows have been assumed because the rotation parts are small, and the airspeed is relatively low.

The CFD simulation results are shown in Figure 3, and . The traces of fluid particles and the velocity profiles in planes have been chosen for a better visualization of the results to show them in a clear and precise manner from both a qualitative and quantitative perspective. Each frame is associated with its simulation time (t) and its propeller angle (α). The evolution of the airflows across the quadrotor can be seen in Figure 3. The traces arise from a plane located over the quadrotor and are attracted to the rotors. Their speed grows gradually as they approach the rotors and rapidly when they pass by them. Their trajectories follow the periphery of the air volume, avoiding the center of the quadrotor.

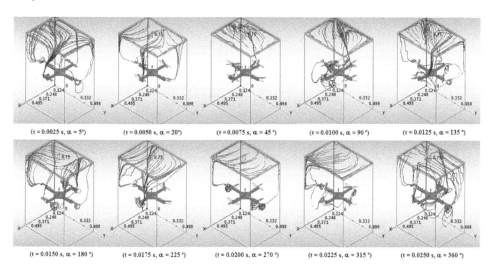

Figure 3. Airflows over the quadrotor.

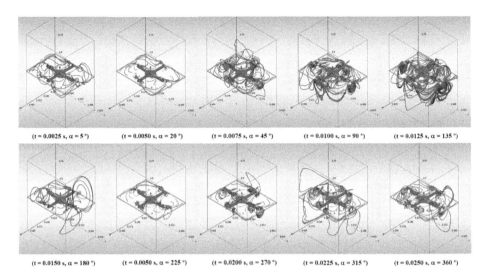

Figure 4. Airflows under the quadrotor.

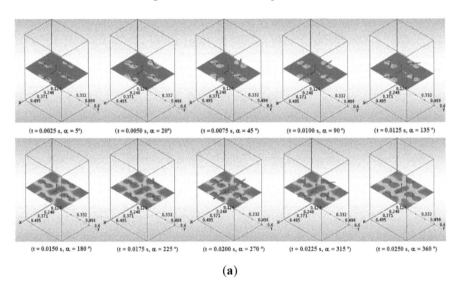

(a)

Figure 5. *Cont.*

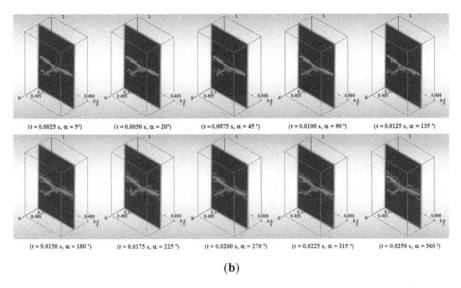

| (t = 0.0025 s, α = 5°) | (t = 0.0050 s, α = 20°) | (t = 0.0075 s, α = 45°) | (t = 0.0100 s, α = 90°) | (t = 0.0125 s, α = 135°) |

| (t = 0.0150 s, α = 180°) | (t = 0.0175 s, α = 225°) | (t = 0.0200 s, α = 270°) | (t = 0.0225 s, α = 315°) | (t = 0.0250 s, α = 360°) |

(b)

Figure 5. (a) Airspeed profile in the horizontal plane; (b) Airspeed profile in the vertical plane.

The evolution of the airflows under the quadrotor can be seen in Figure 4. The traces arise from a plane located immediately under the quadrotor and form a series of vortices around the rotors. Their speed grows rapidly when they pass near the rotors and falls gradually when they pass by them. As in the previous case, the center of the quadrotor is relatively free of airflows.

Figure 5 shows the airspeed profiles in the horizontal and vertical planes. As shown, the maximum speed is obtained within the rotors whereas the minimum speed is located near the center of the quadrotor. These results agree with the results of the previous works [32,33] and the hypotheses assumed in this work.

3.2. Measurement of Quadrotor Airflows

In order to determine the airspeed at different points around the quadrotor, an experiment has been performed to validate the results of the CFD simulations with a real quadrotor in real conditions.

This experiment has been performed using the Parrot AR.Drone 2.0 quadrotor that is shown in Figure 6. The quadrotor was attached in a support with a Cardan joint, which allows its attitude to change while maintaining its location and altitude. This mechanism facilitated the experiment and reduced the risk of an accident. A grid of 24 positions located both under and over the quadrotor was defined to measure the air speed with a digital anemometer. Ten readings were registered at each grid point to calculate an average value at each grid point.

Figure 6. Layout of the experiment.

Figure 7 depicts the results of this experiment. The airflows over the quadrotor are shown on the left side of the figure, and the airflows under the quadrotor are shown on the right. The X and Y axes show the location in centimeters, and the Z axis shows the airspeed in meters per second. The points are the measurements of the anemometer, and the surface is an interpolation of them.

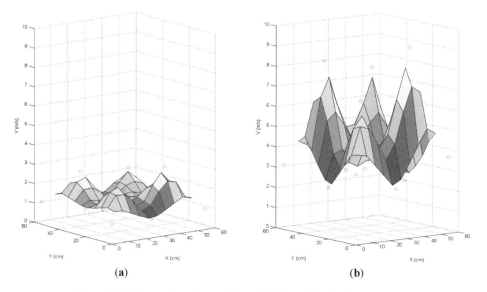

Figure 7. (a) Air speed over the quadrotor; (b) Air speed under the quadrotor.

The results of this experiment are coherent with the results of the CFD simulations. The airspeed presents four maxima within the four rotors and a depression in the center of the quadrotor. In addition, the results show that the velocity of the air repelled by the rotors (Figure 7) is higher than the velocity of the air attracted by them (Figure 7).

Both the simulations and experiments confirm the hypothesis that the optimal location for the sensors is on the center of the top side of the quadrotor. A proposal of the most adequate areas for allocating the different sensors is shown in Figure 8.

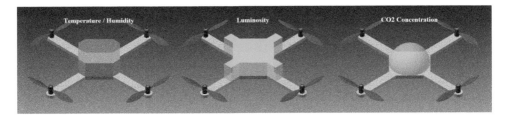

Figure 8. Proposal of sensor location.

3.3. Measuring on-Board a Quadrotor

The previous simulation and experiments have determined the optimal placement for the sensors in the quadrotor, but they have not provided any conclusions about the feasibility of performing measurements with the sensors in this location. Therefore, an additional experiment has been conducted to validate that the rotors' airflows have no significant influence in the sensors' measurements. This experiment consisted of taking measurements at a series of points under two different conditions: with the propellers stopped and with the propellers active.

Three sources of temperature, humidity and carbon dioxide were used to create gradients of these variables in the workspace. As in the previous experiment, a Parrot AR.Drone 2.0 quadrotor was used to transport the sensor system.

Measurements were taken from a distance of 5 m to the sources at intervals of 1 m. In the first test, the quadrotor was moved by hand, and in the other, it flew autonomously. In both experiments, the quadrotor was at a height of 0.5 m, and the sources were located on the ground. The results of this experiment are shown in Figure 9. As can be seen, there are differences between the measurements obtained with the rotors stopped and when they are moving. However, the average relative errors in temperature (3.71%), humidity (1.65%) and carbon dioxide concentration (3.84%) can be considered to be negligible. These errors can be associated with multiple factors apart from the influence of the propellers, including possible changes in the environment over time, particularly with regard to temperature and humidity, and the response time of the sensors, particularly with regard to carbon dioxide.

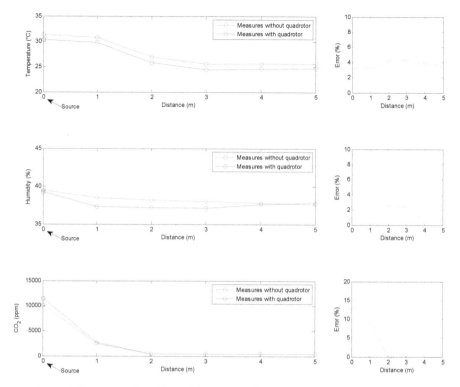

Figure 9. Temperature, humidity and CO_2 measured with propellers stopped and working.

4. Experiments, Results and Discussion

In order to validate the developments made in the laboratory, a series of field experiments was carried out in a greenhouse located in Almeria (Andalucia, Spain), an area with massive use of greenhouse farming.

In these experiments, maps of temperature, humidity, luminosity and carbon dioxide concentration in a greenhouse have been built using the mini-UAV based sensory system. The quadrotor followed a pre-planned path to avoid collisions with obstacles and obtained its location using visual odometry (*i.e.*, following a line and stopping at squares printed on the ground) and measurements from its Inertial Measurement Unit (IMU). In order to avoid possible errors due to the response times of the sensors, the quadrotor stopped at the waypoints until their measurements were stable.

The greenhouse was rectangular (106 m × 47 m) and had a height of 3 m; there were two doors on the front side of the building and two windows along its roof. The experiments were performed on 2 June 2014, starting at 9:00 a.m. and finishing at 10:00 a.m. During the experiments, the greenhouse was fallow, a tractor was working inside, and the doors and windows were open for ventilation.

Different perspectives of the greenhouse (e.g., outside, inside, front and top) are shown in Figure 10. The covered surface, the measurement points and the path followed in the experiments are detailed in the top view (Figure 10). The maps of temperature, humidity, luminosity and carbon dioxide concentration obtained in the greenhouse are shown in Figure 11. The points show the measurements of the sensors, and the surfaces show interpolations between these points.

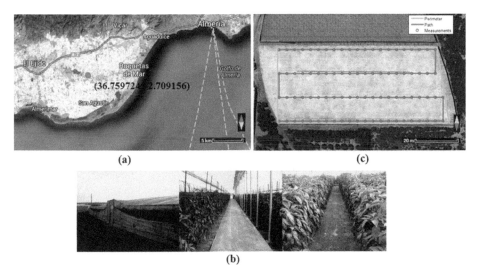

Figure 10. (a) The "sea of plastics" in Almería (Andalucía, Spain). Map data ©2014 Google, based on BCN IGN Spain; (b) Inside and outside of the greenhouse; (c) Top view of the greenhouse. Map data ©2014 Google, based on BCN IGN Spain.

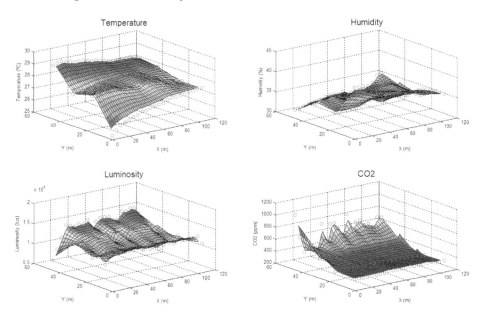

Figure 11. Maps of temperature, humidity, luminosity and CO_2 concentration of the greenhouse.

As shown, the temperature grew from the first measurement (25.3 °C), located at (1,1), to the last measurement (29.6 °C), located at (46,1). This fact is explained by considering the time differential of these measurements, which began at 9:00 and ended at 9:22; these measurements corresponded to the transition between nighttime and daytime temperatures and the overall warming of the greenhouse.

In contrast, the humidity declined from 43% to 33% from the beginning to the end of the experiment at the same locations as the temperature measurements. This behavior is also justified by considering the differences between nighttime and daytime humidities. Additionally, the values of humidity (e.g., 30%–50%) were lower than is typical in greenhouses (e.g., 70%–90%); this was likely because the greenhouse was not in production during these experiments, and a supply of humidity from the evapotranspiration of plants was absent.

The luminosity map is shown to be more regular, but it presents some shadowed locations. It is noticeable that the greenhouse cover filters luminosity: the sensor measurements were approximately 40,000 lux outside and 14,000 lux inside.

Finally, the carbon dioxide concentration shows spatial variation; specifically, this variable increased more in the Y-axis, which was likely due to the tractor mentioned before, which was working in that area, and the ventilation, which was worse on that side of the greenhouse.

The expected correlation among some variables and time is shown in Figure 12. This fact highlights one of the limits of the mini-UAV based sensory system—its inability to take simultaneous measurements at different points. However, when considering a steady state, the coverage time required is small enough to still monitor the complete greenhouse and obtain valuable information. Depending on the size of the greenhouse, a fleet of mini-UAVs instead of a single mini-UAV could be used to obtain more homogeneous measurements and build maps more efficiently.

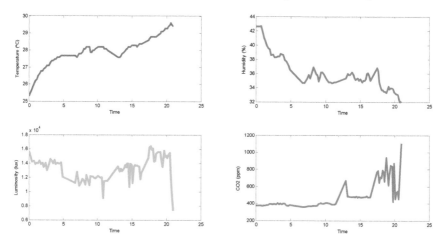

Figure 12. Variable dependence with time.

The results of these experiments demonstrate that the mini-UAV sensory system is able to measure the temperature, humidity, luminosity and carbon dioxide concentration in a greenhouse, allowing maps of the distribution of these variables to be built, and to capture the spatial and temporal variation of these variables.

5. Conclusions

This paper proposes a quadrotor-based sensory system for measuring environmental variables of a greenhouse. In contrast to Wireless Sensor Networks (WSNs), Unmanned Ground Vehicles (UGVs) and other solutions, Unmanned Aerial Vehicles (UAVs) are able to obtain measurements at nearly any point in the three dimensional space of the greenhouse, which facilitates activities such as local climate control and crop monitoring. The primary contributions of this paper are the determination of the optimal location of sensors on the quadrotor and the validation of a quadrotor as a platform for measuring environmental variables.

First, an exhaustive study of quadrotor aerodynamics was performed in order to determine the optimal allocation for sensors in the quadrotor. This study was supported by Computational Fluid

Sensors **2015**, *15*, 3334–3350

Dynamics (CFD) simulations and experiments and has concluded that the optimal location for the sensors is the central part of the top side of the quadrotor. The results of this study can be applied to different contexts, including the design of a high-efficiency quadrotor and the location of other sensors and actuators.

Second, a set of field experiments was performed in a greenhouse to validate the mini-UAV sensory system. These experiments have shown that the system can collect the environmental variables of the greenhouse, including the gas concentrations together with their spatial and temporal variability and possible disturbances. Differences in the sensor measurements that can be attributed to the rotors' influence were bounded; relative errors were lower than 4%. The system allows for climate control, crop monitoring and failure detection in a greenhouse and can be implemented in other industries and infrastructures. Finally, the system can incorporate other sensors for measuring other gases such as CO, CH_4, SO_2 or NO_2, if required.

Acknowledgments: This work has been supported by the Robotics and Cybernetics Research Group at Technical University of Madrid (Spain) and has been funded under the projects "ROTOS: Multi-robot system for outdoor infrastructures protection", sponsored by the Spanish Ministry of Education, Culture and Sport (DPI2010-17998); the "ROBOCITY 2030 Project", sponsored by the Autonomous Community of Madrid (S-0505/DPI/ 000235); and the SAVIER Project, sponsored by Airbus Defence & Space.

Author Contributions: Juan Jesus Roldan conceived the sensory system, designed and performed the experiments, analyzed the results and wrote the paper, Guillaume Joossen was involved in performing the experiments and reviewing the paper, David Sanz collaborated in the design of some experiments and the review of the paper, Jaime del Cerro and Antonio Barrientos analyzed the experimental results and assisted in the experimental work and the review of the paper.

Conflicts of Interest: The authors declare no conflict of interest.

References

1. Akyildiz, I.F.; Su, W.; Sankarasubramaniam, Y.; Cayirci, E. Wireless sensor networks: A survey. *Comput. Netw.* **2002**, *38*, 393–422.
2. Ruíz-García, L.; Lunadei, L.; Barreiro, P.; Robla, I. A review of wireless sensor technologies and applications in agriculture and food industry: State of the art and current trends. *Sensors* **2009**, *9*, 4728–4750.
3. Wang, N.; Zhang, N.; Wang, M. Wireless sensors in agriculture and food industry—Recent development and future perspective. *Comput. Electron. Agric.* **2006**, *50*, 1–14.
4. D'Oleire-Oltmanns, S.; Marzolff, I.; Peter, K.D.; Ries, J.B. Unmanned Aerial Vehicle (UAV) for Monitoring Soil Erosion in Morocco. *Remote Sens.* **2012**, *4*, 3390–3416.
5. Khan, A.; Schaefer, D.; Tao, L.; Miller, D.J.; Sun, K.; Zondlo, M.A.; Harrison, W.A.; Roscoe, B.; Lary, D.J. Low Power Greenhouse Gas Sensors for Unmanned Aerial Vehicles. *Remote Sens.* **2012**, *4*, 1355–1368.
6. Antonio, P.; Grimaccia, F.; Mussetta, M. Architecture and Methods for Innovative Heterogeneous Wireless Sensor Network Applications. *Remote Sens.* **2012**, *4*, 1146–1161.
7. Pawlowski, A.; Guzman, J.L.; Rodríguez, F.; Berenguel, M.; Sánchez, J.; Dormido, S. Simulation of greenhouse climate monitoring and control with wireless sensor network and event-based control. *Sensors* **2009**, *9*, 232–252.
8. Park, D.H.; Kang, B.J.; Cho, K.R.; Shin, C.S.; Cho, S.E.; Park, J.W.; Yang, W.M. A study on greenhouse automatic control system based on wireless sensor network. *Wirel. Pers. Commun.* **2011**, *56*, 117–130.
9. Zhang, Q.; Yang, X.L.; Zhou, Y.M.; Wang, L.R.; Guo, X.S. A wireless solution for greenhouse monitoring and control system based on ZigBee technology. *J. Zhejiang Univ. Sci. A* **2007**, *8*, 1584–1587.
10. Correll, N.; Arechiga, N.; Bolger, A.; Bollini, M.; Charrow, B.; Clayton, A.; Dominguez, F.; Donahue, K.; Dyar, S.; Johnson, L.; *et al.* Building a distributed robot garden. Proceedings of the IEEE/RSJ International Conference on Intelligent Robots and Systems, St. Louis, MO, USA, 10–15 October 2009.
11. Mandow, A.; Gómez de Gabriel, J.; Martínez, J.L.; Muñoz, V.F.; Ollero, A.; García Cerezo, A. The Autonomous Mobile Robot AURORA for Greenhouse Operation. *IEEE Robot. Autom. Mag.* **1996**, *3*, 18–28.
12. Sánchez-Hermosilla, J.; González, R.; Rodríguez, F.; Donaire, J.G. Mechatronic Description of a Laser Autoguided Vehicle for Greenhouse Operations. *Sensors* **2013**, *13*, 769–784.

13. Spiess, T.; Bange, J.; Buschmann, M.; Vorsmann, P. First application of the meteorological Mini-UAV "M^{2}AV". *Meteorol. Z.* **2007**, *16*, 159–169.

14. Buschmann, M.; Bange, J.; Vörsmann, P. MMAV-A Miniature Unmaned Aerial Vehicle (Mini-UAV) for Meteorological Purposes. Proceedings of the 16th Symposium on Boundary Layers and Turbulence, Portland, OR, USA, 9–13 August 2004.

15. Berman, E.S.; Fladeland, M.; Liem, J.; Kolyer, R.; Gupta, M. Greenhouse gas analyzer for measurements of carbon dioxide methane and water vapor aboard an unmanned aerial vehicle. *Sens. Actuators* **2012**, *169*, 128–135.

16. White, B.A.; Tsourdos, A.; Ashokaraj, I.; Subchan, S.; Zbikowski, R. Contaminant cloud boundary monitoring using network of UAV sensors. *Sens. J.* **2008**, *8*, 1681–1692.

17. Neumann, P.P.; Asadi, S.; Lilienthal, A.J.; Bartholmai, M.; Schiller, J.H. Autonomous gas-sensitive microdrone: Wind vector estimation and gas distribution mapping. *IEEE Robot. Autom. Mag.* **2012**, *19*, 50–61.

18. Primicerio, J.; di Gennaro, S.F.; Fiorillo, E.; Genesio, L.; Lugato, E.; Matese, A.; Vaccari, F.P. A flexible unmanned aerial vehicle for precision agriculture. *Precis. Agric.* **2012**, *13*, 517–523.

19. Zarco-Tejada, P.J.; González-Dugo, V.; Berni, J.A. Fluorescence, temperature and narrow-band indices acquired from a UAV platform for water stress detection using a micro-hyperspectral imager and a thermal camera. *Remote Sens. Environ.* **2012**, *117*, 322–337.

20. Valente, J.; Sanz, D.; del Cerro, J.; Barrientos, A.; de Frutos, M.A. Near-optimal coverage trajectories for image mosaicking using a mini quadrotor over irregular-shape fields. *Precis. Agric.* **2013**, *14*, 115–132.

21. Valente, J.; Sanz, D.; Barrientos, A.; del Cerro, J.; Ribeiro, A.; Rossi, C. An air-ground wireless sensor network for crop monitoring. *Sensors* **2011**, *11*, 6088–6108.

22. Valavanis, K.P. *Advances in Unmanned Aerial Vehicles: State of the Art and the Road to Autonomy*; Springer: Berlin, Germany, 2008.

23. Hernández, L.; Pestana, J.; Casares, D.; Campoy, P.; Sanchez-Lopez, J.L. Identification and cascade control by nonlinearities reversion of a quadrotor for the Control Engineering Competition CEA IFAC 2012. *Rev. Iberoam. Autom. Inform. Ind.* **2013**, *10*, 356–367.

24. Henten, E.J.V. Greenhouse Climate Management: An Optimal Control Approach. Doctoral Dissertation, Landbowuniversiteit te Wageningen, The Netherlands, 1994.

25. Linker, R.; Seginer, I. Greenhouse temperature modeling: A comparison between sigmoid neural networks and hybrid models. *Math. Comput. Simul.* **2004**, *65*, 19–29.

26. Nachidi, M.; Benzaouia, A.; Tadeo, F. Temperature and humidity control in greenhouses using the Takagi-Sugeno fuzzy model. Proceedings of the IEEE International Symposium on Intelligent Control, Munich, Germany, 4–6 October 2006.

27. Stanghellini, C.; de Jong, T. A model of humidity and its applications in a greenhouse. *Agric. For. Meteorol.* **1995**, *76*, 129–148.

28. Lieberman, M.; Baker, J.E.; Sloger, M. Influence of plant hormones on ethylene production in apple, tomato, and avocado slices during maturation and senescence. *Plant Physiol.* **1977**, *60*, 214–217.

29. Mellis, D.; Banzi, M.; Cuartielles, D.; Igoe, T. Arduino: An open electronic prototyping platform. Proceedings of 2007 Computer/Human Interaction Conference (CHI), San Jose, CA, USA, 28 April–3 May 2007.

30. Upton, E.; Halfacree, G. *Raspberry Pi User Guide*; John Wiley & Sons: Hoboken, NJ, USA, 2013.

31. Quigley, M.; Conley, K.; Gerkey, B.; Faust, J.; Foote, T.; Leibs, J.; Wheeler, R.; Ng, A.Y. ROS: An open-source Robot Operating System. Proceedings of ICRA Workshop on Open Source Software, Kobe, Japan, 12–17 May 2009; p. 5.

32. Aleksandrov, D.; Penkov, I. Optimal gap distance between rotors of mini quadrotor helicopter. Proceedings of the 8th DAAAM Baltic Conference, Tallinn, Estonia, 19–21 April 2012.

33. Poyi, G.T.; Wu, M.H.; Bousbaine, A.; Wiggins, B. Validation of a quadrotor helicopter Matlab/Simulink and Solidworks models. Proceeding of the IET Conference on Control and Automation 2013: Uniting Problems and Solutions, Birmingham, UK, 4–5 June 2013.

Article

Inspection of Pole-Like Structures Using a Visual-Inertial Aided VTOL Platform with Shared Autonomy

Inkyu Sa [1,*], Stefan Hrabar [2] and Peter Corke [1]

[1] Science and Engineering Faculty, Queensland University of Technology, Brisbane 4000, Australia;
   E-Mail: peter.corke@qut.edu.au
[2] CSIRO Digital Productivity, Brisbane 4069, Australia; E-Mail: Stefan.Hrabar@csiro.au
* E-Mail: i.sa@qut.edu.au; Tel.: +61-449-722-415; Fax: +61-7-3138-8822.

Academic Editors: Felipe Gonzalez Toro and Antonios Tsourdos
Received: 14 July 2015 / Accepted: 26 August 2015 / Published: 2 September 2015

Abstract: This paper presents an algorithm and a system for vertical infrastructure inspection using a vertical take-off and landing (VTOL) unmanned aerial vehicle and shared autonomy. Inspecting vertical structures such as light and power distribution poles is a difficult task that is time-consuming, dangerous and expensive. Recently, micro VTOL platforms (*i.e.*, quad-, hexa- and octa-rotors) have been rapidly gaining interest in research, military and even public domains. The unmanned, low-cost and VTOL properties of these platforms make them ideal for situations where inspection would otherwise be time-consuming and/or hazardous to humans. There are, however, challenges involved with developing such an inspection system, for example flying in close proximity to a target while maintaining a fixed stand-off distance from it, being immune to wind gusts and exchanging useful information with the remote user. To overcome these challenges, we require accurate and high-update rate state estimation and high performance controllers to be implemented onboard the vehicle. Ease of control and a live video feed are required for the human operator. We demonstrate a VTOL platform that can operate at close-quarters, whilst maintaining a safe stand-off distance and rejecting environmental disturbances. Two approaches are presented: Position-Based Visual Servoing (PBVS) using an Extended Kalman Filter (EKF) and estimator-free Image-Based Visual Servoing (IBVS). Both use monocular visual, inertia, and sonar data, allowing the approaches to be applied for indoor or GPS-impaired environments. We extensively compare the performances of PBVS and IBVS in terms of accuracy, robustness and computational costs. Results from simulations and indoor/outdoor (day and night) flight experiments demonstrate the system is able to successfully inspect and circumnavigate a vertical pole.

Keywords: aerial robotics; pole inspection; visual servoing, shared autonomy

---

1. Introduction

This paper presents an inspection system based on a vertical take-off landing (VTOL) platform and shared autonomy. The term "shared autonomy" indicates that the major fraction of control is accomplished by a computer. The operator's interventions for low-level control are prohibited but the operator provides supervisory high-level control commands such as setting the goal position. In order to perform an inspection task, a VTOL platform should fly in close proximity to the target object being inspected. This close-quarters flying does not require global navigation (explorations of large known or unknown environments) but instead requires local navigation relative to the specific geometry of the target, for instance, the pole of a streetlight. Such a system allows an unskilled operator to easily and safely control a VTOL platform to examine locations that are otherwise difficult to reach. For example, it could be used for practical tasks such as inspecting for bridge or streetlight defects.

Sensors **2015**, *15*, 22003–22048

Inspection is an important task for the safety of structures but is a dangerous and labor intensive job. According to the US Bureau of Transportation Statistics, there are approximately 600,000 bridges in the United States and 26% of them require inspections. Echelon, an electricity company, reported that there are 174.1 million streetlights in the US, Europe, and UK [1]. These streetlights also require inspections every year. These tasks are not only high risk for the workers involved but are slow, labour intensive and therefore expensive. VTOL platforms can efficiently perform these missions since they can reach places that are high and inaccessible such as the outsides of buildings (roof or wall), high ceilings, the tops of poles and so on. However, it is very challenging to use these platforms for inspection because there is insufficient room for error and high-level pilot skills are required as well as line-of-sight from pilot to vehicle. This paper is concerned with enabling low-cost semi-autonomous flying robots, in collaboration with low-skilled human operators, to perform useful tasks close to objects.

Multi-rotor VTOL micro aerial vehicles (MAVs) have been popular research platforms for a number of years due to advances in sensor, battery and integrated circuit technologies. The variety of commercially-available platforms today is testament to the fact that they are leaving the research labs and being used for real-world aerial work. These platforms are very capable in terms of their autonomous or attitude stabilized flight modes and the useful payloads they can carry. Arguably the most common use is for the collection of aerial imagery, for applications such as mapping, surveys, conservation and infrastructure inspection. Applications such as infrastructure inspection require flying at close-quarters to vertical structures in order to obtain the required images. Current regulations require the MAV's operator to maintain visual line-of-sight contact with the aircraft, but even so it is an extremely challenging task for the operator to maintain a safe, fixed distance from the infrastructure being inspected. From the vantage point on the ground it is hard to judge the stand-off distance, and impossible to do so once the aircraft is obscured by the structure. The problem is exacerbated in windy conditions as the structures cause turbulence. The use of First-Person View (FPV) video streamed live from the platform can help with situational awareness, but flying close to structures still requires great skill and experience by the operator and requires a reliable low-latency high-bandwidth communication channel. It has been found that flight operations near vertical structures is best performed by a team of three people: a skilled pilot, a mission specialist, and a flight director [2]. For small VTOL MAVs to truly become ubiquitous aerial imaging tools that can be used by domain experts rather than skilled pilots, their level of autonomy must be increased. One avenue to increased autonomy of a platform is through shared autonomy, where the majority of control is accomplished by the platform, but operator input is still required. Typically, the operator is relieved from the low-level relative-control task which is better performed by a computer, but still provides supervisory high-level control commands such as a goal position. We employ this shared autonomy approach for the problem of MAV-based vertical infrastructure inspections.

It is useful for an operator to be able to "guide" the MAV in order to obtain the required inspection viewpoints without the cognitive workload of "piloting" it. We provide the additional autonomy needed by implementing visual plus inertial-based pole-relative hovering as well as object circumnavigation shown in Figure 1. By tracking the two edges of the pole in the image and employing Position-Based Visual Servoing (PBVS) or Image-Based Visual Servoing (IBVS), the platform is able to maintain a user specified distance from the pole and keep the camera oriented towards the pole. The operator is also able to control the height and yaw of the platform. Since the pole is kept centred in the image, a yaw rate control command results in an orbit about the pole. A cylindrical workspace around the pole is therefore available to the operator for manoeuvres.

Figure 1. The vertical take-off and landing (VTOL) platform used for our pole inspection experiments. It includes a front-facing camera, downward-facing ultrasonic sensor and an onboard inertial measurement unit (IMU) for attitude control. All processing occurs onboard using a quad-core Acorn Risc Machine (ARM) Cortex-A9 processor.

1.1. Related Work

1.1.1. Climbing Robots for Inspection Tasks

As mentioned before, inspecting structures, such as light and power distribution poles is a time-consuming, dangerous and expensive task with high operator workload. The options for inspecting locations above the ground are rather limited, and all are currently cumbersome. Ladders can be used up to a height of 10–15 m but are quite dangerous: each year 160 people are killed and 170,000 injured in falls from ladders in the United States [3]. Cherry pickers require large vehicle access, sufficient space to operate and considerable setup time.

Robotics and mechatronics researchers have demonstrated a variety of climbing robots. Considerable growth in sensor and integrated circuit technology has accelerated small and lightweight robotics development. Typically, these robots are inspired by reptiles, mammals and insects, and their type of movement varies between sliding, swinging, extension and jumping.

The flexible mechatronic assistive technology system (MATS) robot has five degrees of freedom (DOF) and a symmetrical mechanism [4]. The robot shows good mobility features for travel, however, it requires docking stations that are attached to the wall, ceiling, or anywhere the robot is required to traverse. The bio-mimicking gekko robot, StickyBot [5], does not require docking stations since it has hierarchical adhesive structures under its toes to hold itself on any kind of surface. It has, however, limitations for payload and practical applications. A bridge cable inspection robot [6] is more applicable than the StickyBot in terms of its climbing speed and payload carrying ability. It climbs the cables by means of wheels which remain in contact with the cable for traction. A climbing robot with legged locomotion was developed by Haynes *et al.* [7]. This robot was designed for high-speed climbing of a uniformly convex cylindrical structure, such as a telephone or electricity pole. NASA's Jet Propulsion Laboratory recently demonstrated a rock climbing robot utilizing a hierarchical array of claws (called microspines) to create an attachment force of up to 180 N normal to the surface [8]. This robot also can drill a hole with a self-contained rotary percussive drill while it is attached to the surface.

Since climbing robots are in contact with the surface they can perform contact-based high-precision inspection with high performance sensors. They are also able to perform physical actions on the surface, not just inspections [9]. These climbing robots could not only replace a worker undertaking risky tasks in a hazardous environment but also increase the efficiency of such tasks. Climbing robots, however, require complex mechanical designs and complicated dynamic analysis. Their applications are also limited to structures with specific shapes and surface materials. They require setup time and climb slowly, so the inspection task can be time-consuming.

1.1.2. Flying Robots for Inspection Tasks

VTOL platforms on the other hand offer a number of advantages when used for infrastructure inspection. They have relatively simple mechanical designs (usually symmetric) which require a simple dynamic analysis and controller. VTOL platforms can ascend quickly to the required height and can obtain images from many angles regardless of the shape of the structure. Recent advanced sensor, integrated circuit and motor technologies allow VTOL platforms to fly for a useful amount of time while carrying inspection payloads. Minimal space is required for operations and their costs are relatively low. The popularity of these platforms means that hardware and software resources are readily available [10].

These advantages have accelerated the development of small and light-weight flying robotics for inspection. Voigt *et al.* [11] demonstrated an embedded stereo-camera based egomotion estimation technique for the inspection of structures such as boilers and general indoor scenarios. The stereo vision system provides a relative pose estimate between the previous and the current frame and this is fed into an indirect Extended Kalman Filter (EKF) framework as a measurement update. The inertial measurements such as linear accelerations and rotation rates played important roles in the filter framework. States, (position, orientation, bias, and relative pose) were propagated with IMU measurements through a prediction step and the covariance of the predicted pose were exploited to determine a confidence region for feature searching in the image plane. This allowed feature tracking on scenes with repeating textures (perception aliasing), increased the total number of correct matches (inliers), and efficiently rejected outliers with reasonable computation power. They evaluated the proposed method on several trajectories with varying flight velocities. The results presented show the vehicle is capable of impressively accurate path tracking. However, flights tests were performed indoors in a boiler mock-up environment where disturbances are not abundant, and using hand-held sequences from an office building dataset. Based on this work, Burri *et al.* [12] and Nikolic *et al.* [13] show visual inspection of a thermal power plant boiler system using a quadrotor. They developed a Field Programmable Gate Array (FPGA) based visual-inertial stereo Simultaneous Localization and Mapping (SLAM) sensor with state updates at 10 Hz. A model predictive controller (MPC) is used for closed loop control in industrial boiler environments. In contrast to their work, we aim for flights in outdoor environments where disturbances such as wind gusts are abundant and the scenes include natural objects.

Ortiz *et al.* [14] and Eich *et al.* [15] introduced autonomous vessel inspection using a quadrotor platform. A laser scanner is utilized for horizontal pose estimation with Rao-Blackwellized particle filter based SLAM (GMapping), and small mirrors reflected a few of the horizontal beams vertically downwards for altitude measurement. These technologies have been adopted from the 2D ground vehicle SLAM solution into aerial vehicle research [16] and often incorporated within a filter framework for fast update rates and accurate state estimation [17]. While such methods are well-established and optimized open-source software packages are available, one of the main drawbacks is the laser scanner. Compared to monocular vision, a laser scanner is relatively heavy and consumes more power, which significantly decreases the total flight time. Instead, we propose a method using only a single light-weight camera, a geometric model of the target object, and a single board computer for vertical structure inspection tasks.

1.2. Contributions and Overview

This paper contributes to the state-of-the-art in aerial inspections by addressing the limitations of existing approaches presented in Section 1.1 with the proposed high performance vertical structure inspection system. In this paper, we make use of our previous developed robust line feature tracker [18] as a front-end vision system, and it is summarized in Section 2.2. A significant difference to our previous works [19–21] in which different flying platforms had been utilized is the integration of both PBVS and IBVS systems on the same platform. By doing so, we are able to compare both systems quantitatively. We also conduct experiments where a trained pilot performs the same tasks using manual flight and with the aid of PBVS and IBVS and demonstrate the difficulty of the tasks. For evaluation, motion capture systems, a laser tracker, and hand-annotated images are used. Therefore, the contributions of this paper are:

- The development of onboard flight controllers using monocular visual features (lines) and inertial sensing for visual servoing (PBVS and IBVS) to enable VTOL MAV close quarters manoeuvring.
- The use of shared autonomy to permit an un-skilled operator to easily and safely perform MAV based pole inspections in outdoor environments, with wind, and at night.
- Significant experimental evaluation of state estimation and control performance for indoor and outdoor (day and night) flight tests, using a motion capture device and a laser tracker for ground truth. Video demonstration [22].
- A performance evaluation of the proposed systems in comparison to skilled pilots for a pole inspection task.

The remainder of the paper is structured as follows: Section 2 describes the coordinate system definition used in this paper, and the vision processing algorithms for fast line tracking. Sections 3 and 4 present the PBVS and IBVS control structures which are developed for the pole inspection scenario, and with validation through simulation. Section 5 presents the use of shared autonomy and we present our extensive experimental results in Section 6. Conclusions are drawn in Section 7.

2. Coordinate Systems and Image Processing

2.1. Coordinate Systems

We define three right-handed frames: world $\{\mathcal{W}\}$, body $\{\mathcal{B}\}$ and camera $\{\mathcal{C}\}$ which are shown in Figure 2. Note that both $\{\mathcal{W}\}$ and $\{\mathcal{B}\}$ have their z-axis downward while $\{\mathcal{C}\}$ has its z-axis (camera optical axis) in the horizontal plane of the propellers and pointing in the vehicle's forward direction. We define the notation ${}^{a}\mathbf{R}_{b}$ which rotates a vector defined with respect to frame $\{b\}$ to a vector with respect to $\{a\}$.

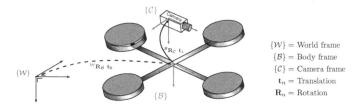

$\{\mathcal{W}\}$ = World frame
$\{\mathcal{B}\}$ = Body frame
$\{\mathcal{C}\}$ = Camera frame
$\mathbf{t}_{n}$ = Translation
$\mathbf{R}_{n}$ = Rotation

Figure 2. Coordinate systems: body $\{\mathcal{B}\}$, world $\{\mathcal{W}\}$, and camera $\{\mathcal{C}\}$. Transformation between $\{\mathcal{B}\}$ and $\{\mathcal{C}\}$ is constant whereas $\{\mathcal{B}\}$ varies as the quadrotor moves. ${}^{\mathcal{C}}\mathbf{R}_{\mathcal{B}}$ rotates a vector defined with respect to $\{\mathcal{B}\}$ to a vector with respect to $\{\mathcal{C}\}$.

2.2. Image Processing for Fast Line Tracking

Our line tracker is based on tracking the two edges of the pole over time. This is an appropriate feature since the pole will dominate the scene in our selected application. There are many reported line extraction algorithms such as Hough transform [23] and other linear feature extractors [24] but these methods are unsuitable due to their computational complexity. Instead we use a simple and efficient line tracker inspired by [25]. The key advantage of this algorithm is its low computation requirement. For 320 × 240 pixel images every iteration is finished in < 16 ms and uses only 55% of the CPU quad-core ARM Cortex-A9.

2.2.1. 2D and 3D Line Models

A 3D line can be described using various parameterizations including two 3D points, the intersection of two 3D planes, closest point with direction or two projections. These representations vary in terms of their properties including completeness, reprojection characteristics with a perspective camera and the number of internal constraints [26]. *Plücker coordinates* [27] have been widely used in the computer vision and the robotic community for 3D line reconstruction [28], line based visual servoing [29] and SLAM [30]. Plücker coordinates describe a line joining the two 3D points $^W\mathbf{A}$ and $^W\mathbf{B} \in \mathbb{R}^3$ in the world frame according to

$$^W\mathbf{L} = {}^W\tilde{\mathbf{A}}{}^W\tilde{\mathbf{B}}^T - {}^W\tilde{\mathbf{B}}{}^W\tilde{\mathbf{A}}^T \tag{1}$$

where $^W\mathbf{L}$ is a Plücker matrix $\in \mathbb{R}^{4\times4}$. The tilde denotes the homogeneous form of the point ($\in \mathbb{P}^3$).

Consider a perspective projection represented by a camera matrix (intrinsic and extrinsic) $\mathbf{C}(\xi_C) \in \mathbb{R}^{3\times4}$

$$\mathbf{C}(\xi_C) = \begin{bmatrix} f_x & 0 & u_0 \\ 0 & f_y & v_0 \\ 0 & 0 & 1 \end{bmatrix} \begin{bmatrix} 1 & 0 & 0 & 0 \\ 0 & 1 & 0 & 0 \\ 0 & 0 & 1 & 0 \end{bmatrix} \xi_C^{-1} \tag{2}$$

$$= \mathbf{K}\mathbf{P}_0\xi_C^{-1}$$

where $\xi_C \in SE(3)$ is the camera pose with respect to the world coordinate frame, f_x and f_y are focal lengths, u_0 and v_0 are the coordinates of the principal point.

The 3D line $^W\mathbf{L}$ is projected to a 2D line on the camera image plane by

$$[\ell]_\times = \mathbf{C}(\xi_C) {}^W\mathbf{L} \, \mathbf{C}(\xi_C)^T \tag{3}$$

where $[\ell]_\times$ is a skew-symmetric matrix and $\ell = (\ell_1, \ell_2, \ell_3)$ is the homogeneous line equation on the image plane

$$\ell_1 u + \ell_2 v + \ell_3 = 0 \tag{4}$$

where u and v are the horizontal and vertical image plane coordinates respectively (see Figure 3). We reparameterize the line as

$$\ell = [\alpha, \beta]^T, \text{ where } \alpha = \frac{\ell_1}{\ell_2}, \beta = \frac{-\ell_3}{\ell_2} \tag{5}$$

and α is the slope and β is the x-axis intercept (in pixels), see Figure 3b. Note that this parameterization is the $\frac{\pi}{2}$ rotated form of the conventional 2D line equation, in order to avoid the singular case for a vertical line. There is a singularity for a horizontal line ($\ell_2 = 0$) but we do not expect this in our application.

2.2.2. Line Prediction and Tracking

We use a linear feature velocity model for line prediction

$$\hat{\ell}_{k+1} = \ell_k + \Delta\dot{\ell}_k \tag{6}$$

where k is the timestep, $\hat{\ell}_{k+1}$ is the predicted line in the image plane, $\dot{\ell}_k$ is the feature velocity, ℓ_k is the previously observed feature and Δ is the sample time. In order to calculate feature velocity, we compute an *image Jacobian*, $\mathbf{J}_l$, which describes how a line moves on the image plane as a function of camera spatial velocity $v = [^C\dot{x}, {}^C\dot{y}, {}^C\dot{z}, {}^C\omega_x, {}^C\omega_y, {}^C\omega_z]^T$ [31].

$$\dot{\ell}_k = J_{1k} v_k \tag{7}$$

This image Jacobian is the derivative of the 3D line projection function with respect to camera pose, and for the line parameterization of Equation (5) $J_1 \in \mathbb{R}^{2\times 6}$.

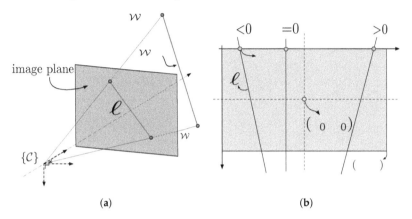

Figure 3. (**a**) Perspective image of a line ^{W}L in 3D space. a and b are projections of the world point and ℓ is a line on the image plane; (**b**) Image plane representation of slope (α) and intercept (β).

The line tracker has two phases: bootstrapping and tracking. The computationally expensive Canny edge detection and Hough transform are utilized only once for bootstrapping. The tracking phase is invoked while the vehicle is flying. There are two steps in the tracking phase: line searching and line model fitting. Horizontal gradient (Sobel kernel) images are computed which emphasise vertical lines in the scene. We sample at 60 points uniformly distributed vertically along the predicted lines. We then compute maxima along a fixed-length horizontal scan line centred on each of these points, see Figure 4a. The horizontal scan line length is empirically set to 24 pixels. These maxima are input to a line fitting algorithm using RANSAC [32], to update the line model for the next iteration.

For vision-based control methods it is critical to have feature tracking that is robust to agile camera motion and lighting condition changes. To handle agile motion we make use of inertial measurements, acceleration and angular velocity in the body coordinate frame, to predict where the feature will be in the image plane for the next frame. Figure 4 shows an example of the prediction result. At this moment, the camera had an acceleration of $1.7\,\text{m/s}^2$ and rotation rate of $19\,^{\circ}/\text{s}$. The yellow and red lines in Figure 4a denote the cases without and with prediction respectively, and shows qualitatively that the red line is closer to the true edge than the yellow line.

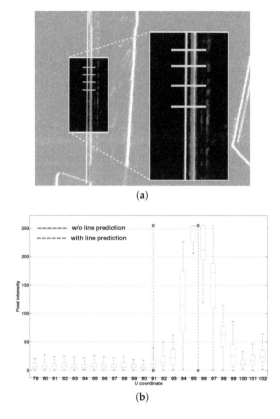

(a)

(b)

Figure 4. (**a**) The image at time k is shown with the tracked line from time $k - 1$ (without prediction case in yellow) and the predicted line from time $k - 1$ (with prediction case in red). We search for maxima along 60 horizontal search lines (cyan), and each is 24 pixels wide; (**b**) The predicted line is close to the maxima whereas there is 4.5 pixel offset without prediction.

Figure 4b shows the statistical result over multiple search lines in a single frame. We measure pixel gradient magnitude along fixed length horizontal search lines (the cyan lines in Figure 4a) and then plot them against image u coordinates in Figure 4b. The red predicted line is closer to the maxima whereas there is an offset in the yellow line. This offset varies with motion of the camera. More details and experimental results for the line tracker and prediction is presented in [18].

Although we implemented enhancements such as sub-pixel interpolation and feature prediction to improve tracking performance, the line tracker still suffered (tracking failures and noisy tracking) in both indoor and outdoor environments for a variety of reasons. In some cases, man-made structures caused tracking of the pole edges to fail because of other strong vertical edge features in the scene. In other cases the tracker was still able to track the pole edges but the tracking was noisy due to the background scene complexity (for example because of trees in the background).

This reveals the limitations of a naive gradient magnitude-based tracker. Robust and accurate object detection algorithms can be utilized to address this challenge. The tracker only searches within the region-of-interest (ROI) determined by the algorithms. However, we have to scarify update rates or agility of the flying robot due to onboard computational limits.

466

3. Position Based Visual Servoing (PBVS)

PBVS uses measured visual features, camera calibration parameters, and prior knowledge about the target in order to determine the pose of a camera with respect to the target. We use an Extended Kalman Filter for state estimation with a Plücker line representation as shown in Figure 5. State estimation and control are performed in SE(3). This section presents details of the horizontal and vertical Kalman Filter frameworks shown in Figure 6 and simulation results.

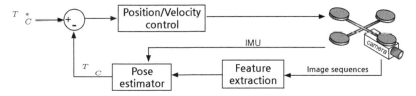

Figure 5. Position-based visual servoing diagram. f is a feature vector. $^T\hat{\xi}_C$ and $^T\xi^*_C$ are the estimated and the desired pose of the target with respect to the camera.

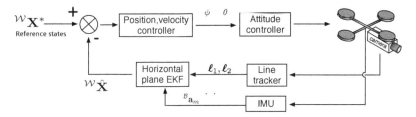

Figure 6. Block diagram of horizontal plane state estimator and control used for the PBVS approach. u^ϕ and u^θ denote control inputs for roll and pitch commands. ℓ_1 and ℓ_2 are tracked 2D lines and $^B\mathbf{a}_m$ is onboard inertial measurement unit (IMU) acceleration measurement.

3.1. Horizontal Plane EKF

The position and velocity of the vehicle in the horizontal plane is estimated using monocular vision and inertial data. These sensor modalities are complementary in that the IMU outputs are subject to drift over time, whereas the visually acquired pole edge measurements are drift free and absolute with respect to the world frame, but of unknown scale.

3.1.1. Process Model

Our discrete-time process model for the flying body assumes constant acceleration [33].

$$^W\hat{\mathbf{X}}(k+1|k) = \mathbf{A}\,^W\hat{\mathbf{X}}(k|k) + \mathbf{Bb}_k + \mathbf{v} \tag{8}$$

where $^W\mathbf{X}_k = \left[^W x_k, {}^W y_k, {}^W \dot{x}_k, {}^W \dot{y}_k, \phi_k, \theta_k\right]^T$. There is an ambiguity for $^W y$ and yaw angle (ψ), as both result in the target appearing to move horizontally in the image. Although these are the observable states by both the camera and the IMU, it is a challenge to decouple them with our front-facing camera configuration and without using additional sensors. Therefore, we omit yaw (heading) angle estimation in the EKF states and assume it is controlled independently, for example using gyroscope and/or magnetometer sensors.

$\hat{\mathbf{X}}\langle k+1|k\rangle$ is the estimate of $\mathbf{X}$ at time $k+1$ given observations up to time k. $\mathbf{b}_k = \left[{}^{W}\ddot{x}_k, {}^{W}\ddot{y}_k, \dot{\phi}_k, \dot{\theta}_k\right]^T$ represents the sensor-observed motion of the vehicle. $\mathbf{A}$ and $\mathbf{B}$ describe the evolution of a state vector and are given by

$$\mathbf{A} = \begin{bmatrix} 1 & 0 & \Delta t & 0 & 0 & 0 \\ 0 & 1 & 0 & \Delta t & 0 & 0 \\ 0 & 0 & 1 & 0 & 0 & 0 \\ 0 & 0 & 0 & 1 & 0 & 0 \\ 0 & 0 & 0 & 0 & 1 & 0 \\ 0 & 0 & 0 & 0 & 0 & 1 \end{bmatrix}, \mathbf{B} = \begin{bmatrix} \frac{1}{2}\Delta t^2 & 0 & 0 & 0 \\ 0 & \frac{1}{2}\Delta t^2 & 0 & 0 \\ \Delta t & 0 & 0 & 0 \\ 0 & \Delta t & 0 & 0 \\ 0 & 0 & \Delta t & 0 \\ 0 & 0 & 0 & \Delta t \end{bmatrix} \tag{9}$$

It is worth mentioning that accelerometers measure the difference between the actual acceleration of a robot and the gravity vector in $\{B\}$ [34,35]. Therefore, accelerations in $\{W\}$ are

$${}^{W}\mathbf{a} = \begin{bmatrix} {}^{W}\ddot{x} \\ {}^{W}\ddot{y} \\ {}^{W}\ddot{z} \end{bmatrix} = {}^{W}\mathbf{R}_B {}^{B}\mathbf{a}_m - \mathbf{g} \tag{10}$$

where $\mathbf{g}$ is gravitational acceleration, $[0,0,g]^T$ and ${}^{B}\mathbf{a}_m$ is the accelerometer measurement. Process noise $\mathbf{v}$ is assumed to be Gaussian in nature:

$$\mathbf{v} \sim \mathcal{N}(0, \mathcal{Q}) \tag{11}$$
$$\mathcal{Q} = \text{diag}\begin{bmatrix} \sigma_{W_x}^2 & \sigma_{W_y}^2 & \sigma_{W_{\dot{x}}}^2 & \sigma_{W_{\dot{y}}}^2 & \sigma_{\phi}^2 & \sigma_{\theta}^2 \end{bmatrix}$$

where $\mathcal{Q}$ is the covariance matrix of the process noise. $\mathcal{N}(0, \mathcal{Q})$ denotes a zero-mean Gaussian noise process, and σ is the standard deviation of the corresponding states. The covariance propagation step follows the standard Kalman Filter procedure.

3.1.2. Measurement Model

Four points $\in \mathbb{R}^3$ that lie on the two sides of the pole are defined in $\{W\}$. Two Plücker lines, ${}^{W}\mathbf{L}^1$ and ${}^{W}\mathbf{L}^2$, are formed and projected onto the image plane as shown in Figure 7.

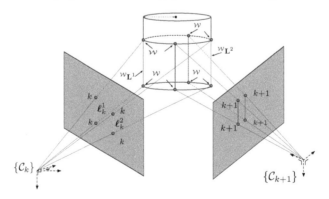

Figure 7. Projection model for a cylindrical object. ${}^{W}A, {}^{W}B, {}^{W}C, {}^{W}D \in \mathbb{R}^3$ denote points in the world frame with $a_k, b_k, c_k, d_k \in \mathbb{R}^2$ denoting their corresponding projection onto a planar imaging surface at a sample k. Although we actually measure different world points between frames, they are considered to be the same point due to the cylindrical nature of the object and the choice of line representation.

We partition the measurement into two components: visual $\mathbf{z}_{cam}$ and inertial $\mathbf{z}_{IMU}$. The measurement vector is

$$
\mathbf{Z}_k = \begin{bmatrix} \ell_k^1 \\ \ell_k^2 \\ \phi_k \\ \theta_k \end{bmatrix} = \begin{bmatrix} \mathbf{z}_{cam} \\ \mathbf{z}_{IMU} \end{bmatrix}
\tag{12}
$$

where $\ell_k^i \in \mathbb{R}^2$ are the 2D line features from the tracker as given by Equation (5). The projected line observation is given by the nonlinear function of Equation (3)

$$
\mathbf{z}_{cam} = \begin{bmatrix} h_{cam}(^{\mathcal{W}}\mathbf{L}^1, {}^{\mathcal{W}}\hat{x}_k, {}^{\mathcal{W}}\hat{y}_k, \hat{\phi}_k, \hat{\theta}_k, \mathbf{w}) \\ h_{cam}(^{\mathcal{W}}\mathbf{L}^2, {}^{\mathcal{W}}\hat{x}_k, {}^{\mathcal{W}}\hat{y}_k, \hat{\phi}_k, \hat{\theta}_k, \mathbf{w}) \end{bmatrix}
\tag{13}
$$

$$
= \begin{bmatrix} \mathbf{C}(^{\mathcal{W}}\hat{x}_k, {}^{\mathcal{W}}\hat{y}_k, \hat{\phi}_k, \hat{\theta}_k) {}^{\mathcal{W}}\mathbf{L}^1 \mathbf{C}(^{\mathcal{W}}\hat{x}_k, {}^{\mathcal{W}}\hat{y}_k, \hat{\phi}_k, \hat{\theta}_k)^T \\ \mathbf{C}(^{\mathcal{W}}\hat{x}_k, {}^{\mathcal{W}}\hat{y}_k, \hat{\phi}_k, \hat{\theta}_k) {}^{\mathcal{W}}\mathbf{L}^2 \mathbf{C}(^{\mathcal{W}}\hat{x}_k, {}^{\mathcal{W}}\hat{y}_k, \hat{\phi}_k, \hat{\theta}_k)^T \end{bmatrix}
\tag{14}
$$

Note that the unobservable states $^{\mathcal{W}}z_k$ and ψ are omitted. $\mathbf{w}$ is the measurement noise with measurement covariance matrix, $\mathcal{R}$

$$
\mathbf{w} \sim \mathcal{N}(0, \mathcal{R})
\tag{15}
$$

$$
\mathcal{R} = \text{diag} \begin{bmatrix} \sigma_{\alpha1}^2 & \sigma_{\beta1}^2 & \sigma_{\alpha2}^2 & \sigma_{\beta2}^2 & \sigma_\phi^2 & \sigma_\theta^2 \end{bmatrix}
$$

We manually tune these parameters by comparing the filter output with Vicon ground truth. We generated the run-time code for Equation (14) using the MATLAB Symbolic Toolbox and then exporting the C++ code. This model is 19 K lines of source code but computation time is just 6 µs.

The update step for the filter requires linearization of this line model and evaluation of the two Jacobians

$$
H_x = \frac{\partial h_{cam}}{\partial \mathbf{x}}|_{\hat{\mathbf{x}}(k)}, \quad H_w = \frac{\partial h_{cam}}{\partial \mathbf{w}}
\tag{16}
$$

where H_x is a function of state that includes the camera projection model. We again use the MATLAB Symbolic Toolbox and automatic code generation (58 K lines of source code) for H_x. It takes 30 µs to compute in the C++ implementation with the onboard CPU quad-core ARM Cortex-A9.

The remaining observations are the vehicle attitude, directly measured by the onboard IMU ($\mathbf{z}_{IMU}$) and reported at 100 Hz over a serial link. The linear observation model for the attitude is

$$
\mathbf{z}_{IMU} = \begin{bmatrix} \phi_k \\ \theta_k \end{bmatrix} = \mathbf{H}_{IMU} {}^{\mathcal{W}}\hat{\mathbf{x}}
\tag{17}
$$

$$
\mathbf{H}_{IMU} = \begin{bmatrix} 0 & 0 & 0 & 1 & 0 & 0 \\ 0 & 0 & 0 & 0 & 1 & 0 \end{bmatrix}
\tag{18}
$$

The measurements $\mathbf{z}_{cam}$ and $\mathbf{z}_{IMU}$ are available at 60 Hz and 100 Hz respectively. The EKF is synchronous with the 60 Hz vision data and the most recent $\mathbf{z}_{IMU}$ measurement is used for the filter update. Inputs and outputs of the horizontal plane EKF are presented in Figure 5.

3.1.3. Simulation Results

In order to validate the EKF line model and Jacobian, we create a virtual camera observing four points, $\mathbf{P}_i \in \mathbb{R}^3$ and move the camera with sinusoidal motion in 4 DOF ($^{\mathcal{W}}x, {}^{\mathcal{W}}y, \phi, \theta$) using the simulation framework of [36]. To emulate the errors in line measurements we set the measurement uncertainties to be $\sigma_\alpha = 1°$ and $\sigma_\beta = 4$ pixels in the line parameters, α and β, from Equation (5).

Estimation results, their confidence boundary and noise parameters are shown in Figures 8–10. Most of the states are within 3σ confidence level. The total simulation time is 10 s, with a sampling rate of 100 Hz. We see good quality estimates of position and velocity in the horizontal plane, whilst decoupling the effects of attitude on the projected line parameters. We see that the x-axis forward estimation is noisier than the y-axis since image variation due to change in camera depth is much less than that due to fronto-parallel motion.

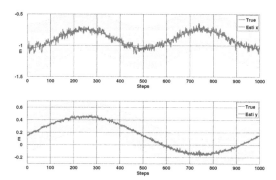

Figure 8. Simulation results for Position Based Visual Servoing (PBVS) tracking: position estimation. $^{W}\hat{x}$ and $^{W}\hat{y}$ with 3σ confidence boundary.

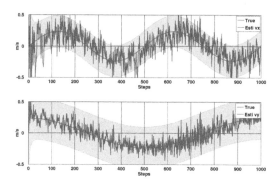

Figure 9. Simulation results for PBVS tracking: velocity estimation. $^{W}\hat{x}$ and $^{W}\hat{y}$ with 3σ confidence boundary.

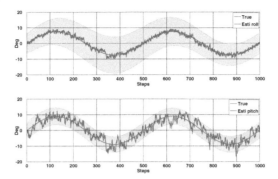

Figure 10. Simulation results for PBVS tracking: angle estimation. $\hat{\phi}$ and $\hat{\theta}$ with 3σ confidence boundary.

3.2. Kalman Filter-Based Vertical State Estimation

Various options exist to determine altitude. For instance a downward-looking camera with an object of known scale on the ground and/or vertical visual odometry on the target object (pole) using the forward-facing camera. Due to onboard computational limits we opt, at this stage, to use a sonar altimeter. We observe altitude directly using a downward-facing ultrasonic sensor at 20 Hz, but this update rate is too low for control purposes and any derived velocity signal has too much lag. Therefore we use another Kalman Filter to fuse this with the 100 Hz inertial data which includes vertical acceleration in $\{\mathcal{B}\}$. The sonar sensor is calibrated by least square fitting to ground truth state estimates. The altitude and z-axis acceleration measurement in $\{\mathcal{B}\}$ are transformed to $\{\mathcal{W}\}$ using $\hat{\phi}$ and $\hat{\theta}$ angles and Equation (10). $^{\mathcal{W}}\mathbf{X}^{\text{alt}}$ is the vertical state, $\left[^{\mathcal{W}}z, {}^{\mathcal{W}}\dot{z}\right]^{T}$ and the process model is given by

$$^{\mathcal{W}}\hat{\mathbf{X}}^{\text{alt}}_{\langle k+1|k\rangle} = \mathbf{A}^{\text{alt}} \left[\begin{array}{c} ^{\mathcal{W}}\hat{z}_{\langle k|k\rangle} \\ ^{\mathcal{W}}\hat{\dot{z}}_{\langle k|k\rangle} \end{array} \right] + \mathbf{B}^{\text{alt}} {}^{\mathcal{W}}\ddot{z} + \mathbf{v}^{\text{alt}} \tag{19}$$

where

$$\mathbf{A}^{\text{alt}} = \left[\begin{array}{cc} 1 & \Delta t \\ 0 & 1 \end{array} \right], \quad \mathbf{B}^{\text{alt}} = \left[\begin{array}{c} \frac{1}{2}\Delta t^2 \\ \Delta t \end{array} \right] \tag{20}$$

and where $\mathbf{v}^{\text{alt}}$ is the process noise vector of $^{\mathcal{W}}z$ and $^{\mathcal{W}}\dot{z}$. The covariance matrices of the process and measurement noise, $\mathcal{Q}^{\text{alt}}$ and $\mathcal{R}^{\text{alt}}$, are defined as Equations (12) and (16). The observation matrix is $\mathbf{H}^{\text{alt}} = \left[\begin{array}{cc} 1 & 0 \end{array} \right]$.

4. Image Based Visual Servoing (IBVS)

Image based visual servoing (IBVS) omits the pose estimation block of Figure 5 and the control is computed directly from image-plane features as shown in Figure 11. It is a challenging control problem since the image features are a non-linear function of camera pose, and the controller generates desired velocities which the non-holonomic platform cannot follow. In this section we present the relation between camera and image motion, the image Jacobian for line features and an IMU-based de-rotation technique. Simulation results are also presented.

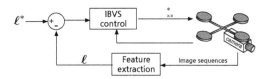

Figure 11. Image-based visual servoing diagram. We model an ordinary camera which has 3 mm focal length, 320 × 240 image resolution. v_{xz}^* is the computed desired translational velocity and ω is used for de-rotation in the Imaged-Based Visual Servoing (IBVS) control block.

4.1. Line-Feature-Based IBVS

IBVS has been exploited in a wide range of robotic applications mainly due to its simplicity and robustness to control error [31,37,38]. Point-feature-based IBVS systems are used commonly because point features are fundamental, general and visually distinct in the image. State-of-the-art scale and rotation invariant feature tracking techniques have been used to demonstrate robust and accurate IBVS. By comparison line-feature-based IBVS implementations are relatively rare, yet lines are distinct visual features in man-made environments, for examples the edges of roads, buildings and power distribution poles.

4.1.1. Image Jacobian for Line Features

The homogeneous equation of a 2D line is $au + bv + c = 0$ with coefficients (a, b, c). Although any line can be represented in this form it does not have a minimum number of parameters. The standard *slope-intercept* form $v = mu + c$ where m is slope and c is intercept is problematic for the case of vertical lines where $m = \infty$. We therefore choose (ρ, θ) parameterization as the 2D line representation as shown in Figure 12

$$u \sin \theta + v \cos \theta = \rho \tag{21}$$

where $\theta \in [-\frac{\pi}{2}, \frac{\pi}{2})$ is the angle from the u-axis to v-axis in radians, and $\rho \in [-\rho_{\min}, \rho_{\max}]$ is the perpendicular distance in pixels from the origin to the line. This form can represent a horizontal line ($\theta = 0$) and a vertical line ($\theta = -\frac{\pi}{2}$).

For a moving camera, the rate of change of line parameters is related to the camera velocity by

$$\dot{\ell} = \begin{bmatrix} \dot{\theta} \\ \dot{\rho} \end{bmatrix} = J_l v \tag{22}$$

where $\dot{\theta}$ and $\dot{\rho}$ are the velocity of a line feature, and are analogous to optical flow for a point feature. These line parameters are simply related to the line parameters introduced earlier by

$$\theta = \tan^{-1} \alpha, \ \rho = \beta \cos \theta \tag{23}$$

The matrix J_l is the *Image Jacobian* or *Interaction matrix* and given by Equation (29) [39]. The lines lie on the equation of a plane $AX + BY + CZ + D = 0$ where (A, B, C) is the plane normal vector and D is the distance between the plane and the camera. The camera spatial velocity in world coordinates is

$$v = \begin{bmatrix} v_x, v_y, v_z & | & \omega_x, \omega_y, \omega_z \end{bmatrix}^T \tag{24}$$

$$= \begin{bmatrix} v_t & | & \omega \end{bmatrix}^T$$

where v_t and ω are the translational and angular velocity components respectively.

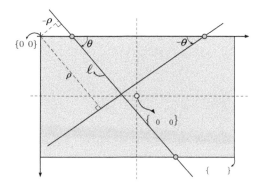

Figure 12. An example of the (ρ-θ) representation for two lines, ℓ. Signs of the two parameters, ρ and θ, are shown for the corresponding positive or negative quantities. $\{u_0, v_0\}$ is the principle point and $\{W, H\}$ denotes the width and the height of the image plane. The origin of the pixel coordinate frame is at the top-left of the image plane by convention.

For the case of N line features we can stack these equations. The left hand side is a $2N \times 1$ matrix, while the stacked Jacobian is $2N \times 6$. In this paper, we consider the $N-2$ case where the two lines are the vertical edges of the pole, which yields

$$\begin{bmatrix} \dot{\theta}_1 \\ \dot{\rho}_1 \\ \dot{\theta}_2 \\ \dot{\rho}_2 \end{bmatrix} = \begin{bmatrix} J_{l_1} \\ J_{l_2} \end{bmatrix} v \tag{25}$$

We can solve for the camera motion required in order to move the image features to the desired value

$$v = J_l^+ \dot{\ell} \tag{26}$$

where J_l^+ denotes the pseudo-inverse of J_l. Given the desired feature vector, ℓ^*, the desired feature velocity is

$$\dot{\ell}^* = \lambda(\ell^* - \ell) \tag{27}$$

where ℓ represents the two tracked line features, ℓ^* is the desired feature positions in the image plane, and λ is a positive scalar for a simple linear controller. Substituting Equation (27) into Equation (26) yields the desired camera velocity:

$$v^* = \lambda J_l^+ (\ell^* - \ell) \tag{28}$$

It is important to note that we do require some *a priori* Cartesian knowledge about the scene: the distance from the camera origin to the plane in which the vertical lines lie and the approximate orientation of that plane. This information is encoded in the parameters of the plane which is required to compute the image Jacobian in Equation (29). We know A, B, C because the plane is vertical and orthogonal to the camera x-axis ($A=0$, $B=0$, and $C=1$). Since we are interested in flying close to the target, we choose a reasonable value for D. We will discuss this point further in Section 4.2.

$$J_l = \begin{bmatrix} \lambda_\theta \sin\theta & \lambda_\theta \cos\theta & -\lambda_\theta \rho & -\rho\sin\theta & -\rho\cos\theta & -1 \\ \lambda_\rho \sin\theta & \lambda_\rho \cos\theta & -\lambda_\rho \rho & -(1+\rho^2)\cos\theta & (1+\rho^2)\sin\theta & 0 \end{bmatrix} \tag{29}$$

$$\text{where} \quad \lambda_\rho = \frac{A\rho\sin\theta + B\rho\cos\theta + C}{D}, \lambda_\theta = \frac{A\cos\theta - B\sin\theta}{D}$$

4.1.2. Unobservable and Ambiguous States with Line Features

Depending on the number of lines and their orientation it may not be possible to recover all camera velocity elements. Some velocities may be unobservable, that is, camera motion in that direction causes no change in the image. Some observed motion may be ambiguous, that is, the same image motion might be caused by two or more different camera velocities. In order to recover all elements of the camera velocity we need to observe at least 3 non-parallel lines. These limitations can be found from examining the null-space of the Jacobian and its dimensions give the number of ambiguous states [40], and are summarised in Table 1. The unobservable velocities could be estimated by alternative sensors such as gyroscopes, magnetometers, or perhaps a downward looking camera that served as a visual compass. These alternative estimates could also be used to resolve ambiguities.

For the case of two vertical lines considered in this paper, the vertical velocity is unobservable and there is ambiguity between a sideways motion (camera x-axis) and a rotation about camera y-axis. Another manifestation is the case where a change in more than one state causes the same feature motions in the image. For example a sideways motion (camera x-axis) and a rotation about camera y-axis.

Table 1. Unobservable and ambiguous velocity components.

# of lines	Rank	Unobservable	Ambiguities	Condition
1	2	v_y	$v_x \sim v_z \sim \omega_y, \omega_x \sim \omega_z$	Line not on the optical axis
2	4	v_y	$v_x \sim \omega_y$	—
3	6 (Full)	—	Lines are not parallel	

4.1.3. De-Rotation Using an IMU

VTOL platforms such as a quadrotor or a hexarotor are underactuated and cannot translate without first tilting the thrust vector in the direction of travel. This rotation immediately causes the image features to move and increases the image feature error, causing poor performance with a simple linear controller like Equation (28). Instead we use IMU measurement of this rotation which we subtract from the observed feature motions, often called *image de-rotation* [41]. The displacements of line features in θ and ρ are a function of a camera rotation about the x, y and z axes in the world coordinate: roll, pitch and yaw. We rewrite Equation (25) in partitioned form [37] as

$$\dot{\ell} = \begin{bmatrix} \frac{1}{D}J_t & | & J_\omega \end{bmatrix} \begin{bmatrix} v_{xz} \\ \omega \end{bmatrix}$$

$$= \frac{1}{D}J_t v_{xz} + J_\omega \omega \tag{30}$$

where $\dot{\ell}$ is a 4×1 optical flow component, $\frac{1}{D}J_t$ and J_ω are 4×2 translational and 4×3 rotational components. They are respectively columns $\{1,3\}$ and $\{4,5,6\}$ of the stacked Jacobian, $\begin{bmatrix} J_{l_1}, J_{l_2} \end{bmatrix}^T$. Note we omit column $\{2\}$ which corresponds to the unobservable state, v_y. The reduced Jacobian is slightly better conditioned (smaller condition number) and the mean computation time is measured at $30\,\mu s$ which is $20\,\mu s$ faster than for the full size Jacobian. Thus v_{xz} contains only two elements,

$\begin{bmatrix} v_x, v_z \end{bmatrix}^T$, the translational camera velocity in the horizontal plane. This is input to the vehicle's roll and pitch angle control loops. The common denominator, D, denotes target object depth which is assumed, and ω is obtained from an IMU. We rearrange Equation (30) as

$$v_{xz} = DJ_t^+ (\dot{\ell} - J_\omega \omega) \tag{31}$$

and we substitute Equation (27) into Equation (31) to write

$$v_{xz}^* = DJ_t^+ (\lambda(\ell^* - \ell) - \underbrace{J_\omega \omega}) \tag{32}$$

The de-rotation term is indicated, and subtracts the effect of camera rotation from the observed features. After subtraction, only the desired translational velocity remains.

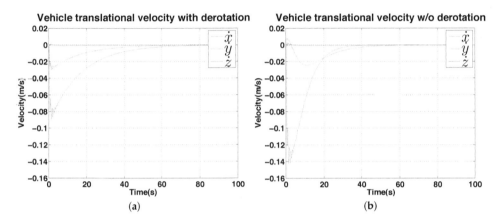

Figure 13. Simulation results for IBVS regulation: velocity response. (**a**) with de-rotation; (**b**) without.

4.2. IBVS Simulation Results

In order to validate the IBVS system and de-rotation method, we use a simulation framework from [36] for a quadrotor dynamic model equipped with a front-facing perspective camera. We implement the mentioned IBVS system with two vertical lines and the de-rotation block in the IBVS control block which yields the required translational velocity, v_{xz}^*.

The vehicle is initially located at $(x, y) = (-1.0, 0.5)$ m and the goal position is $(-2.5, 0)$ m in the world coordinate frame. Figure 14 shows the results of the position changes in x and y and the normalized feature errors over time. There are two parameters, $\lambda = 0.3$ and $D = 1$ which are respectively a positive gain and a depth for this simulation.

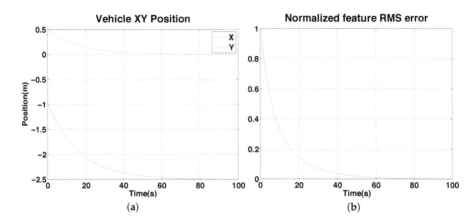

Figure 14. Simulation results for IBVS (with de-rotation): position response. (**a**): vehicle position (x, y) *versus* time showing the vehicle achieves its goal (x = −2.5, y = 0) at around 60 s; (**b**) the normalized root mean squared feature error $(\ell^* - \ell)$ for the same period.

We ran simulations with and without the de-rotation block for comparison. Although the vehicle converges to the goal position slower with de-rotation enabled ($\approx$38 s *versus* $\approx$60 s) as shown in Figure 13, the de-rotation yields smoother velocity commands and less oscillation of the y velocity.

Computing Equation (29) requires knowledge of the plane normal vector and D, the distance from the camera image plane to the plane on which the lines lie [42]. In the literature, many approaches have been demonstrated for depth estimation, e.g., 3D reconstruction of a scene using vision techniques [43], a Jacobian matrix estimator using the kinematics of an arm-type robot and image motion [44] or an Unscented Kalman Filter based estimator using point features and inertial sensor data [45]. These approaches are applicable to our problem however we lack sufficient computational power on our onboard computer to implement them in real time.

Instead we used a fixed value of D (denoted D_{fixed}), and set this to a value that is reasonable for the pole inspection task (for example 1.5 m). IBVS uses D_{fixed} in Jacobian calculation Equation (29) and it is neither the true depth nor current camera pose.

To investigate the control sensitivity to incorrect values of D_{fixed}, we ran simulations with different values of D_{fixed} and the results are plotted in Figure 15. The figure shows how the camera Euclidean distance error (in SE(3), between the goal and current position) changes over time for a variety of values of D_{fixed}. The true value of D at t = 0 was 1.5 m. The plot shows that D_{fixed} effectively acts as a proportional gain, with higher values of D_{fixed} causing faster convergence (for example when the true value of D is 1.5 m, a value of D_{fixed} = 5 m results in convergence after 10 s compared to 70 s for D_{fixed} = 1 m). Since D_{fixed} acts as a proportional gain, there is the potential for the system to become unstable if the difference between D_{fixed} and D is too large. Figure 15, however, shows that the system is stable for a relatively large variation in D_{fixed} values, indicating that using a fixed value for D instead of estimating it online is appropriate for our application.

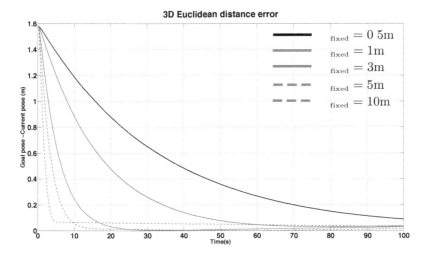

Figure 15. Simulation results for IBVS regulation: error response as a function of time for different assumed constant target object depths.

5. Shared Autonomy

A typical attitude-stabilized multi-rotor has four user-controllable degrees of freedom (DOF), namely horizontal position (x,y), height (z), and heading (ψ). These are usually controlled indirectly with joysticks where the stick positions are mapped to rates (e.g., the "throttle" stick position is mapped to climb rate), or to angular velocity in the case of yaw. These commands are in the body coordinate frame, making it hard for an operator to control position in the 3-dimensional Cartesian world coordinate frame.

We propose reducing the operator's cognitive load and level of skill required by reducing the DOFs that the operator must control, see Figure 16, and letting the system control the remaining DOFs automatically. Additionally, some of the DOFs are controlled in a more direct, intuitive manner rather than indirectly via rate or velocity commands. Since the proposed system can self-regulate the stand-off distance and keep the camera pointed at the target, the operator is left with only two DOFs to control, namely height and yaw rate. Height control is simplified: from the operator providing rate control commands to providing height set-points.

Yaw rate commands are used to induce a translation around the pole, allowing the operator to inspect it from different angles. Changing yaw angle makes the quadcopter circle around the pole (red bar indicates the *front* rotor). References for the x and y position controllers are d_x and 0 respectively. The robot hovers by keeping d_x distance at time t. The operator sends a yaw command and the vehicle rotates by the angle γ which induces a lateral offset d_y at time $t+1$. The vision-based controller moves the robot to the right to eliminate d_y and keeps d_x distance at time $t+2$—the result is motion around the target object.

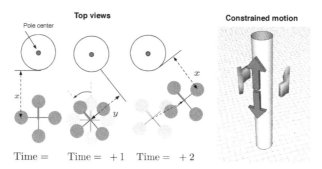

Figure 16. (**left**) illustration of how vehicle induces yaw motion. γ is an angle for the yaw motion and d_x and d_y are distances between the pole and the robot in x-, and y-axis; (**right**) reduced dimension task space for operator commands which is sufficient for inspection purposes.

6. Experimental Results

The experiments we present are summarised in Figure 17 and can be considered with respect to many categories: autonomous or manual, PBVS or IBVS control, hovering or circumnavigation, indoor or outdoor, day or night. The manual pilot experiments pit two human pilots, with different skill levels, against the autonomous system for the tasks of hovering and circumnavigation. Figure 18 shows some sample images from the onboard front-camera captured during various experiments. The demonstration video is available from the following link [22].

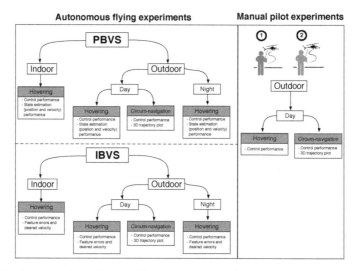

Figure 17. Overview of experiments. There are two categories: autonomous flying with shared autonomy (**left**) and manual piloting with only attitude stabilization (**right**); Autonomous flying consists of PBVS (**top**) and IBVS (**bottom**). Two pilots were involved in the manual piloting experiments. Each box with grey denotes a sub-experiment and describes key characteristics of that experiment.

Figure 18. Sample onboard images from the three hovering test environments: indoor (**left**), day-time outdoor (**mid**) and night-time outdoor (**right**) respectively. The top row contains the raw images while the bottom row contains the corresponding gradient images with overlaid lines corresponding to the goal line positions (green) and the tracked pole edges (red).

6.1. Hardware Configuration

Our research hexarotor platform, an Ascending Technologies Firefly, is fitted with a front-facing camera and a quad-core 1.7 GHz ARM Cortex-A9 computer which performs all the processing onboard. It runs Ubuntu Linux, and the Robot Operating System (ROS) [46] is used as the underlying software framework.

The front-facing camera is a low-cost high-speed Playstation EyeToy connected via USB. This CMOS camera has a rolling shutter which is problematic on a moving platform [47]. We thus set essential camera options (such as selecting low-resolution 320 × 240 images, using fixed exposure and gain and the fastest frame rate available) in order to minimize rolling shutter effects. The IMU provides angular velocity as well as orientation (roll, pitch, yaw) and the 3-axis acceleration through the High-Level-Processor (HLP) and Low-Level-Processor (LLP) of the hexarotor platform [48]. Altitude measurements are obtained from a downward-facing ultrasonic sensor. For night time flying a high-powered LED is mounted to the front to illuminate the scene for the onboard camera.

All our experiments are ground truthed. The indoor flights are conducted inside a Vicon motion capture environment and we attach reflective markers to the vehicle which allows us to measure position and attitude at 100 Hz. Outdoors we attach a single corner reflector and use a Leica TS30 laser tracking theodolite which provides position only measurments at 5 Hz shown in Figure 19. Each ground truth system can provide position measurements with sub-millimeter accuracy [49]. We take considerable care to synchronise the clocks of the flying vehicle and the tracking systems to avoid time skew when comparing onboard estimates with ground truth.

Figure 19. Experimental setup for outdoor flights. The VTOL platform is shown pointing towards the pole. An actuated surveying laser tracks a reflective beacon on the aircraft for position ground truth.

Software Configuration

We implemented PBVS and IBVS using ROS and Figures 20 and 21 show the sensor and software configurations for PBVS and IBVS control respectively. Different colors denote different sampling rates and arrows denote data flow from the sensors to the software components where processing occurs. Each box is an individual ROS node implemented using C++. Precision Time Protocol (PTP) is utilized for time synchronization between the onboard computer and the external ground truth data logger.

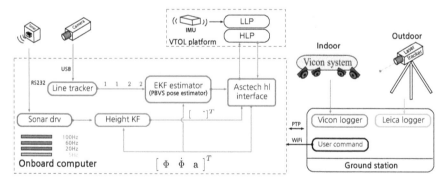

Figure 20. Sensor and software system diagram for PBVS experiments. Different colors denote different sampling rates. The software is implemented using ROS and all computation for a full update cycle happens within 16.6 ms (60 Hz) on average. The line tracker utilizes 55% of the CPU. The Extended Kalman Filter (EKF) includes line model and Jacobian calculations and these steps take 6 μs and 30 μs respectively. Note that only one ground truth source is used at time. The Vicon system is used for indoor ground truth while the Leica laser tracker is used outdoors.

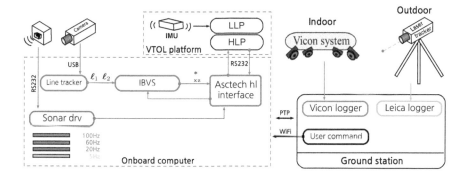

Figure 21. Sensor and software system diagram for IBVS experiments. Different colors denote different sampling rates. All software is implemented using a Robot Operating System (ROS). Note that only one ground truth source is used at time. The Vicon system is used for indoor ground truth while the Leica laser tracker is used outdoors.

6.2. Position-Based Visual Servoing (PBVS)

In this section, hovering and circumnavigation experimental results are presented with 4 sub-experiments. During the hovering experiments, no human inputs are provided. Only yaw rate commands are sent to the flying robot for the circumnavigation experiments. We evaluate control performance by comparing the goal and ground truth position. The evaluation of state estimation is presented with ground truth states and onboard estimator outputs. Figure 22 summarizes the position control and state estimation performance for the indoor and outdoor environments.

State	Control performance			State estimator performance			Unit
	Indoor	Outdoor (day)	Outdoor (night)	Indoor	Outdoor (day)	Outdoor (night)	
x	0.048	0.038	0.047	0.016	0.019	0.019	m
y	0.024	0.028	0.043	0.01	0.024	0.03	m
z	0.011	0.022	0.016	0.008	0.015	0.013	m
$\dot{x}$	—	—	—	0.038	0.12	0.152	m/s
$\dot{y}$	—	—	—	0.015	0.053	0.108	m/s
$\dot{z}$	—	—	—	0.042	0.04	0.031	m/s
ϕ	—	—	—	0.294	—	—	deg
θ	—	—	—	1.25	—	—	deg
Time interval	15~55	10~50	10~50	15~55	10~50	10~50	s

Figure 22. PBVS hovering (standard deviation).

6.2.1. PBVS Indoor Hovering

The control performance is evaluated by computing the standard deviation of the error between the goal position (x, y and z) and ground truth as measured by the Vicon system. The performance of the controller is shown in Figure 23. Interestingly, although the x-axis velocity estimation is noisy, the control performance for this axis is not significantly worse than for the y-axis, the quadrotor plant is effectively filtering out this noise.

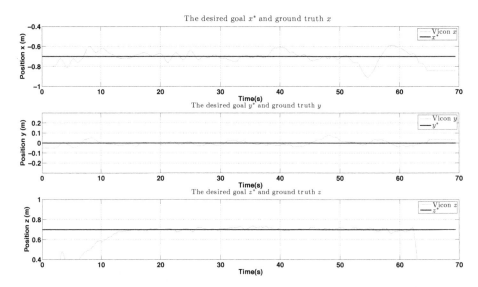

Figure 23. Experimental results for PBVS-based indoor hovering: control performance. Goal (black) and ground truth states (blue). The desired position for $^{W}\hat{x}$, $^{W}\hat{y}$ and $^{W}\hat{z}$ is -0.7 m, 0 m and 0.7 m. We compute standard deviation of errors for each state over the interval 15 s~63 s: $\sigma_x = 0.048$ m, $\sigma_y = 0.024$ m, and $\sigma_z = 0.011$ m.

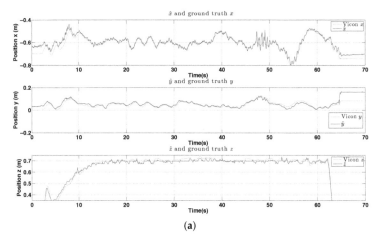

(a)

Figure 24. *Cont.*

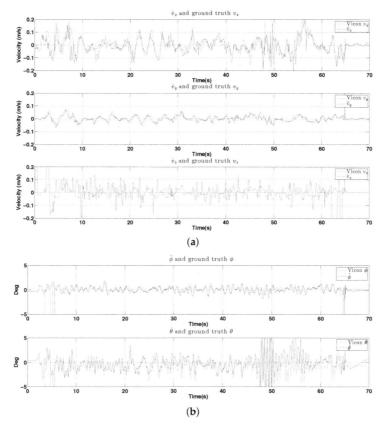

Figure 24. Experimental results for indoor PBVS-based hovering. All states are shown at the same scale. The performance evaluation of the state estimator is summarised in Figure 22. (**a**) Position estimation (red) with ground truth (blue). Note that z-axis is inverted for visualization; (**b**) Velocity estimation (red) with ground truth (blue), $^{W}\hat{x}$, $^{W}\hat{y}$ and $^{W}\hat{z}$; (**c**) Attitude estimation (red) with ground truth (blue), roll $\hat{\phi}$ and pitch $\hat{\theta}$.

We estimate position and orientation except heading angle as shown in Figure 24. The robot oscillates around 48 s–50 s when the line tracker was affected by the noisy background leading to errors in the position estimate $^{W}\hat{x}$.

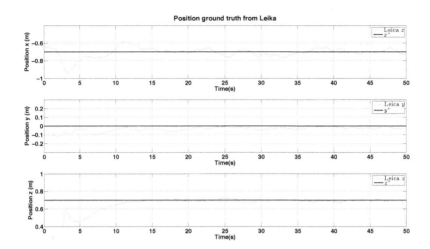

Figure 25. Experimental results for outdoor (day) PBVS-based hovering: control performance. Goal (black) and ground truth states (blue) . The desired position for $^{\mathcal{W}}\hat{x}$, $^{\mathcal{W}}\hat{y}$ and $^{\mathcal{W}}\hat{z}$ is -0.7 m, 0 m and 0.7 m. We compute standard deviations of errors for each state over the interval 10 s~50 s: $\sigma_x = 0.038$ m, $\sigma_y = 0.028$ m, and $\sigma_z = 0.022$ m.

6.2.2. PBVS Day-Time Outdoor Hovering

The VTOL platform was flown outdoors where external disturbances such as wind gusts are encountered. In addition, background scenes were nosier as shown in Figure 18. Figures 25 and 26 show control performance and state estimation during day-time hovering outdoors. The proposed system was able to efficiently reject disturbances and maintain a fixed stand-off distance from a pole (see accompanying video demonstration 2.2). Position and velocity estimation results are shown in Figure 26 and are noisier than for the indoor case due to more complex naturally textured background scenes (see Figure 18). Controller performance is consistent with that observed indoors. All results are within a ± 0.02 m variation boundary.

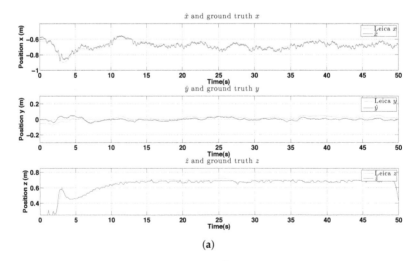

(a)

Figure 26. *Cont.*

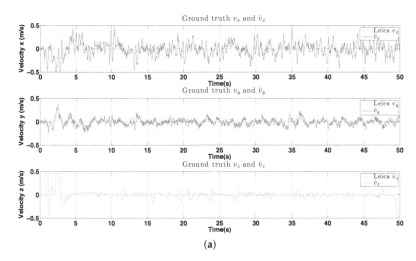

(a)

Figure 26. Experimental results for outdoor (day) PBVS-based hovering: estimator performance. Angle estimation are omitted because the laser tracker can only provide position ground truth of the moving target. All states are shown at the same scale. The performance evaluation is presented in Figure 22. (**a**) Position estimation (red) with ground truth (blue). Note that z-axis is inverted for visualization; (**b**) Velocity estimation (red) with ground truth (blue), ${}^W\hat{x}$, ${}^W\hat{y}$ and ${}^W\hat{z}$.

6.2.3. PBVS Night-Time Outdoor Hovering

We performed night-time PBVS-based hovering experiments and experienced the best control and state estimation performance shown in Figures 27 and 28 respectively. At night there was less wind (average wind speed was less than 1 m/s) and the pole edges were clear in the image since only the pole in the foreground was illuminated by the onboard light. However, the EKF used for horizontal state estimation was extremely sensitive to the measurement noise parameters, $\mathcal{R}$ from Equation (16). We didn't adapt $\mathcal{R}$ for night-time flights (same values as for day-time outdoor and indoor hovering) and this led to poor state estimation and oscillation in the x and y-axes—the worst results among the 3 experiments. This is a potential limitation of the deterministic Extended Kalman Filter (Filter Tuning). [50,51] exploited stochastic gradient descent in order to learn $\mathcal{R}$ with accurate ground truth such as motion capture or high quality GPS. We are interested in this adaptive online learning for filter frameworks; however, this is beyond the scope of this work.

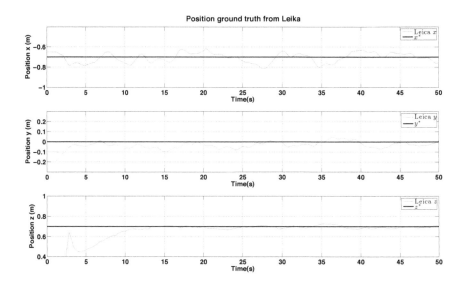

Figure 27. Experimental results for outdoor (night) PBVS-based hovering: control performance. Goal (black) and ground truth states (blue). The desired position for $^W\hat{x}$, $^W\hat{y}$ and $^W\hat{z}$ is -0.7 m, 0 m and 0.7 m. We compute standard deviations of errors for each state over the interval 10 s~50 s. $\sigma_x = 0.047$ m, $\sigma_y = 0.043$ m, and $\sigma_z = 0.016$ m.

6.2.4. PBVS Day-Time Outdoor Circumnavigation

The pole circumnavigation experiment is performed by placing the VTOL platform on the ground with the camera facing the pole to be inspected and at the desired stand-off distance. The line-tracking algorithm is initialized and the operator then commands only goal height and yaw rate to move the VTOL platform around the pole at different heights. The system keeps the camera oriented towards the pole and maintains the stand-off distance. Figure 29 displays different views of the trajectory for a flight where the pole was circumnavigated. A circle with the goal radius is shown with a dashed line, and we see the system tracks the desired stand-off distance well. At the time the average wind speed was about 1.8 m/s blowing from left to right (See the demonstration video 2.4). The trajectory was within ±0.15 m of the goal radius for most of the circumnavigation but the error increased at around x = −0.7 and y = −0.4 due to wind gusts. Note that the laser tracker lost track of the reflective prism on the vehicle when it was occluded by the pole at (x = −0.8~−0.9) and (y = −0.6~−0.2). We computed the standard deviation of the control performance for the flight period (0–75 s) to be 0.034 m. The height change near (x = −0.9, y = 0) is due to the box that was placed on the ground as a takeoff and landing platform. Since the aircraft maintains a fixed height above the ground beneath it, it climbs when flying over the box.

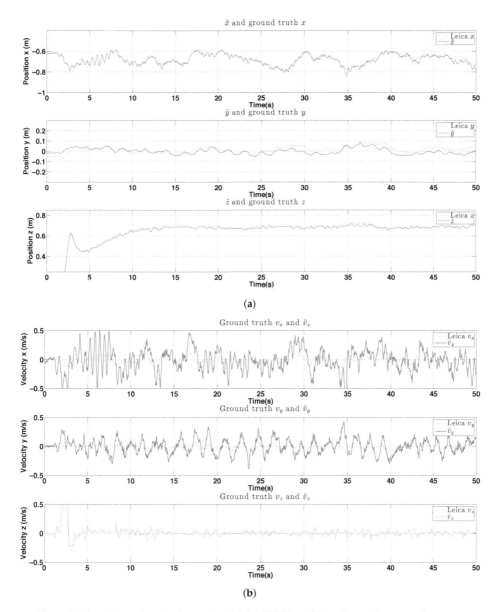

Figure 28. Experimental results for outdoor (night) PBVS-based hovering: estimator performance. Estimated angles are omitted and all states are shown at the same scale. The performance evaluation summary of these plots is presented in Figure 22. (**a**) Position estimation (red) with ground truth (blue). Note that z-axis is inverted for visualization; (**b**) Velocity estimation (red) with ground truth (blue), $^{W}\hat{x}$, $^{W}\hat{y}$ and $^{W}\hat{z}$.

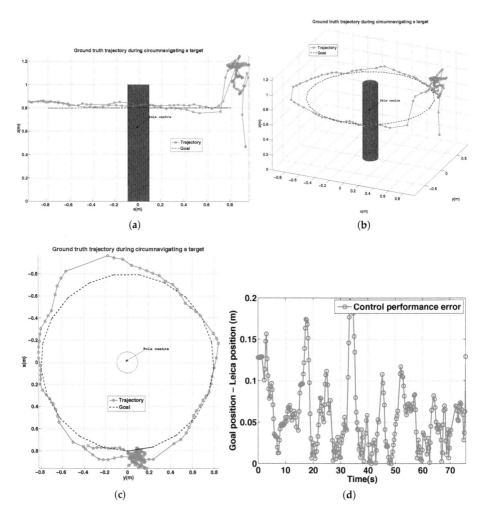

Figure 29. Experimental results for outdoor (day) PBVS-based circumnavigation: control performance. (**a**–**c**) 3D views of the trajectory: Side view, Perspective view and Top view; (**d**) Euclidean error (goal minus actual) trajectory *versus* time.

6.3. Imaged-Based Visual Servoing (IBVS)

We performed pole-relative hovering in 3 environments as shown in Figure 18: indoor (controlled lighting), day-time outdoor and night-time outdoor. For each flight test the platform was flown for approximately a minute and no human interventions were provided during the flight. We set λ and D to 1.1 and 0.8 m respectively for all experiments presented in this and the following section. A summary of the results is presented in the Table 2, while Figures 30–32 show the position results with respect to $\{\mathcal{W}\}$.

Table 2. Hovering standard deviation performance summary of Figures 30–32.

State w.r.t $\{W\}$	Indoor	Outdoor (Day)	Outdoor (Night)	Unit
x	0.084	0.068	0.033	m
y	0.057	0.076	0.050	m
z	0.013	0.013	0.013	m
Duration	15~60	15~55	15~70	s
Wind speed	—	1.8~2.5	less than 1	m/s

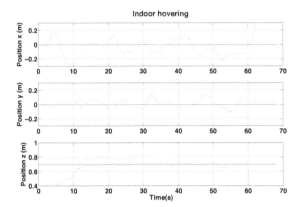

Figure 30. Experimental results for IBVS-based hovering: control performance for indoor IBVS-based hovering. Goal positions shown in red.

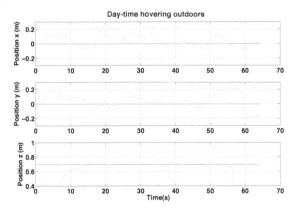

Figure 31. Experimental results for IBVS-based hovering: control performance for outdoor day-time IBVS-based hovering. Goal positions shown in red.

As shown in the Table 2, hovering performance for indoors is similar to that for outdoors despite the fact that there are no wind disturbances indoors. This can be explained by the fact that the hexarotor platform uses a yaw angle estimated from a gyro and magnetometer. Gyros are subject to drift due to biases and vibration noise, while magnetometers are strongly influenced by magnetic perturbations produced by man-made structures indoors. Poor yaw estimates indoors therefore yields a yaw rotation of the vehicle, which in turn causes a y-axis controller error in $\{W\}$. The platform moves in the body y-axis in order to keep the camera oriented towards the target, and this maneuver also causes an x-axis controller error in practice. Furthermore, salient vertical edges in the man-made indoor environment affect hovering performance as well.

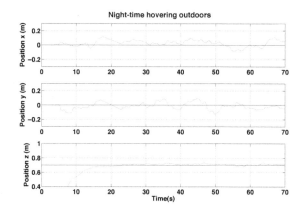

Figure 32. Experimental results for IBVS-based hovering: control performance for night-time IBVS-based hovering. Goal positions shown in red.

6.3.1. IBVS-Based Hovering

Control performance of the IBVS-based controller is shown in Figures 30–32 for indoors, outdoor day-time and outdoor night-time flight. For the day-time outdoor hovering test the average wind speed was 1.8 m/s with gusts of up to 2.5 m/s (See the demonstration video 1.2). The computed velocity demand to the vehicle is shown in in Figure 33.

The yaw estimation is better outdoors but there are wind disturbances. Also, the line features are noisier due to varying light conditions, shadows, and strong gradients from the background which make the edge of the pole weaker. The best performance was achieved for the outdoor night-time flights. Unlike PBVS, IBVS does not require a pose estimator and is therefore not subject to the EKF sensitivity issue described in Section 6.2.3. As for the PBVS night-time flights, there was less wind at night (average wind speed was less than 1 m/s) and the pole edges were well defined in the image since only the pole in the foreground was illuminated by the onboard light.

6.3.2. IBVS Day-Time Outdoor Circumnavigation

Outdoor circumnavigation experiments were conducted using IBVS in a similar manner to the PBVS experiments. The line-tracking algorithm was initialized and the pole edge features that were found became the desired feature positions for the flight. Figure 34 displays the top, side, and perspective views of the trajectory for a flight where the pole was circumnavigated twice. A circle with the goal radius is shown with a dashed line, and we see the system tracks the desired stand-off distance well. At the time the average wind speed was 1.8 m/s~2.5 m/s (See the demonstration video 1.4). The stand-off distance was maintained within 0.17 m error boundary for the entire flight as shown in Figure 34d. For comparison with PBVS, We also computed a standard deviation of the control performance error for the same length of flight time (0–75 s) and obtained 0.034 m. IBVS and PBVS show similar performance as shown in Table 3. The height change near ($x = -0.6$, $y = -0.5$) is due to the box that was placed on the ground as a takeoff and landing platform. Since the aircraft maintains a fixed height above the ground beneath it, it climbs when flying over the box.

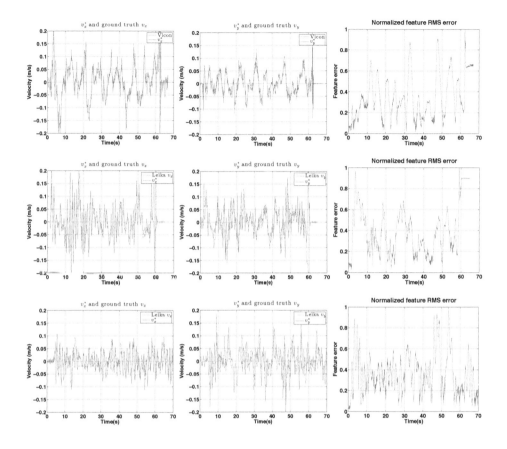

Figure 33. Experimental results for IBVS-based hovering: control demand. v_x^* (**left column**) and v_y^* (**middle column**) compared to ground truth. The first row is for indoor hovering while the second and the third rows are for day and night-time outdoor hovering. Normalized root mean squared (RMS) image feature errors for each are shown in the (**right column**).

6.4. Manually Piloted Experiments

The aim of these experiments was to determine how well a human pilot could perform the inspection tasks, hovering and circumnavigation, that we have demonstrated autonomously. Manual piloting requires great skill and people with this skill are quite scarce. The key skill is hand eye coordination, adjusting the vehicle's position by controlling roll and pitch angle joysticks. These joysticks effectively control vehicle acceleration which is more difficult for humans to master than the more common rate, or velocity, control inputs. The controls are effected with respect to the vehicle's coordinate frame, and pilots of moderate skill level are only able to fly with a constant vehicle heading angle, typically the x-axis away from the pilot. Circumnavigation requires the heading angle to change continuously and this requires high-order piloting skills.

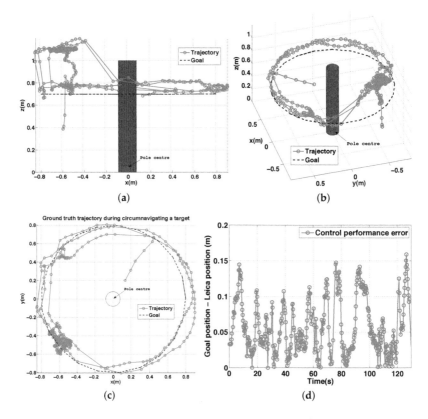

Figure 34. Different views of ground truth trajectory with respect to $\{T\}$ for a pole inspection flight (a–c); (d) is control performance error plot, *i.e.*, 3D Euclidian distance error between the goal and the trajectory. An operator only commands yaw rate using the RC transmitter during the experiment. The constant height, 0.7 m is maintained by the system.

Table 3. Circum-navigation performance comparison.

	PBVS	IBVS	Unit
Max error margin	0.024	0.017	m
Standard deviation	0.038	0.034	m
Duration	0~75	0~125	s

In these experiments we use two pilots with differing skill levels. Pilot 1, one of the authors, has strong skills for manual hovering but is unable to achieve circumnavigation. Pilot 2 is a licensed professional UAV pilot who is able to perform manual circumnavigation flights. In both cases the pilot is making use of the builtin attitude stabilization capability of the vehicle, and the manually piloted experiments were conducted outdoors during the daytime.

Figure 35 is a summary of the results. It compares the performance of the two pilots with that of the shared-autonomy system (PBVS and IBVS control) operated by the weaker of the two pilots. The different colors in the columns denote different experiments. We use a number of performance metrics:

Ground truth is derived from ground truth data from the laser tracker or Vicon to compute statistics of the error with respect to goal position as we have done in earlier parts of this paper. For example, Pilot 1 hovered for around 100 s and we computed $\sigma_x = 0.079$, $\sigma_y = 0.069$, and $\sigma_z = 0.093$ for the period 25–80 s of the first trial shown in Figure 36. For the case of circumnavigation we compute error with respect to the path's circle and do not penalize uneven motion around the circle.

[A] is the percentage pole detection rate in the onboard camera images. If the task is performed correctly the pole will be visible in 100% of the images.

[B] is the standard deviation of the horizontal pole centre position (pixels) in the onboard camera images. This is a more graduated performance measure than A, and says something about the quality of the translational and heading angle control of the vehicle. If the task is performed well this should be 0. Note this statistic is computed over the frames in which the pole is visible.

[C] is the standard deviation of the pole width (pixels) in the onboard camera images. It says something about the quality of the control of the vehicle position in the standoff direction, and if the task is performed well this should be 0. Note this statistic is computed over the frames in which the pole is visible.

Measures *A*, *B* and *C* are computed using a semi-automated line picking software tool (shown in Figure 37) from the recorded image sequences (subsampled to 10 Hz). As shown in Figure 35, autonomous flight outperformed all manual flights. Moreover, pole detection rates (*A*) were 100% for shared autonomy circumnavigation however it decreased to 70%–80% for manually piloted flights. This is due to the fact that the pilot had to control all 4 DOFs and the heading angle had to be constantly changed during circumnavigation in order to keep the camera pointed at the pole. The cognitive load on the pilot was very high for the circumnavigation task and the task was considered to be quite stressful.

Another interesting result was the difference in *B* and *C* for the manual hovering and circumnavigation experiments. These increased significantly from hovering to circumnavigation for manually piloted flights. For example, *B* was 13.33 pixels for Pilot 2's first hovering trial and this increased to 76.5 pixels for circumnavigation (See Figure 35). For Pilot 1's shared autonomy flights the results remained fairly consistent however (*B* = 12–17 pixels for PBVS, and *B* = 30 pixels for IBVS). Figure 38 also illustrates this point as we see the pole width and position in the image remain far more consistent for the shared autonomy flights (Figure 38b,c) compared to the manually piloted flight (Figure 38a). Figure 35 shows the best hovering results were achieved by IBVS (night-time) and PBVS (day-time) and best circumnavigation results were achieved by PBVS (day-time).

Evaluation method	Task	Pilot1 manual			Pilot2 manual			Pilot1 shared autonomy using PBVS			Pilot1 shared autonomy using IBVS			Unit
		1st trial	2nd trial	3rd trial	1st trial	2nd trial	3rd trial	Indoor	Outdoor (day)	Outdoor (night)	Indoor	Outdoor (day)	Outdoor (night)	
Leica laser based	Hovering	$\sigma_x=0.079$ $\sigma_y=0.069$ $\sigma_z=0.093$	$\sigma_x=0.167$ $\sigma_y=0.106$ $\sigma_z=0.086$	$\sigma_x=0.152$ $\sigma_y=0.083$ $\sigma_z=0.153$	$\sigma_x=0.074$ $\sigma_y=0.066$ $\sigma_z=0.049$	$\sigma_x=0.120$ $\sigma_y=0.099$ $\sigma_z=0.049$	$\sigma_x=0.102$ $\sigma_y=0.073$ $\sigma_z=0.034$	$\sigma_x=0.048$ $\sigma_y=0.024$ $\sigma_z=0.011$	$\sigma_x=0.038$ $\sigma_y=0.028$ $\sigma_z=0.022$	$\sigma_x=0.047$ $\sigma_y=0.043$ $\sigma_z=0.016$	$\sigma_x=0.084$ $\sigma_y=0.057$ $\sigma_z=0.013$	$\sigma_x=0.068$ $\sigma_y=0.076$ $\sigma_z=0.013$	$\sigma_x=0.033$ $\sigma_y=0.050$ $\sigma_z=0.013$	Metre (Standard deviation)
Onboard camera based		A=100% B=12.55px C=4.15px	A=100% B=16.64px C=7.41px	A=100% B=16.73px C=4.74px	A=100% B=13.33px C=5.79px	A=100% B=23.12px C=6.98px	A=100% B=13.18px C=6.22px	A=100% B=9.56px C=6px	A=100% B=12.29px C=2.69px	A=100% B=11.06px C=4.97px	A=100% B=20.34px C=8.19px	A=100% B=31.73px C=8.07px	A=100% B=20.37px C=5.17px	
	Time duration	25-80(55s)	25-80(55s)	25-80(55s)	36-73 (37s)	36-73 (37s)	36-73 (37s)	15-55 (40s)	15-55 (40s)	15-55 (40s)	15-55 (40)	15-55 (40)	15-55 (40)	second
	# images	110	110	110	74	74	74	80	80	80	80	80	80	
Onboard camera based	Circumnavi	—	—	—	A=72.5% B=76.5px C=10.83px	A=85.26% B=60.4px C=10.29px	A=71.79% B=65.67px C=11.67px	—	A=100% B=17.17px C=6.13px	—	—	A=100% B=30.28px C=4.69px	—	
	# images	—	—	—	240	154	154	—	110	—	—	110	—	
	Time duration	—	—	—	20-140	20-97	20-97	—	5-60	—	—	5-60	—	

A= Successful pole detection rate
B= Standard deviation of pole centre position in image coordinate
C= Standard deviation of pole width in image coordinate

Figure 35. Performance evaluation of manual pilot and autonomous flights.

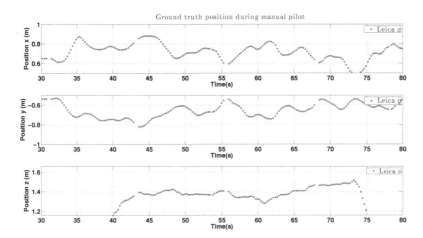

Figure 36. Experimental results for Pilot 2 during manual hovering: position *versus* time.

Figure 37. User interface of tool for generating performance metrics. The four yellow points are manually picked to define the pole edges (two green lines). The red point and star denote the calculated pole centre and width in the image coordinates respectively.

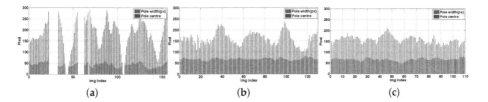

(a) (b) (c)

Figure 38. Experimental results for circumnavigation: pole width (blue) and offset of the pole centre from the image edge (red) *versus* time. Comparison of Pilot 2 manual flight (**a**), and Pilot 1 shared autonomy flights using IBVS and PBVS (**b,c** respectively). Since the image width is 320 pixels, a value of 160 for "Pole centre" indicates the pole was centred in the image. Zero values are for frames where the pole didn't appear in the image. (**a**) Pilot 2; (**b**) IBVS; (**c**) PBVS.

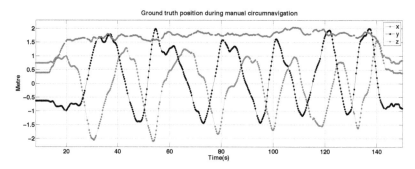

Figure 39. Experimental results for Pilot 2 during manual circumnavigation: position *versus* time.

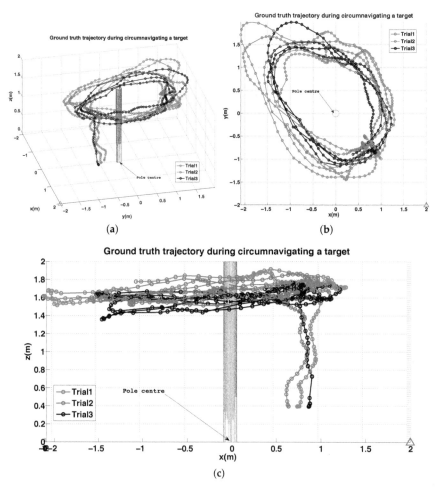

Figure 40. Experimental results for Pilot 2 during manual circumnavigation, 3D views. The magenta triangle indicates where the pilot was standing. (**a**) Perspective view; (**b**) Top view; (**c**) Side view.

Figures 39 and 40 shows 3D position *versus* time for three circumnavigation trials by Pilot 2. The magenta triangle denotes the standing position of Pilot 2 during the circumnavigation experiments. The trajectory is elliptical in shape with its major axis passing through the standing position. We believe there are two reasons for this: firstly, it is most challenging to estimate distance along the optical axis when the only visible cue is slight change in apparent size, and secondly, the pole occluded the vehicle from the pilots's line-of-sight at this position.

In summary the shared autonomy system allowed a less skilled pilot to achieve better task performance than our best human pilot, and at a much lower level of cognitive load and stress.

6.5. Limitations and Failure Cases

The proposed system has limitations which we plan to address. For example, the line tracker failed on occasion when the user commanded a large yaw rate causing the pole to leave the camera field of view (FOV). This can be addressed by either using a wider FOV lens (currently 75°) or by limiting the yaw rate.

Another limitation is the susceptibility of the line tracker to the real-world lighting effects when operating outdoors. If the sun appears in the image it leads to severe flaring and failure of the line tracker. Laser scanners are however also adversely affected when looking at the sun. Shadowing and uneven illumination in both indoor and outdoor environments, (see for example Figure 41), can create an intensity gradient on the surface of the pole and this may be falsely detected and tracked as the pole edge. To avoid these challenges our experiments were conducted in the absence of direct sunlight (early morning, late afternoon or cloudy), and we will improve the robustness to these effects in the future.

Figure 41. Challenging images for indoor (**a**) and outdoor (**b**–**d**) scenes.

We also experienced failures when strong lines were present in the background as shown in Figure 41d. A telecommunication tower produced strong lines and the line tracker failed to track the pole.

The sonar only works reliably up to 2 m and is very noisy on grass or gravel outdoors. We therefore plan to develop a height estimator which combines other sensing modalities such as a barometer, scale from a downward looking camera or vertical visual odometry from the front camera on the pole being inspected.

Figure 42 shows real-world vertical structures such as power line and street light poles. They are not white in colour or have the homogeneous uniform shape that we assume in this paper. Perceptual aliasing caused from identical poles placed in a row represents a common real world problem. The proposed method does not present a straightforward way of resolving these issues however the results obtained from the simplified setup we use do demonstrate its potential.

| (a) | (b) | (c) | (d) |

Figure 42. Sophisticated (**a**), curved (**b**), perceptual aliased real world power line (**c**) and street light pole (**d**) variants.

7. Conclusions

We have presented a VTOL platform-based pole-relative navigation system using PBVS, IBVS and shared autonomy. The target application is aerial inspection of vertical structures such as poles. The pole-relative navigation increases the autonomy of the system and facilitates the use of shared autonomy where the operator is relieved of the cognitive load of controlling all degrees of freedom. By self-regulating its stand-off distance from the pole, height, and keeping the camera facing the pole the system only requires height set points and yaw rate commands (which induce an orbit around the pole).

A common element of all the systems is an efficient and high-performance line tracker which provides estimates of the Plücker coordinate parameters of the observed lines. A key to high performance tracking on such an agile platform is feature prediction which we achieve using an image feature Jacobian and IMU measurements. For PBVS control these tracked features and IMU measurements are fed into a pose estimator of PBVS (Extended Kalman Filter) and we designed a controller based on the estimated states. IBVS control is performed directly using information from only two vertical lines (the pole edges) which leads to some unobservable and also ambiguous vehicle motions. We presented a line-feature based IBVS system which uses IMU data to eliminate the effect of body rotation and directly commands velocity in the horizontal plane.

The IBVS and PBVS systems demonstrated good pole-relative hovering and circumnavigation performance, maintaining position to within 20 cm of the goal position even in the presence off light wind.

The controllers formed part of a shared autonomy system in which the operator is no longer flying the vehicle but providing setpoints in a low DOF object-relative coordinate frame. Experimentally we showed that this allows a less skilled pilot to achieve better task performance than our best human pilot, and at a much lower level of cognitive load and stress. This sets the scene for operation of small VTOL platforms to perform cost-effective single person inspection jobs, rather than the three person crews that are currently the norm.

Finally, even though both systems demonstrate good performance, we prefer IBVS over PBVS for two reasons. Firstly, PBVS requires a pose estimator which takes image features as input and computes the metric pose of a camera. Development of a robust vision-based pose estimator that can be used in varying environments is difficult. Secondly, IBVS is relatively easy to implement since it omits the

Sensors **2015**, *15*, 22003–22048

pose estimation step, and utilize the image features directly. Although IBVS can degrade observability and be poorly conditioned due to a linearization of a highly non-linear model, it works on our pole tracking application.

Acknowledgments: The authors would like to thank Aaron Mcfadyen for assistance while using the VICON system, Navinda Kottege for providing the Leica ROS software and Kyran Findlater for supplying the LED light. In addition we thank to Peter Reid for manual pilot experiments.

Author Contributions: Inkyu Sa contributed to the development of the systems including implementation, field tests, and the manuscript writing. Stefan Hrabar contributed to the manuscript writing, providing significant suggestions on the development, and supporting field tests. Peter Corke provided the manuscript writing and guided the whole study.

Conflicts of Interest: The authors declare no conflict of interest.

References

1. Echelon. Monitored Outdoor Lighting. Available online: http://info.echelon.com/Whitepaper-Monitored-Outdoor-Lighting.html (accessed on 27 August 2015).
2. Pratt, K.S.; Murphy, R.R.; Stover, S.; Griffin, C. CONOPS and autonomy recommendations for VTOL small unmanned aerial system based on Hurricane Katrina operations. *J. Field Robot.* **2009**, *26*, 636–650.
3. Consumer Reports Magazine. Available online: http://www.consumerreports.org/cro/magazine-archive/may-2009/may-2009-toc.htm (accessed on 27 August 2015).
4. Balaguer, C.; Gimenez, A.; Jardon, A. Climbing Robots' Mobility for Inspection and Maintenance of 3D Complex Environments. *Auton. Robot.* **2005**, *18*, 157–169.
5. Kim, S.; Spenko, M.; Trujillo, S.; Heyneman, B.; Santos, D.; Cutkosky, M. Smooth Vertical Surface Climbing With Directional Adhesion. *IEEE Trans. Robot.* **2008**, *24*, 65–74.
6. Xu, F.; Wang, X.; Wang, L. Cable inspection robot for cable-stayed bridges: Design, analysis, and application. *J. Field Robot.* **2011**, *28*, 441–459.
7. Haynes, G.C.; Khripin, A.; Lynch, G.; Amory, J.; Saunders, A.; Rizzi, A.A.; Koditschek, D.E. Rapid Pole Climbing with a Quadrupedal Robot. In Proceedings of the IEEE International Conference on Robotics and Automation, Kobe, Japan, 12–17 May 2009; pp. 2767–2772.
8. Parness, A.; Frost, M.; King, J.; Thatte, N. Demonstrations of gravity-independent mobility and drilling on natural rock using microspines. In Proceedings of the IEEE International Conference on Robotics and Automation, Saint Paul, MN, USA, 14–18 May 2012; pp. 3547–3548.
9. Ahmadabadi, M.; Moradi, H.; Sadeghi, A.; Madani, A.; Farahnak, M. The evolution of UT pole climbing robots. In Proceedings of the 2010 1st International Conference on Applied Robotics for the Power Industry (CARPI), Montreal, QC, Canada, 5–7 October 2010; pp. 1–6.
10. Kendoul, F. Survey of advances in guidance, navigation, and control of unmanned rotorcraft systems. *J. Field Robot.* **2012**, *29*, 315–378.
11. Voigt, R.; Nikolic, J.; Hurzeler, C.; Weiss, S.; Kneip, L.; Siegwart, R. Robust embedded egomotion estimation. In Proceedings of the IEEE International Conference on Intelligent Robots and Systems, San Francisco, CA, USA, 25–30 September 2011; pp. 2694–2699.
12. Burri, M.; Nikolic, J.; Hurzeler, C.; Caprari, G.; Siegwart, R. Aerial service robots for visual inspection of thermal power plant boiler systems. In Proceedings of the International Conference on Applied Robotics for the Power Industry, Zurich, Switzerland, 11–13 September 2012; pp. 70–75.
13. Nikolic, J.; Burri, M.; Rehder, J.; Leutenegger, S.; Huerzeler, C.; Siegwart, R. A UAV system for inspection of industrial facilities. In Proceedings of the IEEE Aerospace Conference, Big Sky, MT, USA, 2–9 March 2013; pp. 1–8.
14. Ortiz, A.; Bonnin-Pascual, F.; Garcia-Fidalgo, E. Vessel Inspection: A Micro-Aerial Vehicle-based Approach. *J. Intell. Robot. Syst.* **2013**, *76*, 151–167.
15. Eich, M.; Bonnin-Pascual, F.; Garcia-Fidalgo, E.; Ortiz, A.; Bruzzone, G.; Koveos, Y.; Kirchner, F. A Robot Application for Marine Vessel Inspection. *J. Field Robot.* **2014**, *31*, 319–341.
16. Bachrach, A.; Prentice, S.; He, R.; Roy, N. RANGE–Robust autonomous navigation in GPS-denied environments. *J. Field Robot.* **2011**, *28*, 644–666.

17. Shen, S.; Michael, N.; Kumar, V. Autonomous Multi-Floor Indoor Navigation with a Computationally Constrained MAV. In Proceedings of the IEEE International Conference on Robotics and Automation, Shanghai, China, 9–13 May 2011; pp. 20–25.
18. Sa, I.; Corke, P. Improved line tracking using IMU and Vision for visual servoing. In Proceedings of the Australasian Conference on Robotics and Automation, University of New South Wales, Sydney, Australia, 2–4 December 2013.
19. Sa, I.; Hrabar, S.; Corke, P. Outdoor Flight Testing of a Pole Inspection UAV Incorporating High-Speed Vision. In Proceedings of the International Conference on Field and Service Robotics, Brisbane, Australia, 9–11 December 2013.
20. Sa, I.; Hrabar, S.; Corke, P. Inspection of Pole-Like Structures Using a Vision-Controlled VTOL UAV and Shared Autonomy. In Proceedings of the IEEE International Conference on Intelligent Robots and Systems, Chicago, IL, USA, 14–18 September 2014; pp. 4819–4826.
21. Sa, I.; Corke, P. Close-quarters Quadrotor flying for a pole inspection with position based visual servoing and high-speed vision. In Proceedings of the IEEE International Conference on Unmanned Aircraft Systems, Orlando, FL, USA, 27–30 May 2014; pp. 623–631.
22. Video demonstration. Available online: http://youtu.be/ccS85_EDl9A (accessed on 27 August 2015).
23. Hough, P. Machine Analysis of Bubble Chamber Pictures. In Proceedings of the International Conference on High Energy Accelerators and Instrumentation, Geneva, Switzerland, 14–19 September 1959.
24. Shi, D.; Zheng, L.; Liu., J. Advanced Hough Transform Using A Multilayer Fractional Fourier Method. *IEEE Trans. Image Process.* **2010**, doi:10.1109/TIP.2010.2042102.
25. Hager, G.; Toyama, K. X vision: A portable substrate for real-time vision applications. *Comput. Vis. Image Understand.* **1998**, *69*, 23–37.
26. Bartoli, A.; Sturm, P. The 3D line motion matrix and alignment of line reconstructions. In Proceedings of the IEEE Conference on Computer Vision and Pattern Recognition, Kauai, USA, 8–14 December 2001; Volume 1, pp. 1–287.
27. Hartley, R.; Zisserman, A. *Multiple View Geometry in Computer Vision*; Cambridge University Press: Cambridge, UK, 2003.
28. Bartoli, A.; Sturm, P. Structure-from-motion using lines: Representation, triangulation, and bundle adjustment. *Comput. Vis. Image Understand.* **2005**, *100*, 416–441.
29. Mahony, R.; Hamel, T. Image-based visual servo control of aerial robotic systems using linear image features. *IEEE Trans. Robot.* **2005**, *21*, 227–239.
30. Solà, J.; Vidal-Calleja, T.; Civera, J.; Montiel, J.M. Impact of Landmark Parametrization on Monocular EKF-SLAM with Points and Lines. *Int. J. Comput. Vis.* **2012**, *97*, 339–368.
31. Chaumette, F.; Hutchinson, S. Visual servo control. I. Basic approaches. *IEEE Robot. Autom. Mag.* **2006**, *13*, 82–90.
32. Fischler, M.; Bolles, R. Random sample consensus: A paradigm for model fitting with applications to image analysis and automated cartography. *Commun. ACM*, **1981**, *24*, 381–395.
33. Durrant-Whyte, H. *Introduction to Estimation and the Kalman Filter*; Technical Report; The University of Sydney: Sydney, Australia, 2001.
34. Mahony, R.; Kumar, V.; Corke, P. Modeling, Estimation and Control of Quadrotor Aerial Vehicles. *IEEE Robot. Autom. Mag.* **2012**, doi:10.1109/MRA.2012.2206474.
35. Leishman, R.; Macdonald, J.; Beard, R.; McLain, T. Quadrotors and Accelerometers: State Estimation with an Improved Dynamic Model. *IEEE Control Syst.* **2014**, *34*, 28–41.
36. Corke, P. *Robotics, Vision & Control: Fundamental Algorithms in MATLAB*; Springer: Brisbane, Australia, 2011. ISBN 978-3-642-20143-1.
37. Corke, P.; Hutchinson, S. A new partitioned approach to image-based visual servo control. *IEEE Trans. Robot. Autom.* **2001**, *17*, 507–515.
38. Malis, E.; Rives, P. Robustness of image-based visual servoing with respect to depth distribution errors. In Proceedings of the IEEE International Conference on Robotics and Automation, Taipei, Taiwan, 14–19 September 2003; pp. 1056–1061.
39. Espiau, B.; Chaumette, F.; Rives, P. A new approach to visual servoing in robotics. *IEEE Trans. Robot. Autom.* **1992**, *8*, 313–326.

40. Pissard-Gibollet, R.; Rives, P. Applying visual servoing techniques to control a mobile hand-eye system. In Proceedings of the IEEE International Conference on Robotics and Automation, Nagoya, Japan, 21–27 May 1995; pp. 166–171.
41. Briod, A.; Zufferey, J.C.; Floreano, D. Automatically calibrating the viewing direction of optic-flow sensors. In Proceedings of the IEEE International Conference on Robotics and Automation, St Paul, MN, USA, 14–18 May 2012; pp. 3956–3961.
42. Corke, P.; Hutchinson, S. Real-time vision, tracking and control. In Proceedings of the IEEE International Conference on Robotics and Automation, Saint Paul, MN, USA, 14–18 May 2012; pp. 622–629.
43. Feddema, J.; Mitchell, O.R. Vision-guided servoing with feature-based trajectory generation [for robots]. *IEEE Trans. Robot. Autom.* **1989**, *5*, 691–700.
44. Hosoda, K.; Asada, M. Versatile visual servoing without knowledge of true Jacobian. In Proceedings of the IEEE International Conference on Intelligent Robot Systems, Munich, Germany, 12–16 September 1994; pp. 186–193.
45. Omari, S.; Ducard, G. Metric Visual-Inertial Navigation System Using Single Optical Flow Feature. In Proceedings of the European Contol Conference, Zurich, Switzerland, 17–19 July 2013.
46. Quigley, M.; Conley, K.; Gerkey, B.; Faust, J.; Foote, T.B.; Leibs, J.; Wheeler, R.; Ng, A.Y. ROS: An open-source Robot Operating System. In Proceedings of the IEEE International Conference on Robotics and Automation Workshop on Open Source Software, Kobe, Japan, 12–17 May 2009; pp. 5–11.
47. O'Sullivan, L.; Corke, P. Empirical Modelling of Rolling Shutter Effect. In Proceedings of the IEEE International Conference on Robotics and Automation, Hong Kong, China, 31 May–7 June 2014.
48. Achtelik, M.; Achtelik, M.; Weiss, S.; Siegwar, R. Onboard IMU and Monocular Vision Based Control for MAVs in Unknown In- and Outdoor Environments. In Proceedings of the IEEE International Conference on Robotics and Automation, Shanghai, China, 9–13 May 2011.
49. Leica TS30. Available online: http://www.leica-geosystems.com/en/Engineering-Monitoring-TPS-Leica-TS30_77093.htm (accessed on 27 August 2015).
50. Bachrach, A.G. Autonomous Flight in Unstructured and Unknown Indoor Environments. Master's Thesis, MIT, Cambridge, MA, USA, 2009.
51. Abbeel, P.; Coates, A.; Montemerlo, M.; Ng, A.Y.; Thrun, S. Discriminative Training of Kalman Filters. In Proceedings of the Robotics: Science and Systems, Cambridge, MA, USA, 8–11 June 2005; pp. 289–296.

Article

Dual-Stack Single-Radio Communication Architecture for UAV Acting As a Mobile Node to Collect Data in WSNs

Ali Sayyed [1], Gustavo Medeiros de Araújo [2], João Paulo Bodanese [1,3] and Leandro Buss Becker [1,*]

[1] Department of Automation and Systems, Federal University of Santa Catarina, Florianópolis 88040-900, Brazil;
   E-Mails: ali.sayyed@hotmail.com (A.S.); joao.bodanese@gmail.com (J.P.B.)
[2] Campus Araranguá, Federal University of Santa Catarina, Araranguá 88906-072, Brazil; E-Mail:
   gustavo.araujo@ufsc.br
[3] Instituto Senai de Sistemas Embarcados, Florianópolis 88040-900, Brazil
* E-Mail: leandro.becker@ufsc.br; Tel.: +55-4837217606.

Academic Editor: Felipe Gonzalez Toro
Received: 16 July 2015 / Accepted: 7 September 2015 / Published: 16 September 2015

Abstract: The use of mobile nodes to collect data in a Wireless Sensor Network (WSN) has gained special attention over the last years. Some researchers explore the use of Unmanned Aerial Vehicles (UAVs) as mobile node for such data-collection purposes. Analyzing these works, it is apparent that mobile nodes used in such scenarios are typically equipped with at least two different radio interfaces. The present work presents a Dual-Stack Single-Radio Communication Architecture (DSSRCA), which allows a UAV to communicate in a bidirectional manner with a WSN and a Sink node. The proposed architecture was specifically designed to support different network QoS requirements, such as best-effort and more reliable communications, attending both UAV-to-WSN and UAV-to-Sink communications needs. DSSRCA was implemented and tested on a real UAV, as detailed in this paper. This paper also includes a simulation analysis that addresses bandwidth consumption in an environmental monitoring application scenario. It includes an analysis of the data gathering rate that can be achieved considering different UAV flight speeds. Obtained results show the viability of using a single radio transmitter for collecting data from the WSN and forwarding such data to the Sink node.

Keywords: mobile nodes; unmanned aerial vehicles; communication architecture; data collection; wireless sensor networks

1. Introduction

Over the last 15 years, Wireless Sensor Networks (WSNs) have been studied extensively and applied in a large number of real world applications, ranging from environment monitoring (e.g., pollution, agriculture, volcanoes, structures and buildings health), to event detection (e.g., intrusions, fire and flood emergencies) and target tracking (e.g., surveillance). WSNs usually generate a large amount of data by sensing their environment and detecting events, that must be managed and forwarded to a Sink node in an efficient and reliable manner.

The introduction of mobility in WSN has attracted significant interest in recent years [1]. Mobility can be introduced in any component of a sensor network, including the regular sensor nodes, relay nodes (if any), data collectors, sink or any combination of these. From nodes deployment and localization to data routing and dissemination, mobility plays a key role in almost every operation of sensor networks. For instance, a mobile node can visit other nodes in the network and collect data directly through single-hop transmissions [2]. Similarly, a mobile node can move around the sensor network and collect messages from sensors, buffer them, and then transfer them to the Sink [3]. This significantly reduces not only message collisions and losses, but also minimizes the burden of data forwarding by nodes, and as a result spreads the energy consumption more uniformly throughout the network [1]. Mobility can also improve WSNs in terms of extending lifetime [4], coverage [5], reliability [6], and channel capacity [7]. A detailed survey on the usage and advantages of mobility in different phases of the WSN operation is given in [8].

Unmanned Aerial Vehicles (UAVs) are increasingly being used in WSNs, due to the cost-effective wireless communication and surveillance capabilities they provide [9]. UAVs play a prominent role in a variety of real world applications like homeland defense and security, natural disaster recovery, real-time surveillance, among others. UAVs and WSNs can cooperate in a number of different ways to improve network efficiency and performance [10–14]. Benefits can be identified in both directions. For instance, UAVs that have more resources for sensing and computing, can move to particular locations to perform more complicated missions. In addition, the WSN could provide extended sensory capabilities to guide the navigation of UAVs.

In order to support simultaneous interaction with WSN and Sink, the UAV must have at least two types of communication protocols. As sensor nodes have constrained resources, their communications protocol should be efficient and have a small footprint. This holds for communication among the WSN nodes as well as WSN-to-UAV and UAV-to-WSN. When considering communication between the UAV and the Sink, more robust protocols can be adopted, given that both sides have much more computing power when compared to the nodes in a WSN. In order to support UAV-to-Sink communication, there are two possibilities: either the UAV carries a second radio (with different physical layer) or it uses the same radio for both purposes. Most related works analyzed so far make use of two or even three radios for such purposes, as further discussed.

In this paper, we detail the Dual-Stack Single-Radio Communication Architecture (DSSRCA), which was first presented in [15]. DSSRCA allows a mobile node (e.g., a UAV) to communicate in a bidirectional manner with a WSN and a Sink node using the same radio. This architecture is composed of a dual stack protocol designed specifically to meet the different network QoS requirements present in UAV-to-WSN and UAV-to-Sink communications, such as best-effort and more reliable communication. DSSRCA suits situations where the mobile node acts as data collector for the WSN.

Obtained results confirm the viability of using a single radio transmitter for collecting data from the WSN and flushing such data back to a Sink node, or even for sending telemetry data to a Base Station. The designed prototype allowed a real UAV to collect data from heterogeneous ground sensor devices using best-effort communication and send data back to the Sink using reliable communication. This paper also includes a simulation analysis that observes the bandwidth consumption in an environmental monitoring application scenario.

The remaining parts of this paper are organized as follows. Section 2 presents and discusses some related works, including different schemes proposed for using mobile data collectors in WSNs. In Section 3, the proposed Dual-Stack Single-Radio Communication Architecture (DSSRCA) is explained, along with the hardware and software components, and some experimental results with a real UAV. Next, in Section 4, a detailed simulation study related to bandwidth consumption in an environmental monitoring application scenario is presented and discussed. Finally, conclusions and future work directions are addressed in Section 5.

2. Related Works

A considerable number of approaches exploiting mobility for data collection in WSNs have been proposed in recent years. Mobility can be introduced in any component of the WSN, including regular sensor nodes, relay nodes (if any), data collectors, sink or any combination of these. For instance, in the case of *mobile sensor nodes*, as shown in [16,17], animals / people with attached sensors, moving in the network field, not only generate their own data, but also carry and forward data coming from other nodes that they have been previously in contact with. These mobile sensors eventually transfer all their data when in contact with the sink node. Similarly in case of *mobile sink*, as shown in [4,18,19], the overall energy consumption of the network is minimized by changing the position of the sink node which collects data from sensor nodes either directly (*i.e.*, by visiting each of them) or indirectly (*i.e.*, through relays or other nodes). Finally, *mobile data collectors*, which are neither producers nor consumers of messages in a sensor network, perform the specific tasks of collecting data from sensor nodes when in their coverage range and eventually passing it in to the Sink. In this paper we focus mainly on approaches based on mobile data collectors (MDCs), because in our application scenario the UAV acts as a MDC that moves in the network area with predictable / deterministic mobility patterns while collecting data from the underlying ground nodes.

2.1. Data Collection Schemes Based on Mobile Data Collector

In this case, data is buffered at the source nodes until a mobile data collector visits and collects data over a single-hop wireless transmission. Existing proposals in this category can be further classified in three categories according to the mobility patterns of the mobile data collector [20,21]. *i.e.*, *Random mobility*, *Predictable mobility* and *Controlled mobility*.

In the rest of this sub section, we present a brief overview of the state-of-the-art in this regard. Interested readers should refer to [21] for a detailed survey.

2.1.1. Mobile Data Collectors with Random Mobility

The concept of mobile data collector was first introduced in [3], in which MDCs are referred to as Data Mules. In this proposal, generated data is buffered at source sensors until MDCs move randomly (using a Markov model based on a two-dimensional random walk) and collect data opportunistically from sensors in their direct communication range. The MDCs then carry the collected data and forward it to a set of access points. In this case, static nodes periodically wake up and listen for advertisements from mobile node for a short time. If it does not hear any beacon message from a mobile node, it return to sleep. Otherwise it starts transferring data to the mobile node.

An efficient Data Driven Routing Protocol (DDRP) was proposed in [22] where mobile sinks periodically broadcast beacon messages to their one-hop neighbors as they move around the network area. Each node maintains a variable called Dist2Sink (Distance to Sink), which stores the shortest number of hops to the sink. Beyond a certain value, Dist2Sink is equal to infinity, which means the node has no route to the sink. When a node's Dist2Sink variable is equal to infinity, it waits for a certain amount of time to overhear about a new valid route to the sink. However, if unsuccessful within a time bound, nodes adopt a random-walk strategy until the packet finds a valid route to the sink or is timed out and dropped. The main problem with DDRP is that with each movement of the mobile sink, subsequent topological changes occur that are propagated in the entire network in the form of the overhearing mechanism.

Safdar *et al.* proposed an improvement in [23] for Routing Protocol for Low Power and Lossy Networks (RPL) to efficiently support sink mobility. In case of basic RPL, mobile sink frequently broadcasts its presence which is propagated in the entire sensor field and each node in the network makes a Directed Acyclic Graph (DAG) to the sink. This promotes extensive network traffic and high energy consumption with each movement of the sink. However, in [23] the sink's topological update is restricted to a confined zone, consisting of a few hops around the sink. Hence, nodes within the confined zone can immediately send data to the sink using the known DAGs. The size of confined zones is increased for low sink mobility and decreased for high sink mobility. Nodes outside the confined zones implement on demand sink discovery. With each unsuccessful attempt, the zone size for broadcasting the route discovery request is increased. This procedure is repeated until a network wide broadcast for route discovery is initiated.

2.1.2. Mobile Data Collectors with Predictable or Deterministic Mobility

In this case, the static and mobile nodes agree on a specific time at which the data transfer may initiate. Mobile nodes follow a very strict schedule and other nodes know exactly when mobile nodes will be in their communication range. For instance, in [24], mobile nodes are assumed to be on board of public transportation shuttles that visit sensor nodes according to a schedule. In this way the sensor nodes calculate the exact active time and wake up accordingly to transfer data.

In [25] a scheme called Multiple Enhanced Specified-deployed Sub-sinks (MESS) for WSNs is proposed for data collection using multiple sub-sinks. The sub-sinks are considered as enhanced nodes having more resources and deployed at equal distances along the accessible path, creating a strip in the sensing field and providing a set of meeting points with the mobile sink. Each sub-sink working as an access-point to the mobile sink notifies the underlined network segment about the service it is offering.

Similarly in [26] Oliveira *et al.* proposed a greedy algorithm called Wireless HIgh SPEed Routing (Whisper) to forward data to a high speed sink. The scheme is based on the assumption that all nodes know their own locations, their neighbors' locations, and the path and displacement of the mobile sink. In this case, the sink, moving at high speed, does not stay permanently at the location of interest, and therefore refreshes its future location based on its speed and trajectory. The interested sensor nodes cannot directly send message to the fast moving sink and therefore forward their data towards the estimated meeting point with the sink. The meeting point is estimated on the basis of various delays in message transmission together with the node's own location, the neighbors' locations, and the estimated sink location.

2.1.3. Mobile Data Collectors with Controlled Mobility

In some WSNs data collection is performed only when events of interest occurs. In this case, controlled mobility is desired since it allows triggering mobile nodes on demand to visit sensors and avoiding sensor buffer overflows. For instance, in [9] the problem is proved to be NP-complete and a heuristic solution called Earliest Deadline First (EDF) and its two variants are presented. With the EDF solution, the next node to be visited by the mobile node is chosen as the one that has the earliest buffer overflow deadline.

Banerjee *et al.* proposed a scheme in [27], in which multiple resource-rich mobile cluster heads (CHs) are employed to prolong network lifetime and ensure delay requirements of real time applications are met. In this case, mobile cluster heads cooperatively collect data from different network segments and deliver it to a central Sink. All CHs move in a manner ensuring their connectivity with the base station while covering most interested areas at the same time. The authors propose three different movement strategies for CHs in order to reduce multi-hop communication and enhance network lifetime.

Similarly, authors in [7] used different controlled mobility approaches to optimize the speed of mobile nodes while collecting data in a sensor network. The first approach is called stop and communicate because as the mobile node enters the communication range of a static node that has some data to send, it stops there until all buffered data has been collected. The duration of the stop depends on the data generation rate of the source node. The second way to optimize speed is called adaptive speed control in which the speed of mobile node is changed according to the number of encountered nodes and the percentage of collected data with respect to buffered messages. Different groups of nodes are made according to the amount of data collected from them (low, medium or high). The mobile node moves slowly in a group with a low collected data percentage, while it moves faster when it is in the communication range of nodes with a high percentage of collected data.

2.2. Preliminary Conclusions

There are numerous other mobility based data collection approaches that target one aspect of data collection or another and have different pros and cons. After analyzing all the state-of-the-art solutions regarding mobile data collectors it is possible to conclude that:

- Approaches using a single radio in the mobile data collector only perform data collection, *i.e.*, the mobile node does not communicate with the sink node while on the quest for data collection.

– Approaches targeting simultaneous communication with sensor nodes and sink or base station use at least two different radios on the mobile data collector.

This motivated us to design an architecture that can communicate with both the WSN and the sink node using the same radio. Our work is different in the way that it focuses on using a dual stack that allows a mobile node, in our case a UAV, with deterministic mobility patterns, to collect data from heterogeneous ground sensor nodes and send telemetry or collected data back to the Sink node using the same communication hardware and without compromising the data collection rate. The details about this architecture are presented in the next section.

3. Dual-Stack Single-Radio Communication Architecture (DSSRCA)

The proposed Dual-Stack Single-Radio Communication Architecture (DSSRCA) aims to provide a flexible communication infrastructure that supports UAV-to-WSN and UAV-to-Sink communications using a single radio. It is able to commute between best-effort and reliable communication strategies according to the different network QoS requirements present in UAV-to-WSN and UAV-to-Sink communications. An overview of DSSRCA organization in terms of partitions, layers, and sub-components is illustrated in Figure 1.

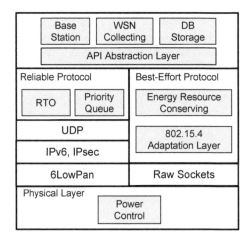

Figure 1. Dual-Stack Single-Radio Communication Architecture (DSSRCA) overview: customized transport layer on top of 6LoWPAN (**left**) and the small-footprint Wireless Sensor Network (WSN) protocol (**right**).

The top layer in Figure 1 refers to the application level, which could be either at the Sink node or at the mobile data collector (MDC). On the left, it is the customized transport layer on top of 6LoWPAN that constitutes the reliable protocol used in UAV-to-Sink communication. Finally, the small-footprint best-effort protocol used for UAV-to-WSN communication is shown on the right side.

The following subsections detail the two main partitions that compose DSSRCA: Reliable and Best-Effort protocols. Additionally, details regarding the implementation and some experimental results are also presented.

3.1. UAV-to-Sink Communication

DSSRCA provides a partition for performing a reliable communication between the UAV and the Sink. This layer makes use of 6LoWPAN [28] in conjunction with our customized transport layer, which aims at improving the reliability in terms of message transmissions.

The proposed protocol also allows to distinguish messages among three different classes of priorities: (i) low; (ii) normal; and (iii) high. The messages are stored in a priority queue implemented using binary heaps. For each priority class there is a maximum number of re-transmission attempts associated. High priority messages have an unlimited number of re-transmission attempts. Messages with medium priority have n attempts (typically 5), and messages with low priority have no re-transmissions.

To improve reliability, the proposed approach provides basic mechanisms for message re-transmission, duplicated packet detection, and error-checked delivery of a stream of octets between applications running on both the UAV and the Sink. It follows the philosophy of connection-oriented protocols, meaning that applications must first establish a connection with each other before they send data in full-duplex mode.

In order to guarantee proper data delivery, the protocol uses a technique known as positive acknowledgment, in which the sender transmits the message with its own sequence number, and waits for the receiver to acknowledge the message reception by sending an ACK message with the sequence number received. At the moment the sender transmits the data, a timeout mechanism is initiated to wait for the acknowledgment. If the timer expires before the message has been acknowledged, the sender re-transmits the message.

A very important point to ensure the data transmission efficiency, is the tuning of the re-transmission timer (RTO). If the value is too high, the protocol will be slow in reacting to a segment loss, thus reducing the throughput. On the other hand, if the value is too small, it will result in unnecessary re-transmissions that lead to exaggerated consumption of network resources. The goal is to find an RTO value that balances throughput degradation between both cases.

The proposed solution uses a dynamic algorithm to continuously adjust the timeout value based on monitoring the packet transmission delay. The round trip time variance estimation algorithm introduced by [29] was chosen and frequently applied in the TCP congestion control. The algorithm maintains a *SmoothedRTT* variable to represent the best estimated timeout for the packet round-trip time. When a message is sent, a timer is triggered to measure how long it takes to get the confirmation and stores the value in the *MeasuredRTT* variable. Then, the algorithm updates *SmoothedRTT* as shown in Equation (1).

$$SmoothedRTT = \alpha.SmoothedRTT + (1 - \alpha).MeasuredRTT \tag{1}$$

where α is a smoothing factor with a recommended value of 0.9: in each measure, *SmoothedRTT* is updated using 10% from the new estimate and 90% from the previous one.

In order to respond to large fluctuations in the round-trip times, RFC793 recommends using the re-transmission timeout value (RTO), as defined in Equation (2):

$$RTO = min[UBOUND, max[LBOUND, \beta.SmoothedRTT]] \tag{2}$$

where β is a delay variance factor with a value of 1.3. The constant UBOUND is an upper bound on the timeout with a minimum value of 100 ms. The LBOUND is a lower bound on the timeout with a value of 30 ms. The values of β, LBOUND and UBOUND were obtained by conducting empirical tests on the UAV-to-BS communication. This mechanism can lead to a poor performance since the sender does not transmit any message until the acknowledgment of the last message is received. To solve this problem, a new solution based in the TCP sliding window concept [30] is proposed in this work. Sliding window is the number of messages the sender transmits before waiting for an acknowledgment signal from the receiver. For every acknowledgment received, the sender moves the window one position to the right and sends the next message. If the receiver receives a message outside the range of the window, the message is discarded. However, if the message is in the range, but out of order, it is stored. For each message, the sender starts a timeout as previously described.

When transmission losses occur, the proposed algorithm does not reduce the window size as is done by the traditional TCP congestion control algorithms. This is justified since the message loss is related to transmission errors and not to network congestion. Reducing the window size decreases the data delivery rate, and therefore the window size should be maintained or even increased. This technique is discussed in different studies, such as [31], in an effort to adapt TCP to wireless networks or other high error rate mediums.

The multiple segment losses in the window can cause a degradation effect in the data rate. Traditional TCP implementations use cumulative acknowledgment scheme, in which the receiver sends an acknowledgment meaning that the receiver has received all data preceding the acknowledged sequence number. This mechanism forces the transmitter to wait for the timeout to find out that the message has been lost, which causes a flow decrease. The proposed TCP SACK (Selective Acknowledgment) [32] increases TCP performance in networks with high loss rate. The SACK aims to recover multiple lost segments within one RTT. To achieve this, the receiver provides enough information about the segments lost in its sliding window. With this information, the sender knows exactly which segments were lost and re-transmits them before the timeout expires.

For implementing the sliding window algorithm with SACK, the sender maintains a set of sequence numbers corresponding to frames it is permitted to send, which are represented by two variables: S_L (sender lower limit), *i.e.*, the number of the oldest message sent but not acknowledged yet; and S_U (sender upper limit), *i.e.*, the number of the next message to send. Similarly, the receiver also maintains a receiving window corresponding to the set of messages it is permitted to accept, which are also represented by two variables: R_L (receiver lower limit), *i.e.*, the lowest numbered message that the receiver is willing to accept; and R_U (receiver upper limit), *i.e.*, one more than the highest numbered message the receiver is willing to accept. The implemented sliding window algorithm with SACK is presented in Algorithm 1.

Algorithm 1 Sliding Window Algorithm with SACK

1: Transmit all messages in the sender's window (S_L to S_U-1);
2: **if** *Message arrives in the window and in correct sequence (R_L)* **then**
3:     Receiver acknowledges the message
4:     Receiver copy R_L to the application and advances its window
5: **else if** *Message arrives in the window but is out of sequence (sequence $> R_L$ and $<= R_U$)* **then**
6:     Receiver acknowledges the message for the highest message correctly received
7:     Attach a SACK list with the message(s) number(s) missing
8: **end if**
9:
10: **if** *Sender receives an ACK for a message within its window* **then**
11:     Mark this message as correctly sent and received
12:     If message number is S_L then increment S_L and S_U and transmit S_U-1
13: **else if** *Sender receives an ACK with a SACK list* **then**
14:     Re-transmit the correspondent messages contained in the SACK list and restart the timeout
15: **end if**
16:
17: **if** *Timeout Occur* **then**
18:     Re-transmit the corresponding message
19: **end if**

3.2. UAV-to-Sensor Communication

The second partition of DSSRCA aims at providing an efficient and small-footprint protocol to support the communication between the UAV and the WSN. To reduce the protocol footprint it is necessary to bypass the 6LoWPAN stack, which can be done by using Linux *raw sockets*.

Such protocol is based on the frame format shown in Figure 2. *Sensor ID* is a number that identifies the message sender. *Data Type* specifies what type of phenomenon the sensor is observing. *Time Stamp* is used to identify the sensors local time of observation. Finally, the payload is the sensed data. The sensor header requires 6 bytes of the 802.15.4 payload size using short address mode, leaving 110 bytes for the data payload.

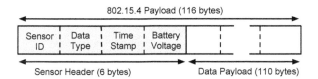

Figure 2. Message structure for Unmanned Aerial Vehicle (UAV)-to-WSN communication.

To collect data from the WSN, it is suggested that the network is organized in clusters [33], *i.e.*, there are cluster head (CH) nodes which collect and aggregate data from other sensor nodes in range. At the proper moment, the UAV triggers the CH with a *request message* for transmitting its data. In fact, this moment depends on the application under consideration.

Whenever a CH receives the *request message* and has data to transmit, it does so using unslotted CSMA-CA in the frame format previously described. When the UAV receives a message, an acknowledgment frame (*ack*), at the MAC level, is transmitted to confirm receipt. The CH continues to transmit until it has no more data or up to the moment it stops receiving the *ack*. Three attempts of re-transmissions are performed until the CH gives up transmitting. If this happens, the CH must wait for a new *request message* to re-start the transmission.

An important detail about this protocol regards the need for a modification in the UAV transmission power if compared to UAV-to-Sink communication. At the maximum UAV transmission power of +19 dBm, the CH would receive the *request message* even at a distance of hundreds of meters away from the UAV. In this case, the sensor would try to perform transmissions but would fail to reach the UAV due to the long distance. To tackle this issue, a mechanism named *Power Control* in the PHY level was implemented to switch the transmission power of the UAV radio to a lower level (−2 dBm) when communicating with the WSN.

3.3. Hardware Prototype

In order to experiment with DSSRCA in real scenarios, a hardware prototype was developed and embedded in our UAV, as depicted in Figure 3. The core of the proposed solution is the MRF24J40MC transceiver (center-left part of the UAV), which implements an extended range version (and the standard version) of the IEEE 802.15.4 protocol. It was selected because the great majority of the WSNs are implemented using the IEEE 802.15.4 standard. It is a protocol suitable for connecting different devices at a short/mid-range, focusing on low-cost, low-speed (250 kbps), and low-power applications. The adopted radio transceiver contains a Power Amplifier (PA) and a Low Noise Amplifier (LNA) that allows communication ranges up to 1200 m. The adopted external antenna (at the center of the UAV) is a full wave +5 dBi antenna manufactured by Aristotle Enterprises. The antenna perpendicular, to the ground, is used for the 72 MHz RC.

Figure 3. UAV equipped with the communication prototype.

The transceiver is connected to the embedded computing platform through a 4-wire SPI interface. As a computing platform, a low-cost credit-card-sized Linux computer equipped with a 1 GHz super-scalar ARM Cortex-A8 processor was selected (Beaglebone black [34]). The Ångstrom Linux distribution [35] is used as the OS. This embedded platform, located in the center-right of the UAV, also interfaces with a GPS and additional sensors/actuators.

3.4. Experimental Evaluation

This section aims to present practical experiments that make it possible to validate the proposed architecture, including its main ideas and the developed hardware prototype. Different experimental setups were created, in order to separately evaluate the two partitions that compose DSSRCA: Best-effort and reliable communication protocols.

The experiments were performed using two different types of mobile data collectors (MDCs): (i) a remotely piloted UAV and (ii) a bicycle. The main reason of using a bicycle equipped with our communication prototype is safety, as long-range experiments-related to the reliable protocol-were performed on a long and busy seaside walkway. On the other hand, the tests of the best-effort communication were performed in an open field using both the UAV and the bicycle.

The physical layer settings and other hardware characteristics of the experimental setups are presented in Table 1.

Table 1. Physical layer settings and other aspects of the experiments.

Parameters	Sink Node/UAV	Sensor
Radio Mode	Microchip MRF24J40MC	MICAz MPR2400CA
Antenna	Full-wave dipole	Half-wave dipole
Tx Power	+19 dBm (UAV-BS) −2 dBm (UAV-WSN)	−1 dBm
Automatic HW Re-transmission	3	3
Channel	11	11

3.4.1. Best-Effort Protocol Experimentation

These experiment were performed using one wireless sensor (MICAz) deployed 70 cm above the ground level and a MDC. The MICAz sensor was configured to transmit 20 bytes of payload per message, with an unlimited number of messages in the buffer, whenever it was in the communication range of the MDC. Both, a remotely piloted UAV and a bicycle were used as MDC. Let us start by addressing the experiments using the UAV. The UAV was supposed to perform a rectilinear flight, approaching the sensor and departing from it at constant speed and altitude (see [36]).

Figure 4 shows the flight data collected in one of the experiments, with distance at the top, altitude in the middle, and speed at the bottom. While the distance (to the sensor) is calculated by the UAV, altitude and speed are obtained from the GPS. As one can see, the altitude and the speed presented significant variations (while the original intention was to keep it constant). This implied changing the distance of the UAV to the sensor in a non-linear manner, making it impossible to obtain a proper estimate of the message reception rate with respect to the distance.

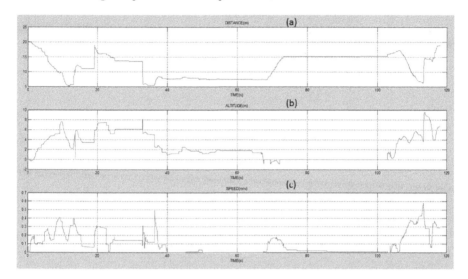

Figure 4. UAV experiment flight log with respect to (**a**) distance; (**b**) altitude; and (**c**) speed.

An interesting phenomenon to be observed is the action of the re-transmissions protocol at the sensor side. As stated before, it makes three re-transmissions attempts before stopping sending messages to the UAV. At this point, the sensor waits for a new *request message*, so that it can start transmitting again. Figure 5, presents the total number of messages received by the UAV during the 110 s of the experiment. The pie graph is used to divide the messages according to the number of re-transmissions necessary to deliver these messages. The results show that 91% are delivered in the first transmission attempt, which represents a very good transmission quality for the experimented distance. It is important to notice that 453 re-synchronization messages (*request message*) were transmitted by the UAV along the experiment. We consider this number to be higher than desired, so improving it will be the focus of further investigation.

These experiments allowed us to conclude that it is very difficult for an amateur pilot to perform a smooth flight trajectory. Only very experienced pilots or autonomous aircrafts would be able to achieve this goal, which would allow testing of the proposed communication prototype in a more suitable manner. Therefore, it was decided that further experiments use the bicycle.

In respect to the experiments with the bicycle, as previously mentioned, they make it easier for one to control both the bicycle path and its speed. The bicycle traveled on a rectilinear path so that it first approached the MICAz and then distanced itself away. The bicycle started at 125 m away from the sensor and stopped 125 m past it. The target speed was 7 m/s (see [37]).

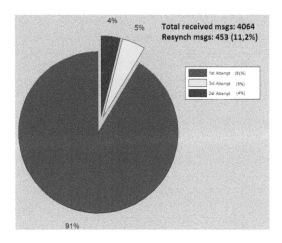

Figure 5. Messages received by the UAV along the experiment.

Figure 6 shows the number of messages collected by the bicycle with respect to the sensor distance. This analysis is interesting in this experiment because the bicycle moves at a constant speed, without stopping. As expected, the distance directly affects the link quality. When the distance between bicycle and sensor increases, the number of messages collected by the bicycle decreases. Therefore, this experiment helps evaluate the distance that is most suitable to start collecting data.

The graph in Figure 7 shows the latency of the messages transmitted by the sensor. The measurement consists in the time difference between the moment the packet is sent to the transceiver upon receiving the ACK at MAC level. The dark line in the graph denotes the polynomial trend. It is possible to observe that the average message latency is about 15 ms from 5 to 80 m. Moreover, the latency increases at distances greater than 80 m since the link quality decreases and more re-transmissions are necessary. This experiment is important in order to highlight the boundaries of a good data collection mechanism. With this information, one can tune the protocol to optimize data gathering performance.

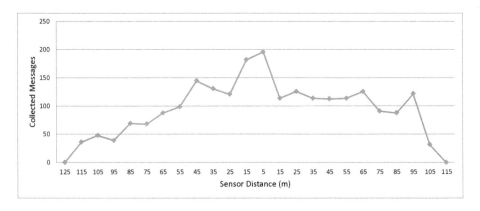

Figure 6. Histogram of messages collected by the bicycle.

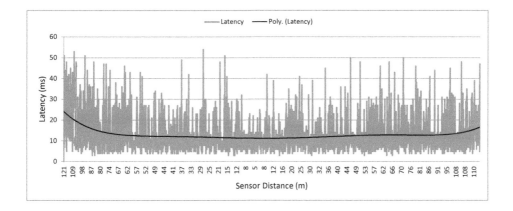

Figure 7. Sensor messages latency in the bicycle experiment.

3.4.2. Reliable Protocol Experimentation

As mentioned, these experiments were performed on a long seaside walkway. Besides natural obstacles such as poles and trees, it is in an urban area and the environment poses possible interference from other wireless technologies (e.g., WiFi)-which is good for testing the proposed re-transmission mechanism.

In this experiment, the MDC took about 9 min, at speeds between 4 and 8 m/s, to move 1400 m away from the sink and then return to it. It was configured to transmit one message per second. The message was composed of telemetry information with a payload of 32 bytes: 128 bits for the geographical coordinates and 128 bits for speeds north and west. The sink not only records the received telemetry data, but also transmits a ping message to the MDC to monitor the link.

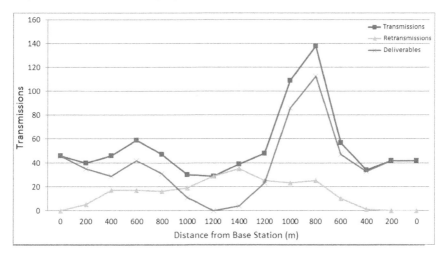

Figure 8. data collected from the reliable protocol evaluation.

During the experiment, 489 messages were transmitted by the MDC and received by the Sink. Figure 8 plots the three variables that are directly associated with the performance of the proposed re-transmission algorithm. The *Transmission* variable denotes the total number of messages that were transmitted, including those delivered and the re-transmissions attempts. The *Deliverables* variable is the number of messages successfully delivered and *Re-transmissions* is the number of failed re-transmissions that did not receive the ACK. The graph shows the average of these three variables at every 200 m.

It can be observed in the graph that in the first 900 m the number of delivered messages is considerably satisfactory. Then it started dropping until it reach its lowest point, contrasted with the increase in the re-transmissions (which reaches its highest level). When the MDC starts moving back there is a peak in the total number of delivered messages. This large number of messages includes those previously generated (and buffered), but that could not have been delivered due to the long distance from the sink.

It is important to note that the re-transmissions mentioned above are in the application layer, not in the link layer. The link layer re-transmissions can be performed automatically by the MRF24J40MC hardware using an *ack* at the MAC level. Such feature remained active during all of the experiments.

In order to measure the data rate, in another experiment we placed the MDC close to the sink and performed a 10 MB mission file download. This experiment was repeated several times using different block sizes. Whenever the 6LowPan receives a message (block) from the application to transmit (in this case at the Sink node), the message is typically divided into multiple fragments in order to fit the 127 bytes of the 802.15.4 maximum transmission unit (MTU). The MTU has 75 bytes available for payload in the first fragment and about 85 bytes for the remainder ones. A MAC-ACK is sent for any fragment correctly received in the destination node. If any single fragment is lost, it will not be re-transmitted, and all others already received will be discarded.

It can be observed in Figure 9 that the data rate initially increases, since using bigger message blocks imply less confirmation messages at the application level. It is possible to notice a slight drop in the data rate from 500 to 600 bytes before it saturates. This comes from increasing the number of required fragments from 6 directly to 8, and increasing the number of fragments enhances exponentialy the probability of transmission failures. The tendency for saturation in the long run with transmission blocks bigger than 500 bytes can be observed in the tendency curve (the thin line that is not in the legend). In addition, it can also be observed in the figure that the use of windows size higher than one has a negative effect on the data rate. This occurs because when the UAV tries to transmit the application-level acknowledgment, the Sink tries to transmit the next message that is allocated in the transmission window. Multiple occurrences of such events increases the collision probability, which occasionally results in transmission losses.

The experiment was repeated using a link with a 40% probability of loss. It is apparent in Figure 10 that the data rate has behavior opposite to that of the previous experiment. The more the message size increases, the smaller the data rate becomes. This happens because bigger message blocks increase the probability of losing a single frame, which results in discarding all other frames already received by the 6LoWPAN. The same drop from 500 bytes to 600 bytes observed in Figure 9 is again observed here.

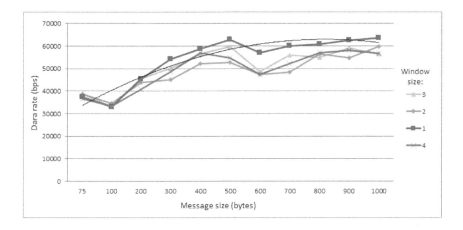

Figure 9. Data flow between mobile data collectors (MDC) and Sink.

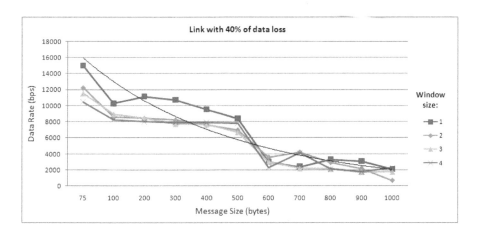

Figure 10. Data rate in the communication between the Sink and MDC, in a link with 40% loss probability.

It should also be observed that if there is no loss in the network, it is desirable for the application to send bigger blocks because fewer acknowledgment messages at the application level are required. However, when there are losses in the network, it is more advantageous for the application to send messages of a smaller size, because the loss of a single frame does not discard the frames already received.

In the next experiment, the message size was fixed at 75 bytes and the size of sliding window was changed with a link of 20% loss probability, in a network with 50 ms of latency. The results obtained are shown in Figure 11. It can be observed that increasing the window size up to the limit of seven, the data rate increases. Thus, it allows us to conclude that it is always advantageous to use a window size larger than one when there is latency in the network; otherwise, the number of collisions causes the data rate to decrease.

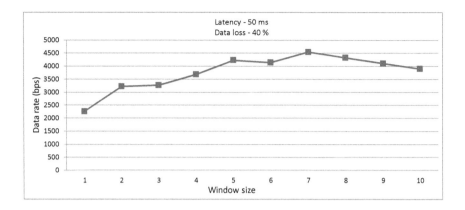

Figure 11. Data rate in the communication between the Sink and MDC, in a link with 40% loss probability and 50 ms network latency.

4. Extended Simulation Analysis

The simulation analysis presented in this section intends to validate one of DSSRCA's features, which is the fact that it uses only one radio for communication. Therefore, the experiment simulates a UAV acting as mobile data collector (MDC), which collects data from a WSN and then forwards it to a Sink node.

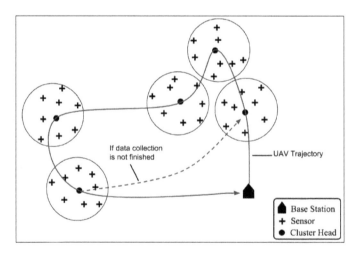

Figure 12. Simulation scenario: UAV flies over 45 nodes organized into five clusters.

It considers a practical example of an environmental monitoring system where sensors, organized into non-overlapping clusters, are deployed in a remote area. The sensor nodes sense their environment and save data in their buffers. Each sensor transmits data to its corresponding cluster head (CH), which in turn saves it and eventually send it to the UAV whenever it is in communication range. The cluster heads are responsible for coordinating the operation of the cluster members and also for communication with the UAV. A simulated environment of this scenario was developed in Omnet++ 4.41 with Inetmanet-2.0. The simulated environment was 1000 m × 1000 m. The altitude of UAV was

Sensors **2015**, *15*, 23376–23401

kept constant at 15 m above the ground. We used 45 sensor nodes, organized into five clusters on the basis of distance from each others. Such simulation scenario is depicted in Figure 12.

To perform this experiment the following reasonable assumptions were made:

- A Sink node knows the location and propagation model (communication range) of all the ground sensors.
- The UAV is equipped with GPS, so that it can calculate its current position.

The transmission range of each node, including the CH is approximately a circle with 90 m radius. The radio model used for all nodes including UAV is 802.15.4. The UAV can make use of the extended transmission power (+19 dBm) to send data (telemetry or collected) back to the Sink, while traversing the network area. Before starting data collection, the Sink generates a flight plan for UAV, as it has all the necessary information regarding locations and radio ranges of sensors. The length of each data collection lap is approx. 1650 m. It should be noted that when the UAV starts flying, it already knows what trajectory to follow. The UAV can always estimate the start and end points of each CH zone. It is also assumed that the UAV trajectory passes over the CH node according to the flight plan generated by the Sink. The mechanism of this initial flight plan generation and its optimization is not considered here and is out of the scope of this work. We also do not consider how the sensors within each cluster behaves and communicate with CH. We assume that the sensor nodes remain silent (go into sleep mode) whenever the UAV enters the communication range of a CH. These sensor nodes remain in sleep mode until the UAV goes out of the coverage zone of that particular CH. This helps to avoid message collisions and disturbance in CH-to-UAV communication. This sleep/wake-up coordination is simply achieved by cooperation between UAV and CH. Whenever a UAV is about to enter a CH coverage zone it announces its speed while contacting the CH. Knowing his coverage area, the CH can easily calculate the time, the UAV is about to spend in communication range and instruct the sensors to go to sleep mode for at least that duration.

After the generation of the flight plan, the UAV begins traversing the network area and starts collecting data. As the UAV knows the location of each CH, when it enters the communication range of CH1, it asks CH1 to start sending its data. In case, the UAV does not know the location of each CH, a three-way handshake could be used for identifying the underlying CH.

In this experiment, the UAV was tested at different speeds, starting from the Sink on a pre-defined path, covering a total distance of 1650 m and returning back to the Sink again. In first case, the UAV is configured to only collect data from ground sensors and does not communicate with the Sink at all. In later cases, the UAV is configured to send one message of 70 bytes periodically after 200 ms, 500 ms, 1 s, 2 s, 3 s, 4 s, and 5 s to the Sink using its extended transmission power. The message sent by the UAV could be a telemetry packet or data collected from ground sensors.

All along the path of one complete lap of 1650 m at speeds of 5, 10, and 15 m/s, different numbers of messages were transmitted by UAV to the Sink, depending on the size of period the UAV used (e.g., 200 ms, 500 ms, 1 s, 2 s, 3 s, 4 s and 5 s) which we call the **Flush Back Period**. In this case, we present in Figures 13–15, the averages of the *transmitted*, *delivered*, and *re-transmitted* variables. The *transmitted* denotes the total number of messages that were transmitted, including delivered and re-transmissions attempts. The *delivered* variable is the number of messages successfully delivered and *re-transmitted* is the number of failed re-transmissions that did not receive the ACK.

Similarly, the re-transmission rate of telemetry data (sent by UAV to BS), with respect to the speed of UAV and Flush Back Period, is shown in Figure 16.

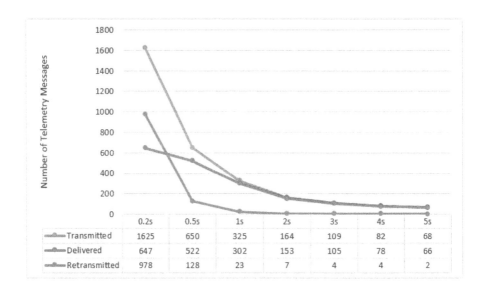

Figure 13. UAV to BS Communication at 5 m/s with respect to different Flush Back Periods.

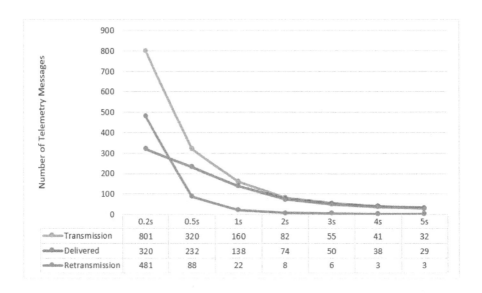

Figure 14. UAV to BS Communication at 10 m/s with respect to different Flush Back Periods.

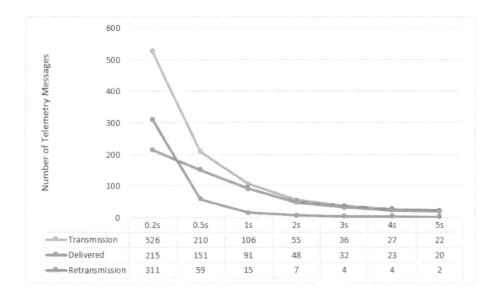

Figure 15. UAV to BS Communication at 15 m/s with respect to different Flush Back Periods.

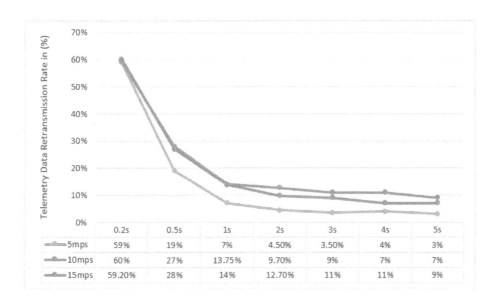

Figure 16. Re-transmission Rate of UAV to BS Communication for different UAV speed and Flush Back Period.

It can be noted here that the successful delivery rate (from UAV to BS) increases with decreasing the UAV Speed and increasing Flush Back Period. The majority of the re-transmission attempts are in areas where the UAV is furthest from the BS. This behavior is also confirmed by our practical experiment shown in Figure 6. It must be noted that we do not consider how cluster members communicate with

CH, although the communication between cluster members and CH is very important and it can affect the communication between UAV and CH. As described before, we use a simple and intelligent way for the CH, knowing the UAV speed and hence its approximated contact period, to instruct all non CH nodes to remain in sleep mode until the UAV is gone.

To measure the goodput, we repeated the simulation experiment several times with different UAV speeds and with a different Flush Back Period for one complete lap of 1650 m. It is assumed that each of the cluster Heads has infinite data to transmit to the UAV. In the first case, the UAV is configured to only collect data from the underground sensors and not to communicate with the Sink at all. In later cases, the UAV is configured to send one message of 70 bytes periodically after 200 ms, 500 ms, 1 s, 2 s, 3 s, 4 s, and 5 s to the Sink.

A very slight increase in the network goodput can be seen in Figure 17, when the UAV also sends data back to the Sink. This is due to the additional messages the UAV sends to Sink. In the case of short Flush Back Periods, the data rate is slightly decreased because the UAV spends more time sending messages to the Sink. It can be seen here that the data rate with and without sending telemetry data is roughly same. This shows the viability of using a single transmitter for a mobile data collector in a WSN. Furthermore it can also be noticed that the larger the distance or gap between clusters of sensors, the higher the goodput. The gap between sensor clusters, in which the UAV is not busy communicating with the underlying CH, can be used to flush collected data back to the Sink, and hence increase the goodput. In this way, an intelligent mechanism can also be implemented to vary the Flush Back Period so as to use the available bandwidth wisely.

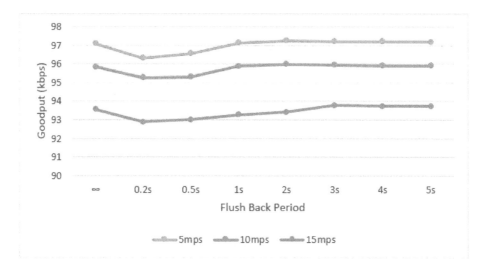

Figure 17. Network goodput for different UAV speeds and Flush Back Periods.

It should be noted that the UAV and Sink talk on a communication channel other than the one used for communication between UAV and sensor network. This helps reduce the number of message collisions and disturbance in other part of the sensor network, as the UAV uses increased power for communicating with the sink.

In our simulation experiment, we assumed that the UAV can estimate its current location along its trajectory and can calculate when it will enter communications range of each CH while traversing the network area. This knowledge can be exploited by configuring the UAV to flush back its collected data when it is not in the range of any underlying CH. In this case, the UAV does not simply send telemetry data periodically, but uses full-fledged transmission to send back collected data to the Sink. The same simulation was repeated with this configuration for the UAV using Algorithm 2. In this case, the Network Goodput increases to 158.5 kbps when the UAV moves at 5 m/s, to 156.16 kbps when the UAV moves at 10 m/s and to 153.9 kbps when the UAV moves at 15 m/s.

Algorithm 2 Full-Fledged Transmission from UAV to BS when free (not in the range of any cluster head)

1: **procedure** CONTINUOUSLY ESTIMATE UAV CURRENT LOCATION
2:    **while** *UAVCurrentLocation* does not lies in a CH Range **do**
3:       Send collected data back to BS
4:    **end while**
5: **end procedure**

A marginal increase in the bit rate can be seen here. This is due to the fact that in this case the UAV is busy communicating throughout its trajectory. In this particular example, more than 75% of the collected data is sent back to the Sink before completing an entire round. As pointed out earlier, the larger the gap between sensor clusters, the more spare time the UAV has, and the more data it will send back to the Sink. In dense sensor networks or WSNs in which the sensors clusters are overlapped or nearly overlapped, we would not see this significant increase in data rate, because the UAV would find very little time to flush data back to the BS. Nevertheless, in this case, the UAV could still communicate with the BS using an appropriate flush back period.

5. Conclusions and Future Work

In this paper, we presented a dual stack single radio architecture that can be used in mobile sinks, specifically an Unmanned Aerial Vehicle (UAV), to address the problem of data collection in static, ground, wireless sensor nodes. We discussed how a low cost, dual stack protocol can be used by a short range UAV (or any other mobile device) to collect data from a WSN. Moreover, the viability of using a single radio transmitter for simultaneously collecting data from the WSN and communicating with the Sink node or Base Station was shown. Our architecture design allows the UAV to collect data from heterogeneous ground sensor devices and send telemetry or collected data back to the Sink using the same communication hardware and without compromising the data rate. This prototype design has clearly established the feasibility of using a single communication hardware for mobile nodes for data gathering in sensor networks.

This paper also detailed the proposed communication protocol, which is capable of dealing with two different requirements: a best effort communication between the UAV and the WSN and a more reliable communication to be used between the UAV and Sink. This paper presented the successful use of the proposed architecture within an environmental monitoring application scenario. It can also be observed that the proposed dual stack architecture is not only valid for the UAV, it is also generic enough to be used for ground mobile nodes as well.

In this work, we did not focus on how sensor nodes communicate with the CH, although in a real application this needs to be configured in a way that sensor-to-CH communication does not collide with CH-to-UAV communication. Different sleep/wake-up pattern can be used for such purposes. It is also possible to exploit cooperation (information sharing) between UAV and ground nodes to avoid messages collisions. Similarly, it is also possible for static nodes, particularly the CH, to learn and predict the mobility pattern of UAV to further enhance UAV detection. This will not only decrease message collisions, but also enhance network lifetime and allow more accurate sleep/wake-up patterns for sensor nodes. This along with effect of single radio transmitter on the lifetime of network needs to be further addressed.

Future work also includes removing the UAV remote pilot by designing and implementing a dynamic speed control for the UAV, in which speed is adaptively and autonomously adjusted according to the application needs. Similarly, a dynamic Flush Back Period is also under consideration so as to use available bandwidth wisely. Finally, additional practical experimentation alternating best-effort and reliable transmissions should be performed.

Acknowledgments: The authors would like to thank the Brazilian funding agencies CAPES, CNPq, and FINEP. Special thanks also goes to our project partners. Carlos Pereira and Ivan Muller, from project FINEP E3/SA-WH, who helped with the wireless nodes. Guilherme Raffo, Fernando S. Gonçalves, Patrick J. Pereira, Richard Andrade, and Vinicius D. Woyakewicz, from Provant team, who helped in the UAV construction and in the experimental flights.

Author Contributions: The proposed communication architecture was developed within the scope of the MSc Thesis from João Paulo Bodanese, advised by Leandro Buss Becker and Gustavo Medeiros de Araújo. Ali Sayyed is under the same advisors that researchs the use of UAVs as a mobile node to collect data in WSNs. All authors contributed equally in the formulation and preparation of this manuscript.

Conflicts of Interest: The authors declare no conflicts of interest.

References

1. Wang, G.; Cao, G.; La Porta, T.; Zhang, W. Sensor relocation in mobile sensor networks. In Proceedings of the INFOCOM 2005 24th Annual Joint Conference of the IEEE Computer and Communications Societies, Miami, FL, USA, 13–17 March 2005; Volume 4, pp. 2302–2312.

2. Chatzigiannakis, I.; Kinalis, A.; Nikoletseas, S. Sink Mobility Protocols for Data Collection in Wireless Sensor Networks. In Proceedings of the 4th ACM International Workshop on Mobility Management and Wireless Access, Torremolinos, Spain, 2–6 October 2006; pp. 52–59.

3. Shah, R.; Roy, S.; Jain, S.; Brunette, W. Data MULEs: Modeling a three-tier architecture for sparse sensor networks. In Proceedings of the First IEEE International Workshop on Sensor Network Protocols and Applications, Anchorage, AK, USA, 11 May 2003; pp. 30–41.

4. Gandham, S.; Dawande, M.; Prakash, R.; Venkatesan, S. Energy efficient schemes for wireless sensor networks with multiple mobile base stations. In Proceedings of the 2003 GLOBECOM '03. IEEE Global Telecommunications Conference, San Francisco, CA, USA, 1–5 December 2003; Volume 1, pp. 377–381.

5. Liu, B.; Brass, P.; Dousse, O.; Nain, P.; Towsley, D. Mobility Improves Coverage of Sensor Networks. In Proceedings of the 6th ACM International Symposium on Mobile Ad Hoc Networking and Computing, Urbana-Champaign, IL, USA, 25–28 May 2005; pp. 300–308.

6. Anastasi, G.; Conti, M.; di Francesco, M. Reliable and Energy-Efficient Data Collection in Sparse Sensor Networks with Mobile Elements. *Perform. Eval.* **2009**, *66*, 791–810.

7. Kansal, A.; Somasundara, A.A.; Jea, D.D.; Srivastava, M.B.; Estrin, D. Intelligent Fluid Infrastructure for Embedded Networks. In Proceedings of the 2nd International Conference on Mobile Systems, Applications, and Services, Boston, MA, USA, 6–9 June 2004; pp. 111–124.

8. Sayyed, A.; Becker, L.B. A Survey on Data Collection in Mobile Wireless Sensor Networks (MWSNs). In *Cooperative Robots and Sensor Networks 2015*; Koubaa, A., Martínez-de Dios, J.R., Eds.; Springer International Publishing: Gewerbestrasse, Switzerland, 2015; Volume 604, pp. 257–278.

9. Somasundara, A.A.; Ramamoorthy, A.; Srivastava, M.B. Mobile element scheduling for efficient data collection in wireless sensor networks with dynamic deadlines. In Proceedings of the 25th IEEE International Real-Time Systems Symposium, Lisbon, Portugal, 5–8 December 2004; pp. 296–305.

10. Martinez-de Dios, J.; Lferd, K.; de San Bernabé, A.; Núñez, G.; Torres-González, A.; Ollero, A. Cooperation between UAS and wireless sensor networks for efficient data collection in large environments. *J. Intell. Robot. Syst.* **2013**, *70*, 491–508.

11. Culler, D.; Estrin, D.; Srivastava, M. Guest editors' introduction: Overview of sensor networks. *Computer* **2004**, *37*, 41–49.

12. Ho, D.T.; Shimamoto, S. Highly reliable communication protocol for WSN-UAV system employing TDMA and PFS scheme. In Proceedings of the GLOBECOM Workshops (GC Wkshps), Houston, TX, USA, 5–9 December 2011; pp. 1320–1324.

13. Sotheara, S.; Aomi, N.; Ando, T.; Jiang, L.; Shiratori, N.; Shimamoto, S. Effective data gathering protocol in WSN-UAV employing priority-based contention window adjustment scheme. In Proceedings of the Globecom Workshops (GC Wkshps), Austin, TX, USA, 8–12 December 2014; pp. 1475–1480.

14. Shih, C.Y.; Capitán, J.; Marrón, P.J.; Viguria, A.; Alarcón, F.; Schwarzbach, M.; Laiacker, M.; Kondak, K.; Martínez-de Dios, J.R.; Ollero, A. On the Cooperation between Mobile Robots and Wireless Sensor Networks. In *Cooperative Robots and Sensor Networks 2014*; Springer International Publishing: Berlin, Germany, 2014; pp. 67–86.

15. Bodanese, J.P.; Araujo, G.M.D.; Steup, C.; Raffo, G.V.; Becker, L.B. Wireless Communication Infrastructure for a Short-Range Unmanned Aerial. In Proceedings of the 2014 28th International Conference on Advanced Information Networking and Applications Workshops, Washington, DC, USA, 5–10 July 2014; pp. 492–497.

16. Yan, H.; Huo, H.; Xu, Y.; Gidlund, M. Wireless sensor network based E-health system-implementation and experimental results. *IEEE Trans. Consum. Electron.* **2010**, *56*, 2288–2295.

17. Ehsan, S.; Bradford, K.; Brugger, M.; Hamdaoui, B.; Kovchegov, Y.; Johnson, D.; Louhaichi, M. Design and Analysis of Delay-Tolerant Sensor Networks for Monitoring and Tracking Free-Roaming Animals. *IEEE Trans. Wirel. Commun.* **2012**, *11*, 1220–1227.

18. Chatzigiannakis, I.; Kinalis, A.; Nikoletseas, S. Sink Mobility Protocols for Data Collection in Wireless Sensor Networks. In Proceedings of the 4th ACM International Workshop on Mobility Management and Wireless Access, Torremolinos, Spain, 2 October 2006; pp. 52–59.

19. Choi, L.; Jung, J.K.; Cho, B.H.; Choi, H. M-Geocast: Robust and Energy-Efficient Geometric Routing for Mobile Sensor Networks. In Proceedings of the 6th IFIP WG 10.2 International Workshop on Software Technologies for Embedded and Ubiquitous Systems, Capri Island, Italy, 1–3 October 2008; pp. 304–316.

20. Ekici, E.; Gu, Y.; Bozdag, D. Mobility-based communication in wireless sensor networks. *IEEE Commun. Mag.* **2006**, *44*, 56–62.

21. Khan, A.W.; Abdullah, A.H.; Anisi, M.H.; Bangash, J.I. A Comprehensive Study of Data Collection Schemes Using Mobile Sinks in Wireless Sensor Networks. *Sensors* **2014**, *14*, 2510–2548.

22. Shi, L.; Zhang, B.; Mouftah, H.T.; Ma, J. DDRP: An efficient data-driven routing protocol for wireless sensor networks with mobile sinks. *Int. J. Commun. Syst.* **2013**, *26*, 1341–1355.

23. Safdar, V.; Bashir, F.; Hamid, Z.; Afzal, H.; Pyun, J.Y. A hybrid routing protocol for wireless sensor networks with mobile sinks. In Proceedings of the 2012 7th International Symposium on Wireless and Pervasive Computing (ISWPC), Dalian, China, 3–5 July 2012; pp. 1–5.

24. Chakrabarti, A.; Sabharwal, A.; Aazhang, B. Using Predictable Observer Mobility for Power Efficient Design of Sensor Networks. In Proceedings of the 2nd International Conference on Information Processing in Sensor Networks, Palo Alto, CA, USA, 22–23 April 2003; pp. 129–145.

25. Tang, B.; Wang, J.; Geng, X.; Zheng, Y.; Kim, J.U. A novel data retrieving mechanism in wireless sensor networks with path-limited mobile sink. *Int. J. Grid Distrib. Comput.* **2012**, *5*, 133–140.

26. Oliveira, H.; Barreto, R.; Fontao, A.; Loureiro, A.; Nakamura, E. A Novel Greedy Forward Algorithm for Routing Data toward a High Speed Sink in Wireless Sensor Networks. In Proceedings of the 19th International Conference on Computer Communications and Networks (ICCCN), Zürich, Switzerland, 2–5 August 2010; pp. 1–7.

27. Banerjee, T.; Xie, B.; Jun, J.H.; Agrawal, D.P. Increasing lifetime of wireless sensor networks using controllable mobile cluster heads. *Wirel. Commun. Mob. Comput.* **2010**, *10*, 313–336.

28. IETF, L. 6LowPan-IPv6 Over Low Power WPAN. Available online: http://tools.ietf.org/wg/6lowpan (accessed on 25 May 2015).

29. Jacobson, V. Congestion Avoidance and Control. In Proceedings of the SIGCOMM '88, Stanford, CA, USA, August 1988; pp. 314–329.

30. Tanenbaum, A. *Computer Networks*, 4th ed.; Prentice Hall Professional Technical Reference, Prentice Hall: Upper Saddle River, NJ, USA, 2003.

31. Balakrishnan, H.; Padmanabhan, V.; Seshan, S.; Katz, R. A comparison of mechanisms for improving TCP performance over wireless links. *IEEE/ACM Trans. Netw.* **1997**, *5*, 756–769.

32. Mathis, M.; Mahdavi, J.; Floyd, S.; Romanow, A. TCP Selective Acknowledgment Options. Avaliable online: http://www.rfc-editor.org/info/rfc2018 (accessed on 2 April 2015).

33. Kwon, T.; Gerla, M. Clustering with power control. In Proceedings of the Military Communications Conference Proceedings, 1999 (MILCOM 1999), Piscataway, NJ, USA, 31 October–31 November 1999; Volume 2, pp. 1424–1428.

34. BeagleBoard.org. BeagleBone Black. Available online: http://beagleboard.org/black (accessed on 4 April 2015).

35. The Angstrom Linux Distribution. Available online: http://www.angstrom-distribution.org/ (accessed on 22 April 2015).
36. Goncalves, F.S. UAV as a Mobile Data Collector (Flight Conducted to Test the Use of UAV to Collect Data from a WSN) (Video File). Available online: https://youtu.be/ZWbt9MEtU4Y (accessed on 23 April 2015).
37. Becker, L. Bicycle as a Mobile Data Collector (Using Bicycle As a Mobile Data Collector in WSN) (Video File). Available online: https://youtu.be/sH2frl1_hzo (accessed on 4 April 2015).

Article

Development and Evaluation of a UAV-Photogrammetry System for Precise 3D Environmental Modeling

Mozhdeh Shahbazi [1,*], Gunho Sohn [2,†], Jérôme Théau [1,†] and Patrick Menard [3,†]

[1] Department of Applied Geomatics, Université de Sherbrooke, 2500 Boulevard de l'Université, Building A6, Sherbrooke, QC J1K 2R1, Canada; jerome.theau@usherbrooke.ca
[2] Department of Earth and Space Science and Engineering, York University, 4700 Keele Street, Petrie Science & Engineering Building, Toronto, ON M3J 1P3, Canada; gsohn@yorku.ca
[3] Centre de géomatique du Québec, 534 Jacques-Cartier Est, Building G, Chicoutimi, QC G7H 1Z6, Canada; pmenard@cgq.qc.ca
* Author to whom correspondence should be addressed; mozhdeh.shahbazi@usherbrooke.ca;
  Tel.: +1-418-698-5995 (ext. 1625).
† These authors contributed equally to this work.

Academic Editor: Felipe Gonzalez Toro
Received: 15 September 2015; Accepted: 20 October 2015; Published: 30 October 2015

Abstract: The specific requirements of UAV-photogrammetry necessitate particular solutions for system development, which have mostly been ignored or not assessed adequately in recent studies. Accordingly, this paper presents the methodological and experimental aspects of correctly implementing a UAV-photogrammetry system. The hardware of the system consists of an electric-powered helicopter, a high-resolution digital camera and an inertial navigation system. The software of the system includes the in-house programs specifically designed for camera calibration, platform calibration, system integration, on-board data acquisition, flight planning and on-the-job self-calibration. The detailed features of the system are discussed, and solutions are proposed in order to enhance the system and its photogrammetric outputs. The developed system is extensively tested for precise modeling of the challenging environment of an open-pit gravel mine. The accuracy of the results is evaluated under various mapping conditions, including direct georeferencing and indirect georeferencing with different numbers, distributions and types of ground control points. Additionally, the effects of imaging configuration and network stability on modeling accuracy are assessed. The experiments demonstrated that 1.55 m horizontal and 3.16 m vertical absolute modeling accuracy could be achieved via direct geo-referencing, which was improved to 0.4 cm and 1.7 cm after indirect geo-referencing.

Keywords: UAV; modeling; photogrammetry; calibration; georeferencing; ground control point; mine

1. Introduction

1.1. Background

Unmanned aerial imagery has recently been applied in various domains such as natural resource management, spatial ecology and civil engineering [1–3]. Most of these unmanned aerial vehicle (UAV) applications require geospatial information of the environment. Consequently, three-dimensional (3D) environmental modeling via UAV-photogrammetry systems (UAV-PS) has become a matter of growing interest among both researchers and industries. However, a surveying-grade UAV-PS has critical differences from traditional photogrammetry systems, which should be considered carefully in its development and application. The following paragraphs discuss the background of UAV-PSs and the efforts made to evaluate their capacities.

Typically, development of a UAV-PS starts with selecting the platform as well as the imaging and navigation sensors compatible with it. Regarding the platform, the payload capacity, endurance, range, degree of autonomy must be considered. In some studies, pre-packaged UAVs are used, e.g., AscTec Falcon8 [4], Aeryon Scout [5], SenseFly eBee [6]. Such systems offer safety and ease of operation. However, they offer less flexibility regarding sensor selection and adjustment.

Navigation sensors play two roles in a UAV-PS: auto-piloting the platform and determining the exterior orientation (EO) parameters of images. High-grade inertial navigation systems (INS) can be used in order to eliminate the requirement for establishing ground control points (GCPs) and to achieve enough spatial accuracy via direct georeferencing (DG) [7]. However, consumer-grade systems are preferred considering the costs and limitations of access to base stations for differential or real-time-kinematic (RTK) global-positioning-system (GPS) [8,9]. In such systems, different strategies might be taken for increasing the positioning accuracy—e.g., replacing poor-quality GPS elevation data with height measurements from a barometric altimeter [10]. Accuracy of DG depends on the performance of INS components and the accuracy of platform calibration. Moreover, the system-integration scheme is important since it controls the synchronization between imaging and navigation sensors. Depending on flight speed and accuracy of INS measurements, the delay between camera exposures and their geo-tags can cause serious positioning drifts [7,11].

When indirect georeferencing is performed, considerable care should be given to several factors such as the accuracy of multi-view image matching, on-the-job self-calibration and GCP positioning. Discussed briefly in few studies [10,12,13], the method used to locate GCPs on the images and the configuration of the GCPs are also important factors in determining the final accuracy of indirect georeferencing. Accordingly, the optimum configuration of GCPs required to achieve a certain level of accuracy is a significant concern in the field of UAV-PS. In most of UAV applications, only a minimum number of GCPs in a special configuration and with a limited positioning accuracy can be established. In order to ensure that the results, based on these conditions, can satisfy the accuracy requirements of the application, it is important to have an *a priori* knowledge of the final accuracy.

In terms of imaging sensors, a high-resolution digital visible camera is the key element for photogrammetric mapping. Despite the benefits of non-metric digital cameras such as low price, light weight and high resolution, the instability of their lens and sensor mounts is still a concern in unmanned aerial mapping. Therefore, intrinsic camera calibration must be performed to determine the interior orientation (IO) and distortion parameters of the camera. When metric accuracies are required, offline camera calibration is suggested [14]. However, offline calibration parameters change slightly during the flight due to platform vibrations and instability of camera components [15]. A solution to this problem is to calibrate the camera by adding its systematic errors as additional parameters to aerial block bundle adjustment (BBA), which is known as self-calibration. However, inaccuracy of image observations may influence the calibration parameters as they are all adjusted together with completely unknown parameters such as object-space coordinates of tie points and EO parameters [16]. Thus, motion blur and noise, which are inevitably present in unmanned aerial images, affect the accuracy of calibration. Besides, the numerical stability of self-calibration decreases highly depending on the aerial imaging configuration. Therefore, careful solutions are required to address the issues of on-the-job self-calibration for unmanned aerial imagery.

1.2. Environmental Application

Regarding the environmental application, the system of this study was applied for gravel-pit surveying and volumetric change measurement. This environment was selected because of two reasons. Firstly, open-pit mines provide a challenging environment for 3D modeling. That is, considerable scale variations are introduced to the images due to the low altitude of platform in comparison with the terrain relief [17]. Secondly, there are several mining and geological applications which require high-resolution accurate 3D information of open-pit mines, e.g., geotechnical risk assessment. Previous studies have shown that the topographic data must provide a ground resolution of 1–3 cm in order to predict hazardous events such as ground subsidence, slope instability and landslides [18]. Furthermore, mining companies have to quantify the amount of extracted mass and stocked material regularly. The map scale required for volumetric measurement in earthworks is usually between 1:4000 and 1:10,000 [19]. Considering the requirements of mining applications, including spatial and temporal resolution, speed of measurement and safety criteria, unmanned aerial systems can be better solutions for mine mapping in comparison with traditional terrestrial surveying techniques. This can be noticed by the significant increase in use of UAV-PSs in mining applications during the last few years [20–22].

1.3. Objectives and Assumptions

This paper presents the details of development and implementation of a UAV-PS. In addition to general aspects of the development, the main focus of this study is to discuss the issues and to perform the experiments that are usually ignored or not thoroughly addressed for UAV-PSs. First, the paper concentrates on the procedures for camera and platform calibration as well as system integration. Instead of discussing the regular aspects of calibration, the main focus is on the design of the test-field and automatic target detection assuming that these elements impact the efficiency of calibration significantly. Regarding the system integration, it is assumed that the developed software solution is able to integrate the navigation and imaging sensors accurately without needing any additional mechanism.

Afterwards, the photogrammetric processing workflow is presented. Some aspects of image pre-processing are discussed and their impacts on the accuracy of modeling are investigated. Then, assuming that the accuracy of on-the-job self-calibration is affected by the imaging network, a BBA strategy is suggested to control this adverse effect. This assumption and efficiency of the BBA strategy are also verified. Furthermore, several experiments are designed to assess the effect of GCPs configuration on modeling accuracy. The main assumption that these experiments verify is that a minimum number of GCPs can provide an accuracy level equivalent to the one achievable with redundant number of GCPs under two conditions. First, they are distributed over the whole zone and their visibility in images is maximized. Second, the imaging configuration is proper. That is the imaging configuration ensures scale consistency of the network.

The rest of the paper is structured as follows: first, the equipment is presented. Then, the procedure of system development, including camera calibration, platform calibration and system integration, are discussed in Section 3. Afterwards, Sections 4 and 5 describe the methodology of data acquisition and data processing. The experiments performed to evaluate the system are presented in Section 6, and the results are discussed in Section 7. At the end, the conclusions and final remarks are presented in Section 8.

2. Equipment

2.1. Platform

The platform used in this project is a Responder helicopter built by ING Robotic Aviation Inc. (Ottawa, ON, Canada) (Figure 1a). Responder is a vertical take-off & landing UAV which is equipped with a lightweight, carbon-fiber gimbal. This platform has 12 kg payload capacity and cruise operational endurance of 40 min. With our whole independent package of sensors, computer and batteries weighing about 3 kg, the platform could safely fly for 25 min in a day with wind speed of 19 km/h. The platform is equipped with an open-source autopilot—ArduPilot Autopilot Suite. It comes with a portable, compact ground control station to visualize, plan and control autonomous flights (Figure 1b).

2.2. Navigation Sensor

The navigation sensor is a GPS-aided INS, MIDGII from Microbotics Inc. (Hampton, VA, USA) (Figure 1c, stacked on the top of the camera). The unit measures pitch and roll with 0.4° and heading (yaw) with 1–2° of accuracy. Its positioning accuracy is 2–5 m depending on availability of wide area augmentation system (WAAS). The output rate of the unit can be extended up to 50 Hz.

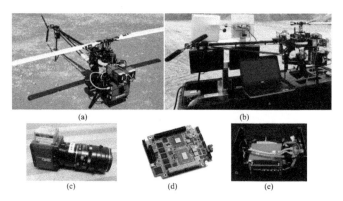

Figure 1. Equipment (**a**) Aerial platform; (**b**) Ground control station; (**c**) INS attached to camera; (**d**) Computer board; (**e**) Computer board and power supply stacked together.

2.3. Imaging Sensor

The imaging sensor is a GE4900C visible camera (Prosilica, Exton, PA, USA) (Figure 1c). It has a 36.0528 mm × 24.0352 mm sensor at pixel size of 7.4 μm. It is equipped with a 35 mm F-mount lens and supports minimum exposure time of 625 μs. The fact that the camera has a global shutter and charge-coupled-device (CCD) progressive sensor makes the imaging more robust against motion blur, interlacing artifact and read-out delay, which are all essential for UAV-PS [23]. Global shutter controls the incoming light all over the image surface simultaneously. Thus, at any time instance, all photo detectors are either equally closed or equally open. This is in contrast with rolling shutters where exposures move row by row from one side to another side of the image. In the CCD architecture, only one-pixel shift happens to move the charge from image to storage area. Therefore, the readout time and energy consumption decrease considerably. Progressive scanning is also strongly preferred for grabbing moving images, since the images are free of interlacing artifacts caused by the time lag of frame fields.

2.4. Onboard Computer

The computer applied in this study is an ultra-small, single-board system (CoreModule 920, ADLINK, San Jose, CA, USA), which is based on a 1.7 GHz Intel Core™ i7 processor (Figure 1d). The board is stacked together with a PC/104 power supply (Figure 1e). The power supply receives 12.8-volt DC input from a 3200 mAh LiFePO4 battery pack. In return, it provides +5 V regulated DC voltage to the computer, +12 V to the camera and +5 V to a fan for cooling the processing unit. With this configuration, the embedded system is capable of acquiring, logging and storing images with a rate of 3 frames per second and navigation data with a rate of 50 Hz during approximately 70 min.

3. System Development

3.1. Camera Calibration

In this study, offline camera calibration is performed using a test-field (Figure 2) via a conventional photogrammetric calibration method known as inner-constrained bundle adjustment with additional parameters [24]. In this study, Brown's additional parameters are applied to model the systematic errors of the camera [25]. Therefore, the camera IO parameters, radial and decentering lens distortions as well as in-plane scale and shear distortions of the sensor [26] are modeled via calibration. As digital camera calibration is a well-studied topic in photogrammetry, detailed theories are avoided here. Instead, other important aspects, including our methodology for test-field design and target detection, are discussed.

Figure 2. Camera calibration test-field.

3.1.1. Design of the Calibration Test-Field

Two camera parameters determine the size and depth of a test-field: focus distance and field of view. For aerial imaging, camera focus distance is set to infinity so that different altitudes can be covered in the depth of field (DoF) of the camera. However, the focal length extends slightly during focusing (When the focus distance is changed, a small group of elements inside the lens, instead of the whole barrel, is moved to provide the focus. Thus, the focal length changes slightly.). This means that the focus distance should remain fixed all the time. Therefore, it should be ensured that the calibration test-field can provide focused photos at short distances while the focus distance is still set to infinity. Considering Equation (1), the far and near limits of DoF (H_f, H_n) depend on F-number (d), focal length (f), circle of confusion diameter (c) and focus distance (h). That is all the objects located between H_n and H_f from the camera can be imaged sharply:

$$H_f = h/1 - (h-f)c\frac{d}{f^2}, \quad H_n = h/1 + (h-f)c\frac{d}{f^2} \quad (1)$$

If the focus distance is set to infinity $(h \to \infty)$, then the F-number should be increased largely to provide focus at short ranges. By setting the F-number to its maximum value (d_{max}), the minimum focus distance can be determined ($H_{n_{min}}$). Thus, the distance of the test-field from the camera, namely the test-field depth, should be larger than $H_{n_{min}}$. Notice that by maximizing the F-number, the aperture opening (A) decreases (Equation (2)), and calibration images become very dark. To compensate this, the exposure time should be increased according to Equation (3). Let t_1 be the minimum exposure time of the camera which is usually selected for aerial imaging to reduce motion blurring artifacts. Accordingly, A_1 is the aperture opening which provides proper illumination for outdoor acquisition in combination with t_1. If A_2 is the aperture opening when maximizing the F-number, then t_2 is the exposure time that should be set to avoid either underexposure or overexposure:

$$A = \pi f^2 / 4d^2 \qquad (2)$$

$$A_1 t_1 = A_2 t_2 \qquad (3)$$

Afterwards, the width and height of the test-field should be determined. It is essential to model the systematic errors based on the distortions observed uniformly across the whole image [24]. Therefore, it should be ensured that the test-field is large enough to cover approximately the whole field of view (FoV) of the camera. The horizontal and vertical angles of FoV (α_h, α_v) can be calculated from the sensor size (W, H) and the focal length (f) as in Equation (4). Therefore, the minimum size required for the test-field ($W_t \times H_t$) can be determined via Equation (5)

$$\alpha_h = 2\tan^{-1}(W/2f) \quad , \quad \alpha_v = 2\tan^{-1}(H/2f) \qquad (4)$$

$$W_t = 2H_{n_{\min}} \tan(\alpha_h/2) \quad , \quad H_t = 2H_{n_{\min}} \tan(\alpha_v/2) \qquad (5)$$

3.1.2. Target Detection

Figure 3 demonstrates the approach of this study for detecting the targets. Unless otherwise indicated, the procedures mentioned in the diagram are fully automatic.

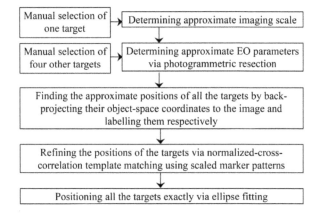

Figure 3. Diagram of the target detection method.

In this study, the targets are designed as black and white rings with crosshairs passing through the circles' centers. The reason for using circular targets is that once the image of a circle is deformed under any linear transformation, it appears as an ellipse. Then, the techniques of ellipse detection can be applied to position it accurately. The following paragraphs explain the ellipse fitting method developed to determine the accurate positions of the targets.

Assuming that (x_a, y_a) is the approximate position of a target, a rectangular window is centered at (x_a, y_a). The window is transformed to binary format, and its connected black components are detected. Each closed component represents a candidate region. Let B denote the set of all the pixels belonging to the boundary of a candidate region. If the region is actually a target, then its boundary can be modeled as an ellipse with Equation (6):

$$(x - x_0)^2 + K(y - y_0)^2 = R^2 \; ; \forall (x,y) \in B \qquad (6)$$

where (x_o, y_o) is the center, R is the flattened radius and K is the aspect ratio of the ellipse. A small percentage of the points belonging to B are reserved as checkpoints. Other points are served as observations to determine the ellipse parameters with least squares fitting technique. Once the ellipse is defined mathematically, the checkpoints are validated against the ellipse model. If the fitting error is less than a given threshold, then the candidate region is recognized as a target, and the ellipse center denotes the exact position of the target.

.

3.2. Aerial Platform Calibration

The goal of platform calibration is to make sure that EO parameters of images are represented in an earth-centered, earth-fixed (ECEF) reference system, in which the navigation positioning data are represented as well. To this end, the vector between the perspective center of the camera and the center of the INS body-fixed system—known as lever-arm offset—as well as the rotations of the camera coordinate system with respect to the INS system—known as bore-sight angles—should be determined. In this study, the lever-arm offset is ignored. This is due to the fact that the offset between the INS and the camera never exceeds a few centimeters, which is far below the precision of GPS measurements (a few meters).

Attitude outputs from the INS are presented as Euler angles, also known as Cardan. The Cardan consists of three rotations: roll (ϕ), pitch (θ) and yaw (ψ). The rotation matrix R_b^n—composed of Euler angles—rotates vectors from the INS body-fixed coordinates system (b) to the local geodetic system (n) as in Equation (7). Likewise, the rotation matrix R_n^e—composed of geodetic latitude (φ) and longitude (λ)—rotates vectors from the local geodetic system to the ECEF system (e) as in Equation (8). Therefore, the rotation matrix R_b^e rotates vectors from the INS body-fixed coordinate system to the ECEF system as in Equation (9) [27]:

$$R_b^n = R_z(\psi)R_y(\theta)R_x(\phi) \tag{7}$$

$$R_n^e = R_z(\pi - \lambda)R_y(\pi/2 - \varphi) \tag{8}$$

$$R_b^e = R_n^e R_b^n \tag{9}$$

The required rotation matrix for image georeferencing is R_i^e, which describes the rotations from the camera coordinate system (i) to the ECEF one. The rotation matrix R_i^e can be calculated using the rotation matrix R_b^e and the bore-sight matrix R_i^b:

$$R_i^e = R_b^e R_i^b \tag{10}$$

To determine the bore-sight matrix R_i^b, first, a network of targets is established in the ECEF coordinate system (Figure 4). Then, the targets are photographed using the camera which is firmly installed on the platform with the INS. Simultaneously, the INS data (R_b^n) is logged. At the post-processing stage, the position ($\vec{X}_i^e$) and orientation (R_i^e) of the camera center are calculated via photogrammetric resection. Using the geodetic coordinates of the camera center (φ_i^e, λ_i^e)—derived from Cartesian coordinates $\vec{X}_i^e$—the rotation matrix R_n^e is calculated. Then, the rotation matrix R_b^e from the INS body-fixed system to the ECEF system is determined via Equation (9). Finally, by substituting R_i^e and R_b^e to Equation (10), the unknown bore-sight matrix R_i^b is determined.

Figure 4. Test-field for platform calibration.

Notice that, in this study, the camera and the INS were stacked together, in a fixed status as in Figure 1c. Consequently, the platform could be calibrated before installing the sensors on the UAV.

3.3. System Integration

A UAV-PS consists of a platform, camera, navigation system and control system. The control system is responsible for various tasks including power control, setting the data-acquisition parameters, data logging, data storage and time synchronization. In this study, the hardware of the control system simply includes the computer and the power supply as described in Section 2.4. The software solution developed for this control system contains three main classes: INS, camera, and clock.

The main functionality of the clock class is to get the time up to nanoseconds from the system-wide clock and assign it to any acquisition event in a real-time manner. The INS class is responsible for communication with the INS and recording the navigation messages. Each navigation message contains the information of position, rotation and GPS time. The system time at the moment of receiving each navigation message is also assigned to that message by the clock class. The GPS time of the messages is assigned to a shared variable as well, which has external linkage to the camera class. The camera class is responsible for communication with the camera and setting its attributes including triggering time interval, exposure and gain value, and frame size. Although several methods of acquisition are available for Prosilica cameras, software triggering mode is used to facilitate the synchronization process. That is the camera is triggered automatically based on defined intervals, e.g., every 500 ms. The end of camera exposure is set as an event, and a callback function is registered to this event. The functionality of this callback is to save the acquired frame and tag the navigation information to it. This information includes the GPS time and the navigation data received from the INS as well as the system time observed at the epoch of the exposure-end event. Finally, the software makes these classes operate together. It starts two threads in the calling process to execute the main functions of the classes simultaneously.

Notice that the GPS time, tagged to each image, is determined by the INS class and the frequency of INS data is 50 Hz. Therefore, the GPS timestamp is theoretically less than 20 ms different from the exact time of the exposure. Assuming a flight speed of 20 km/h, the time-synchronization error of 20 ms can cause 11 cm of shift between the true position and the tagged position of an image. When a navigation-grade GPS is used, such a shift is quite below the precision of GPS measurements and can be ignored [7]. However, if differential GPS measurements are used, then this error must be systematically handled [11]. To do so, in a post-processing step, the difference between the system time and the GPS time that are tagged to each image is used to derive the exact navigation data corresponding to that image via a linear interpolation over the two INS messages, between which the image is acquired.

4. Data Acquisition

In this study, the data were acquired from a gravel pit at Sherbrooke, QC, Canada. The extent of the gravel-pit mine is shown in Figure 5. Two series of data over a period of two months were acquired—August and October 2014. Two main zones were considered for the experiments (Figure 5). The red zone represents one part of the gravel pit which was covered by stockpiles, and the green zone represents the zone covered by cliffs and rocks.

The Study Area

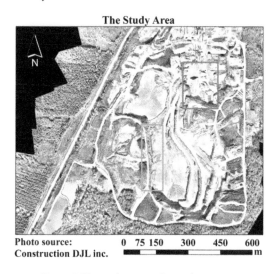

Photo source: 0 75 150   300   450   600
Construction DJL inc.   m

Figure 5. The study area and mapping zones.

4.1. Data-Acquisition Planning

In order to perform flight planning, there exist several software packages. However, in this study, a simple software solution is developed to satisfy the specific needs of the project for both flight and fieldwork planning. The interface of the software is shown in Figure 6. The main inputs of the software are the platform and sensor characteristics as well as the desired overlap and ground resolution for imagery. In order to calculate the position of the sun, the flight time is needed too. Knowing the position of the sun helps to minimize shadow effects; the larger the solar elevation angle, the shorter the shadows. The software allows users to either load or graphically choose the predicted positions of GCPs—red triangles in the display panel of Figure 6. It is, then, possible to determine the flight zone—blue polygon and the flight home—yellow lozenge. The software designs and logs the flight plan afterwards. One of the significant applications of this software is to design the approximate spatial distribution of GCPs considering two conditions. First, GCPs should be installed at stable locations distributed over the whole imaging zone. Second, their visibility in the images should be maximized.

Figure 6. Interface of the flight-planning software.

4.2. Fieldwork

The first task of the fieldwork was to initialize the GPS base receiver for collecting RTK measurements. The absolute coordinates of the base point were determined with 2–5 mm accuracy. The next step was to install the targets at locations predicted during the flight planning stage (see Section 4.1). Then, their positions were measured using the R8 GNSS System (Trimble, Sunnyvale, CA, USA)—a high-precision, dual-frequency RTK system. Following the same concept as camera calibration (Section 3.1.2), GCPs were marked as circular targets (Figure 7a). Once the GCPs were established, the flights for image acquisition started. Table 1 presents the flight conditions. Labels are given to the acquired datasets for further use in this paper.

Table 1. Information of the data-acquisition sessions.

Characteristic	Flight Date		
	August 2014	August 2014	October 2014
	Dataset A	Dataset B	Dataset C
Weather temperature (°C)	22	26	10
Wind speed (Km/h)	8	19	8
Zone structure	Stockpiles	Cliffs	Cliffs
Approximate flight altitude (m)	80	90	90

Upon termination of image acquisitions, the terrestrial surveying for gathering check data started. The check data included sparse 3D point clouds measured by Trimble VX laser scanner over the whole mapping area and a dense point cloud measured by a FARO Focus laser scanner (FARO, Lake Mary, FL, USA) over a small pile. Figure 7b,c show the configuration of the control points, the stations where VX scanner was installed as well as the zone where the dense 3D point cloud was measured by FARO scanner (blue polygon). In order to facilitate the accurate registration of individual FARO scans, several targets with checkerboard pattern and reference spheres were installed at different levels of the pile (Figure 7d).

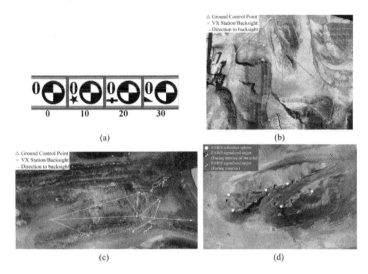

(a)   (b)

(c)   (d)

Figure 7. Surveying plans (**a**) Markers for GCPs; (**b**) Configuration of GCPs and laser-scanner stations for dataset A; (**c**) Configuration of GCPs and laser-scanner stations for dataset C; (**d**) Configuration of targets for FARO scanner over one pile.

5. Data Processing Workflow

The main steps of data processing in aerial photogrammetry are illustrated in Figure 8. The ordinary methods to perform these steps are not discussed here. Instead, the methodology of this study to improve some of these procedures is presented.

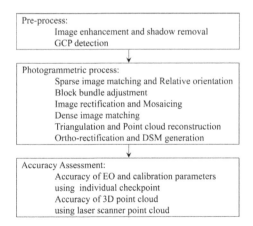

Figure 8. Photogrammetric workflow to produce topographic data from images.

5.1. Image Pre-Processing

5.1.1. Intensity Enhancement

As in the system of this study, in most of UAV-PSs, small-format cameras are used because of weight limitations. One of the main problems caused by such cameras is their small ground coverage. This characteristic makes the sequence of images vulnerable to photometric variations, even though the flight time is usually quite short [28]. Noticeable radiometric changes among adjacent images make both sparse matching and pixel-based dense-matching more difficult [29,30]. An example of such situation is given in Figure 9. These images are from a test dataset not listed in Table 1. They were taken with a Prosilica GT1920C camera with sensor size of 8.7894 mm × 6.6102 mm and focal length of 16 mm. As shown in Figure 9a, some images are considerably darker than their neighboring images since a patch of dark clouds had passed through the zone. Some of the images additionally suffer from lack of texture diversity (Figure 9c). As a result, matching such images becomes difficult. Therefore, their relative orientation parameters cannot be determined, and the ortho-mosaic cannot be generated either (Figure 9e).

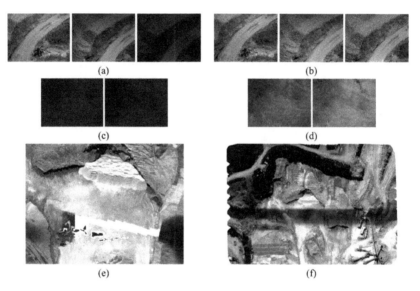

Figure 9. Intensity enhancement (**a**) A sequence of illuminated and dark images; (**b**) Images of Figure 9a after correction; (**c**) Two adjacent dark images with low texture diversity; (**d**) Images of Figure 9c after correction; (**e**) Failure in ortho-mosaic generation; (**f**) Correct mosaic after intensity enhancement.

When thematic applications are required, sophisticated techniques using spectral observations should be used for radiometric correction of images. Otherwise, simple image enhancement methods can be used to reduce the relative photometric variations. In this study, a combination of white balancing and histogram matching is proposed. Starting by the image with proper illumination as the reference image, its dark neighboring image is first white-balanced. Then, the intensity histograms of both images are calculated. If the correlation of the cumulative distribution functions of the histograms is more than 0.5, then no further enhancement is needed. Note that a cumulative histogram represents the intensity rank of each pixel with respect to other pixels regardless of general radiometric changes [31]. If the correlation of two distributions is less than 0.5, then histogram matching is performed to adjust the intensity values in the dark image. With this method, each image is relatively corrected only with respect to the images immediately adjacent to it. Therefore, the trace of dark images is still visible globally. However, this intensity enhancement makes the matching performable, and a correct ortho-mosaic can be generated (Figure 9f).

5.1.2. Shadow Removal

Another radiometric effect on aerial images is caused by shadows. In thematic applications such as atmospheric correction and classification, shadow regions lead to inevitable errors [32]. Shadows can also cause spatial errors in 3D modeling. This usually happens when the sun direction changes slightly during the flight and the shadow edges move as a consequence [33]. In our experiments, this problem was observed in dataset B (Table 1). In order to investigate the effects of shadows on the quality of 3D modeling, a simple technique is proposed to detect and remove the shadow regions from single images. The summary of this technique is presented in Figure 10. In Section 7.3, the effects of shadow removal on both the photometric appearance and the accuracy of the 3D point clouds are analyzed.

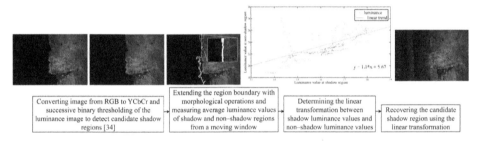

Figure 10. Workflow for automatic shadow detection and removal.

5.1.3. GCP Detection

Once the image intensities are enhanced, ground control points can be detected. Automatic detection of GCPs in images is important from two aspects. Firstly, detecting GCPs manually in large sets of UAV images is a cumbersome task. Secondly, the accuracy of target detection directly affects the accuracy of georeferencing and calibration. Therefore, more attention has recently been paid to this process [10,35]. The method applied to position GCPs is presented in Figure 11. It is mainly based on localization of GCPs using direct EO parameters, color-thresholding and ellipse detection as described in Section 3.1.2. Although this process is automatic, the results should manually be verified to remove incorrect detections.

Figure 11. Workflow for automatic detection of GCPs.

5.2. Photogrammetric Processing

Photogrammetric processes are applied to aerial images in order to generate various types of topographic products such as 3D point cloud, geo-referenced mosaic and digital surface model (DSM). In this study, the main photogrammetric processes are performed using Pix4D software [36]. This software has recently been popular among researchers and commercial users for UAV-photogrammetry [33]. Several experiments were designed using this software in order to evaluate the performance of the developed UAV-PS (Section 6.2).

Furthermore, a BBA strategy is suggested for on-the-job self-calibration. This strategy is able to control the adverse effects of noisy aerial observations as well as correlation of IO and EO parameters on the accuracy of self-calibration. The main solution of this strategy is to transform the intrinsic camera calibration parameters from unknowns to pseudo-observations. That is, they should be considered as semi-unknowns with known weights. The experimental way to determine these weights is presented in Section 6.1, and the results are discussed in Section 7.6. In the following paragraphs, the principals of self-calibration with pseudo-observations are presented. Since BBA and least-squares adjustment are well-studied topics, the details are avoided here. Readers are referred to [16,37,38] for more theoretical information.

The mathematical model of bundle adjustment with additional parameters can be presented as in Equation (11), where the observation equations (*F*) are based on co-linearity condition. These equations are functions of measurements (*L*), unknowns (*Y*) and pseudo-observations (*X*). The measurements are image coordinates of tie points. The unknowns include ground coordinates of tie points and EO parameters of images. The pseudo-observations include intrinsic calibration parameters—both IO parameters and distortion terms:

$$F(X,Y,L) = 0 \tag{11}$$

The linear form of Equation (11) is obtained using a Taylor series first-order approximation:

$$W + A\delta K + B\delta L = 0 \tag{12}$$

where *K* is a concatenated vector by *X* and *Y*, *W* is the miss-closure matrix, and *A* and *B* are the matrices of first-order partial derivatives of *F* with respect to *K* and *L*, respectively. Assuming that P_L is the weight matrix of the measurements, P_X is the weight matrix of the pseudo-observations, and *D* is the matrix of datum constraints, then the least-squares solution for δK can be obtained as in the following equation:

$$\Sigma_K = \left(P_X + A^T(BP_L^{-1}B^T)^{-1}A + D^TD\right)^{-1} - D^T(DD^TDD^T)^{-1}D$$

$$\delta\hat{K} = -\Sigma_K(A^T(BP_L^{-1}B^T)^{-1}W) \tag{13}$$

The vector of residuals, δL , is also estimated as follows:

$$\delta\hat{L} = -P_L^{-1}B^T(BP_L^{-1}B^T)^{-1}(A\delta\hat{K} + W) \tag{14}$$

These partial solutions, $\delta\hat{K}$ and $\delta\hat{L}$, are successively added to the initial estimations of the unknowns and values of the measurements and pseudo-observations until reaching convergence. The role of the weight matrix P_X in Equation (13) is to control the changing range of pseudo-observations.

6. Experiments

In this section, the experiments which were performed to assess different aspects of the developed system are presented. The results obtained from these experiments are, then, discussed in Section 7.

6.1. Laboratory Experiments

6.1.1. Calibration

The camera was calibrated several times during a period of few months before starting the data acquisition. The final parameters were obtained as the average of the parameters from these tests. The stability of each calibration parameter was also determined as its variance at these tests. In the BBA strategy of Section 5.2, these variances were used as the weights of pseudo-observations.

In order to verify the accuracy of offline calibration parameters obtained from these test, 10 check images were captured. In these images, the targets were detected using the method of Section 3.1.2. Some of them were reserved as checkpoints, and others were served as control points. Using the control points and the calibration parameters, the EO parameters of the images were determined via space resection. Then, the 3D object-space coordinates of the checkpoints were back-projected to the images. The difference between the back-projected position of a checkpoint and its actual position on an image is called the residual. The residuals show how accurate the calibration parameters are modeled. The results obtained from this test are presented in Section 7.1.

In order to analyze the efficiency of automatic target detection, compared with manual target detection, similar calibration and assessment tests were performed using the targets that were detected manually. Positions of the manual targets were different from those of the automatic targets with an average of 1.3 pixels and maximum of 2.4 pixels. The results obtained from this experiment are also presented in Section 7.1.

6.1.2. Time Synchronization

In order to verify how precisely the camera exposures could be tagged via INS messages, a simple experiment was performed. For each image, the INS log file was searched, and the INS message whose GPS time was exactly equal to the GPS timestamp of the image was detected. If the system time tagged to that INS message were adequately close to the system time tagged to the image, then it could be concluded that the GPS timestamp tagged to the image was accurate too. The results of this test, performed on more than 2500 images, are discussed in Section 7.2.

6.2. Photogrammetric Tests

Initially, the images were processed using Pix4D. The 3D triangulated mesh objects for dataset A, dataset B after shadow removal and dataset C are presented in Figure 12. To evaluate the accuracy of dense reconstruction, cloud-cloud comparison was performed between the image point clouds and terrestrial laser-scanner point clouds. To this end, the CloudCompare open source software was applied [39]. For each point in the laser cloud, the closest point in the image cloud was found, and the distance between them was calculated. Then, the distances were analyzed to measure the spatial accuracy of image point clouds. The results of these analyses are presented in Section 7.3. Once the accuracy of individual point clouds was assessed, they were used to produce other topographic data such as slope maps. Using the DSMs of two different dates, volumetric changes within the site were measured as well. The results are discussed in Section 7.7.

(a)   (b)   (c)

Figure 12. Triangulated mesh for (**a**) Dataset A; (**b**) Dataset B after shadow removal; (**c**) Dataset C.

The second series of the experiments were performed to determine how the number and spatial distribution of GCPs affect the accuracy of 3D modeling. Traditionally, it is known that having more than enough GCPs with good geometrical configuration improves the accuracy of the results [37]. However, in most of UAV-mapping applications, only a minimum number of GCPs can be established. Therefore, it is important to have an *a priori* knowledge of how the final accuracy of 3D modeling would be affected by the GCPs. To this end, several experimental tests were designed. These tests are described through Table 2. In each test, the initial photogrammetric processing was performed using Pix4D, which included tie point generation, block bundle adjustment and self-calibration. Then, the accuracy of the results was evaluated against checkpoints. The checkpoints in these experiments (Figure 13) were either the GCPs not used as control points and/or some of the individual laser-scanner points that were transformed to checkpoints. To do this, the ground coordinates of each laser point were back-projected to the images via the accurate indirect EO parameters. Then, SURF feature descriptors were calculated over the projected pixels in all the images [40]. If such pixels represented salient features, and if the distances between their descriptors were smaller than a threshold, then that laser point was considered as a checkpoint. In addition, all the checkpoints were manually verified to avoid errors.

Table 2. Description of experimental tests for verifying the effect of number/distribution of GCPs.

Test Label	Figure	Descriptions	
GCPTest 1		Number of GCPs	22
		Distribution	Covering the whole imaging zone
GCPTest 2		Number of GCPs	3
		Visibility *	9, 12 and 21 images
		Distribution	Evenly distributed over the imaging zone
GCPTest 3		Number of GCPs	3
		Visibility	4–6 images
		Distribution	Well distributed over the imaging zone
GCPTest 4		Number of GCPs	3
		Visibility	19, 20 and 22 images
		Distribution	Positioned near the ends of flight strips
GCPTest 5		Number of GCPs	3
		Visibility	5, 12 and 21 images
		Distribution	Established at the flight home due to inaccessibility to the rest of the imaging zone
GCPTest 6		Number of GCPs	3
		Visibility	15, 17 and 22 images
		Distribution	Established along a hypothetical road due to inaccessibility to other areas

* Number of images, in which every GCP is visible.

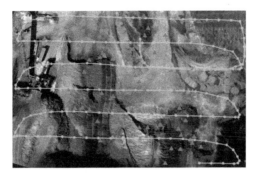

Figure 13. Flight trajectory and distribution of checkpoints in dataset A.

Moreover, we were interested in assessing the results of these tests not only with GCPs that were measured by RTK GPS, but also with GCPs whose positions were measured by other techniques. Since the GCPs were originally measured only by accurate RTK GPS, it was decided to simulate the measurements for other techniques. To this end, three reference points were established outdoor, and their exact coordinates with RTK GPS system were measured. Then, other GPS devices were used to re-measure their coordinates. Using the observations made by each device, the positioning errors of GCPs were simulated. First, a Garmin GLO-GPS was used, and more than 10,000 observations were recorded over the reference points. This device is WAAS-enabled and receives position information from both GPS and GLONASS satellites. The root mean square (RMS) positioning error for this device was 2.40 m horizontally and 6.04 m vertically. Similarly, a series of 2000 observations were made with a SXBlueII GPS. This device is also WAAS-enabled and performs additional code-phase measurements and multi-path error reduction. The RMS positioning error with this device was 0.65 m horizontally and 0.69 m vertically. The results obtained in different tests using these types of GCPs are presented in Section 7.4.

To analyze the effect of imaging configuration in absence/presence of GCPs on the accuracy of 3D modeling, a sequence of nine images was considered (Figure 14a). Ground control points A, B and C were visible in images 1–4, while point D was visible in images 5–6. The following situations were, then, designed and tested.

i.   OverlapTest 1: Each image was overlapped with at least three connected images. For example, image 5 had common tie points with both images 3 and 4. Such a connection is illustrated via the connectivity matrix in Figure 14b.

ii. OvelapTest 2: The situation was the same as OverlapTest 1. However, image 5 did not have any common tie points with both images 3 and 4; *i.e.*, it was only overlapped with image 4. The connectivity matrix of Figure 14c shows this situation.

The objective of these tests was to find out whether the lack of overlap in OverlapTest 2 could cause problems in BBA, and how important the role of GCPs was to solve those problems. The results obtained from each test and the issues involved with each situation are assessed in Section 7.5.

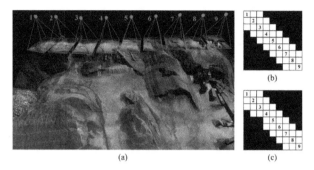

Figure 14. (a) Experiment to assess the effect of imaging configuration; (b) Connectivity matrix in OverlapTest 1; (c) Connectivity matrix in OvelapTest 2.

The final series of photogrammetric experiments were performed to analyze the effect of on-the-job self-calibration on the accuracy of IO and EO parameters. Also, the proposed self-calibration strategy of Section 5.2 was evaluated, and the results were compared with those of traditional self-calibration. To this end, the following situations were considered, and correlation analysis was performed at each situation in order to determine the dependency of IO parameters to EO parameters.

i.   CalibTest 1: Offline camera calibration was performed using a well-configured imaging network (Figure 15a).
ii.   CalibTest 2: On-the-job calibration was performed using typical aerial images, which were all acquired from almost the same altitude (Figure 15b).
iii. CalibTest 3: On-the-job calibration was performed using aerial images, which were acquired from varying altitudes (Figure 15c).

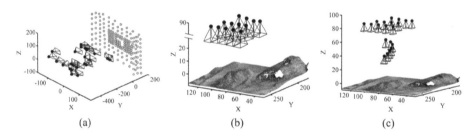

Figure 15. Self-calibration experiments (a) CalibTest 1; (b) CalibTest 2; (c) CalibTest 3.

7. Results and Discussion

7.1. Calibration Results

Figure 16a shows the residual vectors for the checkpoints after camera calibration. The mean and standard deviation (StD) of the residuals on the checkpoints at x- and y-directions are 0.32 ± 0.18 pixel and 0.20 ± 0.16 pixel, respectively. Figure 16b presents the residual vectors based on the manual target detection. Notice that the targets on check images were detected automatically and, only, the targets used for calibration were measured manually. As a result, the automatic target detection improves the accuracy of calibration 81.25% in comparison with the noisy observations based on manual detection even though the noise level does not exceed 2.4 pixels.

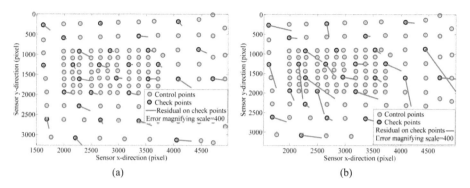

Figure 16. Calibration results (**a**) Residuals on checkpoints based on automatic target detection; (**b**) Residuals on checkpoints based on manual target detection.

7.2. Precision of Time-Synchronization

Figure 17 demonstrates an example of the results obtained from the time-synchronization test. The *x*-axis shows the image number. The *y*-axis shows ΔT, which is the absolute difference between the system time tagged to each image and the system time tagged to its corresponding INS message. It is, indeed, the difference between the real exposure-end time of an image and the GPS timestamp tagged to it. As it can be noticed, these differences are random; however, they do rarely exceed 20 ms. This is due to the fact the INS frequency is 50 Hz (1 message per 20 ms). The main reason why this difference (ΔT) is random is that the camera exposures do not start on very exactly fixed intervals—e.g., every 500 ms. Instead, there is a few milliseconds of random delay/advance from the defined interval—e.g., 502 ms. In average, it can be concluded that the GPS timestamp tagged to any image is approximately 11 ± 7 ms delayed/advanced from the exact time of the exposure.

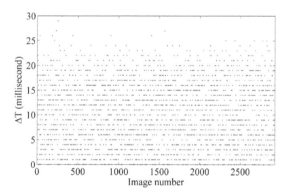

Figure 17. Results of the time-synchronization test.

7.3. Accuracy of 3D Point Clouds

As mentioned in Section 6.2, image point clouds were compared with terrestrial laser-scanner point clouds. Figure 18a,b illustrate the histograms of horizontal and vertical distances between the point cloud of dataset A and that of the laser scanner (see Table 3 as well). The vertical accuracy of this point cloud is 1.03 cm, and its horizontal accuracy is 1.58 cm.

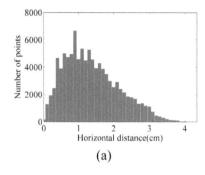
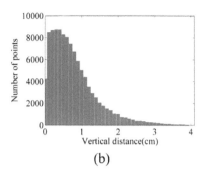

(a) (b)

Figure 18. Histograms of distances between the point clouds from dataset A and laser scanner (**a**) Horizontal distances; (**b**) Vertical distances.

Table 3. Summary of distances between the image-based point clouds and laser-scanner ones.

Dataset	Horizontal Distance (cm)			Vertical Distance (cm)		
	Mean	RMS	StD	Mean	RMS	StD
A	1.38	1.58	0.77	0.80	1.03	0.66
B before shadow removal	1.79	2.03	0.96	1.41	1.72	0.99
B after shadow removal	1.62	1.82	0.83	1.32	1.62	0.95
C	1.88	2.07	0.84	1.63	2.02	1.18

As another test, the dense point cloud measured by FARO laser scanner was transformed to a raster DSM. Then, the absolute difference of the image-based DSM from the laser DSM was calculated. Figure 19 presents the vertical difference between the image-based DSM and laser DSM. In more than 78% of the zone, this vertical difference is less than 1 cm.

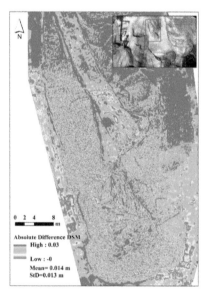

Figure 19. Absolute difference between the DSM from FARO laser scanner and that of dataset A.

For dataset B, as mentioned in Section 5.1.2, shadow regions were removed from the images. Figure 20a,b show the geo-referenced mosaics of the site before and after shadow removal, respectively. As it can be seen, the results are visually improved. The shadow-free mosaic can be used in thematic applications where shadow effects cause errors.

(a)   (b)

Figure 20. Image mosaics of dataset B (**a**) Before shadow removal; (**b**) After shadow removal.

The point clouds obtained before and after shadow removal were also evaluated using the laser-scanner data. The results are presented in Table 3 and Figure 21. According to the results, the vertical accuracy is improved from 1.72 cm to 1.62 cm with shadow removal. Therefore, no noticeable improvement can be observed via this test. This is principally due to the fact that the terrestrial laser-scanner points over the shadow region were not dense enough. However, when the DSMs before and after shadow removal were compared, the differences could be observed more clearly. As shown in Figure 22, large vertical differences, as large as 8 cm, can be observed in edges of the cliffs. These are the zones where more than one shadow was casted on the objects—the shadow from two higher rows of the rocks. It is believed that this type of shadow causes errors in the dense matching and 3D reconstruction process.

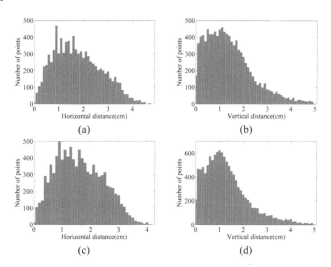

Figure 21. Histograms of distances between the point clouds from dataset B and laser scanner (**a**) Horizontal distances before shadow removal; (**b**) Vertical distances before shadow removal; (**c**) Horizontal distances after shadow removal; (**d**) Vertical distances after shadow removal.

Figure 22. Absolute difference between the DSMs of dataset B before and after shadow removal.

Finally, the point cloud of dataset C was evaluated against the terrestrial laser point cloud. The results are presented in Table 3 and Figure 23. Generally, the accuracy of the point cloud from dataset A is higher than both dataset B and dataset C. This could be explained by the structure of the stockpiles in dataset A, where less occlusion happens in aerial images. For datasets B and C, the vertical and layered structure of the cliffs causes more errors.

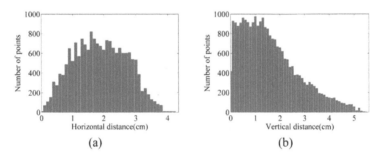

Figure 23. Histograms of distances between the point clouds from dataset C and laser scanner (**a**) Horizontal distances; (**b**) Vertical distances.

7.4. Effects of Ground Control Points

Firstly, the results obtained via direct georeferencing are discussed. In fact, DG may be interpreted in two senses. In the first sense, direct EO parameters of images from the navigation data are directly used for 3D modeling without any photogrammetric refinement applied to them. In this case, offline intrinsic camera calibration parameters should be used as no on-the-job self-calibration is performed. This strategy is mostly used for rapid mosaicking and ortho-photo generation. The results from this test for dataset A are shown as DG type 1 in Table 4. In the second sense, direct navigation data are used as inputs into initial photogrammetric processing and the EO parameters are slightly refined within a free-network adjustment. Then, these refined EO parameters are used for 3D modeling. The results from this test are shown as DG type 2 in Table 4. Notice that in this table and the following ones, the mean error represents the average of absolute errors—not the signed ones. As it can be seen, the horizontal and vertical accuracy is improved 31% and 73%, respectively, after applying the initial processing. The main reason for this improvement is that all the relative-orientation errors between images are corrected within the initial processing. This can be observed from the fact that standard deviations of errors decrease considerably after initial processing.

Table 4. Accuracy of direct georeferencing on checkpoints.

Experiment	X-Direction Error (m)			Y-Direction Error (m)			Vertical Error (m)		
	Mean	RMS	StD	Mean	RMS	StD	Mean	RMS	StD
DG type 1	1.146	1.406	0.837	2.478	3.088	1.892	10.440	11.670	5.359
DG type 2	1.938	1.943	0.137	1.162	1.166	0.090	3.159	3.169	0.268

The results obtained from the experimental tests with GCPs (Table 2, Section 6.2) are presented in Table 5. For each checkpoint, the EO and calibration parameters of images—after indirect geo-referencing at each experimental test—were used to determine its ground coordinates via intersection. Then, the error on the checkpoint was measured as the difference between its ground-truth 3D coordinates and the calculated coordinates.

In order to provide a better understanding of the way each configuration or device affects the results, the relative changes of accuracy are represented in Table 6. These change rates are calculated as the percentage of RMS improvement with regard to the lowest accuracy. Therefore, the improvement rate of 0.0% shows the reference value used for change-percentage measurement.

Table 5. Horizontal accuracy on checkpoints based on different GCP experiments.

Error	Experiment	Trimble R8			SXBlue			Garmin GLO		
		Mean	RMS	StD	Mean	RMS	StD	Mean	RMS	StD
Horizontal Error (cm)	GCPTest 1	0.2	0.4	0.3	61.9	61.9	3.0	180.0	180.7	12.4
	GCPTest 2	0.3	0.4	0.6	68.0	69.0	1.7	158.2	160.8	19.6
	GCPTest 3	0.8	0.9	1.2	73.9	74.1	4.6	216.4	216.6	9.0
	GCPTest 4	0.3	0.4	0.7	63.8	62.9	2.3	160.2	165.3	30.2
	GCPTest 5	0.6	0.8	0.9	74.7	76.3	14.8	227.0	228.0	20.5
	GCPTest 6	0.3	0.5	0.6	72.6	72.8	5.1	189.3	193.5	31.4
Vertical Error (cm)	GCPTest 1	1.2	1.7	1.2	13.8	15.5	7.0	412.9	413.0	10.5
	GCPTest 2	1.6	2.0	1.2	41.1	49.7	28.3	355.5	355.8	12.4
	GCPTest 3	4.1	4.3	1.4	73.6	73.6	2.8	434.9	436.0	32.5
	GCPTest 4	1.4	2.0	1.4	43.2	48.5	22.1	432.1	433.5	35.2
	GCPTest 5	2.4	3.0	1.8	121.6	147.1	83.6	431.2	446.1	115.6
	GCPTest 6	1.4	1.9	1.4	80.2	97.4	55.7	432.1	433.5	35.2

Table 6. Improvement rate of accuracy on checkpoints based on different GCP experiments.

Device	Horizontal-Accuracy Percentage Change						Vertical-Accuracy Percentage Change					
	Experiment						Experiment					
	1 *	2	3	4	5	6	1	2	3	4	5	6
R8 RTK	99.8	99.8	99.6	99.8	99.6	99.8	99.6	99.6	99.0	99.6	99.3	99.6
SXBlue	72.9	69.7	67.5	72.4	66.5	68.1	96.5	88.9	83.5	89.1	67.0	78.2
Garmin GLO	20.7	29.5	5.0	27.5	0.0	15.1	7.4	20.2	2.3	2.8	0.0	2.8

* Reads as GCPTest 1.

As the results show, in order to reach the highest accuracy, it is recommended to provide a large number of well-distributed GCPs (as in GCPTest 1). However, if this is not possible, then the best solution is to install the GCPs at different sides of the imaging zone, where they can also be visible in as many images as possible (as in GCPTest 2). To ensure this condition, the best practice is to install them near the ends of the flight strips so that they are visible in several images from two adjacent strips (as in GCPTest 4). Typically, it is preferred to install GCPs at places with height variation. However, the results from GCPTest 6, where the control points are almost at the same elevation, are much more accurate than those of GCPTest 3, where GCPs have high height variation but low visibility. Finally, the least accurate results are obtained from GCPTest 5, where the GCPs are positioned at the flight home. Therefore, this solution should be avoided unless there is no other possibility. Besides, in this situation, it should be ensured that the GCPs can be commonly visible in at least three images. In order to plan any of these situations, the flight-planning software (Section 4.1) can be used.

7.5. Effects of Imaging Configuration

The above-mentioned experiments prove that careful application of minimum GCPs can also yield a high level of modeling accuracy. However, such accuracy level is only achievable if images provide a stable imaging and network configuration. The importance of this fact is analyzed based on the overlap tests described in Section 6.2.

When performing BBA, the coordinate datum requires seven defined elements to compensate its rank deficiencies, namely scale, position and rotation. These defined elements can be provided with either minimum constraints in controlled networks or inner constraints in free networks. When enough overlap exists among images, both free and controlled network adjustments can be performed correctly without facing any additional rank deficiencies. Figure 24a shows the orientations of cameras and ground coordinates of tie points calculated correctly in a free network based on OverlapTest 1.

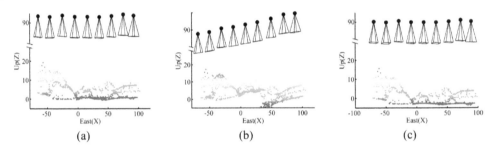

Figure 24. Effect of imaging configuration (**a**) Inner-constrained adjustment based on OverlapTest 1; (**b**) Inner-constrained adjustment based on OverlapTest 2; (**c**) Controlled adjustment based on OverlapTest 2 using four GCPs.

However, only one image not having enough overlap with its adjacent ones can disturb this ideal configuration. As in OverlapTest 2, image 5 does not have any common tie point with image 3. Therefore, there is no tie point to make a connection between one part of the network including images 1–4 and the other part of the network including images 5–9. Notice that this disconnection happens even though image 5 and image 4 have common tie points. As a result, the coordinate datum faces eight rank deficiencies—one additional scale deficiency. In order to resolve this, one more constraint is required. If no ground-truth measurement is available, then the solution is to assign an arbitrary scale factor to one of the unknowns. Figure 24b illustrates the results by assigning a scale factor based on the DG data to one of the tie points between images 4 and 5. In this situation, although the BBA can be solved, a wrong scale change is introduced between the two parts of the network. As a result, this solution must be avoided unless the DG data are very accurate. The practical solution to this problem is to add the ground observations of control points to the adjustment. In this example, control point D provides the additional scale constraint required to solve the 8th rank deficiency of the datum (Figure 24c). It can be concluded, that configurations of both the terrestrial data (GCPs or any other types of ground measurements) and the aerial images decide the final accuracy of 3D modeling.

7.6. On-the-job Self-Calibration Results

As the results in Section 7.1 show, noisy image observations affect the results of self-calibration to a great extent even if the noise level is very low. Similarly, on-the-job self-calibration of aerial images is affected by the noise in images, which is usually inevitable in UAV imagery. Another factor that affects the accuracy of on-the-job self-calibration is the particular configuration of aerial network. That is the images are acquired from a relatively fixed altitude. In fact, this network configuration reduces the numerical stability of calibration in terms of the increase in the correlation between the unknown parameters. Especially, IO parameters—the principal point offset and focal length (x_p, y_p, f)—become strongly correlated with EO parameters—the position of the camera center (C_x, C_y, C_z). As a result, intrinsic calibration parameters become physically meaningless since they become dependent parameters that change relatively with the changes of EO parameters.

This effect can be practically controlled in close-range photogrammetry by providing various orientations and object depth levels as in CalTest 1. Figure 25a presents the correlation analysis of the self-calibration based on these images. As it is noticed, this condition results in very low correlation between IO and EO parameters. However, the same analysis for on-the-job self-calibration based on aerial images of CalTest 2 presents very high correlation between IO and EO parameters, specifically between focal length and imaging depth (Figure 25b).

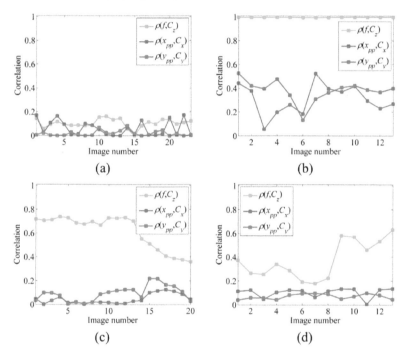

Figure 25. Correlation analysis in self-calibration based on (**a**) CalTest 1; (**b**) CalTest 2; (**c**) CalTest 3; (**d**) CalTest 2 by applying the proposed BBA strategy.

One of the advantages of UAVs is that they can fly obliquely and at very low altitudes. Therefore, it is possible to provide more orientation variations as in CalTest 3. As a result, the average correlation between focal length and imaging depth can be reduced 38% by this new configuration (Figure 25c). However, such maneuvers are not possible in all the UAV mapping applications. Therefore, the solution proposed in Section 5.2 can be applied to improve the self-calibration. This strategy reduces the correlation between the unknowns without the need to change the network configuration (Figure 25d). For instance, the average correlation between focal length and imaging depth is reduced 60%.

7.7. Application-Dependant Results

Figure 26a presents the major cut/fill regions based on dataset B and dataset C that were gathered with an interval of two months. As expected, most places at this zone were excavated. In Figure 26b, the volumetric change per cell is measured for every cell of the DSM. The volumetric change is measured as the difference of elevation in the before-DSM from the after-DSM which is multiplied by the cell area (1.69 cm^2). Therefore, positive values represent excavation or cut, and negative values represent fill. The vertical accuracies of dataset B (before) and dataset C (after) are 1.32 cm and 1.63 cm, respectively. Therefore, the volumetric change measurement at each cell is performed with accuracy of 3.54 cm^3. Figure 27 presents the slope map based on dataset A. As it can be seen, very detailed slope information is extractable from such a map, which can be used in various geological applications.

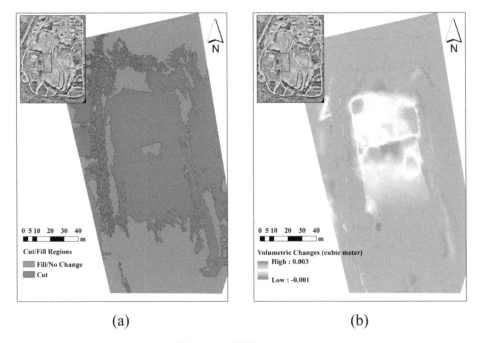

(a)   (b)

Figure 26. (a) Cut/fill regions; (b) Volumetric change measurement.

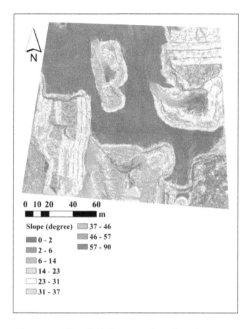

Figure 27. Classified slope map based on dataset A.

8. Conclusions

Various aspects of the development and implementation of a UAV-PS were discussed in this study. These included the camera offline calibration, platform calibration, system integration, flight and fieldwork planning, data acquisition, photogrammetric processing and application in open-pit mine mapping. Based on the experiments, it was concluded that the accuracy of 3D modeling with the system, either in terms of the accuracy of indirect georeferencing or the spatial accuracy of the point clouds, was better than 2 cm.

In addition to general photogrammetric experiments, several tests were performed to analyze the specific issues of UAV-based 3D modeling, and solutions were proposed to address them. It is hoped that the lessons learnt from these experiments give a more clear insight of the capacities of UAV-PSs for the upcoming studies and applications. In brief, the impact of automatic target detection on the accuracy of camera calibration was investigated. It was shown that an improvement of 81% in the accuracy of calibration could be achieved with our target detection technique in comparison with manual target detection. Regarding the system integration, it was validated that the developed software package was capable of synchronizing the navigation and imaging sensors with an approximate delay of 11 ms without requiring any additional mechanism. Moreover, the impacts of high photometric variations among images and shadowed regions on the accuracy of 3D modeling were verified. Besides, the use of a BBA strategy was suggested to improve the accuracy of on-the-job self-calibration by reducing the correlation of intrinsic camera calibration parameters to other BBA elements such as EO parameters. It was shown that, using this strategy, the correlation of IO and EO parameters could be reduced by 60% in an unsuitable imaging network. This strategy can be used in applications where the accurate, on-the-flight intrinsic calibration parameters are required independently. Furthermore, several experiments were performed to assess the effect of GCPs configuration on modeling accuracy. It was shown that a minimum number of GCPs could provide a high accuracy level if they were distributed evenly over the whole zone and their visibilities in images were maximized. However,

Sensors **2015**, *15*, 27493–27524

under such conditions, the scale consistency of the imaging network needed to be ensured by providing high overlap among images.

Acknowledgments: This study is supported by grants from the: Centre de Géomatique du Québec, Fonds de Recherche Québécois sur la Nature et les Technologies, and Natural Sciences and Engineering Research Council of Canada. The authors would also like to thank Kim Desrosiers from Centre de Géomatique du Québec who took time from his schedule to help us optimize the system-integration software.

Author Contributions: Mozhdeh Shahbazi performed the theoretical developments and planning of this system. Unless otherwise mentioned in the paper, she wrote the software packages for implementing the system, performing the experiments and analyzing the results. Furthermore, she prepared the first draft of this manuscript. The manuscript was revised and modified by the other authors as well Patrick Ménard performed the hardware development. All the authors assisted in data acquisition and fieldworks. Gunho Sohn and Jérôme Théau provided the direction and supervision of this study. They provided their constructive comments and consults to perform this study.

Conflicts of Interest: The authors declare no conflict of interest.

References

1. Shahbazi, M.; Théau, J.; Ménard, P. Recent applications of unmanned aerial imagery in natural resource management. *Gisci. Remote Sens.* **2014**, *51*, 339–365. [CrossRef]
2. Anderson, K.; Gaston, K.J. Lightweight unmanned aerial vehicles will revolutionize spatial ecology. *Front. Ecol. Environ.* **2013**, *11*, 138–146. [CrossRef]
3. Liu, P.; Chen, A.; Huang, Y.; Han, J.; Lai, J.; Kang, S.; Wu, T.; Wen, M.; Tsai, M. A review of rotorcraft unmanned aerial vehicle (UAV) developments and applications in civil engineering. *Smart Struct. Syst.* **2014**, *13*, 1065–1094. [CrossRef]
4. Anai, T.; Sasaki, T.; Osaragi, K.; Yamada, M.; Otomo, F.; Otani, H. Automatic exterior orientation procedure for low-cost UAV photogrammetry using video image tracking technique and GPS information. *Int. Arch. Photogramm. Remote Sens. Spat. Inf. Sci.* **2012**. [CrossRef]
5. Bahr, T.; Jin, X.; Lasica, R.; Giessel, D. Image registration of high-resolution UAV data: The new hypare algorithm. *Int. Arch. Photogramm. Remote Sens. Spat. Inf. Sci.* **2013**. [CrossRef]
6. Boccardo, P.; Chiabrando, F.; Dutto, F.; Tonolo, F.G.; Lingua, A. UAV deployment exercise for mapping purposes: Evaluation of emergency response applications. *Sensors* **2015**, *15*, 15717–15737. [CrossRef] [PubMed]
7. Turner, D.; Lucieer, A.; Wallace, L. Direct georeferencing of ultrahigh-resolution UAV imagery. *IEEE Trans. Geosci. Remote* **2014**, *52*, 2738–2745. [CrossRef]
8. Xiang, H.; Tian, L. Method for automatic georeferencing aerial remote sensing (RS) images from an unmanned aerial vehicle (UAV) platform. *Biosyst. Eng.* **2011**, *108*, 104–113. [CrossRef]
9. Chiang, K.W.; Tsai, M.L.; Chu, C.H. The development of an UAV borne direct georeferenced photogrammetric platform for ground control point free applications. *Sensors* **2012**, *12*, 9161–9180. [CrossRef] [PubMed]
10. Turner, D.; Lucieer, A.; Watson, C. An automated technique for generating georectified mosaics from ultra-high resolution unmanned aerial vehicle (UAV) imagery, based on structure from motion (SfM) point clouds. *Remote Sens.* **2012**, *4*, 1392–1410. [CrossRef]
11. Chiang, K.W.; Tsai, M.L.; Naser, E.S.; Habib, A.; Chu, C.H. New calibration method using low cost MEMS IMUs to verify the performance of UAV-borne MMS payloads. *Sensors* **2015**, *15*, 6560–6585. [CrossRef] [PubMed]
12. Ai, M.; Hu, Q.; Li, J.; Wang, M.; Yuan, H.; Wang, S. A robust photogrammetric processing method of low-altitude UAV images. *Remote Sens.* **2015**, *7*, 2302–2333. [CrossRef]
13. Wu, C.T.; Hsiao, C.Y.; Chen, C.S. An assessment of errors using unconventional photogrammetric measurement technology with UAV photographic images as an example. *J. Appl. Sci. Eng.* **2013**, *16*, 105–116.
14. Remondino, F.; Fraser, C. Digital cameras calibration methods: Considerations and comparisons. In Proceedings of the ISPRS Commission V Symposium on Image Engineering and Vision Metrology, Dresden, Germany, 19 September 2006; pp. 266–272.

15. Rieke-Zapp, D.; Tecklenburg, W.; Peipe, J.; Hastedt, H.; Haig, C. Evaluation of the geometric stability and the accuracy potential of digital cameras-comparing mechanical stabilisation *versus* parameterisation. *ISPRS J. Photogramm.* **2009**, *64*, 248–258. [CrossRef]

16. Yang, Y. Robust estimation for dependent observation. *Manuscr. Geod.* **1994**, *19*, 10–17.

17. Zhang, Y.; Xiong, J.; Hao, L. Photogrammetric processing of low-altitude images acquired by unpiloted aerial vehicles. *Photogramm. Rec.* **2011**, *26*, 190–211. [CrossRef]

18. Ivory, J. An Evaluation of Photogrammetry as a Geotechnical Risk Management Tool for Open-Pit Mine Studies and for the Development of Discrete Fracture Network Models. Master's Thesis, UCL Australia, Adelaide, SA, Australia, 2012.

19. Patikova, A. Digital photogrammetry in the practice of open pit mining. *Int. Arch. Photogramm. Remote Sens. Spat. Inf. Sci.* **2004**, *34*, 1–4.

20. Siebert, S.; Teizer, J. Mobile 3D mapping for surveying earthwork projects using an Unmanned Aerial Vehicle (UAV) system. *Autom. Constr.* **2014**, *41*, 1–14. [CrossRef]

21. Bemis, S.P.; Micklethwaite, S.; Turner, D.; James, M.R.; Akciz, S.; Thiele, S.T.; Bangash, H.A. Ground-based and UAV-based photogrammetry: A multi-scale, high-resolution mapping tool for structural geology and paleoseismology. *J. Struct. Geol.* **2014**, *69*, 163–178. [CrossRef]

22. Cryderman, C.; Mah, S.B.; Shufletoski, A. Evaluation of UAV photogrammetric accuracy for mapping and earthworks computations. *Geomatica* **2014**, *68*, 309–317. [CrossRef]

23. Holst, G.C.; Lomheim, T.S. *CMOS/CCD Sensors and Camera Systems*, 2nd ed.; JCD Publishing: Winter Park, FL, USA, 2007.

24. Luhmann, T.; Robson, S.; Kyle, S.; Harley, I. *Close Range Photogrammetry: Principles, Techniques and Applications*; John Wiley & Sons: Hoboken, NJ, USA, 2007.

25. Brown, D.C. Close-range camera calibration. *Photogramm. Eng.* **1971**, *37*, 855–866.

26. Dörstel, C.; Jacobsen, K.; Stallmann, D. DMC—Photogrammetric Accuracy—Calibration Aspects and Generation of Synthetic DMC Images. In Proceedings of Optical 3D Sensor Workshop, Zurich, Switzerland, 6 September 2003; pp. 4–12.

27. Grewal, M.S.; Weill, L.R.; Andrews, A.P. *Global Positioning Systems, Inertial Navigation, and Integration*, 2nd ed.; John Wiley & Sons Publication: Hoboken, NJ, USA, 2007.

28. Honkavaara, E.; Hakala, T.; Markelin, L.; Rosnell, T.; Saari, H.; Makynen, J. A process for radiometric correction of UAV image blocks. *Photogramm. Fernerkund.* **2012**, *2*, 115–127. [CrossRef]

29. Mukherjee, D.; Wu, Q.J.; Wang, G. A comparative experimental study of image feature detectors and descriptors. *Mach. Vis. Appl.* **2015**, *26*, 443–466. [CrossRef]

30. Hirschmüller, H.; Scharstein, D. Evaluation of stereo matching costs on images with radiometric differences. *IEEE Trans. Pattern Anal.* **2009**, *31*, 1582–1599. [CrossRef] [PubMed]

31. Jung, I.L.; Chung, T.Y.; Sim, J.Y.; Kim, C.S. Consistent stereo matching under varying radiometric conditions. *IEEE Trans. Multimed.* **2013**, *15*, 56–69. [CrossRef]

32. Adeline, K.R.M.; Chen, M.; Briottet, X.; Pang, S.K.; Paparoditis, N. Shadow detection in very high spatial resolution aerial images: A comparative study. *ISPRS J. Photogramm.* **2013**, *80*, 21–38. [CrossRef]

33. Sona, G.; Pinto, L.; Pagliari, D.; Passoni, D.; Gini, R. Experimental analysis of different software packages for orientation and digital surface modelling from UAV images. *Earth Sci. Inform.* **2014**, *7*, 97–107. [CrossRef]

34. Chung, K.L.; Lin, Y.R.; Huang, Y.H. Efficient shadow detection of color aerial images based on successive thresholding scheme. *IEEE Trans. Geosci. Remote Sense.* **2009**, *47*, 671–682. [CrossRef]

35. Rumpler, M.; Daftry, S.; Tscharf, A.; Prettenthaler, R.; Hoppe, C.; Mayer, G.; Bischof, H. Automated end-to-end workflow for precise and geo-accurate reconstructions using fiducial markers. *ISPRS Ann. Photogramm. Remote Sens. Spat. Inf. Sci.* **2014**. [CrossRef]

36. Pix4D, UAV Mapping Software. Available online: https://pix4d.com (accessed on 10 August 2015).

37. Wolf, P.R.; Dewitt, B.A. *Elements of Photogrammetry: With Applications in GIS*, 3rd ed.; The McGraw-Hill Companies: Boston, MA, USA, 2000.

38. Vanicek, P.; Krakiwsky, E.J. *Geodesy: The Concepts*, 2nd ed.; Elsevier Science Publishers: New York, NY, USA, 1986; pp. 242–283.

39. CloudCompare-Open Source project. Available online: http://www.danielgm.net/cc (accessed on 10 August 2015).
40. Bay, H.; Ess, A.; Tuytelaars, T.; van Gool, L. Speeded-up robust features (SURF). *Comput. Vis. Image Underst.* **2008**, *110*, 346–359. [CrossRef]

Article

Prototyping a GNSS-Based Passive Radar for UAVs: An Instrument to Classify the Water Content Feature of Lands

Micaela Troglia Gamba [1,*], Gianluca Marucco [1], Marco Pini [1], Sabrina Ugazio [2], Emanuela Falletti [1] and Letizia Lo Presti [2]

[1] Istituto Superiore Mario Boella (ISMB), Via P.C. Boggio 61, 10138 Torino, Italy; E-Mails: marucco@ismb.it (G.M.); pini@ismb.it (M.P.); falletti@ismb.it (E.F.)

[2] Politecnico di Torino - Corso Duca degli Abruzzi 24, 10129 Torino, Italy; E-Mails: sabrina.ugazio@polito.it (S.U.); letizia.lopresti@polito.it (L.L.P.)

* E-Mail: trogliagamba@ismb.it; Tel.: +39-11-2276-447; Fax: +39-11-2276-299.

Academic Editor: Felipe Gonzalez Toro

Received: 10 July 2015 / Accepted: 2 November 2015 / Published: 10 November 2015

Abstract: Global Navigation Satellite Systems (GNSS) broadcast signals for positioning and navigation, which can be also employed for remote sensing applications. Indeed, the satellites of any GNSS can be seen as synchronized sources of electromagnetic radiation, and specific processing of the signals reflected back from the ground can be used to estimate the geophysical properties of the Earth's surface. Several experiments have successfully demonstrated GNSS-reflectometry (GNSS-R), whereas new applications are continuously emerging and are presently under development, either from static or dynamic platforms. GNSS-R can be implemented at a low cost, primarily if small devices are mounted on-board unmanned aerial vehicles (UAVs), which today can be equipped with several types of sensors for environmental monitoring. So far, many instruments for GNSS-R have followed the GNSS bistatic radar architecture and consisted of custom GNSS receivers, often requiring a personal computer and bulky systems to store large amounts of data. This paper presents the development of a GNSS-based sensor for UAVs and small manned aircraft, used to classify lands according to their soil water content. The paper provides details on the design of the major hardware and software components, as well as the description of the results obtained through field tests.

Keywords: UAV; GNSS-reflectometry; GNSS bistatic radar; prototyping

---

1. Introduction

Over the past few decades, Global Navigation Satellite System (GNSS) signals have not been used solely for navigation purposes. Indeed, since satellites can be considered a passive source of radiation, GNSS signals have been used for remote sensing applications, which consist of the processing of GNSS signals reflected back from the ground. Such reflected signals can be used to characterize the Earth's surface, because they have different characteristics from those of the signal directly received from the satellite, in terms of delay, Doppler shift, power strength and polarization. These differences depend on the geophysical properties of the scattering surface; therefore, they potentially carry information about the surface geophysics.

GNSS signals are broadcast over the L-band, and many experiments have successfully demonstrated GNSS-reflectometry (GNSS-R) for the remote sensing of land and ocean surfaces [1] using the GPS L1 at 1575.42 MHz. Wind retrieval and altimetry, mainly from static platforms [2], are the most consolidated applications, while new employments, such as soil moisture sensing, ice monitoring, water level and snow thickness measurements [3], are continuously emerging and are presently under development. More recently, the joint use of GNSS-R data and other sensors, such as optical, infrared, thermal and microwave radiometers, turns out to be promising for accurate soil moisture estimation [4] and sea surface salinity retrieval [5,6]. In addition to better environmental monitoring, it is expected that new GNSS-R data will represent valuable inputs to numerical weather prediction (NWP) systems. Although today's NWPs are capable of predicting many meteorological events, their accuracy is sometimes poor or they have an insufficient lead-time to initiate actions aimed at protecting life and property. For instance, uncertainty in present meteorological forecasts and the lack of integration of currently scattered monitoring networks represent a bottleneck for flood and drought risk assessment at local and regional scales. GNSS-R can be a means to provide additional data at low cost, mainly if new GNSS-R devices are mounted on-board unmanned aerial vehicles (UAVs). Today, UAVs offer a broad range of solutions for many civilian applications and can be equipped with several types of sensors for environmental monitoring; some UAVs are also more cost effective with respect to manned light aircraft. Until now, many instruments for GNSS-R have been proposed, and several algorithms have been developed for the estimate of geophysical properties of the scattering materials (e.g., [1,7]) and for altimetry (e.g., [2,3,8,9]). The hardware of traditional GNSS-based passive radars consists of custom GNSS receivers, based on application-specific integrated circuits (ASIC) or field programmable gate arrays (FPGA). In most cases, they require a personal computer (PC) and sufficient memory to store a large amount of data [10,11]. GNSS-based passive radars use a right-hand circular polarized (RHCP) antenna pointing toward the zenith for the reception of the direct signals from satellites and a second left-hand circular polarized (LHCP) antenna pointing towards the nadir for the reception of the reflected signals. Some devices (e.g., [12]) have been designed to collect the LHCP-only reflected component, because most of the reflected power has this polarization. More advanced versions (e.g., [13]) enable the reception of both LHCP and RHCP polarizations, because even weak RHCP reflected signals carry valuable information for precise measurements, like for the estimate of the soil moisture. It is worth noticing that other configurations making simultaneous use of horizontally- and vertically-polarized antennas, such as in [14], are possible, but their use on-board a small UAV would pose several mounting and reception issues.

The main drawback of many GNSS-based bistatic radars proposed so far is the heavy and bulky set up, which prevents the use of such devices on-board small and light UAVs. To overcome this problem, some researchers spent effort to design more compact and portable instruments. For example, Esterhuizen [15] proposed a software receiver over a Nano-ITX single board computer combined with two radio frequency (RF) front-ends featuring a common clock, connected to a universal serial bus (USB) bridge for high-speed data transfer. In [16,17], a prototype of an FPGA-based real-time GPS reflectometer is presented, which computes the full two-dimensional delay Doppler maps every 1 ms and performs coherent and incoherent averaging. Other remarkable examples are the designs of Starlab: Oceanpal® [12] and the SAMGNSS reflectometer [18,19] are two instruments that they developed. While the former collects the LHCP reflected GNSS signals from the sea surface, the latter enables the reception of both polarization components of the reflected signal for soil moisture retrieval [13,18–21].

This paper presents the design and prototyping of a GNSS-based bistatic radar for small UAVs to be employed in environmental monitoring campaigns, for the water content classification of land and for the detection of water surfaces. Section 2 introduces the major requirements that guided the first phase of the design: among all, the light weight and the reduced size of the sensor, as well as the need for a GNSS antenna able to receive the LHCP and RHCP components of the reflected signals over two separate channels. Section 3 provides an overview of the hardware architecture, with a functional block diagram and the layout of the components, which were integrated into a case with an airfoil shape. It also explains some details of the software running on the microprocessor that controls the overall system, developed under a software radio paradigm to introduce flexibility. The remarkable hardware feature consists of the capability to simultaneously collect both polarizations' data streams, synchronized with the same clock. This makes the proposed sensor different with respect to the reflectometer presented in [18], which, on the contrary, switches among the reflected RHCP and LHCP RF signals and, therefore, processes them in a sequential way. Although the focus of this work is not on the reflectometry data post-processing, but on the GNSS sensor design and implementation, Section 4 briefly discusses the background of such a discipline to give better evidence of the overall process that starts from GNSS measurements and ends with moisture-related information. Thus, Section 5 proceeds with the description of some results obtained in the field with a small aircraft. Finally, Section 6 concludes the paper with some open issues and the expected developments and exploitation of this work.

2. Rationale and Requirements

The objective of this work is the design and prototyping of the on-board sensor to collect measurements of GNSS reflected signals suitable to enable the estimate of some soil parameters, in particular the soil moisture, using the GNSS sensor mounted on-board a small, possibly unmanned, aircraft. To implement the radar capabilities, the direct signal coming from the satellite is received for positioning purposes, in order to evaluate and geo-reference the specular reflection point on the ground, as described in Section 4; furthermore, the characteristics of the direct signal are used as a reference for the processing of the reflected one, to enable the remote sensing of the soil features.

The design flow of the on-board sensor has been distributed over three main layers, depicted in Figure 1:

- the hardware platform,
- the GNSS signal processing and
- the signal processing for soil parameter retrieval.

The first step is the definition of the hardware architecture, *i.e.*, the RF front-end, the microprocessor board and the antennas. Then, the design of proper GNSS signal processing techniques follows, to detect and estimate the relative delay and the amplitude of the reflected GNSS signals. The major concern at this stage is the extreme weakness of the reflected signals, which may lose around 13 dB for the LHCP and 23 dB for the RHCP with respect to the direct signal [18], received at a nominal power on the order of −160 dBW. Finally, proper remote sensing algorithms post-process the raw GNSS data.

The focus of this paper is explicitly on the hardware architecture of the prototype. For this reason, only a few details are given about the signal processing and soil parameter retrieval algorithms; the interested reader may refer to [22,23]. Nonetheless, Section 5 shows some examples of the results obtained from processing the data recorded by the prototype, during one of the test flights.

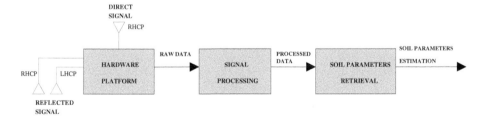

Figure 1. High-level information flow and the three layers of the prototype design.

The design of the GNSS-based bistatic radar was guided by a set of requirements dictated by the target application. The subset of functional requirements applicable to the hardware platform discussed here is constituted by six propositions:

Advanced antenna configuration. The sensor shall be able to handle three antennas, implying three RF chains and three digital streams, as depicted in Figure 1.

Storage capability. The sensor shall be able to store raw measurements, observed during a flight mission.

Direct and reflected signals' synchronization. The three signal streams (one direct, two reflected) shall be synchronized in sampling, storage and processing. Since the reflected signals are in general very weak, their processing can obtain significant benefits if aided by the direct signal, but aiding procedures require the synchronization among the three streams.

Flexibility. The radar shall be programmable and reconfigurable, at least in terms of the receiving bandwidth, signal conditioning and digital signal processing parameters (e.g., the number of correlation points).

Processing capability. The radar should be equipped with enough computing resources to allow the implementation of some on-board digital signal processing algorithms.

Size and weight. The sensor must be lightweight, *i.e.*, <3 kg, and small, *i.e.*, $\leq$200 mm $\times$ 250 mm $\times$ 250 mm (length $\times$ width $\times$ height), to be mounted on-board UAVs and light aircraft.

3. Prototype Design

From the high level application requirements listed above and from the indications received by the UAV manufacturer during the phases of development, the fundamental features of the sensor's hardware platform were derived. They are presented in the next subsections, organized by hardware components, software components and functional validation.

3.1. Hardware Components

The essential hardware components of the on-board sensor are:

1. the GNSS antennas,
2. the commercial off-the-shelf (COTS) RF front-ends (FEs) and
3. the digital signal processing (DSP) stage.

Figure 2 shows the functional block diagram of the sensor, with the major hardware components highlighted and their connections.

While a conventional low-cost hemispherical GNSS L1 RHCP patch antenna, properly mounted to point to the zenith and normally available on-board, is enough to receive the direct GNSS signals, the reflected ones require an *ad hoc* antenna oriented toward the nadir. Since one of our purposes was the reception of the reflected signals with both polarizations (LHCP + RHCP), we preferred a single dual-polarization antenna instead of two separate single-polarization ones, in order to limit weight and volume. However, very stringent requirements were set against the level of cross-polarization isolation, which represents a measure of the cross-talk between the two nominal polarizations. The work in [18] suggests a value lower than −24 dB, in particular for the measurement of the very weak RHCP component of the reflected signals against the stronger LHCP component. Unfortunately, commercial products typically do not meet both requirements of weight and cross-polarization isolation. Nonetheless, we decided to adopt the dual-polarization L1/L2 GNSS Antcom antenna 1G1215RL-PP-XS-X RevA [24], whose cross-polarization isolation declared by the manufacturer is −17 dB; despite its suboptimal performance in polarization separation to perform precise GNSS-R polarimetric measurements [18], it is light, small and has a quite flat profile. Another custom Antcom device based on the G8ANT-52A4SC1-RL model, whose cross-polarization rejection specification was set at −24 dB, showed RF compatibility problems with the front-end and cross-polarization issues, making its use more difficult and even having lesser performance.

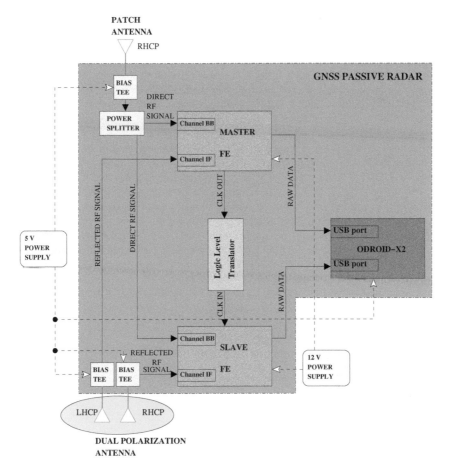

Figure 2. Hardware scheme of the sensor.

The second stage of the hardware platform is the RF front-end. It includes the stages of signal conditioning, RF down conversion, filtering and analog-to-digital conversion (ADC). The number of RF chains in the FE must be equal to the number of separate signals and polarizations: one RF chain is employed for the direct signal and connected to the zenith-pointing antenna, while two RF chains are devoted to the two LHC- and RHC-polarized reflected signals and connected to the two ports of the nadir-pointing antenna. The same clock reference must be distributed on the three chains. To implement this setup, two FEs of the "Stereo" family commercialized by Nottingham Scientific Ltd (NSL) were selected [25], configured in a master/slave architecture. Each FE embeds two full and synchronized receiving chains, implemented in two separate chipsets: the MAX2769B, covering the GNSS upper L-band and indicated as L1, and the MAX2112, covering both the upper and lower L-bands and indicated as LB. In Figure 2, we identified the two chains as "Channel IF" and "Channel BB" respectively, to indicate the different down conversion schemes applied in the two chains. The FE contains one shared clock (TXC 26 MHz TCXO) with interfaces for alternative oscillators and external frequency input. Slaving two FEs to the same clock guarantees four synchronized channels. To do this operation, a logic level translator, from low voltage positive emitter coupled logic (LVPECL) to low voltage complementary metal oxide semiconductor (LVCMOS) levels was specifically designed and manufactured as a printed circuit board (PCB). In Figure 2, the exact connection between antennas and FE chains is indicated: the direct signal is split and sent to Channel BB of both the FEs as a reference, while the reflected LHCP and RHCP are connected to the Channel IF of the master and slave boards, respectively.

Finally, a DSP stage is necessary to process the digital data after the ADC. For our purposes, the software-defined radio (SDR) is the preferred technology over other solutions like FPGA or ASIC-based platforms, thanks to its flexibility, re-configurability and reduced development time in the prototype integration. The DSP stage was required to support a memory of at least some tens of GBytes, e.g., in a secure digital (SD) card or an embedded multimedia card (eMMC) card, for fast storage of the raw data produced during a flight mission. A number of I/O USB ports was also necessary to handle digitalized data streams and to give access to the sensor configuration parameters. The chosen platform is the Open-Android (ODROID)-X2 [26]. It is an open development 1.7 GHz ARM Cortex-A9 Quad Core platform with 2 GB RAM memory and PC-like performance. The ODROID-X2 was one of the most powerful boards available on the market at the time the activity began. It provides 2 GB RAM memory and a number of peripherals, like a high-definition multimedia interface (HDMI) monitor connector and six USB ports. It is able to support the input from the two FEs streams, thanks to the real-time management of two USB ports and the fast memory storage. This board hosts an Ubuntu Linaro Operative System (OS) distribution, booting from a 64 GB eMMC. In the current version, the sensor serves as data grabber: data are received, sampled and stored in the memory. Further developments will address the implementation of some more advanced processing directly on-board.

As indicated in Figure 2, a power supply of 5 V is employed for the ODROID-X2 and all of the antennas, while 12 V is used for the FEs in order to guarantee a proper functioning of the device, in particular the stability of the master clock. The bias-tees (BTs) allow a stable power supply to the antennas, while decoupling the DC from the RF signal entering the FE.

A summary of the fundamental hardware components of the radar prototype is reported in Table 1.

Table 1. Summary of the selected principal hardware components. eMMC, embedded multimedia card.

Hardware Component	Selected Device
Antenna (towards zenith):	Aircraft's hemispherical L1 patch
Antenna (toward nadir):	Antcom dual-polarization L1/L2 1G1215RL-PP-XS-X RevA
RF front-end:	NSL Stereo (2 boards mutually synchronized)
DSP (μ-processor board):	ODROID-X2, 1.7 GHz ARM Cortex-A9 Quad Core platform, 2 GB RAM
Memory:	64 GB eMMC

3.2. Hardware Assembly

After the definition of the functional architecture, the system components were assembled inside a proper case. The authors already showed a preliminary assembled prototype in [27], but such a configuration implied a parallelepiped-shaped case, which was not the best in terms of aerodynamic performance. For this reason, the final sensor case was designed *ad hoc* in carbon fiber with a neutral wing profile, and the internal components were mounted accordingly. Some computer-aided design (CAD) views are reported in Figure 3. In particular, Figure 3A depicts the carbon fiber case in light violet, double-ended with two aluminum plates in grey. The bottom plate serves as the support for the nadir-oriented antenna, depicted by the dark violet cylindrical disk, while the top lid is designed to be screwed to a rectangular plate, called the trolley unit, which connects the sensor mechanically and electrically to the aircraft body. The trolley is specifically designed to be hosted in a structure (bay) composed of two rails and fastened externally to the lower part of the aircraft, for rapid boarding of the sensors. The bay can host 3–4 sensor carts boarded with a "plug-and-play" mechanism. The yellow object, which stands out from the case, acts as heat sink of the internal parts: in fact, it terminates with a cylindrical part directly in contact with the microprocessor of the ODROID-X2 board, depicted in orange. The two FEs in green are just behind the ODROID-X2. All of the boards are attached to the internal faces of the case by means of small resin supports. The chosen airfoil is better visualized in

Figure 3B, where it is possible to observe how the internal components are placed, in order to optimize the available space.

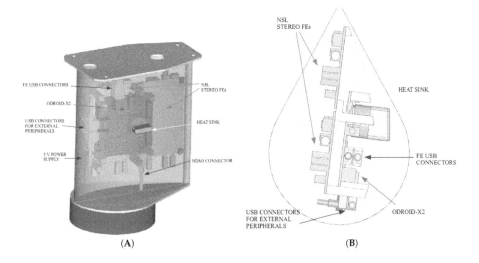

Figure 3. Three-quarter (**A**) and top (**B**) views of the sensor case 3D and 2D CADs, with the main internal components visible.

Once the carbon fiber case has been manufactured, the assembly of the sensor was completed. Figure 4 shows some pictures of the final version of the prototype. In particular, the carbon fiber case is well visible in Figure 4A,B: it is screwed to the nadir-oriented antenna at the bottom side, by means of a circular aluminum plate with the function of both support and ground plane, and to the trolley unit at the top side. Figure 4B shows the sensor under lab testing connected to an HDMI monitor, mouse, keyboard and external power supply. The case has been specifically designed in such a way as to be easily opened, for fast checks and maintenance service, as illustrated in Figure 4C,D, where internal components and connections are visible.

Figure 4. Photos of the sensor prototype, closed in its carbon fiber case (**A**), opened (**C,D**) showing internal components and equipped with a keyboard, mouse, HDMI monitor and power supply for the in laboratory tests (**B**).

Designed and assembled in such a way, the sensor prototype dimensions resulted in being 200 mm × 200 mm × 250 mm (length × width × height) with a weight of less than 3 kg, including the nadir-pointing antenna with its ground plane, so as to be sufficiently light and compact to be mounted on-board UAVs and small aircraft. In particular, the target UAV belongs to the civilian category of remotely-controlled (RC) light UAVs, or more in general, light unmanned aircraft systems (UASs), with a maximum take-off mass (MTOM), including the fuel, of less than 150 kg and an autonomy of 1–2 h. Additionally, the manned airplane is a two-seat ultralight aircraft with fixed wings.

3.3. Software Components

As said in Section 3.1, the ODROID-X2 has PC-like performance: it features several peripherals and hosts an Ubuntu Linaro Operative System (OS). Consequently, the entire development work was done directly on the target platform, with no need for another machine for cross-compiling. Nevertheless, the implementation of the master-slave configuration required not only the integration of the logic level translator, described in Section 3.1, at the hardware level, but also some additional work at the software level: the original FE drivers required to be modified and recompiled on the target platform, in order to manage the connection of the two FEs unambiguously.

Software components include routines for FE configuration, usage modes and grabbing functionality. Thanks to the flexibility of the selected FEs, based on the SDR paradigm, the user is allowed to create his or her own setup, configuring the parameters listed in the first column of Table 2. In our setup, the parameter configuration is the same for both FEs in the sensor. In order to manage two USB data streams, the default sampling frequency chosen for the prototype was the lowest possible permitted by the manufacturer; this choice allows for storing more than 30 min of data on the ODROID-X2 eMMC, which is sufficient for our purposes.

Table 2. Configuration options for the RF front-ends. Channel IF and Channel BB are the two RF chains of each front-end. The rightmost column contains default values used during the tests. The admissible ranges are derived from the examples presented in the Stereo front-end (FE) user manual and have not been completely tested by the authors.

Configurable Parameter	Admissible Range	Default Value
Sampling frequency	13 ÷ 40 MHz	13 MHz
Channel IF, carrier frequency	{L1, E1, G1}	1575.42 MHz
Channel IF, intermediate frequency	Not specified	3.55 MHz
Channel IF, double-sided bandwidth	2 ÷ 9.66 MHz	4.2 MHz
Channel BB, carrier frequency	{L1, E1, G1, L2, G2, L5, E5a, E5b}	1575.42 MHz
Channel BB, intermediate frequency	0 MHz	0 MHz
Channel BB, single-sided bandwidth	1.39 ÷ 10.09 MHz	4.0 MHz
Channel BB, filter gain	0 ÷ 15 dB	6 dB

The sensor can be used in two modes, based on the number of signals the user desires to process:

1. Basic mode: direct channel + one LHCP reflected channel (only the master FE enabled);
2. Advanced mode: direct channel + two reflected channels (LHCP and RHCP).

Each mode is implemented via software, using proper shell scripts for the FE configuration, which are executed as startup applications. In this way, at power up, the ODROID-X2 automatically boots the OS, configures the FEs based on one of the above-described usage modes and launches the data grabbing, which uses the eMMC module as the storage unit. The start and stop commands and the duration of data grabbing are parameters to be defined based on the flight plan. Note that, with a 13 MHz sampling frequency, the two modes have a different impact on the necessary amount of memory: the advanced configuration requires 1.56 GB/min, allowing one to save more than 30 min of data, while the basic configuration halves the rate, thus doubling the total amount of storable data. Since the raw data are stored to memory, ready for off-line processing, the sensor is fully enabled for all of the available GNSS signals, and the information of all visible satellites is fully preserved: this approach facilitates a thorough validation of the prototype and an accurate interpretation of the soil parameter estimation.

3.4. Functional Tests for the Validation of the Sensor

For the functional validation of the prototype, an intensive test campaign was conducted in the laboratory. Such tests were divided into two categories: first, we validated the sensor in a controlled environment, generating GNSS signals through a hardware signal generator; then, we tested the signal conditioning with live GNSS signals.

First, we were able to verify all hardware components, from RF to IF, as well as the software routines implementing the usage modes and the grabbing functionality. The use of a professional GNSS hardware generator [28] allowed for excluding effects due to phenomena related to a real environment, such as multipath and interferers. Several tests were performed on each single RF receiving chain of the two FEs, which were first validated separately. Then, all of the channels were tested simultaneously, with the sensor configured such that the master FE provided the reference clock to the slave one. The same GNSS signal was split and sent to the four channels by means of a four-way power splitter, as reported in the simplified scheme of Figure 5. The signals at the RF input were digitalized, and the samples were stored in the ODROID-X2 eMMC memory, then post-processed by a software receiver. Several test metrics were considered in the analysis and validation process: the power spectral density (PSD) of the digitalized signal, the amplitude of the main peak of the cross ambiguity function (CAF) computed during the acquisition of the GNSS signals, the quality of the tracking loop lock through the mean and variance of the correlators and the estimate of the carrier-to-noise power density ratio (C/N_0). The sensor successfully passed all of the tests and demonstrated the ability to process the signal properly in all cases. The FEs resulted in being well synchronized through the master-slave configuration, whereas the ODROID-X2 was able to handle the two FEs' streams, thanks to the real-time management of two USB ports and the fast memory storage. The post-processing analysis revealed that the receiver is able to successfully acquire and track all generated satellite signals in all performed tests. As an example, Figure 6 shows the estimated C/N_0 for two tracked satellites, processing the streams of samples at the output of the four channels. The C/N_0 is estimated at the tracking loop stage and provides a valid measure of the quality of the received signal [29]. Looking at Figure 6, it can be noticed that the values of C/N_0 measured on samples out of the slave FE after the initial transient (*i.e.*, the pink line for the first channel and green for the second) are on average 1 dB lower with respect to the C/N_0 estimated on the corresponding chains of the master FE, depicted in blue and black, respectively. The reason for this small power loss is explained considering that the slave FE receives the clock from the master, via a logic translator circuit, which introduces noise to the reference signals of the slave board.

In Figure 7, another example of the in-lab test results is reported. Here, the estimated PSDs of the digitalized signals entering the FEs are shown: Channel IF signals in Figure 7A and Channel BB signals in Figure 7B. In particular, the black plot represents the master spectrum, while the pink one is the slave spectrum. The GPS signal is strong and well evident in the bandwidth center. From these results, a strong interferer at ± 2.4 MHz respectively for $IF = 0$ Hz (BB) and $IF = 3.55$ MHz (IF) appears. This is probably due the FE clock, but being far from the GPS main lobe, no performance degradation is produced in the acquisition and tracking loop.

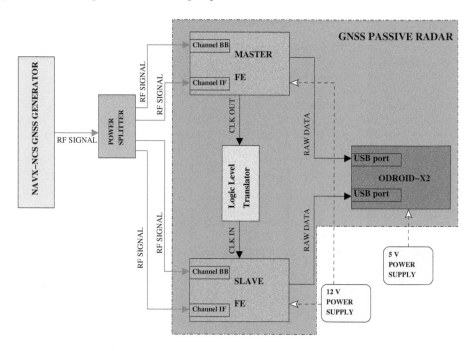

Figure 5. In-laboratory test setup with the simulated scenario.

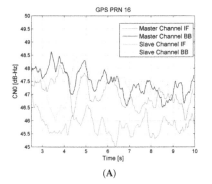
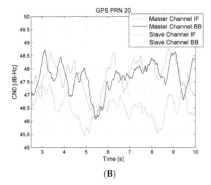

(A)   (B)

Figure 6. C/N_0 evaluation for PRN16 (**A**) and PRN 20 (**B**), obtained during a test in a simulated scenario.

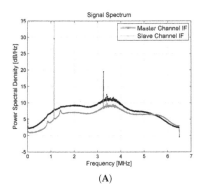
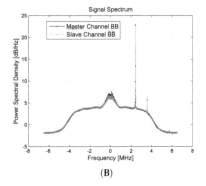

(A)   (B)

Figure 7. Comparison of the estimated power spectral densities of the signals at the two channels of both FEs, generated through a hardware signal generator and supplied to the FEs input ports through a wired RF connection. (**A**) Channel IF; (**B**) Channel BB.

In the second part of the validation, we processed real signals received from the antennas, although limited to live RHCP signals, because it was not possible to replicated in-lab LHCP reflected signals. Anyway, from a functional perspective, such tests were necessary to check the performance of the FEs when connected to the antennas and to detect any distortions on the received signals. The test setup corresponds to the scheme of Figure 2, with the only difference that the dual-polarization antenna is pointed towards the sky, as well as the antenna for the reception of direct signals, and only the RHCP channel is evaluated. The results of these tests, using the same performance metrics mentioned before, allowed for an accurate calibration of the hardware platform parameters, in particular at the signal conditioning stage, to find the best match between antennas and RF receiving stages. Using the calibrated setup, the sensor showed good performance, and no anomalies on the received signals were detected.

4. Soil Moisture Retrieval from Reflection Measurements: A Background on the Discipline

In this section, we briefly review the basic principles of the soil moisture estimate using GNSS reflectometry. A review of the theoretical background of this discipline is necessary in order to clearly motivate the implementation choices made in the prototype design.

In order to quantitatively estimate the soil moisture, the soil dielectric constant has to be evaluated, applying models that take into account the soil characteristics, including in particular the soil composition [18,30,31]. Different methods exist, which either exploit the LHCP reflected signal only or both the LHCP and the RHCP reflections. The soil dielectric constant depends in particular on two parameters related to the GNSS signal: the soil reflection index and the incidence angle (which directly depends on the satellite elevation angle).

In the following paragraphs, the main equations are provided, explaining the physical principles at the basis of the soil moisture retrieval through the measurement of the reflected GNSS signal. The principle is that the soil dielectric constant changes depending on the soil water content, which has an impact on the soil reflectivity properties; it is highly dependent on the soil composition, as well. In particular, the more the soil is moist, the less an incising electromagnetic wave penetrates it in depth, which translates into a higher reflected power [32]. However, a role is played also by the surface roughness, which decreases the specular reflected power.

The two basic principles summarized hereafter exploit, on the one hand, the LHCP measurements only, on the other one, the joint processing of the LHCP and RHCP measurements.

4.1. LHCP-Based Soil Moisture Retrieval

In the simplified hypothesis of a specular reflection, the soil reflection index can be estimated from the ratio between the reflected and the directly incident signal power. This quantity can be evaluated as the ratio of the estimated signal-to-noise ratios (SNR) on the reflected and on the direct channel, respectively. In this way, the differences in the hardware receiving chains of the reflected and the direct signals can be compensated. Once the soil reflection index is known, the dielectric constant can be retrieved. Thus, in order to estimate the soil dielectric constant from the GNSS reflection measurement, it is needed to *a priori* know the satellite elevation (incidence angle) and to have at least a rough knowledge of the soil composition. Then, the reflection index must be estimated, which is a function of the reflected power percentage after the incidence. In what follows, the concepts explained so far will be provided in formulas.

If the reflecting surface can be well approximated as a perfectly smooth surface, then a specular reflection can be assumed. In such a case, neglecting any surface roughness and, therefore, any non-coherent components in the reflection, the reflected GNSS signals results in being mainly LHCP, in particular considering the satellites with close to the zenith elevation. For the direct signal propagating in free space, the SNR is directly proportional to the transmitted power P^t and the transmitter gain G^t, the receiver gain for the direct signal chain G^r_{dir}, the signal wave-length λ and the processing gain G_D. Then, it is inversely proportional to the transmitter-receiver distance R and to the noise power, which, for the direct signal receiving chain, is indicated as $P^r_{N,dir}$. Thus:

$$SNR_{dir} = \frac{P^t G^t}{4\pi R^2} \cdot \frac{G^r_{dir}\lambda^2 G_D}{4\pi P^r_{N,dir}} \tag{1}$$

Similarly, for the LHCP reflected channel, the SNR can be expressed as the power ratio between the reflected signal and the noise related to that channel. It can be written replacing in Equation (1) G^r_{dir} with the receiver gain through the appropriate chain $G^r_{refl,l}$ and the traveled distance R with the sum of the distances from the satellite to the reflection point (R_A) and back to the receiver (R_B). Furthermore, it is necessary to account for the additional path loss due to the reflection, which can be written as $\frac{1}{4}\left(|\Gamma_{vv}| + |\Gamma_{hh}|\right)^2$ [33], where $|\Gamma_{vv}|$ and $|\Gamma_{hh}|$ are the reflection indexes for the vertical and the horizontal polarizations, respectively, which combine together in the case of circular polarization. The noise power in the reflected signal chain is $P^r_{N,l}$. Thus:

$$SNR_{refl,l} = \frac{1}{4}\frac{P^t G^t}{4\pi(R_A + R_B)^2} \cdot \frac{G^r_{refl,l}\lambda^2 G_D}{4\pi P^r_{N,l}} \cdot \left(|\Gamma_{vv}| + |\Gamma_{hh}|\right)^2 \tag{2}$$

where the subscript l in $SNR_{refl,l}$ and $P^r_{N,l}$ refers to the LHCP reflected polarization.

As said above, the reflection index is a function of both the reflecting surface characteristics and the incidence angle; therefore, it can be expressed as a function of the soil dielectric constant ϵ_r and the satellite elevation angle θ:

$$\Gamma_{vv}(\epsilon_r, \theta) = \frac{\sin\theta - \sqrt{\epsilon_r - \cos^2\theta}}{\sin\theta + \sqrt{\epsilon_r - \cos^2\theta}} \tag{3}$$

$$\Gamma_{hh}(\epsilon_r, \theta) = \frac{\epsilon_r \sin\theta - \sqrt{\epsilon_r - \cos^2\theta}}{\epsilon_r \sin\theta + \sqrt{\epsilon_r - \cos^2\theta}} \tag{4}$$

In order to evaluate the soil dielectric constant, from which the soil moisture can be retrieved if there is some knowledge of the soil composition, the ratio between the reflected SNR in Equation (2) and the direct SNR in Equation (1) is computed. It results in being:

$$\frac{SNR_{refl,l}}{SNR_{dir}} = \frac{R^2}{(R_A + R_B)^2} \cdot (|\Gamma_{vv}(\epsilon_r, \theta)| + |\Gamma_{hh}(\epsilon_r, \theta)|)^2 \cdot C \tag{5}$$

where $C = \frac{G^r_{refl,l}}{P^r_{N,l}} \cdot \frac{P^r_{N,dir}}{G^r_{dir}}$ depends on the hardware differences in the receiving chains, mainly due to antennas and RF filtering gains. The actual value of C must be determined with a calibration.

One of the more robust ways to calibrate the system for soil moisture purposes is the on-water calibration, used for example in [34], through multiple over-water overflights. This is because the expected reflected power over water is well known given the incidence angle, while over the terrain the uncertainty is higher, due to the imperfect knowledge of the soil composition and its inherent dis-homogeneity. In order to have a more accurate calibration, a measurement campaign should be done *in situ* with other sensors (hygrometers), for different soil types in different moisture conditions. This would involve the need of performing measurements for a long time, in order to have reliable measurements, and to compare all of the obtained results with the other sensors in the terrain. However, for the application at hand, the on-water calibration is proven to be quite an effective low-cost solution [34].

After the calibration, the dielectric constant ϵ_r in Equation (5) is solvable via numerical routines, given the knowledge of R, R_A, R_B and θ. It has to be noted that from Equation (5), only $|\epsilon_r|$ can be evaluated: in order to get the full soil moisture information, the real and imaginary parts of ϵ_r need to be separated, which is possible considering empirical dielectric models, such as the one proposed in [31].

4.2. LHCP + RHCP-Based Soil Moisture Retrieval

The retrieval algorithm described above is based on the assumptions of having a smooth reflection surface. Nonetheless, in order to better take into account the effects of the soil roughness, which makes the reflection different from specular, another approach is needed, which exploits the availability of both the LHCP and the RHCP SNR measurements.

The roughness of a surface impacts its capability of reflecting an incident electromagnetic field along a principal direction (reflection angle); this capability is typically quantified in terms of the so-called radar cross-section RCS [35]. The RCS of an object is in turn a function of: (i) the object dimensions and shape; (ii) the electromagnetic wave incident angle; and (iii) the reflecting material (through the so-called normalized radar cross-section (NRCS), σ^0). The NRCS is a function of the dielectric properties of the material and separates into a horizontal and a vertical polarization component.

For these reasons, we expect it to be possible to extract the dielectric constant of the soil by estimating the NRCS for the two circular polarizations of the reflected GNSS signals (it is well known that the two LH and RH circular polarizations of an electromagnetic wave can be written as combinations of the two linear polarizations) [33].

The SNR of the reflected signals, which was expressed in Equation (2) for the LHCP reflection, can be expressed for both the LHCP and the RHCP reflections as:

$$SNR_{refl,l} = \frac{G^t G^r_{refl,l} \lambda^2 |\sigma_{lr}|}{(4\pi)^3 R_A{}^2 R_B{}^2} \frac{P^t}{P^r_{N,l}} \tag{6}$$

$$SNR_{refl,r} = \frac{G^t G^r_{refl,r} \lambda^2 |\sigma_{rr}|}{(4\pi)^3 R_A{}^2 R_B{}^2} \frac{P^t}{P^r_{N,r}} \tag{7}$$

where the notation is the one adopted in Equation (2), while the parameters σ_{ij} represent the RCS for the circular polarized components of the incident and the reflected waves, for which the subscripts are such that i indicates the polarization of the reflected signal and j indicates the polarization of the incident wave. Furthermore:

$$\sqrt{\sigma_{lr}} = \frac{\sqrt{A}}{2} \left(\sqrt{\sigma^0_{hh}} + \sqrt{\sigma^0_{vv}} \right) \tag{8}$$

$$\sqrt{\sigma_{rr}} = \frac{\sqrt{A}}{2} \left(\sqrt{\sigma^0_{hh}} - \sqrt{\sigma^0_{vv}} \right) \tag{9}$$

where σ^0_{hh}, σ^0_{vv} are the horizontal and vertical polarization components of the NRCS and A is the total illuminated area, or glistening zone, which depends on the reflection geometry [22,35]. The NRCS is a key parameter in reflection theory: through its estimation, the characteristics of the reflecting surface can be retrieved, but to do that, a good model of the reflecting system is required. A detailed analysis of the NRCS and of the effects of the geometry (incidence angle) is available in [36], where also the soil inhomogeneity is taken into account. The more accurate the model applied is, the more accurate the soil estimate will be. However, at this stage of the work, a simple approximated model has been considered, without a thorough study of the soil characteristics, as for instance the soil roughness.

Combining Equations (6) and (7) with Equations (8) and (9), the ratio between Equations (6) and (7) can be written as:

$$\frac{SNR_{refl,l}}{SNR_{refl,r}} = \frac{\left| \sqrt{\sigma^0_{hh}} + \sqrt{\sigma^0_{vv}} \right|}{\left| \sqrt{\sigma^0_{hh}} - \sqrt{\sigma^0_{vv}} \right|} \cdot C' \tag{10}$$

where $C' = \frac{G^r_{refl,l}}{P^r_{N,l}} \cdot \frac{P^r_{N,r}}{G^r_{refl,r}}$ is a calibration constant similar to C in Equation (5). The parameters σ^0_{hh} and σ^0_{vv} are functions, in particular, of the soil dielectric constant and the incidence angle (the satellite elevation); it can be stated that:

$$\begin{cases} \sigma^0_{hh} = f_1(\epsilon_r, \theta) \\ \sigma^0_{vv} = f_2(\epsilon_r, \theta) \end{cases} \tag{11}$$

The functions f_1 and f_2, can be described through proper scattering models that take into account various other physical parameters involved in the reflection phenomena, other than the satellite elevation angle θ, assumed to be known, and the dielectric constant ϵ_r to be estimated. Different scattering models have been proposed in the literature [31,35,37]; for instance, applying the so-called small perturbation method (SPM), the RCS components can be expressed as a function of ϵ_r and θ [35]. To do that, for simplicity, the variables α_{hh} and α_{vv} can be introduced as:

$$\begin{cases} |\alpha_{hh}| = \sqrt{\sigma^0_{hh}} \\ |\alpha_{vv}| = \sqrt{\sigma^0_{vv}} \end{cases} \tag{12}$$

Then, the relationship between α_{hh}, α_{vv} (*i.e.*, σ_{hh}^{o}, σ_{vv}^{o}) and ϵ_r, θ can be expressed as follows [35]:

$$\begin{cases} \alpha_{hh} = \dfrac{1 - \epsilon_r}{\left(\cos\theta + \sqrt{\epsilon_r - \sin^2\theta}\,\right)^2} \\ \alpha_{vv} = \dfrac{(1 - \epsilon_r)(\epsilon_r - \sin^2\theta - \epsilon_r \sin^2\theta)}{\left(\epsilon_r \cos\theta + \sqrt{\epsilon_r - \sin^2\theta}\,\right)^2} \end{cases} \tag{13}$$

Thus, combining the expressions in Equations (12) and (13) with Equation (10) and determining the value of C' through the calibration phase, the dielectric constant ϵ_r can be numerically solved.

5. Signal Processing and Results of an In-Field Test

Several in-flight data collection campaigns have been executed in order to test the performance of the prototype in different configurations. The aim of the test campaigns was to demonstrate the capability of the prototype to provide the GNSS measurements necessary to implement a soil moisture retrieval algorithm, such as one of those mentioned in the previous section. In this section, we first review the signal processing principles at the basis of our project (Section 5.1), then some results of the reflectometry measurement campaign are shown (Section 5.2).

5.1. Signal Processing Principles

With the scope of implementing the reflectometry functionalities, a MATLAB®-based software receiver has been modified here to make it able to properly process the data from the four channels of the sensor. The principle of this architecture comes from the GPS software receiver described in [38], and it has been chosen for the many advantages that the software-defined paradigm includes, in particular for its flexibility. Particular attention was paid to the study of the algorithms to detect the reflected signal to cope with the major challenges presented by the reflected signals, namely the extremely low power and the very short phase coherence.

In fact, the reflected signal is not a single specular reflection from the so-called specular point, but it is the sum of several contributions (scattering) from a reflecting area, namely the glistening zone, whose size depends on the incidence angle, the receiver altitude and the surface roughness (models exist that allow one to find a suitable approximation).

This scattering effect causes a much shorter phase coherence compared to the direct signal, in particular in dynamic environments, such as in flight; for this reason, an open-loop strategy is in general advisable to detect the reflections.

Furthermore, the scattering effect, reducing the reflected power reaching the nadir-pointing antenna, worsens for lower incident angles, in particular for the LHCP components. Thus, although Equation (5) takes the incidence angle into account, the accuracy of the estimate decreases when the satellite elevation reduces [34]: the SNR diminishes, meaning that the impact of the noise becomes heavier on the measurement. For this reason, the surface scattering effect needs to be mitigated in order to estimate the actual reflected signal power; this can be done by averaging over time the measurements.

Furthermore, in order to detect a low-power signal, the integration time needs to be increased as much as possible, even if this means getting lower spatial resolution on the measurements. However, the coherent integration time cannot be longer than the signal coherence interval; therefore, a trade-off solution needs to be found. A key strategy introduced in the software scheme is the channel aiding, which means that information from the direct signal processing is exploited to detect the reflected signal, since the Doppler frequency of the two signals is expected to differ by only a few tenths of Hertz, and the delay is expected to be within an interval depending on the satellite elevation and the aircraft altitude.

The effects of the secondary multipath are neglected here. Concerning the reflected signal, the interest is on the principal reflection; it might occur that some unexpected (and undesired) reflections from some targets, including buildings, are received together with the reflection from the considered reflection point on the terrain, as a multipath signal. Such cases are not predictable, but they are expected to be rare, and such an error is considered acceptable for this kind of application. Concerning the direct signal, undesired multipath signals may occur due to the reflections from the aircraft or the UAV. However, given the small dimensions of the aircraft on which the sensor is designed to be mounted and given the position of the antenna on the wing, the multipath effects are expected not to be significant with respect to the noise [23]. Anyway, a better analysis of the antenna gain together with the multipath effects should be included in the future developments of the sensor.

5.2. Test Campaign Results

Some in-flight tests were executed to assess the prototype performance. The sensor was mounted on two different platforms, as shown in Figures 8 and 9: a manned ultra-light aircraft (Digisky's Tecnam P92) and a UAV (Nimbus' CFly). Some results are shown here from a flight test with the P92 aircraft, which flew over a countryside nearby Turin (Italy), also overflying two small lakes (the Avigliana lakes). The overflown area has been chosen because it includes test scenarios of interest. In fact, this area is in the countryside north of Turin, not far from the airport from where the aircraft used for the tests can take off and land, and it includes water basins, such as lakes and rivers. Moreover, a swampy area is present around the lakes, which looks particularly interesting in the framework of this work, since the evaluation of the soil moisture is the main goal. In that region, different cultivated areas are also present, which are interesting from the perspective of a future agriculture application. Forest zones are present, as well, which are characterized by weaker reflections, due to the higher scattering effects; on this topic, several studies have been presented in the literature to address the analysis of the vegetation characteristics through the GNSS reflections, as, for instance, in [33] and later in [39]. Similarly, the inhabited zones, including buildings, roads or bridges, give a different reflection depending on the surface composition, roughness and inclination. However, the detection of these kinds of targets is not the focus of this work.

(A)   (B)

Figure 8. Sensor prototype mounted on the manned Digisky Tecnam P92 aircraft ready to take off (**A**) and during the flight (**B**).

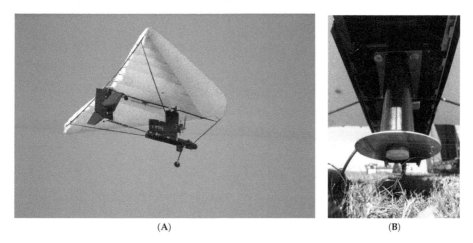

(A)   (B)

Figure 9. Sensor prototype mounted on the Nimbus C-Fly UAV during the flight (**A**) and on the ground after landing (**B**).

The prototype was used in the advanced mode, as detailed in Section 3.3, collecting data from both the RHCP and the LHCP channels of the nadir-pointing antenna. The collected signals were then processed to get the aircraft and the satellite position and to compute the specular points for each satellite. Then, the direct and reflected SNR were estimated from both the RH and the LH circular polarizations, so as to enable the post-processing algorithms presented in Sections 4.1 and 4.2. It is important to highlight that the data post-processing necessary to convert the reflectometry measurements in Equation (5) or Equation (10) in estimates of the soil moisture is highly sensitive to the accuracy of either the terrain composition model or the terrain scattering model used in the conversion process, as well as to their sensitivity to the signal incidence angle (satellite geometry). Even if the focus of this paper is on the prototype design, both of the hardware and software parts, some results are presented here of the soil parameter retrieval process, in order to validate the system in terms of the final output.

Figures 10 and 11 show the plot on the map of the estimated specular points for two satellites (PRN30 and PRN 13, respectively); the color is proportional to the ratio between the reflected LHCP SNR and the direct GPS signal SNR, which is related to the geophysical quantity of interest, the soil dielectric constant, through Equations (3) and (4). As explained above, if a suitable model of the soil composition is given, then the full soil moisture information can be retrieved through the measurement of the LHCP reflections only, assuming the approximation of a smooth surface.

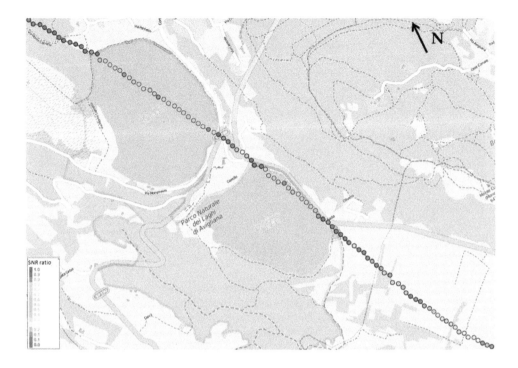

Figure 10. Specular points of PRN 30 over the Avigliana lakes. The color is proportional to the SNR ratio between the left-hand circular polarized (LHCP) reflection and the direct GPS signal.

As expected and indicated by the red points in Figures 10 and 11, the reflection from the water surface is much higher than from the terrain [34], where weaker and different values correspond to different lands, such as forest or fields, with different moisture levels. Furthermore, the strength of the LHCP reflections is more intense as the satellite elevation increases. In this case, while the elevation of PRN 13 is around 45°, PRN 30 has an elevation lower than 10°, showing reduced power values on the reflected signal.

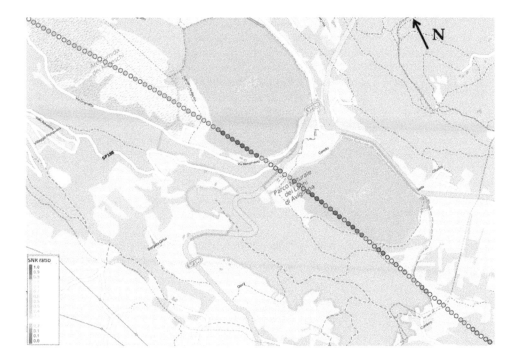

Figure 11. Specular points of PRN 13 over the Avigliana lakes. The color is proportional to the SNR ratio between the LHCP reflection and the direct GPS signal.

Quantitative measurements validate this expectation: on the water, the SNR ratio assumes values between 0.7 and 1 (the differences are mainly due to the elevation angle and the noise effects), whereas when the specular point is on the terrain, the measured values are very different. Figure 10 shows that for PRN 30, the reflection points out of water are mainly on a forest area (northern points) and on irrigated fields (southern points), showing SNR ratios below 0.2 and between 0.2 and 0.4, respectively. As shown in Figure 11, the reflection points for PRN 13 are southern with respect to PRN 30, and they fall also in the so-called "Area umida dei Mareschi", *i.e.*, the "Mareschi humid zone". In that region, as expected, the SNR ratios oscillate in a range between 0.45 and 0.65, sometimes very close to the values assumed on water basins, due to the swamp effect. Then, in the southern region of irrigated fields, again, the measurement is similar to PRN 30, with values between 0.2 and 0.45, with small differences due to the satellite elevation and specular point positions.

Figure 12 shows a comparison between some in-flight real measurements and the expected values of the reflection coefficient for different satellite elevations, *i.e.*, different incidence angles, and three types of surface: water (blue), wet soil (green) and dry soil (brown). The expected values of the reflection coefficient, represented by dotted lines, are available in the literature [18,40] and, being the result of accurately calibrated test campaigns, have become the theoretical reference for this kind of measurement. Note that in the literature, the qualitative expressions dry soil and wet soil are largely used, to indicate poor and abundant water content in the soil, respectively. Since it is correct to think that the expressions wet and dry soil are related to a range of water content values, in Figure 12, a region of values is indicated for the dry and the wet soil, by the dashed bars, brown and green respectively. The circle dots in the figure indicate the measured values, for PRN 13 and 15 respectively. The measurements, as for Figure 11, are taken at a rate of 1 Hz. From Figure 12, it can be seen how the values obtained on the water surface are different for two satellites at different elevation, as expected, as well as in the other regions. Note that different colors are used to plot the circle dots, depending on the region in which the reflection point lays, known from the map. For instance, for the PRN 13, when the reflection point is in the Mareschi humid zone, the reflection index is plotted using the light green color. As is visible from Figure 12, the reflection coefficient over that region has a value that matches the expected one. However, the variance of the measurements, including those obtained on the water surface, which should be expected to be fairly homogeneous, is due to different factors. First, it has to be noted that, in general, the estimate of the SNR of the direct GPS signal, even in static conditions, has a certain variance, in the order of a few dB-Hz, due to several reasons, including the high noise present in the signal. In this environment, in-flight, more effects contribute to the variance, as, in particular, the antenna gain, not being omnidirectional. The use of different antennas being non-co-located, for the direct and the reflected signal, increases the effects of these phenomena. However, the level of accuracy reached is as expected from the project requirements, given the low-cost devices, which cause several residual errors, not including accurate hardware calibration of the antennas (using inertial systems) and not calibrating other effects, such as the system vibrations.

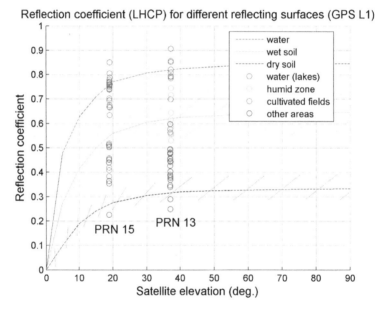

Figure 12. Reflection coefficient: comparison between expected and measured values, for different reflecting surfaces and satellite elevations.

A further test was done by comparing the retrieved values of soil moisture, for the same area, computed using the reflections from different satellites. This test can be a good proof of the goodness of the system, allowing a comparison between different measurements (signals from different satellites) of the same quantities (same area). This was possible thanks to the multiple passing of the aircraft close to the same area during the flight test. In this way, it happens that different reflections, corresponding to different satellites and then to different incidence angles and to different reflection coefficients, must theoretically give the same result in terms of dielectric constant and moisture, since the reflection points lay in the same area. Applying the soil moisture retrieval algorithm as explained in Section 4.1 to the real in-flight measurements, it is found that when the reflection point is on the lakes' surface, the average estimated value of the so-called volumetric soil moisture, *i.e.*, the estimate of the soil water content, is as expected $m_v \simeq 1$ for all of the PRNs, 13, 15 and 30. Over other regions, for instance cultivated fields, when the reflection points of different satellites lay on the same field (visible from the map, with the satellite view), the mean value of the estimated soil parameters match for different satellites. In the Mareschi humid zone, for instance, the measured mean value is $m_v = 0.75$ for PRN 13, while it is $m_v = 0.78$ for PRN 15. The matching between these measurements, together with the comparison between the expected and the obtained measured reflection coefficients in Figure 12, represents a good test of the measurement system, when *in situ* measurements with other sensors are not available.

As said before, the results described above are obtained from the processing of the LHCP reflection. Although it is expected that the processing of the RHCP measurement can improve the accuracy of the overall results, as discussed in Section 4.2, an insufficient cross-polarization separation at the antenna stage is likely responsible for the little improvement observed in our data collections. For this reason, we limit the present discussion to the results of the LHCP-only approach mentioned in Section 4.1. Nonetheless, this test proved the prototype to be effective in order to provide GPS reflection measurements useful to retrieve soil parameters, such as its moisture. The overall accuracy of the methodology depends on several parameters, including the antenna performance, particularly in terms of cross-polarization separation, the applicability of the model used to estimate the soil parameters and the accuracy of the knowledge of the terrain composition.

6. Conclusions

This paper presents the design and development of a GNSS passive radar for the classification of lands, based on the water content feature, and the detection of water surfaces' extent and scattering objects on the ground. Such GNSS passive radar is intended for small UAVs; therefore, size and weight constrained the design of the whole system from the beginning. The sensor features four synchronized RF channels, which are used to receive the direct and the reflected GNSS signals separately over RHCP and LHCP polarizations. The RF part is connected to a commercial embedded micro-processor, which hosts the software routines to control the flow of the digital samples of all channels. The sensor guarantees the storage of more than 30 min of data, if the sampling frequency of the signals is set to 13 MHz. Although the sensor has been integrated with low-cost COTS components, the design followed the software radio paradigm and, for this reason, allows for a significant level of flexibility of the system settings, e.g., the possibility to use only a subset of the four channels, custom frequency plan and variable bandwidths. The sensor has been intensively tested in-lab and validated through some flight tests. These served to assess the performance in a real environment, including the electromagnetic compatibility with other UAV devices, the sensor reliability to store data in an automatic fashion and the mechanical resistance of the sensor's case during take-off and landing stress. The sensor successfully demonstrated its ability to receive reflected signals, both LHC and RHC polarized. This result is comparable to others presented in the literature, but on the one hand, it allows the simultaneous grabbing of RHCP and LHCP reflections and, on the other hand, has been obtained with a prototype much lighter and smaller with respect to those used in other experiments.

Among all of the results, it is important to underline the lesson learned from the analysis performed over some of the collected datasets. The cross-polarization isolation between the RHCP and LHCP channels of the antenna pointing at the nadir is critical for the system performance. In fact, if the cross-polarization rejection is lower than the minimum required, a portion of the LHCP power obscures the RHCP reflected signals, which cannot be correctly measured. This limits the fine computation of correct soil moisture parameters and identifies the nadir-pointing dual-polarization antenna as the most critical system element, as it requires very high cross-polarization isolation, typically unavailable as COTS.

Furthermore, considering the extreme weakness of the reflected signals, another critical point is the accurate characterization, calibration and control of the electromagnetic environment during the tests and on-board the aircraft during the data collections: the effect of the electromagnetic interference

Sensors **2015**, *15*, 28287–28313

from the surrounding electronic systems during the data collections, especially on-board unmanned vehicles, can be destructive for the GNSS-R processing and, therefore, must be carefully controlled.

Acknowledgments: The authors would like to thank the SMAT–F2 (System of Advanced Monitoring of the Territory–Phase 2) Project partners, which supported this work.

Author Contributions: Micaela Troglia Gamba was the principal developer of the prototype. She was responsible of the design of the architecture and of the software components. Gianluca Marucco was responsible for the hardware architecture and setup. Marco Pini was the coordinator and responsible for all of the activities at ISMB on the design, development and testing of the prototype. Sabrina Ugazio was in charge of the signal post-processing. Emanuela Falletti coordinated the in-lab test campaign. Letizia Lo Presti was the coordinator and responsible for the activities in Polito related to signal post-processing.

Conflicts of Interest: The authors declare no conflict of interest.

References

1. Garrison, J.; Katzberg, S.; Hill, M. Effect of sea roughness on bistatically scattered range coded signals from the Global Positioning System. *Geophys. Res. Lett.* **1998**, *13*, 2257–2260.
2. Dampf, J.; Pany, T.; Falk, N.; Riedl, B.; Winkel, J. Galileo altimetry using AltBOC and RTK techniques. *Inside GNSS* **2013**, *8*, 54–63.
3. Rodriguez-Alvarez, N.; Aguasca, A.; Valencia, E.; Bosch-Lluis, X.; Camps, A.; Ramos-Perez, I.; Park, H.; Vall-llossera, M. Snow thickness monitoring using GNSS measurements. *IEEE Geosci. Remote Sens. Lett.* **2012**, *6*, 1109–1113.
4. Sánchez, N.; Alonso-Arroyo, A.; Martínez-Fernández, J.; Piles, M.; González-Zamora, A.; Camps, A.; Vall-Llosera, M. On the Synergy of Airborne GNSS-R and Landsat 8 for Soil Moisture Estimation. *Remote Sens.* **2015**, *7*, 9954–9974.
5. Camps, A.; Bosch-Lluis, X.; Ramos-Perez, I.; Marchán-Hernández, J.F.; Rodríguez, N.; Valencia, E.; Tarongi, J.M.; Aguasca, A.; Acevo, R. New Passive Instruments Developed for Ocean Monitoring at the Remote Sensing Lab-Universitat Politècnica de Catalunya. *Sensors* **2009**, *9*, 10171–10189.
6. Bosch-Lluis, X.; Camps, A.; Ramos-Perez, I.; Marchan-Hernandez, J.F.; Rodriguez-Alvarez, N.; Valencia, E. PAU/RAD: Design and Preliminary Calibration Results of a New L-Band Pseudo-Correlation Radiometer Concept. *Sensors* **2008**, *8*, 4392–4412.
7. Nogués-Correig, O.; Cardellach Galí, E.; Campderrós, J.S.; Rius, A. A GPS-Reflections Receiver That Computes Doppler/Delay Maps in Real Time. *IEEE Trans. Geosci. Remote Sens* **2007**, *45*, 156–174.
8. Martin-Neira, M.; Caparrini, M.; Font-Rossello, J.; Lannelongue, S.; Vallmitjana, C.S. The PARIS concept: An experimental demonstration of sea surface altimetry using GPS reflected signals. *IEEE Trans. Geosci. Remote Sens.* **2001**, *39*, 142–150.
9. Ribot, M.A.; Kucwaj, J.-C.; Botteron, C.; Reboul, S.; Stienne, G.; Leclère, J.; Choquel, J.-B.; Farine, P.-A.; Benjelloun, M. Normalized GNSS Interference Pattern Technique for Altimetry. *Sensors* **2014**, *14*, 10234–10257.
10. Vinande, E.; Akos, D.; Masters, D.; Axelrad, P.; Esterhuizen, S. GPS bistatic radar measurements of aircraft altitude and ground objects with a software receiver. In Proceedings of the 61th Annual Meeting of the Institute of Navigation, Cambridge, MA, USA, 27–29 June 2005; pp. 528–534.
11. Esterhuizen, S.; Masters, D. Experimental characterization of land-reflected GPS signals. In Proceedings of the 18th International Technical Meeting of the Satellite Division of the Institute of Navigation (ION GNSS 2005), Long Beach, CA, USA, 13–16 September 2005; pp. 1670–1678.
12. Caparrini, M.; Egido, A.; Soulat, F.; Germain, O.; Farres, E.; Dunne, S.; Ruffini, G. Oceanpal®: Monitoring sea state with a GNSS-R coastal instrument. In Proceedings of the IEEE International Geoscience and Remote Sensing Symposium (IGARSS), Barcelona, Spain, 23–28 July 2007; pp. 5080–5083.
13. Pierdicca, N.; Guerriero, L.; Caparrini, M.; Egido, A.; Paloscia, S.; Santi, E.; Floury, N. GNSS reflectometry as a tool to retrieve soil moisture and vegetation biomass: Experimental and theoretical activities. In Proceedings of the International Conference on Localization and GNSS (ICL-GNSS), Turin, Italy, 25–27 June 2013; pp. 1–5.
14. Ceraldi, E.; Franceschetti, G.; Iodice, A.; Riccio, D. Estimating the soil dielectric constant via scattering measurements along the specular direction. *IEEE Trans. Geosci. Remote Sens.* **2005**, *43*, 295–305.
15. Esterhuizen, S. The Design, Construction, and Testing of a Modular GPS Bistatic Radar Software Receiver for Small Platforms. Master Thesis, University of Colorado, Boulder, CO, USA, 2006.

16. Camps, A.; Marchan-Hernandez, J.F.; Ramos-Perez, I.; Bosch-Lluis, X.; Prehn, R. New Radiometer Concepts for Ocean Remote Sensing: Description of the Passive Advanced Unit (PAU) for Ocean Monitoring. In Proceedings of the IEEE International Geoscience and Remote Sensing Symposium (IGARSS), Denver, CO, USA, 21 July–4 August 2006; pp. 3988–3991.

17. Marchan-Hernandez, J.F.; Camps, A. ; Rodriguez-Alvarez, N. ; Bosch-Lluis, X. ; Ramos-Perez, I.; Valencia, E. PAU/GNSS-R: Implementation, Performance and First Results of a Real-Time Delay-Doppler Map Reflectometer Using Global Navigation Satellite System Signals. *Sensors* **2008**, *8*, 3005–3019.

18. Egido, A. GNSS Reflectometry for Land Remote Sensing Applications. Ph.D. Thesis, Universitat Politècnica de Catalunya, Barcelona, Spain, 7 May 2013.

19. Egido, A.; Caparrini, M.; Ruffini, G. ; Paloscia, S.; Santi, E.; Guerriero, L.; Pierdicca, N.; Floury, N. Global Navigation Satellite Systems Reflectometry as a Remote Sensing Tool for Agriculture. *Remote Sens.* **2012**, *4*, 2356–2372.

20. Paloscia, S.; Santi, E.; Fontanelli, G.; Pettinato, S.; Egido, A.; Caparrini, M.; Motte, E.; Guerriero, L.; Pierdicca, N.; Floury, N. Grass: An experiment on the capability of airborne GNSS-R sensors in sensing soil moisture and vegetation biomass. In Proceedings of the IEEE International Geoscience and Remote Sensing Symposium (IGARSS), Melbourne, Australia, 21–26 July 2013; pp. 2110–2113.

21. Egido, A.; Paloscia, S.; Motte, E.; Guerriero, L.; Pierdicca, N.; Caparrini, M.; Santi, E.; Fontanelli, G.; Floury, N. Airborne GNSS-R Polarimetric Measurements for Soil Moisture and Above-Ground Biomass Estimation. *IEEE J. Sel. Top. Appl. Earth Obs. Remote Sens.* **2014**, *7*, 1522–1532.

22. Zavorotny, V.U.; Voronovich, A.G. Bistatic GPS signal reflections at various polarizations from rough land surface with moisture content. In Proceedings of the IEEE International Geoscience and Remote Sensing Symposium, Honolulu, HI, USA, 24–28 July 2000; pp. 2852–2854.

23. Zavorotny, V.U.; Larson, K.M.; Braun, J.J.; Small, E.E.; Gutmann, E.D.; Bilich, A.L. A Physical Model for GPS Multipath Caused by Land Reflections: Toward Bare Soil Moisture Retrievals. *IEEE J. Sel. Top. Appl. Earth Obs. Remote Sens.* **2010**, *3*, 100–110.

24. Dual Polarization ANTCOM Antennas Catalog. Available online: http://www.antcom.com/ documents/catalogs/RHCP-LHCP-V-H-L1L2GPSAntennas.pdf (accessed on 5 November 2015).

25. Nottingham Scientific Ltd (NSL) GNSS SDR Front End and Receiver. Available online: http://www.nsl.eu.com/primo.html (accessed on 27 April 2015).

26. Hardkernel Co. Ltd ODROID-X2 Platform. Available online: http://www.hardkernel.com/ main/products/prdt_info.php?g_code=G135235611947 (accessed on 5 November 2015).

27. Troglia Gamba, M.; Lo Presti, L.; Notarpietro, R.; Pini, M.; Savi, P. A New SDR GNSS Receiver Prototype For Reflectometry Applications: Ideas and Design. In Proceedings of the 4th International Colloquium Scientific and Fundamental Aspects of the Galileo Programme, Prague, Czech, 4–6 December 2013; pp. 571–579.

28. Navx-Ncs Professional. Available online: http://www.ifen.com/products/navx-gnss-test-solutions/ ncs-gnss-rf-signal-generator/professional.html#c3 (accessed on 5 November 2015).

29. Falletti, E.; Pini, M.; Lo Presti, L. Low complexity carrier to noise ratio estimators for GNSS digital receivers. *IEEE Trans. Aerosp. Electron. Syst.* **2011**, *1*, 420–437.

30. De Roo, R.D.; Ulaby, F.T. Bistatic specular scattering from rough dielectric surfaces. *IEEE Trans. Antennas Propag.* **1994**, *2*, 220–231.

31. Hallikainen, M.T.; Ulaby, F.T.; Dobson, M.C. Microwave dielectric behavior of wet soil -part 1: Empirical models and experimental observations. *IEEE Trans. Geosci. Remote Sens.* **1985**, *GE-23*, 25–34.

32. Rodriguez-Alvarez, N.; Bosch-Lluis, X.; Camps, A.; Vall-llossera, M.; Valencia, E.; Marchan-Hernandez, J.F.; Ramos-Perez, I. Soil Moisture Retrieval Using GNSS-R Techniques: Experimental Results over a Bare Soil Field. *IEEE Trans. Geosci. Remote Sens.* **2009**, *47*, 3616–3624.

33. Sigrist, P.; Coppin, P.; Hermy, M. Impact of Forest Canopy on Quality and Accuracy of GPS Measurements. *Int. J. Remote Sens.* **1999**, *20*, 3595–3610.

34. Masters, D.; Axelrad, P.; Katzberg, S. Initial results of land-reflected GPS bistatic radar measurements in SMEX02. *Remote Sens. Environ.* **2004**, *92*, 507–520.

35. Ulaby, F.T.; Moore, R.K. ; Fung, A.K. *Microwave Remote Sensing: Active and Passive*; Addison-Wesley Reading: Boston, MA, USA, 1982.

36. Clarizia, M.P. Investigating the Effect of Ocean Waves on GNSS-R Microwave Remote Sensing Measurements. Ph.D. Thesis, University of Southampton, Southampton, UK, 7 October 2012.

37. Ticconi, F.; Pulvirenti; L.; Pierdicca; N. Models for scattering from rough surfaces *Electromagn. Waves* **2011**, *10*, 203–226.
38. Borre, K.; Akos, D. ; Bertelsen, N.; Rinder, P. ; Jensen, S.H. *A Software—Defined GPS and Galileo Receiver: A Single-Frequency Approach*; Birkhauser: Boston, MA, USA, 2007.
39. Rodriguez-Alvarez, N.; Bosch-Lluis, X.; Camps, A.; Ramos-Perez, I.; Valencia, E.; Park, H.; Vall-llossera, M. Vegetation Water Content Estimation Using GNSS Measurements *IEEE Geosci. Remote Sens. Lett.* **2012**, *9*, 282–286.
40. Larson, K.M.; Braun, J.J.; Small E.E.; Zavorotny V.U. Environmental Sensing: A Revolution in GNSS Applications. *Inside GNSS* **2014**, *9*, 36–46.

Article

Enabling UAV Navigation with Sensor and Environmental Uncertainty in Cluttered and GPS-Denied Environments

Fernando Vanegas * and Felipe Gonzalez

Australian Research Centre for Aerospace Automation (ARCAA), Queensland University of Technology (QUT), 2 George St, Brisbane, QLD 4000, Australia; felipe.gonzalez@qut.edu.au
* Correspondence: fernando.vanegasalvarez@hdr.qut.edu.au; Tel.: +61-7-3138-1772

Academic Editor: Assefa M. Melesse
Received: 21 December 2015; Accepted: 5 May 2016; Published: 10 May 2016

Abstract: Unmanned Aerial Vehicles (UAV) can navigate with low risk in obstacle-free environments using ground control stations that plan a series of GPS waypoints as a path to follow. This GPS waypoint navigation does however become dangerous in environments where the GPS signal is faulty or is only present in some places and when the airspace is filled with obstacles. UAV navigation then becomes challenging because the UAV uses other sensors, which in turn generate uncertainty about its localisation and motion systems, especially if the UAV is a low cost platform. Additional uncertainty affects the mission when the UAV goal location is only partially known and can only be discovered by exploring and detecting a target. This navigation problem is established in this research as a Partially-Observable Markov Decision Process (POMDP), so as to produce a policy that maps a set of motion commands to belief states and observations. The policy is calculated and updated on-line while flying with a newly-developed system for UAV Uncertainty-Based Navigation (UBNAV), to navigate in cluttered and GPS-denied environments using observations and executing motion commands instead of waypoints. Experimental results in both simulation and real flight tests show that the UAV finds a path on-line to a region where it can explore and detect a target without colliding with obstacles. UBNAV provides a new method and an enabling technology for scientists to implement and test UAV navigation missions with uncertainty where targets must be detected using on-line POMDP in real flight scenarios.

Keywords: unmanned aircraft; UAV target detection; Partially-Observable Markov Decision Process (POMDP); path planning; Robotic Operating System (ROS); uncertainty; robust navigation

1. Introduction

Ground, underwater and aerial robots are widely used for environmental monitoring and target detection missions [1–5]. Among these robots, UAVs use ground control stations to plan a path to a goal before flying, using Global Positioning System (GPS) sensors as their source of localisation [6–10]. However, reliable GPS localisation is not always available due to occlusions or the absence of the satellite signal. The accuracy of such localisation systems also decreases for low cost UAVs. Flying in GPS-denied environments and with only on-board sensors as the source of localisation is challenging, particularly when there are obstacles in the airspace that need to be avoided or if there is uncertainty in the goal location, which requires the UAV to fly and explore the area until the target is detected.

POMDP is a mathematical framework to model decision making problems with different sources of uncertainty [11–14], which makes it useful for implementing UAV navigation in cluttered and GPS-denied environments. The performance of POMDP solvers has improved, especially in the ability to cope with larger numbers of states $|S| > 2000$ and observations $|O| > 100$ [15,16], which is the case in UAV navigation missions. UAV navigation has been formulated previously as a Partially-Observable Markov Decision Process (POMDP) [17–19]. However, most of these studies only consider simulated scenarios [20] and use off-line solvers to compute the policy [21]. Another research study [22] modelled the uncertainty in target localisation using a belie state and planned a series of actions to solve the mission; however, this study relied on accurate localisation of the UAV in its environment.

In a POMDP, it is desirable to calculate a large number of possible sequences of actions and observations in order to increase the quality of the planning, but these must be done within a limited time, which depends on the system dynamics. This paper presents a system in which the actions that the UAV executes are designed based on the capabilities of the motion control system. This approach provides a method that enables us to model the POMDP transition function using characteristic step responses of four states in the UAV, and it also allows us to define a maximum computation time for each iteration of the on-line POMDP algorithm.

We developed *UAV Uncertainty-Based Navigation* (UBNAV) for UAVs to navigate in GPS-denied environments by executing a policy that takes into account different sources of uncertainty. This policy is calculated on-line by a POMDP path planning algorithm. Two on-line POMDP solvers, *Partially-Observable Monte Carlo Planning* (POMCP) [14] and *Adaptive Belief Tree* (ABT) [16], will be compared and integrated into a modular system architecture along with a motion control module and a perception module. The system uses a low cost multi-rotor UAV flying in a 3D indoor space with no GPS signal available. UBNAV updates its motion plan on-line while flying based on sensor observations taking into account motion and localisation uncertainties in both the UAV and the target.

UBNAV incorporates different sources of uncertainty into a UAV motion plan on-line in order to navigate challenging scenarios using only on-board sensors. The system uses motion commands instead of waypoints and updates a policy after receiving feedback from a perception module. This approach provides ease in modelling the decoupled system dynamics by using time step responses for a set of holonomic actions in four states of the UAV. The system also guides the UAV towards regions where it can localise better in order to reduce the uncertainty in localisation of both the UAV and the goal target. UBNAV enables researchers to implement and flight test on-line POMDP algorithms for the purpose of UAV navigation in GPS-denied environments or where perception has high degrees of uncertainty.

2. UAV Navigation as a Sequential Decision Problem

2.1. Markov Decision Processes and Partially-Observable Markov Decision Processes

A Markov Decision Process (MDP) is an effective mathematical framework to model sequential decision problems affected by uncertainties [23]. When MDP is used for UAV or robotic missions, the objective is to generate a policy that allows the UAV or robot to decide what sequence of actions it should take, taking into account the uncertainties in motion, in order to maximise a return or cost function.

MDPs assume that the robot states are completely observable. However, a UAV or robot that has limitations in perception due to its sensors and the environment in which it is moving has partial observability of its states. The perception of the UAV or robot is not completely accurate or is insufficient, which means there are deviations from the real state, creating partial observability.

Partially-Observable Markov Decision Processes (POMDP) incorporate the uncertainties in sensors and the partial observability of the UAV and target locations in the environment [24]. Formally, a POMDP consists of the following elements $(S, A, O, T, Z, R, \gamma)$. S represents the set of states in the environment; A stands for the set of actions the UAV or robot can execute; O is the set of observations; T is the transition function for the state after taking an action; Z is the distribution function describing the probability of observing o from state s after taking action a; R is the set of rewards for every state; and γ is the discount factor. POMDP relies on the concept of belief, b, or belief state, which is a probability distribution of the system over all of the possible states in its state-space representation at a particular time.

POMDP enables us to represent the observations that a UAV or robot receives $o \in O$ using observation functions Z that map probability distributions to states and actions. A policy $\pi : \mathcal{B} \to A$ allocates an action a to each belief $b \in \mathcal{B}$, which is the set of possible beliefs. Given the current *belief b*, the objective of a POMDP algorithm is to find an optimal policy that maximizes an accumulated discounted return when following a sequence of actions suggested by the policy π. The accumulated *discounted return R_t* is the sum of the discounted rewards after executing every action in the sequence from time t onwards $R_t = \sum_{k=t}^{\infty} \gamma^{k-t} r_k$, where r_k is the immediate reward received at particular time step t for taking action a_t, and γ is a discount factor that models the importance of actions in the future. Choosing a value lower than one for γ signifies that actions performed in short-term steps are more important than those in long-term steps. The *value function V^π* is the expected return from belief b when following policy π, $V^\pi(b) = \mathbb{E}[\sum_{t=0}^{\infty} \gamma^{k-t} r_k | b, \pi]$. An optimal policy for solving the POMDP problem maximizes the value function $\pi^*(b) = \arg\max_\pi V^\pi(b)$.

Uncertainty in UAV localisation can be represented using an observation function Z. This observation function models the UAV localisation according to the accuracy of the GPS signal in the scenario and the precision of the onboard odometry system. The objectives of the UAV mission can be represented using a reward function R, and the motion plan is the calculated policy π^* that maximises an expected discounted accumulated return for a sequence of motion commands or actions. The UAV dynamics and the motion uncertainty are incorporated into the POMDP using the transition function T, which enables one to predict the next state s' of the UAV after an action a is executed.

2.2. POMCP and ABT

In this study, we implemented and tested two of the fastest on-line POMDP algorithms (to the authors knowledge), *Partially-Observable Monte Carlo Planning* (POMCP) [14] and *Adaptive Belief Tree* (ABT) [16], in hardware and software in order to test the system for UAV navigation missions.

POMCP [14] uses the Monte Carlo Tree Search (MCTS) algorithm [25] to produce a search tree of possible subsequent belief states b and proved to be successful in problems with large domains. The algorithm reduces the search domain by concentrating only on the states that can be reached by the UAV or robot from its initial belief state b_0 after performing actions and receiving observations. In order to calculate the transition to the next states, the algorithm uses a black box simulator that has the transition functions (T) for the UAV. In our case, the UAV dynamics are the transition function for the UAV and are modelled as motion equations in four degrees of freedom using a continuous state space. The POMCP algorithm samples states from an initial belief state and performs thousands of Monte Carlo simulations applying the set of actions that the UAV can perform; see Table 1. POMCP uses the MCTS algorithm to guide the search of the sequences of actions in the reachable belief state space and builds a search tree according to the observations received after the UAV performs the actions.

Table 1. Summary of actions in navigation and target finding problem.

Action a	Forward Velocity V_f^* (m/s)	Lateral Velocity V_l^* (m/s)	Altitude Change Δz_a^* (m)	Heading Angle Ψ_a^* (deg)
Forward	0.6	0	0	90
Backward	-0.6	0	0	90
Roll left	0	0.6	0	90
Roll right	0	-0.6	0	90
Up	0	0	0.3	90
Down	0	0	-0.3	90
Hover	0	0	0	90

POMCP initially runs a planning stage, where it generates a policy π that is stored as a tree with nodes containing belie states that are represented by particles. The algorithm then outputs the action a that maximises an expected accumulated return. Afterwards, this action a is executed, and then, an observation o is received. With this observation, POMCP performs an update of the belief state b by updating the tree and selects the node that matches the observation received in the tree search. The algorithm incorporates new particles to avoid particle deprivation and initiates a new search round from the matched node. The search depth of the planning stage is controlled by the planning horizon h parameter. The longer the planning horizon, the more time the algorithm will take to build the search tree.

ABT is an on-line POMDP solver that also uses Monte Carlo simulations to predict future belief states and a set of state particles to represent the belief state [16]. Both algorithms, POMCP and ABT, generate a search tree to store the policy. The root of the tree is a node containing the state particles representing the initial belief of the environment. The tree branches out according to the probability of selecting actions and receiving observations.

ABT updates the policy π after receiving an observation o by keeping the search tree without deleting it, which increases the number of possible sampling states in the tree search. On the other hand, POMCP only keeps the belief node that matches the received observation. The rest of the search tree is deleted, and a new round of calculations for building the policy takes place after every iteration.

3. System Architecture

UBNAV is developed as a modular system that consists of four modules that interact with a multi-copter UAV. A diagram with the system architecture is shown in Figure 1. The system contains an on-line POMDP module that runs the POMDP on-line solver and outputs actions that are executed by the motion control module. It also includes a formulation of the navigation mission as a POMDP. The on-line POMDP module updates the belief state b according to received observations o. We use existing open-source code for the POMCP [14] and ABT algorithms [26] as the base code and modified the codes to integrate all of the modules using the Robotic Operating System (ROS) [27]. The motion control module controls four states in the UAV using PID controllers. Actions in the formulated POMDP are a combination of reference values for each of the four controllers. The observation module uses the on-board sensors information and calculates and updates the UAV position. The final module is the AR Drone Autonomy Lab driver for ROS [28], which reads multiple UAV sensors and receives commands to control the roll, pitch, yaw and thrust of the UAV.

All modules execute in parallel in different threads with different update rates. The motion control module, the observation module and the AR Drone driver module execute at 100 Hz in order to control the dynamics of the UAV and to compute the odometry and estimate the location of the UAV. On the other hand, the POMDP solver executes at 1 Hz in order to guarantee that the UAV reaches a steady state after executing actions and to have a policy that enables the UAV to accomplish the mission. The system source code is written in C++, is available as ROS packages and can be provided as open-source code upon request.

3.1. On-Line POMDP Module

We did several modifications and additions to the POMDP solvers' source code in order to allow them to work as ROS nodes and to integrate them into the modular system. We also created a visualisation scenario in order to examine the performance of the solvers in simulation.

The on-line POMDP module runs the on-line POMDP planning algorithms. This module initialises the parameters for planning horizon h, the map of the environment, target location, odometry uncertainty and an initial belief state b_0. This module produces a policy π based on the initial belie state b_0 and outputs the action a to be executed according to the calculated policy. The action a is then executed by the motion control module, and the observation module calculates the UAV position using the on-board sensed velocity, altitude and heading angle. The observation module also checks whether the target is detected by the downward-looking camera and detects obstacles located within approximately 1 m in front of the UAV with an on-board front-looking camera.

Once the observation is received, the POMDP node updates the belie state b, to match the received observation and replenishing particles until a time-out is reached. Based on the current belie state b, the POMDP solver calculates a new policy π (POMCP) or updates the policy π (ABT) and outputs the subsequent action a based on the updated policy π.

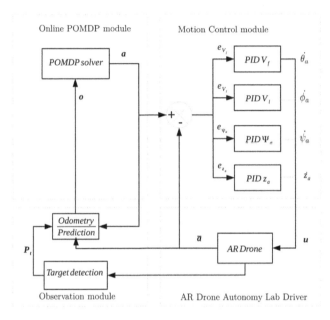

Figure 1. POMDP ROS system architecture.

3.2. Motion Control Module

The UAV motion control module is composed of four independent controllers that actuate on the following states of the UAV: Forward velocity V_f, lateral velocity V_l, yaw angle (heading angle) Ψ_a and altitude z_a. The motion control module receives reference states $a = \{V_f^*, V_l^*, \Psi_a^*, z_a^*\}$ from the POMDP solver (see Table 1) and subtracts the actual states $\bar{a} = \{V_f, V_l, \Psi_a, z_a\}$ from the references sates a to generate error signals that are used by each of the PID controllers.

The output of the PIDs is a control vector $u = \{\theta_a, \phi_a, \psi_a, z_a\}$, where θ_a, ϕ_a and ψ_a are pitch, roll and yaw rates, respectively, and $\dot{z}_a$ is the rate of climb. These outputs are sent to the AR Drone Autonomy lab driver, which transforms them into control signals that are sent to the UAV.

The UAV on-board navigation system calculates forward V_f and lateral V_l velocities using optical flow obtained from the downward-looking camera. The UAV obtains the yaw angle Ψ_a from the IMU and magnetometer readings and the altitude z_a from on-board ultrasonic and barometric pressure sensors that are fused using a proprietary Kalman filter.

The duration of an action execution T_a is chosen to guarantee that the PID controller reaches a steady state when transitioning from different actions for all states, as shown in Figure 2. Therefore, we set the POMDP solver time to to be equal to the action duration, that is $T_P = T_a$.

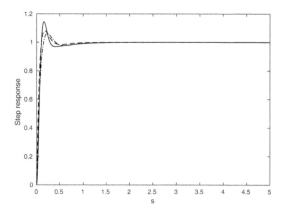

Figure 2. PID time responses to a unit step input for V_f and V_l (dashed line), Ψ (dash-dot line) and z (solid line). All PID controllers reach steady state within 1 s.

3.3. Observation Module

The observation module calculates the current multi-rotor position and heading angle based on the sensed forward and lateral velocities, accelerations and the measured yaw angle Ψ_a. The forward and lateral velocities in the UAV's frame are transformed to the fixed world frame and are integrated to calculate the UAV coordinates x_a and y_a in the world frame. The UAV altitude z_a is also read from the AR Drone ROS driver and is calculated on-board.

The UAV position $P_a = \{x_a, y_a, z_a, \Psi_a\}$ is calculated based on the actual states $\tilde{a}$ obtained from the on-board sensors from the AR Drone ROS driver. The POMDP solver calculates and updates the policy based on the position and heading angle that the UAV will have by the time the previous action is finished, *i.e.*, $t - T_a$. Thus, a prediction of the UAV position and heading angle $P_{a_{t+1}}$ is calculated based on the current UAV position and yaw angle in the world frame, the action currently being executed and the action duration time T_a. This prediction is calculated using the characteristic step responses for the four degrees of freedom shown in Figure 2, the commanded state reference values a and the actual state values $\tilde{a}$.

The observation module also detects and calculates the target position P_t by using a specific tag figure and the AR Drone Autonomy Lab driver to determine whether the tag is present in the downward looking camera image and its position within the image. The last type of observation in the module identifies obstacles. In this case, we use Augmented Reality (AR) tags that are placed on the obstacles and which can be detected with the front camera using the ROS package Arsys for reading Aruco-type tags [29].

The use of the AR tags enables the system to have a global source of localisation since the tags can give an accurate indication of the UAV camera pose with respect to the world frame by using coordinate frame transformations. However, these tags can only be detected when the AR tag is located within the UAV front camera Field Of View (FOV) and within a distance of approximately 1 m. The model of the downward looking camera FOV is described in Section 4.5.

4. Navigation and Target Detection

4.1. Problem Description and Formulation

Consider a multi-rotor UAV flying in a 3D space in a GPS-denied environment filled with obstacles. An example of such an environment is the indoor scenario shown in Figure 3. The UAV has the mission to navigate within a limited region of the airspace with boundaries x_{lim}, y_{lim} and z_{lim} in the x, y and z

coordinates, respectively. The UAV's task is to fly from the initial hovering position to a region where the target is believed to be located in order to explore and detect it using an onboard downward-looking camera, avoiding obstacles whose location is known by the UAV navigation system. After taking off, the UAV hovers for five seconds before starting the mission. This initial hovering incorporates uncertainty into the UAV position due to the UAV drifting around the take off position. The target is stationary, and its location is known by the system with uncertainty.

(a)   (b)

Figure 3. Two typical scenarios for the navigation and target detection problem. (**a**) Scenario with five obstacles; (**b**) scenario with four obstacles with Augmented Reality (AR) tags.

The navigation and target detection must be done using only on-board sensors, whilst overcoming uncertainties in the UAV states information and uncertainty in target location. We use continuous state representation to model the transition function and a discrete set of observations O as described in Section 4.5.

The problem is formulated as a POMDP with the following elements: the state of the aircraft in the environment (S), the set of actions that the multi-rotor can execute (A), the transition function describing the state transition after applying a specific action (T), the observation model (O) that represents the sensed state and uncertainty of the UAV after taking an action and the reward function (R).

4.2. State Variables (S)

The state variables considered in the POMDP formulation are the position and yaw (heading) angle of the UAV in the world frame $P_a = (x_a, y_a, z_a, \Psi_a)$, the target's position $P_t = (x_t, y_t, z_t)$ and the UAV's forward V_f and lateral velocity V_l.

4.3. Actions (A)

There are seven possible actions $a \in A$ that the UAV can execute: an action to go forward at speed V_f, an action to go backwards at speed $-V_f$, an action to roll left at speed V_l and an action to roll right at speed $-V_l$. Actions up and down increase or decrease the altitude by 0.3 m, with UAV's forward V_f =0 m/s and lateral velocity V_l = 0 m/s. The hover action maintains the UAV's velocity at zero. The set of actions is summarised in Table 1.

4.4. Transition Functions (T)

In order to represent the system dynamics, a kinematic model is described as in Equation (2) to Equation (3). The position of the aircraft is determined by calculating the change in position due to its current velocity, which is controlled by the motion control module and is selected according to the commanded action. A transformation from the UAV's frame to the world frame is also calculated in these equations.

A normal probability distribution with mean value and standard deviation <1 m around the take off position, as shown in Equation (1), is used to model this uncertainty on the UAV initial position.

The orientation of the UAV is determined by its heading angle Ψ_a. The uncertainty in motion is incorporated in the system by adding a small deviation to the heading angle σ_a using a normal distribution with mean value equal to the desired heading angle and restricted to the range $-2.0° < \sigma_a < 2.0°$, which represents the uncertainty on the yaw angle control system.

A discrete table with the characteristic values of a step input response was obtained experimentally for each of the controllers for every degree of freedom in the motion controller module. These step responses (see Figure 2) are included in the transition function in order to incorporate the transient changes in the UAV speed when it transitions from action to action after receiving the command from the on-line planner. The uncertainty in UAV position due to the initial hovering is modelled as a Gaussian probability distribution, described by Equation (1), with mean value μ and standard deviation $\sigma < 1$ m.

$$P(x) = \frac{1}{\sigma\sqrt{2\pi}} e^{-(x-\mu)^2/2\sigma^2} \tag{1}$$

$$\begin{bmatrix} \Delta x_{a_t} \\ \Delta y_{a_t} \\ \Delta z_{a_t} \end{bmatrix} = \begin{bmatrix} \cos(\Psi_{a_t} + \sigma_{a_t}) & -\sin(\Psi_{a_t} + \sigma_{a_t}) & 0 \\ \sin(\Psi_{a_t} + \sigma_{a_t}) & \cos(\Psi_{a_t} + \sigma_{a_t}) & 0 \\ 0 & 0 & 1 \end{bmatrix} \begin{bmatrix} V_{a_{f_t}}\Delta t \\ V_{a_{l_t}}\Delta t \\ \Delta z_a \end{bmatrix} \tag{2}$$

$$\begin{bmatrix} x_{a_{t+1}} \\ y_{a_{t+1}} \\ z_{a_{t+1}} \end{bmatrix} = \begin{bmatrix} x_{a_t} \\ y_{a_t} \\ z_{a_t} \end{bmatrix} + \begin{bmatrix} \Delta x_{a_t} \\ \Delta y_{a_t} \\ \Delta z_{a_t} \end{bmatrix} \tag{3}$$

where x_{a_t}, y_{a_t} and z_{a_t} are the x, y and z aircraft coordinates at time t, $V_{a_{f_t}}$ and $V_{a_{l_t}}$ are forward and lateral velocities in the multi-rotor's frame at time t and Ψ_{a_t} and σ_{a_t} are heading and heading deviation at time t.

4.5. Observation Model (O)

An observation for the POMDP model is composed of: (1) the UAV position in the world frame with onboard odometry as the source; (2) the UAV position with reduced uncertainty in the world frame if obstacles are detected; and (3) the target location if it is detected by the downward-looking camera. The UAV odometry calculation has an uncertainty that is approximated using a Gaussian distribution, as in Equation (1). The mean value of this distribution is the UAV position calculated by the odometry system, and the variance is equal to the error in the odometry calculation. This error is bigger for UAV positions calculated based only on optical flow sensor and smaller if the AR tags are detected in any of the obstacles by the front camera.

Equation (4) is used to calculate the x_a and y_a positions of the UAV.

$$\begin{bmatrix} x_{a_{t+1}} \\ y_{a_{t+1}} \end{bmatrix} = \begin{bmatrix} x_{a_t} \\ y_{a_t} \end{bmatrix} + \begin{bmatrix} \cos(\Psi_{a_t}) & -\sin(\Psi_{a_t}) \\ \sin(\Psi_{a_t}) & \cos(\Psi_{a_t}) \end{bmatrix} \begin{bmatrix} V_{a_{f_t}} \\ V_{a_{l_t}} \end{bmatrix} \Delta t \tag{4}$$

If the target is detected by the onboard downward-looking camera, the AR Drone ROS driver provides the target position within the image. This position is transformed to a position in the world frame.

The FOV of the downward-looking camera, shown in Figure 4, is modelled and defined experimentally as follows: the image width depends on the UAV altitude and is defined as $w_c = z_a \alpha_w$;

the image height is defined as $h_c = z_a \alpha_h$; and there is a shift from the UAV frame to camera frame that is defined as $shift = z_a \alpha_s$, where $\alpha_h = 0.56$, $\alpha_w = 0.96$ and $\alpha_s = 0.2256$ are intrinsic parameters of the camera and are obtained by camera calibration and experimental tests.

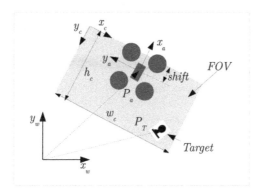

Figure 4. Field of view parameters for the UAV onboard camera, target and world frames.

4.6. Reward Function (R)

The objectives of the UAV navigation mission can be represented in the POMDP using a reward function. The UAV receives a high reward of 300 if it detects the target within the downward-looking camera FOV. Hitting an obstacle or going out of the scenario incurs a penalty of -70, and every other movement carries a cost of -2. The values of the reward and penalties were selected based on existing test cases of POMDP problems. These values were tuned by experimentation on a large number of simulations (≈ 500) and flight tests (≈ 50). A summary of the reward function values is shown in Table 2.

Table 2. Summary of rewards and costs in the UAV navigation problem.

Reward/Cost	Value
Detecting the target	300
Hitting an obstacle	-70
Out of region	-70
Movement	-2

4.7. Discount Factor (γ)

A discount factor γ of 0.97 was selected by experimentation on a large number of simulations (≈ 500) and flight tests (≈ 50) and taking into account the distance travelled by the UAV in every step at the selected speed and the size of the flying area.

5. Results and Discussion

We conducted simulation and real flight experiments for four tests cases to compare the performance of ABT and POMCP.

5.1. Simulation

We tested the UAV navigation formulation in simulation with both solvers, POMCP and ABT, by running each algorithm, using the model dynamics presented in Sections 4.4 and 4.5. We used boost C++ library random generators in order to have high quality normal distributions as described in Section 4.1.

The target was placed at four different locations inside the flying area, and the UAV was initially located around $(3.0, 0.55, 0.7)$, with uncertainty in each of the locations, as described by Equation (1). Each target location was tested in simulation by running POMCP and ABT 100 times. We created a simulated 3D model of the environment to visualise the UAV path in the scenario to inspect the evolution of the belief state of the system and for visualising the UAV position and the target location as clouds of white and red particles, respectively (Figure 5). Initially, the UAV position, represented by the white particle cloud in Figure 5a, has an uncertainty around the starting position at the bottom of the image. The spread of the particles represents the drift caused by UAV take off and initial hovering. This uncertainty increases as the UAV flies towards the goal (Figure 5b). As the UAV gets closer to one of the obstacles and detects the AR tag attached to the obstacle with the front camera, the uncertainty in the UAV position is reduced, as seen in Figure 5c. The UAV then flies towards the target until it detects it, shown in Figure 5d, and the uncertainty in its position is reduced by the knowledge that it acquires from the position of the target in the image taken by the downward-looking camera.

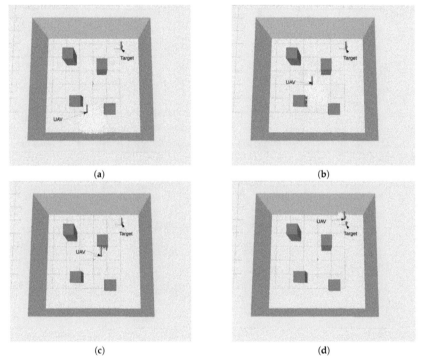

Figure 5. Belief evolution for the navigation problem using the POMCP solver. Target's position (red); UAV position (white). (**a**) Initial belief, UAV location (white particles) and target location (red particles); (**b**) belief after some steps; (**c**) UAV in front of obstacle; uncertainty is reduced; (**d**) target within UAV downward camera FOV.

A summary of the results for the discounted return and the time to detect the target for each test in simulation and in real flight is shown in Table 3.

Table 3. Simulation and real flight results for Adaptive Belief Tree (ABT) and POMCP.

Solver	Number of Obstacles in Scenario	Target Location (x, y, z)	Flight Time to Target (s) (Simulation)	Flight Time to Target (s) (Real Flight)	Success (No Collision) % (Real Flight)
POMCP	5	$(5.0, 6.0, 0.15)$	14	25	100
	4	$(2.65, -2.33, 0.15)$	20	21	100
ABT	5	$(5.0, 6.0, 0.15)$	16	25	90
	4	$(2.65, -2.33, 0.15)$	19	27	80

The results indicate that for all cases in the simulation, both POMDP solvers were able to reach and detect the target in less than 20 s. Results also show that paths produced by ABT are shorter than the ones produced by POMCP (Figures 6 and 7). This can also be seen in the flight time to detect the target, which is shorter for ABT in three of the four cases; see Table 3. In the case of the target located at $(5, 6, 0.15)$, the path computed by ABT shows that the UAV takes a trajectory flying between the obstacles, whereas with POMCP, the UAV flies over the obstacles to avoid the risk of collision.

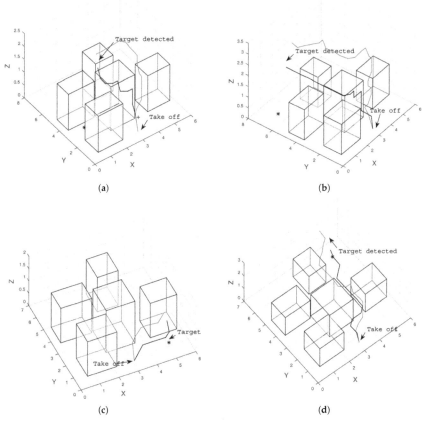

Figure 6. Example of trajectories to four different target locations produced by ABT (black), and POMCP (red) in simulation. (**a**) Target located at (1.0, 3.0, 0.15); (**b**) target located at (1.0, 6.5, 0.15); (**c**) target located at (5.0, 1.0, 0.15); (**d**) target located at (5.0, 6.0, 0.15).

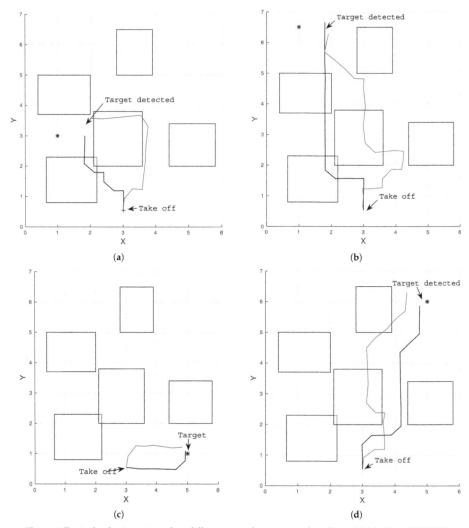

Figure 7. Example of trajectories to four different target locations produced by ABT (black) and POMCP (red) in simulation. (**a**) Target located at (1.0, 3.0, 0.15); (**b**) target located at (1.0, 6.5, 0.15); (**c**) target located at (5.0, 1.0, 0.15); (**d**) target located at (5.0, 6.0, 0.15).

5.2. Real Flight Tests

We conducted experiments in real flight scenarios with the target located at different positions. We ran the system 10 times for each location of the target for both solvers to compare their performance. We also compare two scenarios, one with five obstacles and another with four taller obstacles with AR tags on each of the obstacles, which enable the UAV to correct the uncertainty in the onboard odometry once the tags are in the FOV of the front camera.

The UAV has to take off from an initial position and hover for 4 s to initialise its on-board sensors. The forward (pitch) and lateral (roll) velocity controllers start to actuate when taking off, and the altitude controller sets an initial value for the UAV to hover at 0.7 m above the ground.

The initial POMDP navigation policy is computed once the UAV is airborne, and the POMDP module outputs an action after 4 s of hovering. The system computes a new policy at each step of 1 s of duration.

Figure 8a shows a comparison of the UAV position estimation performed by the observation node and using the VICON positioning system. Even though there is an error in the estimated path, the system is robust enough to account for this drift, is able to find a path using the UAV on-board odometry and finds the target without colliding with the obstacles.

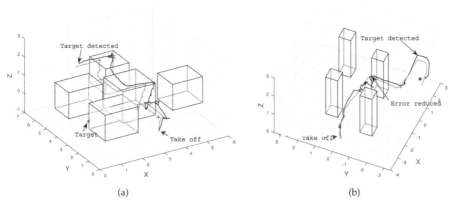

(a)   (b)

Figure 8. Comparison of onboard computed position (red) *vs.* the VICON positioning system (black). (**a**) Without resetting position error; (**b**) resetting position error with AR tags.

The system was also tested incorporating the detection of the obstacles by the UAV, using AR tags that allow for a more reliable source of positioning in places where the UAV is closer to the obstacles than the onboard odometry. A comparison of the UAV position estimation against the VICON system using the AR tags to reduce the uncertainty is shown in Figure 8b. Results indicate that the system is able to reduce the uncertainty and the error in the position estimation using AR tags as beacons in the environment.

Results in Table 3 indicate that the performance of both POMDP solvers is affected by real flight conditions, and the flight time to detect the target increases in all of the cases. The UAV can accomplish the mission in 100% of the cases using POMCP for both scenarios and 85% of the cases using ABT in real flight, as show in Table 3. A comparison of the paths flown by the UAV to four different target locations produced by POMCP and ABT algorithms are shown in Figures 9 and 10.

Figure 9a–c show the trajectories taken by the UAV in a scenario with five obstacles. In this case, both solvers show that it is safer to fly over the obstacles to reach the target. On the other hand, Figure 9d shows that the algorithms find a safe route to the target by flying between obstacles in the scenario with four obstacles. However, in the last case, the UAV flies close to the obstacle first in order to reduce its uncertainty by detecting the beacon or landmark represented by the AR tag on the obstacle and then continues flying towards the target.

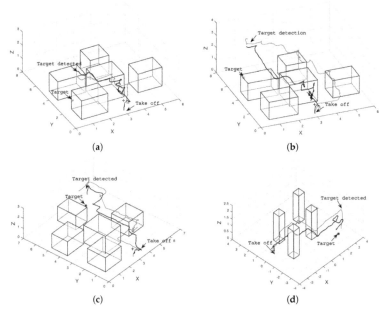

Figure 9. Example of trajectories to four different target locations produced by ABT (black) and POMCP (red). (**a**) Target located at (1.0, 3.0, 0.15); (**b**) target located at (1.0, 6.5, 0.15); (**c**) target located at (5.0, 6.0, 0.15); (**d**) target located at (2.65, −2.33, 0.15).

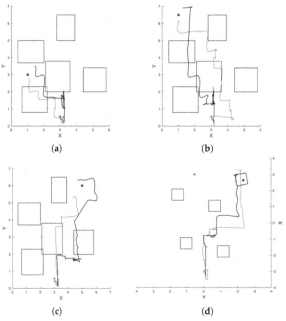

Figure 10. Example of trajectories to four different target locations produced by ABT (black) and POMCP (red) in simulation. (**a**) Target located at (1.0, 3.0, 0.15); (**b**) target located at (1.0, 6.5, 0.15); (**c**) target located at (5.0, 6.0, 0.15); (**d**) target located at (2.65, −2.33, 0.15).

5.3. Recommendations for Implementation of UAV Navigation Using On-Line POMDP in a Real Flight

Implementation of UAV navigation in cluttered and GPS-denied environments formulated as a POMDP is a complex task. Some considerations should be taken into account for the implementation of the system for real flights.

The first consideration is to perform a thorough analysis of the dynamic capabilities of the UAV and, in our case, the design of motion controllers on four decoupled states. The design and tuning of these motion controllers takes time, repeated flight testing and varies according to UAV specifications. This step is indispensable and enables us to model the characteristic response of the UAV, which facilitates modelling the POMDP transition function based on the set of actions chosen.

The next consideration is to characterise the controllers' time response by using system identification software and the data collected from the real flight tests. The transition function uses this characteristic time response to calculate the next state of the UAV after performing an action, and it is also used to simulate the performance of UBNAV. This step is also required for the implementation of navigation in other path planning algorithms, which require low level motion control that outputs a series of waypoints [3,4,10] or velocity commands that a lower level motion controller executes [30].

A third important consideration for a POMDP implementation is the observation function, which models the characteristics, ranges, accuracy, disturbances and noise measurements. Having an exact knowledge of all of these characteristics is not always possible, but some approximations can be made. If a camera is used as one of the sensors, then camera calibration and a model for the FOV that depends on the UAV altitude is needed. Several flight tests measuring the drift in the initial hovering and the yaw angle measurements are needed in order to approximate the model using Gaussian distributions.

A fourth consideration is the discount factor γ that is used to balance the importance of actions during the planning stage. Selecting a discount factor of one means that there is no discount applied to the return received after performing an action, which in turn gives the same importance to actions that are executed in the short term and in the long term. Conversely, assigning a value much lower than one discounts the return for actions executed in the long term and in turn gives more importance to short-term actions. In this study, the discount factor was selected by testing in 100 Monte Carlo simulations with values for the discount factor ranging from 0.95–1, taking into account that the UAV should not collide with obstacles (short-term goal) and should also reach and detect the target (long-term goal).

Finally, the reward function (R) was tuned by performing multiple Monte Carlo simulations. Values of the reward and penalties were tuned focusing first on reaching the primary goal, which is to detect the target. Afterwards, values of penalties for colliding with obstacles and going out of the flying area were tuned by fixing the reward for reaching the goal and running multiple simulations with different values for the penalty. Finally, in order for the UAV to choose shorter paths, values in the interval [0,20] for the motion penalty were tested running multiple simulations.

6. Conclusions

This study developed and tested UBNAV for UAV navigation in GPS-denied and cluttered environments where there is uncertainty in sensor perception and target localisation. UBNAV performed full navigation and target detecting missions using a low cost platform with only on-board sensors in GPS-denied environments and in the presence of obstacles. UBNAV allows researchers to implement, simulate and test different on-line POMDP algorithms for navigation with uncertainty and also permits the implementation of different motion controllers and perception modules into the system.

The system executes a set of holonomic actions instead of a classic waypoint approach. This approach facilitates the execution of the motion planning by modelling the UAV dynamics as a decoupled system with a set of actions that execute in four states of the UAV. It models the uncertainties in the motion and perception by including deviations in the states using Gaussian distributions.

The system also allows for selecting the planning time in an on-line POMDP solver for a UAV navigation mission, taking into account the duration of the actions. These actions are based on the time response of the motion controllers of the UAV.

Experimental results show that the system is robust enough to overcome uncertainties that are present during a flight mission in a GPS-denied and cluttered environment. UBNAV guides the UAV towards regions where it can localise better in order to reduce the uncertainty in the localisation of both the UAV and the goal target. Results indicate that the system successfully finds a path on-line for the UAV to navigate to a location where it could detect a target with its onboard downward-looking camera without colliding with obstacles for 100% of the time for the simulation and 96.25% of the time in 80 different real flight experiments.

The system has also the potential to be used in an outdoor environment with additional perception information by modelling a Global Positioning System (GPS) where the uncertainty can be introduced depending on the resolution and the quality of the GPS module. Adding the GPS module uncertainty and the wind disturbances as another source of uncertainty into the POMDP model along with real flight testing is currently being explored.

Acknowledgments: This work is supported by Queensland University of Technology, The Australian Research Centre for Aerospace Automation and the COLCIENCIAS 529 scholarship. We also would like to thank the anonymous reviewers for their comprehensive reviews and comprehensive feedback.

Author Contributions: This work is part of the doctoral studies of Fernando Vanegas supervised by Felipe Gonzalez. Fernando Vanegas developed the system, controllers, implemented the simulation and flight test environments and wrote this manuscript. Felipe Gonzalez provided general ideas about the work, supervision for simulation flight testing and wrote sections of the manuscript.

Conflicts of Interest: The authors declare no conflict of interest.

References

1. Gonzalez, L.F. *Robust Evolutionary Methods for Multi-Objective and Multidisciplinary Design Optimisation in Aeronautics*; University of Sydney: Sydney, Australia, 2005.
2. Lee, D.; Gonzalez, L.F.; Periaux, J.; Srinivas, K.; Onate, E. Hybrid-Game Strategies for multi-objective design optimization in engineering. *Comput. Fluids* **2011**, *47*, 189–204.
3. Roldan, J.; Joossen, G.; Sanz, D.; del Cerro, J.; Barrientos, A. Mini-UAV Based Sensory System for Measuring Environmental Variables in Greenhouses. *Sensors* **2015**, *15*, 3334–3350.
4. Everaerts, J. The use of unmanned aerial vehicles (UAVs) for remote sensing and mapping. *Int. Arch. Photogramm. Remote Sens. Spat. Inf. Sci.* **2008**, *37*, 1187–1192.
5. Figueira, N.M.; Freire, I.L.; Trindade, O.; Simoes, E. Mission-Oriented Sensor Arrays And UAVs; A Case Study On Environmental Monitoring. *ISPRS Int. Arch. Photogramm. Remote Sens. Spat. Inf. Sci.* **2015**, *XL-1/W4*, 305–312.
6. Lupashin, S.; Schollig, A.; Sherback, M.; D'Andrea, R. A simple learning strategy for high-speed quadrocopter multi-flips. In Proceedings of the IEEE International Conference on Robotics and Automation, Anchorage, AK, USA, 3–7 May 2010; pp. 1642–1648.
7. Muller, M.; Lupashin, S.; D'Andrea, R. Quadrocopter ball juggling. In Proceedings of the IEEE/RSJ International Conference on Intelligent Robots and Systems, San Francisco, CA, USA, 25–30 September 2011; pp. 5113–5120.
8. Schollig, A.; Augugliaro, F.; Lupashin, S.; D'Andrea, R. Synchronizing the motion of a quadrocopter to music. In Proceedings of the IEEE International Conference on Robotics and Automation, Anchorage, AK, USA, 3–8 May 2010; pp. 3355–3360.
9. Hehn, M.; D'Andrea, R. A flying inverted pendulum. In Proceedings of the IEEE International Conference on Robotics and Automation, Shanghai, China, 9–13 May 2011; pp. 763–770.
10. Haibin, D.; Yu, Y.; Zhang, X.; Shao, S. Three-Dimension Path Planning for UCAV Using Hybrid Meta-Heuristic ACO-DE Algorithm. *Simul. Model. Pract. Theory* **2010**, *18*, 1104–1115.
11. Smith, T.; Simmons, R. Heuristic search value iteration for POMDP. In Proceedings of the 20th Conference on Uncertainty in Artificial Intelligence, Banff, AB, Canada, 7–11 July 2004; pp. 520–527.

12. Pineau, J.; Gordon, G.; Thrun, S. Anytime point-based approximations for large POMDP. *J. Artif. Intell. Res.* **2006**, *27*, 335–380.

13. Hauser, K. Randomized belief-space replanning in partially-observable continuous spaces. In *Algorithmic Foundations of Robotics IX*; Springer: Berlin, Germany, 2011; pp. 193–209.

14. Silver, D.; Veness, J. Monte-Carlo Planning in large POMDP. In Proceedings of the 24th Annual Conference on Neural Information Processing Systems, Vancouver, BC, Canada, 6–9 December 2010.

15. Kurniawati, H.; Du, Y.; Hsu, D.; Lee, W.S. Motion planning under uncertainty for robotic tasks with long time horizons. *Int. J. Robot. Res.* **2011**, *30*, 308–323.

16. Kurniawati, H.; Yadav, V. An Online POMDP Solver for Uncertainty Planning in Dynamic Environment. In Proceedings of Results of the 16th International Symposium on Robotics Research, Singapore, 16–19 December 2013.

17. Pfeiffer, B.; Batta, R.; Klamroth, K.; Nagi, R. Path planning for UAVs in the presence of threat zones using probabilistic modeling. *IEEE Trans. Autom. Control* **2005**, *43*, 278–283.

18. Ragi, S.; Chong, E. UAV path planning in a dynamic environment via partially observable Markov decision process. *IEEE Trans. Aerosp. Electr. Syst.* **2013**, *49*, 2397–2412.

19. Dadkhah, N.; Mettler, B. Survey of Motion Planning Literature in the Presence of Uncertainty: Considerations for UAV Guidance. *J. Intell. Robot. Syst.* **2012**, *65*, 233–246.

20. Al-Sabban, W.H.L.; Gonzalez, F.; Smith, R.N. Wind-Energy based Path Planning For Unmanned Aerial Vehicles Using Markov Decision Processes. In Proceedings of the IEEE International Conference on Robotics and Automation, Karlsruhe, Germany, 6–10 May 2013.

21. Al-Sabban, W.H.; Gonzalez, L.F.; Smith, R.N. Extending persistent monitoring by combining ocean models and Markov decision processes. In Proceedings of the 2012 MTS/IEEE Oceans Conference, Hampton Roads, VA, USA, 14–19 October 2012.

22. Chanel, C.P.C.; Teichteil-Königsbuch, F.; Lesire, C. Planning for perception and perceiving for decision POMDP-like online target detection and recognition for autonomous UAVs. In Proceedings of the 6th International Scheduling and Planning Applications woRKshop, São Paulo, Brazil, 25–29 June 2012.

23. Thrun, S.; Burgard, W.; Fox, D. *Probabilistic Robotics*; MIT Press: Cambridge, MA, USA, 2005; Volume 1.

24. Pineau, J.; Gordon, G.; Thrun, S. Point-based value iteration: An anytime algorithm for POMDP. In Proceedings of the Eighteenth International Joint Conference on Artificial Intelligence, Acapulco, Mexico, 9–15 August 2003; Volume 18, pp. 1025–1032.

25. Levente, K.; Szepesvari, C. Bandit Based Monte-Carlo Planning. In *Machine Learning*; Springer: Berlin, Germany, 2006; pp. 282–293.

26. Klimenko, D.; Song, J.; Kurniawati, H. TAPIR: A Software Toolkit for Approximating and Adapting POMDP Solutions Online. In Proceedings of the Australasian Conference on Robotics and Automation, Melbourne, Australia, 2–4 December 2014.

27. Quigley, M.; Conley, K.; Gerkey, B.; Faust, J.; Foote, T.; Leibs, J.; Wheeler, R.; Ng, A.Y. ROS: An open-source Robot Operating System. In Proceedings of the ICRA Workshop on Open Source Software, Kobe, Japan, 12–17 May 2009; Volume 3.

28. Krajnik, T.; Vonasek, V.; Fiser, D.; Faigl, J. AR-drone as a platform for robotic research and education. In *Research and Education in Robotics-EUROBOT*; Springer: Berlin, Germany, 2011; pp. 172–186.

29. Garrido-Jurado, S.; Munoz-Salinas, R.; Madrid-Cuevas, F.J.; Marin-Jimenez, M.J. Automatic generation and detection of highly reliable fiducial markers under occlusion. *Pattern Recognit.* **2014**, *47*, 2280–2292.

30. Garzon, M.; Valente, J.; Zapata, D.; Barrientos, A. An Aerial-Ground Robotic System for Navigation and Obstacle Mapping in Large Outdoor Areas. *Sensors* **2013**, *13*, 1247–1267.

Article

UAV-Based Estimation of Carbon Exports from Heterogeneous Soil Landscapes—A Case Study from the CarboZALF Experimental Area

Marc Wehrhan [1,*], Philipp Rauneker [2] and Michael Sommer [1,3]

[1] Leibniz Centre for Agricultural Landscape Research (ZALF), Institute of Soil Landscape Research, Eberswalder Straße 84, Müncheberg 15374, Germany; sommer@zalf.de

[2] Leibniz Centre for Agricultural Landscape Research (ZALF), Institute of Landscape Hydrology, Eberswalder Straße 84, Müncheberg 15374, Germany; philipp.rauneker@zalf.de

[3] University of Potsdam, Institute of Earth and Environmental Sciences, Karl-Liebknecht-Str. 24-25, Potsdam 14476, Germany

* Correspondence: wehrhan@zalf.de; Tel.: +49-33432-82-109; Fax: +49-33432-280

Academic Editors: Felipe Gonzalez Toro and Antonios Tsourdos
Received: 15 December 2015; Accepted: 15 February 2016; Published: 19 February 2016

Abstract: The advantages of remote sensing using Unmanned Aerial Vehicles (UAVs) are a high spatial resolution of images, temporal flexibility and narrow-band spectral data from different wavelengths domains. This enables the detection of spatio-temporal dynamics of environmental variables, like plant-related carbon dynamics in agricultural landscapes. In this paper, we quantify spatial patterns of fresh phytomass and related carbon (C) export using imagery captured by a 12-band multispectral camera mounted on the fixed wing UAV Carolo P360. The study was performed in 2014 at the experimental area CarboZALF-D in NE Germany. From radiometrically corrected and calibrated images of lucerne (Medicago sativa), the performance of four commonly used vegetation indices (VIs) was tested using band combinations of six near-infrared bands. The highest correlation between ground-based measurements of fresh phytomass of lucerne and VIs was obtained for the Enhanced Vegetation Index (EVI) using near-infrared band b_{899}. The resulting map was transformed into dry phytomass and finally upscaled to total C export by harvest. The observed spatial variability at field- and plot-scale could be attributed to small-scale soil heterogeneity in part.

Keywords: UAV; multispectral; VI; agriculture; carbon export; soil landscape

1. Introduction

The application and enhancement of remote sensing methods and sensors have led to a better understanding of how leaf properties (age, shape, nutrient and water status) affect leaf reflectance and leaf emittance. Research on the contribution of plant canopy architecture, solar illumination conditions and soil reflectance to total canopy reflectance resulted in improved estimates of canopy parameters such as leaf area, standing phytomass, crop type and yield [1,2]. These techniques and methods can provide a valuable contribution to the analysis of processes controlling the carbon budget of agricultural landscapes.

Recently, Leaf Area Index (LAI) derived from multi-temporal broad-band optical and microwave remote sensors, has been assimilated successfully in carbon cycle modeling. LAI values of winter wheat were used to update the simulated LAI of a process-based model of cereal crop carbon budgets to improve daily net ecosystem exchange (NEE) fluxes and at-harvest cumulative NEE at the field-scale of different European study sites [3]. Time-series of Landsat TM and ETM+ data were used for leaf chlorophyll (Chl) retrieval to improve model simulations of gross primary productivity (GPP) of corn.

The satellite-based Chl estimates were used for parameterizing the maximum rate of carboxylation (V_{max}), which represents a key control on leaf photosynthesis within C_3 and C_4 photosynthesis models [4]. These approaches of data assimilation are promising but rely on the availability of remote sensing data. In the case of required time-series or time-sensitive applications such as management decision support, crop stress or erosion related events, most remote sensing platforms provide unfavorable revisit times (>5 days). In addition, the quality and finally the exploitability of satellite imagery depends on current weather conditions during image acquisition. On the local scale, the use of optical and thermal remote sensing sensors mounted on satellites and manned airborne platforms is either limited in spatial and spectral resolution or suffer from high operational costs [5].

The development of Unmanned Aerial Vehicles (UAVs) is an opportunity to overcome some of these limitations. The technical progress in the development of sensors, embedded computers, autopilot systems and platforms enables the construction of lightweight remote sensing systems with user defined flight intervals. For multitemporal data acquisition with high spatial resolution, these systems are much more economical than manned aircraft and allow a more flexible mission design than with the use of satellites. However, the challenge of using UAVs for environmental research is less their operational use but (i) the generation of radiometric and geometric corrected imagery; and (ii) the conversion of the spectral information to vegetation biophysical parameters. Previous studies focused on these two essential aspects for different sensors mounted on UAVs. Numerous procedures were applied to reduce the impacts of noise [6], radial illumination fall-off (vignetting) [6–8], lens distortions [5] and bidirectional reflectance effects [7] on imagery captured by different non-calibrated charge-coupled device (CCD) or complementary metal-oxide-semiconductor (CMOS) sensors. For the conversion of the pre-processed digital numbers (DN) into at-surface reflectance the above-mentioned authors applied an empirical line approach using a linear transformation derived from ground-based reflectance measurements of calibration targets and the respective image DNs.

Several UAV-based studies performed over agricultural fields were conducted with the aim to estimate vegetation biophysical parameters either by object-based image analysis [9] or by using vegetation indices (VIs) [7,10,11]. VIs are linear, orthogonal or ratio combinations of reflectance calculated from different wavelengths of the visible (VIS) and near-infrared (NIR) part of the electromagnetic spectrum [12] and widely used proxies for temporal and spatial variation in vegetation structure and biophysical parameters of agricultural crops [13]. Numerous VIs have been developed in the last decades to improve the relationships between the spectral response and the respective characteristics of vegetation canopies such as net primary production, LAI, vegetation fraction, chlorophyll density or the fraction of absorbed photosynthetic active radiation (FAPAR) [14–16]. Most VIs suffer from strong non-linearity and sensitivity to external factors such as solar and viewing geometry, soil background and atmospheric effects [17]. Apart from limitations caused by insufficient spectral resolution of available sensors, most improvements refer to the reduction of these external factors. Jiang *et al.* [18] give a brief review of different VIs, their limitations and proposed improvements.

Lelong *et al.* [7] implemented different VIs in generic relationships to successfully estimate LAI and nitrogen uptake of wheat varieties grown on micro-plots. Zarco-Tejada *et al.* [10,11] demonstrated that narrow-band VIs derived from multi- and hyperspectral imagery captured during different UAV missions enable the detection of chlorophyll fluorescence emission variability as a stress status indicator of olive, peach and citrus trees. Although there is a growing demand on quantitative and spatially consistent data of major components of the carbon budget [19,20], to our knowledge, UAV-based remote sensing have not been used for the estimation of the C export by harvest of crops to date. Apparently easy to achieve, considerable uncertainties arise from the spatio-temporal variation of environmental factors (soils, weather conditions) and harvesting techniques (crop residues) [19].

In this case study we examined the potential of narrow-band multispectral imagery to estimate the C export of lucerne (Medicago sativa) from plots in different terrain positions and soil types. The output of a workflow for data pre-processing, a calibrated orthorectified image mosaic, was used to examine the predictive accuracy of the normalized difference vegetation index (NDVI) [21], the transformed soil-adjusted vegetation index (TSAVI) [22], the two-band vegetation index (TBVI) [23] and the enhanced vegetation index (EVI) [24] for fresh phytomass of lucerne. The total C export was then calculated from relationships between ground-based measurements of phytomass and carbon content of lucerne. Finally, yield data collected at each harvest date were used to estimate the yearly C export.

2. Methods

2.1. Study Area

UAV imagery was acquired in the Federal State of Brandenburg (NE-Germany) at the CarboZALF experimental area (53.3793N, 13.7856E) (Figure 1). The subcontinental climate is characterized by a mean annual air temperature of 8.7 °C and a mean annual precipitation of 483 mm (1992–2011, ZALF Research Station Dedelow). The 6 ha research area is embedded in a hummocky ground moraine landscape characterized by intense agricultural land use. Past and recent soil erosion by water and tillage leads to a very high spatial variability of soils and related growth conditions for crops. Only 20% of the region are unaffected by soil erosion. The 12 plots are extensively instrumented with the aim to conduct a long-term study (>10 years) on the impact of climate change, management and different soil types on gas exchange, carbon budget and carbon stocks of arable land in glacial landscapes [25]. The soils of the 6 ha experimental area represent a full gradient in erosion and deposition, namely a non-eroded Albic Cutanic Luvisols (plots 1–6), strongly eroded Calcic Cutanic Luvisols (plots 11–12), extremely eroded Calcaric Regosols (plot 7), and a colluvial soil, *i.e.*, Endogleyic Colluvic Regosols (Eutric) over peat (plots 9–10). In 2014, lucerne was grown on eight plots while corn and sorghum were grown on the remaining plots.

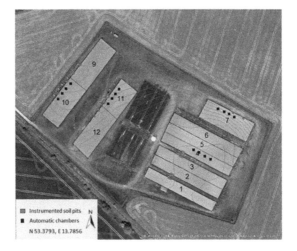

Figure 1. The CarboZALF experimental area near Dedelow (NE Germany): plot design and instrumentation.

2.2. UAV-Platform Carolo P360

The Carolo P360 is a fixed wing construction, developed by the Institute of Aerospace Systems of the Technical University Braunschweig (Figure 2). With a wingspan of 3.6 m and a takeoff weight of almost 22.5 kg including the complete battery set, the UAV is capable of carrying an additional payload of approximately 2.5 kg. The battery set consists of two Lithium-Polymer (LiPo) batteries (each 10 cells and 10 Ah) for the electric drive motor, two LiPos (each 2 cells and 3.55 Ah) for the autopilot system including servo actuators and one LiPo (4 cells, 3.55 Ah) for the payload (sensors and control unit). The payload, mounted inside the fuselage, is protected during takeoff and landing by landing gear doors, opened and closed via radio control. Due to the size of the opening, the dimension of the sensor optics is limited to 10 cm width and 24 cm length. The battery set allows flight durations of approximately 40 min at ground speeds between 20 and 30 m·s^{-1} including the time for climbing and landing. However, for security reasons a recommended ground speed should not fall significantly below 24 m·s^{-1}.

The combination of a strong electric motor (9.5 KW), large wheels and a rigid, spring mounted landing gear enables the UAV to takeoff and land on short airstrips (length: 70 m, width: 30 m including reserve) with rough surfaces. Takeoff, climbing and landing have to be performed manually by a pilot via radio control. During the autonomous flight, the task of the backup pilot is to observe and, in case of an emergency, abort the autonomous flight and land manually. The backup pilot is mandatory due to regulations of the national civil aviation authorities (CAA).

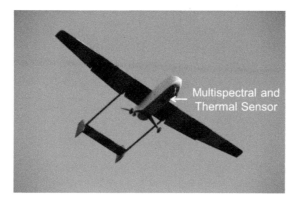

Figure 2. The Carolo P360 unmanned aerial vehicle (UAV) during a mission with open landing gear doors.

The autonomous flight of the UAV is controlled by a MINC autopilot. Navigation filter and algorithms utilize data from the inertial measurement unit (IMU) and GPS measurements. IMU measurements are combined with long time stable and precise but less frequent GPS measurements in a discrete error state Kalman filter. This algorithm requires the GPS signal of less than four satellites [26]. Extensions of the MINC system include a flight data recorder that stores collected IMU and GPS data on a 256 Mb MM card in a 2 s interval for further processing.

2.3. Mounted Sensors

Sensors mounted on UAVs are limited in their dimensions and weight. Despite the use of lightweight materials, a reduction in manufacturing quality, data storage capacity and on-board processing features is inevitable [6]. However, the development of miniaturized electronic components in the last decades enables scientist to mount a variety of sensors on UAVs for small-scale remote sensing applications. The Carolo P360 is equipped with two sensors, a multispectral and a thermal sensor. However, thermal sensor data were not used in this study.

The core of the sensor equipment is a 12-band miniature multi-camera array Mini-MCA 12 (MCA hereafter) (Tetracam Inc., Chatsworth, CA, USA). The compact modular construction of the MCA integrates two basic modules into one rugged chassis. Each module consists of an array of six individual CMOS sensors (1280 × 1024 pixels; pixel size 5.2 μm), lenses (focal length 8.5 mm) and mountings for user definable band-pass filters (Figure 3). Images can be stored on 2 GB CF cards for every sensor either in DCM or RAW format. The PixelWrench2 software (Tetracam Inc., Chatsworth,

CA, USA), shipped with the camera, enables the conversion of those images into single/multiband TIF or JPG format. Camera settings can be modified and enable the user to select camera exposure times between 1 ms and 20 ms (1 ms increment) and store images at a dynamic range of either 8 or 10 bit. In order to find an adequate exposure time, several ground-based tests prior to this study were performed at blue sky conditions over vegetation plots. A high dynamic range for most bands and no saturation at the same time was found for a fixed exposure time of 4 ms, which was then used in this study. The dynamic range was set to 10 bit.

Figure 3. Tetracam Inc. miniature multi-camera array Mini-MCA 12 with mounted narrow-band (10–40 nm) filters that cover the spectral range between the visible and the near-infrared light (470–953 nm; both center wavelengths).

The interchangeable band-pass filters, manufactured by Andover (Andover Corp., Salem, NH, USA), were selected prior to delivery, based on their center wavelengths and bandwidths. The 12 narrow-band filters cover, almost equally distributed, the spectral range from visible to near infrared wavelengths with focus on the characteristic reflectance features of healthy vegetation including the chlorophyll absorption band around 650 nm, the red-edge region between 680 nm and 730 nm and one of the water absorption bands around 950 nm. The bandwidth (full width at half maximum FWHM) varies from 9.1 nm to 40.8 nm with increasing bandwidths towards larger wavelengths. The filter configuration and characteristics are given in Table 1.

Table 1. Filter configuration of the Mini-MCA 12 and optical properties of the mounted filters. For band 2 (b), no fact sheet has been provided (N/A).

Band	Center Wavelength (nm)	FWHM * Coordinates (Bandwidth) (nm)	Bandwidth (10%) (nm)	Peak Transmission (%)
b_{471}	471	466.0–475.1 (9.1)	12.8	68.3
b_{515}	515	N/A ($\approx$10.0)	N/A	N/A
b_{551}	551	545.5–555.6 (10.1)	14.8	56.4
b_{613}	613	607.7–617.8 (10.2)	14.2	67.6
b_{658}	658	653.4–662.9 (9.5)	13.6	69.2
b_{713}	713	708.1–717.7 (9.6)	13.4	63.0
b_{761}	761	756.2–766.7 (10.5)	14.7	71.9
b_{802}	802	797.3–807.3 (10.1)	14.5	56.3
b_{831}	831	826.3–835.8 (9.5)	13.1	55.3
b_{861}	861	856.4–866.4 (10.1)	14.0	64.2
b_{899}	899	891.3–907.7 (16.4)	22.9	63.6
b_{953}	953	933.0–973.8 (40.8)	58.2	69.6

* FWHM = Full width at half maximum.

2.4. Ground Control Software

The ground control software MAVCDesk enables the operator to plan and control a mission. The planning of an autonomous flight encompasses the definition of waypoints, flight altitude, ground speed and a waypoint sequence mode. All required mission settings are finally transmitted to the autopilot via telemetry. During the mission, the telemetry antenna serves as a receiver for important UAV status information, which enables a visual position control (Figure 4). In the case of irregular behavior of any important flight parameter, the operator instructs the backup pilot to switch over to manual control.

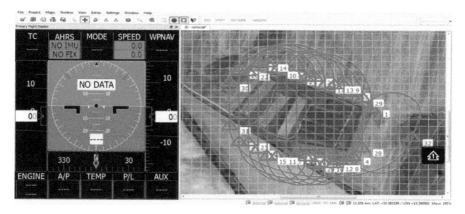

Figure 4. Screenshot of the MAVCDesk software. (**Left**) Primary flight display (not active) showing UAV status information; (**Right**) Visualization of the flight path across a map of the CarboZALF experimental area.

2.5. Mission Settings

The UAV mission was conducted on 27 August 2014 under clear sky conditions and low wind speeds. The ground sampling distance (GSD) of 0.1 m was chosen as a compromise of flight altitude, cruising speed, shutter interval and high spatial resolution. The recommended cruising speed of the UAV is 25 m·s^{-1} and the camera shutter interval is fixed at a rate of 2 s. Sufficient overlap of consecutive images is necessary for post-processing of the imagery. Thus, we chose a flight altitude of 163 m, resulting in an overlap of approximately 50% in flight direction. To achieve a sufficient across flight overlap of at least 60%, we selected a distance of 40 m between the flight paths. Twenty-six waypoints were predefined, each marking a start- and endpoint of 13 parallel flight paths with a total length of 5.8 km excluding the loop lines.

2.6. Image Processing

The final goal of post-processing of recorded MCA imagery is the conversion of measured digital numbers (DN) into georeferenced at-surface reflectance images. This multistage procedure consists of three major components: (i) radiometric image correction; (ii) transformation of sensor coordinates into a geographic coordinate system and image alignment; and (iii) absolute radiometric calibration. The radiometric image correction includes noise reduction, correction of sensor-based illumination fall-off (vignetting) and lens distortion. The transformation of sensor coordinates includes the fusion of recorded GPS measurements with collected images, band-wise automated aerial triangulation (AAT), the minimizing of remaining geometric distortions and the alignment of the 12 single bands to one multispectral image using ground control points (GCPs). Radiometric calibration is the conversion of measured DNs into at-surface reflectance. For image correction we followed a practical approach proposed by Kelcey and Lucieer [6], developed for a single image captured by a Mini-MCA 6. They used dark offset imagery to create average noise corrections and images of a homogeneous illuminated near-Lambertian white surface to create a per-pixel flat-field correction factor. The procedure incorporates correction techniques proposed by Mansouri *et al.* [27].

2.6.1. Noise Reduction

Noise is defined as unwanted electrical or electromagnetic energy that degrades the quality of signals and data. In the case of a CMOS-based camera, noise complies with all temporal and systematic errors added to a recorded signal during image acquisition. Noise is introduced both by the sensor (e.g., non-uniform pixel responses) and the electronics (e.g., electrical interferences) that amplify the output signal of the sensor for digitization [27]. Due to a random component, it is impossible to calculate the precise proportion of sensor noise to sensor signal within an image. Prominent examples are periodic noise, checkered patterns and horizontal band noise caused by the progressive shutter of CMOS sensors [6].

The dark offset subtraction technique is a statistical image based approach, which reduces the noise component of an image by subtracting a dark offset image. A dark offset image represents the average per-pixel noise and is generated by multiple repetitions in a completely darkened environment. The dark offset imagery was created for each sensor of the MCA in a darkened room with black painted walls. In addition, the MCA lenses were covered with black cardboard. For each of the 12 sensors, a total of 120 images at exposure level of 4 ms were taken to calculate the per-pixel average dark offset. The examples in Figure 5 show the dark offset images for bands 1 and 11 with strong periodic noise features in b_{471} (Figure 5a), a global checkered pattern in b_{899} (Figure 5b) and the overlapping horizontal progressive shutter band noise visible in both images.

(a) Dark offset image – b_{471} (b) Dark offset image – b_{899}

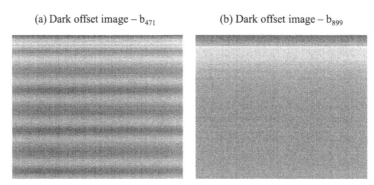

Figure 5. (**a**) Dark offset image of b_{471} showing periodic noise and progressive shutter band noise; (**b**) Dark offset image of b_{899} showing a global checkered pattern and progressive shutter band noise.

2.6.2. Vignetting Correction

The effect of radial fall-off of light intensity from the center towards the periphery in photographed images is known as vignetting. Different sources of vignetting contribute to a progressive reduction of irradiation across the image plane and may cumulate up to 60% at the periphery of an image. Although lens manufacturers go the limit to what is technically feasible, the geometry of the sensor optics contributes most to this effect [6,28]. In order to minimize the vignetting, an image based correction method was applied to each of the 12 bands of every single image captured during the mission. The method basically uses a look-up table (LUT) for each band, composed of correction factors for each pixel derived from flat field imagery. The proper generation of flat field imagery requires an evenly illuminated white surface with Lambertian properties and constant spectral characteristics. However, for practical considerations, a white surface with near Lambertian properties may serve to generate acceptable flat field imagery.

Numerous artificial white and black materials (various types of paper, rubber and plastic) were examined regarding their spectral properties. A halogen lamp was used for an even illumination of the targets and an ASD FieldSpec 4 Wide-Resolution spectrometer (ASD Inc., Boulder, CO, USA) was used to collect spectral reflectance data. The instrument measures the spectral radiance ($W \cdot m^{-2} \cdot sr^{-1} \cdot nm^{-1}$) over the wavelength range of 350–2500 nm with a spectral resolution of 3 nm at 700 nm and 30 nm at 1400 nm and 2100 nm, respectively. In order to yield the relative reflectance of a target (vegetation, calibration panels, soil), the measurement of the spectral radiance from a reference panel with Lambertian characteristic and constant spectral properties over VIS and NIR wavelengths was required. Therefore, a Spectralon® reference panel was mounted on a tripod for collecting reference spectra prior to every single measurement of a target. Without fore optics, the bare fiber has a 25° field of view.

A matt white Bristol Cardboard with a density of 625 $g \cdot m^{-2}$ was found to show the highest and most uniform reflectance over the same wavelength range from 466 nm to 978 nm that is covered by the filter configuration of the MCA.

Vignetting imagery was created under diffuse illumination conditions using the matt white Bristol Cardboard. The camera was operated manually at a distance of approximately 1 m with the sensors pointing downwards to the Cardboard. Between each triggering the orientation of the camera has been changed slightly to minimize possible heterogeneities on the surface of the cardboard. During image acquisition we ensured that the cardboard completely covered the field of view of all 12 bands. In a first step, the per-pixel average was calculated from a total of 10 images for each of the 12 sensors at different exposure levels, followed by a subtraction of the respective dark offset imagery.

To account for the horizontal band noise induced by the progressive shutter of the camera, a shutter correction factor has been calculated in a second step. Each flat field image has been averaged along the y-axis (row-wise). While most profiles showed a behavior that could be approximated by a 3-grade polynomial function, satisfying approximation for b_{761}, b_{899} and b_{953} could only be achieved by a 5-grade polynomial function. For the sake of consistency, we used a 5-grade polynomial for the approximation in all 12 bands. Figure 6 shows the y-axis average and its approximation for two examples, b_{831} and b_{899}, respectively. For each row, the shutter correction factor has then been calculated by dividing the row-wise average by the approximation. Assuming a multiplicative row-wise brightness modification of the shutter, each row of the average flat field image was then multiplied by the respective correction factor.

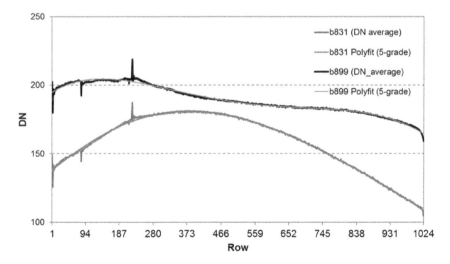

Figure 6. Row-wise average and 5-grade approximation of band-noise affected flat-field images of b_{891} and b_{899}.

Finally, the correction factor LUT for each sensor was then calculated by dividing all pixel values of the flat field imagery by the maximum pixel value that occurred in the respective image assuming the maximum value (the brightest pixel) to be an unaffected representation of the measured radiance. Figure 7 depicts an example for flat field images generated for b_{831} and b_{899}.

(a) Flat field correction image – b_{831} (b) Flat field correction image – b_{899}

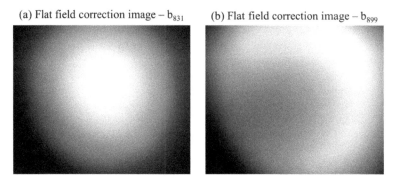

Figure 7. (**a**) Flat field image generated for vignetting correction of b_{831}; and (**b**) for vignetting correction of b_{899}.

The example in Figure 8 demonstrates the effect of the consecutive correction steps (noise and vignetting) to an uncorrected single image recorded in b_{831}. Due to its relatively small contribution, noise reduction is almost invisible in the resulting image. However, the successful correction of the vignetting effect, especially in the upper and lower left corners, is evident.

(a) Uncorrected image – b_{899} (b) Corrected image – b_{899}

Figure 8. (**a**) Example for an uncorrected image (RAW format) recorded in b_{831}; and (**b**) the respective image after noise reduction and consecutive vignetting correction.

2.6.3. Lens Distortion Correction

Lens distortions arise from the symmetry of a photographic lens. The most frequent distortions are radially symmetric and known as barrel or pincushion distortion. In the case of a barrel distortion, image magnification decreases with distance from the optical axis. It increases in the case of a pincushion distortion. Both effects result in a radial displacement of measured per-pixel radiance.

A commonly applied correction technique for both types of distortion is the plumb-line approach described in the Brown–Conrady model [29], which is implemented in the PhotoScan-Pro V.1.2. software (Agisoft LLC, St. Petersburg, Russia). Since no correction factors are provided in the Exif tags of the imagery, the required internal and external orientation of each camera (band) is estimated automatically from the geometry of an image sequence during the image alignment process [30].

This first step in the PhotoScan-Pro workflow only requires the input of the focal length (8.5 mm) and the pixel size (5.2 µm) together with the GPS coordinates recorded for each individual image.

2.6.4. Mosaicking and Georeferencing

Due to the proprietary nature of the software, the underlying algorithms are not known in detail. Anyway, the program workflow involves common photogrammetric procedures in a Structure from Motion (SfM) workflow, including the search for conjugate points by feature detection algorithms used in the bundle adjustment procedure, approximation of camera positions and orientation, geometric image correction, point cloud and mesh creation, automatic georeferencing and finally the creation of an orthorectified mosaic [31]. This workflow (lens distortion correction included) was applied to each of the 12 bands independently. The result of the workflow applied to b_{761} and the reconstructed flight path from the recorded GPS locations used for image alignment is illustrated in Figure 9.

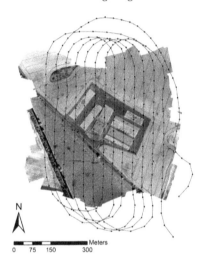

Figure 9. Image mosaic of b_{761}. Overlay: Reconstructed flight path from recorded GPS locations (black dots).

The ERDAS Imagine software (Hexagon Geospatial, Norcross, GA, U.S.) was then used to improve the spatial accuracy and to transform the single bands to the local coordinate system ETRS 89 UTM 33 using precisely measured GCPs. Finally, the 12 bands were stacked to a single multispectral image.

2.6.5. Radiometric Calibration

The retrieval of biophysical parameters of vegetation canopies requires an absolute calibration of the collected imagery because a recorded DN is not only a function of the spectral characteristics of vegetation or soils but also of environmental condition [32]. These include in particular the atmospheric conditions during the flight and the respective illumination geometry (solar zenith and sensor viewing angles). Several approaches exist to calculate the at-surface reflectance either by using radiative transfer models (RTM) or a combination of RTM and ground-based *in situ* measurements of the reflectance of a calibration target, a so-called in-flight calibration [33]. Both approaches require a well-calibrated sensor and on-site measurements of the atmospheric conditions at the date of image acquisition. To overcome these requirements, attempts were made to establish linear relationships between ground-based reflectance measurements and recorded DNs [34]. This empirical line approach accounts for both the influence of illumination geometry and atmosphere [32]. Prerequisite is the availability of low and high reflectance targets, with homogenous spectral characteristics over the wavelength range covered by the sensor. The approach has been applied successfully to different former UAV missions using the MCA [5,6,8] and is well suited for UAV remote sensing applications for several reasons: (i) the spectra of different materials potentially suited as targets can be examined prior to UAV missions; (ii) the spectra of selected calibration targets can be measured close to the time of image acquisition; and (iii) due to the high spatial resolution of images, the dimensions of the targets remain small and easy to carry.

The targets, in the following referred to as calibration panels, were constructed from thin chipboards with a dimension of 0.5 m × 0.5 m. The chipboards were coated with the matt white Bristol cardboard already used for the vignetting correction and with black cardboard with almost uniform low reflectance in the respective wavelength range. Five pairs of black and white calibration panels were placed in the four corners and the center of the study area. The spectral characteristics of each panel were measured during a time period of one hour around image acquisition. The measured reflectance of the white and black calibration panels tends to be lower under laboratory conditions than under clear sky conditions. This may be caused by non-Lambertian reflectance characteristics of the used materials which may also be a reason for the disparate reflectance values of the five black and white calibration panels over the relevant wavelength range (450–1000 nm). Slight angular deviations from the horizontal plane of the calibration panels showed undesired impacts on the reflectance especially of the white ones, partially exhibiting reflectance greater than one. Instead of using an averaged reflectance of all calibration panels, the reflectance of the one pair (P1), which comes closest to the respective laboratory reflectance was used (Figure 10).

The DN of the pixel with the highest value within the white calibration panels and the lowest value within the black calibration panels was plotted against the corresponding ground-based reflectance, calculated by averaging over the rounded FWHM bandwidths, for each of the 12 bands. The resulting 12 empirical lines were finally used to perform a band-by-band conversion of per-pixel DN into per-pixel reflectance.

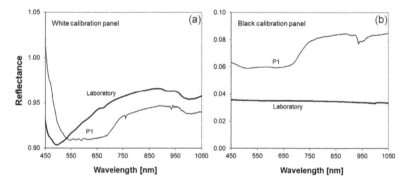

Figure 10. (**a**) Reflectance of the white calibration panel (matt white Bristol cardboard) from laboratory and field measurements at P1; (**b**) Reflectance of the black calibration panel (black cardboard) from laboratory and field measurements at P1.

2.7. Ground-Based Measurements

2.7.1. Fresh, Dry Phytomass and Total Carbon Content of Lucerne

Lucerne belongs to the legume family and is usually grown for fodder production. Within crop rotations, lucerne is frequently grown for soil improvement due to its nitrogen-fixing properties. It was grown on eight equally managed plots, each fertilized with 300 kg· ha^{-1} phosphate and 110 kg· ha^{-1} potash. The date of the UAV mission corresponded with the growth stage of beginning flowering (BBCH—Code 61), and was conducted one day before the fourth harvest.

Fresh phytomass was sampled at 22 permanent observation sites after collecting spectral data. Plants were cut at ground level within an area of 0.25 m^2 from two locations close to the permanent sites. After weighing, the samples were chaffed in order to guarantee uniform drying. Dry phytomass and the corresponding water content were determined after oven drying at 60 °C until constant weight (48 h). This intermediate step is required to estimate the amount of carbon content of each of the samples.

The carbon content of green lucerne was determined in an earlier study (2013, beginning flowering) at the CarboZALF experimental area by the ZALF Institute for Landscape Biogeochemistry (data not published). Samples of lucerne were analyzed in the ZALF central laboratory according to standard methods (spectral elementary analysis, DIN ISO 10694:1995). The total carbon content ranges between 41% and 46%. The mean of 43% (SD = 0.4, N = 100) is in good agreement with a reported mean of 45% used for most diverse crops in regional studies [35,36] and was finally multiplied with the amount of dry phytomass calculated for each pixel.

2.7.2. Total Carbon Content of Lucerne per Vegetation Period

In order to estimate the total C export of the entire growing season, the averaged total carbon per plot calculated from MCA imagery were multiplied by factors derived from time-series of independently collected ground-based measurements of dry phytomass. These samples were routinely collected before each of the four harvest dates at least four representative locations within the individual plots. To determine dry phytomass, 1 m^2 of plants was cut at ground level and oven dried at 60 °C (until constant weight (48 h). The factors were determined by dividing the sum of dry phytomass of all harvest dates by the dry phytomass collected at the fourth harvest date.

2.7.3. Spectral Response of Vegetation and Bare Soil

Spectral reflectance measurements of vegetation, calibration targets and bare soils were collected using an ASD FieldSpec 4 Wide-Resolution spectrometer. Since the distance between fiber optics and canopy was approximately 0.8 m, the collected spectra represented an average reflectance from a circle of 0.36 m in diameter. For compensation of slight movements of the fiber optics introduced by the operator while collecting the spectra, a number of 10 repetitions were set to default.

The sampling of vegetation spectra was performed between 11:00 a.m. and 1:00 p.m. local time, ± 1 h before and after image acquisition. Forty-four spectra were taken at 22 permanent observation sites representing the natural variability of site properties. Each is represented by two spectral measurements for compensation of slight variations of standing fresh phytomass in its surrounding. The mean spectra of the two repetitions were then averaged to a single spectral response curve that represents the characteristics of the fresh phytomass at the site. The spectral measurements of the ten calibration panels (five white and five black panels) were taken during the same period (11:00 a.m. and 1:00 p.m. local time) as the vegetation spectra. The panels were placed in pairs (a white and a black) close to the four corners and the center of the study area.

The spectral response of bare soil was collected at 18 September between 11:00 a.m. and 1:00 p.m. local time under clear sky conditions. Sample locations were identical to those selected for the collection of vegetation spectra. One day after the seeding of winter wheat, the topsoil showed a gentle surface roughness. From this, little influence on light scattering can be assumed. Topsoil conditions varied in moisture (7% to 12%) according to differences in terrain position and soil properties. The spectra were then used to calculate slope and intercept of the soil-line required for the calculation of the TSAVI. The best correlation ($R^2 = 0.99$) between the b$_{658}$ and one of the six NIR bands was found for b$_{756}$ (Figure 11). The resulting parameters of the regression ($a = 1.07$; $b = 0.02$) were then used for the calculation of TSAVI.

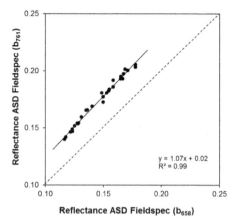

Figure 11. Relationship between ASD Fieldspec measurements of topsoil reflectance in the wavelengths corresponds to Mini-MCA 12 bands b_{658} and b_{756}.

2.8. Description and Calculation of VIs

The NDVI is an intrinsic vegetation index that simply accounts for the chlorophyll absorption feature in the red (R) and the structural information inherent in high NIR reflectance of a green vegetation canopy. It does not involve any external factor other than the measured spectral reflectance.

$$\text{NDVI} = (\text{NIR} - \text{R})/(\text{NIR} + \text{R}) \tag{1}$$

The TSAVI incorporates slope and intercept of the soil line together with an adjusted coefficient to account for first-order soil background variation.

$$\text{TSAVI} = a(\text{NIR} - a\text{R} - b)/[(\text{R} + a(\text{NIR} - b) + 0.08(1 + a^2)] \tag{2}$$

where a is the slope and b is the intercept of the soil line. The value 0.08 is an adjusted coefficient.

The TBVI is a general formulation of the NDVI and was used to examine the predictive accuracy of other provided band combinations than used by the NDVI.

$$\text{TBVI}_{i,j} = (\text{Ref}_j - \text{Ref}_i)/(\text{Ref}_j + \text{Ref}_i) \tag{3}$$

where $i, j = 1, \ldots, N$, where N is the number of narrow bands and Ref is the reflectance measured in a narrow band.

EVI, originally developed as a standard satellite vegetation product for the Moderate Resolution Imaging Spectroradiometer (MODIS), combines atmospherically corrected blue (B), R and NIR reflectance with coefficients of an aerosol resistance term and a soil-adjustment factor. The use of the EVI has been motivated by studies reporting a trend to more linear relationships with vegetation biophysical parameters such as standing biomass and LAI of crops and a wider range of values at the same time [4,37].

$$\text{EVI} = 2.5 ((\text{NIR} - \text{R})/(\text{NIR} + C_1\text{R} - C_2\text{B} + \text{L})) \tag{4}$$

where 2.5 is a gain factor. C_1 and C_2 are coefficients of the aerosol resistance term and L is the soil background reflectance adjustment where $C_1 = 0.06$; $C_2 = 0.08$ and L = 1.

In order to investigate the potential of the six available bands in the NIR, six variations of the NDVI, the TSAVI and the EVI were calculated. Variations of the TBVI were calculated for all band combinations except those covered by the NDVI variations (the R band in combination with the six NIR bands).

3. Results and Discussion

3.1. Radiometric Calibration

The empirical line approach used for sensor calibration produces a set of 12 linear relationships between DNs and ground measured reflectance of a white and black calibration panel. Figure 12 shows the empirical lines for bands b_{471}–b_{713} (Figure 12a) and bands b_{761}–b_{953} (Figure 12b).

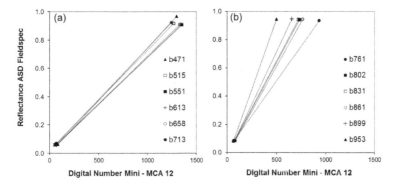

Figure 12. Relationship between ground measured reflectance of black and white calibration panels and the respective digital numbers acquired by Mini-MCA 12. (**a**) Bands 1–6; and (**b**) Bands 7–12.

Raw imagery DNs of the white panels for bands 1–6 showed values close to saturation (DN 1024). After flat-field correction those pixels showed values higher than 1024. This is caused by flat-field correction factors >1, in cases where the white panel was situated close to the periphery of an image. Image DNs of the calibration panels and the corresponding ground measured reflectance are listed in Table 2 together with the respective regression.

Table 2. Image digital numbers and ground measured reflectance of the white and black calibration panels in bands 1–12 and the respective regressions (empirical lines).

Band	DN Mini-MCA 12		Reflectance ASD Fieldspec		
	White Panel	Black Panel	White Panel	Black Panel	Regression
b_{471}	1299.5	87.1	0.968	0.062	R = 0.000748 * DN − 0.003615
b_{515}	1269.1	55.4	0.917	0.059	R = 0.000707 * DN + 0.019753
b_{551}	1355.0	62.6	0.909	0.060	R = 0.000657 * DN + 0.018536
b_{613}	1345.2	74.1	0.910	0.060	R = 0.000669 * DN + 0.009897
b_{658}	1330.0	74.0	0.911	0.060	R = 0.000678 * DN + 0.010134
b_{713}	1247.4	70.4	0.923	0.070	R = 0.000725 * DN + 0.018933
b_{761}	933.6	64.9	0.936	0.080	R = 0.000985 * DN + 0.015794
b_{802}	733.6	76.8	0.941	0.082	R = 0.001307 * DN − 0.018208
b_{831}	722.7	71.6	0.943	0.083	R = 0.001321 * DN − 0.011693
b_{861}	765.1	82.6	0.945	0.084	R = 0.001262 * DN − 0.020314
b_{899}	656.6	75.7	0.947	0.084	R = 0.001485 * DN − 0.028555
b_{953}	499.0	64.6	0.945	0.080	R = 0.001991 * DN − 0.048845

The spatial subset depicted in Figure 13 shows a composite from MCA b_{658} (R), b_{551} (G) and b_{471} (B) after calibration, band alignment, layer stacking and transformation from geographical coordinates (WGS 84) to projected coordinates (ETRS 89 UTM 33).

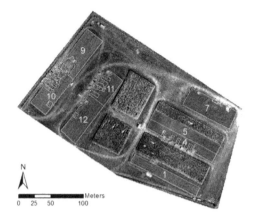

Figure 13. RGB composite image of the CarboZALF experimental area from calibrated Mini-MCA 12 bands b_{658}, b_{551} and b_{471}.

3.2. Empirical Line Quality Assessment

An examination of calibrated MCA spectra and the corresponding ground-based measurement reveals good agreement. Generally, the spectra of both ground-based and calibrated MCA reflectance show the characteristic features of a green vegetation canopy with different amounts of biomass and soil covers (Figure 14). The depicted examples represent sites with high (28), medium (1) and low (5) amounts of fresh phytomass of lucerne.

Reflectance in the VIS is low in the blue and red domain and shows the characteristic peak in the green domain. After the transition from VIS to NIR wavelengths around 712 nm, NIR reflectance varies between 0.33 and 0.65 at 761 nm and between 0.38 and 0.71 at 899 nm. The water absorption band around 953 nm reveals differences between the spectra regarding the relative decline of the reflectance compared to b_{899}. The example of a bare soil reflectance curve in Figure 14 indicates the general ability of calibrated MCA imagery to produce realistic spectra not exclusively for vegetation. The spectrum represents a small area free of vegetation within plot 7 and shows the typical monotonous increase of reflectance from VIS to NIR wavelengths within a realistic range of values.

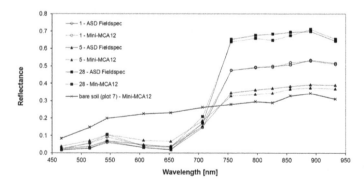

Figure 14. Comparison of the spectral response of lucerne extracted from calibrated Mini-MCA 12 bands with ground measured ASD Fieldspec reflectance and with bare soil reflectance (extracted from calibrated Mini-MCA 12 bands; ASD Fieldspec reflectance not available). The selected sites represent high (28), medium (1) and low (5) amounts of fresh phytomass of lucerne. The bare soil spectrum represents an area free of vegetation within plot 7.

However, the VIS spectral response of lucerne in calibrated MCA imagery is generally higher than the ground based reflectance. In the NIR domain, MCA acquired reflectance closely matches the ground-based measurements. Coefficients of determination reveal high correlations between ground-based and image reflectance for the six NIR-bands b_{761}–b_{953}, poor correlations for the four VIS bands b_{471}–b_{613} and the NIR band b_{713} and a moderate correlation for the red band b_{658} (Table 3). The corresponding root mean square errors (RMSE) are low in terms of absolute reflectance but differ extremely in relation to the range of values in the respective bands. The mean relative error (MRE %) illustrate the discrepancy between the VIS (33%–104%) and the NIR bands (4%–23%). We assume the poor matching in the VIS bands are partially caused by the saturated pixels of the white panels used for the creation of the empirical lines. Additional greyscale panels in the medium reflectance range (0.3–0.7) would have been helpful to make the empirical line relationship more robust [38]. Nonetheless, additional sources of error might be present since other authors reported similar discrepancies for the VIS response of the MCA [39].

Table 3. R^2, RMSE and MRE [%] for the relationships between the reflectance acquired by the 12 Mini-MCA bands and ground-based measurements at the 22 permanent observation sites.

	b_{471}	b_{515}	b_{551}	b_{613}	b_{658}	b_{713}	b_{761}	b_{802}	b_{831}	b_{861}	b_{899}	b_{953}
R^2	0.16	0.10	0.04	0.19	0.40	0.11	0.88	0.91	0.90	0.89	0.88	0.84
RMSE	0.001	0.003	0.007	0.004	0.003	0.016	0.028	0.025	0.026	0.027	0.027	0.027
MRE%	51.2	104.4	58.0	33.0	82.6	22.7	4.0	3.6	4.3	3.8	4.4	4.6

3.3. Ground-Based Measurements of Vegetation

The different amounts of fresh phytomass of lucerne, sampled at 22 locations across the different plots reflect the spatial heterogeneity of naturally occurring site properties (terrain and soil). Although all plots were treated equally, the averaged fresh phytomass of each permanent observation site ranges between 440 g·m^{-2} and 2080 g·m^{-2}. The overall mean is 1469 g·m^{-2} and the coefficient of variation (CV) is 34%. The amount of dry phytomass ranges between 158 g·m^{-2} and 426 g·m^{-2}. The overall mean is 329 g·m^{-2} with a CV of 23%. The variation in the corresponding water content ranges between 282 g·m^{-2} and 1709 g·m^{-2}. With a CV of 38%, the variation is similar to the variation in fresh phytomass. The statistic evaluation of the data shows a strong linear correlation (R^2 = 0.89) between fresh and dry phytomass with a RMSE of 24 g·m^{-2} (Figure 15).

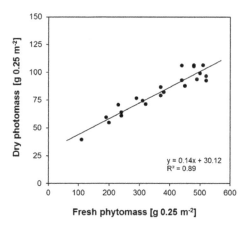

Figure 15. Relationship between fresh and dry phytomass of lucerne measured at the 22 permanent observation sites.

3.4. VI Performance

Coefficients of determination were calculated for the expected exponential relationships between ground measured fresh phytomass and all band combinations of the four VIs, as described above. The results indicate clear differences between the VIs and, with the exception of TBVI, only little differences between the variants using one of the six NIR bands for VI calculation. Regardless of the used NIR band, the best relationships are obtained for EVI, followed by TSAVI using the soil-line and NDVI. Coefficients of determination calculated for EVI range between 0.86 (b_{761} and b_{953}) and 0.88 (b_{802}, b_{861} and b_{899}). Lower R^2s can be observed for TSAVI and NDVI but again, MCA b_{761} and b_{953} (0.80 for TSAVI; 0.71 for NDVI) are less suited than b_{802} and b_{899} (0.82 for TSAVI; 0.72 for NDVI). Due to the construction of TBVI exclusively from NIR bands only five band combinations with b_{953} are possible. Low correlations were obtained when using combinations with b_{761} and b_{861} ($R^2 = 0.42$ and 0.38, respectively), moderate correlations for the combinations with b_{802} and b_{831} ($R^2 = 0.65$) and a high correlation ($R^2 = 0.77$) for the combination with b_{899} ($TBVI_{b899/b953}$ hereafter).

The small differences between the results obtained for EVI, TSAVI and NDVI variants can be explained by the low variation of canopy reflectance in the different NIR bands and the corresponding low reflectance in the R band (and B band in the case of EVI). The high correlation of the $TBVI_{b899/b953}$ may be explained by the relatively large difference in reflectance compared to other NIR band combinations. The distance increases with higher amounts of fresh phytomass due to the maximum of NIR reflectance in b_{899} observed for all measured spectra and the relatively strong decline of the reflectance in b_{953}, caused by the respectively higher absolute water contents. However, the highest correlations between the examined VIs and ground-based measurements of fresh phytomass were found for the variants using NIR band b_{899}. Regardless the mathematical construction of VIs from available broad or narrow band sensors the disadvantage concerning the non-linearity caused by saturation effects, especially in the case of dense vegetation canopies, is still present. Saturation levels are reported in numerous studies using field measurements of biophysical canopy parameters or leaf and canopy radiative transfer model [17,40,41]. The studies compared the predictive power and stability of several narrow- and broad-band VIs for estimation of LAI under different environmental conditions (canopy architecture, soil background and illumination geometry). Values for NDVI of 0.90 and 0.75 for TSAVI when using a single soil line are typical for dense vegetation canopies (LAI > 2). The observed saturation effect for the relationship between VIs and fresh phytomass in this study is caused by the strong linear relationship between LAI and fresh phytomass ($R^2 = 0.88$; LAI was measured simultaneously to fresh phytomass).

In this study, most NDVI values range in a narrow span between 0.89 and 0.91 when fresh phytomass exceeds 1200 g·m^{-2}, which holds true for 15 out of 22 samples (Figure 16a). The same effect can be observed for the TSAVI where the same 15 samples range in a span between 0.70 and 0.74 (Figure 16b). These findings indicate that both NDVI and TSAVI are not reliable estimators for fresh phytomass of a dense green vegetation canopy typical for lucerne. In terms of R^2 s, the $TBVI_{b899/b953}$ performs better than the NDVI but worse than the TSAVI. The lower R^2 compared with TSAVI is caused by a larger scatter in the data. This is probably the result of the small differences in the reflectance in b_{899} and b_{953}. Consequently, results are more sensitive to remaining noise in the data after sensor calibration than the results for VIs calculated from NIR and VIS bands with large differences in reflectance. Nonetheless, $TBVI_{b899/b953}$ exhibits a trend to more linearity than both NDVI and TSAVI (Figure 16c). Although developed for broad MODIS bands with the respective optimized coefficients to reduce impacts of soil and aerosol, EVI is the best predictor for fresh phytomass (Figure 16d). As reported in literature [4,37,42] the relationship is more linear and the range of values is wider (0.43) than observed for NDVI (0.25) and TSAVI (0.22). The RMSE and MRE for fresh phytomass using EVI are 193 g·m^{-2} and 11%, respectively.

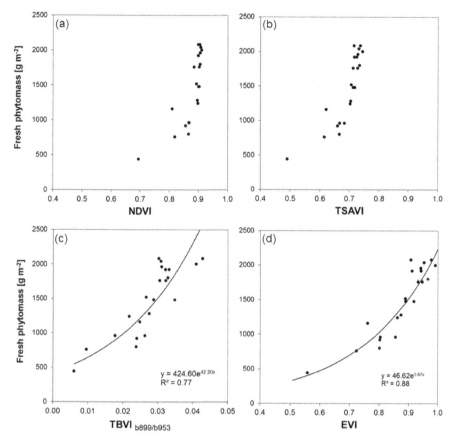

Figure 16. Relationships obtained between (**a**) NDVI; (**b**) TSAVI; (**c**) TBVI$_{b899/b953}$; and (**d**) EVI constructed from VIS bands in combination with NIR band b$_{899}$ (except TBVI$_{b899/b953}$) and fresh phytomass of lucerne at the 22 permanent observation sites.

3.5. Spatial Variability of Fresh Phytomass

The relationship between ground-based measurements of fresh phytomass and EVI calculated from calibrated MCA imagery was used in a first step to produce a map of fresh phytomass of lucerne for the eight plots of the CarboZALF experimental area (Figure 17). The high resolution enables a clear spatial differentiation of areas with extremely low (<250 g·m^{-2}) and high amounts of fresh phytomass (>3000 g·m^{-2}). The spatial patterns across and within the individual plots will be discussed in the context of total C export by harvest. The eye-catching area in the southwest corner of plot 9 with extremely low amounts of fresh phytomass relates to an inundated spot (two months) as a result of high rainfall in spring. This part will be excluded in the further evaluation.

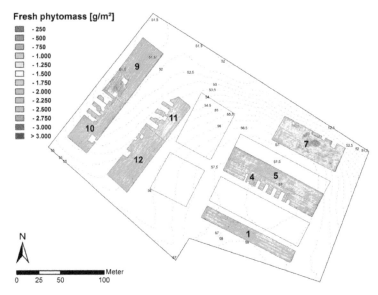

Figure 17. Spatial distribution of fresh phytomass of lucerne within the eight plots of the CarboZALF experimental area. Outclipped areas are disturbed areas due to experimental devices (autochambers, pathways, *etc.*).

3.6. Total C Export by Harvest—Quantities and Spatial Variability

To convert fresh phytomass into a map of total C export, the linear relationship between fresh and dry phytomass depicted in Figure 14 was used. The result was then multiplied by the averaged total carbon content (44%) to quantify the C export. Due to the linear transformation, the spatial patterns across and within the individual plots remain the same (Figure 18). The majority of values (99%) range between 75 $g \cdot m^{-2}$ and 225 $g \cdot m^{-2}$. In order to evaluate effects of terrain and prevalent soil type on the total C export, the median (M) for each individual plot was calculated (observations were normally distributed). The lowest C export is at plot 7 (M = 124 $g \cdot m^{-2}$), which represents an extremely eroded soil at steep slope. The C export from colluvial soils in the hollow (plot 9, M = 164 $g \cdot m^{-2}$; plot 10, M = 163 $g \cdot m^{-2}$) is only slightly higher than from non-eroded soils at the flat hilltop (plot 1, M = 162 $g \cdot m^{-2}$; plot 5, M = 154 $g \cdot m^{-2}$ and plot 4, M = 151 $g \cdot m^{-2}$). The internal spatial heterogeneity is higher at flat hilltop positions (plots 1, 4 and 5) and at steep slopes (plot 7). The plots in other terrain positions (plots 9, 10 11 and 12) appear more homogenous with coincidently higher phytomasses.

Although plots 12 and 11 are located in a similar flat slope position, the difference between the respective exports (M = 157 $g \cdot m^{-2}$ and 146 $g \cdot m^{-2}$ respectively) is higher than from other adjacent plots (9, 10 and 4, 5). This effect may be the result of a manipulation experiment that was conducted in 2010 to simulate landscape-scale erosion processes [25]. However, the interpretation of this effect is beyond the scope of this paper.

Weather conditions in 2014 were ideal for growing since plant growth was not limited by rainfall input. Therefore, differences across the plots are not very distinct. Generally, the observations match the spatial arrangement of soil types in the respective terrain positions, which is characteristic for this hummocky soil landscape. The Calcaric Regosol (plot 7) represents widespread extremely eroded soils with very dense parent material (glacial till) at 30 cm depth. Therefore, the rooting space is rather limited. The Endogleyic Colluvic Regosol (plots 9 and 10) in the hollow has the highest organic matter and nutrient stocks compared to other soils and shows local groundwater level is approximately 80 cm, hence additional water supply for an enhanced plant growth by capillary rise. The strongly eroded

Calcic Cutanic Luvisol (plots 11 and 12) at midslope and the non-eroded Albic Cutanic Luvisols at the flat hilltop are generally fertile and characterized by high available water capacities and good root penetration.

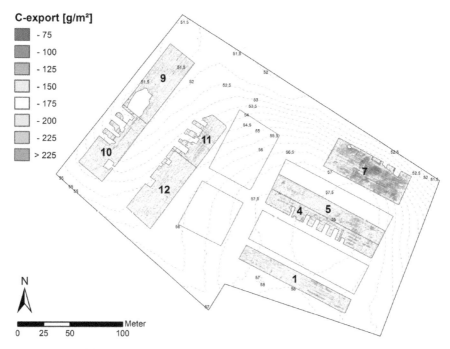

Figure 18. Spatial distribution of total exported carbon by harvest within the eight plots of lucerne at the CarboZALF experimental area.

3.7. Total C Export by Harvest Per Year—Temporal Trends and Spatial Variability

For the estimation of the C export over the entire growing season, the samples collected at the four harvest dates were used (from monitoring program in 2014). While the relationship between terrain position/prevalent soil type and exported carbon by the fourth harvest shows a weak trend, the dependency becomes more obvious when the development of dry phytomass over the entire growing season is taken into account. The amounts generally decrease continuously from the first to the fourth harvest date. However, the differences between soils become obvious over time (Figure 19). The strongest decline can be observed for the extremely eroded soil (plot 7) hereafter referred to as C1. Whereas soils with additional water supply, either by groundwater (plots 9 and 10) or lateral water fluxes in 1.5m depth (plot 12, without manipulation), showed the lowest decline over time (C3). The non-eroded soils at the plateau (plots 1, 4, 5) and the manipulated plot 11 behaved intermediate (C2).

The averaged export of total carbon for the fourth harvest (27 August 2014) ranges between $124\,\text{g}\cdot\text{m}^{-2}$ from C1 and $161\,\text{g}\cdot\text{m}^{-2}$ from C3 (Table 4). With $156\,\text{g}\cdot\text{m}^{-2}$, C2 cannot be distinguished clearly from C3. Multiplying these values with the case specific factors calculated from the summed phytomass divided by the phytomass from the fourth harvest, the total exported carbon per year ranges between $624\,\text{g}\cdot\text{m}^{-2}$ from C1 and $718\,\text{g}\cdot\text{m}^{-2}$ from C2, which is slightly more than the $697\,\text{g}\cdot\text{m}^{-2}$ from C3. This is caused by the higher mean dry phytomass estimated from UAV imagery for the plots belonging to C2 ($363\,\text{g}\cdot\text{m}^{-2}$ *vs.* $316\,\text{g}\cdot\text{m}^{-2}$).

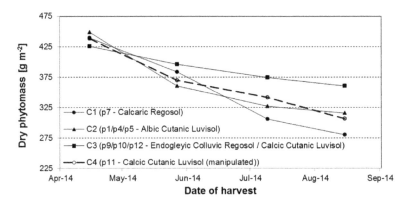

Figure 19. Temporal decline of above ground dry phytomass of lucerne between the first and fourth harvest in 2014.

Table 4. Yearly estimates of dry phytomass production and C export from four different Terrain/Soil type combinations (C1–C4) calculated from UAV imagery and ground based monitoring data.

Case	Terrain position	Soil type (FAO)	Monitoring 2014			UAV Mission (14-08-27)			
			Dry phytomass		factor	Dry phytomass		C export	
			4. harvest [g·m^{-2}]	per year [g·m^{-2}]		4. harvest [g·m^{-2}]	4. harvest [g·m^{-2}]	per year [g·m^{-2}]	CV [%]
C1	Steep slope	Calcaric Regosol	280	1409	5.03	288	124	624	21
C2	Flat hilltop	Albic Cutanic Luvisol	316	1452	4.60	363	156	718	17
C3	Midslope/hollow	Calcic Cutanic Luvisol/Endogleyic Colluvic Regosol	361	1556	4.32	376	161	697	14
C4	Midslope	Calcic Cutanic Luvisol-manipulated	307	1456	4.75	339	146	693	14

Altogether, the differences are relatively small between the groups due to weather conditions in 2014, which were almost optimal for plant growth. Nevertheless, differences in site properties (terrain and soil) are known to result in quite different growth conditions [43,44] and seasonal and intra-annual changes in weather conditions affect the within-field variability of phytomass production [45,46]. Taylor *et al.* [47] reported higher within-field variation of crop yield in dryer years, which was spatially associated with soil properties. The effect was less pronounced in wetter years with adequate water supply, which was most recently confirmed in a study by Stadler *et al.* [48]. In addition, soil related within-field variability was found to be more pronounced and visible (beginning senescence) at the end of the growing season [44,45]. As a consequence, mapping of small-scale variability of crop characteristics and finally C-export is most effective in a narrow time window. UAV-based remote sensing meets all requirements for this purpose and helps to reduce time consuming and expensive ground-based measurement campaigns.

4. Conclusions and Outlook

In this case study, we presented a successful approach of how to use the combination of UAV-based high resolution remote sensing data and ground truth measurements for the estimation of the total C export by harvest of lucerne. The image pre-processing of the 12-band multispectral images has led to a considerable reduction of noise and vignetting effects. Together with the subsequent mosaicking and band alignment, the workflow is operational and will reduce the pre-processing time in future UAV missions. The conversion of recorded digital numbers to at-surface reflectance was only successful for the six NIR bands (b_{761}-b_{953}) and the R band b_{658}. However, these were the most important bands for the calculation of the VIs used in this study. Nevertheless, improvements of the experimental design to obtain the empirical lines for absolute radiometric calibration of the remaining bands in the VIS are necessary.

The strong correlation between fresh phytomass of lucerne and the EVI (R^2 = 0.88) using the NIR band b_{899} demonstrated the power of this VI even in the case of a dense green vegetation canopy. The map of the total C export revealed the potential of high spatial image resolution to: (i) map high small-scale variability within and across the different plots (75 g·m^{-2} and 225 g·m^{-2}); and (ii) identify the spatial pattern as a result of different terrain positions and ascociated soil types.

Future UAV missions should include important annual crops such as winter wheat or corn. The temporal flexibility of the UAV should be exploited for intra- and inter-annual studies of the temporal carbon dynamics, coupled with research campaigns focus on other components of the carbon budget (e.g., gas exchange measurements). The fixed-wing Carolo P360 proved to be a valuable instrument for mapping vegetation parameters over small experimental areas. The high cruising speed and the potential flight endurance of at least 30 min. have the potential to map larger areas which cover the full spectrum of terrain position/soil type combinations in this heterogeneous soil landscape under equal imaging conditions.

Finally, the respective results should be coupled with other available sources of spatially consistent proximal and remote sensing data and assimilated in state of the art models to gain improved insight into the processes controlling the carbon budget of agricultural landscapes.

Acknowledgments: The authors would like to thank Madlen Pohl and the staff of the ZALF Research Station in Dedelow for providing total carbon content data and yield data of lucerne, respectively. Special thanks go to Antje Wehrhan and Ingrid Onasch for carrying out the ground truth measurements, the UAV development team, at TU Braunschweig, for technical support and finally Frank Emmel, our back-up pilot, for ensuring a safe conduction of the UAV mission and technical assistance.

Author Contributions: Marc Wehrhan planned and operated the UAV mission, applied image pre-processing and interpretation (VI calculation) and sampled field data of lucerne. Philipp Rauneker developed and provided the software tool for image correction procedures including noise reduction and vignetting correction. Marc Wehrhan wrote the manuscript with considerable editorial contributions of Philipp Rauneker and Michael Sommer.

Conflicts of Interest: The authors declare no conflict of interest.

References

1. Pinter, P.J.; Hatfield, J.L.; Schepers, J.S.; Barnes, E.M.; Moran, M.S.; Daughtry, C.S.T.; Upchurch, D.R. Remote sensing for crop management. *Photogramm. Eng. Remote Sens.* **2003**, *69*, 647–664. [CrossRef]
2. Hatfield, J.L.; Gitelson, A.A.; Schepers, J.S.; Walthall, C.L. Application of Spectral Remote Sensing for Agronomic Decisions. *Agron. J.* **2008**, *100*, 117–131. [CrossRef]
3. Revill, A.; Sus, O.; Barrett, B.; Williams, M. Carbon Cycling of European Croplands: A Framework for Data Assimilation of Optical and Microwave Earth Observation Data. *Remote Sens. Environ.* **2013**, *137*, 84–93. [CrossRef]
4. Houborg, R.; Cescatti, A.; Migliavacca, M.; Kustas, W.P. Satellite retrievals of leaf chlorophyll and photosynthetic capacity for improved modelling of GPP. *Agric. For. Meteorol.* **2013**, *177*, 10–23. [CrossRef]
5. Berni, J.A.J.; Zarco-Tejada, P.J.; Suárez, L.; Fereres, E. Thermal and Narrowband Multispectral Remote Sensing for Vegetation Monitoring From an Unmanned Aerial Vehicle. *IEEE Trans. Geosci. Remote Sens.* **2009**, *47*, 722–738. [CrossRef]
6. Kelcey, J.; Lucieer, A. Sensor Correction of a 6-Band Multispectral Imaging Sensor for UAV Remote Sensing. *Remote Sens.* **2012**, *4*, 1462–1493. [CrossRef]
7. Lelong, C.C.D.; Burger, P.; Jubelin, G.; Roux, B.; Labbé, S.; Baret, F. Assessment of Unmanned Aerial Vehicles Imagery for Quantitative Monitoring of Wheat Crop in Small Plots. *Sensors* **2008**, *8*, 3557–3585. [CrossRef]
8. Laliberte, A.S.; Goforth, M.A.; Steele, C.M.; Rango, A. Multispectral Remote Sensing from Unmanned Aircraft: Image Processing Workflows and Applications for Rangeland Environments. *Remote Sens.* **2011**, *3*, 2529–2551. [CrossRef]
9. Peña, J.M.; Torres-Sánchez, J.; de Castro, I.A.; Kelly, M.; Lopez-Granados, F. Weed Mapping in Early-Season Maize Fields Using Object-Based Analysis of Unmanned Aerial Vehicle (UAV) Images. *PLoS ONE* **2013**, *8*, e77151. [CrossRef]

10. Zarco-Tejada, P.J.; Berni, J.A.J.; Suárez, L.; Sepulcre-Cantó, G.; Morales, F.; Miller, J.R. Imaging chlorophyll fluorescence with an airborne narrow-band multispectral camera for vegetation stress detection. *Remote Sens. Environ.* **2009**, *113*, 1262–1275. [CrossRef]

11. Zarco-Tejada, P.J.; González-Dugo, V.; Berni, J.A.J. Fluorescence, temperature and narrow-band indices acquired from a UAV platform for water stress detection using a micro-hyperspectral imager and a thermal camera. *Remote Sens. Environ.* **2012**, *117*, 322–337. [CrossRef]

12. Bouman, B.A.M. Accuracy of estimation the leaf area index from vegetation indices derived from drop reflectance characteristics, a simulation study. *Int. J. Remote Sens.* **1992**, *13*, 3069–3084. [CrossRef]

13. Rundquist, D.; Gitelson, A.; Derry, D.; Ramirez, J.; Stark, R.; Keydan, G. Remote Estimation of Vegetation Fraction in Corn Canopies. *Pap. Nat. Resour.* **2001**, *274*, 301–306.

14. Clevers, J.G.P.W. The application of a weighted infra-red vegetation index for estimating leaf area index by correcting for soil moisture. *Remote Sens. Environ.* **1989**, *29*, 25–37. [CrossRef]

15. Myneni, R.B.; Williams, D.L. On the relationship between FAPAR and NDVI. *Remote Sens. Environ.* **1994**, *49*, 200–211. [CrossRef]

16. Elvidge, C.D.; Chen, Z. Comparison of broad-band and narrow-band red and near-infrared vegetation indices. *Remote Sens. Environ.* **1995**, *54*, 38–48. [CrossRef]

17. Rondeaux, G.; Steven, M.; Baret, F. Optimization of Soil-Adjusted Vegetation Indices. *Remote Sens. Environ.* **1996**, *55*, 95–107. [CrossRef]

18. Jiang, Z.; Huete, A.R.; Didan, K.; Miura, T. Development of a two-band enhanced vegetation index without a blue band. *Remote Sens. Environ.* **2008**, *112*, 3833–3845. [CrossRef]

19. Osborne, B.; Saunders, M.; Walmsley, D.; Jones, M.; Smith, P. Key questions and uncertainties associated with the assessment of the cropland greenhouse gas balance. *Agric. Ecosyst. Environ.* **2010**, *139*, 293–301. [CrossRef]

20. Smith, P.; Lanigan, G.; Kutsch, W.L.; Buchmann, N.; Eugster, W.; Aubinet, M.; Ceschia, E.; Beziat, P.; Yeluripati, J.B.; Osborne, B.; *et al.* Measurements necessary for assessing the net ecosystem carbon budget of croplands. *Agric. Ecosyst. Environ.* **2010**, *139*, 302–315. [CrossRef]

21. Rouse, J.W.; Haas, R.H.; Schell, J.A.; Deering, D.W. Monitoring Vegetation System in the Great Plains with ERTS. In *Third Earth Resources Technology Satellite-1 Symposium*; NASA SP-351: Greenbelt, MA, USA, 1974; pp. 3010–3017.

22. Baret, F.; Guyot, G. Potentials and limits of vegetation indices for LAI and APAR assessment. *Remote Sens. Environ.* **1991**, *35*, 161–173. [CrossRef]

23. Thenkabail, P.S.; Smith, R.B.; de Pauw, E. Evaluation of Narrowband and Broadband Vegetation Indices for Determining Optimal Hyperspectral Wavebands for Agricultural Crop Characterization. *Photogramm. Eng. Remote Sens.* **2002**, *68*, 607–621.

24. Liu, H.Q.; Huete, A.R. A feedback based modification of the NDVI to minimize canopy background and atmospheric noise. *IEEE Trans. Geosci. Remote Sens.* **1995**, *33*, 457–465.

25. Sommer, M.; Augustin, J.; Kleber, M. Feedback of soil erosion on SOC patterns and carbon dynamics in agricultural landscapes—The CarboZALF experiment. *Soil Tillage Res.* **2015**. [CrossRef]

26. Scholtz, A.; Krüger, T.; Wilkens, C.-S.; Krüger, T.; Hiraki, K.; Vörsmann, P. Scientific Application and Design of Small Unmanned Aircraft Systems. In Proceedings of the 14th Australian International Aerospace Congress, Melbourne, Australia, 28 February–03 Match 2011.

27. Mansouri, A.; Marzani, F.S.; Gouton, P. Development of a protocol for CCD calibration: Application to a Multispectral Imaging System. *Int. J. Robot. Autom.* **2005**, *3767*, 1–12. [CrossRef]

28. Goldman, D.B.; Chen, J.-H. Vignette and Exposure Calibration and Compensation. *IEEE Trans. Pattern Anal. Mach. Intell.* **2010**, *32*, 2276–2288. [CrossRef] [PubMed]

29. Hugemann, W. *Correcting Lens Distortions in Digital Photographs*; Ingenieurbüro Morawski + Hugemann: Leverkusen, Germany, 2010.

30. Dall' Asta, E.; Roncella, R. A Comparison of Semiglobal and Local Dense Matching Algorithms for Surface Reconstruction. In Proceedings of the ISPRS Technical Commission V Symposium, Riva del Garda, Italy, 23–25 June 2014.

31. Conçalves, J.A.; Henriques, R. UAV photogrammetry for topographic monitoring of coastal areas. *ISPRS J. Photogramm. Remote Sens.* **2015**, *104*, 101–111. [CrossRef]

32. Moran, S.; Bryant, R.; Thome, K.; Ni, W.; Nouvellon, Y.; González-Dugo, M.P.; Qi, J.; Clarke, T.R. A refined empirical line approach for reflectance factor retrieval from Landsat-5 TM and Landsat-7 ETM+. *Remote Sens. Environ.* **2001**, *78*, 71–82. [CrossRef]

33. Chen, W.; Yan, L.; Li, Z.; Jing, X.; Duan, Y.; Xiong, X. In-flight calibration of an airborne wide-view multispectral imager using a reflectance-based method and its validation. *Int. J. Remote Sens.* **2013**, *34*, 1995–2005. [CrossRef]

34. Smith, G.M.; Milton, E.J. The use of the empirical line method to calibrate remotely sensed data to reflectance. *Int. J. Remote Sens.* **1999**, *20*, 2653–2662. [CrossRef]

35. West, T.O.; Bandaru, V.; Brandt, C.C.; Schuh, A.E.; Ogle, S.M. Regional uptake and release of crop carbon in the United States. *Biogeosciences* **2011**, *8*, 2037–2046. [CrossRef]

36. Zhang, X.; Izaurralde, R.C.; Manowitz, D.H.; Sahajpal, R.; West, T.O.; Thomson, A.M.; Xu, M.; Zhao, K.; LeDuc, S.; Williams, J.R. Regional scale cropland carbon budgets: Evaluating a Geospatial Agricultural Modeling System Using Inventory Data. *Environ. Model. Softw.* **2015**, *63*, 199–216. [CrossRef]

37. Wardlow, B.D.; Egbert, S.L.; Kastens, J.H. Analysis of time-series MODIS 250m vegetation index data for crop classification in the U.S. Central Great Plains. *Remote Sens. Environ.* **2007**, *108*, 290–310. [CrossRef]

38. Del Pozo, S.; Rodríguez-Gonzálvez, P.; Hernández-López, D.; Felipe-García, B. Vicarious Radiometric Calibration of a Multispectral Camera on Board an Unmanned Aerial System. *Remote Sens.* **2014**, *6*, 1918–1937. [CrossRef]

39. Von Bueren, S.K.; Burkart, A.; Hueni, A.; Rascher, U.; Tuohy, M.P.; Yule, I.J. Deploying four optical UAV-based sensors over grassland: Challenges and Limitations. *Biogeosciences* **2015**, *12*, 163–175. [CrossRef]

40. Haboudane, D.; Miller, J.R.; Pattey, E.; Zarco-Tejada, P.J.; Strachan, I.B. Hyperspectral vegetation indices and Novel Algorithms for Predicting Green LAI of crop canopies: Modeling and Validation in the Context of Precision Agriculture. *Remote Sens. Environ.* **2004**, *90*, 337–352. [CrossRef]

41. Broge, N.H.; Leblanc, E. Comparing prediction power and stability of broadband and hyperspectral vegetation indices for estimation of green leaf area index and canopy chlorophyll density. *Remote Sens. Environ.* **2000**, *76*, 156–172. [CrossRef]

42. Huete, A.; Didan, K.; Miura, T.; Rodriguez, E.P.; Gao, X.; Ferreira, L.G. Overview of the radiometric and biophysical performance of MODIS vegetation indices. *Remote Sens. Environ.* **2002**, *83*, 195–213. [CrossRef]

43. De Benedetto, D.; Castrignanò, A.; Rinaldi, M.; Ruggieri, S.; Santoro, F.; Figorito, B.; Gualano, S.; Diacono, M.; Tamborrino, R. An approach for delineating homogenous zones by using multi-sensor data. *Geoderma* **2013**, *199*, 117–127. [CrossRef]

44. Rudolph, S.; van der Kruk, B.; von Hebel, C.; Ali, M.; Herbst, M.; Montzka, C.; Pätzold, S.; Robinson, D.A.; Vereecken, H.; Weihermüller, L. Linking satellite derived LAI patterns with subsoil heterogeneity using large-scale ground-based electromagnetic induction measurements. *Geoderma* **2015**, *241–242*, 262–271. [CrossRef]

45. Sommer, M.; Wehrhan, M.; Zipprich, M.; Weller, U.; zu Castell, W.; Ehrich, S.; Tandler, B.; Selige, T. Hierarchical data fusion for mapping soil units at field scale. *Geoderma* **2003**, *112*, 179–196. [CrossRef]

46. Diacono, M.; Gastrinianò, A.; Troccoli, A.; De Benedetto, D.; Basso, B.; Rubino, P. Spatial and temporal variability of wheat grain yield and quality in a Mediterranean environment: A Multivariate Geostatistical Approach. *Field Crops Res.* **2012**, *131*, 49–62. [CrossRef]

47. Taylor, J.C.; Wood, G.A.; Earl, R.; Godwin, R.J. Soil factors and their Influence on Within-Field crop Variability II: Spatial Analysis and Determination of Management Zones. *Biosyst. Eng.* **2003**, *84*, 441–453. [CrossRef]

48. Stadler, A.; Rudolph, S.; Kupisch, M.; Langensiepen, M.; van der Kruk, B.; Ewert, F. Quantifying the effect of soil variability on crop growth using apparent soil electrical conductivity measurements. *Eur. J. Agron.* **2015**, *64*, 8–20. [CrossRef]

Article

Wavelength-Adaptive Dehazing Using Histogram Merging-Based Classification for UAV Images

Inhye Yoon [1], Seokhwa Jeong [1], Jaeheon Jeong [2], Doochun Seo [2] and Joonki Paik [1,*]

[1] Department of Image, Chung-Ang University, 84 Heukseok-ro, Dongjak-gu, Seoul 156-756, Korea;
   E-Mails: inhyey@gmail.com (I.Y.); sukhwa88@gmail.com (S.J.)
[2] Department of Satellite Data Cal/Val Team, Korea Aerospace Research Institute, 115 Gwahangbo, Yusung-Gu, Daejon 305-806, Korea; E-Mails: jjh583@kari.re.kr (J.J.); dcivil@kari.re.kr (D.S.)
* E-Mail: paikj@cau.ac.kr; Tel.: +82-2-820-5300; Fax: +82-2-814-9110.

Academic Editors: Felipe Gonzalez Toro and Antonios Tsourdos
Received: 9 December 2014 / Accepted: 10 March 2015 / Published: 19 March 2015

Abstract: Since incoming light to an unmanned aerial vehicle (UAV) platform can be scattered by haze and dust in the atmosphere, the acquired image loses the original color and brightness of the subject. Enhancement of hazy images is an important task in improving the visibility of various UAV images. This paper presents a spatially-adaptive dehazing algorithm that merges color histograms with consideration of the wavelength-dependent atmospheric turbidity. Based on the wavelength-adaptive hazy image acquisition model, the proposed dehazing algorithm consists of three steps: (i) image segmentation based on geometric classes; (ii) generation of the context-adaptive transmission map; and (iii) intensity transformation for enhancing a hazy UAV image. The major contribution of the research is a novel hazy UAV image degradation model by considering the wavelength of light sources. In addition, the proposed transmission map provides a theoretical basis to differentiate visually important regions from others based on the turbidity and merged classification results.

Keywords: image dehazing; image defogging; image enhancement; unmanned aerial vehicle images; remote sensing images

1. Introduction

Acquisition of high-quality images is an important issue in securing visual information of unmanned aerial vehicle (UAV) platforms. However, most UAV images are subject to atmospheric degradation. Among various factors of image degradation, haze or fog in the atmosphere results in color distortion, which can lead to erroneous analysis of important object regions. In reviewing the literature, dark channel prior-based defogging methods are first analyzed. We then address issues of the limitations and problems of the dark channel prior-based methods and justify the need of the wavelength-adaptive model of hazy image formulation.

A major approach to dehazing utilizes the dark channel prior that decomposes an image into the hazy and haze-free regions. The dark channel prior is a kind of statistics of the haze-free outdoor images. It is based on the assumption that most local patches in a haze-free outdoor image contain some pixels that have very low intensities in at least one color channel [1]. In order to solve the color distortion problem of He's method, Yoon et al. proposed an edge-based dark channel prior and corrected color distortion using gradient-based tone mapping [2]. Xie et al. used a combined bilateral and denoising filter to generate the transmission map [3]. Gao et al. further reduced halo effects using a guided filter and applied the maximum visibility to control the turbidity [4]. Park applied the weighted least squares-based edge-preserving smoothing filter to the dark channel prior and performed multi-scale tone manipulation [5]. Kil et al. combined the dark channel prior and local contrast enhancement to remove haze and correct color at the same time [6]. Yeh *et al.* proposed a fast dehazing algorithm by analyzing the haze density based on pixel-level dark and bright channel priors [7]. He *et al.* extended their original work by introducing the atmospheric point spread function to restore sharper dehazed images [8]. Shi estimated the amount of hazy components by detecting the sky region using the dark channel prior [9]. Although the proposed method shares a similar framework with Shi's method, it detects the sky region using various features instead of the dark channel prior. In addition, wavelength-adaptive enhancement is another contribution of the proposed work. Long *et al.* proposed an improved dehazing algorithm using the dark channel prior and a low-pass Gaussian filter for remote sensing images [10].

Although the dark channel prior has played an important role in various dehazing algorithms, the related methods suffer from color distortion and edge degradation, since they do not consider the wavelength characteristics in the image degradation model. In order to solve these problems, there were various dehazing algorithms without using the dark channel prior. Narasimhan *et al.* acquired two differently-exposed images for the same scene to estimate the amount of light penetration and the depth [11,12]. Shwartz *et al.* used two differently-polarized filters to reduce the polarized hazy component [13]. The use of two images makes these algorithms computationally expensive and impossible to be implemented in real-time. Schechner *et al.* proved that noise is amplified in a distant object region, because of the low transmission ratio [14]. Fattal measured the reflection ratio and performed dehazing under the assumption that directions of reflection should be identical at the same location [15]. Tan observed that the amount of haze or fog and contrast depend on the distance from the camera [16], and Kratz estimated statistically-independent components of image albedo and distance using Markov random fields [17]. Although three-dimensional (3D) geometry provides more reliable intensity of haze or fog, the related algorithms cannot be implemented in real-time video systems, because of the complex geometric transformation steps. Additional dehazing algorithms based on physical characteristics of haze or geometric information of the imaging process were proposed in [18,19].

The third group of dehazing algorithms can be categorized as an application-specific approach. Gibson applied the dehazing algorithm before and after video compression to reduce coding artifacts [20], and Chiang utilized the characteristics of the underwater image to modify the dark channel prior and proposed a dehazing algorithm with color and wavelength correction [21]. Pei proposed a combined color correction and guided filtering to remove blue shift in the night image [22]. Wen computed the scattering and transmission factors of light in underwater images and successfully removed haze using the difference of light attenuation Yoon *et al.* proposed color preserved defogging algorithms by considering the wavelength dependency [23,24].

The common challenge of existing dehazing methods includes color distortion and the associated high computational load to correct colors without considering the wavelength dependency. In order to solve this problem, we first present a wavelength-adaptive hazy UAV image degradation model, and a spatially-adaptive transmission map is generated using geometric classes and dynamic merging according to the model. The proposed wavelength-adaptive transmission removes hazy components without color distortion. As a result, the proposed algorithm needs neither additional optical equipment nor *a priori* distance estimation. The proposed dehazing method can significantly increase the visual quality of an atmospherically-degraded UAV image in the sense of preserving color distortion without the halo effect. Although the proposed model is established primarily for UAV images, it can be used to enhance any aerial hazy photography, video surveillance systems, driving systems and remote sensing systems.

This paper is organized as follows. Section 2 describes the wavelength-adaptive UAV image formation model, and Section 3 presents the proposed single image-based dehazing approach. Experimental results are given in Section 4, and Section 5 concludes the paper.

2. Wavelength-Adaptive UAV Image Formation Model

Rayleigh's law of atmospheric scattering provides the relationship between the scattering coefficient β and the wavelength λ, which is defined as [11]:

$$\beta(\lambda) \propto \frac{1}{\lambda^\gamma} \tag{1}$$

where $0 \leq \gamma \leq 4$ depends on the size of particles distributed in the atmosphere. In the haze-free atmosphere, haze particles can be considered to be sufficiently smaller than the wavelength of the light, and γ takes its maximum value, which is $\gamma = 4$. More specifically, if γ increases, the amount of scattering by haze particles becomes more dependent on the wavelength. On the other hand, in the hazy atmosphere, the size of haze particles is larger than the wavelength of the light, and γ takes its minimum value, such as $\gamma \approx 0$. For a small γ, the amount of scattering becomes less dependent on the wavelength of light.

If there is a sufficiently large amount of fog or haze, the amount of scattering light is assumed to be uniform regardless of the wavelength. This assumption is proved by the simple experiment based on the theory of Narasimhan [11,12]. In order to observe the wavelength-dependent scattering, Figure 1 shows four images acquired from the same indoor scene with different turbidities generated by the different amount of steam using a humidifier.

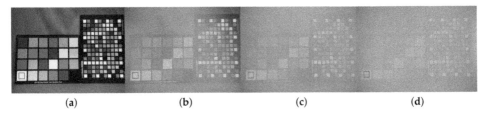

(a) (b) (c) (d)

Figure 1. Test images acquired from the same indoor scene with a different amount of steam generated by a humidifier: (**a**) the haze-free image without steam; and (**b**–**d**) the hazy images with different amounts of steam.

In Figure 1, the white region enclosed by the red rectangles contain the pure white color, and color distributions of the red rectangular regions in the four images were used. Theoretically, pure white regions, as shown in Figure 1a, have unity in all RGB color channels. However, the real values are not exactly the same as unity, but are close to unity because of noise. The color distribution appears as a "point" in the three-dimensional RGB space, as shown in Figure 2a. On the other hand, as the amount of haze increases in the atmosphere, the distribution becomes elongated ellipsoids, as shown in Figure 2b–d, since the turbidity of haze decreases the brightness of the object by scattering.

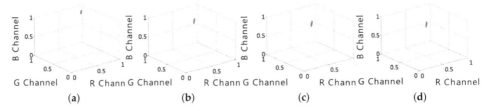

(a) (b) (c) (d)

Figure 2. The distribution of the RGB color of the red rectangular regions in Figure 1.

More specifically, Figure 3 shows RGB color histograms in four white regions enclosed by the red rectangle in Figure 1 and supports the observation in Figure 2.

This simple experiment shows that the amount of scattering depends on the turbidity of haze, which follows the observation by Narasimhan [12]. For this reason, restoration of the original color and brightness of an object is a challenging problem in the hazy environment.

Since the turbidity of a hazy image varies by the size of atmospheric particles and the distance of an object, the dehazing process should be performed in a spatially-adaptive manner. Let a small region with homogeneous color and brightness, or simply a cluster, have mean brightness values in RGB color channels as C_r, C_g and C_b. The effect of scattering can be estimated from the quantity of C_r, C_g and C_b using the color alignment measure (CAM) proposed in [25]:

$$L = \lambda_r \lambda_g \lambda_b / \sigma_r^2 \sigma_g^2 \sigma_b^2 \tag{2}$$

where λ_r, λ_g and λ_b represent eigenvalues of the RGB color covariance matrix of the white region and σ_r^2, σ_g^2 and σ_b^2 the corresponding diagonal elements of the covariance matrix. L represents the amount of correlation among RGB components.

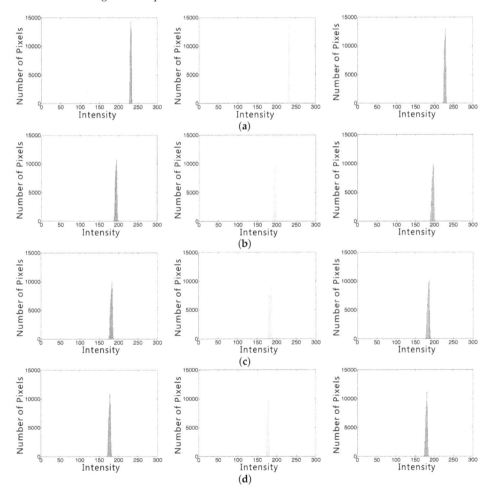

Figure 3. The RGB color histograms of the white regions in Figure 1: (**a**) red, green and blue (from left to right) color histograms of the white regions of the haze-free image shown in Figure 1a; and (**b–d**) red, green and blue color histograms of the hazy images shown in Figure 1b–d.

The CAM represents the degree of dispersion among RGB channels and increases if the dispersion becomes more dependent on the wavelength. Table 1 summarizes mean brightness values C_r, C_g and C_b of RGB components and CAM values of four white regions in Figure 1. As shown in the table, the more turbid the atmosphere, the less the scattering is affected by the wavelength.

A white region is suitable to estimate the CAM parameter, since it contains pan-chromatic light. Table 1 summarizes RGB components of mean brightness values, C_r, C_g and C_b and the corresponding CAM values of four white regions in Figure 1. As shown in the table, as the turbidity increases, the scattering amount becomes independent of the wavelength.

Based on this experimental work, the proposed degradation model of wavelength-dependent hazy unmanned aerial vehicle (UAV) images can be considered as an extended version of Yoon's work in [24], as shown in Figure 4.

Table 1. Mean RGB values of the white region in Figures 1 and the corresponding color alignment measure (CAM) values.

Image Type	C_r	C_g	C_b	CAM
Figure 1a	0.9044	0.9049	0.8884	0.0044
Figure 1b	0.7562	0.7640	0.7648	0.0024
Figure 1c	0.7089	0.7147	0.7199	0.0021
Figure 1d	0.6657	0.6711	0.6743	0.0019

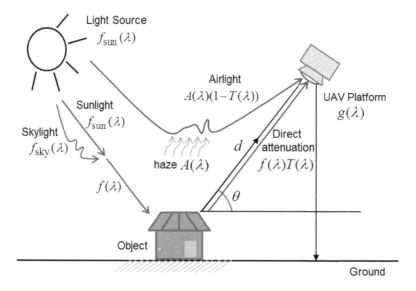

Figure 4. The proposed wavelength-adaptive UAV image formation model in a hazy atmosphere acquired by a UAV platform. $f_{sun}(\lambda)$ and $f_{sky}(\lambda)$ respectively represent the sun and sky light. The sky light $f_{sky}(\lambda)$ represents the light component that is scattered in the atmosphere.

The mathematical expression of the proposed UAV image formation model is given as:

$$g(\lambda) = f(\lambda)T(\lambda) + A(\lambda)(1 - T(\lambda)) \tag{3}$$

where $g(\lambda)$ represents the hazy image of wavelength $\lambda \in \{\lambda_{red}, \lambda_{green}, \lambda_{blue}\}$, $f(\lambda) = f_{sun}(\lambda) + f_{sky}(\lambda)$ is the original haze-free image, $A(\lambda)$ is the atmospheric light and $T(\lambda)$ is the transmission map. As described in Equation (3), $f(\lambda)$ is the combination of the scattering light in the atmosphere $f_{\text{sun}}(\lambda)$ and the unscattered light $f_{\text{sky}}(\lambda)$.

In the right-hand side of Equation (3), the first term $f(\lambda)T(\lambda)$ represents the direct attenuation component and the second term $A(\lambda)(1 - T(\lambda))$ represents the air light component. The former describes the decayed version of $f(\lambda)$ in the atmosphere or the space, while the latter results from scattering by haze and color shifts. The proposed degradation model given in Equation (3) can be considered as a wavelength-extended version of the original hazy image formation model proposed in [1].

Given the UAV image degradation model, the dehazing problem is to restore $\hat{f}(\lambda)$ from $g(\lambda)$ by estimating $T(\lambda)$ and $A(\lambda)$. $\hat{f}(\lambda)$ represents the estimated value of the original haze-free image.

3. The Proposed Single UAV Image-Based Dehazing Approach

The proposed dehazing algorithm consists of image segmentation and labeling, modified transmission map generation, atmospheric light estimation and intensity transformation modules, as shown in Figure 5.

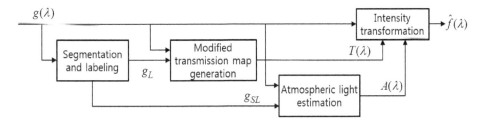

Figure 5. The proposed single UAV image-based dehazing algorithm for enhancing hazy UAV images.

The label image $g_L(\lambda)$ is first generated using histogram merging-based classification in the hazy image $g(\lambda)$. Second, the modified transmission map $T(\lambda)$ is generated based on the wavelength-dependent atmospheric turbidity. The corresponding atmospheric light $A(\lambda)$ is then estimated in the labeled version of the sky image g_{SL}. Finally, the proposed method can significantly enhance the contrast and visibility of a hazy UAV image using the estimated atmospheric light and modified transmission map. As a result, the enhanced image $\hat{f}(\lambda)$ is obtained by adaptively removing hazy components.

Although the wavelength-dependent model was already proposed for underwater image enhancement, the effect of the wavelength may be trivial in the visible spectrum of the air for a UAV image. On the other hand, the proposed work pays attention to the dependency of the wavelength together with the object distance and the amount of scattering. The early work of wavelength-dependent vision through the atmosphere can be found in [26].

3.1. Image Segmentation Based on Geometric Classes

The turbidity of the atmosphere varies by the distance of an object and atmospheric degradation factors. Since the conventional transmission map determines the proportion of the light reflected by the object reaching the UAV camera, Tan *et al.* assumed that light traveling a longer distance is more attenuated, yielding the transmission map defined as [16]:

$$T(\lambda) = e^{-\beta d(x,y)} \tag{4}$$

where β represents the scattering coefficient of the atmosphere by color wavelength and $d(x, y)$ the depth or distance of a point corresponding to the image coordinates (x, y).

However, the depth information is difficult to estimate using a single input UAV image. In addition, existing dehazing methods based on the estimation of the atmospheric light and transmission map exhibit various artifacts, such as color distortion, incompletely removed haze and unnaturally enhanced contrast, to name a few [16].

In order to solve these problems, we present a novel transmission map generation method using geometric classes, such as sky, ground and vertical structures. The proposed transmission map provides a theoretical basis to differentiate important regions from the background based on the turbidity and merged classification results.

Existing transmission map generation methods commonly perform image segmentation by minimizing a cost function, such as the ratio cut [27]. Comaniciu *et al.* proposed an image segmentation algorithm using edge information based on the mean shift [28]. Bao *et al.* detected the edge of the input image using a modified canny edge detector with scale multiplication [29]. Erisoglu *et al.* propose a segmentation method to estimate the initial cluster using the *K*-means algorithm [30]. However, most existing methods could not completely solve the problems of noise amplification and edge blurring.

In order to overcome the above-mentioned limitations in existing segmentation methods, the geometric class-based approach in [24] is used for the three-dimensional (3D) context-adaptive processing. The geometric class-based pre-segmentation approach proposed in [24] was inspired by the Hoiem's work in [31], and the proposed approach is an improved version of Yoon's work in [24], where coarse geometric properties are estimated by learning the appearance model of geometric classes. Although Hoiem *et al.* defined a complete set of geometrical classes for segmenting a general scene in their original works, we present a simplified version by selecting only three classes, sky, vertical and ground. Hoiem uses a complete set of geometrical classes, but it causes extremely high computational complexity and an over-segmentation problem. Figure 6 shows a simple illustration to describe the proposed segmentation algorithm.

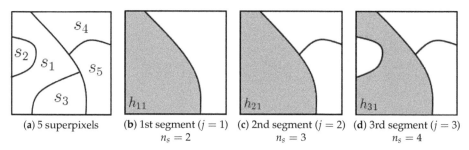

(a) 5 superpixels   (b) 1st segment ($j = 1$)   (c) 2nd segment ($j = 2$)   (d) 3rd segment ($j = 3$)
$n_s = 2$ $n_s = 3$ $n_s = 4$

Figure 6. Illustration of the proposed segmentation algorithm: (a) five superpixels of an input image and (b–d) regions containing s_1 for three hypotheses, such as $h_{j1}, j = 1, 2, 3$.

Figure 6a shows a sample input image decomposed into five superpixels, $s_i, i = 1, \ldots, 5$. In order to assign one of three classes, such as $v \in \{(S)ky, (V)ertical, (G)round\}$, to each s_i, we perform image segmentation multiple times using different hypotheses. For the first superpixel s_1, Figure 6b–d show regions containing s_1 for the j-th hypothesis, such as h_{j1} for $j = 1, 2, 3$. Let b_i and $\bar{b}_{ji} \in \{S, V, G\}$ respectively represent labels of s_i and h_{ji}; then, the most suitable label for s_i maximizes the confidence value, defined as:

$$C(b_i = v|g(\lambda)) = \sum_{j=1}^{3} P(b_{ji} = v|g(\lambda), h_{ji}) P(h_{ji}|g(\lambda)) \qquad (5)$$

where C is the label confidence and P the likelihood.

For example, the confidence that the first superpixel s_1 is labeled as sky is computed as:

$$
\begin{aligned}
C(b_1 = S|g(\lambda)) = &\, P(\tilde{b}_{11} = S|g(\lambda), h_{11})P(h_{11}|g(\lambda)) \\
&+ P(\tilde{b}_{21} = S|g(\lambda), h_{21})P(h_{21}|g(\lambda)) \\
&+ P(\tilde{b}_{31} = S|g(\lambda), h_{31})P(h_{31}|g(\lambda))
\end{aligned}
\tag{6}
$$

The first step of label generation is to estimate superpixels from small, almost-homogeneous regions in the image. Since superpixel segmentation usually preserves object boundaries at the cost of over-segmentation, it can provide accurate boundary information of a subject. We then merge adjacent regions using histogram classification. Next, we uniformly quantize each color channel into 16 levels and then estimate the histogram of each region in the feature space of $16 \times 16 \times 16 = 4096$ bins. The following step merges the regions based on their three classified intensity ranges, such as dark, middle and bright, as shown in Figure 7b.

The similarity measure $\rho(R_A, R_C)$ between two regions R_A and R_C is defined as:

$$
\rho(R_A, R_C) = \max_{k=1,\dots K} \rho(R_A, R_k)
\tag{7}
$$

Let R_C and R_A be respectively a region-of-interest and one of its adjacent regions. If there are K adjacent regions R_k, for $k = 1, \dots, K$, R_C is definitely equal to one of the R_k's. If $\rho(R_A, R_C)$ is the maximum among K similarities $\rho(R_A, R_k)$, $k = 1, \dots, K$, the two regions are merged as [32]:

$$
R'_C = R_C \cup R_A
\tag{8}
$$

where R'_C has the same histogram as that of R_C. This process repeats until there are no more merging regions. The next step computes multiple labels based on simple features, such as color, texture and shape. We generate multiple labels of the hazy image with n_s different regions. We determine the most suitable class for each region using the estimated probability that all superpixels have the same class, which represents each geometric class, such as sky, vertical and ground. Given the hazy UAV image $g(\lambda)$, the superpixel label can be expressed as:

$$
C(b_i = v|g(\lambda)) = \sum_j^{N_h} P(b_j = v|g(\lambda), h_{ji})P(h_{ji}|g(\lambda))
\tag{9}
$$

where C represents the label confidence, b_i the superpixel label, v the possible label value, N_h the number of multiple hypotheses and h_{ji} the region containing the i-th superpixel for the jth hypothesis. $P(h_{ji}|g(\lambda))$ represents the homogeneity likelihood and $P(b_j = v|g(\lambda), h_{ji})$ the label likelihood. Consequently, the sum of the label likelihoods for a particular region and the sum of the homogeneity likelihoods for all regions containing a particular superpixel are normalized.

The proposed segmentation method consists of the following steps: (i) the initial segmentation is performed using superpixels; (ii) the entire histogram is divided into three ranges, as shown in Figure 7b; and (iii) the labeled image is obtained using histogram merging when adjacent segments have similar histograms. In order to simply generate the labeled image, three classes are defined as $v \in \{(S)ky, (V)ertical, (G)round\}$ using color, texture, shape and location and then computing the probability that a new segment has the same class.

Figure 7a shows an input hazy image. Figure 7b shows three classified regions, dark, middle and bright, in the histogram of the input image, and Figure 7c shows the segmentation result using the proposed algorithm. We can obtain the labeled image g_L using the same class of image as shown in Figure 7d.

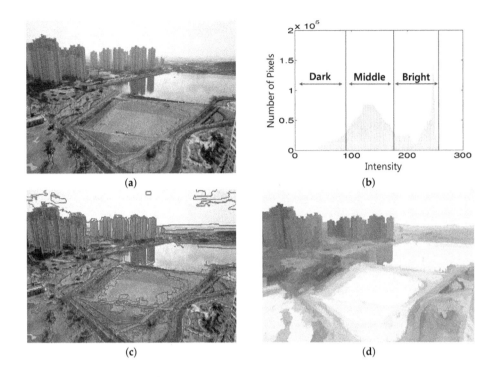

Figure 7. (**a**) A hazy aerial image; (**b**) the corresponding histogram classified into the dark, middle and bright ranges; (**c**) the result of histogram merging-based segmentation; and (**d**) the result of labeling.

3.2. Spatially-Adaptive Transmission Map

The conventional dark channel prior-based dehazing method generates the transmission map by searching for the lowest intensity in the patch centered at (x,y), denoted as $\Omega(x,y)$, from Equation (3), as [1]:

$$
\min_{c\in\{R,G,B\}}\left\{\min_{(p,q)\in\Omega(x,y)}\left(\frac{g_c(p,q)}{A}\right)\right\}
$$
$$
= T(x,y)\min_{c\in\{R,G,B\}}\left\{\min_{(p,q)\in\Omega(x,y)}\left(\frac{f_c(p,q)}{A}\right)+(1-T(x,y))\right\} \tag{10}
$$

where subscript c represents one of the RGB color channels and (p,q) the pixel coordinates of local patch centers.

Since the minimum intensity in a patch of the ideal haze-free image tends to zero, the transmission map can be computed as:

$$
\tilde{T}(x,y) = 1 - \min_{c\in\{R,G,B\}}\left\{\min_{(p,q)\in\Omega(x,y)}\left(\frac{g_c(p,q)}{A}\right)\right\} \tag{11}
$$

The conventional transmission map results in a halo effect and color distortion in the finally dehazed UAV image, since intensity discontinuity across edges is not considered in the reconstruction process [1]. Although an image matting-based halo effect reduction method has been proposed in the literature, it requires extremely high computational load, which is unsuitable for practical applications. To solve this problem, we generate a modified transmission map by incorporating the classification and labeling results into the conventional transmission map given in Equation (4). The modified transmission map is defined as:

$$T(\lambda) = e^{-\beta g_{Gi}(\lambda)^{-\alpha}} \tag{12}$$

where β represents the scattering coefficient of the atmosphere, λ the wavelength in red (700 μm), green (520 μm) and blue (440 μm) and α the wavelength exponent. For the experiment, $\alpha = 1.5$ was used. In Equation (12), since $g_L(\lambda)$ contains sharp edges, the guided filter is used to generate the context adaptive image $g_{Gi}(\lambda)$ that has smoothed edges. The context-adaptive image can be computed by the weighted average of the guided filter as [33]:

$$g_{G_i}(\lambda) = \sum_j W_{ij}(g_L) g_i(\lambda) \tag{13}$$

where i, j are the pixel coordinates, $g(\lambda)$ represents the input hazy UAV image and g_L the guidance label image, and the weighted filter kernel W_{ij} is given as:

$$W_{ij}(g_L) = \frac{1}{|\omega|^2} \sum_{(i,j) \in \omega_{mn}} \left(1 + \frac{(g_{Li}(\lambda) - \mu_{mn})(g_{Lj}(\lambda) - \mu_{mn})}{\sigma_{mn}^2 + \varepsilon} \right) \tag{14}$$

where ω_{mn} is a window centered at (m, n), $|\omega|$ is the number of pixels in ω_{mn}, ε is a regularization parameter, μ_{mn} is the mean and σ_{mn} is the variance of ω_{mn}.

The scattering coefficient is determined by the amount of haze, object distance and camera angle. This work uses a camera angle of 60 degrees [34]. The scattering coefficients are defined as:

$$\beta = \begin{cases} 0.3324, & \lambda = 700 \ \mu m \ \text{(red)} \\ 0.3433, & \lambda = 520 \ \mu m \ \text{(green)} \\ 0.3502, & \lambda = 440 \ \mu m \ \text{(blue)} \end{cases} \tag{15}$$

In order to generate the transmission map depending on the wavelength, the feature-based labeled image is used. Therefore, the wavelength-dependent transmission map is proposed to adaptively represent hazy images. As a result, the proposed spatially-adaptive transmission map can mitigate the contrast distortion and halo effect problems using the context-adaptive image.

3.3. Estimation of Local Atmospheric Light and Intensity Transformation

Conventional dehazing methods estimate the atmospheric light from the brightest pixel in the dark channel prior of the hazy image. To the best of our knowledge, He *et al.* [1] were the first to raise the issue of color distortion when the atmospheric light is incorrectly estimated from an undesired region, such as a white car or a white building. To address this problem, we estimate the atmospheric light using the labeled sky image that can be generated from the 'sky' class, as shown in Figure 8b. As a result, the atmospheric light can be estimated as:

$$A(\lambda) = \max(g(\lambda) \cap g_{SL}) \tag{16}$$

where $g(\lambda)$ represents the hazy image and g_{SL} the labeled sky image. This procedure can mitigate the color distortion problem in conventional dehazing methods.

Given the adaptive global atmospheric light $A(\lambda)$ and the modified transmission map $T(\lambda)$, the dehazed image is finally restored as:

$$\hat{f}(\lambda) = \frac{g(\lambda) - A(\lambda)}{T(\lambda)} + A(\lambda) \tag{17}$$

Figure 8a shows the proposed transmission map of Figure 7a. As a result, the proposed spatially-adaptive transmission map has continuous intensity values in the neighborhood of boundaries. Figure 8b shows the labeled sky image using the proposed segmentation method. Figure 8c shows the dehazed image using the modified transmission map in Figure 8a. As shown in Figure 8c, the proposed method significantly enhanced the contrast of the hazy image without color distortion or unnatural contrast amplification.

(a)   (b)   (c)

Figure 8. Results of dehazing: (**a**) the transmission map using the proposed method; (**b**) the labeled sky image using the proposed segmentation method; and (**c**) the dehazed image using the proposed modified transmission map and atmospheric light.

4. Experimental Results

In this section, we show the experimental results to compare the performance of the proposed dehazing algorithm with conventional methods.

(a)   (b)   (c)

Figure 9. Performance evaluation of color restoration using a simulated hazy image: (**a**) original haze-free image; (**b**) the simulated hazy image; and (**c**) the dehazed image using the proposed method.

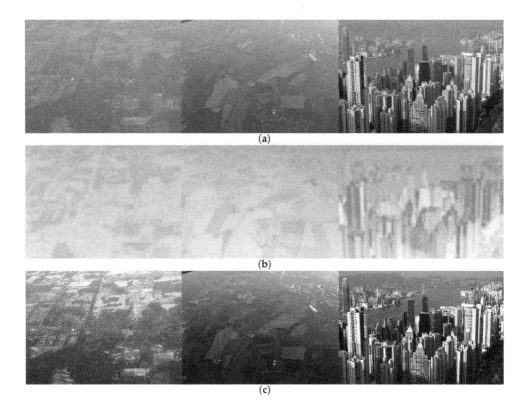

Figure 10. Experimental results of the proposed dehazing method: (**a**) input hazy UAV images; (**b**) the modified transmission maps; and (**c**) the dehazed UAV images using the proposed method.

In order to demonstrate the performance of original color restoration, a set of hazy images are first generated by simulation. The test images are then enhanced using the proposed dehazing algorithm to evaluate the accuracy of color restoration, as shown in Figure 9.

Figure 10a shows three hazy unmanned aerial vehicle (UAV) images; Figure 10b shows the modified transmission map using the proposed method; and Figure 10c shows the dehazed results using the proposed method. As shown in Figure 10c, the proposed dehazing method preserves fine details without color distortion. If there are no sky regions in an image, as shown in Figure 10, the atmospheric light can be simply estimated using the maximum intensity value.

The performance of the proposed dehazing algorithm is compared with three existing dehazing methods. Figure 11a shows two hazy images, and Figure 11b shows the dehazed images using the conventional dark channel prior-based method [1], where the hazy components were removed at the cost of color distortion in the sky region. Figure 11c shows dehazed images using Fattal's method proposed in [15], where the haze is not completely removed in regions far from the camera. Figure 11d shows dehazed results using Tan's method proposed in [16], where color distortion is visible. As shown in Figure 10e, the dehazed image using the proposed method shows significantly improved image quality without color distortion or unnaturally amplified contrast. Furthermore, the proposed method maintains the haze in the sky and removes haze surrounding the buildings.

Figure 11. Experimental results of various dehazing methods: (**a**) input hazy images; (**b**) dehazed images using He's method; (**c**) dehazed images using Fattal's method; (**d**) dehazed images using Tan's method; and (**e**) dehazed images using the proposed method.

In order to justify the performance of dehazing, Tarel *et al.* compared a number of enhancement algorithms [35]. On the other hand, the proposed work is evaluated in the sense of both subjective and objective manners using visual comparison in Figure 11 and quantitative evaluation in Table 2, respectively. We compared the performance of the proposed dehazing algorithm with three existing state-of-the-art methods in the sense of the visibility metric proposed by Zhengguo. The visibility metric is used to calculate the contrast-to-noise ratio (CNR) of dehazed images.

Table 2. Quantitative performance of the proposed and three state-of-the-art dehazing methods in the sense of both contrast-to-noise ratio (CNR) and the measure of enhancement (ME).

Input Image	Dehazing Method	CNR	ME
	Haze Image	53.0588	-
	He's Method [1]	68.8744	15.8156
	Fattal's Method [14]	71.8029	18.7441
	Tan's Method [15]	84.9931	31.9343
	Proposed Method	72.4338	19.3750
	Haze Image	58.8830	-
	He's Method [1]	72.8756	13.9626
	Fattal's Method [14]	67.5625	8.6795
	Tan's Method [15]	88.5869	29.7039
	Proposed Method	81.4254	22.5424

Figure 12. Experimental results of the proposed dehazing method: (**a**) input hazy images acquired by a video surveillance camera, an underwater camera and a vehicle black box camera, respectively; (**b**) the modified transmission maps; and (**c**) the dehazed images using the proposed method.

Figure 12a shows hazy images acquired by a video surveillance camera, an underwater camera and a vehicle black box camera. Figure 12b shows the modified transmission maps using the proposed method, and the results of the proposed dehazing method are shown in Figure 12c. Based on the experimental results, the proposed dehazing method can successfully restore the original color of the scene with a moderately hazy atmosphere, but its dehazing performance is limited with a severe amount of haze in the atmosphere.

Figure 13a shows three satellite images acquired by the Korea Aerospace Research Institute (KARI). Figure 13b shows the modified transmission maps, and Figure 13c shows the dehazed results using the proposed method. Based on the experimental results, the proposed dehazing method can successfully restore the original color of the scene with a moderately hazy atmosphere, except with an excessive amount of haze, such as thick cloud.

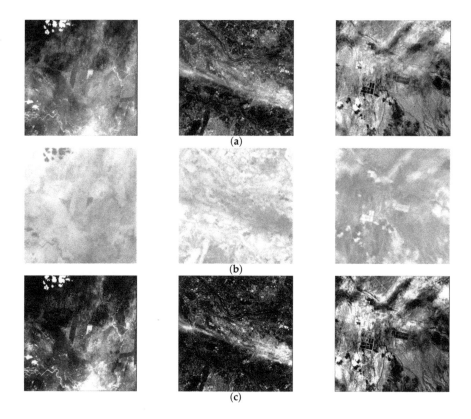

Figure 13. Experimental results of the proposed dehazing method: (**a**) input hazy satellite images courtesy of the Korea Aerospace Research Institute (KARI); (**b**) the modified transmission maps; and (**c**) the dehazed images using the proposed method.

5. Conclusions

In this paper, we presented a spatially-adaptive dehazing algorithm based on a wavelength-adaptive hazy image formation model. As a major contribution, the proposed wavelength-based dehazing method can mitigate the color distortion problem in conventional methods. By incorporating the wavelength characteristics of light sources into the UAV image degradation model, the proposed transmission map removes hazy components in the input image. Another contribution is that the proposed algorithm needs neither additional optical equipment nor *a priori* distance estimation.

Experimental results show that the proposed dehazing algorithm can successfully restore the original color of the scene containing the wavelength-dependent scattering of atmosphere. This proposed algorithm can be used for various applications, such as video surveillance systems, intelligent driver assistant systems and remote sensing systems. The proposed wavelength-adaptive dehazing algorithm is particularly suitable for preprocessing multispectral registration of satellite images for enhancing aerial images with various types of haze, fog and cloud.

Acknowledgments: This work was supported by the Korea Aerospace Research Institute (KARI), the ICT R&D program of MSIP/IITP. [14-824-09-002, Development of global multi-target tracking and event prediction techniques based on real-time large-scale video analysis, and the Technology Innovation Program (Development of Smart Video/Audio Surveillance SoC & Core Component for Onsite Decision Security System) under Grant 10047788.

Acknowledgments: Inhye Yoon initiated the research and designed the experiments; Seokhwa Jeong performed experiments; Jaeheon Jeong and Doochun Seo analyzed the data; Joonki Paik wrote the paper.

Conflicts of Interest: CThe authors declare no conflict of interest.

References

1. He, K.; Sun, J.; Tang, X. Single image haze removal using dark channel prior. In Proceedings of the IEEE Conference on Computer Vision and Pattern Recognition, Miami, FL, USA, 20–26 June 2009; pp. 1956–1963.
2. Yoon, I.; Jeon, J.; Lee, J.; Paik, J. Spatially adaptive image defogging using edge analysis and gradient-based tone mapping. In Proceedings of the IEEE International Conference on Consumer Electronics, Las Vegas, NV, USA, 9–12 January 2011; pp. 195–196.
3. Xie, B.; Guo, F.; Cai, Z. Universal strategy for surveillance video defogging. *Opt. Eng.* **2012**, *51*, 1–7.
4. Gao, R.; Fan, X.; Zhang, J.; Luo, Z. Haze filtering with aerial perspective. In Proceedings of the IEEE International Conference on Image Processing, Orland, FL, USA, 30 September 2012; pp. 989–992.
5. Park, D.; Han, D.; Ko, H. Single image haze removal with WLS-based edge-preserving smoothing filter. In Proceedings of the IEEE International Conference on Acoustics, Speech, and Signal Processing, Vancouver, BC, Canada, 26–31 May 2013; pp. 2469–2473.
6. Kil, T.; Lee, S.; Cho, N. A dehazing algorithm using dark channel prior and contrast enhancement. In Proceedings of the IEEE International Conference on Acoustics, Speech, and Signal Processing, Vancouver, BC, Canada, 26–31 May 2013; pp. 2484–2487.
7. Yeh, C.; Kang, L.; Lee, M.; Lin, C. Haze effect removal from image via haze density estimation in optical model. *Opt. Express* **2013**, *21*, 27127–27141.
8. He, R.; Wang, Z.; Fan, Y.; Feng, D. Multiple scattering model based single image dehazing. In Proceedings of the IEEE Conference on Industrial Electronics and Applications, Melbourne, Australia, 19–21 June 2013; pp. 733–737.
9. Shi, Z.; Long, J.; Tang, W.; Zhang, C. Single image dehazing in inhomogeneous atmosphere. *Opt. Int. J. Light Electron Opt.* **2014**, *15*, 3868–3875.
10. Long, J.; Shi, Z.; Tang, W.; Zhang, C. Single remote sensing image dehazing. *IEEE Geosci. Remote Sens. Lett.* **2014**, *11*, 59–63.
11. Narasimhan, S.; Nayar, S. Vision and the atmosphere. *Int. J. Comput. Vis.* **2002**, *48*, 233–254.
12. Narasimhan, S.; Nayer, S. Contrast restoration of weather degraded images, *IEEE Trans. Pattern Anal. Mach. Intell.* **2003**, *25*, 713–724.
13. Shwartz, S.; Namer, E.; Schecher, Y. Blind haze separation. In Proceedings of the IEEE International Conference on Computer Vision, Pattern Recognition, New York, NY, USA, 17–22 October 2006; pp. 1984–1991.
14. Schechner, Y.; Averbuch, Y. Regularized image recovery in scattering media. *Pattern Anal. Mach. Intell.* **2007**, *29*, 1655–1660.
15. Fattal, R. Single image dehazing. *ACM Trans. Graph.* **2008**, *27*, 1–9.
16. Tan, R. Visibility in bad weather from a single image. In Proceedings of the IEEE Conference on Computer Vision and Pattern Recognition, Anchorage, AK, USA, 24–26 June 2008; pp. 1–8.
17. Kratz, L.; Nishino, K. Factorizing scene albedo and depth from a single foggy image. In Proceedings of the IEEE International Conference on Computer Vision, Kyoto, Japan, 29 September 2009; pp. 1701–1708.
18. Ancuti, C.; Ancuti, C. Single image dehazing by multi-scale fusion. *IEEE Trans. Image Process.* **2013**, *22*, 3271–3282.
19. Fang, S.; Xia, X.; Xing, H.; Chen, C. Image dehazing using polarization effects of objects and airlight. *Opt. Express* **2014**, *22*, 19523–19537.
20. Gibson, K.; Vo, D.; Nguyen, T. An investigation of dehazing effects on image and video coding. *IEEE Trans. Image Process.* **2012**, *21*, 662–673.
21. Chiang, J.; Chen, Y. Underwater image enhancement by wavelength compensation and dehazing. *IEEE Trans. Image Process.* **2012**, *21*, 1756–1769.

22. Pei, S.; Lee, T. Nighttime haze removal using color transfer pre-processing and dark channel prior. In Proceedings of the IEEE International Conference on Image Processing, Orlando, FL, USA, 30 September 2012; pp. 957–960.

23. Yoon, I.; Kimg, S.; Kimg, D.; Hayes, M.; Paik, J. Adaptive defogging with color correction in the HSV color space for consumer surveillance system, *IEEE Trans. Consum. Electron.* **2012**, *58*, 111–116.

24. Yoon, I.; Hayes, M.; Paik, J. Wavelength-adaptive image formation model and geometric classification for defogging unmanned aerial vehicle images. In Proceedings of the IEEE International Conference on Acoustics, Speech, and Signal Processing, Vancouver, BC, Canada, 26–31 May 2013; pp. 2454–2458.

25. Bando, Y.; Chen, B.; Nishita, T. Extracting depth and matte using a color-filtered aperture. *ACM Trans. Graph.* **2008**, *27*, 1–9.

26. Middleton, W. Vision through the atmosphere. *Encyclopedia of Physics, Geophysic II*; Springer-Verlag: Berlin, Germany, 1957; Volume 48, pp. 254–287.

27. Wang, S.; Siskind, J. Image segmentation with ration cut. *IEEE Trans. Pattern Anal. Mach. Intell.* **2003**, *25*, 675–690.

28. Comaniciu, D.; Meer, P. Mean shift: A robust approach toward feature space analysis. *IEEE Trans. Pattern Anal. Mach. Intell.* **2002**, *24*, 603–619.

29. Bao, P.; Zhang, L.; Wu, X. Canny edge detection enhancement by scale multiplication. *IEEE Trans. Pattern Anal. Mach. Intell.* **2005**, *27*, 1485–1490.

30. Erisoglu, M.; Calis, N.; Sakallioglu, S. A new algorithm for initial cluster centers in k-means algorithm. *Pattern Recog. Lett.* **2001**, *32*, 1701–1705.

31. Hoiem, D.; Efros, A.; Hebert, M. Recovering surface layout from an image. *Int. J. Comput. Vis.* **2007**, *75*, 151–172.

32. Ning, J.; Zhang, L.; Zhang, D.; Wu, C. Interactive image segmentation by maximal similarity based region merging. *Pattern Recog.* **2010**, *43*, 445–456.

33. He, K.; Sun, J.; Tang, X. Guided image filtering. In Proceedings of the European Conference on Computer Vision, Crete, Greece, 5–11 September 2010; pp. 1–14.

34. Preetham, A.; Shirley, P.; Smits, B. A practical analytic model for daylight. In Proceedings of the ACM SIGGRAPH, Los Angeles, CA, USA, 8–13 August 1999; pp. 91–100.

35. Tarel, J.; Hautiere, N.; Caraffa, L.; Cord, A.; Halmaoui, H.; Gruyer, D. Vision enhancement in homogeneous and heterogeneous fog. *IEEE Intell. Transp. Syst. Mag.* **2012**, *2*, 6–20.

MDPI

Article

A Space-Time Network-Based Modeling Framework for Dynamic Unmanned Aerial Vehicle Routing in Traffic Incident Monitoring Applications

Jisheng Zhang [1,2], Limin Jia [3], Shuyun Niu [2], Fan Zhang [2], Lu Tong [4] and Xuesong Zhou [5,*]

[1] School of Traffic and Transportation, Beijing Jiaotong University, Beijing 100044, China; zjs2107@163.com
[2] Research Institute of Highway Ministry of Transport, No. 8 Xitucheng Rd., Haidian District, Beijing 100088, China; nsy@itsc.cn (S.N.); zhangfan@itsc.cn (F.Z.)
[3] State Key Laboratory of Rail Traffic Control and Safety, Beijing Jiaotong University, No. 3 Shangyuancun, Haidian District, Beijing 100044, China; lmjia@vip.sina.com
[4] School of Traffic and Transportation, Beijing Jiaotong University, No. 3 Shangyuancun, Haidian District, Beijing 100044, China; ltong@bjtu.edu.cn
[5] School of Sustainable Engineering and the Built Environment, Arizona State University, Tempe, AZ 85287, USA
* Author to whom correspondence should be addressed; xzhou74@asu.edu; Tel.: +1-480-9655-827.

Academic Editor: Felipe Gonzalez Toro
Received: 18 April 2015; Accepted: 4 June 2015; Published: 12 June 2015

Abstract: It is essential for transportation management centers to equip and manage a network of fixed and mobile sensors in order to quickly detect traffic incidents and further monitor the related impact areas, especially for high-impact accidents with dramatic traffic congestion propagation. As emerging small Unmanned Aerial Vehicles (UAVs) start to have a more flexible regulation environment, it is critically important to fully explore the potential for of using UAVs for monitoring recurring and non-recurring traffic conditions and special events on transportation networks. This paper presents a space-time network- based modeling framework for integrated fixed and mobile sensor networks, in order to provide a rapid and systematic road traffic monitoring mechanism. By constructing a discretized space-time network to characterize not only the speed for UAVs but also the time-sensitive impact areas of traffic congestion, we formulate the problem as a linear integer programming model to minimize the detection delay cost and operational cost, subject to feasible flying route constraints. A Lagrangian relaxation solution framework is developed to decompose the original complex problem into a series of computationally efficient time-dependent and least cost path finding sub-problems. Several examples are used to demonstrate the results of proposed models in UAVs' route planning for small and medium-scale networks.

Keywords: unmanned aerial vehicle; traffic sensor network; space-time network; lagrangian relaxation; route planning

1. Introduction

Reliable and timely traffic information is the foundation of network-wide traffic management and control systems. An advanced traffic sensor network needs to rapidly detect non-recurring traffic events and reliably estimate recurring traffic congestion along key freeway and arterial corridors. Most of commonly used traffic sensors are equipped at fixed locations, such as loop detectors, microwave detectors, video cameras, Automatic Vehicle Identification (AVI) readers, *etc.* Fixed traffic sensors can constantly monitor traffic dynamic characteristics of the specific location for a long time horizon, but they cannot provide a full spatial and temporal coverage in a network due to construction and maintenance budget constraints. With the use of movable or mobile traffic sensors, the next-generation transportation sensor network is expected to offer a more reliable and less costly approach to rapidly detect complex and dynamic state evolution in a transportation system.

Our research will focus on how to integrate existing fixed and emerging mobile sensors into a dynamic traffic monitoring system that can significantly improve spatial coverage responsiveness to important events. Specifically, the new generation of small Unmanned Aerial Vehicles (UAVs) now offers outstanding flexibility as low-cost mobile sensors. UAVs can be launched quickly and exchange data with the control center in real time by using wireless transmission systems. While in the last decade UAVs have been widely used in the military field, UAVs are still facing some technical and institutional barriers in civilian applications, for instance, strict airspace and complicated route restrictions. Recently, many countries, such as the United States and China, have begun considering and evaluating flexible air traffic control rules that allow the low attitude space (lower than 1000 m) management for UAV-based civil engineering applications, such as, traffic detection, weather monitoring, disaster response and geological survey. This emerging trend presents a great opportunity for the transportation agencies and operators to explore the full potential of UAVs in road traffic network surveillance and traffic incident monitoring. The common equipped sensors on the UAVs can produce entire images of an investigation area or a special location, which can be further post-processed to monitor semi-continuous traffic state evolution. In this research, we are interested in developing computationally efficient optimization models for using UAVs to improve the effectiveness of traffic surveillance in conjunction with traditional fixed traffic sensors.

To rapidly and reliably capture traffic formation and congestion on the traffic network, a dynamically configured sensor network should be able to recognize time-varying traffic flow propagation that expands to both space and time dimensions. In this research, we adopt a modeling approach from the time-geography field [1,2], in order to systematically take into account both geometry and topology of the road network and time attributes of each event along UAV cruising routes. The particular area of interest for this UAV dynamic route planning application is how to rapidly enable road traffic and incident monitoring. Based on a linear programming model and a space-time network characterization, we develop an UAV routing/scheduling model which is integrated with existing fixed traffic monitoring sites for road segments with various frequencies of incidents. The goal of our model is to minimize the total cost in terms of the detection delay of spatially and temporally distributed incidents by both fixed sensors and UAVs. The total budget and UAVs' feasible routing routes are also considered as practical constraints in our model. To address the computational efficiency issue for real-world large scale networks, a Lagrangian relaxation method is introduced for effective problem decomposition.

The remainder of this paper is organized as follows: a literature review and problem statements are presented first in the next section. In Section 3, a space-time network-based UAV routing planning model is developed to integrate with existing fixed traffic detectors to maximize spatial and temporal coverage. Section 4 further presents the Lagrangian relaxation solution algorithmic framework, followed by several illustrative examples and numerical experiment results to demonstrate the effectiveness of the proposed models in Section 5.

2. Literature Review

2.1. Related Studies

On-line applications of intelligent traffic network management call for the reliable detection, estimation and forecasting of dynamic flow states so that proactive, coordinated traffic information and route guidance instructions can be generated to network travelers for their pre-trip planning and en-route diversion. The problem of how to optimize traffic sensor locations to maximize the spatial coverage and information obtainable has been extensively studied by many researchers. Gentili and Mirchandani [3] offered a comprehensive review for three different sensor location optimization models (sensor type, available a-priori information and flows of interest), and classified them into two main problems: the observability problem and the flow-estimation problem. A partial observability problem is also studied by Viti *et al.* [4]. For origin-destination demand and estimation applications, many sensor location methods are based on the study by Yang and Zhou [5] which focuses on how to maximize the coverage measure in terms of geographical connectivity and OD flow demand volume. For travel time estimation, Sherali *et al.* [6] proposed a quadratic binary optimization model for locating AVI readers to capture travel time variability along specified trips. A dynamic programming formulation was developed by Ban *et al.* [7] to minimize link travel time estimation errors through locating point sensors along a corridor. In the reliable sensor location problem studied by Li and Ouyang [8], an integer programming model is developed to consider random sensor failure events. Based on a Kalman filtering framework, Xing *et al.* [9] extended an information-theoretic modeling approach from Zhou and List [10] for designing heterogeneous sensor networks in travel time estimation and prediction applications. It should be noted that, the water network sensor placement problem is also closely related to the problem of traffic sensor network design, as many studies such as that of Berry *et al.* [11] focus on how to improve spatial coverage and event detectability for water pollution sources, by adapting p-median location models.

UAV systems have been used as emerging mobile monitoring tools to conduct different tasks by loading various sensors, such as high-resolution camera, radar, and infrared camera. There are a wide range of UAVs studies for different transportation domain applications. For instance, by using high-resolution images from UAVs, the Utah Department of Transportation examined how to improve their highway construction GIS databases [12]. The Florida Department of Transportation studied the feasibility of using surveillance video from UAVs for traffic control and incident management [13]. Recently, Hart and Gharaibeh [14] examined the use of UAVs for roadside condition surveys.

It is also widely recognized that, there are still a number of limitations for using UAVs in civilian transportation applications. First, the accuracy of traffic information collection depends on weather conditions and specific types of sensors carried by UAV. The maximum flight distance or flight time of UAV is constrained by the fuel weight and number of battery units. If the number of available UAVs is given, it is important to optimize the cruise route plan of UAVs in order to cover more roads of interest under the UAV capacity constraints. Ryan *et al.* [15] considered this UAV routing problem as a multiple Travel Salesman Problem (TSP) with the objective of maximizing expected target coverage and solved it by applying a Reactive Tabu Search. In the study by Hutchison [16], the monitored roads are divided into several sub-areas first, and then all the selected roads in each sub-area is covered by one UAV, equipped with a simulated annealing algorithm. Yan *et al.* [17] also considered the UAVs routing problem as a multi-vehicle TSP and introduced a generic algorithm to design close-to-optimal routes to consider different flight paths between two target roads.

In the area of collaborative UAV route planning, the method proposed by Ousingsawat and Campbell [18] first finds the shortest path between two points, and then solves the corresponding task assignment problem. Tian *et al.* [19] introduced the time window requirement of reconnaissance mission for each target road, and considered the constraints of maximum travel time of each UAV through a Genetic Algorithm. In the study by Wang *et al.* [20], a multi-objective ant colony system algorithm is used for UAV route planning in military application with both route length and danger exposure being minimized in the cost functions. The multi-objective optimization model by Liu *et al.* [21] aims to minimize the total distance and maximize the total number of monitored points subject to the available number of UAVs and maximum cruise distance constraints. The model is solved by the genetic algorithm to search for satisfactory UAV routes. The studies by Liu *et al.* [22] and Liu *et al.* [23] introduce a time window constraint and examine UAV route planning methods without and with flight distance constraints. They used a K-means clustering algorithm to decompose the UAV cursing area into a number of sub-areas, and further applied a simulated annealing-based solution algorithm. The multi-objective optimization model proposed by Liu *et al.* [24] aims to minimize UAV cruise distance and minimize the number of UAVs being used.

Mersheeva and Friedrich [25] adopted a metaheuristic variable neighborhood search algorithm, and Sundar and Rathinam [26] presented a mixed integer programming model for UAV route planning with refueling depot constraints. A recent study by Ning *et al.* [27] specifically considers the mobility constraints of traffic sensors, and a measure of traffic information acquisition benefits was used to evaluate the surveillance performance. Their proposed hybrid two-stage heuristic algorithms include both particle swarm optimization and ant colony optimization components. Table 1 offers a systematic comparison for the literature reviewed in Section 2.1.

Table 1. Summary of existing UAV route planning studies.

Paper	Model and Formulation	Solution Algorithm	Factors under Consideration
Ryan *et al.* (1998) [15]	Multi-vehicle Traveling Salesman Problem	Reactive tabu search heuristic within a discrete event simulation	Target coverage, time window
Hutchison (2002) [16]	Two-stage model for problem decomposition and single-vehicle TSP problem	Simulated annealing method	Target coverage, UAV flight distance
Yan *et al.* (2010) [17]	Multi-vehicle TSP problems	Genetic algorithm	Flying direction on each link
Ousingsawat and Campbell (2004) [18]	Cooperative reconnaissance problem	A* search and binary decision	Maximum time duration, target coverage, UAV conflicts
Tian *et al.* (2006) [19]	Cooperative reconnaissance mission planning problem	Genetic algorithm	Maximum time duration , UAV conflict, time window
Wang *et al.* (2008) [20]	Multi-objective optimization model	Ant colony system algorithm	Minimum length and threat intensity of the path
Liu, Peng, Zhang (2012) [21]	Multi-objective optimization model	Non-dominated sorting genetic algorithm	Number of UVAs
Liu, Peng, Chang, and Zhang, (2012) [22]	Multi-objective optimization model	Multi-objective evolutionary algorithm	Time window
Liu, Chang, and Wang (2012) [23]	Traveling Salesman Problem	Simulated annealing method	Target coverage, number of UAVs
Liu, Guan, Song, Chen, (2014) [24]	Multi-objective optimization model	Evolutionary algorithm based on Pareto optimality technique	UAV flight distance, number of UAVs
Mersheeva and Friedrich (2012) [25]	Mixed-integer programming model	Variable neighborhood Search	Target coverage, UAV flying time
Sundar and Rathinam (2012) [26]	Single-vehicle UAV routing problem	mixed integer, linear programming	Target coverage, refuel depot
This paper by Zhang *et al.*	Linear integer programming model within space-time network	Problem decomposition through Lagrangian relaxation and least cost shortest path algorithm for subproblems	Detecting recurring, non-recurring traffic conditions and special events

While a large number of studies have been devoted to the UAV route planning problem with general spatial coverage measures, the potential benefits of utilizing UAVs to capture traffic propagation in both space and time dimensions have not been adequately exploited, especially for cases with stochastic non-recurring traffic incidents with large impacts on traveler delay and route choice. Most of the existing research only focuses on the static monitoring coverage measure, while it is critically needed to adopt a systematic space-time coverage measure for the integrated sensor network design problem with both fixed and mobile sensors. For the VRP problem focusing on UAV routing in a traffic network, the corresponding theoretical and algorithmic aspects along this line are still relatively undeveloped, and these challenging questions calls for flexible and systematic modeling methodologies and efficient solution algorithms for large-scale network applications.

2.2. Illustrations of Conceptual Framework for Space-Time Networks

In this paper, a time geography based modeling approach is adopted for the traffic sensor network design problem. This theory is introduced by Hagerstrand [1] to specifically use space-time paths and space-time prisms in accessibility assessment. A space-time path represents the path taken by an individual agent in a continuous space, with a travel time budget constraint. A space-time prism is the set of all points that can be reached by an individual, given a maximum possible speed from a starting point and an ending point in space-time [2,28]. A simple space-time path example is illustrated in Figure 1.

The agent (*i.e.*, UAV in our research) starts from location x_1 (as an UAV depot) at time t_1 and departs from x_1 to x_2 (e.g., incident site) at t_2; at time t_3, the agent arrives at x_2 and stays at x_2 until t_4 (to monitor the traffic delay), then the agent moves towards location x_3 and arrives there at time t_5. The different slopes represent different flying speeds in our example.

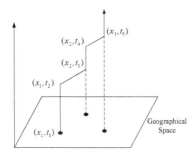

Figure 1. Illustration of a space-time path.

A space-time prism is an important concept for analyzing travelers' accessibility on the transportation network and it is used in many transportation planning studies. Utilizing this space-time prism concept in the UAV routing planning application, we can also clearly examine the relationship between geographic space and time horizon. The accessibility or reachability by an UAV sensor can be represented by the prism volumes. In order to consider more spatial constraints within a real-world road network, such as connectivity, speed range, and geometry, a network-space prism concept can be used. Figure 2 from Kuijpers and Othman [28] shows a network time prism in network-time space (shown in red regions) and its potential paths' projection to road networks (shown in green). The green projection on the physical network represents individual's potential traveling routes from the origin to the destination.

Figure 3 shows an UAV trajectory in the context of network-time paths. With regards to the admissible air space and flying speed considerations, the feasible UAV's routes could be predefined along the physical road network and possible air space discretized in both space and time. In Figure 3, the blue lines are the physical network links; black lines are the UAV route on physical links; purple dash line is the UAV route on admissible air space.

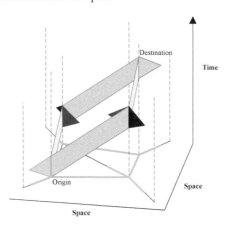

Figure 2. Constrained space–time prism (red) on road networks (green and black) and its spatial projection (green). Adapted from [28].

With a discretized space-time network construct, one of the remaining challenging issues is how to maximize the traffic information obtainable at strategically critical locations and time-sensitive durations within the UAV traveling budget constraint in conjunction with the existing fixed sensor detection infrastructure. We assume that there are probabilistic traffic incident event data, obtainable from historical data or observed directly from a real time environment. With the time as the horizontal axis, Figure 4a shows the space-time feature of traffic incident, where the congestion due to the traffic incident is propagated and dissipated along the corridor as the time advances. Accordingly, one can define a space-time vertex set denoted as $\phi(a)$, for each incident event a at a location. The road segments with very low incident rates typically need to be patrolled once or twice a day in order to find soft time windows with low priority for the UAVs. To better consider the UAV speed and altitude restrictions in further research, one can create a multi-dimensional model, where each vertex is characterized by the longitude, latitude and altitude at different time stamps, and accordingly limit the feasible route search space by considering the vertexes satisfying altitude restrictions and the arcs satisfying speed requirements.

Ideally the entire space-time vertex in a set should be fully observed by either fixed traffic sensors or UAVs at any given time, which means that the incident and its impact area are fully detected. Without loss of generality, this paper assumes that, if a fixed traffic sensor is located on a site inside the vertex set of $\phi(a)$, then this sensor can cover all the space-time vertex on this location at all time. If an UAV flies to site i at time $t + 1$, as shown in Figure 4b then the space-time vertex $(i, t + 1)$ is marked as covered, if it stays at this site for five time periods, then all six out of the seven total space-time vertexes on site i are observed with 1 time unit of detection delay.

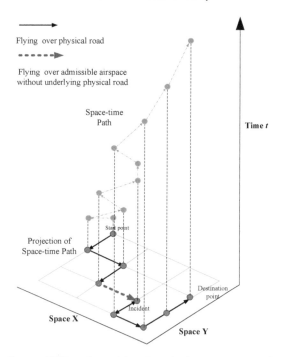

Figure 3. UAV routing path in a discretized space-time network.

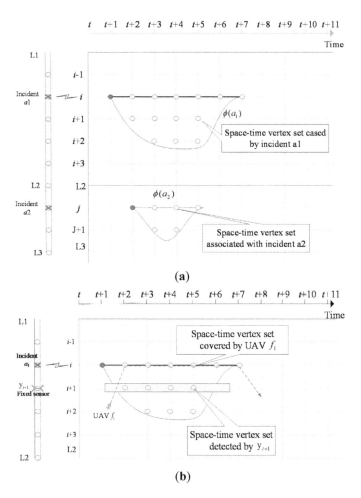

Figure 4. (a) Sets of space-time vertexes affected by different events *a*1, and *a*2; (b) Set of space-time vertexes detected by fixed sensor and UAV.

3. Model Description

3.1. Notations

We first introduce some key notations used in the UAV routing planning problem. Parameters are shown in Table 2:

Table 2. Subscripts and parameters used in mathematical formulations.

Symbol	Definition
A	set of traffic accidents/events
N	set of nodes in transportation network
V	set of space-time vertexes
E	set of space-time traveling arcs that UAV can select
F	set of UAVs
$a, a\prime$	indices of traffic accidents/events, $a, a\prime \in A$
i, j	indices of candidate sensor locations or possible UAV locations, $i, j \in N$
t, s	indices of time stamps
$(i, t), (j, s)$	indices of space-time vertexes
(i, j, t, s)	index of space time traveling arc, $(i, j, t, s) \in E$
f	index of unmanned plane, $f \in F$
o^f, d^f	indices of origin and destination locations of UAV f
EDT^f, LAT^f	earliest departure time and latest arriving time of UAV f
$T(a)$	start time of event a
$\phi(a)$	set of space-time vertexes which can represent the space-time impact area of event a
$\phi\prime(a)$	subset of time-space vertex set $\phi(a)$ which excludes the space time vertexes covered by the fixed detectors. $\phi\prime(a) \subseteq \phi(a)$
$d_{i,j,t,s}$	cost for UAVs to travel between location i and location j, on time period between time t and time s
d_i	cost for constructing a fixed sensor at location i
$c^F_{i,t}(a)$	fail-to-detect cost of event a at location i and time t
B	total budget for constructing fixed sensors
$K(f)$	total distance budget for operating UAV f

Variables are shown in Table 3:

Table 3. Decision variables used in mathematical formulations.

Symbol	Definition
$x_{i,t}(a)$	event detected variable (= 1, if event a is detected at location i, time t; otherwise = 0)
$x^F_{i,t}(a)$	event virtually detected variable (= 1, if event a is virtually detected at location i, time t; otherwise = 0)
y_i	fixed sensor construction variable (= 1, if fixed sensor is allocated at location i; otherwise = 0)
$w_{i,j,t,s}(f)$	UAV routing variable (= 1, if space-time arc (i,j,t,s) is selected by UAV f; otherwise = 0)

In this paper, we do not assume a constant cruising speed of Ubetween a node pair (i, j). Instead, we allow different travel times (denoted as time t to time s) between a link (i, j) to reflect different flying speed, corresponding to various degrees of fuel consumptions. AV can also stay at the node for an extended time period within the regulation constraints, denoted as a staying arc $(i, i, t, t + 1)$ at the same node i. In addition, the virtual sensors are introduced in our model to capture the cost loss of non-coverage. Thus, we assume that every point at a vertex set within the incident impact area could be monitored by either fixed sensors, UAVs or virtual sensors. Accordingly, one should predefine a much higher monitoring cost for virtual sensors compared to fixed sensors UAVs to encourage ehysical coverage as much as possible. The parameter $c^F_{i,t}(a)$ is used to represent the fail-to-detect cost of event a at location i and time t, which could cover generalized cost factors sh as (i) response delay in detecting incidents; and (ii) time-dependent traffic incident impacts for different events and (iii) the time duration spent to monitor the traffic impact area.

W assume the preferred departure time and arrival times of UAVs are given without loss of generality. The finial optimization goal in our model is to minimize the monitoring cost of all incidents by using fixed sensors or UAVs in the space-time network, subject to a number of essential constraints such as flow balance constraints for each flight trajectory, budget constraints for both fixed sensors and UAV routing costs.

3.2. Space-Time Network Construction

The concept of space-time network (STN) is widely used in both transportation geography and transportation network modeling literature and it aims to integrate physical transportation networks with individual time-dependent movements or trajectories. For more detail about conceptual model of STN, please see the references on Hagerstrand [1], Miller [2] and Kuijpers and Othman [28]. In this paper, we have created the UAV flying ty based on STN.

Ge set N (with a set of incidents A) and physical link set G, the next task is to build the STN structure that can model the network-based UAV trajectory. The steps for building a space-time network for an UAV flying trajectory is shown below:

Step 1: Build space-time vertex V

Add vertex (i, t) to V for $i \in N$ and each t.

Step 2: Build space-time arc set E

Step 2.1: Add space-time traveling arc (i,t), $(j, t + TT(i,j,t))$ to E, for physical link$(i,j) \in E$, where $TT(i, j, t)$ is the link travel time from node i to node j starting at time t.
Step 2.2: Add a set of space-time staying/waiting arcs for a pair of vertexes (i,t), $(i, t+1)$ to E, for each time t.

A hypothetic 3-node network is created with time-invariant link travel time in order to illustrate the concept of space-time network construction and reasonable UAV flying trajectory. The detailed information of this hypothetic 3-node network is shown in Table 4.

Table 4. Flying time in hypothetic 3-node network.

Link	Travel Time
$(o, d_1), (d_1, o)$	2
$(d_1, d_2), (d_2, d_1)$	3

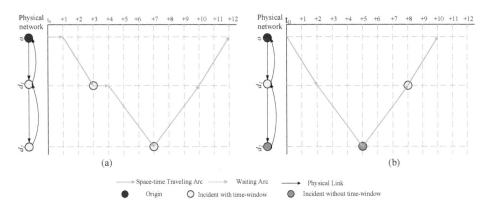

Figure 5. An illustration of flying trajectory space-time network building.

Two cases are shown in Figure 5: (a) two incidents with time-window requirements; and (b) one incident with time-window requirement and one incident without (tight) te-window requirement. Table 5 lists the detailed parameters of incidents for case (a) and case (b), respectively.

Table 5. Description of incidents on Figure 5.

Incident Num.	Location of Vertex	Start Time
Case (a)-1	d_1	3
Case (a)-2	d_2	7
Case (b)-1	d_1	8
Case (b)- 2	d_2	/

First, space-time traveling arcs are constructed only when their corresponding physical links or admissible airspace exist, and the planning time horizon is assumed as 12 time units. We can illustrate how a feasible tour is generated in the space-time network. In case (a) of Figure 5, an UAV first uses waiting arc$(o, o, t_0, t_0 + 1)$, then uses traveling arc $(o, d_1, t_0 + 1, t_0 + 3)$ to reach d_1 and detect no.1 incident. It then uses waiting arc $(d_1, d_1, t_0 + 3, t_0 + 4)$ on d_1 to wait until time $t_0 + 4$, then it will fly to d_2 through arc $(d_1, d_2, t_0 + 4, t_0 + 7)$ to detect incident no. 2. This is followed by a return trip to the departure node o through (d_2, d_1) and (d_1, o).

In case (b) of Figure 5, incident 1 occurs at time $t_0 + 8$, and incident 2 is a minor roadside or non-blocking incident without a specific visiting time requirement. An UAV first travels to d_2 at $t_0 + 5$, then it finishes the monitoring task for no. 1 incident through travelling arc $(d_2, d_1, t_0 + 5, t_0 + 8)$. The UAV will go back to the departure node o through (d_1, o) and arrive o at time $t_0 + 10$.

3.3. Model Description

We now describe the formal problem statement as follows. The general objectives for the route optimization problem could include information processing cost, operational cost, construction cost. Our model specifically aims to maximize the spatial and temporal coverage for the incident road segments, given the total construction budget of fixed sensors and the UAV operational cost constraints. Thus, the equivalent objective function is to minimize the non-detecting cost.

Model:

$$\text{Obj. } min \sum_{a \in A, (i,t) \in \phi(a)} \left[c_{i,t}^F(a) \times x_{i,t}^F(a) \right] \tag{1}$$

Subject to:

Event detection constraint: an event must be detected/virtually detected exactly once.

$$x_{i,t}(a) + x_{i,t}^F(a) = 1 \forall (i,t) \in \phi(a) \tag{2}$$

In Equation (2), for the incidents with time windows, if it is detected by fixed sensors or UAVs, then $x_{i,t}(a) = 1$ and $x_{i,t}^F(a) = 0$; otherwise, it is covered by the virtual sensors where $x_{i,t}(a) = 0$ and $x_{i,t}^F(a) = 1$.

Eetection and sensors coupling constraint: an event can be detected at certain space-time vertex, if this vertex is covered by a fixed sensor or UAVs.

$$x_{i,t}(a) \leq y_i + \sum_{f \in F, j, s} w_{i,j,t,s}(f), \text{ forall } a \in A, (i,t) \in \phi(a) \tag{3}$$

It should be remarked that in Equation (3), $x_{i,t}(a)$ is a variable to represent whether the space time vertex (i, t) in event a is detected by an UAV or fixed sensor. If $x_{i,t}(a) = 1$, then this space time vertex (i, t) is detected by an UAV or fixed sensor, that is, $\sum_{f \in F, j, s} w_{i,j,t,s}(f) = 1$ or $y_i = 1$. Otherwise, $x_{i,t}(a) = 0$ indicates $y_i = 0$ and $\sum_{f \in F, j, s} w_{i,j,t,s}(f) = 0$.

Flow balance constraint: to depict a time-dependent UAV tour in the space-time network, a set of flow balance constraints is formulated below. This model permits more than one depots exist in the network, and also permits more than one UAVs to execute the mission. We define a super origin vertex and super sink vertex for every UAV, and all vertexes follow the flow balance constraints strictly.

$$\sum_{(i,j,t,s)\in V} w_{i,j,t,s}(f) - \sum_{(i,j,t,s)\in V} w_{j,i,s,t}(f) = \begin{cases} 1 & i = o^f, t = EDT^f \\ -1 & i = d^f, t = LAT^f \text{ for all } f \in F \\ 0 & \text{otherwise} \end{cases} \tag{4}$$

UAVs' conflict-free constraint: each vertex at a specific time can only pass one UAV.

$$\sum_{f\in F,j,s} w_{i,j,t,s}(f) \leq 1 \forall(i,t) \tag{5}$$

It should be remarked that UAV should satisfy this conflict-free constraint for all the space-time vertex (i, t).

Fixed sensor budget constraint: the total budget should include fix sensor construction cost.

$$\sum_i y_i d_i \leq B \tag{6}$$

UAV operational constraint: the maximum fly distance or flying time constraint for each UAV.

$$\sum_{(i,j,t,s)\in E} d_{i,j,t,s} \times w_{i,j,t,s}(f) \leq K(f) \quad \forall f \in F \tag{7}$$

Without the loss of generality, our paper considers the total flying time. In Equation (7), $d_{i,j,t,s}$ is fixed for each road segment. There are also binary definitional constraints for variables $x_{i,t}(a)$, $x^F(a)$, y_i and $w_{j,i,s,t}(f)$.

$$x_{i,t}(a) \in \{0,1\}, \text{ for all } a \in A, i \in N$$
$$x_{i,t}^F(a) \in \{0,1\}, \text{ for all } a \in A, i \in N$$
$$y_i \in \{0,1\}, \text{ for all } i \in N$$
$$w_{j,i,s,t}(f) \in \{0,1\}, \text{ for all}(j,i,s,t) \in E, f \in F$$

3.4. Model Simplification

To derive a simple form of the model, we first substitute Equation (2) into the objective function and obtain the following objective function in Equation (8).

$$\min \sum_{a\in A,(i,t)\in\phi(a)} \left[c_{i,t}^F(a) \times (1 - x_{i,t}(a)) \right] \tag{8}$$

In practice, after years of incident detection system construction, fixed sensors have been equipped at important locations in highway networks. Accordingly, in our model the location of fixed sensors are assumed in advance, and the corresponding variable y_i equals to 1 at location i where a fixed sensor is allocated. Then $x_{i,t}(a) = 1$ for $y_i = 1$, so we can construct $\phi\prime(a)$ to exclude the space time vertexes covered by the fixed detectors.

Without loss of generality, we use the time unit (min) as the generalized cost unit so that we can minimize the monitoring cost of all incidents. One can further use the value of time as the coefficient to convert the different degree of non-detection to generalized monetary costs, in conjunction with the other system costs involving fixed sensor operations and UAV energy costs typically expressed as a function of UAV flight distance and speed.

According to Equation (5), any two UAVs cannot arrive at a same space-time vertex (i, t) to avoid conflicts, which means that $\sum_{f\in F,j,s} w_{i,j,t,s}(f)$ can only equal to 1 or 0 this case, if $\sum_{f\in F,j,s} w_{i,j,t,s}(f) = 1$ at space-time vertex (i,t), then $x_{i,t}(a) = 1$; otherwise, if $\sum_{f \in F,j,s} w_{i,j,t,s}(f) = 0$, $x_{i,t}(a) = 0$. Thus, we can derive Equation (9) as:

$$x_{i,t}(a) = \sum_{f \in F,j,s} w_{i,j,t,s}(f) \quad \text{for all } a \in A, (i,t) \in \phi(a), y_i = 0 \tag{9}$$

and then obtain the simplified model with optimization function Equation (10):

$$\min \sum_{a \in A,(i,t) \in \phi\prime(a)} \left\{ -c_{i,t}^F(a) \times \sum_{f \in F,j,s} w_{i,j,t,s}(f) \right\} \tag{10}$$

Subject to, Equations (4), (5) and (7) and the binary variable definitional constraints.

4. Lagrangian Relaxation-Based Solution Algorithms

The Lagrangian relaxation technique is commonly used for solving optimization problems that contain "hard" constraints. Compared to the primal problem, the relaxation problem can often be diverted/decomposed to classic or easy-to-solve sub-problems.

4.1. Lagrangian Function

As constraints in Equations (5) and (7) are considered as hard constraints, they are further relaxed by introducing two sets of multipliers, namely, non-conflict multiplier $\beta_{i,t}$ and UAV budget multiplier $\mu(f)$. The Lagrangian relaxation function $L_{w(f)}$ can be defined in Equation (11) subject to the flow balance constraint (4). For each UAV, the problem can simplified as time-dependent least-cost path sub-problem:

$$\begin{aligned} L_{w(f)} = & \sum_{a \in A,(i,t) \in \phi\prime(a)} \left\{ -c_{i,t}^F(a) \times \sum_{j,s} w_{i,j,t,s}(f) \right\} + \sum_f \mu(f) \times \left[\sum_{(i,j,t,s) \in E} d_{i,j,t,s} \times w_{i,j,t,s}(f) - K(f) \right] \\ & + \sum_{i,t} \left[\beta_{i,t} \times \left(\sum_{j,s} w_{i,j,t,s}(f) - 1 \right) \right] = \sum_{(i,j,t,s) \in E} [c\prime_{i,j,t,s} \times w_{i,j,t,s}(f)] - \sum_f \mu(f) \times K(f) - \sum_{i,t} \beta_{i,t} \end{aligned} \tag{11}$$

where, the generalized cost term $c\prime_{i,j,t,s} = \mu(f) \times d_{i,j,t,s} + \beta_{i,t} - \sum_a \left\{ \theta_{i,t}^a \times c_{i,t}^F(a) \right\}$. Parameter $\theta_{i,t}^a = 1$, if (i,t) is included in event set a. $\mu(f) \times d_{i,j,t,s}$ reflects the use of fuel, and if the total energy is insufficient, then $\mu(f)$ has to be increased to penalize fuel-inefficient routes.

$\beta_{i,t}$ reflects the potential conflict between flights. If there are more than 2 flights at the same space-time vertex, then $\beta_{i,t}$ needs to be increased to prevent the conflict.

Overall, $\sum_a \left\{ \theta_{i,t}^a \times \left[c_{i,t}(a) - c_{i,t}^F(a) \right] \right\}$ reflects the benefit collected by flight routing plan, e.g., early detection and complete space-time coverage for the entire event, as well as potential loss due to non-detected space-time points.

4.2. Solution Procedure

In this section, we further explain the optimization algorithm for solving Lagrangian relaxation-based problem $L_{w(f)}$.

Step 1. Initialization

Set iteration number $m = 0$;

Choose positive values to initialize the set of Lagrangian multipliers $\mu(f)$, $\beta_{i,t}$;

Step 2. Solve simplified problems

Solve subproblem $L_{w(f)}$ using a standard time-dependent least cost path algorithm and find a path solution for each UAV f.

Calculate primal, and gap values of $L_{w(f)}$;

Step 3. Update Lagrangian multipliers

Update Lagrangian multipliers $\mu(f)$, $\beta_{i,t}$ using subgradient method;

Step 4. Termination condition

If m is larger than a predetermined maximum iteration value, or the gap is smaller than a previously given toleration gap, terminate the algorithm, otherwise $m = m + 1$ and one must go back to Step 2.

To generate the upper bound feasible solutions, one can also use a Lagrangian heuristic algorithm by iteratively fixing the feasible routing solution for individual vehicles, while the adjusted multipliers from the lower bound solutions can be used to guide the iterative refinement of the generalized cost for the UAV routing problem. For detailed procedures on time-dependent shortest path algorithms and sub-gradient updating rules in a Lagrangian relaxation solution framework, we refer interested readers to the study by Meng and Zhou [29] for train routing and scheduling applications.

5. Numerical Experiments

This section evaluates the results of the UAV route planning under different conditions, which aims to demonstrate the effectiveness of the proposed method in the context of real-world networks.

5.1. Simple Illustrative Example

Using the simple demonstrative network in Figure 6, we want to illustrate how the proposed model can effectively plan UAV routes.

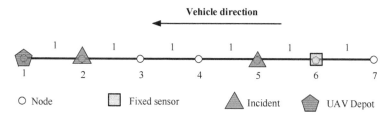

Figure 6. Demonstration of UAV cruise route.

In this example, the UAV depot is located at node 1 and the fixed sensor is installed at node 6. The flying time of UAV for every two successive nodes are 1 time unit and the total allowed flying time duration is 16 time unit. The UAV can be deployed after time 1.

Assume there is only one UAV and two incidents on this network. No. 1 incident on node 2 starts at time 6 and ends at time 8, and it also propagates to node 3 at time 7. No. 2 incident affects node 5, 6 and 7 and the duration for each nodes are time 3 to time 8.5, time 4 to time 7 and time 6 respectively. The space-time influence area of these two incidents is shown below in Figure 7.

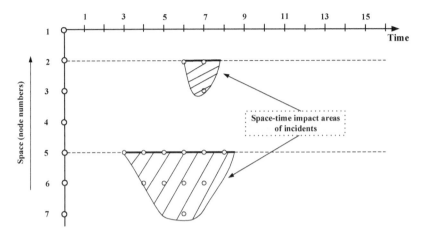

Figure 7. Space-time impact area of these two incidents with first time window of 6–8 and second time window of 3–8.5.

The UAV's route planning problem is implemented and solved by a commercial solution solver, General Algebraic Modeling System (GAMS) using the model we developed in previous sections. The optimal routing plan of UAV is shown in Figure 8, where the vertical axis shows the space dimension with the horizontal axis for the time dimension. Given the total flying time constraint of 15 time units, the optimal route of UAV for this example is: node 1 (time 1) → node 2 (time 2) → node 3 (time 3) → node 4 (time 4) → node 5 (time 5–time 8) → node 4 (time 9) → node 3 (time 10) → node 2 (time 11) → node 1 (time 12–time 16). Due to the flying time constraint, the optimal solution cannot cover No. 1 incident, and the UAV arrives at node 5 after time 2, since No. 2 incident start and it does not reach to node 6 and node 7, while node 6 is covered by the fixed sensor.

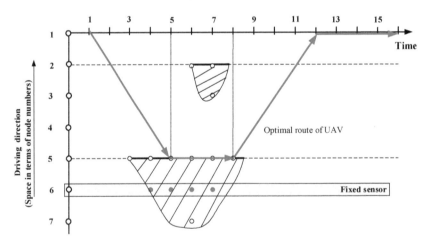

Figure 8. Optimal route of UAV from GAMS with a total flying time constraint of 15 time units.

5.2. Medium-Scale Experiments

A simplified Sioux Falls network consisting of 24 nodes and 76 directional links is shown in Figure 9. We assume the depot of UAV is located at node 16, and fixed sensors are installed at nodes 6,

22 and 24. There is one available UAV for this network and the total allowed flying time duration is 500 min. The flying time durations of the UAV on each road are shown in Table 6. The optimization model is solved on a personal computer with an Intel i7-3630 QM 2.4GHz CPU and 16 GB RAM.

Four incidents are assume to occur on this network on nodes 2, 12, 15 and 23. The detailed propagation and duration time for each incident are listed in Table 7.

There are a total of 161 space-time vertexes covered by these four incidents. Since there are fixed sensors on nodes 6, 22 and 24, thus 46 space-time vertexes of incidents can be detected by fixed sensors. The other 161 − 46 = 115 space-time vertexes of incidents will be considered to be covered by UAV routes. The optimal UAV route is shown in Figure 10.

The optimal UAV cruise route is: node 16 (Depot, 1–76 min) → node 8 (86 min) → node 6 (90 min) → node 2 (incident 1, 100–112 min) → node 1 (124 min) → node 3 (132 min) → node 12 (incident 2, 140–166 min) → node 13 (172 min) → node 24 (180 min) → node 23 (incident 4, 185–212 min) → node 14 (220 min) → node 15 (incident 3, 230–245 min) → node 19 (251 min) → node 17 (255 min) → node 16 (depot, 259–500 min). In this optimal UAV route, 83 space-time vertexes are detected by the UAV, while the others are undetected.

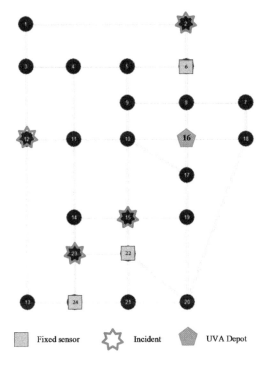

Figure 9. Simplified Sioux Falls network.

Table 6. UAV flying time on all links.

From Node	To Node	Flying Time (min)	From Node	To Node	Flying Time (min)	From Node	To Node	Flying Time (min)	From Node	To Node	Flying Time (min)
1	2	12	8	7	6	13	24	8	19	17	4
1	3	8	8	9	20	14	11	8	19	20	8
2	1	12	8	16	10	14	15	10	20	18	8
2	6	10	9	5	10	14	23	8	20	19	8
3	1	8	9	8	20	15	10	12	20	21	12
3	4	8	9	10	6	15	14	10	20	22	10
3	12	8	10	9	6	15	19	6	21	20	12
4	3	8	10	11	10	15	22	6	21	22	4
4	5	4	10	15	12	16	8	10	21	24	6
4	11	12	10	16	8	16	10	8	22	15	6
5	4	4	10	17	16	16	17	4	22	20	10
5	6	8	11	4	12	16	18	6	22	21	4
5	9	10	11	10	10	17	10	16	22	23	8
6	2	10	11	12	12	17	16	4	23	14	8
6	5	8	11	14	8	17	19	4	23	22	8
6	8	4	12	3	8	18	7	4	23	24	4
7	8	6	12	11	12	18	16	6	24	13	8
7	18	4	12	13	6	18	20	8	24	21	6
8	6	4	13	12	6	19	15	6	24	23	4

Table 7. Incidents information.

	Incident 1	Incident 2	Incident 3	Incident 4
Start node	2	12	15	23
Start time (min)	100	140	230	185
Space-time impact area (node, duration)	2, (100–125) 6, (121–130)	12, (140–165) 13, (160–165)	15, (230–245) 22, (238–247) 21, (245–249)	23, (185–212) 24, (200–225) 21, (222–225)

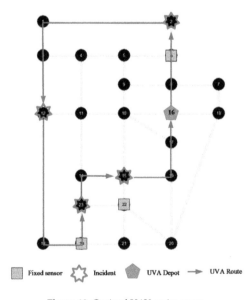

Figure 10. Optimal UAV cruise route.

A quick sensitivity analysis has been also performed by varying the UAV airbase locations on different nodes. The total cost are used as the evaluation index, and the results shown in Figure 11 indicates that the locations at nodes 12, 15, and 23 are more advantageous. As those locations are coincident with the presumed incident locations, so the observations from Figure 11 are expected for this medium scale network. However, the results also indicate that we need to select the UAV depot location more carefully and systematically to minimize the total unexpected cost. The objective value in Figure 11 corresponds to the total penalty cost for undetected space-time vertexes (by the fixed sensors or UAVs). Without loss of generality, we use the time unit (min) as the generalized cost unit. One can further use the value of time as the coefficient to convert the degree of non-detection to generalized monetary costs, in conjunction with the other system costs involving operations and energy costs.

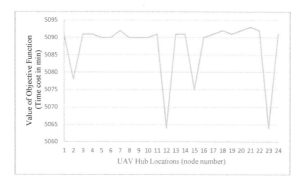

Figure 11. Sensitivity analysis of different UAV's depots.

5.3. Chicago Networks

The above results are provided from a standard optimization solver, which has difficulties in solving large-scale instances on real-world regional networks. Obviously, if we can implement the time-dependent least cost shortest path algorithm directly in high-performance programming languages such as C++ or Fortran, then the proposed Lagrangian relaxation solution procedure can better handle the most computational consuming step for large-scale applications. To this end, we implemented the proposed Lagrangian relaxation algorithm in C++, and test the computational performance of the proposed algorithm on the large-scale Chicago sketch network with 933 nodes, shown in the left plot of Figure 12. In this example, we consider 2 h as the planning horizon (*i.e.*, 120 time intervals with 1 min as temporal resolution), and then randomly generate incident locations. The impact of incidents is simulated through an open-source dynamic traffic assignment simulator [30], DTALite using the time-dependent traffic origin-destination demand tables with reduced capacity due to incidents.

The right plot of Figure 12 demonstrates a sample UAV routing with four UAVs, which could help readers understand the complexity of the problem solved. First, we consider 20 randomly generated incident sites with four UAVs to be scheduled. Figure 13 shows the evolution of lower bound (LB) and upper bound (UB) values in the first 100 iterations, which converges to a significantly small relative solution gap of 5.09%, defined as (UB-LB)/UB. We also observe that, the solution quality gap between upper bound and lower bound starts to reduce steadily after the first 10–15 iterations, as the subgradient algorithm needs to take a few iterations to approximate Lagrangian multipliers with reasonable values. The slow converging pattern afterwards can be explained by the relatively small step size used in the subgradient algorithm when iterative algorithm further proceeds.

Figure 14 further shows the relative solution gap for different numbers of randomly generated incidents, namely 10, 20, 30 and 40 for the same given four UAVs. A small case of 10 incidents converges the optimal solution with the gap of 0% within 20 iterations. When there are a large number of locations to be covered, the algorithm results in larger solution gaps with a slower converging pace. Specifically, 30 and 40 incidents lead to relative solution gaps of 9.7% and 19.4%, respectively, and there are about 8 out of 30 and 19 out of 40 incidents which cannot be covered by UAVs at all. When the number of locations to be covered increases but still with limited UAV resources, the proposed approximation algorithm has to handle the complexity in determining the trade-off of visiting different sites with constrained space-time prism. Overall, the large solution gap is introduced by additional complexity introduced by the number of space-time locations and corresponding LR multipliers. On the other hand, we need to still recognize both theoretical and practical value of the Lagrangian solution algorithm, as it can produce results with exact guarantee on solution accuracy (say 5% or 20%), and the generated upper and lower bound solutions further provide the guidance and benchmark for heuristic algorithms to find close-to-optimal solutions within computational budget.

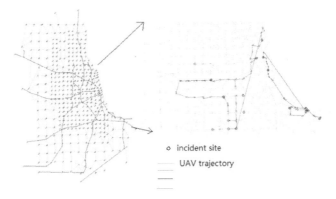

Figure 12. Left: Chicago sketch network with 933 nodes and 2950 links; right: Sameple UAV routing map in the subarea with circles representing incident sites and color lines representing flight routes for four different UAVs.

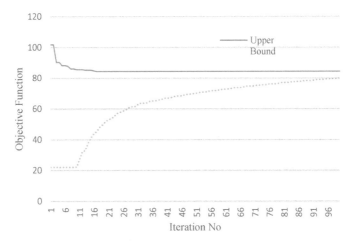

Figure 13. Evolution of Lagrangian relaxation-based upper and lower bound series in the Chicago sketch network with four UAVs and 20 incidents; the lower bound value is generated using Equation (11) and the upper bound is generated by converting possibly infeasible routing solution to satisfy all constraints.

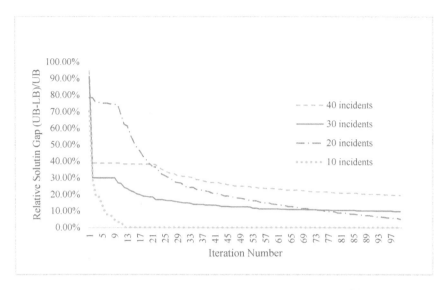

Figure 14. Convering patterns of relative solution gap for four UAVs with different numbers of incidents in Chicago sketch network.

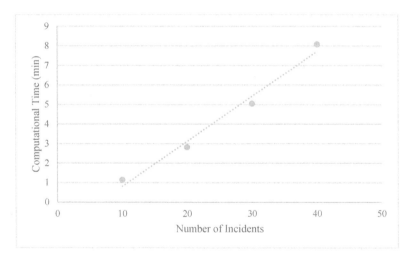

Figure 15. Computaional time with different numbers of incidents in Chicago sketch network.

The computational time of the proposed solution increases almost linearly with an increase of incident numbers, shown in Figure 15. In general, when there are many incidents to be covered, the number of Lagrangian multipliers in Equation (11) also increases, which could require a significant amount of additional computational time for the time-dependent least cost shortest path algorithm to find the optimal solutions in the proposed space-time network. The whole search process with 100 iterations typically takes an average of 1 to 10 CPU min under different numbers of incidents, while the a single iteration of the LR-based lower bound and upper bound generation uses an average of 28.19 milliseconds for the case of 20 incidents. As expected, the most significant amount of time has been spent for constructing a space-time path in the test network with 2950 links and 120 time intervals.

6. Conclusions and Future Research Plan

The fixed sensor-oriented traffic sensor network design problem has been widely studied. With UAVs as a special type of mobile sensors, this paper aims to develop a practically useful and computationally efficient mobile sensor routing model for non-recurring and recurring traffic state detection. Based on the time geography perspective, we present a linear integer programming model to maximize spatial and temporal coverage of traffic state detection under various UAV speed, admissible airspace and operational budget constraints. A Lagrangian relaxation solution framework is developed to effectively simplify the original complex problem into standard time-dependent least cost path problems. Using a number of illustrative and real-world networks, our proposed model offers a unified fixed and mobile sensor network framework and efficient routing/scheduling algorithms for improving road network observability.

With special focus on mobile sensors on daily operations, this paper considers the fixed sensor locations as predefined parameters. In our future research, we will jointly optimize the fixed sensor locations and UAV's route planning under a large number of random traffic conditions, within a stochastic optimization modeling framework. By doing so, it would be interesting to analyze the cost-benefit between mobile and fixed sensors in order to establish a mutually complementary and fully integrated sensor network. Our further research will be also focused on the efficient exact and heuristics algorithms for real-time UAV routing algorithms with unknown or predicted traffic conditions.

Acknowledgments: The first four authors are supported by the project titled "Highway network operation state intelligent monitoring and security services key technology development and system integration" (No. 2014BAG01B02) in China.

Author Contributions: Jisheng Zhang and Limin Jia designed the research framework as well as the UAV routing model, solution algorithm and data analysis used in this paper. Shuyun Niu, Fan Zhang and Xuesong Zhou make the contributions on literature, model and experiment parts. Lu Tong and Xuesong Zhou performed the Lagrangian relaxation algorithm. All the authors prepared the manuscript.

Conflicts of Interest: The authors declare no conflict of interests.

References

1. Hagerstrand, T. What about people in regional science? *Reg. Sci. Assoc.* **1970**, *24*, 7–21. [CrossRef]
2. Miller, H.J. Modelling accessibility using space-time prism concepts within geographical information systems. *Int. J. Geogr. Inf. Syst.* **1991**, *5*, 287–301. [CrossRef]
3. Gentili, M.; Mirchandani, P.B. Locating sensors on traffic networks: Models, challenges and research opportunities. *Transp. Res. Part C: Emerg. Technol.* **2012**, *24*, 227–255. [CrossRef]
4. Viti, F.; Rinaldi, M.; Corman, F.; Tampère, C.M.J. Assessing partial observability in network sensor location problems. *Transp. Res. Part B: Methodol.* **2014**, *70*, 65–89. [CrossRef]
5. Yang, H.; Zhou, J. Optimal traffic counting locations for origin–destination matrix estimation. *Transp. Res. Part B: Methodol.* **1998**, *32*, 109–126. [CrossRef]
6. Sherali, H.D.; Desai, J.; Rakha, H. A discrete optimization approach for locating automatic vehicle identification readers for the provision of roadway travel times. *Transp. Res. Part B: Methodol.* **2006**, *40*, 857–871. [CrossRef]
7. Ban, X.; Herring, R.; Margulici, J.D.; Bayen, A. Optimal sensor placement for freeway travel time estimation. In *Transportation and Traffic Theory 2009: Golden Jubilee*; Lam, W.H.K., Wong, S.C., Lo, H.K., Eds.; Springer US: New York, NY, USA, 2009; pp. 697–721.
8. Li, X.; Ouyang, Y. Reliable sensor deployment for network traffic surveillance. *Transp. Res. Part B: Methodol.* **2011**, *45*, 218–231. [CrossRef]
9. Xing, T.; Zhou, X.; Taylor, J. Designing heterogeneous sensor networks for estimating and predicting path travel time dynamics: An information-theoretic modeling approach. *Transp. Res. Part B: Methodol.* **2013**, *57*, 66–90. [CrossRef]
10. Zhou, X.; List, G.F. An information-theoretic sensor location model for traffic origin-destination demand estimation applications. *Transp. Sci.* **2010**, *44*, 254–273. [CrossRef]

11. Berry, J.; Carr, R.; Hart, W.; Phillips, C. Scalable water network sensor placement via aggregation. In *World Environmental and Water Resources Congress 2007*; American Society of Civil Engineers: Tampa, FL, USA, 2007; pp. 1–11.
12. Barfuss, S.L.; Jensen, A.; Clemens, S. *Evaluation and Development of Unmanned Aircraft (UAV) for UDOT Needs*; Utah Department of Transportation: Salt Lake City, UT, USA, 2012.
13. Latchman, H.A.; Wong, T.; Shea, J.; McNair, J.; Fang, M.; Courage, K.; Bloomquist, D.; Li, I. *Airborne Traffic Surveillance Systems: Proof of Concept Study for the Florida Department of Transportation*; University of Florida, Florida Department of Transportation: Gainesville, FL, USA, 2005.
14. Hart, W.; Gharaibeh, N. Use of micro unmanned aerial vehicles in roadside condition surveys. In *Transportation and Development Institute Congress 2011*; American Society of Civil Engineers: Chicago, IL, USA, 2011; pp. 80–92.
15. Ryan, J.L.; Bailey, T.G.; Moore, J.T.; Carlton, W.B. Reactive Tabu search in unmanned aerial reconnaissance simulations. In Simulation Conference Proceedings, Washington, DC, USA, 1998; Volume 871, pp. 873–879.
16. Hutchison, M.G. A method for Estimating Range Requirements of Tactical Reconnaissance UAVs. In Proceedings of the AIAA's 1st Technical Conference and Workshop on Unmanned Aerospace Vehicles, Virginia, VA, USA, 20–23 May 2002.
17. Yan, Q.; Peng, Z.; Chang, Y. Unmanned aerial vehicle cruise route optimization model for sparse road network. *Transportation Research Board of the National Academies*, Washington, DC, USA, 2011; 432–445.
18. Ousingsawat, J.; Mark, E.C. Establishing trajectories for multi-vehicle reconnaissance. In Proceedings of the AIAA Guidance, Navigation, and Control Conference and Exhibit, Providence, RI, USA, 16–19 August 2004.
19. Tian, J.; Shen, L.; Zheng, Y. Genetic Algorithm Based Approach for Multi-Uav Cooperative Reconnaissance Mission Planning Problem. In *Foundations of Intelligent Systems*; Esposito, F., Raś, Z., Malerba, D., Semeraro, G., Eds.; Springer: Berlin/Heidelberg, Germany, 2006; Volume 4203, pp. 101–110.
20. Wang, Z.; Zhang, W.; Shi, J.; Han, Y. UAV route planning using multiobjective ant colony system. In Proceedings of the 2008 IEEE Conference Cybernetics and Intelligent Systems, Chengdu, China, 21–24 September 2008; pp. 797–800.
21. Liu, X.; Peng, Z.; Zhang, L.; Li, L. Unmanned aerial vehicle route planning for traffic information collection. *J. Transp. Syst. Eng. Inf. Technol.* **2012**, *12*, 91–97. [CrossRef]
22. Liu, X.; Peng, Z.; Chang, Y.; Zhang, L. Multi-objective evolutionary approach for uav cruise route planning to collect traffic information. *J. Central South Univ.* **2012**, *19*, 3614–3621. [CrossRef]
23. Liu, X.; Chang, Y.; Wang, X. A UAV allocation method for traffic surveillance in sparse road network. *J. Highw. Transp. Res. Dev.* **2012**, *29*, 124–130. [CrossRef]
24. Liu, X.; Guan, Z.; Song, Y.; Chen, D. An Optimization Model of UAV Route Planning for Road Segment Surveillance. *J. Central South Univ.* **2014**, *21*, 2501–2510. [CrossRef]
25. Mersheeva, V.; Friedrich, G. Routing for continuous monitoring by multiple micro uavs in disaster scenarios. In Proceedings of the 20th European Conference on Artificial Intelligence (ECAI 2012), Montpellier, France, 27–31 August 2012; pp. 588–593.
26. Sundar, K.; Rathinam, S. Route planning algorithms for unmanned aerial vehicles with refueling constraints. In Proceedings of the American Control Conference (ACC), Montréal, Canda, 27–29 June 2012; pp. 3266–3271.
27. Ning, Z.; Yang, L.; Shoufeng, M.; Zhengbing, H. Mobile traffic sensor routing in dynamic transportation systems. *IEEE Trans. Intell. Transp. Syst.* **2014**, *15*, 2273–2285.
28. Kuijpers, B.; Othman, W. Modeling uncertainty of moving objects on road networks via space-time prisms. *Int. J. Geogr. Inf. Sci.* **2009**, *23*, 1095–1117.
29. Meng, L.; Zhou, X. Simultaneous train rerouting and rescheduling on an N-track network: A model reformulation with network-based cumulative flow variables. *Transp. Res. Part B: Methodol.* **2014**, *67*, 208–234. [CrossRef]
30. Zhou, X.; Taylor, J. DTALite: A queue-based mesoscopic traffic simulator for fast model evaluation and calibration. *Cogent Eng.* **2014**, *1*, 961345. [CrossRef]

MDPI AG

St. Alban-Anlage 66

4052 Basel, Switzerland

Tel. +41 61 683 77 34

Fax +41 61 302 89 18

http://www.mdpi.com

Sensors Editorial Office

E-mail: sensors@mdpi.com

http://www.mdpi.com/journal/sensors

9 783038 427537